OPTICS TODAY

EDITED BY
JOHN N. HOWARD

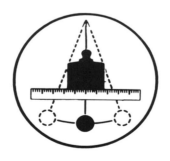

READINGS FROM PHYSICS TODAY
NUMBER THREE

AMERICAN INSTITUTE OF PHYSICS
NEW YORK, NEW YORK
1986

Readings from Physics Today

PHYSICS TODAY, a publication of the American Institute of Physics, provides news coverage of national and international research activities in physics as well as government and institutional activities that affect physics. Both technical and nontechnical developments are covered by scientific articles, news, stories, book reviews, letters to the editor, calendars of meetings, and editorial opinion.

Articles in PHYSICS TODAY are intended to be of interest to—and understandable by—a broad audience of professionals from all subfields of physics as well as people with a general interest in physical science.

Optics Today is the third book in a series of volumes that contains reprinted articles and news material from PHYSICS TODAY in other areas and subfields of physics.

Cover and title design by Charles Grenner.

Printed in the United States of America

Pub. No. R-315.2

Library of Congress Catalog Card No. 86-71346

ISBN 0-88318-499-0

INTRODUCTION

In assembling a volume of optics articles and news items from the last dozen years of *Physics Today*, I had the happy advantage that two earlier reprint volumes have already appeared within the past year—*Astrophysics Today* and *The History of Physics*—so I have adopted a similar format. This has also caused a slight overlap, inasmuch as much of astrophysics is optical astronomy. We did not wish to have major duplication with the astrophysics volume, so we have selected only one article and three or four news items that are specifically optical astronomy. Similarly, in history we have selected only one major historical sketch, celebrating the centennial of the invention of the electric light by Edison in 1879. Those few overlapped items are intended as tantalizing samples to persuade you also to acquire both of those splendid earlier volumes, if you do not already have them.

As in the earlier volumes we also had a problem of selection: budget and size considerations required omitting some of the optics items that had originally appeared in *Physics Today*. We chose to omit those that we felt were too specialized, or which had become dated. For example, in March 1972 *Physics Today* contained an interesting panel discussion by several leading researchers on the Future of Lasers. But, by 1985, a rather large part of those future predictions had already become past accomplishments (and are recounted in this volume in articles by some of the same researchers).

We have tried to organize the articles into clusters of related topics, although some articles cover more than one topic. Similarly, we have here and there written a few introductory words about some topics, although each of the articles contains its own introduction and is complete in itself. The volume is for browsing—a smorgasbord of tasty items—and represents an interesting assortment of the optical activities of the past dozen years.

We mentioned above the necessity of selection of topics and some elimination. We were also faced with the opposite problem: there were a few active areas of recent optical research that are not included in this volume because there simply were no articles on those topics in PHYSICS TODAY in the time period under consideration. For example, the reduction of interferometric and spectroscopic data by Fourier techniques, or techniques of image analysis and processing, or applications of optics to computing are all busy, active areas that are barely mentioned here. Perhaps they will be better represented in the next collection. Even so, the present volume is an excellent representation of the vigor and variety of optics today.

John N. Howard
Editor, *Applied Optics*

TABLE OF CONTENTS

AUTHOR AFFILIATIONS

Robert R. Alfano is Professor of Physics and Herbert Kayser Professor of Electrical Engineering at the City College of the City University of New York.

John M. Anderson is with the General Electric Research and Development Center in Schenectady, New York.

David T. Attwood is a staff member of the Center for X-ray Optics, at Lawrence Berkeley Laboratory in California.

G. C. Baldwin is Emeritus Professor of Nuclear Engineering and Science at Rensselaer Polytechnic Institute, Troy, New York. He is now on the staff of the Physics Division at Los Alamos National Laboratory.

Harrison H. Barrett holds a joint professorship with the Optical Sciences Center and the Department of Radiology, Arizona Health Sciences Center at the University of Arizona, Tucson.

Nicolaas Bloembergen is the Rumford Professor in the Division of Applied Sciences of Harvard University, Cambridge, Massachusetts.

Richard G. Brewer is an IBM Fellow at the IBM Almaden Research Center in San Jose, California.

Walter L. Brown is a member of the Physics Division at AT&T Bell Laboratories in Murray Hill, New Jersey.

Solomon J. Buchsbaum is Vice President, Network Planning and Customer Services, AT&T Bell Laboratories at Holmdel, New Jersey.

George Chapline leads the theoretical studies effort of the University of California Lawrence Livermore Laboratory's x-ray laser program.

Daniel S. Chemla is head of the Quantum Physics and Electronics Research Department at AT&T Bell Laboratories at Holmdel, New Jersey.

Alan G. Chynoweth is Vice President–Applied Research at Bell Communications Research, Morristown, New Jersey.

Timothy Coffey is Director of Research at the Naval Research Laboratory in Washington, D.C.

Esther M. Conwell is a research fellow at Xerox Webster Research Center, Rochester, New York.

Henry Ehrenreich is in the Division of Applied Sciences and the Physics Department of Harvard University.

John L. Emmett is the leader of "Y" Laser Fusion Division at the University of California Lawrence Livermore Laboratory.

Michael Ettenberg is head of Optoelectronic Devices Research, RCA Laboratories, Princeton, New Jersey.

James J. Ewing is a staff scientist in the laser fusion program at Lawrence Livermore Laboratory.

Peter Franken is Director of the Optical Sciences Center at the University of Arizona, Tucson, where he is also Professor of Physics.

Concetto R. Giuliano is the Manager of the Optical Physics Department at Hughes Research Laboratories in Malibu, California.

Theodor W. Hänsch is Professor of Physics at Stanford University, Stanford, California.

John N. Howard (editor) is the Editor of *Applied Optics*.

Erich P. Ippen is Professor of Electrical Engineering at the Massachusetts Institute of Technology, Cambridge, Massachusetts.

Robert J. Keyes, formerly head of the Optics and Infrared Group at the Massachusetts Institute of Technology Lincoln Laboratory, is a private consultant.

R. V. Khoklov (deceased) was Professor of Physics at Moscow State University as well as being Rector of the University.

Robert H. Kingston heads the Optics Division of the Massachusetts Institute of Technology Lincoln Laboratory.

Henry Kressel is Managing Director, E. M. Warburg, Pincus & Co., New York.

Ivan Ladany is a member of the technical staff, RCA Laboratories, Princeton, New Jersey.

Emmett N. Leith is a professor in the Electrical Engineering Department at the University of Michigan, Ann Arbor.

V. S. Letokhov is the head of the Laser Spectroscopy Laboratory and Deputy Director for Research at the Institute of Spectroscopy, of the Academy of Sciences of the USSR, and also a professor at the Moscow Physico-Technical Institute.

H. Richard Leuchtag was an associate editor for PHYSICS TODAY at the American Institute of Physics and is now a staff member of the Department of Biology, Texas Southern University, Houston, Texas.

Marc D. Levenson is on the staff of the IBM Almaden Research Center in San Jose, California.

Barbara G. Levi is a contributing editor of PHYSICS TODAY, and a member of the research staff at the Center for Energy and Environmental Studies, Princeton University.

Tingye Li is head of the Lightwave Systems Research Department at AT&T Bell Laboratories in Holmdel, New Jersey.

Harry Lockwood is Director of Advanced Components Technology Center, GTE Laboratories, Waltham, Massachusetts.

Gloria B. Lubkin is the Editor of PHYSICS TODAY at the American Institute of Physics.

John H. Martin was the executive and technical assistant for The American Physical Society's Study Group on Solar Photovoltaic Energy Conversion.

Aden Baker Meinel is presently at the Jet Propulsion Laboratory, California Institute of Technology, Pasadena, California.

Marjorie Pettit Meinel is presently at the Jet Propulsion Laboratory, California Institute of Technology, Pasadena, California.

Hans Melchior is Professor of Electronics at the Institute for Quantum Electronics at the Swiss Federal Institute of Technology, Zurich, Switzerland.

Aram Mooradian is leader of the quantum electronics group at the Massachusetts Institute of Technology Lincoln Laboratory.

Paul R. Moran is Professor of Radiology at Bowman Gray School of Medicine, Wake Forest University, Winston-Salem, North Carolina and is in the Department of Medical Physics at the University of Wisconsin at Madison.

R. Jerome Nickles is in the Department of Medical Physics at the University of Wisconsin at Madison.

John Nuckolls is Associate Director for Physics at Lawrence Livermore National Laboratory.

John M. Poate is a member of the Physics Division at AT&T Bell Laboratories in Murray Hill, New Jersey.

John S. Saby, now retired, was with the Lighting Research and Technical Services Operation of General Electric in Nela Park, Cleveland, Ohio, when this article was originally published.

Murray Sargent III is Assistant Professor of Optical Sciences at the University of Arizona at Tucson.

Arthur L. Schawlow, who won the Nobel prize for physics for his work on lasers, is Professor of Physics at Stanford University and was president of The American Physical Society in 1981.

Bertram M. Schwarzschild is an associate editor of PHYSICS TODAY at the American Institute of Physics.

Marlan O. Scully is Professor of Physics and Optical Sciences at the University of Arizona at Tucson, and is an Alfred P. Sloan Fellow.

Charles V. Shank is Director of the Electronics Research Center at AT&T Bell Laboratories in Holmdel, New Jersey.

Stanley L. Shapiro (deceased) was a staff member of the Laser Division of Los Alamos Scientific Laboratory.

Phillip Sprangle is head of the Plasma Theory Branch at the Naval Research Laboratory in Washington, D.C.

Yasuharu Suematsu is Professor of Physical Electronics at the Tokyo Institute of Technology.

William Swindell is Head of the Radiotherapy and Radiodiagnostic Physics Section at the Royal Marsden Hospital in Sutton, Surrey, U.K.

Giuliano Toraldo di Francia is Professor of Physics at the Istituto di Fisica Superiore, Università di Firenze, Italy.

James H. Underwood is a staff member at the Center for X-ray Optics, at Lawrence Berkeley Laboratory in California.

Juris Upatnieks is a research engineer of the Institute of Science and Technology at the University of Michigan, Ann Arbor.

E. Joseph Wampler is Professor of Astronomy and Astrophysics at the University of California, Santa Cruz.

Steven Weinberg is Josey Regental Professor of Science at the University of Texas at Austin.

Lowell Wood is a member of the Director's Office and the physics staff of University of California Lawrence Livermore Laboratory.

Eli Yablonovitch is an associate professor in the Division of Applied Sciences of Harvard University, Cambridge, Massachusetts.

James A. Zagzebski is in the Department of Medical Physics at the University of Wisconsin at Madison.

Ahmed H. Zewail is Professor of Chemical Physics at the California Institute of Technology. He has received an Alfred P. Sloan Fellowship and the Camille and Henry Dreyfus Foundation Teacher–Scholar Award.

CHAPTER 1

OPTICS YESTERDAY AND TOMORROW

One of the usual ways to discuss optics is to divide the discussion into the various parts of the optical system: that is, first to discuss light sources—the various ways to produce optical radiation—then to discuss mirrors, lenses, prisms, gratings, interferometers, and so on—the devices and phenomena that bend or disperse or interact with the radiation, and then finally to discuss the optical detectors—the eye, phototubes, or other transducers that convert the optical signal into an image or a meter reading. For convenience, we will follow that general scheme where practical.

In classical physics optics was narrowly defined as relating only to visible light, with the sun or a candle as the light source, and with the eye as the usual detector. Nowadays optics is generally taken to include the entire range of electromagnetic radiation, from very short wavelength ultraviolet radiation and x rays to the far-infrared and submillimeter radiation. Sometimes these limits are pushed even further to include gamma-ray optics, or an "optics" where neutrinos behave analogously to photons; or, at the other end of the spectrum, very long wavelength radio waves. In the present volume we are regarding as optics the various electromagnetic phenomena that are measured and studied by more or less standard optical techniques; we are not really trying to usurp all of physics into optics.

We should mention also that the field of optics has changed over the years as various subfields and fields closely related to optics have matured into independent fields of their own. The study of vision and the physiology of the eye itself, for example, now is considered to be part of the visual sciences, or optometry or ophthalmology. The use of largely optical techniques to study the stars and astrophysical phenomena has long ago made astronomy and astrophysics into a separate discipline. More recently spectroscopy, photography, and various sorts of electro-optics and optoelectronics have all become fields of their own, even though heavily dependent on optics.

In the time period covered by this review there occurred an important anniversary: the centennial of the invention of the electric light by Thomas Edison in 1879. For centuries mankind's illumination had been by flickering candles, torches, and oil lamps. The nineteenth century brought brighter and less smokey kerosene lamps, and chemistry introduced illuminating gas—methane and water gas (steam passed through a mixture of hydrocarbons)—which could be piped into buildings. In 1885 an Austrian chemist, Carl Auer, Baron von Welsbach, invented the incandescent gas mantle, a fine gauze of rare-earth minerals that glowed brilliantly in a flame. Building illumination by incandescent gas mantles was a principal source of illumination until gradually superseded in the 1920s and 1930s by the electric light.

Besides making a practical electric light Edison also had the foresight to provide the generating plant and electrical distribution system that made possible the rapid electrification of the larger cities. Edison (and his competitors) organized teams of scientists and engineers to improve all aspects of the technology, and such newly formed laboratories as the General Electric Laboratory in Schenectady and the Philips Research Laboratory in Eindhoven devoted major efforts to improved light sources. The salute to the centennial of Edison's greatest achievement sketches much of the development of practical light sources over the past century: arc lamps and powerful discharge tube lamps for street and industrial lighting, and filament lamps and fluorescent tubes for home illumination.

John N. Howard

CONTENTS

The electric lamp:
100 years of applied physics

Since Edison's success, progress in physics and in materials science
produced much more efficient and powerful incandescent sources as well
as lamps based on gas discharges and luminescence.

John M. Anderson and John S. Saby

Night-time in 1879, but a scant 100 years ago, was considerably darker than the nocturnal world of today. Artificial illumination was limited to kerosene lamps, illuminating gas, the just-emerging arc lamp for street lighting and a few lingering candles.

Illuminating gas manufacture and distribution had become a well established fact by 1870, having grown rapidly from its commercial origins in about 1800. Even though there was the ever-present danger from explosion and asphyxiation, the gas industry, content only to generate and distribute gas, was lethargic toward research to find better light sources. Efforts in this direction had to come from other quarters. Arc lighting was one such, growing steadily in interest from the first demonstration about 1810 of an electrical arc between charcoal electrodes by Sir Humphrey Davy to the Royal Institution in London. However, a sustained electrical source at high power for commercial arc lighting was not available until the 1870's. It came in the form of dynamos based, in principle, upon a half century of electromagnetic discoveries by Michael Faraday, Joseph Henry, André Ampère, and many others. The arc-lighting industry grew rapidly from the first commercial installations in about 1880, and prospered for 40 years, until better light sources forced its demise.

Another challenge to the illuminating gas industry came in the form of solids made incandescent by electrical current. For their origin we can also go back to Davy who heated platinum to incandescence in 1802. Although the potentialities were apparent, and a number of persons worked in the period up to about 1870 to place platinum, iridium, and carbon rods in vacuum or inert atmospheres, all fell short of a lamp that would burn for an extended period. Probably the biggest problem was attainment of a good vacuum. Another factor, just as for arcs, was the lack of a powerful sustained electrical source.

In 1878 Thomas Edison joined the ranks of those striving for a practical incandescent lamp: Joseph Swan and St. George Lane-Fox in England; William Sawyer, Albon Man, Hiram Maxim and Moses Farmer in the United States. The controversy over who in fact invented the incandescent lamp still rages today, especially the choice between Edison and Swan. Both made contributions—for instance, Swan insisted that carbon was the answer—but we do find it without question that Edison devised both a practical lamp of reasonable size and a complete electrical system, with both

generation and distribution, to exploit his lamp. Because his first carbonized thread lamp ran 40 hours before intentional overvolting destroyed it, and because this happened on 21 October 1879, we take this as the starting point for a century of active lamp research and development, a century rich in interplay between physics and technology.

The understanding of light production in lamps follows from physical principles. By 1879 light was well established as a transverse wave motion, its velocity about 3×10^8 m/s, its spectrum observed, including both ultraviolet and infrared. Kirchoff's cavity radiation defined its laws of total emission and absorption. Radiation from incandescing solids was called "temperature" radiation. To this, Paul Drude (*Theory of Optics,* 1902) further defined "luminescent" radiation resulting from a *change* in the radiating body. What was really meant by this had to await explanation based upon the quantum theory. Luminescence is now universally defined as radiation in excess of thermal, and further defined by the manner of excitation—for example, cathodoluminescence by cathode rays, photoluminescence, and so on.

As the physics of light production became better understood, the level of lamp technology improved. And those in search of better lamps have occasionally discovered new science, including for example, thermionic emission, plasma phenomena and chemical transport theory.

In this article we will follow the historical sequence of lamp development, in incandescent and various kinds of discharge lamps, to the present day. We will give special attention to the efficiency of the different lighting techniques and look to see what improvements in efficiency we might expect in the future.

"Subdividing" light

The arc lamps of the 19th century were noisy; they flickered and gave off noxious gases. Although they were producible in sizes down to 500 candle power, about 6000 lumens, or equivalent to two 40-watt fluorescent lamps, they were hardly usable in the home. The need was clearly recognized for some means to "subdivide" light. Also, arc lamps were operated in series, so that when one was extinguished, all went out unless proper bypass circuits were used. Edison, following the gas-industry practice, imagined a parallel-wire distribution, wherein voltage would be held constant and lamps could be added or dropped at will. To do this he required a high-resistance incandescent conductor in the lamp, starting initially with about 130 ohms. On this point he stood very nearly alone, and it is this vital concept, was well as his practical lamp, that earned him worldwide recognition. Edison, in fact, coined the term "filament" for the incandescent conductor,

which is universally used today. Most workers in the field of electric lighting, including Swan, preferred a low resistance, analogous to the series arc system. Some, like Lord Kelvin and John Tyndall, totally rejected the idea that light could be "subdivided."

Thermal radiation had been well defined experimentally before Max Planck determined the theory in 1900. Scientists therefore recognized that the incandescent wire must have a high temperature to be efficient, that is, to have appreciable radiation in the visible spectrum. The first successful incandescent material was carbon because of its very high sublimation temperature. In practice it operated at about 1700° C, and the life of lamps was limited by evaporation to about 300 hours. Platinum was dangerously close to melting (the melting point is 1773° C) before it gave similar efficiency. On page 36 we show electron micrographs of some early filaments.

After a practical lamp had been secured, work began to measure its characteristics. Following the many experiments of James Joule to find the mechanical equivalent of heat, it seemed natural to find the mechanical equivalent of light, which ultimately was taken to mean the power input required to produce a unit of photometric light—in other words, what we now call the watts per lumen. By 1900 scientists recognized the eye's sensitivity curve, and they had defined a candle as a light source emitting 4π lumens. Herbert Ives[1] measured, in 1910, the emission at the peak of the eye's sensitivity curve, at a wavelength $\lambda = 550$ nm, and found that the emission was equivalent to 0.0012 watts per lumen or about 830 lumens per watt. Today we take this number to be 683 lumens per watt.

Edison's early incandescent lamp was far from ideal. On the lamp's 50th anniversary in 1929, scientists made photometric measurements on a replica of Edison's first carbonized sewing-thread lamp. The lamp gave 1.4 lumens of visible light per watt of total electrical input. By comparison, (see the graph on page 38) a modern incandescent lamp has 17.4 lumens per watt, a figure that is considered "lossy." Lumens per watt, defined as the lamp's "efficacy" (formerly "efficiency"), is a measure of the lamp's efficiency as a light source.

The incandescent lamp industry grew explosively. Edison formed a number of companies to manufacture lamps, dynamos, wire and components. The Pearl Street power station in lower Manhattan was opened in September 1882, and its

---

Edison's laboratory at Menlo Park, New Jersey. This is a reconstruction (built by Henry Ford in the 1920's) at Greenfield Village, Michigan of the laboratory the way it was at the time of Edison's work on the light bulb. It is now part of a museum. (Courtesy of General Electric Co.)

John Anderson is with the General Electric Research and Development Center in Schenectady, New York, and John Saby is with the Lighting Research and Technical Services Operation of General Electric in Nela Park, Cleveland, Ohio.

complete success proved to any lingering doubter that the incandescent lamp was ready to provide a needed service.

Edison continued to experiment with lamps, producing continually improved designs. In 1883 he discovered an effect—the "Edison effect"—that was a precursor of the thermionic tube.

Improving the filament

By 1900 Edison believed there could be no further advancement in the carbon incandescent lamp. In a way he was right. Carbon lamps were mature, but Edison did not anticipate one further development, Willis Whitney's "metallizing" of a flashed filament. This was done by heating the filament to over 3000° C in a carbon resistance furnace; the flashed appearance and temperature coefficient of resistance resembles that of a metal.

In 1897, Walter Nernst, professor of chemistry at Göttingen, devised the "Nernst glower," a variation on the incandescent lamp. The glower was a slender rod of refractory oxides, mostly zirconia (about 85 percent), with yttria and erbia (about 15 percent). Other trace oxides were also included. When indirectly preheated, the rod became an "electrolytic" conductor, typically operating at 0.25 amperes and 200 volts to give 32 candle power. The Nernst lamp was a selective radiator, which accounts in part for its relatively good efficacy. The exact mechanism of conduction was unclear until 1954 when Joseph Weininger and Paul Zemany found that oxide ions were the mobile charge carriers in the rod.[2]

As refractory metals began to appear commercially, it became clear that the days of the carbon lamp were numbered. Osmium, with a melting point of 2700° C, was developed in 1898 by Auer von Welsbach and marketed to a very limited extent, being brittle and quite scarce. Tantalum with a melting point of 2850° C, was not so brittle, more plentiful, and enjoyed considerable popularity from 1905 to 1910.

Tantalum filaments became prone to a failure that later also plagued tungsten filaments: After some hours of being heated, the metal recrystallizes, sometimes with boundaries normal to the wire axis. As these crystals offset, the area of contact between them becomes smaller, and that leads to failure of the lamp.

Tungsten, with a melting point of 3380° C, was the next metal to come, and it remains on the scene to this day. Early processes reduced and sintered the tungsten into filaments. Alexander Just and Franz Hanaman reported, in 1904, the first practical process: they heated tungsten powder in an organic binder under a moist hydrogen and nitrogen atmosphere to remove all carbon, leaving substantially pure tungsten in a matrix (see the electron micrographs on page 36).

Edison. In this photo from around 1883 he is shown with bulbs demonstrating the "Edison effect," which was later shown to be thermionic emission. His other activities kept him from following up on the discovered effect, but he did send sample tubes to other investigators. John Ambrose Fleming, in England, was the first to use the effect to make a diode in 1905. (General Electric Co.)

The efficacy of the tungsten lamp, even at its initial value of 7.8 lumens per watt, was enough to displace all previous incandescent lamps in about ten years, as shown in the graph.

However, the tungsten filament was brittle, and the prevailing wisdom was that it could not be made ductile. William Coolidge, after many years of work, solved the problem by hot working below the annealing temperature. His technique draws out the tungsten crystals into fibers along the wire. But when the filament burns above the annealing temperature in a lamp, it recrystallizes and tends to offset, as we mentioned above for tantalum, as shown in the electron micrographs on page 37. To prevent this problem, Coolidge at first added thoria to inhibit grain growth, so preventing the offset but created another problem: thoriated filaments sag during burning. Later, potassium was added to promote fibrous structure, minimizing transverse grain boundaries, and, consequently, the sagging.

Until 1910, all lamps with oxidizable filaments were operated in vacuum. They all suffered from the evaporation of the filament material, which darkened the glass envelope and progressively reduced the light output.

Perhaps the last major advancement for improvement of efficacy in the common tungsten incandescent lamp was Irving Langmuir's introduction of an inert gas, first nitrogen, and then in 1918, argon, after it became available in volume. This technique retarded evaporation. Langmuir's work in this area led to many other contributions by him—space-charge laws and plasma sheath work, to name but two.

The tungsten lamp story is by no means complete at this point. Elmer Fridrich and Emmett Wiley found, in 1958, that lamps with one or more of the halogens can achieve more lumens per watt than conventional lamps of the same life.[3] This improvement is brought about by a regenerative cycle: Tungsten on the glass (or quartz) envelope reacts with free iodine or bromine and a trace of oxygen to form volatile halides and oxyhalides. These materials are in turn decomposed near the hot tungsten. Thus, in the ideal case, tungsten is kept off the glass wall. One may then use a smaller envelope, to permit a higher pressure of rare gas, and in turn, slower evaporation of the tungsten. One can operate the filament above 3000 K, instead of the usual 2800 K. This raises the efficacy above 20 lumens per watt. Although they are not yet used in the mass home market, these lamps are finding increasing application where their

Replica of the first electric lamp. Its filament is a carbonized sewing thread. Edison himself made the replica in 1929, at a celebration of the fiftieth anniversary of his original achievement organized by Henry Ford in Dearborn, Michigan. Edison's lab was moved to a nearby site and restored for the occasion. (Photo courtesy of US National Park Service, Edison National Historic Site.)

benefits justify their higher cost; examples are film projectors and automobile headlights.

The evolving electric-discharge lamp

Even though Davy demonstrated an electrical arc in air in 1810, the first practical arcs for lighting did not appear until about 1880. For lack of any real scientific understanding about them, electrical arcs were little more than phenomena, albeit with practical application, until at least the beginning of the 20th century. The major discoveries in physics in the last years of the 19th century and up into the 1920's permitted an explanation of the many previous empirical observations both for low- and high-pressure discharges.

Faraday, about 1835, evacuated glass tubes to 3–4 torr and observed the major regions of glow in a low-pressure discharge. Heinrich Geissler, a physicist at Bonn and a skilled glass blower, made discharge tubes and invented a mercury vapor pump that generated a much better vacuum. Geissler's tubes were extensively used in research laboratories throughout Europe. At lower pressures the glowing gas seemed to disappear leaving only a fluorescing of the glass envelope. Julius Plücker (1859) deflected "rays" from the cathode with a magnet;

Johann Hittorf and and Sir William Crookes discovered, in 1869 and 1879, shadow effects due to "cathode rays," and Eugen Goldstein discovered "canal rays" behind holes in the cathode in 1886. In time scientists suspected that the rays were particles and J. J. Thomson undertook, in 1897, to measure their charge and mass. At first he could not believe his results, namely that the ratio e/m was exceedingly large, but soon the concept of an "electron" could not be avoided. The indivisibility of the atom, held generally throughout the 19th century, was shattered. Wilhelm Wien (1889) capped the century by showing that the canal rays were positively charged atoms from the gas in which the discharge took place. The concept of electron ionization of the gas, coupled with the already well-developed kinetic theory of gases, then permitted a serious study of gas-discharge phenomena.

Even though arcs burning in air gave efficacies up to 25 lumens per watt, another promising development was taking place. Geissler tubes were early recognized as good light sources—in essence they are our neon signs today—but their life was short because of gas clean-up. The first practical low-pressure discharge lamp was, in 1895, that of D. McFarlane Moore. He established discharges in air,

nitrogen, or carbon dioxide at a pressure of 0.1 torr. The discharges, at more than 10 kV between cathode and anode, took place in glass tubes up to 60 meters long. The tubes enjoyed some popularity, mostly because of the good color, but by 1912, the technique was on the way out. The Cooper–Hewitt lamp that had become known in 1901 was a more convenient and less expensive lamp with an efficacy up to about 15 lumens per watt. It was a low-pressure discharge in mercury from a liquid cathode to an iron anode. There were many lamps sold for applications where color was not important, for as Maurice Solomon in his book *Electric Lamps,* in 1912, tells us, " . . . the almost complete absence of red and yellow lines in the spectrum causes all colored objects illuminated by mercury vapor lamps to lose their natural colors, and people present the appearance of the dead."

Discharges in gases at the turn of the century were probably more important for helping to establish the evolving quantum physics than for light sources. Line spectra from discharges could now be explained as discrete energy changes in the atom according to Niels Bohr's model of 1913. In the same year, James Franck and Gustav Hertz observed step-wise retardation of electrons by gas atoms. That observation provided direct evidence that there are discrete atomic energy levels and that atoms can be raised to excited levels by collisions with electrons. Einstein's theory of spontaneous emission, absorption, and stimulated emission in 1917 gave a quantitative basis to emission intensity and density of states. Even the more subtle observations of spectroscopy were explained by the wave mechanics of the mid 1920's. A firm theoretical basis now existed for further lamp development.

As material science developed, the many discharge lamps that had been envisioned, some even before 1900, became possible. A major step, for example, was the quartz–molybdenum seal, which became available after World War II. This technique, coupled with the oxide cathode, made a high-pressure mercury lamp with efficacy above 50 lumens per watt practical for widespread use. Today a variety of discharge lamps, both high and low pressure, is used in general lighting.[4]

The low-pressure sodium lamp, one of today's discharge lamps, is useful where geometrical light control is unimportant and where monochromatic illumination is acceptable. First introduced in Europe by the Philips Company in the 1930's, it utilizes a sodium-rare-gas discharge that channels about one third of the total input energy into the yellow sodium resonance "D" lines. Physically, it consists of a long arc tube similar to that of a fluorescent lamp, but lined with sodium resistant glass.

It also shares with the fluorescent lamp

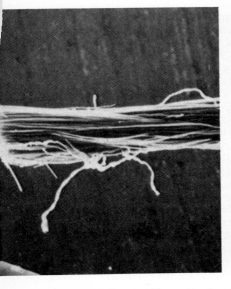
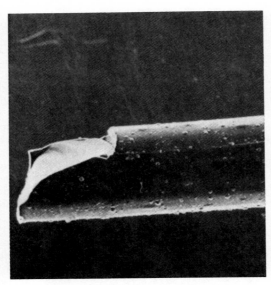
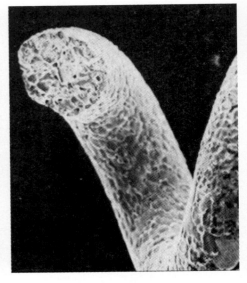

Early lamp filaments. These scanning electron micrographs show (left to right) carbonized thread from a reproduction of Edison's lamp (#8 mercerized cotton); a GEM carbon filament, made by squirting cellulose through a die to make a uniform filament that was then covered with a graphite layer; and a sintered tungsten filament from Langmuir's first successful gas-filled lamp.   (SEM photos by H. F. Webster.)

a high efficiency of generation of resonance radiation. Because the vapor pressure for optimum efficiency requires that the arc tube operate above 200°C, there are energy losses even though the lamp has an elaborate thermal insulation. The primary loss mechanism in this discharge is heat conduction to the walls of the arc tube (see the table on this page). Refinements of this heat-conserving structure are credited with most of the improvements to the present efficacy, which in the larger size lamps (the 180 watt low-pressure sodium lamp is more

than a meter long) exceeds 180 lumens per watt.

Discharge lamps require electrical circuitry called "ballasts" to maintain the required voltage and current. Because of ballast losses the overall efficiency of the lamp-ballast system is lower than that of the lamp alone. Technology exists however, to keep ballast losses below 10 percent in most cases.

Exciting phosphors to fluoresce

At 80 lumens per watt, the fluorescent lamp is one of the most efficient lamps in general use (see table). Introduced in 1938, it consists of a long, phosphor-coated glass envelope, commonly 38 mm in diameter, filled with 1–2 torr of a noble gas plus a small quantity of mercury. Applying a voltage creates an ionized conduction region between the lamp's electrodes. At normal operating temperature, slightly above room temperature, only a few millitorr of mercury are

present; yet mercury with its first resonant line at 254 nm dominates the spectrum. This ultraviolet radiation excites the phosphor coating. In turn, the phosphor creates nearly all the visible light produced by the lamp; only a few percent of the total light comes from visible lines in the mercury spectrum.

Willem Elenbaas studied the mercury–rare-gas discharge and reported his results in the early 1930's. Later, in 1950, Carl Kenty made a careful study of energy-state density in, and radiation from, the fluorescent-lamp discharge.[5] Still later, in 1956–57, John Waymouth and his colleagues showed the way to improve efficacy in lamps with higher loading (watts per unit length).[6] He proposed adding neon to the argon gas.

The ionized conduction region in the fluorescent lamp was the target of a recent study, in 1972, by P. C. Drop and Jan Polman.[7] They calculated the response of the discharge column to signals at up to megahertz frequencies and found that while the behavior in the conduction region at intermediate frequencies deviated in energy states and electron densities from dc behavior, at high frequencies it again approached that under dc fields.

Today, fluorescent lamps produce about two thirds of all the electrically generated light in the world—twice as much as all other lamps combined.

Generating high-intensity discharges

The high-pressure mercury lamp, the first member of a growing family of high-intensity discharge lamps, emerged in the 1930's.[8] It has an inner quartz arc tube surrounded by an envelope that protects its leads from oxidation. The inner tube contains sufficient mercury to produce about 1.5–2 atmospheres when fully vaporized. It also has approximately 20 torr of argon starting gas.

High-pressure mercury lamps waste about 33 percent of their energy as ultraviolet radiation, primarily at 185, 254 and 365 nm (see table).

Power Balance of Existing Lamps

Type	Power W	Efficacy L/W	Radiated Visible	Radiated Not used	Not radiated Conduction	Not radiated Ends
Incandescent, general service	100	17.5	7	80	13	
High-pressure mercury	400	52	15	33	42	10
Low-pressure sodium	180	183	36	4	50	10
Fluorescent	40	80	23	40*	17	20
Metal Halide	400	100	34	20	36	10
High-pressure sodium	400	120	30	20	40	10

*32% lost in the phosphor

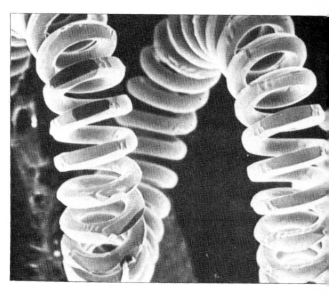

Scientists have been trying to suppress the ultraviolet radiation and improve the visible light output by adding to the inner tube elements with visible atomic spectra. Unfortunately metals, the most desirable additives, are not sufficiently volatile to attain the required concentration in the vapor. Also, many metals form mercury amalgams, which reduce the metal vapor pressure even further. In 1960 Gilbert Reiling discovered that metal halides added to mercury lamps can produce the desired concentration of metal atoms in the vapor. The metal halide lamps that emerged as a result of this discovery are more efficient than high-pressure mercury lamps (see table) but they have somewhat shorter lives. The chemically active species inside attack metal parts and even the quartz tube itself.

High-pressure sodium

In the past few years city streets have been seeing a rapid shift from high-pressure mercury to high-pressure sodium lighting. High-pressure sodium lamps radiate over much of the visible spectrum, giving a golden-white light. The discharge mechanism of high-pressure sodium, whereby resonance "D" lines are broadened and reversed by self absorption, is mostly responsible for the very broad spectrum.

The key to the achievement of this useful discharge was the development of a special arc-tube material, translucent polycrystalline alumina. This material is cheaper than single-crystal alumina, which would also be chemically suitable and translucent. Not only is polycrystalline alumina chemically resistant to sodium but it is also capable of operating when the temperature at the center of the arc tube reaches 1200° C. At this central temperature the cooler ends are hot enough to maintain adequate sodium pressure for optimum light generation.

High-pressure sodium lamps are small—a 400-W arc tube is 10 cm long and it has a diameter of 8 mm, an advantage

in some applications. One can also trade off the relatively high efficacy (see table) for improved color rendition. This tradeoff entails a higher pressure of sodium vapor over the Na–Hg amalgam and consequently a higher temperature within the arc tube and increased infrared losses.

Lamps as indicators

During an electric discharge in gas at low pressure a region near the cathode glows. This phenomenon has been known since Faraday's time. But in 1928 modern neon-glow indicator lamps became possible when Theodore Foulke combined argon and neon to form a Penning mixture. In such a mixture metastable atoms of the majority gas ionize the minority gas atoms, which leads to a lower starting voltage. This technique allows reliable starting of the glow at voltages below 120 V.

In some fluorescent materials—semiconductor phosphors—one can produce light by radiative recombination of hole–electron pairs. Although these phosphors have not been in general illumination use, they have facilitated a variety of widely used special-purpose electric light sources.

A good example is the television screen. Used today only for information display, the screen employs excitation of a phosphor layer by an electron beam. Given sufficient future improvement in phosphors and electron sources, the technique could potentially be applied to general illumination.

One can also excite electrons within luminescent materials by applying external electric fields. Discovered in 1937 by Georges Destriau, this technique made possible one highly specialized application: the thin fire-safe electroluminescent-illuminated instrument panels used some years ago on all the NASA Apollo command spacecraft and Lunar Excursion Modules. Unfortunately, due to the low efficiency of the electron-impact ex-

Coiled-coil tungsten filament. On the left is a scanning-electron micrograph of a filament from a conventional 100-W lamp, fresh from the factory. On the right is similar filament from a bulb that had burned for about 500 hours. Note the growth of crystal structures, outlined by thermal etching, and the start of a few offset fractures; some of the crystals extend over several coils. (SEM photos by H. F. Webster.)

citation mechanism, the Destriau effect has limited potential for general illumination.

A very efficient way of generating hole–electron pairs is by carrier injection across rectifying junctions. Oleg Loessev discovered light production by this phenomenon in silicon carbide as early as 1923. Efficient light generation here requires high injection ratios, a high proportion of radiative recombinations of hole–electron pairs, and rectifying structures that enable escape of light. Electroluminescent light-emitting diodes embody such structures. These diodes pervade the displays in electronic wristwatches and pocket calculators.

Though not cheap or efficient enough for general illumination, light-emitting diodes have adequate quantum efficiencies for use in indicator lights and in injection lasers in gallium arsenide and similar materials. (Quantum efficiency is the ratio of the number of photons emitted by a material to the number of electrons or ultraviolet photons incident on that material.) Another type of indicator lamp combines phosphors that convert infrared radiation into visible light with gallium-arsenside light-emitting diodes. These devices convert 0.9 micron radiation in the infrared into visible light.

Increasing the efficiency

What are the large challenges for the next century of electric light? The most important shortcoming of electric illumination is its low efficiency. Even the most efficient electric lamps require about

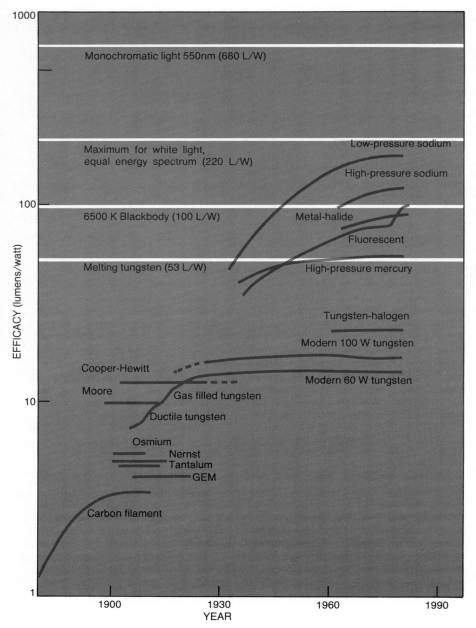

Approximate efficacies of various light sources. We plot efficacies as a function of time for different kinds of electric lamps, showing the evolution within each technology as well as progress from one technology to another. We also indicate efficacies of several standard sources.

rangement. This includes absorption of visible light, as well as the effective emissivity of the coil. Tungsten coils are difficult to design with effective emissivity much greater than 0.5. Geometrical perfection of the reflector, and reasonable accuracy of placement of the coil within the reflector are also important. The Duro Test Corporation has recently announced the development of such a selective-reflector lamp.

Another technique for improving incandescent-lamp efficacy is also available, at least in principle. One can add phosphors to the lamp to absorb radiated infrared and emit visible light. One can excite such phosphors into intermediate energy levels through successive absorption of infrared photons. When the phosphor has absorbed sufficient energy, it radiates a visible photon. Although such phosphors have been used in light-emitting diodes, their efficiency has not been high enough to warrant their use in lamps, especially when absorption of visible light is taken into account. This technique, however, warrants further exploration.

One may ask what would the characteristics of a superior lamp look like? Such a lamp should be compact; it should also be compatible with existing sockets for incandescent lamps; its efficacy should be higher than 40 lumens per watt, and its cost should be low enough to justify adoption by the public. *A priori*, gas-discharge light sources appear to have the best chance of filling these needs. The main hurdle would be the ballast; historically ballasts have added cost and complexity of installation to discharge lamps. However, technology exists today to make sufficiently small electronic ballasts that could conceivably be integrated into the lamp and permit screw-in installation. The General Electric Company recently announced development of such a lamp, the "Electronic Halarc."

Turning to the low-pressure sodium lamp, we find only modest promise for improving this lamp. Attempts to improve its color by additives have extracted energy from the D-lines, decreasing the lamp's efficacy. Improved color without significant penalty in efficacy appears unlikely. Most efficacy improvements to date may be credited to heat conservation by selective infrared reflecting coatings on the glass envelope. The coatings transmit visible light. Improvements in efficacy either by reduction of the loss of heat conduction of the arc tube walls, or by elevating the sodium vapor pressure, are expected to be incremental.

Directions for future improvement of the fluorescent lamp are suggested by the table. Twenty percent of the total input energy goes into so-called "end losses." These include dissipation in the plasma at and near the electrodes, as well as external power used to heat the electrodes. Anode losses can be almost eliminated by

three watts of electric power for every watt of visible light.

Consider the tungsten incandescent lamp, the lamp with the lowest efficiency among all widely used electric lamps (see table). About 80 percent of the tungsten incandescent lamp's energy input is radiated at non-visible wavelengths, primarily in the infrared. The nonradiated losses are primarily due to heat conduction, partly via the electrical leads, but mostly by the inert-gas filling. Even with gas-conduction losses reduced as much as possible by use of heavier and more expensive inert gases, the efficiency of incandescent lamps can be improved by only a few percent.

Thus any significant improvement in the incandescent lamp must be obtained

by reduction of the infrared losses.

One technique under study is to use infrared reflection toward the filament, which could help maintain the high filament temperature at a fraction of the electric energy input of present designs. This technique could double the efficacy. Ironically, this and similar ideas have been occupying researchers' minds during a good part of this century. Approximately 50 US patents on various means of increasing incandescent lamp efficacy by infrared reflection have been issued since 1912.

Reducing the infrared-reflection technique into a practical incandescent lamp design is quite tricky. The energy saving depends critically on the spectral characteristics of the sensitive reflection ar-

Test vehicle for lamps. This 1920 Reo touring car was one of General Electric's "laboratories on wheels" for testing headlamps as well as other lights for automobiles. (General Electric Co.)

operation at high frequency. Recent developments in electronics technology broaden the choice of operating conditions. This should facilitate greatly improved lamp–ballast systems. Reduction of cathode losses requires cathodes that can supply electron emission with less heating.

Fluorescent-lamp phosphors in common use have quantum efficiencies of about 90 percent; yet, in principle, phosphor improvement provides the best opportunity for a truly significant improvement in the fluorescent lamp. One remedy is to put less energy into the ultraviolet photons that are absorbed by the phosphor than the present 5 eV.

One candidate for lower-energy ultraviolet photons is cadmium. Bentley Barnes and Robert Springer have measured low-pressure cadmium discharges.[9] They found that the ultraviolet efficiency of cadmium is comparable to that of mercury. But cadmium has its first resonance line at 326 nm compared to 254 nm for mercury, so that cadmium has 22 percent less energy per quantum. Also, cadmium has nearly 30 percent more visible spectrum than mercury. All in all, up to 28 percent more lumens per watt might be possible with cadmium. But cadmium is toxic. The cadmium lamps require elaborate thermal insulation to permit operation at a high enough wall temperature for optimum discharge conditions. One must also develop phosphors capable of operation at 300° C, in chemical contact with cadmium vapor.

Another remedy for quantum loss lies in developing phosphors that emit two visible photons per excitation. Such phosphors would have two sequential radiative transitions after excitation—the first one to an intermediate energy state, the second one to the ground state. Mercury resonance radiation has sufficient energy to achieve two visible photons per excitation and phosphors with two-step de-excitation do exist. The known two-quantum phosphors, however, generally have very low efficiency.

Can fluorescent lamps be made smaller? Studying this question scientists have been faced with the disadvantages of end losses in short arcs. Anderson showed in 1970 that one can overcome the end-loss problem by ferrite-excited electrodeless discharges.[10] The most severe physical limit to small size is degradation of the phosphors at high intensity.

Turning to high-intensity discharge lamps, we observe that, like fluorescent lamps, they can also be electronically ballasted and they already have the necessary compact form. Because of end effects, however, small high-intensity lamps have traditionally been far less efficient. Also, they warm up slowly after starting, and they are difficult to restart after momentary power interruption. These problems, however, can be overcome by stretching today's technology.

As to metal–halide discharge lamps, researchers expect to arrive at a better understanding of the role of additives, particularly to broaden the choice of suitable additive metals. In turn, this will bring about higher efficacy and more ideal spectra. Mastery of halogen-transport chemistry will facilitate the balancing of the effect of different halides, a key to longer life and better electrical characteristics.

A logical further development of the high-pressure sodium lamp is toward increasing the efficacy by further reducing the conduction losses. Such increases, however, are expected to be small.

Physicists are challenged to generate new ideas for light production. Most of the really major past improvements came by exploiting new combinations of phenomena or technologies. For example, the fluorescent lamp combined the gas discharge with fluorescence, and the high-pressure sodium lamp requires a newly available material.

Gas discharges can be thought of as selective incandescent radiators: How selective can they be made? Several light sources now make use of spectrally selective reflective coatings. How efficient could an incandescent lamp become if its filament were thermally insulated from a reflector? Could a practical arrangement be found whereby the entire radiation outside the visible is reflected onto the filament and the visible radiation is entirely transmitted?

One hundred years ago Thomas Edison would have predicted a number of features of today's incandescent lamps but he would not have foreseen most of the wide variety of electric lamps we take for granted today. Our crystal balls tend to the cloudy beyond the point of estimating improvements of today's lamps. However, we predict with confidence that the innovations of the second century of electric light will require as much applied physics as the first century.

References

The history of electric lighting, particularly of the incandescent lamp, is covered in several books:

T. W. Chalmers, *Historic Researches: Chapters in the History of Physical and Chemical Discovery,* Scribners, New York (1952).

F. A. Lewis *et al., The Incandescent Lamp,* Shorewood, New York (1959).

W. Howell and S. Schroeder, *History of the Incandescent Lamp,* Maqua, Schenectady, (1927).

A. Bright, Jr., *The Electric Lamp Industry,* Macmillan, New York (1949).

Lectures on Illuminating Engineering, (2 vols.), delivered October 1910, The Johns Hopkins Press, Baltimore (1911).

1. H. E. Ives, Trans. Ill. Engr. Soc. **5,** 113, (1910)

2. J. L. Weininger and P. D. Zemany, Chem. Phys. **22,** 1469, (1954)

3. E. G. Fridrich and E. H. Wiley, U.S. Pat No. 2 883 571

4. F. Waymouth, *Electric Discharge Lamps,* MIT Press (1971)

5. C. Kenty, Appl. Phys. **21,** 1309, (1950)

6. J. F. Waymouth and F. Bitter, Appl. Phys. **27,** 122, (1956)

7. P. C. Drop and J. Polman, J. Phys. D **5,** 562, (1972)

8. W. Elenbaas, *The High Pressure Mercury Vapor Discharge,* North Holland, Amsterdam (1951)

9. R. H. Springer and B. T. Barnes, Appl. Phys. **39,** 3100 (1968)

10. J. M. Anderson, US Patents No. 3 521 120 and 4 017 764.

□

Physics looks at solar energy

Large-scale solar-powered "farms," covering thousands of square miles, could produce electricity cleanly and economically using technology we already have.

Aden Baker Meinel and Marjorie Pettit Meinel

PHYSICS TODAY / FEBRUARY 1972

The idea of using the sun as a source of energy has had a long history, but so far it has been a history of bright hope and dismal failure. In the middle 1950's newspaper headlines were full of glowing predictions of what solar energy could do for mankind; the first International Conference on Applied Solar Energy had been held, and solar energy seemed ready to take its place, along with peaceful uses of atomic energy and with interplanetary exploration, on Vannevar Bush's "endless frontier of science." And now in the 1970's nuclear power reactors and spaceflight are realities, yet solar energy, as recently as a year ago, was dismissed by a National Academy of Science–National Research Council committee as of no importance in our future—despite the admitted "energy crisis" looming ahead. Whatever happened to the grand predictions? Our search into the history of solar energy started with this question, because we were curious to know if 1970 technology might yield a different result.

We are all aware of the problems that face the successful transition from the present generation of nuclear power plants to the breeder reactor and the growing problem of disposal of the high-level radioactive wastes. We also read that, in spite of the advance made possible by laser ignition and by the Tokomak configuration, fusion power is still in the uncertain future. Our purpose here is to present our recent studies that may offer a new option—an option based on thermal conversion of solar energy—based upon current technology.

In reading about the history of solar energy it appeared to us that much of the work of the last two decades, with the exception of space applications, had approached solar energy from the wrong basic philosophical perspective. It has always been thought of as something for under-developed countries, those too poor to avail themselves of the world's other energy resources. It was conceived as something cheap and simple—simple enough to be operated by unskilled persons.

Another emphasis has been on solar-energy gadgets built on a scale suitable for the individual user—the dream of each household having a "rooftop" solar collector has been a recurring theme in the literature of the last 20 years or more. Even if such devices were economical, their big drawback would be the maintenance problem. With cen-

tral-station power generation, nearly all maintenance is localized to the central station itself. Individual generators, on the other hand, would need individual servicing, and one would be faced with the need of finding a competent repair man.

The use of solar power by under-developed countries and the "rooftop" units have both been thwarted by the hard facts of economics, even though enterprising inventors can show us ingeneous ways to heat water and houses and operate cookers. We feel that solar energy must be viewed in a new perspective—as something for a technologically advanced nation, utilizing all of the arts of mass manufacturing, and operation on a large scale. Hundreds or thousands of megawatts per power plant begins to make sense when we want to obtain power at today's low electrical energy rates.

Figure 1 shows what such a power plant might look like, according to our proposals discussed in detail below.

Conversion methods

Solar energy can be converted to electricity in several ways. First is direct conversion, wherein a voltage or current is generated by the absorption of photons in a doped semiconductor such as silicon or cadmium sulfide. The second is by thermal conversion,

Aden Meinel is director of the Optical Sciences Center, University of Arizona. He collaborated with his wife, Marjorie Meinel, in the research described here.

Model of a solar-power farm on typical desert terrain. Note that the rows of collectors noticeably darken the desert. This 250-MW power station would use turbine waste heat to desalinate seawater.
Figure 1

1. Sea water 2. Fresh water 3. Desalting plant
4. Cooling tower 5. Thermal storage A
6. Thermal storage B 7. Oil reserve 8. Maintenance
9. 250 MW(E) Turbine 10. Boiler

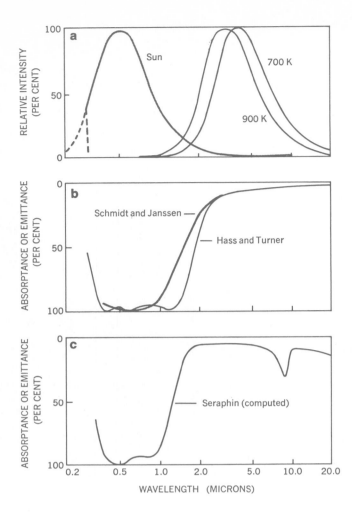

Frequency response of selective surfaces.
Approximately 90% of the solar spectrum
(a) is at wavelengths shorter than 1.3
microns, and the escaping infrared radiation,
even at 900 K, overlaps it very little.   The
selective surface must be black for
wavelengths shorter than 1.3 microns and
mirrorlike for the longer wavelengths.
Interference stacks have the characteristics
shown in b, and the bulk-absorber stack of
figure 4 is expected to have the spectral
performance shown in c.
Figure 2

wherein heat is converted into electrical
energy by way of thermionic, thermo-
electric or magnetohydrodynamic devices, or by ordinary steam power turbines.   The third is through biological
processes by the growing of crops and
subsequent power generation by the
thermal system.

Direct conversion via silicon solar
cells is a well developed technology with
wide application in spacecraft systems.
The cost of silicon cells is high—several
hundred dollars per *watt*—and it is this
high cost that led to solar energy being
dismissed as a resource for the future.
Some new efforts are being made to
lower costs and solve degradation under
terrestrial weather conditions, but the
prospects are clouded for the cost reduction of a thousand-fold that would be
required to get the cost close to the few
hundred dollars per *kilowatt* typical
of bulk electrical power systems.

When we examine biological conversion, we find that each step in the
process from the growing of the crop,
such as algae or good Iowa corn, its
harvest and conversion to methane,
and its burning to provide steam power,
is technologically feasible.   A problem
is apparent in a low net quantum efficiency, 1 or 2 percent, which requires
that extensive land areas be devoted

to the crop.   The basic problem however, is economics: It appears impossible to approach a cost of the order
of a dollar per million BTU's.   Conventional fuels cost about 15 cents for
coal and 35 cents for natural gas.   A
dollar per million BTU's of a crop means
a cost of about ½ cent per pound of the
harvested crop or 10 dollars a ton, far
below the best cost figures for algae
obtained to date.

The history of thermal conversion
shows attempts to use steam engines or
turbines driven by various working
fluids but with very low efficiencies,
of the order of 1 to 2 percent of the
incident energy.   The reason for this
low efficiency is clearly assignable to
the low operating temperatures of the
devices.   The obvious answer is to increase the operating temperature.   In
fact, why not operate them at the temperature and pressure of modern steam
power turbines?   They convert fossil
fuels into electrical power at efficiencies
of approximately 40%.   Here we will
adopt a provisional goal of a system
operating at 500 deg C.

A search for the necessary technology
for achieving high temperatures by an
incident radiation field must start with
the basic physics.   Figure 2a shows the
situation both for the incident solar

flux and for the escaping infrared radiation, which is the dominant cooling
mechanism when the absorbing surface
is located in a vacuum enclosure.
Approximately 90% of the solar spectrum is at wavelengths shorter than 1.3
microns.   We want a surface that is
black over this wavelength interval.
Now consider the escaping radiation.
Even at 900 K there is little overlap of
the heat spectrum with the solar spectrum.   Therefore it is possible to imagine a surface that is black for wavelengths shorter than 1.3 microns and
low   emitting—mirrorlike—at   wavelengths longer than 1.3 microns.   Such
a surface will get hot because it inhibits
the escape of the heat photons.   The
figure-of-merit for this selective behavior is the ratio of the absorptivity in
the visible, a, to the emissivity in the
infrared, e.

Selective surfaces

The peculiar spectral characteristic
of the collector surface can be generated
in a variety of ways.   We have explored
two avenues, both based on technologies
that were still in their infancy a decade
ago.   The first is based on the principle
of the interference filter, in particular,
interference effects in metal–dielectric
multilayer stacks.   If sufficient complexity is permitted, such stacks can
generate almost any spectral profile.
Solar application, however, calls for
longtime stability under high-temperature operation.   Degradation of the
delicate phase balance between the
various   component   layers   through
evaporation   or   interfacial   diffusion
must be kept to a minimum.   Figure 3
shows a promising interference stack
made by Roger Schmidt and John E.

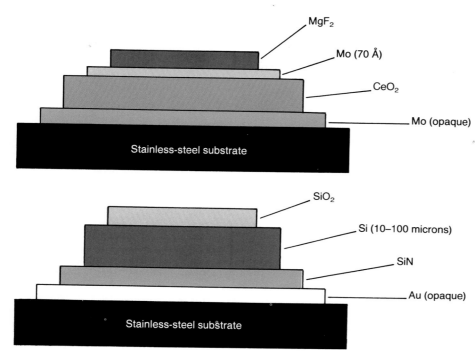

Interference stack made by Roger Schmidt and John E. Janssen, which has the optical characteristics shown in figure 2b.
Figure 3

Bulk-absorber stack proposed by Bernhard Seraphin, which is expected to have the characteristics shown in figure 2c.
Figure 4

Janssen at the Honeywell Laboratories. Its optical characteristic is shown in figure 2b. The classic stacks by George Hass and A. Francis Turner, although superior in their room-temperature profile, do not meet the requirements of high-temperature operation.

An alternate approach suggested by Bernhard O. Seraphin at the Optical Sciences Center, University of Arizona employs the chemical vapor-deposition methods of modern semiconductor device technology for the fabrication of the solar collector surfaces. The double function of the surface—high absorption over the solar emission band and low emissivity over the black-body emission range—is divided between two different components. The high infrared reflectance of a nobel-metal base layer suppresses the emittance. A thin semiconductor film deposited on top of the reflector catches the solar radiation through its intrinsic absorption. The steep absorption edge typical for semiconductors renders the absorber transparent in the infrared, so that the reflector "looks through." A silicon layer between 10 and 100 microns thick is a suitable candidate for the absorber. The absorption edge near 1.4-micron wavelength separates the spectral bands of figure 2a fairly well, and its technology is sufficiently established. The cross section of such an absorber stack is shown in figure 4. Note that the performance of the system relies on the intrinsic properties of the component materials with no phase-match between interfaces required. The reduced sensitivity to degradation, together with convenience and economy of the manufacture, recommend this approach. A projection of the expected

spectral performance, calculated on the basis of available data for the optical constants of the component materials, is shown in figure 2c.

The basic uncertainty regarding these selective surfaces is their lifetime at high temperatures; this must become known for economic reasons. Some interesting physics may lie along the path to this answer.

Complete systems

Today we can get selective films with an a/e of 10 at 500 deg C, and we have reasonable expectations of reaching higher values. The resulting temperatures are shown in figure 5, where X is the optical concentration of sunlight on the absorbing surface. On looking into the system engineering we find that selective coatings alone cannot reach the required temperature, since it must be higher than a stagnation temperature of 500 deg C if we are to extract energy from the surface. Setting a reasonable goal of being able to extract approximately 90% of the energy at 500 deg C we need a surface with $X\,a/e = 100$.

There are various ways to make a unit containing a selective surface, but we prefer the one shown in figure 6 as an engineering model. The flux concentration is provided by either a cylindrical Fresnel lens (or the equivalent parabolic reflector) injecting the energy through a window slot into an evacuated pipe that has a highly reflective inside surface. The selective surface is placed on the exterior of a steel pipe suspended in the vacuum enclosure, and the heat is extracted via a thermal fluid flowing inside the pipe. Engineering considerations point to liquid sodium as the fluid, but some organic fluids and salt eutec-

tics could also be considered. A basic design that meets the requirements of $X\,a/e = 100$ is a surface that has $a/e = 10$ and $X = 10$, where X includes both the effect of flux concentration of the lens or mirror and the reduction in the effective emissivity due to the reflective cavity. These numbers represent modest values obtainable with current technology.

To understand how our proposed system extracts energy let us look at the operating parameters along a long collector element of the type shown in figure 6. The liquid sodium enters cool and increases in temperature until it exits at 500 deg C. The percentage of energy that is extracted as a function of temperature is shown in figure 5. The average energy extracted is therefore larger than that at the system operating temperature. A better way of estimating the effectiveness of the collector might be to multiply P by the Carnot potential at each temperature, since one is interested in the ability to extract power from the heat collected.

One must take a basic capability represented by the engineering model and make a meaningful system. Figure 7 shows a system consisting of three subsystems. The top portion is the energy-gathering subsystem, connected to the adjoining subsystem by means of a liquid-metal loop. This intermediate subsystem brings us face-to-face with the fundamental problem of solar energy; people like to have power 24 hours a day, and the sun only shines part of the day—and some days are cloudy.

Once energy has been converted into electricity it becomes awkward to store. This is a basic problem of direct con-

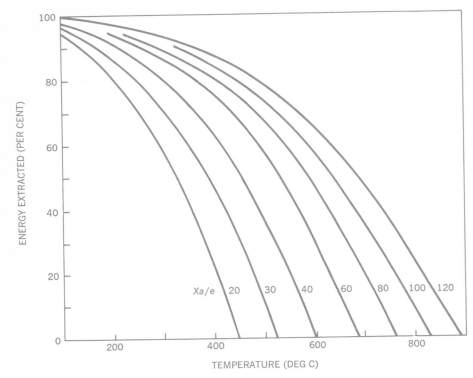

Energy that can be extracted from a
selective surface, expressed as a
percentage varying with surface
temperature and surface quality Xa/e. X
is the optical concentration factor, a the
absorptivity in the visible and e the
emissivity in the infrared.
Figure 5

version devices such as the silicon solar
cell. One possible answer is storage
batteries. In fact, in reading about
solar-power inventions at the turn of
the century, we often find that the
inventor dismissed the problem because
"cheap long-lived batteries are just
around the corner." We are still wait-
ing to turn that corner. Storage batteries are feasible only on spacecraft
and in other special situations where
cost is secondary and the required lifetime short.

Electrical utilities store some electrical energy for "peak-shaving" by
hydrostorage, where water is pumped
into an elevated storage dam during
hours of excess power. This energy is
collected later by letting the water flow
down through hydroelectric generators.
The problem for solar energy is one of
magnitude, however. To store overnight energy for the US would require
Lake Mead (Hoover Dam) to be filled
and emptied every day. Also, people
do not appear to like hydrostorage dams
any more than they like a nearby power
plant, as evidenced by the opposition
to the "Storm King" project in New
York.

Physicists store energy from their
synchrotron magnets between pulses
in mechanical form, using rotating flywheels. Imagine an iron flywheel 30
meters in diameter and 310 meters long;
if they could be kept together at 6000
rpm, one would need 100 such flywheels
to store enough energy for overnight
for the entire US!

Conversion of energy into a chemical
fuel such as hydrogen is also possible,
but storage remains a major problem

except in liquid form. The recent paper
by Lawrence W. Jones in *Science*[1] points
to some attractive consequences of using
liquid hydrogen as a fuel in the future.
Reconversion of a chemical fuel by
burning is not particularly attractive,
because it entails a second Carnot loss
and reduces the total system efficiency
by 60%. If cheap and long-lived fuel
cells were available, this problem would
be greatly reduced, but such devices
are also still "around the corner."

The solution in the system that we
have proposed is to store the energy
needed for overnight and cloudy days
in the form of thermal energy, prior to
conversion. Figure 7 shows a thermal-
storage subsystem with a salt eutectic
wherein heat is stored as heat of fusion
of the eutectic as well as bulk heat
capacity of the material. Other material could be used as the thermal-
storage medium, even liquid sodium
itself, thus avoiding one heat-transfer
stage. One needs a large mass of
eutectic, about 320 000 metric tons per
day of reserve energy for a 1000 MW
power plant, but costs are low enough
per kilowatt-hour to be reasonable;
in fact, they are lower than hydrostorage
of energy, as with the Luddington project of Consumers Power and Detroit
Edison.

The lower part of figure 7 shows the
"using" subsystem, in this case a standard high-pressure steam-turbine system. The turbine draws energy from
thermal storage at the rate demanded
by the utility, thus completely decoupling the rate of solar input from the
use rate.

This diagram is highly simplified,

and when engineers translate it into a
practical system it can get quite a bit
more complicated. For example, one
does not simply dump heat into a tank
and then attempt to remove it. Also,
in actual operation various modes must
be anticipated. The system can be
made to collect energy on a bright
cloudy day if a portion of the collector
field is made with $X \leq 2$, but a subsequent clear day is needed to upgrade
the stored, partially heated, eutectic.
One also wants to minimize energy
losses at night and the consequent
thermal inertia in the morning. This
can be done by purging the collectors of
their hot sodium and storing it overnight. The empty system heats rapidly
in the morning sun and the sodium is
pumped back into the collectors.

Land use

A solar-power farm, according to our
concept, is shown in figure 1. This
illustration shows rather dramatically
that solar energy collection requires
large areas of land, even at 30% conversion efficiency. To produce 1 000 000
MW (electrical) of power, a substantial
fraction of our probable need in the
year 2000 (if one ignores the "exponential idiocy" currently used to make
long-range predictions and instead
substitutes a more rational *per capita*
consumption prediction) our concept
will require the area equivalent to a
square 75 miles on a side, a bit over
5000 square miles of land, of which only
3000 square miles is actually covered
with collectors.

In discussing these matters some
time ago before a student gathering we
got an immediate response from some
of these students who thought that
5000 square miles was an intolerably
large use of land. We realize that environment has become a special area of
concern to many students, and sometimes perspective gets lost in enthusiasm. Our reply has been two-fold.
First, we point out that the US has
a long-established "energy project"
that uses not the 5000 square miles that
we propose but 500 000 square miles—
and it only produces 1% of our energy
needs. We are referring, of course, to
farms and farming. Actually solar
farms need less than one per cent as
much land as is used by agriculture to
produce a major fraction of the remaining energy needs.

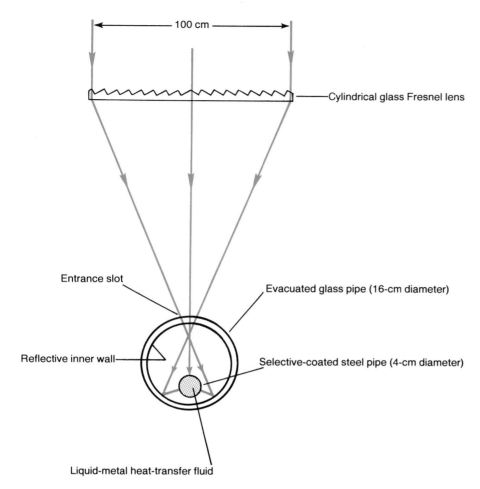

Linear energy-collecting element, shown in schematic cross section. A "solar module" would consist of six to eight of these elements mounted on a single supporting structure.
Figure 6

thermal balance that painting the ironwork of the collectors black or white makes a difference. Solar power is the only source of power conversion where careful design makes an exact local thermal balance possible. Even the use of geothermal power changes the local heat balance, not to mention the highly mineralized effluent that must be safely disposed of. Collection of solar energy does, however, increase the net thermal balance of the earth. The only way to balance this term is to "paint" other areas near the power-consuming areas white, to offset the blackness of the solar collectors. Then we could truly have "alabaster cities gleam . . ."!

The most suitable locations for solar-energy farms like these is in the great western deserts—particularly the area around the conjunction of California, Arizona and Nevada, and also the area along the Rio Grande valley in Texas and New Mexico. These regions are all very far from the major energy-consuming areas of the mid-west and the eastern seaboard; clearly, new technology will be required to develop efficient power-transmission lines for such great distances as these. Cryogenic alternating-current aluminum lines and superconducting direct-current lines are two possibilities already proposed. For the immediate future, transmission will not limit development of solar energy because the power needs of the southwestern and Pacific states would absorb any new energy they could produce.

Economics

Early in our studies it became clear that the central problem associated with solar energy would not be technical feasibility. We have identified the relevant technology as being available today. We are lacking only the verification of this technology and the construction of demonstration units. The central problem is economics. This hard fact will undoubtedly eliminate most of the contenders now being proposed as the new look at solar energy proceeds.

Let us assume that one can convert solar energy at 30% efficiency, a possible goal via thermal conversion following our concept. Assume also: The power must cost no more than 5.3 mills/ KWh (energy cost), which is what the 1971 fuel cost is for generation using

Farms appear to be generally acceptable as one price we must pay for civilization. With power plants it seems to be different. In our talks to various groups (many with environmental interests), we note that there is a definite negative response to the words "power plant." We find no one really upset with "farms," so we take the visual appearance shown in figure 1 and call our concept "solar-power farms." This term is in fact correct, because the solar-power farm is producing a crop—an energy crop—out of incoming sunshine.

The second point is that we must establish a relative scale for land usage. An important fact was pointed out to us by Arizona Public Service Co, the operators of the negatively famous "Four Corners" power plant, the archetype of the strip-mined coal power plant. They noted that their engineering staff had reported they already have under lease more land for strip mining around that plant than would be required for its operation by solar energy—and with solar energy the Navajo's "sheep may safely graze" the area used. As a matter of fact, almost 2 000 000 acres of land are currently leased by the Department of the Interior and the Bureau of Indian Affairs for strip mining of coal for power plants and coal gasification, about the amount

that would be needed for solar generation of the 1 000 000 MW (electrical) needed by the US in the year 2000.

If this is so, why not utilize solar power? The problem is that power is needed now and the strip-mined coal plants are required to meet that need. Even an intensive program, such as the "Apollo" program, could scarcely make solar power a significant contributor a decade from today. Our need for power dwarfs even the space program; however, it is not visible because private industry takes care of most of the task of building new power-generating facilities. In the past year approximately $17 billion will have been committed for new power plants. In that context even an average of a billion dollars a year for the next decade would be reasonable development costs, and such an effort could make solar power a reality because no technological or science breakthroughs appear to be needed.

Note also a second aspect of figure 1. People ask us: "Won't solar collectors alter the climate of the desert?" The solar collectors do noticeably darken the average color of the desert. This is good: Otherwise the extraction of energy in the form of electrical power and the delivery of it elsewhere would cool the desert. One finds that the solar-power farm is so close to an exact local

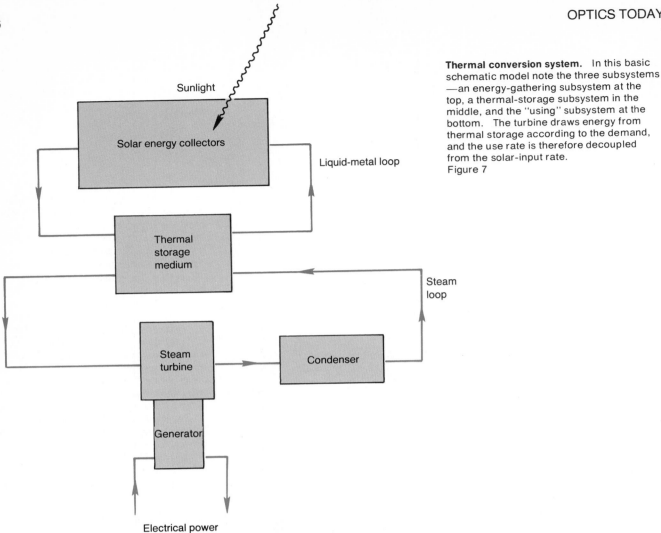

Sunlight

Solar energy collectors

Liquid-metal loop

Thermal storage medium

Steam loop

Steam turbine

Condenser

Generator

Electrical power

Thermal conversion system. In this basic schematic model note the three subsystems —an energy-gathering subsystem at the top, a thermal-storage subsystem in the middle, and the "using" subsystem at the bottom. The turbine draws energy from thermal storage according to the demand, and the use rate is therefore decoupled from the solar-input rate.
Figure 7

natural gas; the interest rate is 10%; the capital amortization is done in 15 years; the site has 330 clear days a year (for example Yuma, Arizona), and the lifetime of the plant is 40 years. One then has a budget of $60 per square meter for the entire collection and energy-storage system. This cost represents a tight constraint but we feel it can probably be met.

The waste heat from the turbines of our solar energy farms, which must be returned to the environment to preserve the thermal balance, can constitute a new energy source if it is properly handled. The critical need of the Southwest at present is not power—it is water. Why not use the waste heat from solar energy farms to desalinate seawater? If one were able to use the waste

heat from 1 000 000 MW (electrical) of farms one could produce 50 billion gallons of fresh water per day, enough for 120 million people. There are, however, a few problems in finding that much saline water and in safely disposing of the resulting brine.

We feel that the quest for solar power can represent a great domestic research goal for the next decade. There are many questions to be explored, and some interesting physics must be done. For example, understanding and predicting the diffusion phenomena that will be the determining factors in the lifetimes of the thin films is essential in accelerated testing of these thin films. There will be much engineering to be done, such as determing the best way to inject and extract heat from the thermal

storage media. We hope that public awareness of the importance and potential reality of solar energy will soon make it possible to begin this effort with the necessary resources. We are convinced that some day, perhaps within the next century, mankind will view the deserts of the earth as one of earth's greatest resources.

* * *

This article is adapted from talks presented at the 1971 Fall meeting of the American Physical Society's Division of Nuclear Physics, in Tucson, Arizona, and at the January 1972 joint APS–AAPT meeting in San Francisco.

Reference

1. L. W. Jones, Science **174,** 367 (1971). □

Solar photovoltaic energy

The study conducted for
The American Physical Society
investigated general systems
questions, solar cell
technologies and directions
for future research.

PHYSICS TODAY / SEPTEMBER 1979

Henry Ehrenreich and
John H. Martin

A large concentrator array. The photograph above shows the space between the Fresnel lenses on the right and the solar cells on the left. (Sandia Laboratories)   Figure 1

Although the Sun is frequently labeled as an "alternative" energy source, it has in fact produced almost all of man's energy throughout history. The world's energy economy now runs primarily on fossil fuels, a form of "solar capital" saved over geological time scales. Because this capital is apparently rather modest in amount and is not being renewed by nature at a rate comparable to our demands, we may eventually exhaust it. We will therefore be forced to turn either to the use of larger non-solar capital stocks such as primordial methane (if it exists), uranium-238 and deuterium, or to the use of regular "solar income" derived from the Sun's daily radiation.

The solar income is about 200 W/m^2, averaged over time, on a ground-level surface tilted to the latitude, and peaks at about 1000 W/m^2 near noon on clear days. It can be used in two ways: directly (for example, heating, photovoltaic conversion

and photochemical conversion) or indirectly (for example, wind, hydroelectricity, ocean thermal gradients and biomass). These techniques have a number of advantages:

▶ They are based on an inexhaustible source of energy

▶ Sunlight and its effects, such as wind and rain, are free and ubiquitous. (Although the distribution of direct and indirect solar resources is not uniform on the Earth's surface, it is far more even than the distribution of mineral resources such as coal, oil and uranium.)

▶ The solar income is large. The US consumes about 80 Quads of primary energy annually (1 Quad = 10^{15} Btu, 1 Btu = 1054 J). The insolation (the energy carried by the light flux from the Sun) on the US is about 5×10^4 Q per year. Accordingly, trapping and converting incoming solar radiation with about 10% end-use efficiency on about one percent of US land would satisfy our energy demands.

▶ Solar-income technologies are flexible and varied. Energy end-use can be matched to a supply of equal quality. For

example, a house can be heated by a low-temperature solar collector rather than electricity. The effect of this is an increased national "second-law efficiency",[1] in which low entropic energy forms, for example those resulting from photovoltaic and photochemical conversion, would be used only for appropriate tasks.

▶ Most of these technologies have no major *inherent* ecological drawbacks. They do have environmental effects, as do all energy technologies, and some of the consequences may be quite serious—firewood in the Third World is a particular problem at present—but these effects can be controlled by wise choice and application of the appropriate form of solar energy.

Some solar-income energy-technologies, such as hydroelectricity, are already in wide use. Others, such as wind power, were once important but have been eclipsed, at least temporarily, in the era of cheap fossil fuel. There are also new technologies, such as photovoltaics (figure 1) and ocean thermal energy conversion, which are now being developed in the laboratory or being deployed for special

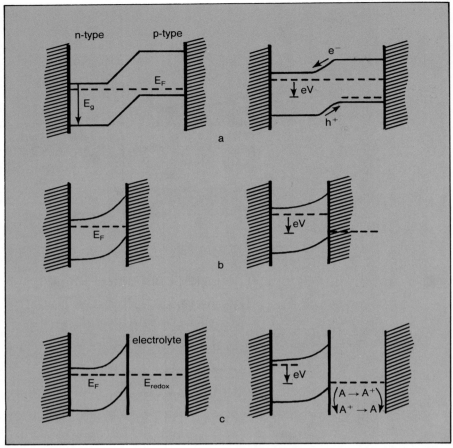

Types of solar cell. We show schematic plots of the energy of the bottom of the conduction band and the top of the valence band across the junction in (**a**) a p–n junction cell, (**b**) a Schottky-barrier cell and (**c**) a semiconductor–liquid junction cell. Plots on the left are for the cells in the dark; on the right we show the situation for illuminated cells. Shading indicates metal electrodes, and the Fermi levels are labelled E_F. Illuminating the cell with sufficiently energetic photons ($h\nu > E_g$) produces a photovoltage V when the cell is in an open circuit.   Figure 2

purposes. These last are so costly that they do not yet represent a significant entry on the US or world energy balance sheet.

The APS Study

The US, through the Department of Energy and its antecedents, has recently increased sharply its efforts to develop and commercialize new energy technologies, including the solar-income technologies, because of the clear necessity to move away from our heavy dependence on oil. The funding level for solar energy in the Department of Energy, for example, has been increased about 25% in the fiscal year 1980 budget and is more than ten times the total government budget for solar energy in 1974. It is gratifying to note that The American Physical Society Study on Solar Photovoltaic Energy Conversion is contributing to the climate

Henry Ehrenreich is chairman of The American Physical Society's Study Group on Solar Photovoltaic Energy Conversion, and John H. Martin is its executive and technical assistant. Both are in the division of applied sciences at Harvard University.

that is shaping such changes in our energy policy. By and large these developments are proceeding in technically sound and carefully considered steps despite the politically based urgings of polemicists.

The American Physical Society's study on photovoltaics was begun in 1977. The White House Office of Science and Technology Policy asked Herman Feshbach, chairman of the MIT physics department and at that time chairman of the Panel on Public Affairs of The American Physical Society, to form a group to study the promises and problems of solar photovoltaic energy conversion as a significant source for electrical energy generation in the United States and, in particular, to delineate an optimal program of research and development. To obtain conclusions that would be substantially free of preconceived opinions, the study group was to be chosen primarily from scientists who possessed appropriate disciplinary backgrounds but were not involved in photovoltaics in a major way. This condition posed no particular problems because the semiconductor field is broad and interdisciplinary.

After he was selected to be Chairman of

the panel, Henry Ehrenreich and Feshbach began to select the Study Group and the equally important Review Committee that would assess the progress of the study and review its conclusions. Our ability to obtain distinguished scientists who were willing to devote substantial time and effort to this task was principally due to its importance in national energy problems. The Department of Energy's program for photovoltaic research, development, demonstration, and commercialization was entering a period involving major decisions regarding its future directions and budget. Everyone recognized that many factors would combine to determine the ultimate direction and funding level of the program, but it was also apparent that as a group we might have a substantial opportunity to influence the development of photovoltaic technology constructively.

The first phase of the study lasted somewhat over one year, from November 1977 to January 1979. During the months preceding the summer of 1978, the Study Group met at length with more than forty experts in photovoltaic science, technology and manufacturing methods to inform itself about key issues. The group also talked extensively with economists concerned with the evaluation of energy technologies. Finally it had the opportunity of meeting with various members of the Department of Energy, the Solar Energy Research Institute, and the Office of Science and Technology Policy, who posed questions, supplied perspective and background, and freely shared their own expertise.

These discussions, backed by literature researches, particularly of directly relevant previous studies, formed the background for a month's session during the summer of 1978 at which the issues were discussed and preliminary position papers were drafted.

The remainder of the fall was devoted to refining the drafts and assembling them into a document, the *Principal Conclusions of The American Physical Society Study Group on Solar Photovoltaic Energy Conversion*. After extensive review by the Review Committee, the Panel on Public Affairs and the APS Executive Council, this report was published in February 1979. Another report emphasizing the scientific and technological opportunities for research and development is in preparation. Its intent is to stimulate greater interest in photovoltaic science on the part of the scientific community. This article will be published in due course in *Reviews of Modern Physics*.

The study was restricted to the problem of terrestrial photovoltaic energy conversion. Because the study's charter was limited to photovoltaics and because it became apparent that just doing justice to the subject of terrestrial photovoltaics would require all the time available to the group, the panel did not address related options such as photothermal generation, wind power, biomass, oceanic thermal energy conversion and the space power satellite. While there are some allusions to funding formats, particularly for long-range research programs with no prospective near-term pay-off, the Study Group felt it inappropriate to comment in any way on the management of DOE or to make specific recommendations concerning funding levels.

Photovoltaic basics

A photovoltaic device absorbs photons by the production of electronic excitations that result in electron–hole pairs. An electric field gradient is built into the device to separate these pairs. The separated carriers can then be used to produce a current through an external circuit because part of the energy absorbed remains as potential energy of the separated carriers. There are several ways to produce this field, among them:

▶ p–n junctions
▶ Schottky barriers
▶ semiconductor–liquid-electrolyte interfaces

These mechanisms are shown schematically in figure 2.

In a p–n junction (figure 2a) a slab of semiconductor is doped to make one side into p-type material and the other into n-type material. The Fermi levels on the two sides must be equal, and this requires that electrons in the conduction band on the p side have a higher potential energy than on the n side. The resultant electric field is located in the junction, or "depletion," region. The thickness of this region depends on the doping. In solar cells it is usually a few microns.

A Schottky barrier (figure 2b) can arise from the transfer of electrons from a n-type semiconductor (or holes from a p-type) to a metal layer deposited on it; the effect is mostly seen in junctions between n-type semiconductors and high-work-function metals. When a metal–semiconductor sandwich containing such a barrier is illuminated, electrons flow into the semiconductor and holes into the metal, producing a photovoltage—or a current in an external circuit.

The same situation prevails in semiconductor–liquid-electrolyte interfaces (figure 2c), in which electrolytes play very much the same role as metals do in Schottky barriers. The holes and electrons transferred into the electrolyte produce chemical reactions, and for appropriate combinations of semiconductors and electrolytes, these reactions can be used to produce fuels directly. This feature represents one of the most attractive incentives for the further development of this type of system, because liquid fuels are difficult to replace in many uses, such as transportation, and relatively difficult to obtain from sources other than oil.

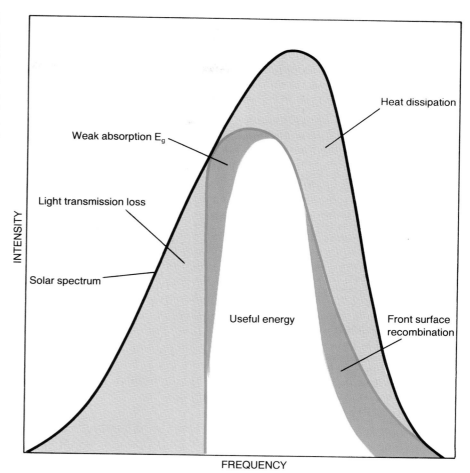

Solar-cell collection efficiency. The outer curve represents the solar power as a function of frequency (or, equivalently, the number of photons as a function of their energy). The light-colored region represents inherent losses in any solar cell based on a semiconductor with bandgap energy E_g, so that the colored curve marks the performance of an ideal single-junction cell. The darker colored region represents two further loss mechanisms present in any real cell. The white area in the center is the remaining useful energy available from the cell. **Figure 3**

The theoretical solar efficiency of a single junction photocell is significantly less than 100%. (See figure 3.) Much of the solar spectrum cannot be used by any given cell: Photons whose energy is less than the band gap of the semiconductor are not absorbed, and while photons whose energy is considerably higher than the band gap are absorbed efficiently in the junction region, a significant portion of the absorbed energy is dissipated by conversion into phonons or heat as the electron and hole dribble respectively to the conduction and valence band edges. In practical devices, moreover, light just above the absorption edge is absorbed only weakly while photons well above the band gap are absorbed so near the surface that the electron–hole pairs often recombine. These opposing effects result in an optimum band gap (1.2–1.4 eV at room temperature) for which a semiconductor is best matched to the solar spectrum. In silicon (band gap of 1.1 eV) the ideal efficiency is 29%. In gallium arsenide (band gap about 1.4 eV) the ideal efficiency is about 36%.

Some of the effects that limit the efficiency of a single-junction cell can be minimized by using a more complex cell design. For example, multijunction cells can be made of layers of materials with different band gaps connected in series, with the material of the largest band gap facing the incident light. Such cells can in principle make more efficient use of the solar spectrum—the theoretical efficiencies can be greater than 50% in cells with three or more layers. Cells of this type are just beginning to be developed in the laboratory. It is too soon to tell whether stable cells of high efficiency can be made this way. There are however a variety of designs of this and other types that hold considerable promise for improvements in cell efficiency.

Currently cell performance is limited not by the theoretical efficiency but rather by cell imperfections such as those associated with the metal grid (figure 4) and surfaces that produce recombination of the light-produced electron–hole pairs. These difficulties have been most nearly surmounted for silicon and gallium arsenide. Certain cell designs eliminate the front surface gridding entirely, but at some increase in complexity. Recombination at the front and back surfaces is often reduced by building into the cell electric fields to reflect the carriers—the use of GaAlAs layers on GaAs cell surfaces is an example.

Reflection of light by the solar cell can be minimized, but not eliminated entirely, through the use of antireflection coatings and textured surfaces. Texturing the surface can also help to increase efficiency by causing light to traverse the cell obliquely, so the absorption path length for a cell of given thickness increases.

Clearly, GaAlAs–GaAs and other more sophisticated cell designs that require elaborate fabrication procedures or rare materials are likely to be expensive (for that reason they are called "jewel cells") and as such they are intended primarily for use in concentrator systems, such as those shown in figures 1 and 5, where the premium on efficiency is particularly high and cell cost is effectively reduced by the concentration ratio.

Rationale for the Study

The Study Group delineated the question of the future value of photovoltaics in the following way: What are the technical problems, and do their solutions lie within the capabilities of present technology? Given this information, what are the nature, magnitude, and proper priority of the technical issues that require attention?

The sequence of questions can be put more specifically as follows: First, what price and performance goals need to be achieved by a photovoltaic technology that can contribute in a major way to the country's future energy mix? Second, what is achievable on the basis of optimistic but realistic extensions of current technological practice? Third, are these technological improvements without substantial modification likely to yield an economically competitive technology within a specified time? Finally, what steps are necessary to achieve a photovoltaic technology producing significant amounts of power in a way that is both time and cost effective?

Cost goals

The description of the setting implied by the answer to the first question requires some form of economic assessment based on a definition of an economically competitive technology. It quickly became apparent to the Study Group that the technical conclusions would be relatively insensitive to the details of the economic assumptions because the required cost reduction turned out to be something of the order of a factor of twenty. This requirement is sufficiently stringent as to require major technical advances, if not breakthroughs, before extensive deployment can become a reasonable course of action. We estimated that module costs would have to be reduced to something like 10–40¢/W_p in 1975 dollars (the subscript denotes the

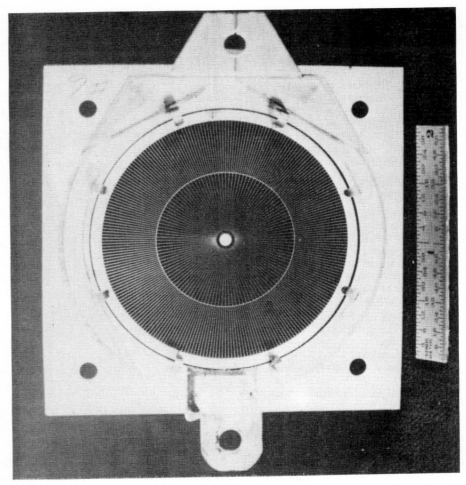

In this solar cell the front (illuminated) surface is covered with a network of metallic electrodes to collect the current produced by the cell.
Figure 4

"peak" wattage resulting from an insolation of 1000 W/m^2) in order to become economically competitive. The estimate of the costs of a photovoltaic system was based on information obtained from several systems-design studies of central power systems. The basis for comparison is the projected cost of coal-generated central-station electrical production in the period 2000–2030, using systems analyses of the likely future capital and operating costs of such plants.

Because our use of the term "central power" in this context has led to a great deal of misunderstanding, a caveat concerning it is in order. "Central power" is used here to mean that the generating capacity is utility-owned and is part of the electrical grid that is the present basis of US electrical distribution. It is not implied that any photovoltaic generation facility would be comparable in size to the largest conventional plants now in use. A typical large coal or nuclear unit is on the order of 1000 MW in capacity; it is probable that insurance costs against damage from local storms, as well as other considerations, will limit photovoltaic plant sizes to at most 100 MW_p or so. In fact, most of the economies of scale associated with installation, maintenance, power

conditioning (that is, conversion of dc to regulated ac), and the like can probably be realized for installations of a few MW_p, which is on the scale of the electric power needs of a small community.

The much debated issue of whether in the long term photovoltaic systems should be deployed in utility-owned central stations or in a purely residential fashion should be regarded as one having low priority at this time. Studies of central power and residential deployment[2] arrive at quite similar allowed costs for the year 2000 assuming consistent energy price escalation rates and balance of systems costs. Even though small-scale residential applications may constitute one of the early uses of solar cells in this country, the primary present problem is to lower the cost of photovoltaic modules by an order of magnitude or more. While exploration of the technical issues involved in all applications via systems tests is necessary to ensure that all necessary parts of the eventual solar–electric systems are being developed, no choice needs to be made now. This is so since photovoltaic arrays can, by and large, be deployed in any fashion. (High-concentration-ratio systems may form an exception.) Thus the use of a central power model in the study

A solar-concentrator array of three units mounted on a tracking device to maintain orientation with the Sun. (This photograph and that of figure 4 are by Sandia Laboratories.) Figure 5

to establish cost targets does not imply any determination on the Study Group's part concerning the ultimate mode of deployment.

The detailed systems studies have been in general agreement that the economics for residential photovoltaic systems is best when there is no residential energy storage, and excess power from the photovoltaic array is sold back to the utility at about half the price paid for the electricity bought. Combined photovoltaic–thermal flat-plate arrays do not appear to possess any advantage over side-by-side electric and thermal solar arrays. This makes the residential system just a very small version of a central power station that may not differ appreciably from its larger counterpart.

The existing electric generation system and the reluctance of homeowners to incur high capital expenses, particularly in a period of rapidly rising house prices, favor the use of photovoltaic systems in central power generation, but local conditions may exist that make purely residential deployment advantageous.

Systems issues

In the *Principal Conclusions* the Study Group addressed five areas of importance.

The first concerns systems questions. Several significant points emerged from this discussion besides those concerning cost goals. One of the more controversial of these was that photovoltaic systems are unlikely to exceed 1% of US electrical energy production in the year 2000. This number represents an order-of-magnitude estimate: Under the most favorable circumstances—high continuing demand for electricity, rapid technological progress, and a national decision to maximize the use of photovoltaic systems by, for example, offering substantial incentives and subsidies—it might be increased to several percent. For photovoltaics to provide more than about 5% of the total electric power would be extremely difficult because storage and other systems problems require resolution at about that level of deployment. (These numbers represent averages for the country as a whole. The Southwest and the Northeast are, respectively, the areas of highest insolation and highest energy cost, so the fraction of solar-electric power in these two areas would be expected to be much higher than the national average.)

It is instructive to examine the kind of commitment of capital and materials that is required to produce a 1% deployment

during a ten-year period extending say from 1990 to the year 2000. One would expect, on the basis of the present rate of progress and the Department of Energy's projections as outlined in its present program plan, that a satisfactory technology might well be ready for commercialization and large-scale deployment at about that time. A build-up at a uniform rate to the 1% level in ten years would require producing photovoltaic systems at a rate of 2000 MW_p per year, more than 1000 times our present level of production.

One percent of total generation corresponds to 5×10^{10} kWh (assuming that electricity demand roughly doubles by 2000; in the energy future of reduced electricity consumption envisioned by Amory Lovins and others, new electricity sources are less important). The thermal fuel displacement would be about 1/2 Quad per year at a thermal-to-electric conversion efficiency of one third after completion of the total deployment. The total capital investment at $1/W_p$ system cost is about twenty billion dollars (in 1975 dollars).

Assuming 10% conversion efficiency from sunlight to electricity leads to an array area of 200 km^2. If the arrays were all built as ground structures for central power or intermediate-sized installations, the total land area required including access roads would be about 400–600 m^2, a relatively modest amount. If the arrays were put on rooftops, about two million houses would be required with a 10-kW_p array on each house.

The materials requirements are more demanding. For example, the steel and concrete used each year to build the structures needed to support ground-based flat-plate solar-cell arrays of a typical recent design would amount to roughly 5 and 17% respectively of the 1974 US annual production of these materials. The use of lightweight concentrators made of novel materials such as plastics would reduce such requirements significantly. Materials savings also result when flat-plate arrays are installed on rooftops if the arrays do not need to be tilted with respect to the roof. Because electrical generation consumes about 20% of all private capital investment, large materials demands may be supportable without major dislocations. Materials usage will have to be considered carefully in system design, however.

The amount of active cell material required may also pose significant limitations. For example, the amount of silicon required to produce 1% of the US electricity in this way would be roughly the same as the present total production of metallurgical (not semiconductor) silicon. Because of the abundance of a high-grade silicon ore (sand) this level of production poses no intrinsic problems, although there is the necessity of building the requisite facilities for low-cost production of high-purity material. This situation is somewhat different for materials like gallium arsenide or cadmium sulfide, which use elements with present production levels far below those required for large-scale use in photovoltaic systems. The present annual production of gallium, for example, is about 7 tons, which is about an order of magnitude less than would be required to use a GaAs concentrator system as the basis for the model deployment above. The gallium problem is one of cost—gallium is abundant in the Earth's crust but there is no source of concentrated ore. Substantial amounts could be recovered as by-products in zinc and aluminum ore processing, but the cost might be prohibitive.

These rather substantial numbers should serve to illustrate that a 1% penetration, corresponding to roughly 0.1% of the 1975 total US energy demand, is far from being insignificant. This point is illustrated more vividly by noting that the amount of electrical power in question is equivalent to the power that would be produced by using 1/3 of the output of the Alaska pipeline to generate electricity. The total capital investment over 10 years equals roughly 2/3 of total 1975 US automobile factory sales. Thus, even this seemingly modest electrical energy source requires build-up of a very substantial industry. While there is nothing that limits in any absolute sense our ability to achieve these goals, or perhaps even goals two or three times as large, the scale of the required effort must not be underestimated.

What of the times required to pay back the amount of energy consumed in creating this industry? Because the production of electricity by other means also involves capital investment of the same general magnitude as photovoltaic systems, the most useful measure is probably the energy payback time of the solar cell itself. This can be estimated to be about one year for silicon cells, which are more highly energy-intensive than most other cells. Out of a projected lifetime of twenty to thirty years, this appears to be quite reasonable and is in line with the payback times estimated for present electrical power plants.

An economically competitive photovoltaic industry would require sufficiently cheap cells, cell encapsulants, support structures and other system components (power conditioning, installation, and so forth). The premium on cell efficiency is particularly high. The cost of energy from a photovoltaic system decreases about 40% for an efficiency increase from 10 to 20% if the area-related system costs (including cell cost) remain constant. Because of costs other than those due to the cells, "free" cells of less than 10% efficiency are likely to be uneconomical in grid-competitive applications.

Before the grid-connected US applications become competitive, it is highly probable that significant intermediate markets involving export and a number of specialized US uses in remote locations will develop. However, a premature entry into large-scale US deployment before the technology has reached fruition might lock us into an overly costly technology. Indeed the national interest may well be served optimally by an emphasis on research and development accompanied by measured and technologically appropriate progress in Government-assisted commercialization. The naive impatience of some solar advocates to deploy solar cells immediately and on a large scale appears to be associated with an inadequate appreciation of the technical problem.

Silicon

The second area discussed in the *Principal Conclusions* concerns crystalline-silicon-based photovoltaics. This is an established and proven technology because of its long and successful use in space vehicles. It thus serves as a standard of reference for the field. However, currently manufactured cells are far too expensive for most terrestrial uses. Because it is the best understood technology, it represents the best short-term hope for a reliable system of at least moderately low cost. For this reason there has been a vigorous development effort by the Jet Propulsion Laboratory and other organizations, largely under the auspices of the Department of Energy.

Crystal-silicon devices are capable of high efficiency and reliable performance, but the rather elaborate manufacturing technology makes the ultimately attainable price per watt difficult to estimate.

The need to understand the demands that cost requirements place on the manufacturing technology has led to some very detailed analyses of the steps in solar cell fabrication. In order to reach the neighborhood of the price goal of $50¢/W_p$ (in 1975 dollars), which DOE has set itself for 1986, it is necessary to design and build factories that are essentially totally automated. Two stages of manufacturing are involved, the first producing purified solar-grade polycrystalline silicon from sand, and the second utilizing that output to fabricate finished solar cell modules with minimal labor cost and a near-perfect yield of functional finished cells.

Some feeling for the procedure is conveyed by considering the steps and their pricing in one approach to fabricating modules from the starting polycrystalline silicon material. In the single-crystal cell technology, the polycrystalline silicon must first be converted into a single crystal material. There are many ways of doing this, the most popular at the present being the Czochralski technique, which grows a cylindrical single crystal of at least several kg from a crucible of molten silicon. According to the estimates analyzed in the APS report this can be accomplished at a projected cost of about $10¢/W_p$ for a module of 16% efficiency. The single crystal boule then must be sliced. While this sounds like an entirely straightforward procedure (one popular writer has suggested that the Italian marble industry has had this art well in hand for many years!), it is in fact a difficult technical problem to find a method for producing about 20 acceptable wafers per centimeter of silicon crystal at a projected cost of 12 $¢/W_p$ (20 ¢ per cut). Saws under development cut up to 1000 wafers simultaneously.

The slices then must be converted into solar cells by doping and other procedures, which are expected to cost $13¢/W_p$. Finally these cells must be interconnected and assembled into modules containing on the order of 100 cells, at an additional cost of $18¢/W_p$. The total projected cost

A silicon ribbon. The usual process for making solar cells involves sawing cells from cylindrical ingots of single-crystal silicon, which wastes much purified silicon. This single-crystal ribbon was grown for experiments to test other, less wasteful, manufacturing processes. (IBM Fishkill Laboratory)   Figure 6

for the module and cells is 62¢/W$_p$, somewhat higher than the DOE goal. Other methods for producing silicon-cell modules (figure 6, for example, shows a single-crystal ribbon) involve process steps that are lower in cost per unit area but result in less efficient modules, so that the final costs in our estimates turn out somewhat higher per watt. We regard these numbers as optimistic projections that will be difficult to surpass in the absence of major technological advances, but JPL, which manages DOE's highly effective silicon research and development effort, has produced lower cost estimates for some of these processes on the basis of somewhat different assumptions.

Major advances in silicon-cell design have been made in recent years. For example, electron–hole recombination losses at the rear surface of the cell have been reduced by building into the cell a back-surface field that repels minority carriers. Light-reflection losses have been nearly eliminated by anti-reflection surface texturing to trap the incident radiation. Losses associated with the heavily doped layer that forms the front surface have been reduced by decreasing the thickness of this layer, as in the "violet cell." The development of other junction configurations, such as the vertical multijunction and interdigitated back contact structure, has been pushed forward rapidly. Indeed, silicon technology is sufficiently advanced that it should be possible to demonstrate with suitable cleverness the achievement of designs yielding nearly 100% of theoretical efficiency under manufacturing conditions. Because of the premium on achieving high efficiencies, this avenue is being actively explored.

Concentrators and thin films

The third area is concerned with concentrator systems that present solar cells with focused sunlight. Concentrators are attractive in principle because they substitute large areas of simple devices such as mirrors and lenses for the large areas of sophisticated electronic devices required by flat plates. However, the savings realized by this strategy must be balanced against the cost for devices that focus and track the sunlight. Structural stability under wind loading is a more severe problem for trackable arrays than for flat plates, and most early concentrator designs were so massive that materials use was prohibitive. Some designs that are much more materials-conservative have recently appeared, but they are still in the early stages of development. Cells with efficiency greater than 20% have been demonstrated in the laboratory (in fact, a device using two cells and a beamsplitter has achieved 28%). Development of practical devices with an efficiency this high would give a strong impetus to the concentrator program.

The fourth area is concerned with thin-film devices or, in a more general sense, with novel active materials other than single-crystal silicon. Most of these have higher absorption coefficients and hence require smaller thicknesses. This is a double advantage: Little material is used and its purity becomes less critical because charge carriers travel shorter distances before macroscopic separation occurs at the barrier or junction. There are many candidate materials, relatively few of which have yet been investigated with any thoroughness. This field offers many research opportunities, some of which are outlined in the *Principal Conclusions.*

Cells with an efficiency of 5–10% can be made from disordered, impure, thin layers of active material. Two common examples are the heterojunction cells made from Cu$_2$S and CdS and cells made from doped hydrogenated amorphous silicon. The advantages of such systems are that they use little active material, that their band gaps can be well-matched to the solar spectrum, and that their absorption coefficients can be much higher than that of crystal silicon. However, there are also disadvantages. Thin films tend to be mechanically fragile, they may undergo

---

The members of the panel

The members of the Study Group, their scientific interests, and their affiliations, at the time of the study, were:

David DeWitt—semiconductor manufacturing and engineering, IBM San Jose

Jerry P. Gollub—physics of turbulence and superconductivity, Haverford College

Robert N. Hall—semiconductor-device science, General Electric Research and Development

Charles H. Henry—semiconductor physics, Bell Laboratories

John J. Hopfield—condensed-matter theory and biophysics, Princeton University and Bell Laboratories

Thomas C. McGill—semiconductor physics, Caltech

Albert Rose—semiconductor-device science, Boston University and University of Delaware

Jan Tauc—semiconductor and metal physics, Brown University

Robb M. Thomson—materials science and metal physics, National Bureau of Standards

Mark S. Wrighton—photo and electrochemistry, MIT

The Review Committee was chaired by Herman Feshbach; its other members were:

N. Bruce Hannay, Vice President and Director of Research, Bell Laboratories

Robert N. Noyce, Vice Chairman of the Board, Intel Corporation

J. Robert Schrieffer, Mary Amanda Wood Professor of Physics at the University of Pennsylvania and Nobel Laureate

Peter A. Wolff, Professor of Physics and Director of the Research Laboratory of Electronics at MIT

---

chemical degradation with time, and they are often difficult to make reproducibly.

Nonetheless, the variety of thin-film materials that could be useful in solar cells is very large indeed. Many promising systems have not yet been explored and characterized using well-known semiconductor techniques. Our knowledge of semiconductors is probably sufficiently extensive at this point to permit an intelligent choice of such systems. The potential payoff would be immense if a truly inexpensive technology based on a thin-film system were to be developed.

Basic research

The final area concerns the fundamental, long-term research programs that are necessary for the development of a significant photovoltaic or photochemical energy technology. The problems are numerous and their scope usually multi-disciplinary. Significant contributions to these programs will be made by a variety of disciplines, including solid-state physics and chemistry, materials science, surface physics, inorganic and organic chemistry, electrochemistry, and biophysics. We can only give a flavor of the types of appropriate activities here.

The photovoltaic process is in principle closely related to the photosynthetic process. One of the most intriguing questions concerns the possibility of man-made photosynthesis with an efficiency of 10 to 30%, more than ten times higher than that typical of natural photosynthesis. Programs in molecular science, directed toward characterizing and synthesizing molecules capable of producing fuel directly from sunlight, and in biophysics, examining natural photosynthesis with a view towards creating artificial analogs, would represent useful beginnings.

Another relevant problem area concerns the properties of interfaces. Despite the crucial importance of interfaces in all aspects of semiconductor electronics this area is only now beginning to receive the attention it deserves. For example, despite the great practical importance of silicon devices, little is yet known about why the SiO$_2$–Si interface is electrically inactive and stable. Equally to the point is the question of why we have been unable to use our empirical knowledge of this system to find appropriate passivation layers for other semiconductors such as the group III–group V compounds.

There are many materials suitable for photovoltaic systems, and in consequence there are many types of interfacial structures. The atomic and chemical structures of interfaces will have to be characterized and perhaps modified using the approaches and techniques of materials science. The investigation of the associated electronic properties will involve electrical engineering and physics. Because different methods of materials preparation strongly affect the character of the interface, solid-state chemistry and crystal-growth science must contribute to this research.

Many interesting problems arise, such as the energetics and kinetics of semiconductor–liquid interfaces, the structural and electronic properties of amorphous semiconductors containing hydrogen and other additives, and the electronic properties of polycrystalline materials. These problems are sufficient to keep condensed-matter scientists happily and fruitfully occupied for some time. Solutions or insight would obviously be of great value in the development of solar photovoltaic energy.

General observations

The time required for photovoltaics to become a mature component of the US energy mix is likely to be some 30 years or more. This assessment is based on the time scale that has been associated historically with the transfer from one dominant energy technology to another (for example, wood to coal and coal to oil) and on the time necessary to build a large industry and a large market. There are many worthwhile shorter-range applications for photovoltaics, some of which we have mentioned in this article. Nonetheless, in the formulation of a program that will best serve the national needs, the central emphasis must be on the eventual establishment of photovoltaics as a major energy technology.

Even though the implementation of a photovoltaic technology that will contribute appreciably to the nation's energy mix may be a long-term proposition, the ultimate prospects for photovoltaic energy conversion are bright in view of the current ferment and rapid rate of progress in both the science and technology relevant to the field. None of the present photovoltaic technologies (with the possible exception of silicon) represents a clearcut choice, because the price of solar cells needs to be reduced by more than an order of magnitude in order to become competitive with the cost of generating electricity by conventional means. These technical problems need to be resolved before the appropriate photovoltaic technology can be determined.

It is in fact not at all clear that the winner will be a single technology based on a single type of cell made from one specific set of materials. During the past year silicon technology has made rapid progress both in the area of materials preparation and cell design. The Cu/CdS system is also undergoing rapid commercialization: At least two US manufacturers, and possibly several others here and abroad, are exploring the commercial fabrication of these systems.

In these somewhat tentative attempts to forecast the development of a new energy technology, it is important to realize that there are no scientific barriers to the success of photovoltaics. While some major technological advances in the next few decades are needed in order to make the technology economically competitive, there is also little doubt that the necessary technological ingenuity to accomplish this is available.

However, it is necessary to ensure that highly talented scientists be attracted to the field. This may require some examination and revision of DOE funding formats: The management styles and funding formats planned for the photovoltaic research and development program must take into account that parts of the program are intrinsically long-range and that there is a real need to compete with other condensed-matter research fields for the talent essential to a successful program. The program must use diverse approaches addressing both near and long-range goals. On the one hand, the candidate technologies available now should be developed at an optimal rate; on the other hand due attention must also be given to the kind of fundamental research leading to new ways in which the Sun's energy may be used efficiently. We have already mentioned the idea of man-made photosynthesis in this regard.

Because none of the present photovoltaic options has shown sufficient promise that its price is reducible to the range needed for grid-connected applications, large-scale deployment or major Federal procurement plans appear to be premature. Certainly something would be learned from such guaranteed buys, and perhaps the large expense may even be justified in view of the prevailing energy crisis. However, the danger that a too primitive technology becomes frozen into the industrial system prematurely is one that is far more serious.

As a result of these considerations, the *Principal Conclusions* emphasizes that an intensive, imaginative, well-funded, and well-managed program of research and development offers the greatest hope for the long-term success of photovoltaic technology as a way of effectively utilizing our solar income.

References

Much of the material of this article is drawn from the *Principal Conclusions of the APS Study Group on Solar Photovoltaic Energy Conversion,* which is available for $5.00 postpaid from The American Physical Society, 335 E. 45th St., New York, N.Y. 10017.

1. *Efficient Use of Energy,* APS Studies on the More Efficient Use of Energy, edited by K. W. Ford et al. (New York, A.I.P. 1975)

2. In addition to the references given in Section 1 of the *Principal Conclusions,* see: (a) Regional Conceptual Design and Analysis Studies for Residential Photovoltaic Systems, Final Report, GE Space Division for Sandia Laboratories, DOE report #SAND78-7039 (2 vols., 1979) and (b) Application of Solar Technology to Today's Energy Needs, Office of Technology Assessment reports OTA-E-66 (vol. 1) and OTA-E-77 (1978). □

Infrared interferometer to measure size and shape of stars

Gloria B. Lubkin

PHYSICS TODAY / JULY 1972

A new high-resolution device for infrared astronomy is being developed at the University of California at Berkeley by Charles Townes and his collaborators, Michael Johnson, Albert Betz and Daniel Galehouse. Applying the new techniques of quantum electronics to a concept originated by Albert A. Michelson, they have built the first infrared stellar interferometer. Townes expects its angular resolution to be orders of magnitude greater than that of an ordinary telescope. It will be used to measure the size and shape of stars and other astronomical objects.

Michelson had used a stellar interferometer with two small receiving telescopes separated by about 20 feet and became the first person to measure the diameter of a star other than our sun. He had an angular resolution of about 10^{-7} radians. The art essentially died until its rejuvenation came with long-baseline microwave interferometry, which has now progressed to intercontinental experiments. Angular resolutions of 10^{-9} have been achieved. Another technique related to Michelson's, developed over the past two decades by Hanbury-Brown in Australia, uses two separate light telescopes and measures the fluctuation in light. By cross correlation of these fluctuations he gets high angular resolution in the visible range.

With intercontinental distances, microwave interferometry has essentially achieved its ultimate resolution, Townes says, unless we put one of the telescopes into space, possibly on the moon. But in the infrared, because of the difference in wavelength, a baseline of only 1 km would give the same resolution obtainable with microwaves—10^{-9}. And an even longer baseline, say 100 km, should be possible.

In the Berkeley interferometer (see figure), a CO_2 laser serves as a local oscillator, whose output is mixed with the signal received by each telescope in a chip of photoconductive copper-doped germanium. The beat frequencies resulting from the mixing preserve the phase and amplitude information of the signal and are amplified electronically. The infrared receiver is sensitive to radiation in a bandwidth of 10^9 Hz, centered at a wavelength of 10 microns. The signals from each heterodyne detector, which lie in the radiofrequency range, are brought together by cable; then their interference is detected by a square-law detector.

As the star swings overhead the difference in path length through the two telescopes changes. To keep the path difference fixed, the new device simply switches in various lengths of cables; the precision required for the 10^9-Hz bandwidth is a few centimeters.

Tests will soon be underway at Kitt Peak with two existing telescopes separated by 20 feet. This interferometer will give a factor of two improvement in resolution over the 200-inch telescope. If all goes well the Berkeley group will then use the three large Kitt Peak telescopes, two at a time; these are arranged in a triangle whose legs are 500 to 800 feet. The telescopes are 60,

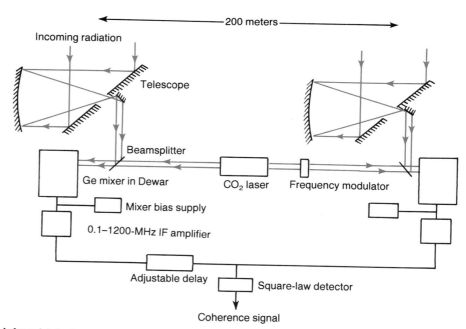

Infrared interferometer. Laser output is mixed with signal from each telescope in germanium mixer. Signals from each heterodyne detector are brought together by cable; their interference is detected by the square-law detector. Telescopes are of Pfund type.

84 and 90 inches. The big interferometer will improve resolution by a factor of 25 or more over the small interferometer, putting the experimenters in a whole new realm of observation.

It should then be possible to measure the sizes of infrared stars and possibly other infrared sources, such as Seyfert galaxies. With improved sensitivity one might eventually be able to resolve quasistellar sources. One can also measure proper motions or the rate of the earth's rotation—how well this can be done depends on properties of the atmosphere that are still not known.

One of the very common beliefs these days is that the stars are surrounded by dust clouds that produce most of the infrared radiation, Townes explained. The actual shape of the radiating region may not be spherical. Instead there may be a ring of clouds bigger than the star itself. "But we won't know until we measure it," he said.

Townes is now trying to raise money to build a pair of 60-inch telescopes, mounted on trailer trucks so that their separation can be varied and so that they can be moved to follow the good weather or even to the Southern Hemisphere. In actual use the telescopes would rest on concrete piers. The telescopes are of the Pfund type. In such a telescope light comes from the star to a plane mirror (see figure), which reflects it into a fixed parabola. The parabola then reflects the light through a small hole in the center of the flat mirror to a focal point beyond it. The focus would be fixed, a feature important for infrared work, Townes says, because one needs to install complex apparatus, such as the heterodyne detectors. Mobility would also allow complex new instrumentation to be installed on the telescopes at a convenient laboratory before they are taken to a remote observing site. The Pfund design minimizes the amount of infrared radiation from the telescope itself that reaches the detector.

The Berkeley telescopes would also be able to be pointed in an absolute fashion so that one can do observations in the daytime without locating the object first, thus enabling the experimenters to work the way radio astronomers do. The angular motion needed for the flat mirror will be more complicated than it would be with a Cassegrain-type telescope, but a modern computer can do the job simply, Townes said.

The parabolic mirrors would be 60 inches and the flat mirrors around 85 inches. Proposals have been submitted to NASA, NSF and to some private donors. The two telescopes would cost $700 000.

Correcting for atmospheric distortion in telescopes

Barbara G. Levi

PHYSICS TODAY / DECEMBER 1974

Optical astronomers are frequently hampered by atmospheric turbulence that distorts telescopic images by causing random phase shifts in the incoming light. These distortions often reduce the resolution of a telescope far below its diffraction limit. But the outlook may be brightened by new techniques now being developed to detect and correct for the phase error in real time. Until now most techniques to correct for atmospheric distortions have been post-detection compensation techniques: The effects of turbulence are extracted after the image data has been recorded. Three of the groups that are conducting major development efforts on real-time compensation described their work at a conference on Optical Propagation Through Turbulence that was sponsored by the Optical Society of America at the University of Colorado, 9–11 July. These three groups are from

Hughes Research Laboratories, from Itek Corporation and from Lawrence Berkeley Laboratory[1] and the Institute for Advanced Studies.

The pre-detection compensation schemes all feature a component for detecting the errors introduced by turbulence, some active optical elements to correct for the errors and an on-line data processing system to calculate the desired response of the optical elements. All components must have extremely fast response times because the pattern of atmospheric disturbances changes typically every 20 msec. Beyond these general features lie interesting differences in approach.

At Itek, John Hardy, Julius Feinlieb and James Wyant use a shearing interferometer[2] to detect the phase errors in the wavefront arriving at the telescope aperture. The phase error introduced by turbulence changes over a

region of roughly 10-cm square, so the interferometer essentially generates a wavefront map by measuring the phase error in each of these regions. The map is measured in about 1 msec, independent of the size of the telescope. The phase corrector is a monolithic piezoelectric mirror.[3] This solid slab is divided into regions, each driven electrically by a separate actuator. The measuring and data processing between the interferometer and the mirror are done in parallel.

The Itek group is the only group that has demonstrated its system in the laboratory. Working with thermal turbulence, they obtained a resolution close to the diffraction limit. Hardy told us that the Itek design is an add-on system that can be relatively small and has a high-frequency response.

Richard Muller and Andrew Buffington of LBL have developed an idea for a

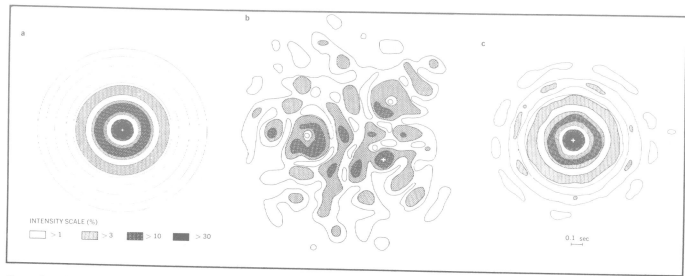

INTENSITY SCALE (%)

☐ > 1   ▦ > 3   ▨ > 10   ■ > 30

0.1 sec

Computer simulation of a ring-shaped telescope aperture (a), the speckle pattern caused by random atmospheric phase distortion (b) and the restored image after one iteration with 25 phase-shifter segments (c). The sharpness criterion was maximization of light through a small hole. These results from LBL resemble other work being done at Hughes Research Laboratory. (From reference 1.)

feedback system that continuously monitors and improves the image quality.[1] By contrast to the above two techniques, they do not directly measure the atmospherically introduced phase errors. Instead their feedback system continuously adjusts the positions of an array of separately movable mirrors in order to maximize a quantity that they call the "sharpness" of the image. They have proved analytically that maximization of the image sharpness will necessarily lead to a completely restored image. Muller and Buffington have investigated the effects of several different definitions of sharpness by computer simulations. A good definition would be the integral of the square of the intensity across the image, but a definition that is easier to implement would be simply maximizing the light through a small hole in a mask.

Freeman Dyson of the Institute for Advanced Studies collaborated with Muller and Buffington by laying the theoretical ground rules for general schemes of optical image improvement, establishing that diffraction-limited resolution in a large, ground-based telescope is in principle possible, and setting forth various expectations and limitations.

At Hughes Research Laboratory, Thomas O'Meara, Wilbur Brown and Larry Miller are applying some of the experience gained in work with transmitter control systems to the problem of optical image improvement, with modifications appropriate to the difference between coherent light and white light from astronomical objects. Like Muller and Buffington, they use image quality as a measure of atmospheric distortions, but differ in some of the definitions of image quality. O'Meara feels that they consider a different variety of maximization systems by using different control algorithms.

The Hughes team is working with analog as well as digital interface systems. For this purpose they are modifying a process of dithering, which is used in transmitter control to maximize the power arriving at a point target. So far they have found no intrinsic advantage favoring either analog or digital processors.

What is the reason for the sudden surge of progress in this field? A general scheme was proposed by Babcock as early as 1953[4] but was presumably never implemented because the technology was not then available. According to Hardy the Itek group first pointed out several years ago that the technology was ready and encouraged development of such systems. O'Meara felt that in addition to the great recent improvement in technology, a motivating factor has been the demonstration that similar systems work for transmitter control.

References

1. R. A. Muller, A. Buffington, J. Opt. Soc. Am. **64**, 1200 (1974).
2. J. C. Wyant, Appl. Optics **13**, 200 (1974).
3. J. Feinlieb, S. G. Lipson, P. F. Cone, Appl. Phys. Lett. **25**, 311 (1974).
4. H. W. Babcock, Publ. Astron. Soc. Pac. **65**, 229 (1953).

Solar Optical Telescope will orbit on space shuttle

Bertram M. Schwarzschild

PHYSICS TODAY / SEPTEMBER 1982

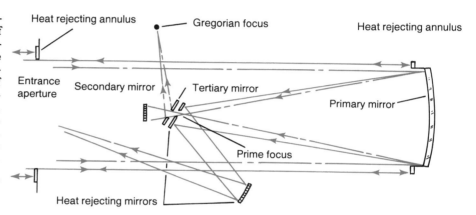

The Solar Optical Telescope will have a folded Gregorian optical design because this provides better heat-rejection capability than does the Cassegrainian configuration more common in night-sky telescopes. The primary mirror focuses the solar image through a hole in a heat-rejection mirror that allows only 3 arc minutes of the Sun to be seen by the concave secondary mirror. A flat tertiary folding mirror directs the image to the off-axis Gregorian focus. The Gregorian optical path is shown as broken colored lines; the heat rejection paths are solid color.

To the naive outsider it seems surprising that solar astronomers complain of being hampered by the inadequate spatial resolution of their telescopes. The Sun is, after all, only eight light minutes away, and most of us don't think of it as possessing small-scale surface features beyond our powers of resolution. Sunspots were studied by Galileo almost four hundred years ago. But in fact, magnetohydrodynamic mechanisms with characteristic scales of 70 km or less appear to be crucial to our understanding of the astrophysics of our nearest star.

The turbulence of the Earth's atmosphere, however, prevents one from resolving solar surface phenomena smaller than about 700 km (one arc second) with any ground-based telescope, except for occasional brief respites of unusually good "seeing." For the past decade, therefore, solar astronomers have been giving serious consideration to the idea of a solar telescope orbiting above the Earth's atmosphere with an aperture diameter of at least a meter—corresponding to a visible-light diffraction limit of about a tenth of an arc second. Besides removing the problem of atmospheric blurring, such an orbiting telescope would be free to look at the ultraviolet part of the solar spectrum, to which our atmosphere is largely opaque.

In 1980, a long-standing study group headed by Richard Dunn (Sacramento Peak Observatory) submitted to NASA its recommendation that a 1.25-meter-diameter, visible–ultraviolet Solar Optical Telescope be built to fly aboard a Space-Shuttle orbiter. This June NASA affirmed its commitment to the SOT project by selecting Perkin–Elmer Corporation as the primary contractor for the detailed design and construction of the telescope. At the same time, Caltech and Lockheed were chosen to build the two principal focal-plane

scientific instruments for the initial SOT flights. NASA plans to fly the SOT aboard one of the existing Shuttles for repeated week-long orbital flights beginning in 1988 or 1989.

Ground-based solar telescopes can achieve $\frac{1}{3}$-arc-second resolution under unusually favorable atmospheric conditions a few times a year. But such conditions persist for ten or fifteen minutes at best. This is particularly unfortunate for observing solar structures because these are dynamically evolving phenomena of brief duration rather than fixed objects like lunar mountains. To understand the dynamics of the solar plasma and its interaction with the Sun's magnetic fields one really wants hours of continuous observation with a resolution of 0.1 arc second. The corresponding distance—70 km—is regarded as a rather fundamental length for solar surface phenomena because it is the mean free path for photons in the photosphere,

the outer layer of the visible Sun.

Since the late 1950s, balloons have occasionally carried solar telescopes aloft. But size limitations have generally kept their diffraction limits to resolutions of no better than $\frac{1}{3}$ arc second. Furthermore, balloons don't get high enough to escape the ultraviolet absorption of the atmosphere, and their useful time on any flight is limited to about eight hours. Richard Fisher (NCAR High Altitude Observatory, Boulder), the Telescope Scientist of the SOT project, told us that it would probably cost as much to build a balloon-borne telescope of sufficient size and pointing stability to achieve 0.1-arc-second resolution as to build the orbiting SOT. Rocket-borne telescopes suffer similar limitations of size and pointing stability, and their flight duration is extremely short.

Despite the instrumental limitations, solar astronomers have in the past two decades discovered intriguing small-scale structures that have generally posed more questions than they answer. These observations strongly suggest further detail below the present resolution limits. "Until the late 1950s, many astronomers had a brain fixation that the Sun, and indeed all celestial sources, had no interesting structure smaller than about one arc second," we were told by Robert O'Dell (now at Rice University), Project Scientist of the Space Telescope (PHYSICS TODAY, March 1981, page 59). "But then the flights of the 12-inch, balloon-borne Stratoscope at the end of the decade revealed that the solar granules possess structure at least down to ⅓ arc second." These granules are closely packed, short-lived bright structures (typically a thousand kilometers across) which give the photosphere a cellular appearance. They are believed to indicate the tops of convection cells, where columns of rising hot gas emerge from inside the photosphere.

In the early 1970s Dunn and his colleagues at the Sacramento Peak vacuum-tower telescope, which is designated to minimize atmospheric turbulence, discovered "filigree" in the dark intergranular borders, where the cooler gas resubmerges. In brief periods of ⅓-arc-second seeing, this filigree looks like tiny bright dots, but it is believed that with higher resolution it will show linear structure. It is not yet clear what sort of convective or magnetic mechanism this filigree is manifesting. To understand the granular structure and filigree one will have to observe their evolution at high resolution for extended, continuous periods.

The chromosphere, the solar layer just outside the photosphere that manifests itself in Fraunhofer absorption and emission lines, also exhibits a wealth of small-scale structure that will require better resolution to clarify. The chromosphere is where one finds solar flares, catastrophic energy releases whose mechanism for converting magnetic to thermal energy is not well understood. Data from the x-ray telescope on the 1980–81 orbiting Solar Maximum Mission suggest that the generating mechanism for these flares involves a spatial scale of less than 70 km. The SOT will be well suited, Fisher suggests, to elucidate their origin.

Imaged in the light of the Balmer α line of hydrogen (Hα), the chromosphere looks somewhat like grass growing on a lawn full of large bald spots. The bald spots are "supergranules"—convective cells an order of magnitude larger than the photosphere granules. They are outlined by a grasslike border of "spicules," fine vertical structures whose thickness is smaller than our present limits of resolution. These spicules appear to result from the interaction of the magnetic flux lines and the

upward flow of material from the photosphere. The magnetohydrodynamic interaction of the plasma with the magnetic field, Fisher told us, is a central problem of solar physics. One will need to look at the spicules with a resolution three to five times finer than their thickness, he explained. Furthermore, because the lifetimes of chromospheric structures range from tens of minutes to hours, one needs long periods of continuous high-resolution seeing to understand their aggregate behavior.

A dramatic temperature rise—from 1.5×10^4 K to 1.5×10^6 K—takes place in the extremely narrow "transition zone" from the chromosphere to the corona, the outermost layer of the Sun.

Chromospheric structure is visible in this photograph of the solar surface with an 0.22 Å-bandwidth birefringent filter tuned $+ \frac{7}{8}$ Å from the center of the Hα line. Grasslike borders of fine spicules outline supergranular cells tens of thousands of kilometers in diameter. The high-contrast system near the bottom is an active sunspot region. Photograph taken at the Sacramento Peak Observatory.

The high temperature of the corona—more than two hundred times hotter than the photosphere—is one of the major puzzles of solar physics. Recent ultraviolet observations by a Naval Research Laboratory group with rocket-borne uv instruments indicate the existence of hitherto unsuspected vertical-flow structures in the chromosphere, with detail smaller than the limits of resolution. Fisher suggests that they may play an important role

in coronal heating. But rocket observations last only a few minutes, and ultraviolet emission-line imaging is particularly slow because one must scan with moving spectroscopic slits; the birefringent narrow-band filters that facilitate visible spectroscopic imaging are not available in the ultraviolet.

The Dunn study group, formally referred to as the Facility Definition Team, made its final report to NASA in January 1980. Their report recommended the construction of a 1.25-meter-diameter Gregorian reflecting telescope with wavelength coverage from 115 nanometers in the uv to 1100 nm in the near infrared, with a diffraction-limited resolution of 0.1 arc second at 500 nm in the visible. Infrared observation is given less emphasis because much of it can be done more conveniently with ground-based telescopes. At longer wavelengths resolution is limited by aperture size rather than turbulence, and the atmosphere has generous windows of transparency in the infrared.

The Gregorian telescope design was preferred to the Cassegrainian scheme more common in night-sky telescopes because it facilitates heat shielding. A major problem with solar telescopes is the enormous heat concentration that tends to distort the optics at the secondary mirror and focus. This heat load is a particular problem for the designers of the SOT because the solar image encountered in the orbital environment cannot easily be simulated in the laboratory. Whereas the convex secondary mirror of a conventional Cassegrainian telescope lies before the focus of its primary mirror, the *concave* Gregorian secondary lies beyond it, permitting the placement of a field stop and heat-rejection mirror at the prime focus to protect subsequent optics from concentrated out-of-field solar energy.

The Dunn group recommended that the Gregorian focus be designed to accommodate a visible-light, universal-filter polarimeter employing birefringent crystals, a visible-light spectrograph and a uv spectrograph. In response, NASA has chosen a Lockheed group led by Alan Title to develop the principal photoelectric imaging instrument for the first SOT flights—a visible–ultraviolet, combined filtergraph/spectrograph. Harold Zirin and his colleagues at Caltech have been chosen to develop the principal photographic instrument. The 65-cm solar telescope at the Big Bear Observatory, built by Zirin's group at the Jet Propulsion Laboratory in 1972, has served as something of a prototype for the SOT design.

Pointing stability. The 7-meter-long, 4000-kg SOT will be mounted on a pointing system in the payload bay of the Shuttle orbiter. When the bay doors are opened in orbit, the front end of the telescope will rise like a cannon. A resolution limit of 0.1 arc seconds would be of little use if one could not hold the telescope steady with corresponding precision. One can hold the Shuttle itself stable to within a few minutes of arc by aligning its principal axis along the gravitational field gradient, so that it experiences no torque. For finer pointing stability (two arc seconds), the Shuttle bay will house a double-gimbal pointing system. The final 0.1-arc-second stability must be provided by the telescope itself. This is to be accomplished by a system that provides information about the movement of the solar image to a compensating servomechanism. This feedback system will involve a major design effort on the part of Perkin–Elmer, Fisher suggests. He points out that one is requiring a significantly better pointing stability of this orbiting instrument than one asks of a telescope sitting firmly on a mountain top.

"The Solar Optical Telescope is the cornerstone of our solar physics program for the rest of the decade," we were told by Edmund Reeves, NASA Program Manager for the SOT. Stuart Jordan (Goddard Space Flight Center) NASA's Project Scientist, will organize periodic working-group meetings of potential SOT users to further the implimentation of the project's scientific objectives. Estimated to cost an order of magnitude less than the $750-million Space Telescope, the SOT will be funded as part of the general Shuttle payload-development program; it will not need the separate Congressional approval a budgetary line item would require. The SOT is expected to fly on one of the presently existing Shuttles, but the first launch date, 1988 to 1989, depends on funding levels, Reeves told us.

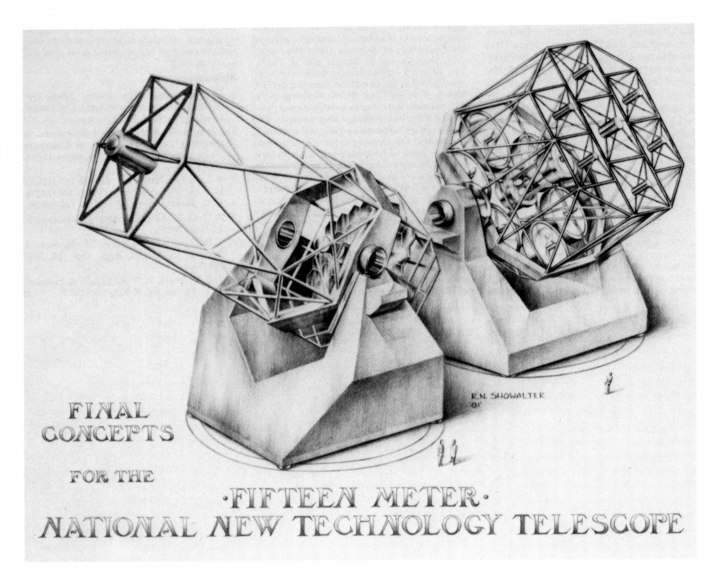

FINAL CONCEPTS FOR THE ·FIFTEEN METER· NATIONAL NEW TECHNOLOGY TELESCOPE

R.N. SHOWALTER '81

Ultraviolet, optical and infrared astronomy

A 15-m telescope on the ground and a far-ultraviolet spectrograph and 10-m infrared telescope in orbit would complement projects already planned and give us totally new insight into the nature of the universe.

E. Joseph Wampler

PHYSICS TODAY / NOVEMBER 1982

While it is possible to trace the history of ultraviolet, optical and infrared astronomy to an era before that of the pre-Christian Greek astronomers at Alexandria, the field is not as static as this maturity might suggest. The flowering of space astronomy and the orders-of-magnitude improvement in detector technology give us exploratory capabilities that are as revolutionary as the invention of the telescope. Decades from now the 1970s will be remembered as a turning point in our understanding of the universe.

With anticipated increases in instrumental capability we will, for the first time, be able to penetrate obscuring dust clouds to study the details of the birth of stars. We will be able to resolve the nuclei of active galaxies in the infrared, observe the distribution of dust clouds and study the processes involved in the conversion of ultraviolet energy to heat. We will be able to

Two ideas for the construction of a 15-meter telescope. Although plans have evolved since this drawing was made, the scale and main features of the two concepts are as shown. (Courtesy Kitt Peak National Observatory.)

obtain low-noise spectra of galaxies and quasars at the edge of the universe and study the conditions of matter at a time that corresponds to the era when our own galaxy was formed. These expectations are based on a modest extrapolation of present understanding and a conservative forecast of the capabilities of proposed new instruments. But it would be astounding if the anticipated enormous increase in instrumental capability throughout a frequency interval corresponding to 13 octaves would not yield totally new insight into the nature of physical reality.

In this article I will discuss the principal scientific currents in ultraviolet, optical and infrared astronomy. I will describe in some detail the major recommendations of the recent Panel on Ultraviolet, Optical and Infrared Astronomy of the National Academy of Sciences' Astronomy Survey Committee. This panel had the broadest charge of any of the committee panels: It was asked to survey progress and capabilities and to set priorities for new instrumentation in those branches of astronomy devoted to collecting and analyzing the information carried by cosmic photons with wavelengths between about 100 angstroms and 1 mm. In this energy range lie the electronic transitions of the outer electrons of neutral and ionized atoms and molecules, the vibrational and rotational transitions of molecules, the peak of the blackbody spectra of materials with temperatures between 3 K and 10^6 K, and "exotic" processes such as synchrotron radiation and the upgrading of photon energy by inverse Compton scattering.

A great variety of telescopes, spectrographs, photometers and cameras work in space and on the ground to measure the temperature, density, chemical composition, luminosity, dynamics and distance of the objects and gas that fill our universe. With the myriad celestial sources to classify and study, and with powerful telescopes equipped with advanced detectors, it is perhaps not surprising that the majority of astronomers work in fields addressed by the

E. Joseph Wampler is professor of astronomy and astrophysics at the University of California, Santa Cruz. He was chairman of the Astronomy Survey Committee's Panel on Ultraviolet, Optical and Infrared Astronomy.

panel on ultraviolet, optical and infrared astronomy.

Where we stand

Today, for the first time, we are appreciating the full complexity of the evolution of galaxies. Radio and optical studies of velocity have not shown the expected decrease in orbital velocity at the optical edges of spiral galaxies. This suggests that galaxies possess giant halos with as much as ten times the visible mass in some form of unseen matter. The nature of this material is unknown. Current observations exclude bright stars and gas as possible constituents of the halo, but very faint stars of low mass are hard to detect and could be present in large numbers.

Studies of the distribution of galaxies in space show that they are not Hubble's "island universes." Rather, they interact in complex ways that have a number of interesting consequences. Collisions between galaxies are believed to trigger the formation of stars, and to disrupt and perhaps even generate spiral arms. Intergalactic space is littered with debris from past collisions and material that has never condensed into galaxies. Gravity has pulled galaxies into stringy associations, as the figure on page 49 shows, where knotty clusters are sites for galaxy mergers. Here the intergalactic gas is heated to temperatures over 100 000 000 K and glows brightly at x-ray wavelengths. The scale of the giant clusters and the volume of seemingly empty space that exists between the strings is much larger than imagined in the early 1970s.

The proof of the existence of neutron stars, first predicted in the 1930s, is one of the major breakthroughs of modern astronomy. Discovered as pulsing radio stars in the late 1960s, it was soon realized that these "pulsars" were rotating collapsed stars. In the early 1970s a second group of pulsars, active at x-ray wavelengths, was discovered by the Uhuru satellite. In this group, collapsed stars are paired with ordinary stars in close binary systems. Gas flows from the ordinary star onto the magnetic poles of the collapsed object; collision processes in the falling gas generate x rays. Because x-ray pulsars are members of binary systems, we can estimate their mass. We can understand most such pulsars as very dense objects of solar mass—that is, neutron stars. But for some systems the derived mass is so high that the collapsed object is thought to be a black hole. Thus astronomers have uncovered a rich zoo of collapsed stellar systems ranging from white dwarfs with central densi-

ties of 10^5 g/cm^3 to neutron stars with densities of 10^{15} g/cm^3 to black holes.

Quasars remain an enigma two decades after their discovery, despite intensive observational and theoretical work. Recent observations couple quasars to distant galaxies and support the belief that their large redshifts are caused by the expansion of the universe. If they are at the edge of the observable universe they are generating energy at the rate of 10^{14} Suns in a volume comparable in size to the solar system. Understanding the detailed nature of this source of energy is the central observational and theoretical problem. We need more light-gathering power throughout the ultraviolet, optical and infrared range to improve the signal-to-noise ratio in the spectrophotometry from which we derive the chemical composition, dynamics and ionization structure of the plasma clouds associated with quasars that were active at an epoch that is twice as old as our Sun.

Technical advances in infrared and submillimeter-wave astronomy have given us the tools to probe, for the first time, the interstellar clouds where new stars are being created. In these dense collapsing clouds, thick concentrations of submicron dust grains shield the interior from ultraviolet starlight and provide a catalyst for the formation of complex molecules that are ubiquitous in these regions. This same dust prevents optical photons from escaping the interior regions. However, the dust is nearly transparent to infrared photons. Furthermore, typical temperatures in the clouds are much lower than stellar temperatures. Thus it is natural to study these complexes at infrared wavelengths. We are now beginning to understand the relative importance of the various processes, but we do not yet understand how star formation is triggered, the details of cloud chemistry or the evolution of a cloud as it collapses to form a new star. To attack these problems we require infrared and submillimeter instruments with better spatial resolution and sensitivity.

Until recently it was believed that the interstellar medium outside these dense clouds was characterized by a temperature of 10^4 K. Observations by the Copernicus satellite show that in

34

addition to a component at 10^4 K there is a 10^6-K component that exists in pressure equilibrium with the lower-temperature gas. This hot gas probably merges with the gas in the galactic halo and eventually with a gaseous wind flowing out of our Galaxy. We need a new telescope to extend the pioneering observations of the now-defunct Copernicus, to chart the physical and chemical nature of this material and to compare the situation in our Galaxy with that of others.

From this very brief description of ultraviolet, optical and infrared astronomy, one may appreciate that the observational data base is collected with a wide assortment of instruments and techniques. The space program is now lifting the opaque atmospheric curtain that hid the vacuum ultraviolet from our view. In the past decade, improvements in detectors have increased the efficiency of optical spectrophotometers by a factor of 100; the sensitivity of infrared detectors has increased by a factor of over a thousand. Proposed cryogenically cooled satellites will be free of the 300 K greybody glow of the Earth's atmosphere. Unhampered by the atmospheric thermal background, and carrying broadband photometers, the satellites will be capable of detecting objects a thousand times fainter than can be seen from the ground.

Instrumentation available to ultraviolet, optical and infrared astronomers is still in a period of rapid improvement. Such improvement is necessary to complement the new facilities in radio, x-ray and γ-ray astronomy. Observations outside the wavelength bands covered by ultraviolet, optical and infrared astronomy are important not only because they provide additional insight into astronomical phenomena but because objects that are powerful sources at extreme wavelengths are usually exotic and astronomically interesting. The observations at ultraviolet, optical and infrared wavelengths give information on distances, morphology and the chemistry and physics of the associated gas. Thus the discovery of each new object creates a new demand for ultraviolet, optical and infrared facilities.

Not an astronomical budget

Sacramento Peak Observatory, Kitt Peak National Observatory and Cerro Tololo Interamerican Observatory are national centers dedicated to ground-based optical and infrared astronomy. Each of these facilities is now in a critical position. Charged with providing observing facilities to American astronomers, they are rapidly falling behind in their ability to fulfill the demand for time. The number of meritorious proposals now far exceeds the available time. The necessary rejection of good proposals constrains scientific creativity. For instance, it is not possible to give large blocks of telescope time to any one project, regardless of merit. Observing runs on the larger telescopes typically last only a few nights. Researchers cannot undertake major statistical programs within these constraints. And the necessity for frequent changes in instruments creates problems in reliability.

The traditionally very strong observatories that are operated by universities and private foundations are also experiencing difficulties. The older observatories are struggling with the problems of outdated equipment, increasing light pollution from nearby growing cities and the erosion of operating funds by inflation. Some of the newer state-owned observatories expanded rapidly during the past two decades and are now finding it increasingly difficult to obtain operating funds as Federal and state governments are reducing their support of science in an attempt to balance budgets. These major observatories, located in good climates, gave America a preeminent position in observational astronomy. Now their decline is reducing the diversity of American astronomy. And the loss of university facilities only makes the problem worse by increasing the pressure on the national centers.

The American program is not as strong as it clearly could be, given the popular support indicated by amateur groups, the renown of astronomical popularizers, and the cultural impact of the more spectacular discoveries such as quasars, neutron stars and black holes. Both the Greenstein (1972) and Whitford (1964) committees recommended a very large ground-based telescope, which has not yet been constructed. Instead, two four-meter

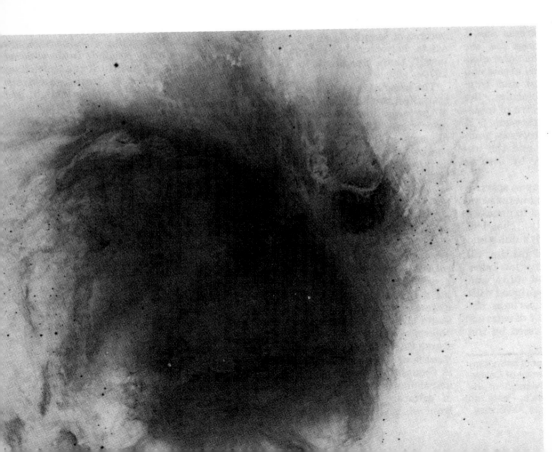

The Orion nebula shows its complexity when photographed through an H_α interference filter. Hot stars are thought to be driving a shock front of ionized gas into a dense molecular cloud where new stars are forming. (European Southern Observatory photograph.)

telescopes, one for each hemisphere, were built by the national centers. Both are extremely productive, as they are characterized by excellent technology and instrumentation and are located at good sites. However, they don't match in size the Russian 6-meter telescope. Australia, Japan and the nations of Europe are mounting ambitious programs of construction that will soon eclipse the US ground facilities. The best astronomical site in the United States, the 14 000-foot-high summit of Mauna Kea, Hawaii, now has more square meters of telescope aperture operated by French, Canadian and English groups than by US groups. Counting the NASA 3-meter infrared telescope in Hawaii (shown in the photograph on page 57), the US national centers have four telescopes with apertures exceeding 2 meters, and no firm commitment to increase this number. The Common Market countries, with a comparable population and smaller gross national product than the United States, have five large telescopes operating on good sites, two more under construction and an additional two funded and in their initial phase of engineering.

Because ground-based optical and infrared observations are inexpensive and flexible compared to space observations, they will remain a central component of astronomy into the twenty-first century. The ground-based observatories therefore must remain healthy. It was evident to the panel on ultraviolet, optical and infrared astronomy that the lack of adequate ground-based facilities is currently a major limiting factor in American astronomy. This deficiency will become more critical as other instruments are built that can identify exciting objects but are incapable of the complementary optical work.

The required optical facilities are not expensive when compared to the overall astronomical budget. A monumental modern telescope is a high technical achievement, yet with modern engineering it is affordable by a large university, a private foundation or even wealthy individuals. Thus uv-to-ir astronomy is unique among the major physical sciences in that it is still possible for a single university to construct and operate a research tool that can undertake work in the forefront of a highly popular scientific field. While most of the telescopes constructed by the Federal government should be operated at the national centers for the benefit of all astronomers, the government can strengthen the base of astronomy by helping private and university observatories that obtain major support from other sources. Healthy private and university groups are an important source of innovative astronomers, new instrumentation, cost-saving construction ideas and diversity of

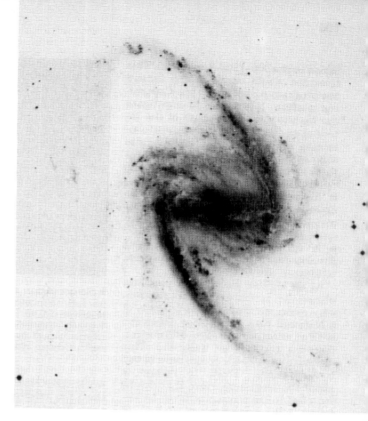

The galaxy NGC 1365 is in the southern constellation of Fornax. In this negative photograph, the dark irregular knots are regions of active star formation that are much larger than the Orion nebula. The light filaments are caused by dust clouds. (European Southern Observatory photograph.)

scientific styles.

In addition to the major new facilities described below, the committee stressed the need to exploit existing facilities fully and to complete the major facilities that have been given the highest scientific ratings by past review groups. The panel on ultraviolet, optical and infrared astronomy identified two instruments in particular as major components of the program for the 1980s. These are the Space Telescope, a 2.4-meter-aperture telescope that is now due to be launched in 1985, and the Shuttle Infrared Telescope Facility (depicted on the cover of this issue), a 0.85-meter-aperture telescope with optics cooled to cryogenic temperatures. Both instruments provide needed capabilities that are not duplicated by other telescopes. The Space Telescope will cover the important wavelength band from 1200 angstroms in the vacuum ultraviolet to the near infrared. In orbit above the distortions, glow and absorptions of Earth's atmosphere, the telescope will be diffraction limited and will detect objects 100 times fainter than can be seen from the ground. The Shuttle Infrared Telescope will also be free of atmospheric limitations. Because the optics are cooled it will be limited by detector noise. For broad-band photometry it will be a thousand times more sensitive than a ground-based telescope and is expected to be able to detect galaxies in formation at the beginning of the universe. Neither of these programs was rated by the pan-

el; rather, they were considered extremely valuable programs from the 1970s that have not yet been completed.

A 15-meter telescope

The excitement of astronomy lies in the study of extreme phenomena—those objects and places where the conditions of temperature, density, scale and dynamics are completely unusual. Radio, x-ray and other space facilities will identify huge numbers of exciting but optically faint sources. Rapid progress in the understanding of the processes involved in the birth of new stars or the formation of our universe requires the will to achieve the technological thrust needed to build the instruments that can collect and analyze the faint gleams of light from key sources.

Assuming that the Space Telescope and the Shuttle Infrared Telescope are flown, remaining major needs are:
▶ A substantial improvement in ground-based facilities capable of observing faint objects with high spectral resolution at optical and infrared wavelengths. There is an urgent need for small telescopes, and a very large telescope is required for the frontier observations.
▶ A large orbiting telescope designed to complement the Shuttle Infrared Telescope by providing high angular and spectral resolution in the wavelength band between 20 microns and 1 mm.
▶ A satellite telescope that combines high efficiency with high spectral reso-

lution in the 900–1200-Å interval. This telescope would continue and extend the program pioneered by the Copernicus satellite. It is needed to investigate the high-temperature phase of the interstellar medium, other regions containing hot gas, and cloud constituents such as H_2, which have major spectral features below the 1200-Å cut-off of the Space Telescope mirrors.

▶ An orbiting solar observatory, as described by Arthur Walker in his article on solar astronomy in this issue (page 60).

▶ A strong program of instrument development to ensure that expensive telescopes are used efficiently.

As its highest priority the panel on ultraviolet, optical and infrared astronomy recommended the construction of a National New Technology Telescope with an effective aperture of 15 meters. (See the sketches on page 44.) This aperture was chosen on the basis of two considerations: First, 15 meters is about the largest aperture that can be matched to the atmospheric conditions found at the best sites and, second, 15 meters is about the largest telescope that can be built without a very costly program of technical development.

The principal optical justification for building a 15-m telescope is its capability for spectroscopy of faint objects. At the darkest sites the background illumination from a square arc second of the night sky corresponds to a blue flux of about magnitude 23 (2.3×10^{-29} ergs/sec cm^2 hz). For a given angular entrance aperture, spectrographs scale in size with telescope aperture. Even at good sites, atmospheric turbulence usually smears star images over at least an arc second of sky. For spectrographs that are not severely limited by slit width, grating technology limits the resolving power R (that is, $\lambda/\Delta\lambda$) to about 20 000 for a 15-m telescope. Such a large telescope collects sufficient light so that with efficient designs it will be possible in only a few hours to obtain high-quality spectra at this resolution for 23rd magnitude stars. This matches well the time it will take the star to cross the meridian. For objects fainter than 23rd magnitude the background of noise from the sky limits the achievable ratio of signal to noise.

A giant telescope able to collect from very faint objects data with a high ratio of signal to noise would allow us to study matter that is now essentially invisible. High priority would be given to studies of galactic halos, the binding mass of clusters and superclusters, studies of the formation of stellar systems and protoplanets, and the chemical composition, dynamics and ionization state of the intergalactic medium. While modern and well-instrumented existing large telescopes can give us exciting glimpses of the

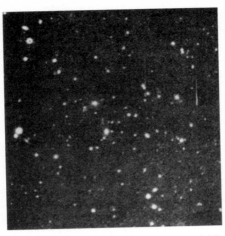

A distant cluster of galaxies with a redshift $\Delta\lambda/\lambda$ of 0.53, corresponding to a velocity of recession of 0.4c. This image is the result of a 5-minute integration by a 500×500-element charge-coupled device on the 200-inch Palomar telescope. (Caltech photo.)

worlds that lie just beyond their grasp, the 15-m telescope is needed to study these frontiers effectively.

At the diffraction limit. It appears possible to use speckle interferometry to obtain some information at the diffraction limit of the 15-meter telescope. Studies using the Multiple Mirror Telescope have achieved speckle information for sources as faint as 17th magnitude. For the 15-meter telescope this would be angular information corresponding to a resolution of 5–7 milliseconds of arc, depending on the choice of telescope design. This is approximately 100 times better than the best seeing conditions found on the surface of the Earth. The techniques, which are still difficult, give very little field and do not produce unique maps for sources with complex optical structure. Nevertheless, their potential is so great that an important part of the justification of the 15-meter telescope is its capacity to resolve objects at its diffraction limit.

In the infrared there are numerous windows in the atmosphere between 2 and 30 microns. Due to the fact that image distortions caused by the atmosphere decrease with increasing wavelength, large telescopes become diffraction limited rather than seeing limited in the infrared. Because the atmosphere and telescope optics are warm, even a 15-meter ground-based telescope could not compete with the Shuttle Infrared Telescope for broad-band photometry over most of the atmospheric windows. But for high-resolution spectroscopy, the instrument is limited by detector noise and its sensitivity increases as the square of the aperture of the telescope. High angular resolution is required to map complex objects such as protostars and galactic nuclei. By combining high spectral resolution

with high spatial resolution, the 15-meter telescope will be unrivaled in its ability to study the chemistry, physics and dynamics of molecular clouds through atmospheric windows in the 2–30 micron interval. It complements the broad-band photometric capabilities of the Shuttle Infrared Telescope and extends to the shorter infrared wavelengths the deep infrared and submillimeter capabilities of the Large Deployable Reflector described below.

Thus the 15-meter telescope will extend our capabilities for optical spectroscopy by an order of magnitude and our capabilities for infrared spectrophotometry by two orders of magnitude. This exceeds the improvement in resolution brought by Galileo's telescope over the naked eye. However, astronomy does have threshold problems. With only a slight improvement in resolution, Galileo discovered the satellites of Jupiter, craters on the moon and spots on the Sun—observations that shook European culture. With our current spectroscopic and infrared instruments, we fall short of being able to make definitive observations of astronomical objects sufficiently far away to study the aging of the universe. Similarly, we do not have the spatial and spectral resolution needed to study the details of the collapse of molecular clouds and the formation of stars, as the nearest examples are too far away to be resolved adequately. The 15-meter telescope, equipped with modern instruments and located on an excellent site, will be capable of attacking such problems in detail.

The cost of constructing the 15-meter telescope is expected to be less than 100 million 1980 dollars, a figure very much smaller than would have been predicted a decade ago. The principal reason for the low cost is the saving introduced by the technology of lightweight mirrors. Also, modern structural design using finite-element analysis, and the practical experience gained in the construction of the 4-meter-class telescopes over the past decade and a half, permit accurate calculation of the structural and dynamical performance of the mirror support system, the telescope mount and the telescope drive before construction begins.

Two competing designs. At present there are two principal competing designs, each with its own advantages and disadvantages. One possibility is to use the design of the lightweight and successful Multiple Mirror Telescope. This telescope was constructed by the University of Arizona and the Smithsonian Astrophysical Observatory on Mt. Hopkins, Arizona. With six 1.8-m mirrors on a common mount feeding a common focus, it achieves an effective aperture of 4.5 m. This innovative concept uses well-established technology, and by now there is considerable observing experience. The major disadvantage of this design is that the individual mirrors for a 15-m telescope are very large and the optics required to produce a cofocal image plane are complicated and introduce additional reflections, which cause the loss of light. Advantages include a large overall size that gives high resolution for speckle and infrared imaging, a comparatively simple mirror support system, and simple pupil masks for infrared work.

A group at the University of Arizona is currently developing technology to cast large, lightweight, cellular Pyrex mirror blanks up to 7.5 meters in diameter that could be used in a 15-meter instrument like the Multiple Mirror Telescope. The group's successes in casting small mirrors are encouraging and within a year they will attempt to cast an intermediate-sized blank. Because the mirrors for a multiple mirror telescope are so large, it is important to find a successful design that combines rigidity, low weight and low thermal inertia.

A competing proposal is to build the telescope using an array of hexagonal mirrors mounted together to simulate a single large mirror with a 15-meter aperture. Because the deformations of a plate bending under its own weight scale as the fourth power of the diameter and inversely as the square of the thickness, small mirrors can be made relatively thin and light. Here the advantage is that the telescope pupil is filled and only a few surfaces are required to reach a stigmatic focal point. Because numerous relay mirrors are not required, this design probably gives the maximum saving of the light that was so laboriously collected. Major disadvantages include the technical difficulties of producing and maintaining the primary mirror surface. Also, there is no smaller telescope, analogous to the Multiple Mirror Telescope, operating to give actual experience for critical, untried subsystems.

Using a small mirror, workers at the University of California, Berkeley, have successfully demonstrated an innovative technique for producing the individual mirror segments required for the primary mirror. Opticians at Kitt Peak are now attempting to use this technique to produce two matched full-sized (2-meter) segments. Laboratory tests show that servo components for sensing and controlling the positions of the segments work at better than required precision. The Berkeley group working on the 10-meter telescope is now attempting a full-scale demonstration of the control system using two mirrors.

Because the segmented mirror approach and the Multiple Mirror Telescope design are the principal contenders, the final choice of the design for the 15-meter telescope depends on the outcome of the current experiments.

One result of the studies leading to a low-cost 15-meter telescope is that the cost of smaller telescopes is dropping substantially. Two university groups, at the University of Texas and the University of California, are hoping to build large telescopes with mostly internal funds, that is, without Federal funds. The Texas group hopes to build a telescope with an aperture of 7.5 m, using a single lightweight mirror. This mirror may come from industry or from the University of Arizona if the group there succeeds in casting very large Pyrex mirror blanks. The California group plans to build a 10-meter telescope using segmented mirrors.

Several universities, including the Universities of California, Michigan and Washington, are seeking funds to build smaller, 2-meter-class telescopes. The panel on ultraviolet, optical and infrared astronomy believed that these instruments, if located on good sites, would not only be effective in their own right but would be valuable adjuncts to the national 15-meter telescope and the space program. To show their strong support they coupled their endorsement of the 15-meter telescope with that for the smaller telescopes, calling for a single program designed to upgrade the capital facilities for ground-based astronomy.

A 10-meter orbiting reflector

The panel gave second priority to developing for launch in the early 1990s a 10-meter-class Large Deployable Reflector. This telescope would be designed to work in the spectral region between 20 microns and 1 mm, a region in which atmospheric absorption very effectively blocks ground observations. Present design goals include diffraction-limited operation at 30 microns, deployment without extensive extravehicular activity by astronauts, and a long orbital lifetime during which instruments could be changed or upgraded. The instrument would be a national facility.

As noted above, the Large Deployable Reflector and the 15-meter telescope would complement each other in giving high spectral and angular resolution throughout the region from 2

Projection of galaxies within 100 million parsecs of the Sun as they would be seen from outside the region—one of a stereoscopic pair of images. An examination of the three-dimensional structure of this system shows that the galaxies are concentrated into string-like associations with large open holes between the groups. (Image courtesy of Richard Miller.)

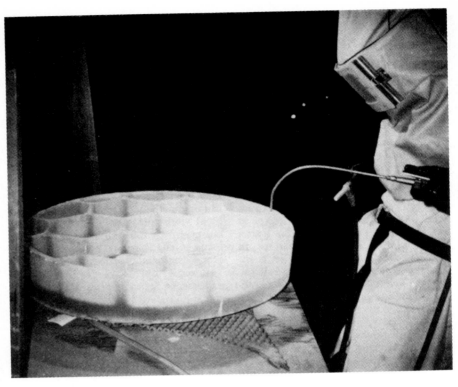

Sandblasting an experimental 30-inch Pyrex mirror blank. (University of Arizona photograph.)

microns to 1 mm of the infrared spectrum. Passive cooling will keep the optical surfaces of the orbiting reflector below 200 K. The technology for constructing this telescope is not yet available, but industry studies now under way should advance technology to the point that construction could begin in the late 1980s. By requiring diffraction-limited performance only at wavelengths longer than 30 microns, the accuracy of the mirror surface and its supporting structure is greatly relaxed compared with the accuracy required for optical and near-infrared telescopes. Nevertheless, it appears that the experience gained in using the Large Deployable Reflector will guide the design of very large optical telescopes that could be orbited in the 21st century.

As is the case for the 15-m telescope, the Large Deployable Reflector is not competitive with the cryogenically cooled Shuttle Infrared Telescope for broad-band, low-spatial-resolution work. But its very large mirror gives it a tremendous advantage for problems that require high spatial and spectral resolution. Many rotational transitions of molecules lie in the submillimeter wavelength band. For instance, the S(0) rotational line of H_2 lies at 28 microns, and rotational lines of HD lie at 29, 38, 56 and 112 microns. Comparison of data obtained in the far ultraviolet with data from the infrared can lead to a firm value for the ratio of hydrogen to deuterium. Because deuterium is believed not to be formed by stars, this ratio is important to the understanding of the nuclear processes in the early history of the universe.

Molecular transitions involving the upper rotational levels of OH, CO and H_2O are important in the studies of the hot, dense regions of molecular clouds. Far-infrared fine-structure transitions in the ground state of O^{++}, N^+ and Si^+ ions are important in studies of the abundance, excitation and ionization of hydrogen, while fine-structure lines from O^0 and C^+ are useful probes of transition regions where atomic hydrogen is not ionized.

The initial size of protostars is believed to be about 2×10^{15} m for an object of solar mass. The angular size of such a protostar is about 20 arc seconds at the distance of the Orion nebula. The Large Deployable Telescope would be capable of studying the density and temperature structure within such an object. At a later evolutionary phase the star would have collapsed to an angular diameter of about 6×10^{13} m. This is comparable in size to the solar system, and at the distance of the Orion nebula is comparable in angular size to the resolution of the Large Deployable Telescope. The angular resolution of the Large Deployable Reflector, the capability of the 15-meter telescope at shorter infrared wavelengths and the high sensitivity of the Shuttle Infrared Telescope will give us a detailed history of the birth of stars. Because it is very hard to detect planetary systems that might be associated with even the nearest stars, our Sun is the only star known to have planets. However, it might be possible to infer the existence of planets through the infrared emission that some theories predict would accompany the formation of planetary and stellar systems.

The angular resolution of the Large Deployable Reflector corresponds to lengths of 300 light-years or less at the distances of the nearer galaxies. These lengths correspond to the extent of galactic nuclei and to the extent of the active regions of star formation in spiral arms. Most of the energy released by galactic nuclei and by collapsing protostars emerges as far-infrared radiation. To compare these phenomena with our own Galaxy, and to understand the physical processes involved, will require the spectroscopic and angular capabilities of the Large Deployable Reflector.

Far-ultraviolet spectrograph. The third priority of the panel on ultraviolet, optical and infrared astronomy was a Far-Ultraviolet Spectrograph. The mirror coatings used on the Space Telescope are designed to be very efficient for wavelengths longward of about 1200 Å but have rapidly vanishing reflectivity for shorter wavelengths. The inability of the Space Telescope to work effectively below 1200 Å led the panel to propose that a high-resolution spectroscopic satellite be launched to observe this important spectral interval. It could be an Explorer-class instrument whose primary mission is to cover the wavelength interval between the interstellar hydrogen absorption edge at 912 Å and the Space Telescope threshold of efficiency at 1200 Å.

The resonance lines of many astrophysically important ions, including O^0, N^+ and O^{+5}, and the electronic transitions of H_2 and HD, occur in this wavelength interval. The densities of H_2, HD and other molecular species found in interstellar clouds are correlated with the density of cosmic rays. Thus we can use information on molecules to trace the density of cosmic rays throughout the Galaxy.

Studies of atomic resonance lines will lead to improved models of the interstellar medium. For instance, the discovery of the resonance lines of O^{+5} at 1032 Å and 1037 Å led to the recognition that some of the interstellar gas in our Galaxy is at a temperature of a million degrees. The sharp absorption lines seen in many quasar spectra are often intepreted as the result of intergalactic clouds or galactic halos along the line of sight. Spectroscopy of the nuclei of nearby Seyfert galaxies will allow us to compare the conditions in the halo of our Galaxy with that found in the Seyfert galaxies. We can compare the results from the sample of nearby of galactic halos with the results from the distant quasars to estimate the amount of evolution that has occurred over cosmological time scales.

Other instrumentation

Telescopes can only collect light and form images. Analysis requires spectrographs, photometers and detectors. Because telescopes are large and expensive, they naturally attract keen interest and provoke much discussion. However, the acquisition of useful data requires both an efficient telescope and efficient instrumentation. Thus the panel on ultraviolet, optical and infrared astronomy listed support for instrument development as a major recommendation.

Astronomical instrumentation is not easy to build. Mechanical tolerances are very tight. The required optical precision is high. The need for efficient systems translates into a requirement for coatings that are efficient over a wide range of wavelengths; and the quest for efficiency also pressures the designer to reduce to an absolute minimum the number of optical surfaces. The resulting designs are compromises between optical aberrations, wavelength coverage, mechanical rigidity, detector format and efficiency. As new detectors are developed, new optical compromises are required to exploit their advantages. For example, an improvement in the noise figure of infrared detectors has required a reduction of the thermal emission from infrared instrumentation. Instrumentation development is a continuing requirement as technology and research fields change.

Because improvements in sensitivity open new fields for research, the improvement of detectors is rightfully an obsession among experimental astronomers. Requirements for detectors include high efficiency at very low light levels, wide dynamic range, gain stability, large numbers of independent detector elements (pixels) and a format that can be matched to telescope or spectrograph image planes. For optical imaging and spectroscopy the new solid-state "charge-coupled device" detectors, with many thousands of pixels, are approaching the theoretical limits. And there is rapid development of infrared detectors. Two-dimensional infrared detectors with low background and high quantum efficiency are becoming available for the first time. These devices will open new vistas for those capable of constructing the matching instrumentation.

Satellite instrumentation needs a substantial improvement in efficiency. There is still no detector for the vacuum ultraviolet that can match the performance of the optical charge-coupled devices. Van Allen belt particles and cosmic rays increase the radiation environment for detectors used in space. This noisy background puts a serious limitation on observing faint objects. The ability to upgrade the instrumentation used with the Space Telescope, the Shuttle Infrared Telescope Facility and the Large Deployable Telescope is an important virtue of their designs. The National Aeronautics and Space Administration and the National Science Foundation are to be commended for their strong programs to develop detectors and instrumentation.

The panel on ultraviolet, optical and infrared astronomy attempted to identify a program that would be broadly balanced, that would take advantage of technical opportunities and would be capable of making rapid advances in promising areas of research. The three major components of the program for research outside the solar system—the 15-meter telescope, the Large Deployable Reflector and the Far-Ultraviolet Spectrograph—are intended to complement the capabilities of Space Telescope and the Shuttle Infrared Telescope Facility. Together these facilities will give us orders-of-magnitude improvement in our present capabilities for imaging and spectrophotometry throughout the wavelength range extending from 900 Å to beyond 1 mm.

The recommendations of the panel were reviewed very favorably by the full Astronomy Survey Committee. The committee ranked the 15-meter-telescope project third among the major programs, saying,[1] "The Committee finds the scientific merit of this instrument to be as high as that of any other facility considered and emphasizes that its priority ranking does not reflect its scientific importance but rather its state of technological readiness." And the large Deployable Reflector was ranked fourth among major programs. The Far-Ultraviolet Spectrograph was ranked second among the moderate new programs after a call for augmenting the NASA Explorer program. The need for instrumentation and detectors received highest ranking among the prerequisites for new research initiatives.

The anticipated enormous increase in instrumental capability is as profound as the invention of the telescope. It brings us to the edge of a revolution in our understanding of the universe. Nature, unconstrained by the limitations of human imagination, is certain to surprise and educate us if we have the will to inquire.

Reference

1. Astronomy Survey Committee, *Astronomy and Astrophysics for the 1980's, Volume 1: Report of the Astronomy Survey Committee*, National Academy Press, Washington, D.C. (1982).

A look at photon detectors

Whether photomultipliers, photodiodes or photoconductors are "best" in a given case depends on signal frequency, on the kinds of noise present and on the coherence of the radiation.

Robert J. Keyes and Robert H. Kingston

PHYSICS TODAY / MARCH 1972

In the best of all possible worlds, we would have the ideal photon detector, a device that caught a photon, gave an unambiguous meter reading and kept count of the number of events. In the real world, these ideal devices do not exist; competing events both outside and inside the detector confuse the true measure of the photons that we are trying to monitor. Phenomena such as quantum noise, "dark" current and background radiation interfere to a degree that depends on the intensity of the signal being measured and on the photon frequency, to mention only a few of the experimental parameters.

Here we shall describe the processes that are important when keeping track of low-intensity photon "beams," at frequencies ranging from the ultraviolet through the infrared, because this is the situation for which most of us need a photon detector. We shall look at what "detecting a photon" really means in operational terms, and see what the physical bases of the competing noise processes are. We shall also look at the

Robert Keyes, formerly head of the Optics and Infrared Group at the Massachusetts Institute of Technology Lincoln Laboratory, is a private consultant; Robert Kingston heads the Laboratory's Optics Division.

special problem of coherent detection, important to those of us who want to analyze laser signals. Finally, we shall briefly review what all this theory means in terms of the sensitivity of real detectors throughout the spectrum.

Photoelectric processes

The oldest forms of photon detector are the eye and photographic film, but these two are not usually suitable for detecting low-intensity signals; we shall limit the discussion here to those devices with a single electrical output that measures the photon stream at some point in space. What we want is clearly a photoelectric event—the production of an excited electronic state by a photon. The best way to proceed depends largely on the photon frequency: In general, vacuum-tube detectors are more responsive to the ultraviolet and visible regions of the spectrum, and semiconductors work better in the infrared (see figure 1). Thermal detectors, which operate as heat engines, are not sufficiently sensitive to be used for low-intensity signals.

In the visible and ultraviolet region of the spectrum, we are most apt to use the original photoelectric effect, photoemission, in which radiation incident on a metallic surface causes the metal to emit an electron into the surrounding

vacuum. Once the electron is emitted, our problem is to detect it. Fortunately, the electron multiplier, which consists of a photoemissive cathode, a series of electrodes at successively higher positive potentials (dynodes) and a collector plate at the highest positive potential, allows us to amplify the initial single-electron pulse, as in figure 2a, so that the measured current pulse is greater than the noise in the measuring amplifier. The dominant source of noise in photomultipliers is the thermal generation of electrons in the photosurface. Because thermal generation is a steep exponential function of temperature, only moderate cooling is needed to eliminate this source of noise.

The quantum efficiency for photoemission can approach unity in the ultraviolet through visible regions of the spectrum, but falls off rapidly to a few percent in the near infrared (one micron). In contrast, the efficiency of electron or hole creation in semiconductors can approach unity throughout the visible and infrared regions, if the semiconductor is sufficiently cooled. The cooling ensures that collision of the photons with already free carriers are negligible compared with the "across the gap" or "bound–free" excitations that we wish to measure; for this as-

Photomultiplier and photodiode. The eleven-stage "Quantacon" photomultiplier (top) has a gallium-arsenide photocathode and is the type used in the Apollo lunar-ranging experiment. Silicon photodiode (bottom) is sensitive in the infrared region of the spectrum. Avalanche multiplication techniques are making silicon photodiodes competitive with photomultipliers in the one-micron region. (Photos by RCA.)

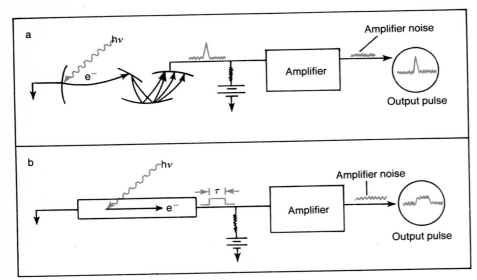

An electron multiplier amplifies the single-electron pulse resulting from photoemission, so that the measured current pulse is greater than amplifier noise (a). In the infrared region of the spectrum, semiconductors (b) must substitute for photomultipliers. The signal to noise ratio is not as great, but "avalanche" multiplication offers some improvement.
Figure 2

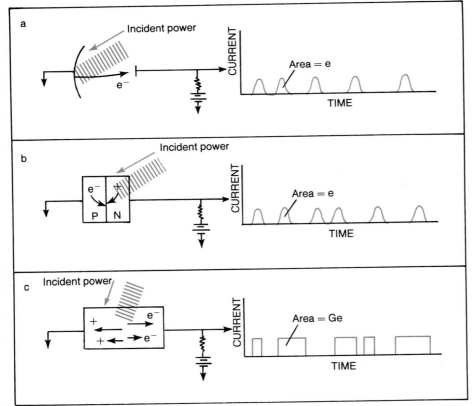

Pulses from a photomultiplier (a) and a photodiode (b) follow a Poisson distribution and have the same unvarying charge content. A photoconductor, on the other hand, emits pulses with charge content that varies from pulse to pulse (c) and depends on the lifetime of the excited carriers.
Figure 3

surance, the temperature must be much less than $h\nu/k$.

If the semiconductor is thick enough, the photon will eventually produce an excited carrier. But production does not ensure detection, and unfortunately the recording of this event by the external circuit is limited by the lifetime of the excited carrier and by the competing noise in the following amplifier (see figure 2b). Some of the noise can be overcome by biasing a semiconductor photodiode so that "avalanche" multiplication of excited carriers occurs. In any case, we see that detection of *individual photons,* as one might picture the constituents of a very weak signal (for example, from the laser-ranging reflector left on the moon by the Apollo astronauts) is very difficult except at short visible wavelengths. But, as we shall see, more serious problems usually limit the use of a photon detector in any practical system.

Signal versus noise

We have posed the problem of the minimum detectable power or energy seen by a photon detector, and we should realize that our definition of a photon is a practical rather than a theoretical one. To measure its existence, we simply state the measurement rule: The average rate of production of photoelectrons (or holes in semiconductors, for that matter) is equal to the product of the quantum efficiency η and the ratio of the incident power P to the quantum energy $h\nu$ corresponding to the photon frequency, so that

$$i_{av} = e\eta P/h\nu$$

where i_{av} is the photocurrent and e is the electron charge.

An essential feature of this law is that the time behavior of the photoelectric events obeys Poisson statistics (random occurrence, see figure 3a), provided that the incident power remains constant. The random time distribution of electron (excited-carrier) production leads to a current fluctuation or noise in the presence of any photocurrent i such that

$$\delta i = e\delta n/T = eN^{1/2}/T$$
$$= e(iT/e)^{1/2}/T = (ei/T)^{1/2}$$

because, for a given measurement time T, the fluctuation δn in the number of events is proportional to the square root of the total number N during the time T.

In a photoemitter, we measure the electron events directly, and the noise current (also known as "shot noise") may be shown to be

$$i_{rms} = (2eiB)^{1/2} \qquad (1)$$

where B is the bandwidth of the detector circuit.

We see then that even a "perfect" photon counter would produce a "noisy" output in the presence of a constant input signal, and, in fact, this quantum noise contribution, along with the thermal noise generated in the detector amplifier, are the two limiting factors in the sensitivity of any photon detector, no matter how high the quantum efficiency.

If only the desired signal were present, we could easily calculate the minimum power detectable by any given photoemissive device, assuming we knew the quantum efficiency and the noise-current contribution of the amplifier. But in many applications the sensitivity is more severely limited by the quantum noise generated by sources of radiation or electron production other than the true "signal." In particular, the "dark" current or thermionic-emission component in a photoemitter adds its share of current fluctuation, and even more important is the unwanted background radiation that emanates from the vicinity of the signal source or from the intervening path or optical components. This background radiation may be perfectly constant in time, yet it produces a quantum or "photon" noise that competes with the signal current. We can make the dark current negligible by cooling the photocathode, but the background-induced noise is strictly controlled by the environment of the signal source and the design of the optics and detector system.

Noise in semiconductors

Shot noise in a photoemitter detector is, as we have seen, a fundamental limitation. What is the equivalent noise mechanism in semiconductor devices? Here we must distinguish between the photodiode and the photoconductor. The photodiode is almost an exact analog of the photoemitter, except that minority carriers (rather than electrons) are produced near a reverse-biased junction and are "collected" after they diffuse to the barrier (see figure 3b), just as in a transistor. As each carrier is collected, it produces an electron-sized pulse at the output, and the noise current is the same as that in a photoemitter.

The photoconductor, in contrast, is a bulk device whose conductance is controlled by the photoinduced generation and eventual recombination of free carriers, either holes or electrons. Photoconductors may be either "intrinsic" or "extrinsic." In an intrinsic photoconductor, incident radiation produces an "across the gap" transition, creating an electron–hole pair by giving a bound electron enough energy to cross to a conduction band. In an extrinsic (impurity-doped) device, a free carrier is excited from a bound state. As we have noted, the device must be cool enough that kT is much lower than the photon energy—otherwise the fluctuations of

thermally excited carriers will overwhelm the contribution of the photoinduced events.

When we apply a voltage to a photoconductor, we get a series of current pulses in the presence of radiation, as we did with the other devices. Here, however, the pulses, although Poisson distributed, will each have an effective charge that is neither equal to the charge on an electron nor constant (see figure 3c). Instead, as each free carrier is randomly produced, it will drift in the internal field of the photoconductor, producing a current at the output electrodes for a finite time until it recombines. Recombination is a probabilistic process, so that there will be a distribution of pulsewidths. The distribution will be exponential, with a mean width equal to the average lifetime of the carrier.

The average charge per pulse may be simply related to the ratio of the carrier lifetime to the time needed for the carrier to travel completely across the photoconductor. If, for example, the applied voltage is such that the carrier drifts *completely* across the device between the time it is excited and the time it recombines, the charge transfer would be one electron. This average charge induced per excited carrier is defined as the gain G, and it may be greater or less than unity, depending on the lifetime and drift velocity of the carriers.

A gain greater than unity seems to imply that the excited carrier drifts out at one end of the semiconductor and reappears magically at the other terminal to drift again. In actuality, it does not, but charge neutrality requires that a new carrier appear at the opposite terminal to replace the old one, and the recombination probability remains the same, so that the effective lifetime may be many transit times.

As for the noise, we now have a series of random pulses with random charge content of average value Ge. It turns out that we can use the shot-noise formula of equation 1 by the simple artifice of replacing the electron by a G-sized electron and, to take into account the random fluctuation in size, multiplying by an extra factor of two. The resulting noise current is then

$$i_{\rm rms} = (4GeiB)^{1/2} = [4Ge(\eta GP/h\nu)B]^{1/2}$$
$$= Ge(4\eta PB/h\nu)^{1/2} \quad (2)$$

We see that the noise behavior of the entire family of photon detectors is markedly similar, if we are willing to deal with random-size electrons and to account for factors of two.

Do not conclude from equation 2 that a small G in a photoconductor is a fruitful way to decrease the noise; remember that the signal current is directly proportional to G so that trying to reduce G (by varying the semiconductor material, for example) is a losing game,

because amplifier noise is the dominant limiting noise in photoconductors.

Limits to photodetection

We are now prepared to look at the way the wavelength and the source of the radiation limit photon detection. Three categories of detection exist: photon-noise limited, amplifier-noise limited and background limited.

Photon noise is the limiting factor only in the visible and near-infrared regions of the spectrum. Here appropriate filtering and field-of-view restrictions can reduce background radiation, and electron-multiplier structure can overcome amplifier noise. The limit, then, is the quantum efficiency, which ranges from a high of about 30 percent in the blue region (4000Å) down to a few percent (in experimental tubes) at one micron in the infrared.[1]

A useful quantity, when discussing photon detection, is noise-equivalent power (NEP), which is defined as the signal power that produces a detector output voltage (or current) equal to the rms noise current. As a benchmark, photon-noise limited detection yields a noise-equivalent power equal to $2h\nu B/\eta$ for the photoemitter or semiconductor photodiodes and $4h\nu B/\eta$ for photoconductors. For a 1-Hz bandwidth, the normal reference for evaluating noise-equivalent power, this value corresponds to 8×10^{-19} watts for unity quantum efficiency at 5000Å.

Beyond one micron (10 000 Å) in the infrared, amplifier noise limits detection capabilities in the presence of low background, because vacuum electron multiplication is no longer possible (photoemission does not occur with these low-frequency photons), and we must resort to semiconductor detectors. Silicon photodiodes that use avalanche multiplication are becoming competitive in the one-micron region; their high quantum efficiency partly offsets the noisy internal amplification and compensates for the poor quantum efficiency of photoemitters in this region. For any semiconductor device, the noise-equivalent power increases precipitously, by several orders of magnitude, as the wavelength increases beyond about one micron. Either amplifier noise or photon noise caused by the high background radiation at the longer wavelengths produces an effective noise current proportional to the square root of the bandwidth. The noise-equivalent power then becomes proportional to the square root of the bandwidth, because the signal current needed to compete with the noise is directly proportional to the input signal power.

In many infrared-detection systems, the background radiation sets the limit, so that improved detectors would be of no advantage. Background radiation is particularly important beyond about

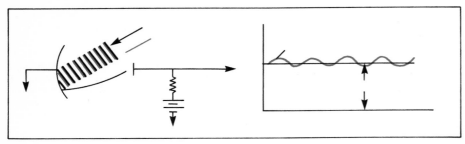

Coherent detection. The photocurrent in coherent detection has two components: a constant induced current, proportional to the incident power P_{LO} of the local oscillator, and a cross-product current that varies sinusoidally at the difference frequency between the local-oscillator field and the signal field.
Figure 4

five microns; in this region the emission from the atmosphere is enough to set the noise limit if one is detecting thermal radiation from a distant source. In this case of background-limited detection, the noise-equivalent power becomes equal to $(4P_B h\nu B/\eta)^{1/2}$ for a photoconductor, where P_B is the power of the background radiation striking the detector.

Detecting coherent signals

Our analysis so far bodes poorly for the detection of laser radiation in all but the visible and near-infrared regions of the spectrum: Although the narrow linewidth of laser beams allows optical designs with narrow spectral filters, which could eliminate all background and theoretically allow photon-noise limited detection far into the infrared, this extension is apparently not feasible because of amplifier noise. Fortunately, heterodyne or coherent detection helps solve this problem, with the help of a local laser source. Just as we do in radio or television reception, we shall mix the incoming weak signal wave with a strong oscillator beam at the detector.[2]

Because we want the signal and local oscillator waves to add coherently (that is, the electric fields should add constructively or destructively), we arrange that the detector only "sees" the effect of a signal wave that arrives within an angle less than λ/d of the oscillator beam, where d is the beam diameter. Otherwise, the two wavefronts will not remain in phase across the detector surface. The induced photocurrent is proportional to the incident power P_{LO} of the local oscillator, that is, to the square of the incident electric field, so that we have a constant current $e\eta P_{LO}/h\nu$ due to the local oscillator (see figure 4). In addition, there is a cross product of the local-oscillator field and the signal field that varies sinusoidally at the difference frequency between the two waves. This term is equal to $2e\eta(P_{LO}P_S)^{1/2}/h\nu$, where P_S is the signal power. A much smaller current component, due to the signal wave alone, may be ignored.

The noise current is now set by the "photon" noise of the local oscillator and is equal to $[2e^2(\eta P_{LO}/h\nu)B]^{1/2}$.

We find that the noise-equivalent power becomes $h\nu B/\eta$ for the photoemitter or the semiconductor photodiode. For the photoconductor, we have $2h\nu B/\eta$. In either case we must make the local-oscillator power so large that its noise current, rather than amplifier noise, is the dominant contribution.

Particularly significant in this mode of detection is the coherent nature of the process: The detector–local-oscillator system only "sees" radiation coming from the diffraction-limited field of view of the illuminated surface, and the output wave from the detector is a frequency-shifted replica of the input waveform. The current output is proportional to the incident electric field. The detection process here, then, is linear, as contrasted with normal photon detection, in which the output current is proportional to input power. Because of this coherence, the bandwidth B required in the detector circuit is the true spectral bandwidth of the incoming signal. Heterodyne detection, then, is limited to narrow spectral lines, typically laser radiation, because detector bandwidths are seldom larger than 10^9 hertz. (A typical blackbody-radiation source is about 10^{13} hertz.) Under these restrictions of spectral purity and diffraction-limited field of view, we see that the minimum detectable signal is effectively the same as the ideal photon-noise limited case for incoherent detection. Coherent detection is effectively limited to laser systems in a more restrictive sense as well, because local-oscillator power of the required frequency purity and intensity is available only from lasers.

Choosing a detector

How does one select a detector from among the available types? A major factor in the decision is usually the photon wavelength. In figure 5 we see the reciprocal of the noise-equivalent power plotted against photon wavelength, under standard conditions—detector area equal to one cm^2 and electrical bandwidth of one hertz. Under these conditions, the reciprocal of noise-equivalent power is known as the "detectivity." Indicated on the figure are the photon-noise limited and coherent

cases, the 290-K background limitations on a sensor exposed to a full hemisphere, the lower limits of noise-equivalent power for a semiconductor detector with insufficient gain to overcome pre-amplifier noise and the response curve for a photomultiplier.

Through the visible region and out to one micron, photomultiplier tubes most closely approach the ideal photon counter. The data plotted in figure 5 are for commercial tubes with alkali-metal cathodes. Recently developed GaAs–Cs$_2$O and InAsP–Cs$_2$O emitters could increase photomultiplier performance by a factor of 10 to 30 in the one-micron region; the improvement is a direct result of increased quantum efficiencies at these wavelengths.

Our present understanding of photoemitter band structures and vacuum work functions indicates that photoemission out to wavelengths of 2.4 microns is possible. A unique characteristic of photoemitters is their high-frequency capability: In the form of simple diodes, gigahertz responses are possible, and for photomultipliers, 100 MHz is the limit. The bandwidth limitations of photomultipliers results from transit-time spreads in the secondary-emission process.

In the spectral region from 1 to 40 microns, photoconductors have the highest sensitivity (about 10^{-15} watts) among the detectors.[3] Photoconductors characteristically have high quantum efficiencies, often as high as 0.6. When sufficiently cooled, their major limitation is preamplifier noise. To reach the photon-noise-limited ($h\nu B/\eta$) performance, internal gains of from 10^3 to 10^4 would be needed. This goal is possible, but we must remember that high gain in photoconductors comes only at the cost of frequency response.

Although avalanche photodiodes do not appear in figure 5, they warrant brief discussion. When avalanche multiplication was first observed in silicon p-n junctions, hopes were raised that the solid-state equivalent of photomultipliers was emerging. Unfortunately, this hope has not been realized. The best avalanche silicon diodes are about equivalent to standard photomultipliers at one micron because of the

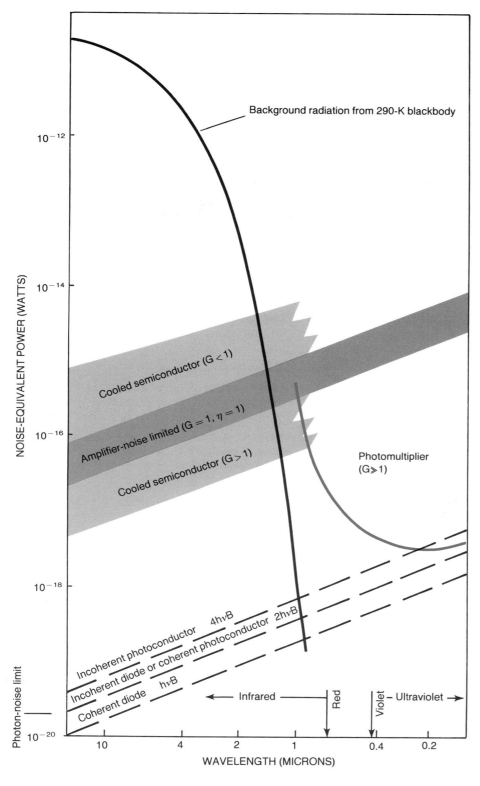

Background radiation from 290-K blackbody

Cooled semiconductor (G < 1)

Amplifier-noise limited (G = 1, η = 1)

Cooled semiconductor (G > 1)

Photomultiplier (G ≫ 1)

4hνB

Incoherent photoconductor

2hνB

Incoherent diode or coherent photoconductor

Coherent diode

hνB

← Infrared

Red

Violet

Ultraviolet →

Photon-noise limit

NOISE-EQUIVALENT POWER (WATTS)

WAVELENGTH (MICRONS)

Noise-equivalent power for photon detectors varies with radiation wavelength. In the region above the colored curve, photomultipliers are more sensitive. Within the shaded area, semiconductor detectors are preferable. Note how background radiation begins to dominate at longer wavelengths. All data shown are for a 1-Hz bandwidth and a 1-cm² detector area, except that the photon-noise limit and photomultiplier curves are minimum values, independent of detector area.
Figure 5

noise limit but only at the expense of an extremely small detector area or an increased response time.

It should be apparent from our discussion that photon detectors in the infrared region are seriously limited in their sensitivity except for measurements of broadband thermal sources, where the background already limits the detection capability. For laser detection, we can resort to the coherent mode, but, just as in a radar receiver, the diffraction-limited behavior of the detection process means that in this situation large collection apertures will lead to extremely narrow fields of view for the system. In addition, the local oscillator must be tuned to within a detector bandwidth of the received signal frequency, so that precise frequency control and fine tuning of the laser are required.

Alternative photon-detection schemes are presently being pursued. One possibility is photon conversion to a shorter wavelength, as in the parametric upconverter, where the output photon may be detected more efficiently in the visible-frequency range. Extension of photo-emitter response into the infrared also shows great promise, as does avalanche multiplication in semiconductors, in which new materials and techniques may allow "noiseless" amplification out to the ten-micron region.

* * *

This work was sponsored in part by the US Department of the Air Force.

extra noise produced in the avalanche process.

Remember that, in figure 5, an area of 1 cm² and an electrical bandwidth of 1Hz were assumed. When background or amplifier noise is the limiting factor, the noise-equivalent power is proportional to $(AB)^{1/2}$, where A is the detector area. For the case of background noise, the dependence is obvious, because background power is proportional to area. In the amplifier-limited situa-

tion, the dependence may be understood when we note that, for a given bandwidth requirement, the maximum usable load resistance R is inversely proportional to detector capacitance, which is proportional to area. The effective amplifier noise current is $(4kTB/R)^{1/2}$, so that the noise-equivalent power is again proportional to $(AB)^{1/2}$. In the limit, we can reduce the bandwidth and the detector area to such values that we approach the photon-

References

1. R. L. Bell, W. E. Spicer, Proc. IEEE **58**, 1788 (1970).

2. M. C. Teich, Proc. IEEE **56**, 37 (1968).

3. R. J. Keyes, T. M. Quist in *Semiconductors and Semimetals* vol. 5 (R. K. Willardson, A. C. Beer, eds.), Academic, New York (1970), chapter 8.

Quantum wells for photonics

New techniques for growing semiconductors with alternating ultrathin layers allow one to produce materials with made-to-order electro-optic properties and potential uses such as optical modulators and solid-state photomultipliers.

Daniel S. Chemla

PHYSICS TODAY / MAY 1985

The crystalline and electronic structures of semiconductors reflect a delicate balance of very large electromagnetic forces, and consequently minute compositional variations or small perturbations can induce large changes in the properties of these materials. For several decades now, research scientists and device designers have exploited this exceptional flexibility to tailor the electronic and optical properties of semiconductors for a variety of fundamental studies and applications. Semiconductor technology has made its most apparent impact, of course, in solid-state electronics.

In recent years, however, the field of photonics, which combines laser physics, electro-optics and nonlinear optics, has burgeoned. Modern lightwave communications exemplify photonic systems: Here optical signals are generated, modulated, transmitted and detected before they are transformed to electrical form for final use. Information processing is another example. Optical processing of information has several advantages over electronic processing, which must usually be done serially and is limited in speed by the broadening of pulses in interconnecting wires and is limited in density by "cross talk" between those wires. Optical systems capable of handling very large quantities of data await only the development of convenient digital optical logic elements with low switching energy.

Daniel Chemla is head of the quantum physics and electronics research department at AT&T Bell Laboratories, in Holmdel, New Jersey.

An ideal material for electro-optic applications such as those mentioned above would be able to transform light into current and vice versa for detection and emission. The material would also exhibit large electronic and optical nonlinearities that would allow one to use it as a transistor and optical gate. By taking advantage of both of these nonlinearities at the same time, one can use the material as an optical modulator. In the last decade we have seen the development of new methods for growing materials epitaxially. Techniques such as molecular beam expitaxy[1] and metal–organic chemical vapor deposition[2] combine an ultraclean growth environment and a slow growth rate to produce samples of extremely high quality. In particular, these techniques allow one to produce heterojunctions that are atomically abrupt and planar. With growth rates as low as 1 Å/sec, one can make layered structures with layer thicknesses ranging from a few angstroms to a few microns, as well as microstructures with continuously tuned composition profiles. These artificial media exhibit novel properties not shown by the parent compounds in the bulk.

Structures consisting of stacks of ultrathin layers are called superlattices or quantum-well structures,[3] and those with continuously varying compositions are called graded-gap structures. In this article I discuss the basic opto-electronic properties of quantum-well and graded-gap structures and describe some recent research that has potential applications in photonics.[4,5] I begin with a look at the physics of the

layered structures, which feature bandgap discontinuities, or abrupt spatial changes in the energy gap between the valence band and the conduction band. Explaining the opto-electronic properties of these quantum-well structures will bring me to such topics as room-temperature excitons, optical nonlinearities and the behavior of carriers constrained to move in two dimensions only (figure 1). I will also say a few words about the structures whose compositions—and band-gap energies—vary smoothly. These graded-gap materials exhibit unusual energy-band gradients that imitate electric fields, which device designers can use to adjust the drift velocities of carriers over a large range. Finally, I will consider the wide variety of present and potential applications that discontinuous and graded-gap materials have in photonics. These applications range from mode-locking laser diodes and high-speed optical samplers to solid-state photomultipliers and optical gates with ultralow switching energies.

The materials. The compound III–V semiconductors, which are made from group-III and group-V elements, have the basic properties necessary for fabricating quantum-well and graded-gap materials. These semiconductors have a direct band gap, that is, they can emit or absorb light without the help of lattice vibrations, and thus they are very efficient absorbers and emitters. They also have large carrier mobilities and are easily doped. More importantly, they can form various solid solutions with identical crystal structures and well-matched lattice parameters but

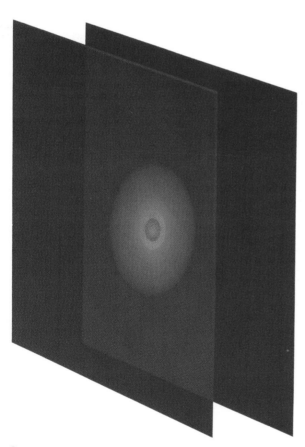

Quantum confined Stark effect, in artist's conception. The drawing on the left shows an electron–hole pair, or exciton, trapped in a quantum well in a superlattice material. The left and right planes represent the walls of the well. When an electric field is imposed, right, the electron (blue cloud) and hole move apart, but the walls of the well are close enough together to prevent the exciton from ionizing.

Figure 1

Quantum-well structure and corresponding real-space energy band structure. The schematic diagram in **a** shows compositional profiling in thin layers. The circle in **b** represents an exciton in the bulk compound, and the ellipse represents an exciton confined in a layer with a low band gap.   **Figure 2**

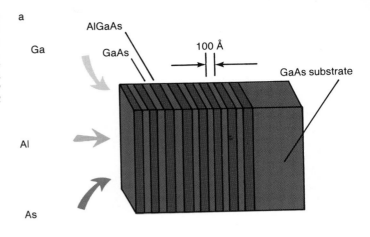

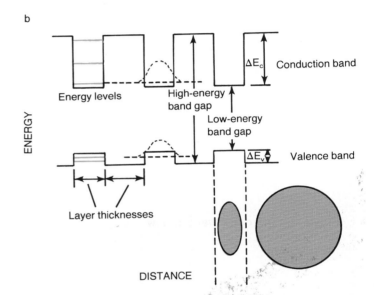

with different energy gaps and refractive indices.

The chemical and physical compatibilities of solid solutions of various III–V compounds make it possible to grow heterostructures involving several compounds. One can make heterostructures with high-quality interfaces, and tailor the optical and electronic discontinuities for specific applications. Although structures have not yet been optimized for more than one function at a time, the III–V alloys could in principle be used for several functions at once. This possibility has motivated much of the tremendous effort over the last 20 years to synthesize III–V semiconductors, study and understand their fundamental characteristics and manipulate their properties. (See Venkatesh Narayanamurti's article on crystalline semiconductor heterostructures, PHYSICS TODAY, October, page 24.)

Physics and structure

Electrons and holes can propagate freely in the periodic potential of semiconductors. The major changes in the dynamics of these charged carriers caused by the semiconductor environment are: replacement of the free-electron masses by much smaller electron and hole effective masses, and a substantial increase in the dielectric constant. As a consequence of these changes, basic physical quantities such as the Bohr radius and the Rydberg constant are drastically modified in semiconductors. In the case of the III–V compounds, the Bohr radius ranges from 10 Å to 500 Å, corresponding to effective Rydberg constants ranging from 100 meV to 1 meV.

The change of scale in these natural units causes a number of processes involving electrons and holes in semiconductors to be different from the free-space atomic processes that they parallel. Thus carriers in semiconductors are more sensitive to small perturbations, a situation that one can exploit for device applications and that one can use to obtain model systems not encountered with free particles. In superlattices and graded-gap structures, modifications of free-particle behavior due to quantum size effects are important. Quantum size effects arise when the dimensions of a quantum system become comparable to the Bohr radius; one can observe these effects in semiconductor microstructures that have dimensions on the order of 100 Å.

The simplest examples of systems where size produces fundamental modifications of optical and electronic properties are quantum-well structures.[6] These structures consist of ultrathin layers of two or more compounds grown one on another periodically, as figure 2a indicates. Because the layers have different band gaps, the energy bands present discontinuities in real space, as shown schematically in figure 2b. Quantization of the carrier motion in the direction perpendicular to the layers produces a set of discrete energy levels. If the energy discontinuities are large enough and the layers with large band gaps are wide enough, then there will be little interaction between adjacent low-gap layers. The carriers confined in each of those layers will behave almost independently. Hence the name quantum-well structures. When the barriers are narrow, or when the energy of a state is comparable to the energy discontinuities, the interaction between layers is important. The wavefunctions of the carriers are extended perpendicularly to the layers, so the behavior of the carriers is modified by the periodic long-range modulation superimposed upon the crystalline potential. Hence the name superlattices.

In quantum-well structures, electrons and holes do not move with their usual three degrees of freedom. They show one-dimensional behavior normal to the layers and two-dimensional be-

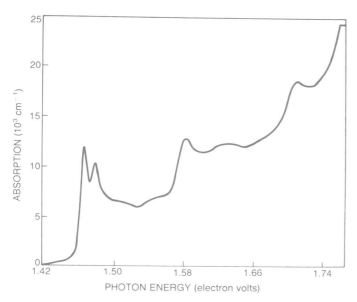

Absorption spectrum of a GaAs–$Al_{0.3}Ga_{0.7}$As quantum-well structure. The steps in this room-temperature spectrum mark transitions between sub-bands. The peaks at the edges of the steps are due to excitons.   Figure 3

havior in the planes of the layers. This reduced dimensionality induces drastic changes in the electric and optical properties of quantum-well materials. For example, one can introduce impurities in the large-gap layers in such a way that the impurity nuclei will be trapped while the carriers that are introduced can migrate toward the low-gap layers and form two-dimensional gases at the interfaces. This technique, known as modulation doping,[7] produces a physical separation between impurities and carriers, leaving the carriers highly mobile. Solid-state physicists have taken advantage of the properties of two-dimensional electron gases in the fabrication of high-speed field-effect transistors and in fundamental studies on the integral and fractional quantum Hall effects.[8]

In layered structures, the conduction and valence bands become sets of two-dimensional sub-bands with step-like densities of states. This increases the number of states that contribute to the optical transitions at the absorption edge. Hence the absorption spectra of quantum-well structures are significantly different from those of three-dimensional semiconductors. Device designers have used this effect to obtain very-low-threshold diode lasers by including quantum wells in the active region of the diode.[9]

Excitons. For undoped quantum-well structures, excitonic effects are further modified by the confinement of carriers. As we will see later, optical effects associated with excitons play a crucial role in many opto-electronics applications of quantum-well structures. When a high-purity semiconductor ab-

sorbs a photon, the electron that is promoted to the conduction band interacts with the hole left in the valence band. The electron and hole can form a bound-state analog of the hydrogen atom called an exciton. This final-state interaction produces a set of discrete and very strong absorption lines just under the band gap. Because the binding energy of the exciton—the energy of its 1s state, or its effective Rydberg constant—is very small, excitons are very fragile. Excitons are sensitive to any kind of defect and are usually observed only at low temperature because they are easily broken apart by thermal phonons.

In a quantum-well structure with layer thicknesses smaller than the Bohr radius, the exciton has to modify its structure to fit into the low-gap layers. It flattens and shrinks. This is illustrated in figure 2b by the colored circle, which represents the 280-Å-diameter exciton of bulk GaAs, and by the ellipse, which represents the exciton when it is confined in a 100-Å quantum well. The electron and the hole are forced to orbit closer to each other, and the binding energy increases by a factor of two to three.

This added stability makes the exciton's resonances observable at room temperature, as demonstrated clearly by the absorption spectrum shown in figure 3. This spectrum was measured at room temperature in a high-quality quantum-well structure that Arthur Gossard grew at AT&T Bell Laboratories. The sample has 65 periods of 96-Å-thick layers of GaAs alternating with 98-Å-thick layers of $Al_{0.3}Ga_{0.7}$As. In figure 3 one can see the steps

associated with the transitions between sub-bands, and one can see exciton peaks before each step. Exciton peaks as clearly resolved as this are usually seen only at very low temperature in bulk semiconductors of ultrahigh purity. The peaks are so apparent in quantum-well structures not only because of the increased exciton binding energy but also because the confinement strongly enhances the contrast with the continuum.[10] In bulk GaAs there is only one exciton resonance. The reduced symmetry of quantum-well structures, however, produces two valence bands and hence two excitons. This results in the double peak seen at the onset of the first transition. Similar exciton resonances, well resolved at high temperature, have been seen in the quantum-well material GaInAs–AlInAs, whose band gap is in the infrared.

It was at first thought that the room-temperature excitonic resonances in quantum-well structures simply offered a convenient way of making use of the large, intensity-dependent absorption and refraction effects usually seen at low temperatures in bulk semiconductors. Indeed quantum-well structures do have giant optical nonlinearities, but these originate from the novel physical properties of the layered structures and present a richer array of potential applications than do the nonlinearities observed at low temperatures in bulk crystals.

When light tuned to the exciton peak illuminates a quantum-well structure, bound electron–hole pairs are first generated and then quickly ionized by the large population of thermal phon-

Excitonic wavefunctions without and with an applied electric field (**a**), and the quantum confined Stark shift in an absorption spectrum (**b**). The wavefunctions illustrate how the walls of a quantum well hold an electron and hole in a bound state, even at applied fields much stronger than the classical ionization field. The absorption spectra are those of a quantum-well structure under three different static electric fields applied normal to the layers. The fields are 10^4 V/cm (bottom curve), 5×10^4 V/cm (middle curve) and 7.5×10^4 V/cm (top curve). Figure 4

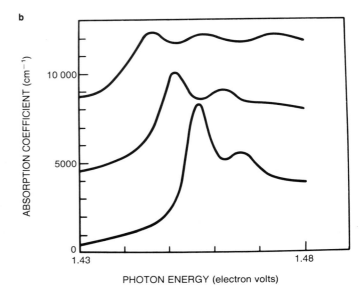

ons present at room temperature. At room temperature the average thermal energy kT is about three times larger than the exciton binding energy. Note for comparison that the ionization temperature for hydrogen atoms is about 2×10^5 K. Recent measurements with femtosecond optical spectroscopic techniques[11] found the ionization time of excitons in quantum-well structures to be 300 fsec. Due to the effects of the electron–hole plasma, the coefficient of absorption and the index of refraction depend strongly on the intensity of the incident light. These dependencies show up in measurements with continuous-wave laser light as well as with picosecond laser pulses. These nonlinearities are several orders of magnitude larger than those observed in normal semiconductors, yet they are still about half as large as the nonlinearities produced by the selective generation of excitons with femtosecond laser sources.[12]

It is commonly believed that the charged electron–hole plasma is more efficient than the neutral bound pairs in shielding electrostatic forces. This is true if the electrons and holes are at the same temperature. However, excitation with ultrashort optical pulses at the resonant frequency generates an exciton gas of temperature near 0 K that relaxes toward a warm electron–hole plasma of temperature around 300 K by absorbing phonons. This most unusual reversed relaxation, and the reduced strength of screening due to the reduced dimensionality in the quantum-well structure, together produce very large optical nonlinearities that have two stages—a stage lasting for subpicosecond times, corresponding to the time it takes for the exciton to ionize, and a stage lasting for nanoseconds, corresponding to the time it takes for the electron–hole pair to recombine.[10]

Excitons in quantum-well materials are also sensitive to electrostatic perturbations. Because the carrier wavefunctions extend to about 100 Å, and

because the confinement or binding energies are only 10–100 meV, moderate electric fields of order 10 mV per 100 Å, or 10^4 V/cm, cause significant perturbations. When an electrostatic field is applied to a three-dimensional exciton, it induces a Stark effect analogous to that seen on atoms: There is a small shift in energy levels that is quickly masked because the energy levels are broadened by the exciton's ionization under the influence of the electric field. One sees this in quantum-well systems when the field is applied parallel to the planes of the layers.

With a perpendicular field, however, an absolutely new process occurs.[13] The field pushes the electron and the hole apart, but the wall of the well prevents ionization by constraining the particles to stay close enough to remain bound, as figure 4a indicates. The ionization can only occur when the particles tunnel out of the well. Consequently, it is possible to apply fields as large as 50 times the classical ionization field, inducing redshifts (figure 4b)

in the absorption peak 2.5 times the binding energy, and still observe exciton resonances! This phenomenon, which figure 1 illustrates, is called the quantum confined Stark effect. Of course, one cannot create a similar situation for atoms, but quantum-well structures provide model systems in which one can study the effects of such extreme conditions and make tests against theory. The observed shifts due to the quantum confined Stark effect are well accounted for by the field-induced variations in the energy of the single-particle state and by pair attraction.[13] This effect also allows one to shift an abrupt and highly absorbing edge into a spectral region where the sample is normally transparent. Such shifts have obvious applications in optical modulation and optical logic, and I will discuss these below.

As figure 2b indicates, the conduction band and the valence band do not contribute equally to the total bandgap discontinuity at a heterojunction. Recent measurements[14] on a GaAs–AlGaAs quantum-well structure, for

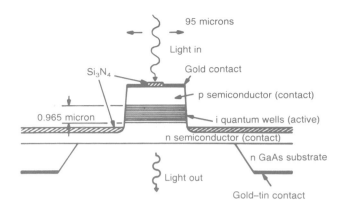

High-speed optical modulator and its optical response to an electrical pulse. The schematic diagram shows a p–i–n diode with a quantum-well optical modulator. The speed of the device is limited by the circuit's RC constant and by the large diameter of the quantum-well structure.　　　Figure 5

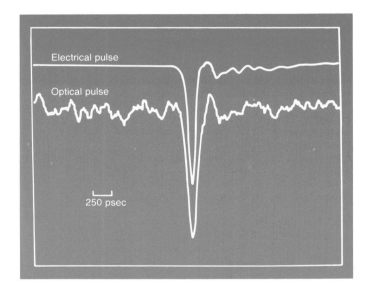

example, indicate that ΔE_c, the change in the energy level of the bottom of the conduction band as one crosses a heterojunction, is about -1.5 times ΔE_v, the change in the energy level of the top of the valence band across the junction. This intrinsic asymmetry between electrons and holes can produce spectacular effects such as impact ionization, which depends critically on energy gains at heterojunctions. As we will see, new types of photodetectors take advantage of this asymmetry.

Graded gaps. So far we have discussed only discontinuous heterojunctions. There is, however, a more subtle way to engineer band structure: by playing with the composition profile. A continuous spatial variation of the composition induces a gradient of the energy bands equivalent to an internal electric field, and because the valence and conduction bands experience different discontinuities, the effective fields seen by holes and electrons are not the same.[5] One can use this new degree of freedom to produce complex internal field profiles, and thus to act locally on the carrier drift velocity. For example, experimenters have measured[15] an equivalent field of order of 10^4 V/cm in a linearly graded gap structure made from $Al_x Ga_{1-x} As$. Such a field can accelerate electrons to velocities as high as 2×10^7 cm/sec.

Microstructures for photonics

The special properties of quantum-well materials suit them to a variety of applications in photonics. Below I consider some of these properties and applications.

The optical nonlinearities of quantum-well structures close to the band edge are the largest measured so far in any semiconductor at room temperature. Because quantum-well structures are easily tuned to laser-diode wavelengths, they have been used in applications where one wants a small amount of excitation to produce a large change in absorption or refraction.

Short-pulse light sources. Passive mode locking of laser diodes allows for compact and efficient short-pulse light sources. However, the design of such mode-locking diodes has suffered from a shortage of adequate saturable absorbers with large enough cross section and short recovery time. Recent experiments on quantum-well structures implanted with light ions have shortened to 150 psec the recombination time of carriers generated by the ionization of excitons. Solid-state physicists have used these fast-recovery saturable absorbers to make stable and reliable mode-locking laser diodes. Such diodes, made from GaAs, have emitted[16] pulses as short as 1.6 psec, the shortest regular pulse trains yet delivered by a diode laser.

Optical switches in cavities. Experiments have demonstrated[17] optical bistability in a sample of quantum-well material held in a Fabry–Perot resonator. Theory indicates[18] that optical switches made from quantum-well materials in precisely tuned etalons will have very low switching energies

High-speed optical modulators. Recent experiments have used the quantum confined Stark effect to achieve the high-speed modulation of light from laser diodes. To build such a modulator, one grows a quantum-well structure in the intrinsic region of a p-i-n diode, as figure 5 shows. By putting a reverse bias on the resulting device, one applies a field without causing current to flow. When the modulator represented in figure 5 was given a 122-psec electrical pulse, it subjected the laser light passing through it to a 131-psec 2.3-dB attenuation. The modulator's response was limited by the effects of capacitance in its 95-micron-diameter structure.[19] The speed of the quantum confined Stark effect is limited only by how quickly the exciton envelope function can follow the applied field. At present, however, the limit is how fast one can change the applied field, so faster operation should be feasible with devices having smaller areas and therefore smaller capacitances. The quantum-well structure that imposed the 2.3-dB modulation

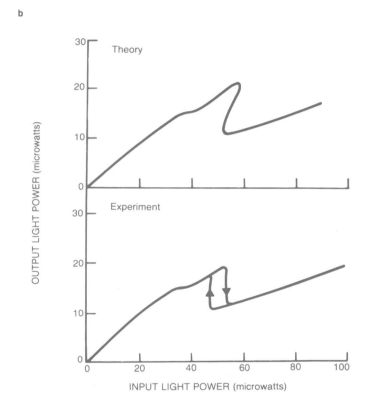

Self-electro-optic device and plots showing its optical bistability. The device in the schematic diagram shows optical bistability when it is connected to a simple resistive load. Figure 6

was only about one micron thick; longer light paths increase the attenuation considerably.

Self-electro-optic devices. When a photon is absorbed in a quantum confined Stark effect p–i–n structure, it generates an electron–hole pair that is separated by the field; here the modulator behaves like a photodetector of unit quantum efficiency. The ability of such a p–i–n device to act as both a modulator and a photodetector provides an internal feedback mechanism when the device is connected to an electronic circuit. This is the basis of a new category of devices, known as self-electro-optic devices, that can operate as optical gates with very low switching energy, self-linearized modulators and optical level shifters.[20]

The optical gate represented in figure 6a operates as follows: One applies just enough voltage to the device to shift the absorption edge so that there is little absorption at the exciton peak; then one directs a light beam of varying intensity onto the device. As the beam's intensity increases, it generates a photocurrent that induces a voltage drop across the resistor. The voltage across the device decreases, the edge shifts back and the absorption increases, increasing the photocurrent. This cycle continues until the device switches to a state of low transmission. The switch back to a highly transmitting state does not occur at the same incident intensity because the self-electro-optic device is now absorbing. Therefore the gate response, shown in figure 6b, is an optical hysteresis loop.

The self-electro-optic devices demonstrated to date typically have total switching energies of 20 fJ per square micron. This switching energy is only one-sixth that reported for any other optically bistable device, despite the fact that self-electro-optic devices operate without resonant cavities. The total switching energy comprises two parts: resonant optical energy, which accounts for about 20% of the total, and electrical energy, which accounts for the other 80%. Self-electro-optic devices are compatible with other III–V semiconductor technologies, and should fit into large-scale integrated arrays.

High-gain avalanche photodetectors can be built from a solid with a large difference between the rates at which electrons and holes create electron–

hole pairs throughout impact ionization. In gallium arsenide, unfortunately, the two ionization rates are about the same. Furthermore, avalanche multiplication is intrinsically a noisy process because of the randomness of the ionization events, and that causes statistical fluctuations in the gain. New concepts based on quantum-well and graded-gap structures have overcome these obstacles. Figure 7a shows the band structure of a p–i–n diode that contains a quantum-well structure in its intrinsic region. When a hot electron enters a gallium arsenide quantum well, it suddenly gains an energy ΔE_c. The ionization threshold for electrons is thus reduced from ΔE_{th} to $\Delta E_{th} - \Delta E_c$, whereas for the holes it is reduced to $\Delta E_{th} - \Delta E_v$. Because the ionization rates depend exponentially

on the thresholds, one can greatly enhance the ratio of electron and hole ionization rates. AlGaAs–GaAs avalanche photodetectors have shown[21] ratios as large as seven. The performances can be improved further by using a sawtooth profile with regions of linear grading followed by abrupt steps, as shown in figure 7b. When one applies a static field to the structure, figure 7c, the impact ionization events occur at each step deterministically, and preferentially for the electrons. The resulting multiplication processes are no longer random, and the gain is almost noise free. The graded-gap structure acts as a solid-state photomultiplier, with the steps in the energy bands corresponding to the dynodes of a traditional photomultiplier tube.[21]

Fast transistors. The transit time of

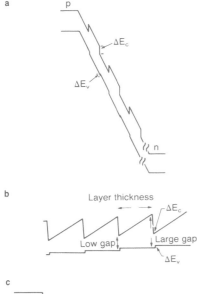

Band structures. a: Band structure of an avalanche photodetector made from a p–i–n quantum-well structure. **b:** Band structure of a sawtooth graded-gap multilayer structure. The composition varies linearly from the low-gap compound to the large-gap compound and then abruptly switches back to the low-gap composition. **c:** Band structure of a graded-gap solid-state photomultiplier with an applied bias voltage. The arrows illustrate how the electrons multiply by impact ionization at discontinuities in the conduction band whereas the holes do not gain enough energy at valence-band discontinuities to participate in impact ionization. Figure 7

the minority carriers in the base of a transistor is governed by diffusion. Almost three decades ago, H. Kroemer of RCA proposed[22] introducing quasielectric fields in the base of a transistor to reduce the carrier transit time. Experimenters recently succeeded in doing this in a phototransistor with a wide-band-gap emitter and a graded-gap base. The 10^4 V/cm quasifield in the base gave an intrinsic response time of 20 psec—less than one-fifth of the diffusion time of carriers in an identical structure without a graded base.[21] Further improvements under study include the ballistic launching of electrons into the base by a conduction-band discontinuity at the emitter–base heterojunction.

The examples I have given of semiconductor properties that are strongly modified by quantum size effects and compositional profiling are just a limited set chosen for their relevance to photonics. In principle, we can use abrupt heterojunctions, selective doping and continuous composition grading to engineer, in almost any arbitrary fashion, the energy-band structure of semiconductor microstructures. The recently developed epitaxial growth techniques already allow us to control the compositions of samples along one dimension. We can envision new techniques that in the near future will give us atomic-scale control over growth in three dimensions. This will enable us to tailor the local band structure of samples according to specific designs. There is no doubt that the resulting materials will present new properties—some unsuspected—and that these new properties will give rise to new applications.

References

1. See, for example, L. L. Chang, K. Ploog, eds., *Molecular Beam Epitaxy and Heterostructures*, NATO Advanced Science Institute Series, Nijhoff, Dordrecht (1985).

2. See, for example, J. B. Mullin, S. J. C. Irvine, R. H. Moss, P. N. Robson, D. R. Wight, eds., *Metal Organic Vapor Phase Epitaxy 1984*, North-Holland, Amsterdam (1984).

3. L. Esaki, R. Tsu, IBM J. Res. Dev. **14**, 61 (1970); for an outline of the history of Esaki's and Tsu's discovery, with Leroy Chang, of artificial semiconductor superlattices, PHYSICS TODAY, March, p. 87.

4. See, for example, D. S. Chemla, D. A. B. Miller, P. W. Smith, *Device and Circuit Applications of III–V Semiconductor Superlattices and Modulation Doping*, R. Dingle, ed., Academic, New York (1985).

5. See, for example, F. Capasso, *Device and Circuit Applications of III–V Semiconductor Superlattices and Modulation Doping*, R. Dingle, ed., Academic, New York (1985).

6. R. Dingle, *Festkörperprobleme* **15**, H. J. Queisser, ed., Pergamon, Braunschweig (1975).

7. R. Dingle, H. L. Stormer, A. C. Gossard, W. Wiegmann, Appl. Phys. Lett. **33**, 665 (1978); H. L. Stormer, Surf. Sci. **132**, 519 (1983).

8. T. Mimura, S. Hiyamizu, T. Fujii, K. Nambu, Japan J. App. Phys. **19**, L225 (1980); D. Delagebeaubeuf, P. Delesclilse, P. Etienne, M. Laviron, J. Cha-

plart, N. T. Linh, Electron Lett. **16**, 667 (1980); H. L. Stormer, *Festkörperprobleme* **24**, P. Grosse, ed., Vieweg, Braunschweig (1984).

9. See Y. Suematsu's article on page 32 of this issue.

10. D. S. Chemla, D. A. B. Miller, J. Opt. Soc. Am. B, to be published July 1985.

11. C. V. Shank, Science **219**, 1031 (1983).

12. W. H. Knox, R. F. Fork, M. C. Downer, D. A. B. Miller, D. S. Chemla, C. V. Shank, Proc. Fourth Int. Conf. Ultrafast Phenomena, Springer-Verlag, Berlin (1984), p. 162; Phys. Rev. Lett. **54**, 1306 (1985).

13. D. A. B. Miller, D. S. Chemla, T. C. Damen, A. C. Gossard, W. Wiegman, T. H. Wood, C. A. Burrus, Phys. Rev. Lett. **53**, 2173 (1984).

14. R. C. Miller, A. C. Gossard, D. A. Kleinman, O. Munteanu, Phys. Rev. **B29**, 3740 (1984); for a review of exciton spectroscopy in quantum-well structures, see R. C. Miller, D. A. Kleinman, Proc. 3rd Trieste IUPAP Semiconductor Symp., J. Lumin. **30**, 520 (1985).

15. B. F. Levine, C. G. Bethea, W. T. Tsand, F. Capasso, K. K. Thornber, R. C. Fluton, D. A. Kleinman, Appl. Phys. Lett. **43**, 769 (1983).

16. Y. Silberberg, P. W. Smith, D. J. Eilenberger, D. A. B. Miller, A. C. Gossard, W. Wiegmann, Optics Lett. **9**, 507 (1984).

17. H. M. Gibbs, S. S. Tarng, J. L. Jewell, D. A. Weinberger, K. Tai, A. C. Gossard, S. L. McCall, A. Pasner, W. Wiegmann, Appl. Phys. Lett. **41**, 221 (1982).

18. P. W. Smith, Proc. Conf. Electro '83, session record 11/1, IEEE, New York (1983).

19. T. H. Wood, C. A. Burrus, D. A. B. Miller, D. S. Chemla, T. C. Damen, A. C. Gossard, W. Wiegmann, IEEE J. Quantum Electron. **QE-21**, 117 (1985).

20. D. A. B. Miller, D. S. Chemla, T. C. Damen, A. C. Gossard, W. Wiegman, T. H. Wood, C. A. Burrus, Appl. Phys. Lett. **45**, 13 (1984); Optics Lett. **9**, 567 (1984); to be published in IEEE J. Quantum Electron. (1985).

21. F. Capasso, Sur. Sci. **513**, 142 (1984).

22. H. Kroemer, RCA Rev. **18**, 332 (1957). □

Optics:
an ebullient
evolution

PHYSICS TODAY / NOVEMBER 1981

Peter Franken

Peter Franken is director of
the Optical Sciences Center at
the University of Arizona,
Tucson, where he is also professor of physics.

Six arms of the 20-beam Shiva laser. *Used in laser-fusion experiments, Shiva can direct a 10 kilojoule, 27 terawatt pulse at a submillimeter-diameter pellet and compress its deuterium–tritium contents to 100 times the liquid density. Flash lamps pump this neodymium-glass laser. (Lawrence Livermore Laboratory, Laser Program.)*

Dye amplifier. *Green beam from a copper-vapor laser (not pictured) optically pumps dye and amplifies red beam. (Photograph from Lawrence Livermore National Laboratory, Laser Program.)*

I did my undergraduate and graduate work in physics right after World War II, as a Columbia victim. It was an era in which optics and electronics enjoyed similarly undistinguished roles in the curriculum, although optics was a required course and electronics was not. But then, electronics was useful, which brought these academic requirements into line with extant pedagogic philosophy. I guess most of us thought optics was useful too, but we didn't really think we had to understand anything more than the thin-lens equation, whereas we really had to know how to do electronics because we were all up to our navels in vacuum tubes and stuff. (Some of you are probably too young to know what a vacuum tube is; imagine a sort of glass cylinder with strange prongs at the base, one of which is always slightly bent. . . .)

It's a different game altogether now. Electronics, of course, retains its position as the premier technology in physics and associated disciplines. Optics has evolved into a far more exuberant adventure: Although the manipulation of the thin-lens equation or arcane discussions of internal conical reflection are still enjoyed, vast new excitements have developed in the optical arena during the last twenty years. This dra-

50 Years of Physics in America

matic evolution of the subject, perhaps hysterical at least as far as the job market or public participation is concerned, not only was stimulated by three dramatic emergences but was literally made possible by them. The three developments I have in mind are the laser, high-speed computation and new materials.

A riot of technology

The laser not only stimulated a grand renaissance in optical physics, it also assumed the role of master tool in experiments and adventures—ranging from the trivialization of alignment to the experimental realizations of incredibly sophisticated nonlinear processes, processes that could not have been effectuated with conventionally bright sources. Along with the almost sexual glee with which the physics community received this new tool, there was the development of vast engineering endeavors that dominate the current employment picture in the physics community and contribute to some budgetary agonies in the private and government sectors. Fusion, communication and medical instrumentation don't come cheap. And, unlike some of the ingredients of Chinese cuisine, there are no substitutes for a laser.

The emergence of computer technology of ever-expanding speed, if not wit, has not only introduced the quantitative improvements one expects from speed itself but has permitted the development of techniques that would be otherwise unrealizable. As a leading example, we have the ultrasophisticated advances in optical testing. This is an era in which mirrors are made up to several meters in diameter and specified to exquisite tolerances with figures far more subtle than spherical sections. Fabricating these beasts without sophisticated testing would be simply impossi-

ble. An obvious rule in the fabrication of optical elements is that an element cannot be made to a precision superior to that with which it can be tested. There were superb tests available before the AIP was even an idea. However, they were by and large rather slow and required substantial skill. The knife-edge test has been familiar to astronomers, amateur and professional alike, for more than a century. But the modern quantitative tests such as the Hartmann test (a geometric yet elegant procedure) and the whole class of interferometers (nongeometric, but still elegant) require ultrafast computational agility if results are to be achieved

Optics institutes. *The Optical Sciences Center (with palm trees) at the University of Arizona, Tucson, and the Institute of Optics at the University of Rochester, New York, are the two major institutions of optics in the Western World. The Rochester building carries the name of the late James P. Wilmot, who was a Rochester business executive and member of the university's Board of Trustees.*

50 Years of Physics in America

in anything like useful time.

I consider the developments of modern optical materials part of the evolution of modern optics; because these materials have permitted startling advances in optical instrumentation—advances that could not have been made with the old standbys. The 200-inch mastodon on Mt. Palomar, one of the few genuine triumphs of American engineering in the 1930s, was fabricated from Pyrex. Ugh. Great for stoves but worse than children when it comes to telescopes. Pyrex expands about three parts per million per degree centigrade, and so the working of a large Pyrex blank into an exquisitely figured mirror re-

quires long delays (many weeks for the Mt. Palomar adventure) between polishing and the initiation of optical tests. Keep in mind that astronomers like to work to specifications of less than a wavelength of light, perhaps 0.1 micron, so that large mirrors push the testing arts to the order of one part in 10^8. With expansion coefficients of a few times 10^{-6} per degree centigrade, we must have thermal homogeneities in the millidegree regime, and that is tough indeed with big chunks of glass that have been attacked by opticians. The new materials, such as Cervit and ultra-low-expansion glass, are endowed with thermal expansion coefficients at room temperature in

the neighborhood of 10^{-9} per degree centigrade, which makes it possible to test almost immediately after the surface is washed down. This new capability, in turn, generates a need for rapid reduction of test data so that the entire fabrication cycle can go virtually uninterrupted. Thus, the emerging testing technology has been totally revolutionized by both the orgiastic inflammation of computer technology and the development of exotic new materials.

The editors asked all of us contributing to this issue to provide an initial "overview" for our articles rather than a terminal summary. Perhaps my comments up to this point serve as an underview, but I have wanted to communicate to you the spirit of the striking retrieval of a discipline from a specialized state—some might even say pedestrian position—to become a premier discipline of physics. Optical science and optical engineering have both participated in this fling forward and could not have enjoyed the surge without the truly revolutionary developments in lasers, computers and new materials.

Lasers

When the first laser was successfully devised in 1960, the event triggered a manic combustion of talent in the government, the physics community and the denizens of Wall Street. In the quieter days of that period, the general atmosphere could best be described as panic. Lasers were going to revolutionize communications, eye surgery, communications, eye surgery and, of course, communications. At the first international meetings there were shrieks from the recognition that a carrier frequency of some 10^{15} hertz would permit, in principle, the transmission of all our television programs across the country on a single light beam, with the con-

50 Years of Physics in America

tents of the Library of Congress thrown in as an occasional hors d'oeuvre. Because xenon-arc coagulators were already being used for some retinal procedures, it was also a trivial extension to realize that lasers might enjoy deployment in the operating theater as well. Interestingly enough, these were just about the only applications that were vigorously touted in 1960, but they did serve to stimulate a lot of folks to make gigantic commitments of time, research facilities and money.

The use of lasers in optics and other technical disciplines has developed even more extensively than most of the evangelists of 1960 imagined. The actual applications, however, were virtually unanticipated during the initial convulsions. It is true that laser scalpels do enjoy an application in eye surgery and even more extensive applications in other surgical endeavors, but the penetration of the communications industry has come about in an entirely unanticipated mode. The original public excitement was attached almost exclusively to communications over very long distances, with light pipes designed primarily to keep weather and dirt out of the light beam. At that time no one I can recall anticipated the real breakthrough, which would come about because of the development of exceedingly low-loss glass fibers.

The basis of extensive optical communications systems today is the use of glass fibers, which are now being fabricated to have losses of less than one decibel per kilometer. In the early 1960s the only manufactured glass fibers were those used for imaging-bundles in various medical endeavors, and these had losses that were typically several decibels per meter.

If it had not been for the then entirely unexpected revolution in materials technology, we would not now be enjoying the incredible revolution in communications

made possible by the laser. A few years ago it was recognized that low-loss fibers would permit a genuine decrease in the cost of intraurban communication; within the last few years the sophistication of couplers and repeaters has permitted realistic speculation about very long optical communication links that might even include a transatlantic cable.

The emergence of extraordinarily high-powered lasers, such as the carbon-dioxide creatures driven by ignition or electric discharge, has given us powers well in excess of 100 kilowatts (continuous!), with an optical fidelity within a factor of 2 of the diffraction limits enunciated long ago by Lord Rayleigh. This remarkable advance in technology, totally unimagined by all folk I knew in the early 1960s, has revolutionized industrial processing as well as avantgarde military technology. High-powered lasers enjoy applications to the fabrication of refractory materials, the cutting of cloth, and a whole host of piercing and annealing operations. In the military amphitheater we have already spent in excess of $1 billion in attempting to build weapon system test facilities for such a variety of straightforward and exotic systems that there appears to be a neverending cacophony in such technical journals as *Aviation Week*, *Fortune* and *Time* magazine.

It is beyond the scope of this article to evaluate the future of high-power laser technology in the military scene, but I think it fair to say that the device might generally accomplish more destruction were it dropped instead of fired. There are, of course, realizable applications such as satellite warfare, because there is little question that current laser power levels are high enough to cause satellite damage, given the current "softness" of many satellites. Whether serious attention should ultimately be given to deployment of these "Star Wars" devices falls outside this ar-

ticle; the question necessarily embraces systems considerations that are tender to the anxieties of the military who are charged with the assessments.

Rather than duck all military issues, however, let me assert without equivocation that the use of lasers as target designators and range finders, just to name two applications, has been thoroughly established and does play an important role in today's armamentarium. It is interesting to note, in retrospect, that the range-finder application was anticipated early in the 1960s, while the possibility of a target designator was recognized quite a bit later.

The applications of medium- and low-power lasers have flared up in existing as well as in unanticipated areas of commerce and technology. In the commercial marketplace we find laser scanners playing an ever more important and useful role in inventory control (at supermarket checkout counters, for example) and in truly mass-market endeavors such as the nascent video disc adventure.

More dear to the AIP heart are the truly fantastic penetrations into a legion of problems in physics. Exquisite ingenuity in spectroscopy, developments in material science, and advances in the whole broad class of nonlinear optical phenomena are yielding extraordinary physics about our world that we could not have discovered otherwise—or at least none of us had been clever enough to discover!

An important challenge for optical science and engineering is the constant demand for optical instruments, ranging from giant telescopes on down to scanning microscopes, which press our testing and evaluation procedures ever more vigorously. The temporal coherence (essentially the monochromaticity) of laser radiation, together with its incredible brightness, permits real-time interferometric procedures that were just

50 Years of Physics in America

about impossible with the old filtered mercury-arc lamps of yesteryear. And then we get to the whole class of devices known as laser scanners. Here the fundamental brightness of the laser permits us to produce exquisitely collimated and intense beams for many applications to scanning processes, printing endeavors and a host of problems in the medical world where challenges and extant capabilities have often not been effectively coupled. Fortunately, the collaboration of qualified physicists and engineers with modern medical chaps has improved vastly over the last decade or two and has led to many genuine and not just glamorous accomplishments.

Image Processing

I prefer the words "image manipulation," but for some people these words have associations that are less than entirely pleasant. I am referring, of course, to the many recent endeavors in which images are manipulated to change their apparent quality and achieve occasionally unbelievable feats of information retrieval. Virtually all the new and powerful techniques of image management today stem from the phenomenal growth of computer capability and the fortunately ever-reduced cost per operation. Let's look at a few examples.

▶ Image reconstruction. Imagine that you take a photograph but manage to swing your camera through some not negligible arc during exposure. The result will be a streaky and blurred image, which enjoys a classification somewhere between unattractive and totally useless. Yet with restoration algorithms that have been created within this decade it would be entirely possible to construct (usually called "reconstruct") an image almost as good as what would have been achieved had you not goofed. Some information about the motion that degrad-

Mirror blanks *made of borosilicate glass, at Schott Optical Company, Duryea, Pennsylvania. The mirrors, 1.2 meters in diameter, will direct beams of the Nova laser toward focusing lenses. (Lawrence Livermore Laboratory photograph.)*

Optical telescope assembly *for the NASA Space Telescope. Light received is reflected by the 2.4-meter primary mirror to the secondary mirror and then back through a hole in the primary mirror for analysis by instruments. (Sketch from Perkin-Elmer Corporation).*

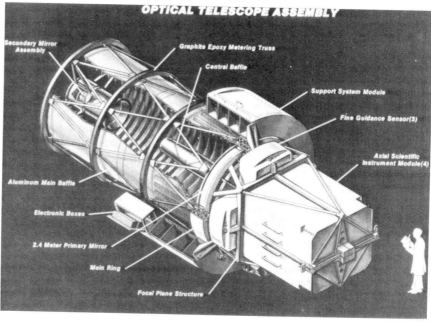

50 Years of Physics in America

ed the original image is necessary, and is often derivable from the picture itself. Although this example is not very exciting in practice, it illustrates one of many kinds of image reconstruction that we can do today but could not have done even as recently as ten years ago because of the enormous computation required.

▶ Pattern recognition. In this game there are problems that have yielded dramatically to the amalgamation of computer and laser technologies. For example, there are several extensive programs for the laser scanning of microscope slides. Such a program gives digital data that we hope can be processed by computer to constitute a cancer screening examination (an automated "pap" smear). Perhaps rather far out, we look forward to the recognition of patterns in mammography and chest radiography in which a computer might recognize textural effects and thus flag early cancers.

I find a very interesting existence theorem question lurking in the pattern-recognition game these days. To wit, there may well be some problems where the total unthinking (albeit fantastically rapid) computational ability of modern devices is fundamentally deficient for the recognition of patterns. One situation where I suspect this might be the case is in the reading of handwritten materials. With the exception of some medical and other professional documents we are exposed to, most of us can read arbitrary handwritten messages. I suspect that we might never be able to "read" these letters and words with a computer even though the image of the handwriting is readily convertible to digital form. I don't pretend to understand how the human mind differs from a digital computer, but people do seem to have a fantastically rapid random access read-and-write memory with an as yet uncalculated limit to its capacity. This clearly plays an

important role in our remarkable abilities of pattern recognition. Millenia ago Socrates enunciated the remarkable fact that it is only necessary for a person to see one cat, or one elephant, to recognize any others he might come across. Socrates, by the way, explained all of this in terms of people "recognizing" the basic image of a cat because the image is part of our original memory store and is triggered by the vision of one particular feline. That was a cop-out, at least partly, but it was also a deep insight to recognize the extraordinary miracle of pattern recognition itself.

Medicine

Much of the modern manipulation of images occurs in the art and science of medical practice. As examples, in this last decade there have been absolutely remarkable advances in the enhancement of diagnostic radiographic images, by subtraction techniques, and there has been a renaissance in novel presentations of nuclear-isotope and computerized tomographic information. The development of these exciting techniques that will make our own clinical lives more satisfactory has, once again, been derived from the extraordinary achievements in computer technology and its adjacent disciplines.

There has been a revolutionary advance in medical imagery, however, that we could have achieved before World War II because it does not depend on either lasers or computers, even though these are sometimes enjoyed. I refer to the application of fiber optics to techniques, formerly very unpleasant to contemplate, for developing images internal to ourselves. A physician can now examine without surgery the entire colon (with a colonoscope), the upper gastrointestinal tract down into the top of the lower intestine (gastroscope),

and the bronchial structure in the lungs (bronchoscope). The general name for these instruments is "endoscope," which, I think, derives from something like "in your end." Having had much of me explored with a variety of these instruments, I can appreciate their flexibility and capability and what this means in clinical terms. As an example, formerly the standard procedure following detection of a lesion some 10 inches or further into the colon (detected, perhaps, by an x-ray examination stimulated by bleeding or some other symptom) was to do full-scale abdominal exploratory surgery. That's a major attack on anybody, and for people who present a surgical risk it is actually an invitation to serious complications, including death. The first 10 inches of the colon could be reached in the old days with a very unyielding and rigid optical instrument called a sigmoidoscope, but beyond that point—where indeed many lesions do occur—exploratory surgery was standard practice. Considering that many of these lesions can be benign, the advent of colonoscopy with fiber-optical instruments that can be snaked a good two meters into our systems permits the surgeon to inspect all the colon and even to perform biopsies and excisions of small lesions. In the event that the reality of cancer is established, the major surgery is called for, of course, but the technique of colonoscopy with the ability to evaluate lesions in a virtually untraumatic fashion has proven to be a godsend to many of us to date and will be for a lot more of us to come.

I would like to digress to comment on an entirely different technology, one that addresses the problem of inserting these devices into a cooperative patient. The key word here is "cooperative," and this is made possible through the miracle of modern pharmacology. Patients who are obliged to receive rather extensive fiber optic

Satellite view of Cape Canaveral, Florida. Man-made structure is clearly visible in this Landsat-3 return beam vidicon image. (Photo courtesy of National Aeronautics and Space Administration/ United States Geological Survey.)

devices, either by swallowing or by less dignified means, are administered, typically, an intravenous blast of a Valium and Demerol mixture that I can only describe as probably the finest entirely legal buzz to be enjoyed in our society today. I don't, however, recommend that any reader develop the symptoms requiring endoscopy just to get this stuff. However, if you ever do face, or face away from, an onslaught of endoscopy, do rest assured that the effect of these drugs is to make the entire experience almost but not quite euphoric.

Classical Optics

To physicists, the adjective "classical" usually means something that was going on before they were born. Classical optics is no exception, encompassing as it does the design and fabrication of a tremendous range of optical instruments as well as the study of phenomena almost all of which were discovered in the 19th century and before. Nevertheless, modern endeavors in classical optics have enjoyed an extraordinary renaissance as a result of the impact of lasers, high-speed computations and new materials. We still make telescopes today but, good heavens, we sure make them differently! The modern telescope is usually a fast instrument with a primary element often endowed with an aspheric figure far more severe than the shallow spheres that underlay the construction of telescopes built around the turn of the century. The basic art of figuring and polishing glass has not really changed over the last century. Opticians still rub glass with much the same materials and methods. However, because of computer technology and the new thermally stable glasses, they now can test their work almost immediately—in "real time"—permitting them to al-

The Refractoscope

The mightiest optical instrument ever conceived by the mind of man, designed to penetrate the vast reaches of inter-stellar space by bending light rays to enable the operator to see around objects.

With the Refractoscope Dr. Huer, super-scientist of the 25th Century, discovers to his horror the attack-camp of the marauding warriors of Mars on the other side of the moon. They are almost ready for battle. Be there when—

MARS INVADES EARTH!!

in

Buck Rogers—25th Century

starting (date) exclusively

in

(Name of Paper)

Newspaper advertisement form for Buck Rogers comic strip series, circa 1929. People on the "Refractoscope" indicate scale. (From Collected Works of Buck Rogers in the 25th Century, A&W Visual Library, New York.)

50 Years of Physics in America

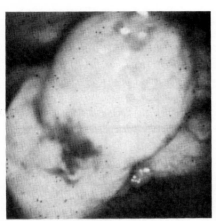

Polyp in descending colon as seen with a colonoscope, a fiber optic instrument. The growth, approximately 2 centimeters in diameter, is about 50 centimeters from the anal verge. Black dots are due to broken fibers. (Photograph courtesy of Hiromi Shinya, Mount Sinai School of Medicine, New York.)

ternately polish and test without the many-week delays that the old ·glasses required. Furthermore, both in testing and alignment, the laser has become just about as essential to a modern optician as the transit is to a surveyor. So the endeavors are "classical," but today's accomplishments and shop practices have changed radially in the past two decades.

One of the most dramatic displays of excitement in classical optics is occurring in the current and somewhat speculative agonies associated with the prayers for a Next Generation Telescope (NGT). The NGT fever permeates the astronomical community and, not surprisingly, titillates the support of our industrial players because building a mammoth is expensive (despite the claims of the devotees) and presents intriguing technical challenges. Current discussions range from a giant multiple-mirror telescope with a 4- to 6-meter primary, all the way to segmented and deep parabolic dishes placed on support structures that

in turn, are in domes beyond the present frontiers of engineering practice. There is a general consensus that the community desire is an effective collection aperture about equivalent to a 15-meter single-primary instrument. The actual design, of course, is far from established, and the "discussions" now being undertaken in several arenas will bring to the ultimate decision some necessary discipline and review, both of technology and astronomical requirements.

As all this goes on, the United States is preparing to launch the first astronomical space telescope, with a parabolic primary 2.4 meters in diameter. I would be astonished if the initial results of this remarkable instrument (expected circa 1984) do not give the research elements of our astronomical enclaves some striking surprises, some of which may lead to modifications of current ambitions to win a behemoth ground-based NGT. I, for one, would be surprised if the space telescope's ability to gather data from the near ultraviolet continuously through to the mid-infrared—and all that at the diffraction limit—fails to provide us with some remarkable observations. It just seems unbelievable that we're not going to find a new pony or two when the atmospheric fence, with its spectroscopic and seeing inhibitions, is finally removed from the observational corral.

Workforce

According to my friends at the American Institute of Physics, American physicists identify optics as their primary discipline in larger numbers than they identify any other subfield except for solid state physics and teaching. This is certainly not inconsistent with what I have seen in academia, industry and federal laboratories. I am also told that some 172 univer-

sity departments in the United States now have active doctoral programs in physics. However, there are only two departments (institutes, actually) granting master's and doctoral degrees in a fairly broad spectrum of optical sciences and optical engineering: The Institute of Optics at the University of Rochester and the Optical Sciences Center at the University of Arizona. Quite a few optical endeavors are being undertaken vigorously at various electrical engineering and physics departments throughout the country, as well as at some special centers, but these do not yet present curricula and graduate options over a broad palette of the modern optical canvas.

As a result of this relatively constricted funnel for graduate education in optics, there is, and has been for the past five or ten years, an intense need for people with graduate training in the subject. Average starting salaries for men and women with doctorates in optical sciences or engineering are approaching $40 000 annually, which is significantly higher than offers currently being tendered to PhDs in physics and engineering. And even the crude predictions I have seen suggest that this situation will obtain for some time to come. Within the field itself the greatest demand from industry and government appears to be for people trained in the engineering aspects of optics, particularly in the design of optical systems, materials technology and the preparation of thin-film structures. The quantum opticians, on the other hand, are competing for employment in an arena that also contains those graduating from the many physics and electrical engineering departments, and so fall into a less panicked marketplace. These comments are not meant to win sympathies, funds or more optical units within universities! I'm just presenting them because they are true.   □

CHAPTER 2

LASER PHYSICS: HIGHER POWER, HIGHER FREQUENCY AND ULTRASHORT

Just after World War II, the field of optics and spectroscopy enjoyed a vigorous renaissance as more sensitive detectors for the visible and near infrared, which had been originally developed for military purposes, became available to the academic and industrial research communities. Improved diffraction gratings of much greater resolving power also became available. The field of microwave spectroscopy similarly was invigorated—both experimentally and in theory—by the wartime development of radar. Then, in 1953, the group under Charles Townes at Columbia University succeeded in obtaining amplification of a microwave signal tuned to the inversion resonance of the ammonia molecule (near 24,000 MHz). They named their device a MASER: Microwave Amplification by Stimulated Emission of Radiation, and the ammonia maser was the stimulus for intense efforts to develop and verify new maser devices. Almost immediately theoreticians began to speculate about how similar processes might work in the visible and infrared portions of the spectrum. Finally, in 1960, several independent efforts all culminated in the successful achievement of the LASER: Light Amplification by Stimulated Emission of Radiation. A paper by C. H. Townes and A. L. Schawlow outlined a practical mechanism; T. H. Maiman operated the first experimental ruby laser; and within months A. Javan had announced the first successful gas laser. (For a very readable history of this entire development see the book *Masers and Lasers* by Mario Bertolotti, Adam Hilger, 268 pp., 1983.)

A veritable revolution then engulfed the field of optics, as everything that had been done before—and many things that had never been done before—could now be done better with lasers. The first lasers were primitive prototypes and not very efficient, and the glorious advantages of this new optical curiosity were mostly speculative daydreams. The first ten or fifteen years of laser research were aimed at improving materials and apparatus and developing practical applications for this new phenomenon. But now the laser is more than 25 years old; we are still very much in the midst of the laser revolution, but we are out of the initial shakedown phase: the papers in the present volume

present a good picture of the current state of laser research and many practical applications.

A laser consists of a resonant cavity containing an active medium that can be excited (by an external energy source—an "optical pump"—or, sometimes, an electric field) into an excited state, and then, triggered by stimulated radiation, emits a cascade of intense, monochromatic, coherent radiation along the axis of the resonant cavity. The radiation is highly monochromatic, with spectral purity typically one part in a million. The radiation beam emitted by the laser is highly directional (i.e., has very little divergence with distance). Laser pulses have been aimed at the moon, where astronauts had left retroreflecting mirrors. It is still possible to detect the return pulse on Earth, even after traversing hundreds of thousands of miles. The intensity of a laser pulse can be very large, as this highly monochromatic, highly directional pulse of light can also be emitted very quickly. In such applications as laser fusion, in which a large amount of energy must be focused on a small target during a short time interval, the momentary power levels can be of the order of tens of terawatts (10^{13} W). And, finally, laser light is coherent spatially or temporally (or both), meaning that the phase and amplitude of the wave at any particular time and position can be calculated from earlier known values. These four properties give laser light huge advantages over ordinary light and make possible the applications discussed in this volume.

The first laser, a ruby rod, emitted bright red light (at 694 nm); other early lasers operated in the near infrared. In the intense research that followed, mechanisms were found for lasers at almost every wavelength from the near ultraviolet to the far infrared. But at wavelengths shorter than the ultraviolet there are new difficulties. Whereas visible light corresponds to ordinary transitions in atoms and molecules, x rays and gamma rays correspond to much more violent phenomena, such as the ejection of an electron from the inner shell of a heavy atom or a change in energy state inside the atomic nucleus. It is therefore much more difficult to produce population inversion in states leading to such energetic photons. Furthermore this energetic radiation passes right through ordinary mirrors. But there is still great interest in this spectral region, because the energies of the photons are greater, and also because the wavelengths of such radiation are so short that images of smaller objects—even molecules themselves—can be resolved.

In the highly competitive environment of the high technology community, whenever a new laser device or effect is described, some other eager group immediately pounces on that idea and then runs with it, trying to extend the laser light conditions to higher peak power, shorter wavelength, faster response, or more efficient operation, and proposing remarkable new applications. The articles in this chapter illustrate many of these developments.

There now exists a rather substantial body of literature on lasers and laser applications, ranging from text and reference books to conference proceedings of the latest symposia in this most active area of research. The Education Committee of the American Association of Physics Teachers has published bibliographies listing much of this material. The magazine *Scientific American* has published many articles at a semipopular level on light and lasers (June 1961, July 1963, February 1968, September 1968, March 1979, to mention just a few). Phillip F. Schewe, an AIP science writer, has written a comprehensive article on lasers for *The Physics Teacher*, pp. 534–547, November 1981, which we have used as the basis for organizing the laser material in the present volume.

John N. Howard

CONTENTS

Frontiers of laser development

Supersonic gas flow, organic dyes and chemical reactions are among the systems exploited in the search for more power, narrower linewidths and greater tunability.

John L. Emmett

PHYSICS TODAY / MARCH 1971

Early lasers had an extremely limited range of operational parameters, and so were of only limited usefulness. Now, however, laser action has been achieved in almost everything from semiconductor diodes to quinine water. What are some of the results of 12 years of laser research?

We can discuss the accomplishments most usefully in terms of five goals: generation of high average power; generation of high peak power; direct conversion of chemical energy to coherent light; extension of the spectral regions in which lasers operate; development of tunable lasers. (The best results to date in each field are listed in the box on page 25.) The projects I have selected are of course only a few examples of the work in progress in each field, and there are entire research areas that fall outside of our five topics.

High average power from lasers

Generation of high average power from lasers has proved to be difficult. Simply scaling up the laser system is rarely successful, because the complex physics associated with many systems requires simultaneous optimization of many technologies (see figure 1). We will concentrate here on two of the most significant problems in achieving high average power: low conversion efficiency and nonuniformity of excitation.

No present laser system is 100% efficient in converting input (excitation energy) into output (coherent radiation). The waste energy may appear as heat, which distorts the optical medium and populates terminal laser levels, or it may populate particular excited metastable levels of the laser system, resulting in absorption of the laser radiation. In either of these cases the result is generally to terminate the laser action or seriously reduce its efficiency, so that we are immediately faced with the problem of removing the waste energy.

Most laser systems are long cylinders or similar shapes with a limited surface-area to volume ratio. In solid-state laser systems the excess energy is conducted to the surface of the rod and is removed by some form of coolant. In gas laser systems the excess energy is usually dissipated by diffusion of the molecules to the walls where cooling or deactivation occurs. Power from both kinds of systems is usually strongly limited by thermal conductivity or diffusion, which are relatively slow processes. The characteristic time associated with energy loss by diffusion in the geometry shown in figure 2 is

$$\tau_{\text{diff}} = (D/\lambda)^2 (\lambda/c) = D^2/\lambda c$$

where λ is the mean free path, D is the tube diameter or a characteristic dimension of the device, and c is the mean particle speed.

In fluids, however, we need not rely on diffusion or thermal conductivity to remove waste energy. Forced convection or high-speed flow can strongly reduce the waste-energy rejection time. The characteristic time associated with this process is

$$\tau_{\text{flow}} = D/V$$

where V is the flow velocity. For typical gas lasers, high-speed flow may increase[1] the average power output by factors of 10^3 to 10^6. In fluids, flow velocity is not reduced by high density, so that we can increase the density of the active medium to increase the power output. This method is not usually possible in diffusion-controlled systems, in which relaxation time increases with increasing density.

Uniform excitation of large volumes of laser material is necessary to produce output fluxes of high optical quality. Whether the pumping mechanism is light, electrical discharges, or any of the many other schemes that have been developed, the achievement of a high degree of volume uniformity has been quite difficult.

Alan E. Hill (Air Force Weapons Laboratory, Albuquerque) is pursuing an interesting solution to both these

John L. Emmett is acting head of the Laser Physics Branch at the Naval Research Laboratory, Washington, D. C.

High-power carbon-dioxide laser installed at the Aero Propulsion Laboratory, Wright-Patterson Field. This device exploits plasma aerodynamics to get uniform excitation (see figure, overleaf). Figure 1

problems. He is working on a high average-power electrically excited carbon-dioxide laser that operates at 10.6 microns.[2] The configuration he uses is shown in figure 3. The entire device, with pumps and heat exchangers not shown in the figure, operates as a closed cycle. The laser channel is $5.6 \times 76 \times 100$ cm and operates at a flow rate of 28 000 cubic feet per minute, providing the waste-energy removal that is essential to operation at high power levels.

Hill's laser exploits plasma aerodynamic concepts to provide uniform electrical excitation of large volumes in high-speed flow streams. The uniform

Best reported results

Average power
60kW at 10.6 microns

Peak power
2.5×10^{13} watts at 1.06 microns

Conversion of chemical energy
4–5% in purely chemical laser

Wavelengths of operation
submillimeter to 1523 Ångstroms

Tunability
many contiguous ranges from 3410 Ångstroms to 3.5 microns; some tunable ranges at longer wavelengths

excitation produced is not dependent on the surface-area to volume ratio of the system or on the proximity of the walls. Discharges at high pressures or in large containers generally collapse in the steady state to a nonuniform discharge with a hot central core; ions from the hot core replace ions lost by diffusion and recombination at the cool periphery. Thus large uniform electrical discharges must be maintained in a transient state.

Discharges can under certain conditions be maintained in a transient state by "freezing" them in a high-speed flow stream; the system in figure 3 accomplishes this task over a useful range of operating conditions. This device has produced a continuous output of 19 kW at an electrical efficiency of 22.6%, corresponding to an output of 14.5 kW/kg/sec and an input of 145 kW/kg/sec. If we add to the output the optical energy lost in the 17 mirrors that are used to provide multiple passes through the laser medium, we have an intrinsic efficiency of more than 27%. The output is obtained in the lowest-order transverse electromagnetic mode (TEM_{00}), attesting to the homogeneity of the active medium.

In the gas dynamic laser invented by Edward T. Gerry, (AVCO),[3] high-speed flow provides excitation of the laser medium and removes the waste energy. Figure 4 is a schematic representation

of the way such a system achieves the necessary excitation and flow velocity; the particular conditions shown are for a CO_2 laser operating at 10.6 microns. This laser starts with the proper gas mixture at a sufficiently high temperature and pressure to provide significant population of the upper laser level.[4] The high-temperature, high-pressure gas is then rapidly expanded through a supersonic nozzle to a high Mach number.

The object here is to lower both the gas pressure and temperature downstream of the nozzle in a time short compared with the vibrational relaxation time of the upper laser level. At the same time water vapor or helium is added to speed the relaxation of the lower laser level, so that the lower level relaxes in a time comparable with or shorter than the expansion time. The result is that several millimeters downstream from the nozzle a population inversion and gain exists between the 00^01 and 10^00 vibrational levels of CO_2, and optical energy can then be extracted from the system.

Note that the problem of uniformity of excitation has been at least partially solved by this approach. In practice the high temperature and pressure can be achieved by burning carbon monoxide or cyanogen with oxygen and adding nitrogen and water vapor. An array of two-dimensional nozzles is usually used to handle the total flow and geometrical requirements. A large device operating at AVCO is capable of approximately 14 kg/sec mass flow. This device, which uses CO as a fuel, has produced a continuous power output of 60 kW in multimode operation, that is with the beam angle greater than the diffraction-limited value, and 30 kW in nearly diffraction-limited operation.[3]

High peak-power lasers

The first ruby laser operated by Theodore H. Maiman[5] produced an array of spikes or pulses with a peak amplitude of about 10 kW; at the time this was an extremely high power level for such a monochromatic source. Soon afterward, "giant pulse" or "Q-switched" operation extended the power levels to the order of 100 MW. We can get even higher power pulses by arranging the phase relations between the longitudinal modes of the laser ("mode locking") so that all these sine waves are superposed. The result is an array of very short pulses, rather than a single pulse as in the Q-switched laser. One of these pulses is then selectively amplified. Mode-locked neodymium-doped glass lasers at 1.06 microns have produced pulses of a few picoseconds. (A one-picosec pulse containing one millijoule has a peak power of 10^9 watts.)

Plasma physicists provide some of the impetus for research in extremely high peak-power lasers with high energy content. Such pulses, focused on small solid targets, can produce clean, high-density, high-temperature plasmas. A group at the Lebedev Institute in Moscow under the direction of Nikolai G. Basov was the first to report neutron emission from solid lithium-deuteride targets irradiated with 30-joule 20-picosec pulses.[6] Others in the US and Europe have published similar results.[7]

In the US high peak-power lasers are being developed at the Lawrence Radiation Laboratory in Livermore, Sandia Corp, Los Alamos Scientific Laboratory, the Naval Research Laboratory (NRL) and the University of Rochester. In all of these programs a single mode-locked pulse or other short pulse is repeatedly amplified in successive Nd:glass stages until the desired pulse energy is obtained. The design of such amplifiers and the development of suitable laser glass has proved to be a difficult problem. The group at the University of Rochester under the direction of Moshe J. Lubin reports the generation of a 600-joule pulse of about 0.2 nanosec duration for a peak power of 1.2×10^{12} watts. The system is presently being adapted to operate at such levels on a repeatable basis. Garth W. Gobeli and his co-workers at Sandia have achieved an output of 75 joules in 3 picosec or a peak power of about 2.5×10^{13} watts, and Dennis Gill of Los Alamos reports 50 joules in pulses of 10–100 picosec.

James E. Swain at Livermore has generated 100 joules in 1 nanosec in an unusual multipass laser system. In one trial, 400 joules was achieved, with destruction of the beam-folding mirrors. The NRL group is developing a system to provide 500-joule pulses with pulse widths adjustable from 20 picosec to 1.5 nanosec; they are presently producing 500-joule, 30-nanosec pulses.

In another wavelength region unusually high-power pulses have been obtained from pulsed CO_2 lasers at 10.6 microns. Hill produced pulses of several joules from CO_2 lasers at 0.1 atmosphere with applied-voltage pulses of 500 kV and obtained peak powers of 200 kW.[8] G. J. Dezenberg and coworkers (US Army Missile Command, Huntsville) extended the work to pulses of

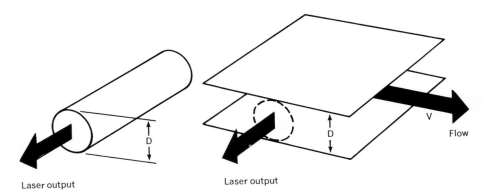

Static and flow lasers. In static lasers (left) waste-energy rejection time depends on diffusion through a tube of diameter **D**. In the high-speed flow lasers the fluid, moving with velocity **V**, can carry off the waste energy faster.   Figure 2

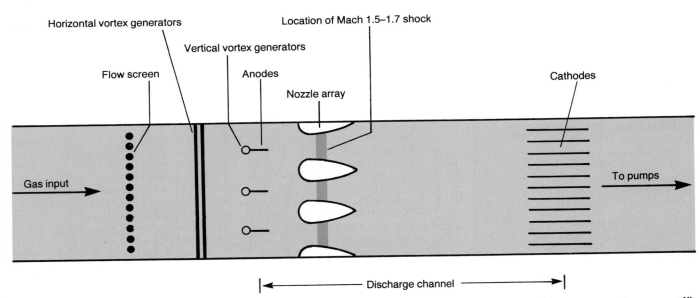

High average-power electrically excited CO_2 laser developed by Alan Hill removes waste energy efficiently and excites the gas uniformly. Screen establishes laminar flow in the gas, which then passes through a series of vortex generators. The plasma, traveling at supersonic speeds through the nozzles, experiences a shock because of the background pressure in the discharge channel.   Figure 3

24 joules and peak powers of 3.6 MW,[9] also at about 0.1 atmosphere.

Early last year, Jacques Beaulieu (Defense Research Establishment, Valcartier, Canada) announced the development of simple pulsed CO_2 lasers operating at atmospheric pressure; these lasers produced pulses in the 10–100 megawatt range with pulse energies of the order of 10 joules.[10] The achievement was made possible by the development of a simple scheme to produce volume excitation of the gas at the high pressure. The excitation consists of numerous fast, simultaneous, individually-ballasted electrical discharges transverse to the laser axis and distributed uniformly along its length. Such short-pulse excitation effectively produces gain-switching operation that leads to high peak powers. By directing the gas flow through the laser cavity, operation at 1000 pulses per sec has been achieved.

The demonstration of CO_2-laser operation at atmospheric pressure has stimulated many other groups to pursue research in this area. At the International Quantum Electronics Conference held in Kyoto, Japan last September, R. Dumanchin (Marcoussis, France) reported achieving 130 joules in a 2-microsec pulse from an electrically pulsed CO_2 laser.[11] Preionization techniques allowed the uniform pumping of 20 liters of gas at atmospheric pressure.

Unlike optically pumped solid-state lasers, these pulsed gas lasers are very efficient—about 5–25%. One atmosphere is clearly not the upper pressure limit, and operation at ten and possibly 100 atmospheres might be achieved if we could develop a suitable method of exciting the gas. Pulsed gas lasers will quite likely compete strongly with optically pumped solid-state lasers for the production of high peak-power energetic pulses.

Chemical lasers

The high energy density that can be released in exothermic chemical reactions has led to the interest in converting chemical energy to coherent optical energy. If we could use the energy of such reactions directly to obtain a population inversion, we could build compact powerful lasers. Chemical lasers may also become a powerful method of studying the kinetics of chemical reactions.

The first chemical-reaction laser was realized by Jerome V. Kasper and George C. Pimentel.[12] In their laser the active molecule was vibrationally excited hydrogen chloride, formed by the reactions

$$Cl + H_2 \rightarrow HCl(v = n) + H$$

$$H + Cl_2 \rightarrow HCl(v = n) + Cl$$

The initial chlorine atoms were produced by flash photolysis according to the reaction

$$Cl_2 + h\nu \rightarrow 2\,Cl$$

The output of this laser is at several frequencies between 3.7 and 3.8 microns, corresponding to several rotational lines associated with the $v = 2 \rightarrow v = 1$ vibrational transition of the HCl molecule.[13]

This laser illustrates two aspects of chemical-laser systems quite well. First, it is generally necessary to have a source of free atoms or free radicals that can react with other atoms or molecules to form an excited-state reactant. The development of chemical lasers is at present dependent on our ability to generate free atoms or free radicals. Second, long-chain reactions, such as the one that produces hydrogen and chlorine atoms, are of significant value because they limit the rate of external production of free radicals. We will need strongly exothermic long-chain reactions to be able to use efficiently the total energy available in a chemical reaction.

Since the original H_2–Cl_2 flash-photolysis laser, many pulsed and continuous lasers based on the hydrogen isotopes and the halogens have been developed, some with average powers of several hundred watts.[14] In most of these systems some external means such as flash photolysis, electrical discharge or shock heating starts the reaction, but Terrill A. Cool and Ronald R. Stephens (Cornell) have developed a purely chemical laser.[15] This laser is based on the reactions:

$$F_2 + NO \rightarrow NOF + F$$

$$F + D_2 \rightarrow DF(v = n) + D$$

$$D + F_2 \rightarrow DF(v = n) + F$$

$$DF(v = n) + CO_2\,(00^00) \rightarrow \\ DF\,(v = n - 1) + CO_2\,(00^01)$$

In the first reaction F_2 is dissociated by a chemical means into free fluorine atoms. The fluorine atoms then mix rapidly with D_2 and CO_2, forming vibrationally excited DF. The excited DF transfers its energy to the CO_2, and laser action occurs at 10.6 microns.

Cool's original laser was a longitudinal-flow device, but present versions operate with fast mixing and flow transverse to the optical axis much as do the other CO_2 flow lasers. Here the value of the flow is to remove the reactants from the laser cavity. Figure 5 shows an operating transverse flow system at NRL; maximum efficiency of present devices is in the 4–5% range (based on the energy available in the $D_2 + F_2$ reactions). The highest-power output obtained to date, 440 watts continuous wave (cw), has been achieved by Theodore J. Falk at the Cornell Aeronautical Laboratory. Since the development of the purely chemical DF–CO_2 transfer laser, other

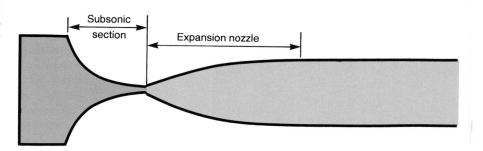

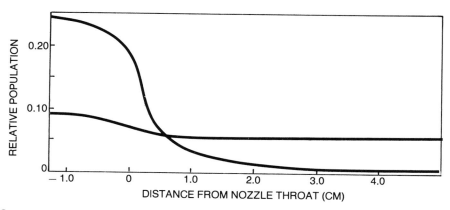

Gas dynamic laser. In this system, developed by Edward Gerry, a mixture of carbon dioxide, nitrogen and water vapor at high temperature and pressure is rapidly expanded through a supersonic nozzle (above), resulting in a population inversion between the upper (color) and lower (black) vibrational levels of carbon dioxide.   Figure 4

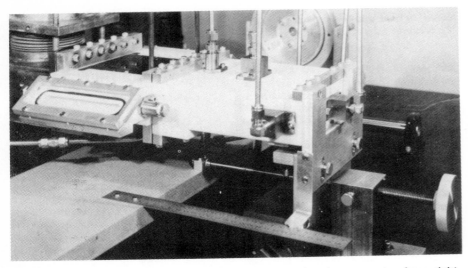

Purely chemical laser at Naval Research Laboratory. Input gases enter from right and flow toward left; laser beam travels in the transverse direction.   Figure 5

purely chemical laser systems have been operated. In the system just discussed, for example, one can omit the CO_2 and get laser action from DF alone.

New wavelength regions

Ever since the realization of the first ruby laser much research has been devoted to extending laser operation to other wavelengths. Laser systems now exist from the near-ultraviolet through the visible and infrared to the submillimeter wavelength region. Significant efforts have been made to develop powerful ultraviolet sources by generating the harmonics of ruby and Nd:glass lasers. In this way lasers were made to operate in the 2300 to 3500 Å region, but still no laser sources existed in the vacuum ultraviolet. The major obstacle to developing vacuum-ultraviolet lasers is that the probability of spontaneous emission increases with the third power of the frequency, so that extremely high pump rates are needed to maintain a population inversion. Last August Rodney T. Hodgson (IBM) announced laser action in the Lyman band of molecular hydrogen.[16] He reported several lines in the 1600-Å region with a total output of about 1.5 kW. Soon after Ronald Waynant and coworkers at NRL reported laser action in the Lyman band of molecular hydrogen with output powers in the 100-kW region.[17] These two almost simultaneous accomplishments followed the initial proposal by P. A. Bazhulin, I. N. Knyazev and G. G. Petrash (Lebedev Institute) for laser action by electron-impact excitation of molecular hydrogen.[18] Abdul W. Ali and Alan C. Kolb[19] followed this proposal with a detailed rate-equation analysis for the Werner and Lyman bands; figure 6 is the relevant energy-level diagram for the system. The Lyman band is the transitions from $B^1\Sigma_u^+ \rightarrow X^1\Sigma_g^+$. The excitation shown is by electron impact. Someone will very likely soon observe stimulated emission in the Werner-band transitions $C^1\Pi_u \rightarrow X^1\Sigma_g^+$.

To provide the extremely fast electron-impact ionization both Hodgson and Waynant used Blumlein pulse generators similar to the one used by John Shipman to obtain 2-MW pulses from N_2.[20] In Waynant's laser system this generator produces a breakdown that starts at one end of the laser channel and propagates with the velocity of light down the channel. Current pulse-rise times of 2.5 nanosec and peak currents of hundreds of kiloamperes are achieved. The gain of these transitions is exceedingly high, so that mirrors are not required, and a single pass of the radiation through the laser channel extracts all the available energy. The output in the forward direction is 10–100 times the output in the reverse direction, amply demonstrating the high stimulated-emission gain achieved in

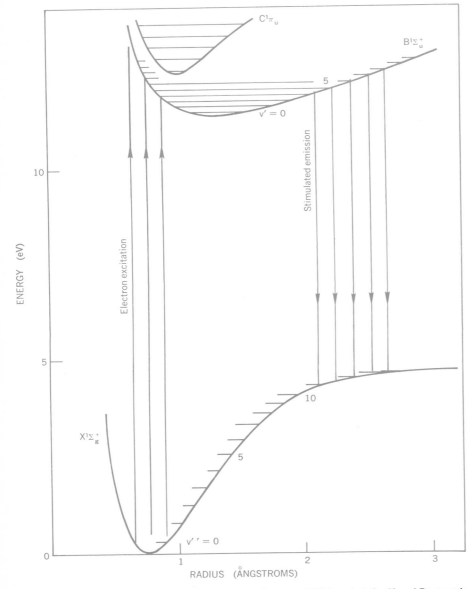

Energy levels in a molecular-hydrogen laser. Groups at IBM and at the Naval Research Laboratory have observed stimulated emission in the Lyman band, corresponding to transitions from $B^1\Sigma_u^+$ to $X^1\Sigma_g^+$. $C^1\Pi_u \rightarrow X^1\Sigma_g^+$ has not yet been seen.   Figure 6

such devices. Hodgson, working with a Blumlein generator without traveling-wave excitation, has to date achieved laser action on 67 lines between 1523 and 1614 Å in H_2, para H_2, HD and D_2 with total output power of about 10 kW. Waynant reports 19 lines between 1523 and 1614 Å in H_2 and D_2 with total output of 1–3 MW. Hodgson also reports laser action on five bands in CO in the 1800 to 2000-Å region. The power output here is quite low—about ten watts.

Development of tunable lasers

Tunable narrow-linewidth sources are now available from the near-ultraviolet to the middle-infrared regions of the spectrum with pulsewidths adjustable between picoseconds and cw operation. A conservative prediction might be that in the next five years, tunable lasers will replace many conventional spectrophotometers and spectrofluorometers, and will become a general-purpose optical measurement tool.

Two types of tunable sources have been extensively demonstrated—dye lasers and parametric oscillators. Each has its own strengths and weaknesses, but the two complement each other; the dyes will probably span the ultraviolet and visible portion of the spectrum with parametric oscillators providing sources tunable from the visible far into the infrared.

The efficient fluorescence of many organic compounds, such as dyes, makes them attractive as possible laser materials. Initial experiments with aromatic hydrocarbons produced disappointing results, but in 1966 Peter P. Sorokin and John R. Lankard (IBM) produced the first unambiguous demonstration of laser action in organic molecules.[21] They obtained laser action in many dye solutions pumped either with lasers or with microsecond-long flashes from high-intensity lamps. Since then dye lasers of all types have been intensively studied in many laboratories.

Optical excitation and emission in an organic dye take place between the rotational–vibrational levels of the ground singlet state S_0 and excited singlet state S_1 (see figure 7). The lifetime of the excited singlet state is typically a few nanoseconds. Only a small excited singlet-state population, however, is needed because the transition is strongly allowed and the laser terminal state is a high-lying vibrational level that is essentially unpopulated. Thus we are essentially dealing with a four-level laser system.

There is, however, a finite intersystem-crossing rate k_{ST}^{-1} from the excited singlet state S_1 to the lowest triplet state T_1. The relaxation of $T_1 \rightarrow S_0$ is generally orders of magnitude slower than the $S_1 \rightarrow S_0$ relaxation, and population can build up in the T_1 states. Because absorptions from T_1 to higher

lying triplet states T_2 generally occur at the same wavelengths as the desired laser emission, a large loss coefficient results, effectively terminating laser action.

In some molecules laser action is also prevented by absorptions from S_1 to other higher-lying singlet states, so that one must choose a dye without $S_1 \rightarrow S_2$ transitions in the region of interest and find ways of deactivating the triplet state. Benjamin B. Snavely (Eastman Kodak, Rochester) and F. P. Shäfer (University of Marburg) found that molecular oxygen dissolved in the dye solution could effectively de-excite or quench the triplet state of the dye rhodamine 6G.[22] Romano Pappalardo, Harold Samelson and Alexander Lempicki (General Telephone and Electronics, Bayside) found similar behavior with cyclo-octatetraene,[23] and other variations are being rapidly developed at many laboratories.

Because the emission bands of these organic molecules are typically several hundred angstroms wide and demonstrate high gain, the addition of a dispersive element in the laser cavity permits tuning over this band. The great interest in the dye laser shown

by many laboratories throughout the world has resulted in the development of many compounds that span the region 3410–11 750 Å in bands of a few hundred angstroms each. Dye lasers have also been mode locked by a wide variety of techniques.

A. Dienes and his coworkers (Bell Telephone Laboratories) have recently extended the tuning range of a few dyes to 1000–1760 Å[25] by the formation of a new molecular complex in the excited state. This "exciplex" is formed of a dye molecule in the excited state and a hydrogen ion from the solution. The ground state of the exciplex is unstable, and fast dissociation takes place after emission of a photon. The exciplex has a different emission band from the simple dye molecule, and a tuning range that encompasses both is possible in a buffered solution. Dienes achieved a tuning range of 3910 to 5670 Å in a buffered solution of the dye 4-methylumbelliferone.

Otis G. Peterson, Sam A. Tuccio and Snavely (Eastman Kodak) have recently developed[26] a cw dye laser. They passed an argon-ion laser beam focused to 10-microns diameter (see figure 8) through a dye cell with a hemi-

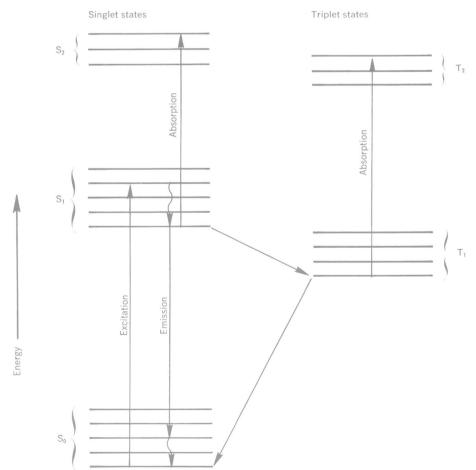

Singlet states   Triplet states

S_2 T_2

Absorption

S_1 T_1

Absorption

Excitation   Emission

Energy

S_0

In an organic dye, optical excitation and emission occur between the vibrational levels of the ground singlet state S_0 and the excited singlet state S_1. Intersystem crossing between singlet and triplet states results in large losses.   Figure 7

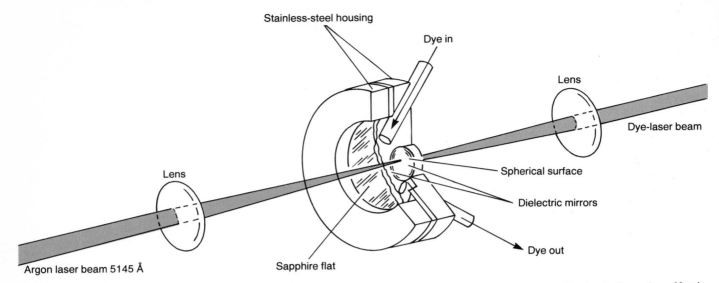

Stainless-steel housing

Dye in

Lens

Dye-laser beam

Spherical surface

Dielectric mirrors

Dye out

Lens

Sapphire flat

Argon laser beam 5145 Å

In continuous-wave dye laser developed by Otis Peterson, Sam Tuccio and Benjamin Snavely, a 10-micron diameter argon laser beam provides the pumping. The group reports observing continuous-wave lasing in three xanthine dyes.   From ref. 30.

Figure 8

spherical resonator. The sapphire flat and the spherical mirror comprising the resonator have high reflectivity at the dye laser wavelength. The dye, rhodamine 6G saturated with air to quench the triplet state, is pumped rapidly through the cell to minimize thermal–optical distortions in the liquid. Peterson originally found an excitation threshold of 200 mW, and he obtained an output of 4 mW with an input of 960 mW. The emission linewidth was about 30 Å centered at about 5965 Å. More recently, with an improved experimental apparatus that has external mirrors and a prism in the cavity, the group reports cw laser action in three of the xanthine dyes.[27] With the prism they could tune the emission of each dye over about 350 Å, so that the three dyes used separately provide a tuning range from 5500 to 6500 Å. The threshold for operation is about 60 mW of argon-ion laser power, and a 20% slope efficiency is achieved. The output of the laser is a spatially coherent TEM_{00} Gaussian mode.

Advances in parametric-oscillator technology became rapid once appropriate crystals with large nonlinear optical coefficients were available. Microwave parametric oscillations and

amplifiers are well known, and the terminology has been carried over to the optical parametric oscillators.[28] Three frequencies are generally involved. The upper frequency ω_p is known as the "pump" frequency, and the signal to be amplified ω_s is known as the "signal" frequency. If these two frequencies are incident on a material possessing a nonlinear interaction that couples the fields, the signal frequency may be amplified with the concomitant generation of a third frequency ω_i known as the "idler" frequency.

In the optical region of the spectrum the nonlinearity that couples the fields is the nonlinear polarizability tensor $\mathbf{X}_{ijk}$ of crystals without a center of symmetry. $\mathbf{X}_{ijk}$ relates the optical-frequency polarization P in the crystal to the incident fields by E_j and E_k by the relation

$$P_i = \sum_{j,k} \mathbf{X}_{ijk} E_j E_k$$

Because $\mathbf{X}_{ijk}$ is almost independent of frequency in regions of the spectrum where the crystal is transparent, optical parametric oscillators can be tuned over a wide range.

As in phase-matched harmonic generation, the polarizability wave must travel at the same velocity as a freely propagating electromagnetic wave if

efficient conversion of energy from the pump to the signal is to be obtained. Thus the **k** vectors of the three waves in the material must satisfy the phase matching condition

$$\mathbf{k}_s + \mathbf{k}_i = \mathbf{k}_p$$

If the pump frequency is fixed, varying the indices of refraction of the material at the signal and idler frequencies will result in tuning of these frequencies. The refractive indices may be easily changed by changing the temperature of the crystal; this may give a range of, say, 5000 Å to 3.5 microns. It is also possible to use the natural birefringence of the crystal, by changing the alignment of the crystal with respect to the signal and idler frequencies. Both methods result in frequency tuning of the oscillator. Either the signal or idler or both can be chosen as the desired output. Figure 9 is a schematic representation of a simple parametric oscillator. It may be resonant at the signal or idler frequencies or both depending on the type of operation desired. Total conversion of the pump energy to signal and idler energy is theoretically possible, and efficiencies well over 50% have been observed.

Joseph A. Giordmaine and Robert C. Miller realized the first tunable optical parametric oscillator in 1965,[29] and since then the field has experienced explosive growth. It is now possible to tune over most of the visible spectrum, a good portion of the near infrared and part of the far infrared. Many crystals, such as potassium dideuterium phosphate, lithium niobate, barium sodium niobate, lithium iodate and proustite (Ag_3AsS_3), have been used successfully as the nonlinear material. Neodymium in glass and in yttrium–aluminum–garnet (YAG), ar-

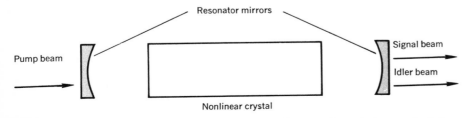

Resonator mirrors

Pump beam

Signal beam

Idler beam

Nonlinear crystal

Parametric oscillator. Input radiation of frequency ω_p goes through resonant cavity formed by the mirrors and the nonlinear crystal. Output radiation is at two frequencies, the "idler" and "signal" frequencies ω_i and ω_s.

Figure 9

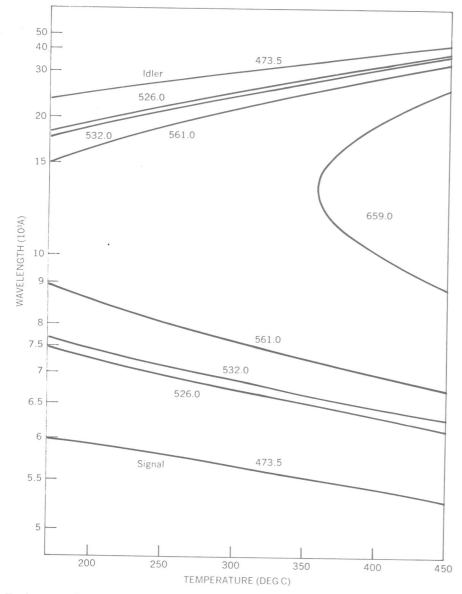

Tuning curve for a parametric oscillator that has lithium niobate as its nonlinear crystal. Each of five curves shows output (signal and idler) for a particular input wavelength (shown in microns) as a function of phase-matching temperature.   Figure 10

gon-ion and ruby lasers have been used as pump sources. Both pulsed and cw operation have been obtained but not always in the same part of the spectrum.

One particular parametric oscillator, manufactured by the Chromatrix Corp, has an exceedingly wide tuning range (see figure 10). The device uses a Nd:YAG oscillator that can oscillate on 13 lines from 9560 to 13 580 Å. A lithium-iodate crystal doubles the output frequency of this laser, to provide the pump source for the parametric oscillator. The nonlinear crystal is lithium niobate, tuned by changing its temperature.

What advances can we expect in the next decade? We hope, of course, for major improvements in power, efficiency and frequency diversity. But, more importantly, we look forward to a period when, with the technology matured, lasers will be increasingly useful as research tools.

*   *   *

I thank all those persons who provided their latest results for inclusion in this article, and I also thank John M. McMahon, who helped me assemble all the material.

References

1. J. Wilson, Appl. Phys. Lett. **8,** 159 (1966).

2. A. E. Hill, Appl. Phys. Lett. (to be published).

3. E. T. Gerry, Bull. Am. Phys. Soc. **15,** 563 (1970).

4. N. G. Basov, A. N. Oraevskii, Sov. Phys.—JETP **17,** 1171 (1963); J. R. Hurle, A. Hertzberg, Phys. Fluids **8,** 1601 (1965).

5. T. H. Maiman, Nature **187,** 493 (1960).

6. N. G. Basov, paper presented at the Fifth International Quantum Electronics Conf., Miami, Fla. (1968).

7. Laser Focus, Oct. 1969, page 14; G. W. Gobeli, J. C. Bushnell, P. S. Perrcy, E. D. Jones, Phys. Rev. **188,** 300 (1969).

8. A. E. Hill, Appl. Phys. Lett. **12,** 324 (1968).

9. G. J. Dezenberg, W. B. McKnight, L. N. McClusky, E. L. Ray, IEEE J. Quantum Electronics, **QE-6,** no. 10, 652 (1970).

10. A. J. Beaulieu, Appl. Phys. Lett. **16,** 504 (1970).

11. R. Dumanchin, J. C. Farcy, J. Michon, J. Rocca Serra, in *Proceedings* of the Sixth International Quantum Electronics Conf., Kyoto, Japan (1970).

12. J. V. Kasper, G. C. Pimentel, Phys. Rev. Lett. **14,** 352 (1965).

13. P. H. Corneil, G. C. Pimentel, J. Chem. Phys. **49,** 1379 (1968).

14. D. J. Spencer, H. Mirels, T. A. Jacobs, R. W. F. Gross, Appl. Phys. Lett. **16,** 235 (1970).

15. T. A. Cool, R. R. Stephens, J. Chem. Phys. **51,** 5175 (1969).

16. R. J. Hodgson, Phys. Rev. Lett. **25,** 494 (1970).

17. R. W. Waynant, J. D. Shipman, R. C. Elton, A. W. Ali, Appl. Phys. Lett. **17,** 383 (1970).

18. P. A. Bazhulin, I. N. Knyazev, G. G. Petrash, Sov. Phys.—JETP **21,** 649 (1965).

19. A. W. Ali, A. C. Kolb, Appl. Phys. Lett. **13,** 259 (1968); A. W. Ali, Catholic University of America, Dept. of Space Science and Appl. Phys. Report 68-009, Oct. 1968.

20. J. D. Shipman, Appl. Phys. Lett. **10,** 3 (1967).

21. P. P. Sorokin, J. R. Lankard, IBM J. Res. Devel. **10,** 162, March 1966.

22. B. B. Snavely, F. P. Shäfer, Phys. Lett. **28A,** 728 (1969).

23. R. Pappalardo, H. Samelson, A. Lempicki, Appl. Phys. Lett. **16,** 267 (1970).

24. W. H. Glenn, M. J. Brienza, A. J. Demaria, Appl. Phys. Lett. **12,** 54 (1968); W. Schmidt, F. P. Shäfer, Phys. Lett. **26A,** 558 (1968); D. J. Bradley, A. J. F. Durrant, Phys. Lett. **27A,** 73 (1968).

25. A. Dienes, C. V. Shank, A. M. Trozzolo, in *Proceedings* of the Sixth International Quantum Electronics Conf., Kyoto, Japan (1970).

26. O. G. Peterson, S. A. Tuccio, B. B. Snavely, Appl. Phys. Lett. **17,** 245 (1970).

27. S. A. Tuccio (to be published).

28. R. H. Kingston, Proc. IRE **50,** 472 (1962); N. M. Kroll, Phys. Rev. **127,** 1207 (1962); S. A. Akhmanov, R. V. Khokhlov, Sov. Phys.—JETP **16,** 252 (1963); J. A. Armstrong, N. Bloembergen, J. Ducuing, P. S. Persham, Phys. Rev. **127,** 1918 (1962); C. C. Wang, C. W. Racette, Appl. Phys. Lett. **6,** 189(1965).

29. J. A. Giordmaine. R. C. Miller, Phys. Rev. Lett. **14,** 973 (1965).

30. B. B. Snavely, Proc. IEEE **57,** 1374 (1969).   □

Rare-gas halide lasers

This new class of ultraviolet lasers, very efficient
and apparently scalable to high power, promises to be of value
in such applications as photochemical separation.

James J. Ewing

PHYSICS TODAY / MAY 1978

The rare-gas monohalides are simple diatomic molecules whose properties and emission spectra were essentially unknown as recently as four years ago. Now they are the active media for gas lasers that could provide overall electrical efficiency as high as 10%. These lasers operate in the ultraviolet and vacuum-ultraviolet spectral regions and are the first lasers outside the infrared region that appear to be scalable to high single-pulse energy (greater that 100 joules per pulse) with high average power and efficiency. Not surprisingly in light of the species' novelty and the laser's promised utility (for example, in photochemical-separation schemes), research and development in this area has grown remarkably since the first laser demonstration three years ago.

The utilization of these intense uv sources in research has already begun. Commercial models are now available that produce 100-mJ pulses of about 30 nanoseconds duration at 100-Hz pulse-repetition frequencies. Lasers with pulse-repetition frequencies of up to 1 kHz have been demonstrated and should be marketed soon. After a brief survey of the short history of rare-gas halide lasers, we shall look at the structure and spectroscopy of these novel and promising lasing species.[1,2] Then we can consider the theory and practical aspects of exciting the lasers, along with the resulting efficiencies.

Rare-gas monohalides belong to a larger class of molecules broadly known as excimers. An excimer is an atomic or molecular aggregate, composed of two atoms or molecules, that is unstable in the ground state, but bound in its excited

state. In figure 1 are schematic potential-energy curves for such a molecule. (Heteronuclear or heteromolecular excimer species such as the rare gas halides are also occasionally called "exciplexes.")

The excited state of the excimer species can radiate in a broad band that is red-shifted from the wavelength of the parent atomic excitation. The size of the band shift depends on the depth of the excited state well (typically 1–5 eV) and on the repulsion of the lower state at the energy minimum of the excimer. Monoatomic gases such as the rare gases and mercury have long been known to radiate broadband continua, emissions that are known to be from the diatomic molecule excimer. These bands are easily observable when rare gases or mercury at high pressures are electrically excited: High pressures favor the excimer formation by increasing the recombination rate of atomic excited states into the molecular upper level.

Excimer emission spectra are useful for lasers because the lower level rapidly dissociates. Since the dissociation time, about 10^{-12} sec, is much less than upper-level radiative lifetimes, 10^{-9} to 10^{-6} sec, population inversions can be readily produced and maintained. However, the broad bandwidth implies low stimulated-emission cross sections and correspondingly low gain for a given excimer density. Excimer cross sections are in the range 10^{-17} to 5×10^{-16} cm^2, whereas similar atomic and molecular transitions at comparable oscillator strengths lie in the range 10^{-15} to 10^{-12} cm^2, in the visible and ultraviolet. To achieve laser action, then, excimer lasers must be vigorously pumped. Typical excitation rate requirements are of order 200 MW per liter for gains of order 10% cm^{-1}. Specific energy densities of about 40 joules per

liter can be expected from excimer lasers. This value compares quite favorably with the extractable energy densities of the high power infrared lasers CO_2, CO and HF.

Before rare-gas halide emissions were discovered, the Xe_2, Kr_2, and Ar_2 excimer bands had been made to lase in the vacuum ultraviolet. These emissions are broad ($\Delta\lambda \approx 200$ Å), with stimulated-emission coefficients[3] of about 10^{-17} cm^2. Problems with mirrors and with competing absorption processes, such as photoionization of the upper level, lead to low efficiencies for these lasers. However, the channeling of atomic ionization and excitation into these excited states was found to be very efficient. Several other known excimer species such as Hg_2, HgXe, and NaXe were also studied but have not yet yielded a laser.

First observations

Rare-gas halide emissions were first broadly characterized in studies of the quenching of metastable atomic states in rare gases by halogen bearing compounds.[4,5] At the low pressures of these kinetic studies, a broad ($\Delta\lambda \approx 100$ Å) undulating continuum appeared. The only plausible mechanism for producing this radiation is the bimolecular reactive quenching of the rare-gas metastable to form electronically excited molecules which then emit. The data suggested the mechanism

$$Xe^* + Cl_2 \rightarrow XeCl^* + Cl$$

$$XeCl^* \rightarrow Xe + Cl + h\nu(300 \text{ nm})$$

where * denotes electronic excitation of either an atom or molecule.

Shortly after this first characterization, production of these novel excimer excited states was found to be highly efficient under the pressure and deposition rate

James J. Ewing is a staff scientist in the laser fusion program at Lawrence Livermore Laboratory.

conditions required for excimer laser action. These high-pressure spectra are considerably sharper than those in the low-pressure regime. The narrow bandwidth and apparent rapid radiative rate implied that these new excimers have stimulated-emission cross sections an order of magnitude larger than do the rare-gas excimers, and the higher stimulated-emission coefficient implies that laser thresholds are easier to reach. Moreover high efficiency behavior is easier to achieve than with several other excimer systems because the stimulated gains well exceed various bulk losses in the excited gases.

The gross kinetic and radiative properties of these systems were rapidly explored, and most of these species have been demonstrated as laser media. At first, excitation was by relativistic electron beams but electric discharge schemes were demonstrated shortly thereafter. Intrinsic medium efficiencies are of order 10%, and overall efficiencies of small-scale, nonoptimized (e-beam and simple discharge) lasers are of order 1%. Single-pulse energies of over 300 J have been reported. In Table 1 are lists of the wavelengths and excitation techniques of the rare-gas halide lasers. Of the lasers demonstrated, ArF (193 nm), KrF (248 nm) and XeF (351, 353 nm) yield the highest efficiencies and highest single-pulse energies. Simple discharge devices have produced energies of 1 J per pulse on the rare-gas fluorides and energies in excess of 100 mJ on XeCl, KrCl and XeBr.

Spectra of rare-gas halides

The spectral properties and most of the important excited-state production and quenching properties of rare-gas monohalides are governed by the simple physical nature of the upper level. The excited states of the rare-gas halides are ion pairs, for example Kr^+F^-. These charge-transfer or ionic states are strongly bound with respect to the lowest atomic excited states. Moreover, we can predict the binding energies of the ionic excited states to within a few percent by noting that the excited state of a rare-gas monohalide ion pair and the ground-state ion pairs of the appropriate alkali halide molecules are virtually isoelectronic. Thus KrF^* is the ion pair Kr^+F^-, which is very similar in its binding energy and formation kinetics to the RbF ground-state molecule.

In figure 2 are potential-energy curves for KrF.[6] The lowest energy curves are essentially repulsive and derive from the collision of a ground state Kr atom with an F atom. The angular momentum symmetries are $Kr:^1S$, and $F:^2P_{3/2,1/2}$. The molecular states that arise are the $X^2\Sigma(\Omega = 1/2)$, where X represents the lowest energy state, and the strongly repulsive $A^2\Pi(\Omega = 3/2, 1/2)$ states. The Π states lie higher in energy because they place two electrons from the outer 2p orbitals of the F atom along the internuclear axis, whereas in the $^2\Sigma$ state only one of these electrons is aligned along the axis. There are three low-lying excited states of

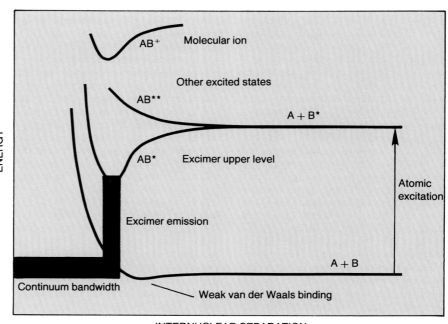

INTERNUCLEAR SEPARATION

Potential energy of excimers. Such species are useful for lasers because the ground state is unstable and the first excited state is bound. Because the lower-level dissociative lifetime in such systems is much less than the upper-level radiative lifetime, population inversions are readily produced and maintained.

Figure 1

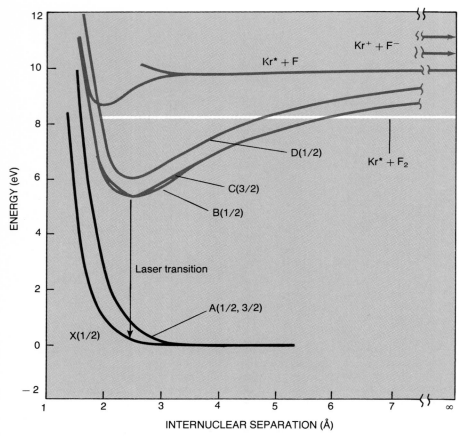

Potential energy curves for KrF.   The laser transition here is from B to X, as it is for all the rare-gas halides.   Note the exceptionally long range of the ion-pair energy (grey curves), resulting in a very large three-body rate coefficient for ion recombination.   (The third body is typically a lighter rare-gas buffer.)   Adapted from calculations published in reference 8.   Figure 2

ionic nature: The F^- ion is a spherically symmetric ^1S state and the Kr^+ ion has $^2P_{3/2,1/2}$ angular momentum, so that these states are $B^2\Sigma(\Omega = \frac{1}{2})$, $C^2\Pi$, $(\Omega = \frac{3}{2})$ and $D^2\Pi(\Omega = \frac{1}{2})$. Here Ω is the sum of electronic orbital and spin angular momenta projected along the molecular axis. Because of spin orbit effects that are especially pronounced in those rare-gas halides containing heavy atoms, such as Kr or Xe, the Ω quantum labeling is preferred to the Σ and Π labeling. The strongest uv emission bands correspond to the $B \to X$, $D \to X$ and, much weaker, $C \to A$ transitions. The lasers all operate on the $B \to X$ transition, which corresponds to an electron hopping from the halide ion to the rare-gas ion, yielding the rare-gas and halogen ground-state atoms. Oscillator strengths for these and analogous charge-transfer bands in other molecules are about 0.1. The radiative lifetimes for uv bands are in the range 5–15 nanosec.

Because the ground state is only weakly repulsive in the region of the $B \to X$ laser transition, the emission band is very sharp for a continuum. In the high-pressure spectrum, fluctuations occur in the high-frequency side of the main band. These fluctuations are due to emission from higher vibrational states of the upper level. Calculations of the various continuum spectra produced by the different

vibrational levels of KrF show that stimulated emission in the peak of the band can extract all of the low lying vibrational levels of the excited B state.

An exception to the general spectroscopic picture given above occurs in the XeF and XeCl systems. Here emission bands are shifted the farthest to the visible, and the excited states are most tightly bound. As a result, significant configuration interaction occurs between ground and ionic excited states, causing weak but quite observable binding in the ground states. The primary emission band at high pressure is a bound-to-bound transition. This band has clearly resolvable vibrational and rotational structure that has not yet been fully analyzed. Because of the bound–bound nature of the main bands in XeF and XeCl, all of the excited B state population can be extracted only if the excited manifold is vibrationally relaxed to the states undergoing stimulated emission. Moreover, the XeF well depth in the ground state is sufficiently large that lower level filling can occur as well. These two features limit the efficiency of room-temperature XeF lasers.

Excitation kinetics

The analogy with alkali halides that governs the basic spectroscopy of these species also governs the formation kinet-

ics: Since the excited state is an ion pair, recombination of positive rare gas ions and readily formed halide ions yields the excited state. A kinetic sequence that can produce the ion pair states is

$$e + Kr \to Kr^+ + 2e$$

$$e + F_2 \to F + F^-$$

$$F^- + Kr^+ + M \to KrF^* + M$$

where M is any third body, typically a lighter rare-gas buffer. For the efficient rare-gas halide laser systems, ion recombination produces the ionic excited states with near unit yield.[7] The only other potential paths for such reactions are predissociation into ground states or excited states of either the rare-gas or halogen atom. These states, however, are apparently not populated in recombinations of Xe^+F^-, Kr^+F^-, Ar^+F^-, Xe^+Cl^-, Kr^+Cl^- and Xe^+Br^- ion pairs. Predissociations can occur in species such as ArI^* where the minimum of the ionic well is above the energy of states that can dissociate into Ar plus highly excited I^*. Such energy channeling has been used to produce other lasers, as we shall see.

Termolecular ion recombination rate constants, as for the last step above, are exceptionally large, about 10^{-25} cm^6 sec^{-1}, about six orders of magnitude larger than termolecular rate coefficients for recombination along covalent potential-energy curves. This magnitude is directly attributable to the tremendously long range of the coulomb potential of these excited states. Ionic termolecular recombination processes effectively become bimolecular processes at pressures above about two atmospheres, with a recombination coefficient of about 2×10^{-6} cm^3 sec^{-1}. In this "diffusion-limited" regime the recombination is governed by the rate of ion–ion approach, which is slow compared to the frequency of collisions with third bodies. The ion recombination path is so important in mixtures containing halogens because the dissociative attachment reaction is exothermic and has a large cross section for many of the typical halogen molecules or halogen-bearing compounds.

A second important excitation path is the reaction of excited states of rare-gas atoms with halogen compounds.[8] Both the metastable states that were first used to generate this class of emission bands and the higher-lying atomic states that can be present in laser plasmas can react to form rare-gas halides. The analogy with alkali–halogen reaction chemistry also holds. Rare-gas excited states have low ionization potentials, similar to those of the metallic alkali atoms. Because of the low ionization potential of these excited atoms and the large electron affinity of halogen molecules, reaction to form ionic rare-gas halide excited states can occur by a mechanism called "harpooning," proposed long ago to explain the large cross sections for reaction of alkali atoms with halogen molecules.[9] The "harpoon" in the mechanism is the easily ionized electron. An intermediate state is involved and is an ionic triatomic species. The "harpoon" electron hops from the rare-gas excited atom or alkali atom to the halogen molecule at an intersection of the potential-energy surfaces. The strong coulomb field thus produced rapidly pulls the diatomic negative ion apart before it can be stabilized. This returns the harpoon electron to the rare-gas or alkali ion, bringing with it a negative halide ion, for example

$$Ar^* + F_2 \rightarrow Ar^+F_2^- \rightarrow Ar^+F^- + F$$

or

$$K + F_2 \rightarrow K^+F_2^- \rightarrow K^+F^- + F$$

For the case of the rare-gas halides, the ion pair state ultimately formed by the rapid decay of the intermediate is an electronically excited state.

The reactive cross sections for such processes are quite large because the potential energy surface intersection takes place at large separations of rare-gas and halogen molecules. To a first approximation, the distance is related to the separation at which the coulombic attraction of the triatomic ionic intermediate equals the infinite-separation energy defect. Thus the crossing distance is given roughly by

$$R_x \approx \frac{e^2}{IP(RG^*) - EA(X_2)}$$

where $IP(RG^*)$ is the ionization potential of the excited state and $EA(X_2)$ is the electron affinity of the halogen molecule. Since ionization potentials for the rare-gas excited states are about 4 eV, and the halogen molecule electron affinities are of order 2 eV, the ion surface becomes degenerate with the initial-reactant energy surface at R_x equal to 5–10 Å or so. Thus the reaction cross section, $\sigma \sim \pi R_x^2$, can be in the range 75–300 Å^2.[8] This general mechanism should apply in a broad range of quenching kinetics between excited atoms and halogen molecules.

Other mechanisms of excited-state production include recombination of excited halogen atoms with rare-gas atoms in the ground state and photodissociation or electron impact dissociation of molecular compounds of xenon and krypton, such as XeF_2 and KrF_2. Finally displacement of one rare-gas atom in an excimer by another may occur, for example

$$ArF^* + Kr \rightarrow KrF^* + Ar$$

These displacement reactions are exothermic, typically by more than 1 eV. From a molecular point of view, the net effect of the reaction is transfer of charge from one rare-gas positive ion to another. These reactions may take place directly at the intersections of potential energy surfaces of those *triatomic* rare-gas ha-

Large KrF laser is excited by electron beam. This laser can produce 100-joule, 1-microsec pulses when operating either as a pure "e-beam" laser or as a discharge laser controlled by an e-beam. Photo from Joseph Mangano, Avco-Everett Research Laboratory, Everett, Mass.   Figure 3

lides that have two rare-gas atoms. Such reactions are not yet completely understood.

The recognition that excited species composed of two rare-gas atoms and one halogen exist was important for understanding the quenching of the excited states of the lasing molecules.[7,10] Diatomic rare-gas ions were known to be bound by 1–2 eV, so that ionic excited-state species composed of rare-gas dimer positive ions, such as Kr_2^+, and halide ions such as F^- are possible, as in $Kr_2F^* = Kr_2^+ + F^-$. Such states are bound relative to the simple excited diatomics and reactions of the form

$$KrF^* + Kr + M \rightarrow Kr_2F^* + M$$

are exothermic. The emission spectra of many of the triatomic rare-gas monohalides have now been characterized. The termolecular rate constants for such excited state quenching processes are about 10^{-31} cm^6 sec^{-1}. At the pressures of rare-gas halide lasers excited by electron beams, about 2 atm, three-body excited-state quenching is a major loss process in the excited state competing with spontaneous and stimulated emission. The importance of three-body quenching processes is another kinetic novelty of this class of laser species. Other excited state loss processes include bimolecular quenching by rare gas atoms, halogen molecules and electrons present in the laser plasma.

The saturation flux for these lasers, that is the stimulating-field intensity at which stimulated decay decreases the gain by a factor of two is determined by the excited-state lifetime $\phi_s = h\nu\sigma\tau$ where $\tau^{-1} = \tau^{-1}$ (radiative) $+ \tau^{-1}$ (collisional). This intensity is of order 1 MW per cm^2 at 2 atmospheres and increases with buffer gas pressure because of the termolecular quenching process.

Aside from production branching ratios for making the excited state, there are two

Table 1. Rare-gas halide laser emission wavelengths

	Peak wavelengths of most intense band (nm)			
	Ne	**Ar**	**Kr**	**Xe**
F	107[a]	193	248	351,353
Cl	b	170	222	308
Br	b	166[a,b]	206[a]	282
I	b	b	185[a,b]	252[a]

[a] Has not demonstrated laser oscillation.
[b] Predissociates, hence emission is weak or unobservable. The helium halides apparently all predissociate.

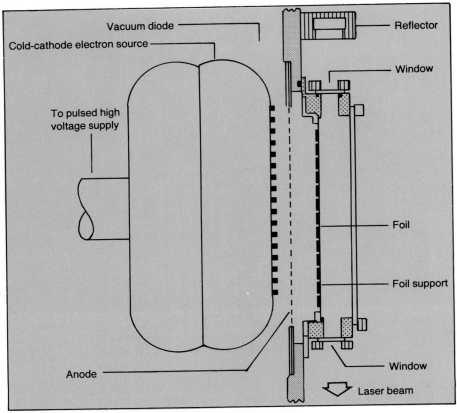

Typical e-beam excited laser uses a cold-cathode electron gun to provide the required beam current density of 10 amps per cm². This cathode is usually an array of metal blades, needles or notched carbon. (Based on laser used at Avco-Everett by James Ewing and Charles Brau.)   Figure 4

principal characteristics that affect the efficiency and scalability of a rare-gas halide laser. These factors are the excitation technique and the optical-extraction efficiency. Losses to absorbing species present in the laser plasma limit the extraction efficiency and we shall see that the amount of absorption relative to gain, as well as the mix of absorbing species in the plasma, depends very much on the excitation approach. The important absorbers are halide ions, rare-gas dimer ions, atomic excited states, diatomic and triatomic rare-gas halides, and parent halogen donor. The best pumping approach for varying applications is still under study. However, the general microscopic properties that dif-

ferentiate pumping techniques are becoming known, and the technology involved with each generic scheme is broadly defined.

Laser excitation methods

Three basic approaches have been used to excite rare-gas halide lasers: direct electron-beam excitation; electric discharge controlled by electron beams; simple discharges with fast-pulse excitation. The pure electron-beam ("e-beam") excited lasers (see figure 3) are volumetrically scalable and are used to produce energies of 100 mJ–300 J per pulse, in pulse lengths from 20 nanosec to 1 microsec. Overall efficiency for this type of excitation is eventually expected

to be about 6%. The e-beam controlled discharge promises to be scalable as well as more efficient and capable of higher pulse-repetition frequencies than the pure e-beam excited systems. The approach has been demonstrated, but it is still far removed from operating at its full potential. The simple discharges are the most widely used laboratory source. Here some external preionization from uv spark sources is usually used to condition, if not control, the discharge. These discharges can produce single-pulse energies of 1 J or so and pulse lengths in the range 10–100 ns, with typical efficiencies of order 1%.[11]

Electron-beam excitation

An excimer laser excited by an e-beam is composed of a high-voltage power supply capable of pulse charging a diode to form a broad area electron beam, a thin foil separating the diode vacuum from the laser medium and the excited medium surrounded by the relevant optics for use as a laser or power amplifier (see figure 4). To produce laser oscillation, the electron-beam current density must be larger than 10 A/cm² for gain lengths longer than about 20 cm. Electron current densities of this or greater magnitude are most conveniently obtained with cold cathode electron guns. The cathode, typically an array of metal blades, needles or notched carbon is pulse charged to a high voltage, ≥300 keV. Electrons are emitted from a plasma that forms near the surface of the cathode. Because the plasma formed at the cathode drifts toward the anode, the anode–cathode spacing is a dynamic quantity. The useful pulse duration at moderately constant beam current is limited by this phenomenon, called diode closure. The efficiency with which pulsed power can be provided to the diode for use in the laser can be as high as 80%, in properly designed systems.

The area that can be irradiated by an individual electron beam is limited by magnetic self-pinching of the beam: The magnetic field induced by the large current flow bends the trajectories of the electrons towards the center of the beam. Control of beam pinch by externally applied magnetic guide fields has been achieved in several large volume lasers; in these, irradiated areas of about one square meter can be achieved. Diode modularization schemes are also being investigated for larger areas.

The electron beam enters the laser medium after passing through a metal or plastic foil, supported on a metal structure, typically an array of holes or slots with 80–90% transmission. The e-beam deposits energy in the foil, and foil heating is a major limitation on the ultimate pulse repetition frequency of pure e-beam excited media. The beam emerges from the foil into the mixture of rare gas and halogen, scatters and loses energy to the

Table 2.   Comparison of efficiency limits for large-scale rare-gas halide lasers

Energy transfer characteristic	Efficiency for e-beam pumped lasers	Corresponding pure discharge efficiency
All processes up to gas deposition	70%	90%
Ratio of energy within usefully excited region to total energy input (fill factor)	80%	90%
Effective quantum efficiency (KrF)	25%	36%
Extraction efficiency	50%	50%
Overall system before gas-flow-power costs	6%	15%

gas. The amount of beam energy deposited in the laser volume depends on the device dimensions and excitation geometry. The ratio of useful uniform deposition to total deposited energy could be 80% in a large device. The excitation results primarily in atomic ionization with some rare-gas excited states being formed. In the Ar–Kr–F_2 mixtures typical of KrF lasers, the beam loses about 20 eV per excitation event. The excitation is in the form of Ar^+ ions and low-energy electrons and a smaller amount, about 20%, of atomic excited states. The rare-gas halide excited states then form by collisional processes as we saw earlier. For the case of KrF* in an Ar buffer, one 5 eV excitation per initial excitation can be achieved. This fourfold loss in efficiency is the first large step in the overall cycle that seriously degrades the potential efficiency, as shown in Table 2.

The second large decrement in efficiency comes from optical loss in the cavity. Electron-beam excitation relies on ionization and recombination to produce upper laser level species. The halide ions, especially the heavier halides, absorb throughout the uv (see figure 5).

At pressures of a few atmospheres, a major fraction of the positive ions are in the form of dimers, such as Ar_2^+. Such ions absorb in the uv also. Note that the XeBr and XeCl laser transitions are subject to substantial absorption due to the ionic species that are present in e-beam excited mixtures. This absorption seriously limits the laser extraction efficiency even though the excited-state production efficiency is very high. For the high efficiency KrF and ArF lasers, significant absorption due to F^- ions, F_2 absorption for KrF, and loss due to photoionization of rare-gas atom and molecule excited states is present in the plasma. This loss limits the length to which a KrF or ArF laser can be built with reasonable extraction efficiency to about 1–2 m. At this length extraction of excited states is about 50% efficient. One of the most significant absorbers is thought to be triatomic species such as Kr_2F^*. Since these species are formed by termolecular excited state quenching, a sufficiently high stimulating field and lower-pressure operation can reduce the loss. Quantitative characterization of this loss and its saturation properties remains as a research problem.

Since extraction efficiencies of large scale devices are limited to about 50%, efficiencies of about 6% are possible for rare-gas halide lasers pumped by electron beams, as shown in Table 2. Single-pulse energies of several kilojoules should be possible, even without significant advances in beam-pinch control.

Controlled electric discharges

Electric discharges have the potential for even higher efficiency. Since the electrical excitation need not be delivered

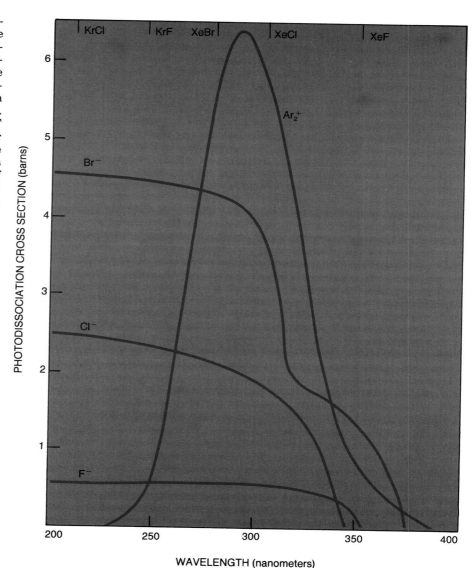

Ionic absorbers can cause optical losses for rare-gas halide lasers. The presence of halide ions, which absorb throughout the ultraviolet, is a result of the excitation process for e-beam lasers. And at pressures above a few atmospheres, dimers such as Ar_2^+ are a major factor. Note particularly the losses for the XeBr and XeCl lasers.

Figure 5

by high-energy electrons passing through a foil, the pulse-repetition frequency at which the laser can operate is potentially higher, implying that higher average power may be realizable. Volume scaling of laser discharges requires a stable volumetric ionization of the laser medium. The most widely used approach so far has been that of e-beam ionization.[12] Excitation and ionization is produced by an electron beam and, one hopes, the bulk of the excitation is produced by low-energy electrons excited by the applied electric field. The excitation sequence is dominated by neutral kinetics rather than ions,

$$e + Kr \rightarrow Kr^* + e$$

$$Kr^* + F_2 \rightarrow KrF^* + F$$

The electrons have a low temperature (about 10 eV) and are secondaries produced by the ionizing relativistic pri-

maries, denoted by e' in the process

$$e' + Kr \rightarrow Kr^+ + e + e'$$

The high atomic densities in the excited state required for producing excimer excited states make reactions such as

$$e + Kr^* \rightarrow Kr^+ + 2e$$

occur rapidly, leading to a discharge instability. Halogen compounds can balance this electron avalanching by attaching to themselves the excess electrons, forming negative ions.[13] However for the rare-gas halide discharge lasers, the ratio of useful excitation produced by the discharge pumping cycle to that produced by the primary ionization is small, in the range 2–5 for discharges that are stable for times of order 1 microsec. In contrast, infrared discharge lasers have discharge enhancements of order 100. The low value of discharge enhancement here is due to the very delicate balance

Electric discharge laser seen here is based on KrF and produces 1 joule. Electrodes in this typical laser are separated by 5 cm vertically; total length is one meter. Photo here is from Julius Goldhar and J. C. Swingle, Lawrence Livermore Laboratory.

Figure 6

between providing sufficient excitation for laser action and not driving the rare-gas mixture into an arc.

The efficiency of such discharges is potentially higher than with e-beam ionization because the effective quantum efficiency is higher. One 10-eV rare-gas excited state can be produced for about 14 eV of discharge-energy deposition under conditions suitable for stably producing about 3% cm^{-1} gain in KrF lasers. Since each rare-gas excited state can ultimately produce one 5-eV photon, the quantum efficiency could be as high as 36%. The power transfer efficiencies and fill factors for the discharge can also be somewhat higher. Assuming a similar optical extraction efficiency, the net overall system efficiency for a pure discharge could be as high as 15% (see Table 2). Since discharge enhancements are low, a realistic upper limit is about 10% for e-beam controlled discharge rare-gas halide lasers. This discharge laser efficiency has not yet been realized, and research is continuing. The achievable extraction efficiency for a given length also remains unknown since the principal absorbers are rare-gas excited states whose absorption is not yet experimentally known.

Simple discharge schemes, the third method, have a demonstrated efficiency of about 1% and are not as readily scalable in volume or in pulse length. They operate in an electric field range where rapid electron multiplication takes place, with the discharge switched off before arc formation. The first simple fast pulse discharges had no source of external ionization, and pulse energies of about 10 mJ

were typical. With preionization schemes to provide some "seed" electrons for the discharge, pulse energies of order 100 mJ are readily achieved, and pressure and discharge-gap scaling has resulted in energies of about 1 J in pulse lengths in the range 20–50 ns. These devices are very similar to simple fast-pulse discharge CO_2 lasers (see figure 6). Although more analysis is needed, it appears that these lasers operate in regimes of lower excited-state production efficiency than those predicted for stable e-beam controlled discharges, thus lowering their overall efficiency. Inefficiencies in power transfer in the discharge circuitry may account for a part of the lowered efficiency. One interesting and important facet uncovered with these devices is the comparably high efficiency with which KrCl, XeCl and XeBr can be made to oscillate. This observation is in marked distinction to early e-beam pumping experiments and is probably due to the preponderance of neutral excitation mechanisms over ion recombination. As we have noted, the ions can absorb strongly if present.

Extensions and applications

Several analogs to the rare-gas halide lasers have been investigated. Molecular diatomic halogens have charge-transfer bands that emit in the ultraviolet and whose upper levels can be efficiently populated in discharges and in e-beam excited media. Laser action on I_2(342 nm), Br_2 (292 nm) and F_2 (150 nm) have been demonstrated.[1,14] Although the fluorescence efficiencies of these species

are quite high, the intrinsic laser medium efficiencies have only been a few percent. Competing loss processes due to overlapping ion or excited-state absorption may be reducing the efficiency. Charge-transfer emissions from diatomic halogens can be readily observed in mixtures that do not initially contain any diatomic halogens. For instance, I_2 emissions and laser action is quite pronounced in mixtures of Ar and HI excited by electron beams. This is quite common behavior if the rare-gas halide species rapidly predissociates into an excited halogen atom rather than exhibiting rare-gas halide emission. All the helium halides, as well as NeCl*, NeBr*, NeI*, ArBr*, and ArI* behave this way. The halogen atoms are formed in high lying excited states with low ionization potentials, and they can react (by the harpooning mechanism) to form ionic excited states of diatomic halogens. Ion recombination reactions can also pump these excited states.

The mercury halides are a new class of lasers that also bear a strong resemblance to the rare-gas halides. HgCl (558 nm), HgBr (504 nm), and HgI (441 nm) lasers have been recently demonstrated by means of e-beams, discharges and photodissociation of HgX_2 vapors.[15] The transitions are not excimer transitions since the lower electronic states are bound by about 1 eV. However, the excited states are basically ionic, and similar production chemistry is involved. Since elevated temperatures are required and surface reactions of ground-state Hg with halogens can destroy the desired laser constituents, the technology development of these lasers may not be as rapid. However, the potential efficiencies are certainly as high as those of rare-gas halides, and the visible wavelengths will propagate through the atmosphere. Developments in the mercury-halide lasers will be closely watched.

Applications of rare-gas halide lasers have already begun, principally in research. These lasers offer a coherent uv source of greater intensity and energy than has been heretofore available. Development of wavelength-conversion processes is underway at several laboratories. Although these lasers are only moderately tunable over a narrow wavelength range, utilization of Raman shifting of these lasers (see figure 7) gives good coverage of most of the near uv. Conversion of KrF and XeF laser light into longer wavelength tunable dye-laser radiation has also been demonstrated. Because of the high efficiency of these lasers, they may prove useful commercially. One possible application of economic significance is in the separation of isotopes. Purification of chemicals by photochemistry at the ArF wavelength has also been demonstrated.

The rare-gas halide lasers operate efficiently in fairly long pulses, greater than

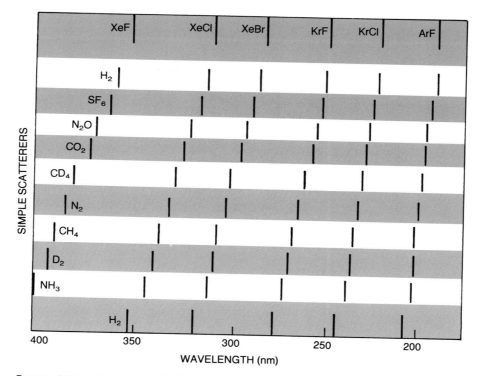

Raman shifting of rare-gas halide lasers improves their coverage of the near-ultraviolet region. Wavelengths seen here are only the first Stokes-shifted Raman frequency   Figure 7

100 ns. Their direct use in laser fusion may thus be precluded since shorter pulses are required. However, a number of schemes exist to produce short pulses with these lasers by pumping various optical converting media that can store energy. For example, the 1.3-micron iodine atom laser has been pumped with XeBr fluorescence. A variety of rare-earth vapor and solid complexes may also be suitable converters of KrF and XeF radiation. Another scheme currently under investigation is the compression of KrF laser pulses by a backward-wave Raman conversion. In this process a long pulse of KrF laser radiation counterpropagates through a Raman active medium with a short pulse at a Raman-shifted frequency. Analysis suggests that a conversion of pump radiation to Raman-shifted radiation can be controllably achieved with time compressions of about 10 at efficiencies in excess of 50%. Combination of this effect with geometric compression schemes (so called angle-multiplexing or pulse-stacking schemes) may result in high-energy pulses in the 1–10-ns regime

from a KrF laser of considerably longer pulse duration: A number of laboratories are investigating this and other applications of these lasers and such applications will define the ultimate utility of the research and development that yielded this new class of lasers.

References

1. J. J.Ewing, C. A. Brau, "High Efficiency UV Lasers" in *Tunable Lasers and Applications* (A. Mooradian, T. Jaeger, P. Stokseth, eds.) Springer, Berlin (1976), page 21. (This article as well as reference 2 contain extensive lists of references.)

2. C. A. Brau, in *Excimer Lasers* (C. K. Rhodes, ed.), Springer, Berlin, to be published.

3. C. K. Rhodes, J. Quant. Elec. **QE10**, 153 (1974).

4. M. F. Golde, B. A. Thrush, Chem. Phys. Lett. **29**, 486 (1974).

5. J. E. Velazco, D. W. Setser, J. Chem. Phys. **62**, 1990 (1975).

6. P. J. Hay, T. H. Dunning Jr, J. Chem. Phys. **66**, 1306 (1977).

7. M. Rokni, J. H. Jacob, J. A. Mangano, Phys. Rev. A **16**, 2216 (1977).

8. J. E. Velazco, J. H. Kolts, D. W. Setser, J. Chem. Phys. **65**, 3468 (1976).

9. K. J. Laidler, *Theories of Chemical Reaction Rates,* McGraw-Hill, New York, NY (1969); J. L. Magee, J. Chem. Phys. **8,** 687 (1940); M. Polanyi, *Atomic Reactions,* Williams and Northgate, London (1932).

10. D. C. Lorents, R. M. Hill, D. L. Huestis, M. V. McCusker, H. H. Nakano, in *Electronic Transition Lasers II* (L. E. Wilson, S. N. Suchard, J. I. Steinfeld, eds.), MIT, Cambridge, Mass (1977), page 30.

11. R. Burnham, N. Djeu, Appl. Phys. Lett. **29**, 707 (1976).

12. J. D. Daugherty, in *Laser Plasmas* (G. Bekefi, ed.), Wiley, New York (1976), page 369.

13. J. D. Daugherty, J. A. Mangano, J. H. Jacob, Appl. Phys. Lett. **28**, 581 (1976).

14. J. K. Rice, A. K. Hays, J. R. Woodworth, Appl. Phys. Lett. **31**, 31 (1977).

15. E. J. Schimitschek, J. E. Celto, J. A. Trias, Appl. Phys. Lett. **31**, 608 (1977); J. H. Parks, Appl. Phys. Lett. **31**, 192 (1977); **31**, 297 (1977); K. Y. Tang, R. O. Hunter, G. Oldenettel, C. Howton, D. Huestis, D. Eckstrom, B. Perry, M. McCusker, Appl. Phys. Lett. **32**, 226 (1978).   □

New sources of high-power coherent radiation

Free-electron lasers and cyclotron-resonance masers show considerable promise for producing previously unattainable levels of power at wavelengths ranging from millimeters to the ultraviolet.

PHYSICS TODAY / MARCH 1984

Phillip Sprangle and Timothy Coffey

Recent progress in novel techniques for generating high-power coherent radiation promises to make available sources with a variety of new and exciting applications. Interestingly, the new techniques have more in common with those used in the earliest sources of coherent radiation—the various microwave generators—than with those used in the more recent optical lasers. Development of new sources based on these techniques is proceeding rapidly at research centers around the world, because the new sources have a great potential for extending the currently available range of wavelengths and levels of power, while maintaining high operating efficiencies. The areas of application that stand to benefit include spectroscopy, advanced accelerators, short-wavelength radar, and plasma heating in fusion reactors.

Conventional sources of coherent radiation, such as the magnetron, the klystron and the traveling-wave devices, have limited power output and efficiency at short wavelengths. To circumvent these limits, researchers have proposed many new concepts and mechanisms, as well as variations on the conventional approaches. Two types of sources that were first demonstrated around 1960 are currently the focus of much attention: free-electron lasers and cyclotron-resonance masers, both of which are powered by relativistic electron beams.

We begin this article with a brief description of the physical mechanism of these and other novel sources of radiation. Then we look at some of the present and future areas of application and give an overview of the relevant experimental programs. Free-electron lasers offer operating efficiencies of

over 20% at millimeter wavelengths, and one can extend their operation to ultraviolet wavelengths while maintaining relatively broad tunability. Figure 1 shows free-electron-laser radiation at visible wavelengths. Cyclotron-resonance masers offer high efficiency and high power at centimeter to millimeter wavelengths.

Physical mechanisms

The terms free-electron laser and cyclotron-resonance maser refer to mechanisms. Each denotes a wide class of coherent sources that operate over a wide range of wavelengths, the words maser and laser having long lost their original limitations to microwaves and light. Although amplification by stimulated emission of radiation is fundamentally quantum mechanical, classical models are sufficient to understand both free-electron lasers and cyclotron-resonance masers.

Electrons radiate when they are accelerated. When an electromagnetic field of proper polarization and phase is imposed on a beam of electrons, the electrons will accelerate in such a way as to radiate coherently. The condition for coherence is that the radiation from the electrons reinforce the original imposed electromagnetic field. The electrons must be moving initially because they radiate at the expense of their kinetic energy. We call a coherent source an amplifier if the imposed field is from an external source, and an oscillator if the imposed field is generated internally by spontaneous radiation from individual electrons.

The free-electron laser, such as the one shown in figure 2, consists of an electron beam, an external "pump field" and the imposed radiation field.[1-4] The pump field, typically a static periodic magnetic field, can be any field that causes the moving electrons to oscillate transversely. Although the basic mechanism of emis-

sion does not rely on relativistic effects, the electrons must be highly relativistic to produce short-wavelength radiation.

The radiation wavelength in the free-electron laser, unlike in most conventional sources, is not fixed by the size of the structure. Furthermore, the lasing medium, being a pump field, cannot break down. Hence, in principle, large structures can generate short wavelengths at high power levels.

We will consider only the common pump field: a static periodic magnetic "wiggler" with its primary field component transverse to the electron and radiation beams, as shown in figure 3. Injected monoenergetic electrons stream through the periodic magnetic field and wiggle, or oscillate transversely, in the same direction as the radiation electric field, and thus can lose energy to the radiation field or gain energy from it. At the point of injection, the electrons are randomly phased and radiate incoherently, generating spontaneous bremsstrahlung radiation. However, the so-called "ponderomotive" wave produced by the beating of the wiggler and radiation fields bunches the electrons within the interaction region and generates coherent radiation in the process, as we will see. The ponderomotive, or trapping wave, originates from the $\mathbf{v} \times \mathbf{B}$ force on the electrons and causes them to bunch in the axial direction. The longitudinal ponderomotive wave, which excites a density wave, acts like the slow-traveling electromagnetic wave in conventional traveling-wave sources. The ponderomotive wave bunches the electrons by decelerating some and accelerating others. If the axial velocity v_0 of the electrons is slightly greater than the velocity of the ponderomotive wave, the average energy of the electrons decreases and the radiation is enhanced. An excessive spread in the velocity of the electrons can greatly

Phillip Sprangle is head of the plasma theory branch, and Timothy Coffey is director of research, at the Naval Research Laboratory, in Washington, D.C.

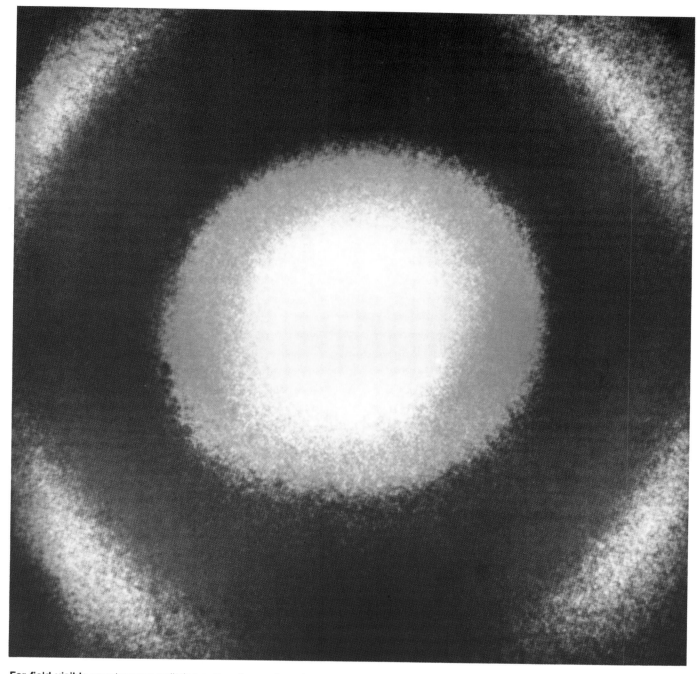

Far-field visible spontaneous radiation pattern from a free-electron laser. The laser used a wiggler magnet and was powered by electrons from the ACO storage ring at Orsay, France. The emission wavelength varies as the point of observation shifts off the beam axis. Figure 1

reduce the bunching and the extraction of energy, especially at shorter wavelengths.

The $\mathbf{v} \times \mathbf{B}$ force that gives rise to the ponderomotive wave involves the electron wiggle velocity v_w, which is typically much less than the axial velocity v_0, and $\mathbf{B}_R$, the strength of the radiation magnetic field. The radiation frequency and wavenumber are related by the vacuum relation $\omega = ck$. The phase velocity v_{ph} of the ponderomotive wave is $\omega/(k + k_w)$, which is about equal to the electrons' axial velocity v_0; here k_w is the wiggler wavenumber. For the bunched streaming electrons, sychronism requires that the radiation frequency satisfy the relation

$$\omega = v_0 k_w /(1 - v_0/c)$$

For highly relativistic electrons the radiation wavelength λ is approximately $\lambda_w / 2\gamma_0^2$, where γ_0 is the relativistic gamma factor $(1 - v_0^2/c^2)^{-1/2}$ and λ_w is the wiggler wavelength shown in figure 3. The radiation wavelength is much smaller than the wiggler wavelength, and can be varied by changing the energy of the electron beam.

One alternative configuration features an additional magnetic field, oriented longitudinally.[5] As we will see, this hybrid-field arrangement is one of the basic features of the cyclotron-resonance maser.

The nature of the electron-beam source and the way the electrons interact divide free-electron lasers into categories.[1-4] Free-electron lasers based on low-current beam sources such as rf linacs, microtrons or storage rings usually operate in the "Compton regime," in which primarily single particles interact, so that we can neglect collective or space-charge effects. In this regime the radiation gain—the growth rate along the interaction length—typically is low, so that practical radiation sources would be oscillators, which do not require high gain. Without enhancement, operating effi-

Free-electron-laser experiment employing a radio-frequency linear accelerator and a tapered wiggler. Here we see equipment that measures the energy spread in the electron beam (right), the tapered wiggler (center) and a time-resolved electron spectrometer (left). (Photograph from Mathematical Sciences Northwest Inc. and the Boeing Aerospace Company.)   Figure 2

ciencies are generally less than 1%. With high-energy electron beams of high quality, or small spread in energy, free-electron lasers in the Compton regime can operate at optical or ultraviolet wavelengths.

Other free-electron lasers are based on intense relativistic electron beams from sources such as induction linear accelerators or pulsed transmission-line accelerators, which Hans Fleischmann has described[6] in these pages. These lasers operate in the "Raman regime," in which collective effects influence the radiation growth rate and the interaction efficiency. Nevertheless, the operating wavelength λ remains at about $\lambda_w/2\gamma_0{}^2$, as in the Compton regime. Pulse-line-generated beams from plasma-induced field-emission diodes typically have relatively flat voltage and current pulses that last a few tens of nanoseconds. The electron beam's low quality and low energy—typically in the MeV range—limit the free-electron laser to millimeter wavelengths. But beam currents in the kiloamperes allow the laser to operate as an amplifier.

A third operating regime, known as the high-gain Compton regime, has features of both the Compton and Raman regimes. Here the wiggler field is so strong that the ponderomotive force on the electrons is dominant over the space-charge forces, and the radiation gain is large.

Cyclotron-resonance masers are far more developed than free-electron lasers and are among the most efficient devices for generating coherent high-power radiation at centimeter and mil-

limeter wavelengths.[7] The mechanism was proposed independently by Richard Q. Twiss, Jurgen Schneider, Andrei Gapanov and Richard Pantell in the late 1950s. Their early theoretical studies demonstrated that relativistic effects associated with monoenergetic electrons gyrating in a magnetic field could result in stimulated cyclotron emission rather than absorption. The first clearly defined experimental confirmation of the mechanism was reported by Jay Hirschfield and Jonathon Wachtel in 1964. Devices based on this mechanism, whether oscillators or amplifiers, are referred to as gyrotrons.

Scientists in the Soviet Union developed the gyrotron concept into a practical source of radiation during the 1960s and 1970s, primarily at Gorkii State University.[8] In the 1970s there were also major advances in the United States at the Naval Research Laboratory as well as at MIT, Yale University, Varian and Hughes. The demonstrated efficiencies and power levels in the millimeter regime are impressive. The Gorkii group, for example, as early as 1975 developed a 22-kW cw oscillator that produced 2-mm radiation with a 22% efficiency.

At the heart of the cyclotron-resonance maser is a beam of nearly monoenergetic electrons streaming along and gyrating about an external magnetic field $B_0\mathbf{e}_z$, as figures 4 and 5 indicate. An imposed electromagnetic field of the form $\mathbf{E} = E_0 \cos(\omega t)\mathbf{e}_y$, closely approximates the field of the transverse-electric mode of a cavity or waveguide when ω is near one of the cutoff

frequencies of the structure. This allows the electrons to radiate coherently. The electrons behave as individual oscillators in the cavity, gyrating about the magnetic field B_0 with a rotation frequency $\Omega_R = \Omega_0/\gamma$, where Ω_0 is the nonrelativistic cyclotron frequency eB_0/m_0c and γ is the usual relativistic mass factor calculated from the transverse electron rotation velocity $v_\perp$.

To understand the process of amplification, we will for simplicity consider only eight electrons, initially distributed uniformly and rotating in clockwise circular orbits as shown in figure 5a. With the radiation field polarized initially as shown, and with the electron orbiting frequency Ω_R slightly lower than the radiation frequency ω, those particles in the upper half plane ($x > 0$) will move closer to resonance with the radiation field and, therefore, lose energy to that field and increase their rotation frequency. Those in the lower half plane will move farther from resonance and hence gain energy and decrease their rotation frequency. The overall result is known as "phase bunching"; viewed after an integral number of wave periods $2\pi/\omega$, most electrons will be in the upper half plane, losing energy and amplifying the field. This mechanism requires only that the radiation frequency slightly exceed the rotation frequency, that the rotation frequency is energy dependent, and that all the electrons have roughly the same transverse rotation velocity $v_\perp$. The cavity length is chosen such that the electrons exit the cavity when their average energy is a minimum.

High-efficiency operation requires a large ratio of transverse to longitudinal electron velocity $v_\perp/v_z$. This ratio is typically between 1 and 3, and demonstrated efficiencies are as high as 60%.

Potential applications

Ultimately, the importance of new sources of coherent radiation will be determined more by their utility than by their novelty. Free-electron lasers and cyclotron-resonance masers have exciting potential applications as sources for spectroscopy, accelerators, radar and plasma heating.

Spectroscopy. A National Academy of Sciences study concludes[9] that the free-electron laser is a promising source for spectroscopy at far-infrared wavelengths greater than 25 microns, and at ultraviolet wavelengths less than 200 nm. This laser's coherence, narrow bandwidth, tunability and stable high power would be especially important in condensed-matter physics and surface chemistry, and in the spectroscopy of atoms, molecules and ions. The short time duration and thus high peak power available from some free-electron-lasers would allow

Table 1. Free-electron-laser experiments

Using rf linacs and microtrons

Laboratory	Class	Wavelength (microns)	Beam energy (MeV)	Peak current (A)
Stanford U.	Amplifier	10.6	24	0.1
Stanford U.	Oscillator	3.3	43	1.3
Los Alamos	Amplifier	10.6	20	10
Los Alamos	Oscillator	10.6	20	30–60
Mathematical Sciences Northwest/Boeing	Amplifier	10.6	20	5
TRW	Amplifier	10.6	25	10
TRW/Stanford U.	Oscillator	1.6	66	0.5–2.5
NRL	Oscillator	16.0	35	5
Bell Labs*	Amplifier	100–400	10–20	5
Frascati*	Amplifier	16	20	0.6

Using pulse-line-generated beams**

Laboratory	Peak power (MW)	Wavelength (mm)	Beam energy (MeV)	Beam current (kA)
NRL	1	0.4	2	30
NRL	35	4	1.35	1.5
NRL/Columbia U.	1	0.4	1.2	25
Columbia U.	8	1.5	0.86	5
Columbia U.	1	0.6	0.9	10
MIT	1.5	3	1	5
Ecole Polytechnique	2	2	1	2

Using electrostatic and induction linacs

Laboratory	Accelerator	Wavelength (mm)	Beam energy (MeV)	Peak current (A)
UCSB	Electrostatic accelerator	0.1–1	6	2
UCSB	Electrostatic accelerator	0.36	3	2
NRL	Induction linac	8	0.7	200
LLNL	Induction linac	3–8	4	400

Using storage-ring beams

Laboratory	Storage ring	Wavelength (microns)	Beam energy (MeV)	Beam current (A)	Gain per pass (%)
Orsay	ACO	0.5	240	2 (peak) 0.03 (average)	0.07 (measured)
Frascati	ADONE	0.5	600	10 (peak) 0.1 (average)	0.02 (measured)
Novosibirsk	VEPP-3	6	340	20 (peak)	0.4 (measured)
Brookhaven	VUV	0.35	500	108 (peak) 1.0 (average)	2 (calculated)
Stanford U.	ARRL (planned)	0.5	1000	200 (peak) 1.0 (average)	—

All entries in tables 1 and 2 represent typical values.
*Microtron beam source.
**Typical pulse times are tens of nanoseconds.

important new applications of radiation ranging in wavelength from 25 to 1000 microns. Pulses as short as tens of picoseconds could probe the dynamics of charge carriers in semiconductors and the dynamics of phonons, plasmons and superconducting gaps. High-power tunable picosecond pulses at wavelengths under 200 nm would substantially strengthen studies of fast chemical kinetics, photochemistry and vibrational relaxation processes that involve more than one photon.

Accelerators. As noted[10] at a recent workshop on laser acceleration of parti-

cles, affordable high-power sources of centimeter waves could lead to shorter high-energy accelerators. Conventional rf accelerators use microwave klystrons that generate about 25 MW of peak power. Recent developments indicate that free-electron lasers and cyclotron-resonance masers could generate gigawatts. These higher powers would mean fewer power tubes and possibly lower total cost. Researchers must resolve practical and scientific questions, however, before they can demonstrate that these new sources would deliver this power with acceptable efficiency and stability.

It may be possible to accelerate particles by reversing the dynamics of the free-electron laser.[10] The electric field of an intense laser beam, such as from a CO_2 laser, together with a wiggler could produce a large-amplitude ponderomotive wave to trap and accelerate electrons. One could energize the trapped electrons by increasing the wiggler field's period or amplitude or both. Accelerating gradients could exceed 100 MeV/m, but laser-beam diffraction would limit the acceleration length, so that electrons would gain at most a few GeV in a single stage. A major question is how to refocus the laser beam for multistage acceleration.

Radar. Most radar operates at microwave frequencies primarily because centimeter-wave power tubes and components are available and atmospheric losses are low. The free-electron laser and cyclotron-resonance maser promise millimeter-wave radar. Atmospheric absorption, although generally higher at millimeter wavelengths, has minima at 35, 94, 220 and 325 GHz. Relative to conventional microwave radar, millimeter-wavelength radar would have narrow beamwidths, large bandwidths and small antennas. Narrow beams permit tracking at low angles of elevation. Large bandwidths enhance resistance to interfering signals—or to electronic countermeasures in the case of military radar—and permit high-range resolution. Millimeter waves are less affected by fog, clouds, rain or smoke than are optical or infrared waves.

A number of issues concerning millimeter-wave radar remain to be resolved. The typical cyclotron-resonance maser uses a high magnetic field and requires superconducting magnets. Free-electron lasers, even at millimeter wavelengths, are now too large for most radar applications, and their high operating voltages are also a problem. The lack of millimeter-wave components has been another practical problem, but these components are developing rapidly.

Fusion power. The problems of plasma heating still prevent practical magnetic-confinement-fusion power reactors. Practical high-power sources at millimeter wavelengths could solve some of these problems.

Recent experiments on the Oak Ridge ISXB Tokamak, using a 35-GHz cyclotron-resonance maser developed at the Naval Research Laboratory, demonstrated large absorption through electron-cyclotron resonance.[7] The ab-

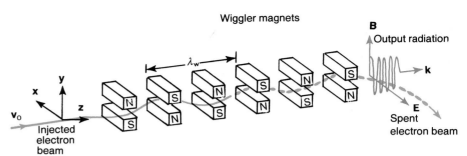

Basic components of a free-electron laser. The pump field is produced by a periodic arrangement of magnets—a "wiggler"—in which the electrons undergo transverse oscillations. The small transverse component of velocity, known as the wiggle velocity, is in the direction of the radiation electric field **E**, and can cause the electrons to lose energy and amplify the radiation field. The interaction between the electrons and the radiation field occurs over the entire length defined by the wiggler magnets. Figure 3

sorption heated the electrons significantly, but because of the low plasma density, the ions, as expected, were not measurably heated. These results imply that high-power cyclotron-resonance masers can heat fusion-reactor plasmas at the required high densities and long confinement times. Free-electron lasers are expected to be less efficient than cyclotron-resonance masers in producing millimeter waves and thus less suitable for plasma heating.

The success of high-power sources of coherent-radiation in any of the potential applications discussed above would be an important development. However, as with any new technology, we can expect the unexpected, implying that we have yet to identify the most important applications.

Enhancing efficiency

In free-electron lasers operating in the Compton regime, a radiation gain per pass of only 0.1 and an intrinsic efficiency of only 1% are typical. In this regime the intrinsic efficiency is given by the reciprocal of twice the number of wiggles in the interaction length. In the Raman regime, high gains are possible, with efficiencies as high as 15%.

One can increase operating efficiencies substantially either by converting the electron kinetic energy to radiation energy with greater efficiency or by recovering a portion of the electron kinetic energy after the electrons interact with the trapping wave. In principle, one can dramatically improve the efficiency with which the electrons transfer their energy to the wave, increasing it from about 1% to 20% in the Compton regime. One approach is to decrease gradually the trapping wave's phase velocity v_{ph}, which is $\omega/(k + k_w)$ or approximately $c(1 - \lambda/\lambda_w)$. By spatially decreasing the wavelength λ_w of the wiggler field, one decreases the phase velocity v_{ph}. In this approach, the electrons remain trapped and lose a large fraction of their kinetic energy to the radiation field.

Instead of decreasing the phase velocity v_{ph}—or in addition to doing so— one can apply a longitudinal accelerating force to the trapped electrons to enhance the efficiency. An external

Table 2. US experimental cyclotron-resonance masers

Oscillators

Laboratory	Frequency (GHz)	Power (kW)	Efficiency (%)	Pulse duration
Varian	28	340	45	Continuous
Varian	60	120	38	Continuous
NRL	35	340	54	1 μsec
MIT	140	180	30	1 μsec
Hughes	60	240	30	100 msec

Amplifiers

Laboratory	Frequency (GHz)	Power (kW)	Efficiency (%)	Pulse duration (microsec)	Bandwidth (%)
NRL	35	10	8	1.5	2–13
Varian	28	65	9	1000	1
Varian	5	120	26	50	6
Yale U.	6	20	10	1	11

uniform axial electric field can provide this accelerating force. A more practical approach is to decrease the spatial amplitude of the wiggler field. In either case, the resulting phase shift of the trapped electrons is such that they perform work on the trapping wave, enhancing the radiation. Norman Kroll of the University of California at San Diego and Marshall Rosenbluth of the University of Texas at Austin have found that instabilities can arise in the trapped-particle mode of operation, resulting in sideband radiation.[4]

In cyclotron-resonance masers, one approach to achieving higher than intrinsic efficiency is to contour the radiation field in the cavity by varying the radius of the cavity wall, as shown in the left half of figure 6. Electrons entering the cavity start phase bunching where the field amplitude is small. The electrons give up relatively little energy until they enter the high-field region. There they radiate more efficiently because they are highly bunched and closer to resonance.

In an alternative approach to efficiency enhancement, experimenters contour the external longitudinal magnetic field axially, as shown on the right side of figure 6. They make the magnetic field near the input smaller than normally required, allowing the electrons to bunch in phase without losing or gaining much energy. Again, as the phase bunching increases, the electrons go into a higher magnetic field and move closer to resonance. In addition, one can recover as much as 90% of the longitudinal energy of the spent electrons.

Experiments

Pioneering in free-electron-laser experiments in the Compton regime is a Stanford University group using the Superconducting Linear Accelerator.[11,12] One such experiment, headed

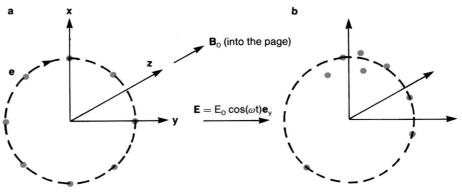

Phase distribution of gyrating electrons. This simplified representation shows the initial phase distribution (at **a** in figure 4) and the distribution after an integral number of wave periods (at **b** in figure 4).
Figure 5

by John Madey,[12] is sketched in figure 7 and listed second in table 1. The 43-MeV electron beam macropulses used in this experiment were 1.5 msec in duration and consisted of 1-mm-long micropulses spaced 25.4 m apart. Spontaneous incoherent radiation from the electrons built up into intense coherent radiation at 3.3 microns, because the gain of the free-electron-laser process peaks near that wavelength. Madey's group carefully separated the optical resonator mirrors so that the round-trip bounce time of the radiation pulses just matched the time between electron micropulses. In the presence of the electron micropulse, the radiation pulse velocity is slightly less than the velocity of light in a vacuum, an effect known as "laser lethargy."[4] In the Stanford experiment, the measured peak output power through a mirror of 1.5% transmittance was 6 kW; hence, the peak radiation power within the resonator was 400 kW. The measured linewidth $\Delta\lambda/\lambda$ of the saturated radiation was about 0.006. The 6% measured gain per pass was in fair agreement with the theoretical value of

about 10%.

At Los Alamos, Charles Brau and his coworkers are developing a highly efficient free-electron-laser oscillator source.[4,12,13] (See PHYSICS TODAY, August 1983, page 17.) This free-electron laser will employ an rf linac accelerator and radiate at 10.6 microns. The experimenters will vary the wiggler wavelength and amplitude spatially and recover part of the energy of the spent electron beam. They anticipate a 20% overall efficiency and an average output power of 100 kW.

At Mathematical Sciences Northwest and Boeing Aircraft, a group led by Jack Slater is developing[12,13] an optical free-electron-laser oscillator that will radiate at 0.5 microns. It employs a radio frequency linac beam with a peak current of 100 A.

Experimenters at the Naval Research Laboratory, in a recent free-electron-laser experiment[13] using an intense relativistic electron beam from a pulse-line generator, produced 35 MW of 4 mm radiation with 2.5% efficiency. (See table 1.) The energy spread of the injected electron beam was uniquely low. Experimental programs at Columbia University, MIT and Ecole Polytechnique are also employing high-current beams generated by pulse lines, as the list in table 1 indicates.[4]

At the University of California, Santa Barbara, Luis Elias and Gerald Ramian are conducting experiments[4,12] with a 6-MeV Van de Graaff accelerator to evaluate a dc energy-recovery scheme. (See table 1.) Their free-electron laser is designed to operate at 200 microns and achieve an output power of 12 kW.

Two free-electron-laser experiments powered by induction linacs are underway in the US, as table 1 indicates. Both operate in the high-gain regime. The Naval Research Laboratory's induction linac experiments, headed by Chris Kapetanakos and John Pasour, feature a uniquely long pulse duration

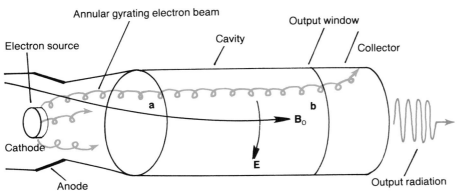

Cyclotron-resonance-maser oscillator, in schematic view. The electron source is a magnetron injection gun. The cathode emits an annular beam that gyrates about an applied magnetic field **B₀** as it propagates through a cavity. The cavity operates in a transverse-electric mode near its cutoff frequency. The spent electron beam is collected, and radiation is emitted through an output window. Figure 5 compares the uniform electron phase distribution at **a** with the bunching at **b**.
Figure 4

of about 2 microsec, enabling the free-electron laser to operate as an oscillator. At present, this free-electron laser operates as a superradiant amplifier generating 4.2 MW at a wavelength of 8 mm and an efficiency of 3%. At Lawrence Livermore National Laboratory, experiments led by Donald Prosnitz and Andy Sessler use the laboratory's 5 MeV Experimental Test Accelerator. Because of this accelerator's short beam pulse of 30 nsec, the laser will operate as an amplifier.

A number of free-electron-laser experiments use electron storage rings (See table 1), with the wiggler being in one of the straight sections. One such storage-ring experiment, headed by Claudio Pellegrini[4] at Brookhaven National Laboratory, will operate at 500 MeV and a peak current of 108 A. The radiation gain should be a few percent at a wavelength of 0.35 microns.

Future direction of research. Charles Roberson, of the Office of Naval Research, and his coworkers suggest[4] the possibility of powering free-electron lasers with intense cyclic electron beams generated by racetrack induction accelerators or modified betatrons. Such sources, however, are still in a proof-of-principle stage of development.

Because wiggler wavelengths are typically at least a few centimeters, optical free-electron lasers require electron beams with energies of at least 50 MeV. Beam energies could be lower with use of a high-frequency electromagnetic pump field, such as an intense laser beam or the output of another free-electron-laser. With a CO_2 laser pump and a 1-MeV electron beam, a free-electron laser could in principle radiate at optical frequencies.

Another interesting possibility for avoiding high beam energies is a two-stage free-electron laser using a single electron beam. The radiation produced in the first stage, which would employ a wiggler field, would become the pump field for the second stage. However, in this scheme, and in the scheme using lasers to generate an electromagnetic pump field, the gain per pass is low, and because beams with extremely low energy spreads are necessary, the trapping efficiency is low.

The electron beam from the Advanced Test Accelerator at Livermore is expected to produce 500-GW pulses of electrons, which could, in principle, generate tunable multigigawatt pulses of radiation at near-optical frequencies.

Maser experiments. Experimenters commonly use magnetron injection guns to produce electron beams for cyclotron-resonance masers (see figure 4). These thermionic sources can generate several amperes and electron energies as high as 100 keV.

To generate millimeter-wavelength radiation with cyclotron-resonance masers, experimenters usually use superconducting sources for the magnetic field. At the fundamental cyclotron harmonic, one needs a 34-kG magnetic field to generate radiation at 94-GHz, or 3 mm. It is possible to overcome the need for superconducting magnets in the generation of millimeter waves by operating at higher cyclotron harmonics, because the required magnetic field is reduced by a factor approximately equal to the harmonic number. The efficiency at the second harmonic remains high and with some designs can be higher than at the fundamental frequency. Generally, however, the efficiency falls sharply beyond the second harmonic.

An example of the state of the art in high-power devices comes from Soviet scientists at Gorkii State University,[14] who have generated 1.25 MW of 45-GHz (6.7 mm) radiation with a pulse duration of 1 to 5 msec and have produced 1.1 MW of 100-GHz (3.0 mm) radiation with a pulse duration of 100 microsec. Both of their oscillators operated at the fundamental cyclotron harmonic with efficiencies of 34%. Another impressive accomplishment of the Gorkii group[15] is a 120-kW cyclotron-resonance maser operating at 375 GHz (0.8 mm) with pulse durations of 0.1 msec. Recently, in the US, Richard Temkin and his coworkers at MIT achieved impressive power levels of over 180 kW at 140 GHz.

To increase the cavity volume and thereby avoid excessive thermal loading and competition among modes, high-power millimeter-wave cyclotron-resonance masers will have to operate in highly "overmoded" cavities, that is, in cavities that support many frequency modes at the same time. Recent results[16] show that one can stabilize highly overmoded cyclotron-resonance masers by adding a small prebunching cavity in front of the large energy-extraction cavity.

Prospects for practical cyclotron-resonance amplifiers are great. Researchers at the Naval Research Laboratory, for example, have achieved[7,17] impressive gains of 18 to 56 dB over large useful bandwidths, this at a frequency of 35 GHz (8.5 mm) and a typical power of 10 kW. Table 2 highlights experimental results in the United States with cyclotron-resonance oscillators and amplifiers.

Replacing the cavity with an open resonator,[18] allows one to operate at submillimeter wavelengths, to select modes and to handle extremely large powers. Preliminary experiments[19] at Yale University on this approach are encouraging.

Other novel sources. Many groups are actively pursuing other concepts for producing high-power radiation. One concept is the nonisochronic reflecting electron system. Here a high-current beam forms a virtual cathode; the emitted electrons oscillate between the

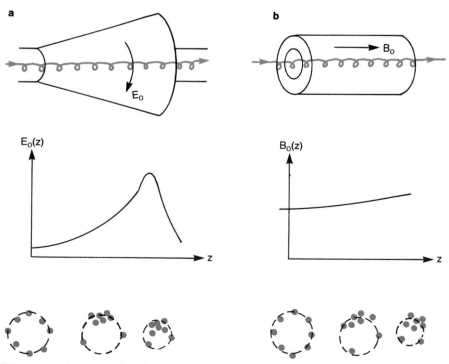

Efficiency enhancement methods used in cyclotron-resonance-maser oscillators. These are the two most common enhancement techniques. In **a**, the radius of the cavity wall increases longitudinally so that the field in the cavity varies as shown. In **b**, the radius of the cavity wall is constant, but the longitudinal external magnetic field varies. The diagrams at the bottom show the electron phase distributions at various points within the cavities.   Figure 6

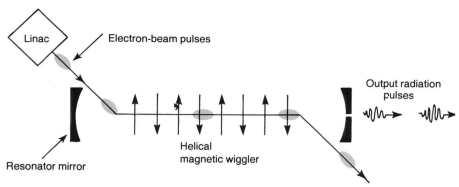

Oscillator using a pulsed electron beam. This is a schematic diagram of a typical free-electron-laser oscillator. The source is a radio frequency linear accelerator. The spacing of electron pulses and the length of the resonator are such that the reflecting pulses of radiation are synchronized with the incoming beam pulses. The mirror on the right is partially transmitting, so that a small fraction of the radiation pulses transiting the resonator can escape. The wiggler magnets are 3.3 cm apart over a total distance of 5.3 m. The magnetic field is 2.3 kG. **Figure 7**

actual and virtual cathodes, bunch in phase and generate radiation copiously. Our own experiments[20] on this source, which is compact, tunable and simple, have produced over 100 MW of 3-cm radiation.

Coherent Cherenkov radiation is a less novel but interesting source of millimeter waves. Some experiments produce this radiation by directing a relativistic electron beam along a dielectric surface. In one such experiment[4] at Dartmouth College, John Walsh, Kevin Felch and their coworkers achieved efficiencies of 10% and power levels of 100 kW at a wavelength of 4 mm.

One novel relativistic magnetron can generate unprecedented levels of coherent radiation at centimeter wavelengths. In experiments[21] headed by George Bekefi at MIT, a relativistic magnetron driven by an electron beam from a pulse line produced microwaves at a power level of 10 GW.

The history of science shows that applications, technology and theoretical concepts usually evolve together. In the case of high-power sources of coherent radiation, we are now witnessing rapid evolution in all these areas. Our ultimate understanding and the most important applications probably remain to be seen.

* * *

We gratefully acknowledge assistance from Cha-Mei Tang in preparing this article.

References

1. N. M. Kroll, W. A. McMullin, Phys. Rev. A **17**, 300 (1978).
2. A. A. Kolomenskii, A. N. Lebedev, Sov. J. Quantum Electron. **8**, 879 (1978).
3. P. Sprangle, R. A. Smith, V. L. Granatstein in *Infrared and Millimeter Waves*, Vol. 1, K. J. Button, ed., Academic, New York (1979).
4. *Free-Electron Generators of Coherent Radiation*, Physics of Quantum Electronics series, S. F. Jacobs, H. S. Pilloff, M. Sargent III, M. O. Scully, R. Spitzer, eds., Addison-Wesley, Reading, Mass. (1980), volumes 7, 8 and 9.
5. R. Davidson, W. McMullin, Phys. Fluids **26**, 840 (1983).
6. H. Fleischmann, PHYSICS TODAY, May 1975, page 35.
7. V. L. Granatstein, M. E. Read, L. R. Barnett in *Infrared and Millimeter Waves*, vol. 5, K. J. Button, ed., Academic, New York (1982).
8. IEEE Trans. Microwave Theory Tech. (special issue) **MTT-25**, No. 6 (1977).
9. *The Free Electron Laser*, the report of the free-electron-laser subcommittee of the Solid State Sciences Committee, National Academy of Sciences, National Academy Press, Washington, D.C. (1982).
10. P. J. Channell, ed., *Laser Acceleration of Particles*, AIP Conf. Proc. No. 91, Am. Inst. Phys., New York (1982).
11. L. R. Elias, W. M. Fairbanks, J. M. J. Madey, H. A. Schwettman, T. I. Smith, Phys. Rev. Lett. **36**, 717 (1976).
12. Bendor Free Electron Laser Conf., Journal de Physique **44**, C1 (1983).
13. IEEE J. Quant. Electron. **QE-19**, a special issue on free-electron lasers (1983).
14. A. A. Andronov, V. A. Flyagin, A. V. Gaponov, A. L. Goldenberg, M. I. Petelin, V. G. Usov, V. K. Yulpatov, Infrared Physics **18**, 385 (December 1978).
15. V. A. Flyagin, A. G. Luchinin, G. S. Nusinovich, Int. J. Infrared and Millimeter Waves **3**, 765 (1982).
16. Y. Carmel, K. R. Chu, M. Read, A. K. Ganguly, D. Dialetis, R. Seeley, J. S. Levine, V. L. Granatstein, Phys. Rev. Lett. **50**, 112 (1983).
17. L. R. Barnett, Y. Y. Lau, K. R. Chu, V. L. Granatstein, IEEE Trans. Electron Devices **ED-28**, 872 (1981).
18. P. Sprangle, J. Vomvoridis, W. Manheimer, Phys. Rev. **A23**, 3127 (1981).
19. N. A. Ebrahim, Z. Liang, J. L. Hirschfield, Phys. Rev. Lett. **49**, 1556 (1982).
20. R. A. Mahaffey, P. Sprangle, J. Golden, C. A. Kapetanakos, Phys. Rev. Lett. **39**, 843 (1977).
21. G. Bekefi, T. J. Orzechowski, Phys. Rev. Lett. **37**, 379 (1976). □

Coherent sources of extreme uv

Bertram M. Schwarzschild

PHYSICS TODAY / NOVEMBER 1983

Light sources in the extreme ultraviolet wavelength regime—from 104 nanometers down to the start of the x-ray region at about 10 nm—are of particular interest for molecular spectroscopy and solid-state surface studies. But this xuv domain poses special difficulties. The beams must be propagated in vacuum, but at these wavelengths there are no longer any solid materials available for vacuum windows or the harmonic generation of xuv light from longer wavelength sources. Crystalline lithium flouride, the last of the solids to maintain its transparency in the ultraviolet, becomes completely opaque at 104 nm. Beams of xuv can of course be generated by synchrotron light sources, but their usefulness is limited: The synchrotron beams do not provide coherent xuv light. Their spectral brightness (power per unit wavelength interval) is at present marginal for high-resolution molecular spectroscopy. Furthermore, synchrotron beams are available only at large storage-ring facilities; they cannot be brought into the experimenter's home laboratory.

Until this year, there have been no lasers capable of producing xuv light. With regard to non-laser sources of coherent xuv light, the production of such beams by nonlinear harmonic generation in gases has required elaborate and cumbersome differential pumping schemes to maintain adequate vacuum in the presence of high gas flow rates.

This year, however, results coming out of several laboratories have brightened the prospects for the convenient generation of coherent ultraviolet beams at wavelengths below 104 nm—the "lithium flouride cutoff." Andrew Kung[1] at Berkeley and a Bell Labs group[2] headed by Jeffrey Bokor, independently of one another, have introduced the use of supersonic pulsed gas jets to generate high-frequency harmonics of longer-wavelength coherent ultraviolet light. With this technique, which circumvents the serious inconveniences posed by high-capacity differential pumping, Kung, collaborating with a Stanford group led by Richard Zare, has developed[3] an easily reproducible, practical system for generating tunable coherent xuv light of extraordinary spectral brightness in laboratories doing molecular spectroscopy. The Bell Labs group, employing a higher pulsed-power system involving mode-locked picosecond dye lasers and excimer laser amplifiers, has set a short-wavelength record for the harmonic generation of coherent xuv light—35.5 nm, the *seventh* harmonic of 248 nm ultraviolet light focused on their pulsed helium jet. Charles Rhodes and his colleagues at the Chicago campus of the University of Illinois have recently reported[4] what appears to be the first true laser emission in the extreme ultraviolet. Their 93-nm stimulated emission came from krypton atoms in a differentially pumped gas cell (not a pulsed jet), optically pumped by the exotic mechanism of *four*-photon excitation with a high-power excimer laser.

Harmonic generation. Because there have been until now been no lasers emitting in the extreme ultraviolet and only a very limited selection in the vacuum ultraviolet regime (the wavelength interval between 200 nm, where oxygen becomes absorbent, and the lithium fluoride cutoff at 104 nm), coherent light below 200 nm has generally been produced by frequency mixing and the generation of harmonics in nonlinear crystals and gases. When subjected to

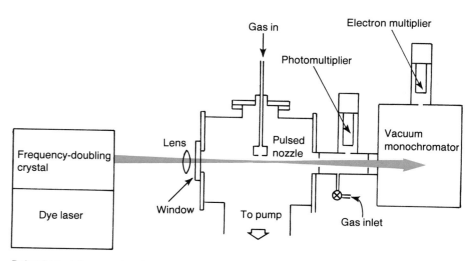

Pulsed gas-jet apparatus developed by the Stanford–Berkeley group as a tunable source of coherent xuv radiation down to a wavelength of 97 nm. Orange light from a tunable dye laser is frequency doubled to about 300 nm in a KDP crystal and then focused near the nozzle that produces supersonic pulsed jets of argon. The gas generates coherent third-harmonic radiation near 100 nm for experiments downstream in the vacuum system. All vacuum-window materials are opaque to uv light below 104 nm.

sufficiently intense light, any material will exhibit nonlinear polarization; its polarization will no longer be simply proportional to the incident electric field. Thus the time-dependent polarization will contain Fourier components at multiples (harmonics) of the incident light frequency, and at sum and difference frequencies when more than one frequency is incident. This nonlinear response of the polarization will radiate light at these harmonic and mixing frequencies.

If a material with sufficiently nonlinear polarizability is subjected to intense laser light, such a scheme provides a practical source of coherent light at shorter wavelengths. KDP (potassium dihydrogen phosphate) crystals, for example, are widely used for frequency doubling of near-ultraviolet laser light. In the case of gases, by contrast, parity conservation permits only the generation of *odd* harmonics; frequencies can be trebled but not doubled. These nonlinear coherent sources should not be thought of as lasers, because they do not exhibit the gain characteristic of true stimulated emission. The efficiency of this kind of harmonic generation is, in fact, well below one percent.

One cannot, of course, use crystalline nonlinear materials like KDP to generate ultraviolet wavelengths below 104 nm; all solids are opaque in the xuv region. Harmonic generation in nonlinear gases, on the other hand, presents its own problems. Coherent harmonic generation increases as the square of the gas density. How does one maintain the necessary high gas pressure at the harmonic generation site while maintaining the required vacuum downstream, when there are no vacuum window materials to be had?

The vacuum is necessary on several counts: Below 200 nm, oxygen is a strong absorber. Below 50 nm (24-eV photons), any gas will be photo-ionized and thus absorbing. The best nonlinear gases, in fact, begin absorbing at considerably longer wavelengths. In any case, the experimental setups on which the xuv beams are intended to impinge generally require a vacuum environment.

Traditionally, experimenters have attempted to achieve the requisite abrupt interface between the region of high nonlinear-gas pressure and the downstream vacuum by successive stages of differential pumping. The gas is confined in a cell with a pinhole at the downstream end. Longer-wavelength uv light is focused through the cell to a point just inside the pinhole, so that the xuv light, which is harmonically generated primarily at the focus (where the light intensity is greatest), can emerge immediately through the pinhole into

an adjoining evacuated chamber. Unfortunately, gas will also escape through this pinhole, making it necessary to pump the vacuum chamber continually with large-capacity pumps.

In practice, one can only get down to a fraction of a Torr with a single stage of such pumping. To achieve the necessary high vacuum (10^{-6} Torr), one needs several stages of differential pumping, with the xuv beam propagating through an aligned row of pinholes. At each pinhole one loses some of the xuv light in the unavoidable tradeoff between solid angle and escaping gas.

This rather demanding technique *does* work. Five years ago, John Reintjes and his colleagues at the Naval Research Lab used differential pumping to produce 38-nm light by seventh harmonic generation—a wavelength not much longer than the recent Bell Labs record. But it requires a very large pumping capacity, which is expensive and gets in the way of the experiments for which one ultimately wants the xuv light. At each pumping stage, furthermore, one loses some of that hard-won light.

Pulsed gas jets. The principal advantage of the newly developed pulsed gas jet alternative to differential pumping, Bokor argues, is that it lets one efficiently collect all the harmonically generated xuv light and direct it into the vacuum experimental setup. It is also much simpler and cheaper than differential pumping, he told us. Bokor minimizes the importance of the few nanometers his group has thus far been able to knock off Reintjes' 1978 short-wavelength record, but he contends that pulsed gas jets will ultimately permit one to get down to much shorter wavelengths than one could practically achieve with differential pumping.

A supersonic pulsed jet of the nonlinear gas to be used for harmonic generation of xuv light is formed by letting the gas flow out of its high-pressure container through a narrow nozzle whose valve opens only for sub-millisecond intervals synchronized with the incident laser pulse. The incident beam, directed at right angles to the gas, is focused at the base of the jet, just above the nozzle, where the region of high gas density is narrowest and most sharply defined. The greater the gas density and the higher the focused laser intensity one brings to bear on it, the stronger will be the harmonic generation of shorter wavelengths.

Because the xuv light in this configuration is generated in the extremely short pathway across the neck of the jet, the loss of xuv light by subsequent absorption in the nonlinear gas is minimal. One can therefore employ molecular gases such as CO and N_2 at short xuv wavelengths where they would otherwise be too absorbing. Furthermore,

there are no pinhole apertures farther downstream to cut off any of the generated light. The beam path is perpendicular to the jet, in contrast to the differential pumping geometry, where the beam propagates parallel to the stream of gas escaping through the pinholes.

Pumping the nonlinear gas away is also much easier in the jet configuration. The jet is aimed directly into the throat of a vacuum pump. Because the nozzle valve is opened only for a brief interval coinciding with the incident laser pulse, one has to deal with only a minimal gas flow and hence a relatively modest pumping capacity.

The backing pressure propagating the jet through the narrow nozzle is sufficient to raise the jet to supersonic velocity. The consequent cooling of the emerging gas is an additional advantage when the nonlinear gas is molecular and twice the incident light frequency corresponds to an intermediate energy level of the molecules. By narrowing the Doppler width of this level, the expansion cooling will then increase the efficiency of the so-called resonant enhancement of the third-harmonic generation.

Zare, Kung and their Stanford colleagues Charles Rettner and Ernest Marinero have concentrated on developing a tunable, narrow-bandwidth source of coherent xuv light for high-resolution molecular spectroscopy that others can reproduce easily and cheaply with commercially available components. Starting with nanosecond pulses of 600-nm orange light from a commercial dye laser, they frequency-double it in a KDP crystal and then direct the resultant near-ultraviolet (300-nm) light at the base of a pulsed jet of argon gas.

With the relatively modest power of the commercial dye laser they are able to generate only the third harmonic of the 300-nm light. Higher harmonics require greater incident laser intensity impinging on the nonlinear gas. By tuning the dye laser around 600 nm, they thus achieve a continually tunable coherent xuv source from 102 nm down to 97 nm.

Using CO as the nonlinear gas in subsequent experiments, the group has obtained coherent fluxes on the order of 10^{12} photons per nanosecond pulse at a repetition rate of 10 Hz. With a bandwidth of 0.01 Å, this represents a much higher spectral brightness than one has available at present-day synchrotron light sources. "Besides," Zare told us, "it's wonderfully convenient to be able to do these experiments in one's own lab." By putting an etalon in the dye laser cavity to narrow its spectral output further, the group expects to achieve narrow-bandwidth xuv beams of still higher spectral brightness.

Zare and his colleagues are particularly interested in high-resolution spectroscopic studies of the hydrogen molecule, most of whose important transitions lie just below the lithium flouride cutoff. By observing the variation in absorption and fluorescence as the xuv beam is tuned through its wavelength range and made to traverse a sample of hydrogen gas, the Stanford group has been able to measure the population densities of individual vibrational and rotational quantum states of the hydrogen molecule down to as few as 10^8 molecules per cubic centimeter. This success encourages Zare to believe that the pulsed jet technique will prove widely applicable for such quantum-specific molecular population surveys.

The first demonstration of high-resolution molecular spectroscopy with harmonicaly generated xuv light was reported in 1980 by Rhodes and his University of Illinois colleagues. They measured absorption line shapes in H_2 with 83 nm light of limited tunability generated in a differentially pumped gas cell by the high-power output of a KrF laser amplifier.

John McTague, chairman of Brookhaven's National Synchrotron Light Source, points out that the installation of free-electron-laser sections in electron storage rings holds out the promise of greatly enhanced xuv spectral brightness from synchrotron sources.

Bokor and his Bell Labs colleagues Philip Buchsbaum, Richard Freeman and Ralph Storz have opted for higher pulsed laser power, to generate higher harmonics and thus shorter wavelengths. Their more elaborate and costly system begins with a mode-locked dye laser producing picosecond pulses of 648-nm red light. Frequency doubling in a KDP crystal followed by nonlinear mixing with infrared laser light in a second crystal yields 248-nm ultraviolet pulses, which are then amplified to 20 millijoules in a KrF excimer laser amplifier.

Excimer lasers, employing stimulated emission from noble-gas molecules that exist only in excited states, are excellent high-power amplifiers of ultraviolet light. But their notoriously

poor spatial and temporal coherence properties, Bokor explains, make it necessary to start with dye-laser light in these experiments rather than using the excimer laser as the primary near-ultraviolet source. The development of this symbiosis of picosecond dye lasers and excimer amplifiers to produce high-power ultraviolet sources for nonlinear generation of shorter wavelength was begun by Rhodes' group at Chicago when Bokor was his student. With excimer amplification and differential pumping, the Chicago group has been able to generation higher harmonics down to 38 nm.

Directing the 248-nm excimer output at the base of a helium gas jet, the Bell Labs group has succeeded in generating its third, fifth and seventh harmonics—with wavelengths down to the record 35.5 nm. The excimer amplifier component in this system, while providing the pulsed power necessary for generating the higher harmonics, makes it much less tunable than the Stanford–Berkeley system. Furthermore, the concentration of power into picosecond pulses precludes the attainment of spectral bandwidths as narrow as those achieved at Stanford with nanosecond pulses. The Chicago group has circumvented the tunability limitations of excimer amplifiers by nonlinear mixing with dye-laser light in a differentially pumped gas cell *downstream* of the amplifier. By this means they have already produced coherent xuv broadly tunable down to 79 nm.

The Bell Labs groups plans to use its short-wavelength coherent source to do photoemission studies of surfaces. By analyzing the energy and angular distribution of photoelectrons liberated by these xuv photons, they will investigate the band structure of the underlying solids. The picosecond harmonic generation technique, Bokor told us, is particularly well suited for time-resolved photoemission studies, where one subjects the surface to an xuv probing pulse at a specific time interval after a different laser pulse has excited the material. The very high repetition rates of synchrotron sources, he argues, are inconvenient for studies of this kind.

The Chicago group, which pioneered the use of excimer amplifiers for the nonlinear generation of coherent vuv and xuv light with differential pumping techniques, has recently been concentrating on the achievement of stimulated emission by means of multiphoton excitation. This work has now resulted in what the group believes to be the first reported xuv laser output. Directing gigawatt pulses of 193-nm light from an ArF excimer amplifier at a differentially pumped krypton gas cell, Rhodes and his colleagues Triveni Srinivasan, Hans Egger, Ting Shang Luk and Herbert Pummer have succeeded in raising inner-shell electrons of the krypton atoms to highly excited levels by the simultaneous absorption of four photons. Returning to a lower state by stimulated emission, these electrons produce laser light at 93 nm. "To our knowledge," the group reports, "this system represents the first inner-shell transition laser and the shortest wavelength reported for stimulated emission." They regard this sort of multiphoton excitation as a particularly promising approach to the production of soft x-ray beams, below 10 nm. Rhodes points out that the conversion efficiency of harmonic generation, the principal alternative, becomes disturbingly minuscule as one goes to higher-order harmonics.

References

1. A. Kung, Opt. Lett. **8**, 24 (1983).
2. J. Bokor, P. Bucksbaum, R. Freeman, Opt. Lett. **8**, 217 (1983).
3. E. Marinero, C. Rettner, R. Zare, A. Kung, Chem. Phys. Lett. **95**, 486 (1983).
4. T. Srinivasan, H. Egger, T. Luk, H. Pummer, C. Rhodes, to be published in Laser Spectroscopy VI, Proc. 6th Int. Conf., Interlaken (1983), eds. H. Weber, W. Luthy, Springer Verlag.

X-ray lasers

George Chapline and Lowell Wood

PHYSICS TODAY / JUNE 1975

The target-positioning system centered in the vacuum chamber and surrounded by diagnostic equipment is used in ultrahigh-intensity laser-irradiation experiments. It is part of the Janus laser system at the Lawrence Livermore Laboratory, which is currently being used for laser fusion experiments and which is also slated for x-ray laser experiments such as those discussed in this article. Figure 1

Experiments with high-power lasers may soon demonstrate stimulated emission of x rays; future devices could have far-reaching impacts on chemistry, biology and crystallography.

One of the great barriers to improving our understanding of the molecular basis of life is that we do not yet have any way of examining individual macromolecules in living tissue. The progress made so far in unravelling the *structures* of biological macromolecules is based on painstaking chemical and x-ray crystallographic analyses of pure, crystalline samples of a given macromolecule.[1] This situation could, however, be completely revolutionized if a source of coherent x rays were available. Although, as we shall show below, the realizaton of this goal is beset with difficulties, experiments aimed towards its accomplishments are under way, as shown in figure 1.

To resolve the structure of molecules one needs radiation of wavelength comparable to the size of atoms. In practice, this means that one must use photons with energies of at least a few kilovolts (that is, x rays) or electrons with energies over a hundred kilovolts (see "Search and Discovery," PHYSICS TODAY, May 1974). The use of x rays as an illuminating source has the advantages that no vacuum enveloping the sample is required, and that the sample suffers much less radiation damage in the process. In particular, by using x rays one should be able to study macromolecules in living tissues, thereby gaining crucial insights into cellular functions. The reason this has not yet been done is that the index of refraction of low-Z elements is close to unity throughout the x-ray region, and therefore targets containing low-Z elements offer poor image contrast when illuminated with ordinary x-ray sources. A source of *coherent* x-rays, on the other hand, would make phase-contrast microscopy possible, which would yield contrasts sufficiently high to allow the

study of biological macromolecules *in situ*. A coherent x-ray source would, in fact, permit one to make three-dimensional holograms of important biological structures, such as the DNA in a cell nucleus (see figure 2). Using this technique one could, for example, make three-dimensional motion pictures—with atom-scale resolution—of the replication of DNA, ribosomal synthesis of proteins and other important biological processes.

Coherent x rays would similarly provide a solution of the famous "phase problem" of x-ray crystallography.[2] To obtain the three-dimensional electron distribution in a crystal one has to know the relative phases as well as the amplitudes of the x rays scattered at different angles. With ordinary x-ray sources one can in general measure only the intensities, that is, the squares of the amplitudes of the scattered x rays. (The phases can often be determined by inspired guesswork and clever tricks such as heavy-atom substitutions, together with very extensive use of digital computers to calculate the implications of the iterated guesses. However, such techniques are in general useful only for structure elucidation of crystals containing molecules of molecular weight less than about 10^5, which is one or two orders of magnitude below that of many biologically crucial molecules such as the large nucleic-acid polymers.) The determination of the atomic structure is now a special problem for every material, and a number of Nobel prizes have been awarded for deciphering particular materials. By contrast, determination of the relative *phases* of the scattered waves would also be straightforward if a coherent source of x rays were available—this would make the determination of the atomic-scale architecture of even the largest molecules a routine matter.

A further possible benefit of using coherent x rays to study materials is that very short (about 10^{-15} sec) x-ray pulses might be attainable with these sources, so that one could "freeze" molecular vibrations. Indeed, the possi-

bility would then exist of watching, in slow motion, thermal vibrations—even shock-wave compressions—in various materials.

What prospects?

It is clear from all this that a coherent source of x rays would have revolutionary implications for many areas of biology, chemistry, physics and materials science. What, then, are the prospects for generating these coherent x rays? There has been a great deal of interest during the past few years in the possibility of generating coherent x rays via an x-ray laser, due in part to several suggestions as to how this might be realized[3-7] and in part to the development of pumping sources that might suffice for pumping an x-ray laser system. This interest was greatly stimulated when workers at the University of Utah reported[8] that coherent x rays were emitted when a pulse from a neodymium-glass laser was focussed onto a thin layer of gelatin containing a dilute solution of $CuSO_4$. This claim of having produced x-ray laser action was, however, based on observations of spots on x-ray film that were subsequently shown to have not been produced by x rays at all.[9] Indeed, it was soon realized that true x-ray laser action was probably impossible to achieve with lasers then (*circa* 1973) existing.

Development of x-ray lasers has not yet occurred because favorable conditions for lasing are extremely difficult to attain at x-ray wavelengths. In the first place, cold matter is highly opaque at all x-ray wavelengths, due to photoelectric absorption. The nearest analog to a transparent optical laser medium (such as glass) at x-ray wavelengths is a completely ionized high-temperature plasma. Secondly, very high pumping powers are required to maintain the population inversions necessary for laser action, because of the very short excited-state lifetimes and the higher transition energies involved. The lifetimes for allowed x-ray transitions are of the order of $10^{-15} \lambda_{\text{Å}}^2$ sec, where $\lambda_{\text{Å}}$ is the wavelength in angstroms. There-

The authors are staff members of the physics department of the University of California's Lawrence Livermore Laboratory. Chapline leads the theoretical studies effort, and Wood is the program manager, of the Laboratory's x-ray laser program.

fore a 10-kilovolt transition ($\lambda = 1.2$ Å) has a radiative lifetime of the order of 10^{-15} sec, so to keep an atom with such a transition in its excited state, one must supply a pumping energy of the order of one watt per atom! Clearly, operation of an x-ray laser will require enormous pumping powers, and it is really only with the ultrahigh-power lasers now being developed that one could hope to produce the power densities required.

Another barrier to the demonstration of lasing action at x-ray wavelengths arises from the fact that an x-ray laser must probably be operated without external mirrors, due to the low reflectivity of materials when $\lambda < 1000$ Å. It has been suggested that instead of using mirrors one might circulate the beam around a circuit, using arrangements of Bragg-angle reflectors[10] but, again, absorption of x-rays at the reflectors strongly erodes the usefulness of this type of scheme. This proposal to run the x-ray pulse around the Bragg-reflector loop even once, also involves times long compared to x-ray radiative lifetimes. In an ordinary laser, the light bounces back and forth between the mirrors, stimulating emission of light from the atoms that have been pumped to the excited state. In the absence of mirrors, one must have a longer laser medium to make up for the lack of round trips between mirrors. This means that substantially larger gains down the length of the laser medium are required if mirrors cannot be used—the exact gain required to produce substantial laser action without mirrors will depend on diffraction losses and the destruction rates of the excited states.[11] For a 1-kilovolt x-ray laser the length, l, of the laser medium must be such that the gain down the length of the laser medium is of the order of a hundred dB; that is,

$$\alpha l \gtrsim 20 \qquad (1)$$

where α is the small-signal gain per unit length. At the same time the gain across the smaller dimensions of the laser must be kept small enough to avoid substantial radial stimulated emission. The diameter of the laser medium therefore must be small compared to its length, and so the medium must be in the form of a thin cylinder.

It turns out the length l needed to satisfy equation 1 will in practice always exceed $c\tau_{spon}$, where c is the speed of light and τ_{spon} is the spontaneous radiative lifetime of the x-ray laser transition. Thus, in general, it will be necessary to use "traveling-wave" excitation to pump the medium, to optimize pumping efficiency. One may picture the operation of an x-ray laser as consisting of the sweep of a suitable pumping excitation along the cylindrical medium at the speed of light; a growing

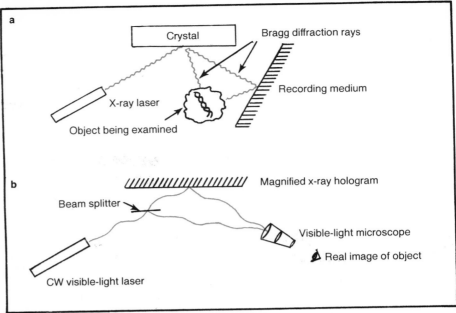

This scheme for obtaining three-dimensional pictures of macromolecules in living matter with x-ray phase-contrast holography is one possible application of an x-ray laser. Diagram **a** shows the recording of the hologram and **b** indicates the way in which a visible image, magnified by the ratio of the wavelength of visible light to that of x rays, is formed.   Figure 2

pulse of coherent x radiation will accompany this traveling excitation wave if a sufficiently large population-inversion density has been attained. A schematic diagram of the form an x-ray laser might take is shown in figure 3.

Pumping requirements

In their classic 1958 paper on the operation of a laser, Arthur Schawlow and Charles Townes showed that a certain threshold inversion density must be achieved to produce laser action. The analogue of their threshold condition for a laser operating without mirrors is $\alpha l \approx 1$. In the absence of losses α is given by

$$\alpha = \left(N_2 - \frac{g_2}{g_1} N_1 \right) \frac{\lambda^2}{8\pi} \frac{1}{\tau_{spon}\Delta\nu} \qquad (2)$$

where N_2 is the population of the upper state, N_1 is the population of the lower state, g_1 and g_2 are the statistical weights of the states and $\Delta\nu$ is the bandwidth of the lasing transition. Since the spontaneous radiative lifetime τ_{spon} of allowed transitions varies as λ^2, it is clear from equation 2 that the threshold inversion density will depend only on the bandwidth $\Delta\nu$ and the length l. Taking into account the fact that one must operate somewhat above threshold to obtain significant stimulated emission (that is, $\alpha l \gtrsim 20$), one finds that the population inversion must satisfy:

$$N^* \gtrsim (\Delta\nu/l)10^{18}\text{cm}^{-3} \qquad (3)$$

where $N^* \equiv N_2 - (g_2/g_1) N_1$ is the effective population-inversion density, and $\Delta\nu$ is in electron volts and l in cm.

In cold matter, the width determined by the Auger effect (about 1 eV) fixes a lower bound on the inversion density needed for laser action. This inversion density is necessarily very high, as is evident from equation 3 and, since the Auger width corresponds to a lifetime of about 10^{-15} sec, this implies that enormous pumping power densities will be needed to maintain required population inversions in cold matter. Furthermore, condition 3 may not be sufficient for laser action in a cold medium because the inversion density must be large enough to overcome the large photoelectric opacity of cold matter.

In hot matter, the minimum inversion density will be determined by the Stark width; Doppler broadening will be less important in most circumstances. Estimates[12] of the Stark width in a solid-density plasma containing identical, almost fully stripped ions leads to $N^*_{min} \approx 5 \times 10^{19}/l$ cm^{-3}. For $l = 1$ cm, this corresponds to a fractional inversion $N^*/N \approx 10^{-3}$, where N is the total atomic density. If the fractional inversion turns out to be smaller than 10^{-3}, then either medium densities higher than solid densities or lengths $l > 1$ cm would be required for lasing. To utilize inversion densities low enough to be comparable to plasma densities available from conventional sources such as ion beams, it is clear that high fractional inversion densities, greater than 0.1, would have to be achieved.

Since the energy needed to produce a population inversion, $h\nu N^* d^2 l$, must be supplied in a time l/c, the pumping power required to operate the x-ray

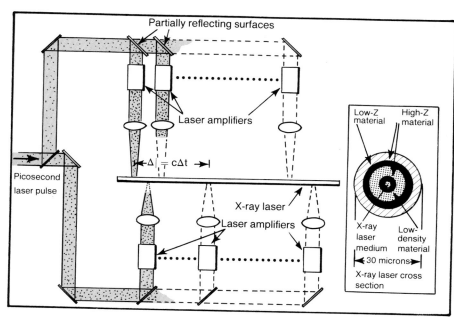

A conceptual design for an x-ray laser. A picosecond laser pulse from a master oscillator-preamplifier chain is fed in portions, via beamsplitters and final amplifiers, onto the x-ray laser target axis. The pulses are sequenced to cause an ultrastrong heating wave to move down axis with the speed of light. The concentric cylindrical target (inset) focusses radial shocks onto the x-ray laser medium to produce the inner-shell vacancies needed.   Figure 3

laser is given by the relationship

$$P_{\text{pump}} = h\nu N^* d^2 c/\epsilon \qquad (4)$$

where ϵ is the pumping efficiency. Obviously one will want to make d as small as possible. One cannot make d much smaller than $\sqrt{(\lambda l)}$, however, because diffraction losses would then become severe. Assuming that $d^2 = \lambda l$, $\Delta\nu = 1$ eV, and $\epsilon = 10^{-3}$ (a value that detailed studies indicate is characteristic), we find that the minimum power required is 1 gigawatt. This power level is well within the capabilities of present-day laser systems. However, before one concludes that existing lasers are capable of pumping an x-ray laser, it should be pointed out that it will not in general be possible to focus a high-power laser pulse onto a target with diameter $\sqrt{(\lambda l)}$; nor is it necessarily possible to sweep a focused beam along an x-ray laser medium in such a way as to produce an optimal traveling-wave excitation. A more realistic appraisal of the pumping requirements in fact suggest that a 10^{12}-watt laser would be needed to pump a 1-kilovolt x-ray laser. Fortunately, 10^{12}-W lasers are just coming to within the state of the art, and therefore meaningful experiments aimed at producing lasing action at $\lambda \approx 10$ Å may soon be possible.

Possibilities

What atomic states will give suitable transitions? To produce x rays with $\lambda < 100$ Å, we require atomic numbers greater than 4 for K_α radiation and greater than 8 for L_α radiation. Transitions between states of high principal quantum number or between subshells, such as $3s \rightarrow 3p$, have the disadvantage that such transitions require high values of Z to produce x rays, and these would lead to relatively large photoelectric losses. Further, states of high principal quantum number and distinct subshells will not exist at the high densities that appear to be required for an x-ray laser medium, because of continuum lowering and Stark mixing of states respectively. From a point of view encompassing oscillator strength, Stark broadening and photoelectric loss, the utilization of inner shell-transitions is favored. On the other hand, population inversions will probably be quenched rather rapidly if the lower lasing state is the ionic ground state, because of the comparatively very large recombination rates into this state. The most promising transitions for x-ray lasers therefore appear to be M-shell $\rightarrow$ L-shell transitions in nearly fully stripped, moderate-Z ions. For example, $3 \rightarrow 2$ transition in hydrogen-like krypton may be a useful transition for demonstrating x-ray lasing action; see figure 4.

Because of the very short lifetimes of allowed x-ray transitions (10^{-13} sec for 1-keV transitions), one might think that it would be more convenient to use metastable states in an x-ray laser. Metastable states of two- and three-electron ions with lifetimes of the order of 10^{-9} sec have been observed in low-density ion beams.[13] However, these metastable states are unlikely to be very long-lived at the relatively high densities required for even a metastable me-

dium, because of Stark mixing into allowed transition states as well as triplet–triplet annihilation processes. Furthermore, even if metastable states could somehow be prepared at the necessary high densities, the gain per unit length attainable with these metastable states would be very low compared to that attainable with allowed transitions—see equation 2. Thus, either impractically high inversion densities or inconveniently great lengths would be needed for metastable-state lasers.

Population-inverting schemes

The problems of producing population inversions in the x-ray region are very different from those at lower photon energies. In the infrared, visible, and near-ultraviolet regions one must deal with the complexities of molecular or solid-state kinetics. While these complexities provide many opportunities for lasing, they also make theoretical calculations difficult and unreliable. X-ray transitions, on the other hand, occur between tightly bound states where the processes affecting level populations are quite well understood. These processes include photoionization and radiative recombination, Auger ejection and dielectronic recombination, electron collisional excitation and de-excitation, ionic collision processes, and charge exchange. Since x-ray transitions take place between states with wavefunctions that are nearly hydrogenic, the cross sections and rate coefficients for these processes can be calculated with reasonable accuracy. These considerations suggest that detailed theoretical calculations of the changes in electronic level populations due to the above processes may be very useful in evaluating the feasibility of x-ray laser pumping schemes, and such calculations are central to our own efforts to produce x-ray lasing action. Although completely detailed calculations have yet to be carried out on any such pumping scheme, it appears that there are a number of situations in which the processes mentioned could lead to population inversions for x-ray transitions.

For example, population inversions could be produced in a straightforward way by exposing a target to an x-ray flash of very high intensity.[3] A population inversion would be created because, for wavelengths in the x-ray region, the probability for photoionization of a tightly bound electron is several times larger than that for a loosely bound electron. Unfortunately, conventional flash x-ray sources are not nearly bright enough to produce the needed high inversion densities on the time scales required. Assuming that the lifetime of the inverted state is 10^{-15} sec (the Auger lifetime), pumping a 1-keV transition requires 10^{16}W/cm^2 for a target 10 microns thick. However,

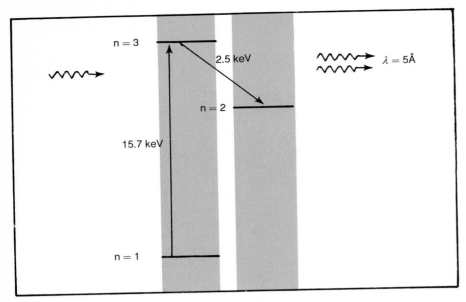

Hydrogenlike krypton provides a possible x-ray laser transition, with a spontaneous radiative lifetime of about 10^{-14} sec and a cross section of about 10^{-18} cm^2 for stimulated emission. Lasers powerful enough to provide the needed pumping are not yet available.   Figure 4

presently available flash x-ray machines cannot produce x-ray intensities much higher than about 10^{10} W/cm^2.

It might be possible, however, to produce an x-ray pumping pulse of the required brightness by focusing a high-energy laser pulse onto a high-Z target.[14] The efficiency of converting laser light into x-rays at appropriate energies is unknown at very high laser intensities, but even it it were as large as 1%, a laser intensity of 10^{18} W/cm^2 would be required. Such an intensity could be generated by focussing a 10^{12}-W laser pulse onto a spot 10 microns across. Such a focussed laser intensity may be achievable in the near future but, as mentioned, we have little idea of what the x-ray conversion efficiencies are at such an intensity.

Another problem with this type of x-ray pumping scheme is the ultrashort rise times required in view of the characteristic decay time for excited states (the Auger lifetime) of 10^{-15} sec. It has been suggested that a quasistationary population inversion can be maintained as long as the lower state of the lasing transition decays faster than the radiative lifetime, but calculations carried out by Timothy Axelrod have shown that, when all Auger processes and the heating of the medium are taken into account, such population inversions do not last longer than about 10^{-14} sec. This time is much shorter than the rise time of any high-power laser pulse that is likely to be available in the near future. Thus the prospects for this type of x-ray laser pumping do not appear promising for the near term.

Selective ionization of a particular inner shell might also be accomplished with energetic ions. For example, when

two atoms of moderate atomic number collide energetically with sufficiently small impact parameter, the Pauli exclusion principle requires the ejection of one or more electrons from one or the other of the atoms of the quasi-molecule transiently formed, at the expense of the kinetic energy of the colliding atoms. When two inner-shell energy levels in the colliding atoms are reasonably well matched, the collision will lead to selective production of vacancies in these shells. The possibility of using this process to achieve lasing at x-ray wavelengths was first discussed by R. A. McCorkle,[7] who proposed sweeping a high-current ion beam along a thin foil at the speed of light. Unfortunately, the gains attainable with this particular scheme are very small, due both to the very limited densities attainable with ion beams and to severe diffraction losses.

It is possible that usefully large gains per unit length could be attained by this process with charge-neutralized beams. For example, David Cheng has demonstrated in experiments at Ames Research Laboratory that charge-neutralized beams of argon ions with kinetic energies exceeding 10 keV and densities greater than 10^{18} cm^{-3} are attainable.[16] A fundamental difficulty with these approaches, however, is that the rise times for ion beams are orders of magnitude longer than typical x-ray-coupled population inversion lifetimes. A way around these difficulties might be to produce a very strong cylindrically imploding shock in a small fiber, with shock velocities of 10^8 cm/sec in condensed media. If the fiber were dense enough ($\rho \approx \rho_{solid}$), then the rise time for inner-shell vacancies due ion–ion

collisions in the shock would be much less than 10^{-12} and could be comparable to the inversion lifetime. One possible approach to attainment of such conditions would be to strongly shock-heat a low-density fiber by supersonically penetrating a surrounding material of higher density with a thermal wave engendered by a short-duration laser pulse.

Rather than attempting to eject electrons preferentially from a given inner shell by quasi-stationary ionization processes, one might consider trying to create population inversions by forcing electrons to recombine with stripped ions into particular preferred states. If, for example, one cooled a dense plasma on a time scale short compared to radiative lifetimes, one might then be able to make use of the fact that collisional recombination of electrons into upper electronic levels is faster at high densities than radiative recombination into lower levels. One possible way to cool a thin x-ray laser medium would be to bring it into contact with a high-Z radiator. However, the pumping powers for this type of scheme are similar to those required for breakeven laser-induced fusion, since radiative cooling rates of the order required (about 1 watt/atom) can only be attained with laser powers $\approx 10^{15}$ W.

Instead of using thermal cooling one might also consider using resonant charge exchange to produce population inversions in ions. A. V. Vinogradov and I. I. Sobel'man of the Lebedev Institute in Moscow have suggested[6] that if a laser-heated plasma of moderate Z is allowed to expand into a cool helium atmosphere, population inversions could result from charge-exchange reactions between the helium atoms and the plasma ions. Another approach, actively being studied by Marlan Scully and co-workers at the University of Arizona, is the production of a 2p–1s population inversion in He$^+$ by shooting an alpha-particle beam at a gas target.[17] The complexity of these schemes makes them difficult to evaluate. Further, these charge-exchange schemes involve the production of relatively very soft x rays ($\lambda \approx 300$ Å), and it is not clear that they can be extended to usefully short wavelengths.

As noted above, a 10^{12}-watt laser pulse might be adequate for pumping an x-ray laser transition of about 1 keV. The duration of the laser pulse will be determined by the requirement that the pumping pulse must be long enough to allow for propagation of the x-ray pulse. With the gains likely to be attainable, the x-ray laser medium will have to be at least 300 microns long, which implies that the effective duration of the pumping laser pulse must be greater than about 1 psec. At least one joule of laser light will be required for such a pulse.

Actually, 10^{12}-watt laser pulses of 100-psec duration are now available as a result of efforts under way around the world to demonstrate the feasibility of laser fusion.[18] Thus x-ray lasers may come into being as a by-product of laser-fusion research.

Free–free x-ray lasers

Most of this discussion has centered on the use of transitions from one bound state to another for producing x-ray laser action, but we might also consider free–free transitions. Free–free transitions differ from bound–bound and free–bound transitions in that they do not have to occur in a charge-neutralized medium. For example, a relativistic beam of electrons could be forced to emit x-rays by passing the beam through a periodic electric or magnetic field.[19] Because of the relativistic contraction effect, the wavelength of the readiation emitted by a beam of electrons, each with energy $\gamma m_e c^2$, where $\gamma = (1 - v^2/c^2)^{-1/2}$ and m_e is the electronic mass, will be the order of $(4\gamma^2)^{-1}$ times the spatial period of the electric or magnetic field. With a magnet period of 1 cm and a 5-GeV electron beam, for example, the wavelength of the emitted radiation would be 1 Å. A first step in this direction is the current effort of John Maday and William Fairbank at Stanford University to generate coherent optical radiation with a 20-MeV electron beam.

The generation of coherent x rays by running an electron beam through a periodic field makes use of the stimulated bremsstrahlung process. Coherent x rays might also, it has been suggested, be generated with stimulated Compton scattering.[20] Indeed, from the point of view of the Weizsäcker–Williams approximation, a stimulated bremsstrahlung laser is the same as a stimulated inverse Compton-scattering laser. The free–free laser analogue of the spontaneous radiative lifetime τ_{spon} is the time for spontaneous Compton scattering $\tau_C = (N_\gamma \sigma_T c)^{-1}$, where N_γ is the density of real photons in a stimulated Compton laser and the density of virtual photons in a stimulated bremsstrahlung laser, and σ_T is the Thomson cross section. The maximum density of either real or virtual photons that is easily attainable is about 10^{24} cm^{-3} so that, in practice, $\tau_C \gtrsim 10^{-10}$ sec.

Since the scattering time is much longer than allowed lifetimes for bound–bound transitions, it would appear that the gain per unit length attainable with free–free transitions must be small. However, the large value of τ_{spon} for free–free transitions is offset by the fact that the effective line width $\Delta\nu$ for free–free transitions can be made fairly small. By using electron beams with well defined electron momenta ($\Delta p/p < 10^{-3}$) and an initial photon distribution with a narrow spectral width, it should be possible to achieve $\Delta\nu \approx 10^{10}$ sec^{-1}. Thus the gain,

$$G_0 \approx N_e(\lambda^2/8\pi)(\Delta\nu\tau_{spon})^{-1}$$

attainable in a free–free laser system can be estimated, in dB/m, as

$$G_0 \approx 10^{-15}N_e\lambda_{\text{Å}}^2 \qquad (5)$$

This relation shows that the principal barrier to a free–free x-ray laser will be the very high electron-beam densities needed. In terms of electron beam's current density, it is clear from equation 5 that beam current densities of the order of 10^7 amp/cm^2 would be required for a free–free x-ray laser of reasonable length. Candidate beams for a stimulated bremsstrahlung laser (such as the 20-GeV SLAC beam) do not have nearly high enough current densities. Pulsed electron-beam machines capable of producing 10^6-amp/cm^2 beams of 10-MeV electrons are, however, now available. These machines were the topic of an article by Hans H. Fleischmann in last month's issue of PHYSICS TODAY (page 34).

Experiments

Unfortunately, these high current densities are achieved in a pinched mode, for which calculations[21] have indicated that the beam is quite "hot"— that is, $\Delta p \approx p$, necessitating beam-current densities of the order of 10^{10} amp/cm^2 for free–free x-ray lasers. There are, however, some indications that effective current densities of this magnitude can be attained with the Bennett fiber technique.[22] If this is the case, a stimulated inverse Compton scattering laser may be within the realm of possibility.

At the present time, x-ray laser ex-periments are being conducted at a number of laboratories around the world. To date, only one group—that of P. Jaegle at the Orsay Laboratories of the University of Paris[23]—has offered even tentative evidence of stimulated emission at x-ray wavelengths. By looking at the x-ray emission of one laser-produced aluminum plasma through another laser-produced aluminum plasma, as shown in figure 5, they claim to have found evidence that a 117.4-Å line of Al^{3+} is amplified in passing through the second plasma. The inferred gain was 10 cm^{-1}, with a substantial uncertainty. The 117-Å line is due to $2p^5 4d^1 \rightarrow 2p^6$ transition. Both plasmas were produced on the same aluminum target by a 20-J, 40-nsec pulse from a neodymium-glass laser.

Since neodymium–glass lasers are the most convenient source of high-power pulses at the present time, most x-ray laser work has utilized these lasers. At the University of Rochester, a group led by James Forsyth and Moshe Lubin is studying the OVII and OVIII lines emitted by oxygen-doped LiD targets irradiated by a 10-psec pulse from a Nd-glass laser. At the Battelle Laboratories in Columbus, Ohio, Philip Mallozzi has studied the x-ray pumping of targets, using a laser-produced high-Z plasma as a source of x-rays. X-ray laser experiments using Nd lasers are also being carried out by Ron Andrews at the Naval Research Laboratories in Washington, and by Michel Duguay at the Sandia Laboratories in Albuquerque. Our own group at the Livermore Laboratory will soon begin x-ray laser work, using high-power Nd-glass lasers developed for laser-induced fusion experiments. The terawatt Cyclops laser system, to be used in these experiments,

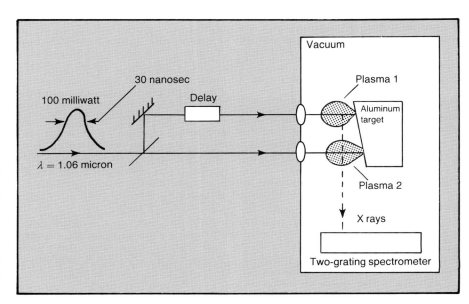

Orsay setup for measuring gain (or loss) of soft x-rays in aluminum plasma. A 20-joule pulse from a neodymium-glass laser produces two plasmas 40 microns apart; plasma 1 acts as the x-ray source, the other as the amplifying medium. Figure 5

This terawatt laser system, Cyclops, may be employed in the irradiation of x-ray laser targets at the Livermore laboratory. Cyclops, the world's most powerful single laser system, was developed in support of the laboratory's laser-fusion program.   Figure 6

is shown under development, in figure 6.

Actually, it is rather difficult to achieve the brightness required for x-ray laser pumping with Nd–glass lasers, as the focal spot from existing high-power lasers of this type cannot be made much smaller than about 50 microns in diameter. From the point of view of focal-spot brightness, it would be desirable to have available a diffraction-limited laser in the visible or near ultraviolet, such as the high-power ruby laser being developed at the Los Alamos Scientific Laboratory by Robert Carman. Indeed, the pumping of lasers operating near $\lambda = 1$ Å may require the development of ultrahigh-power, diffraction-limited ultraviolet lasers.

Applications

The most obvious experimental evidence for lasing is the presence of an intense collimated beam of monochromatic radiation directed along the fiber axis. Even for primitive x-ray lasers, the brightness of such a beam should be extremely high. In fact, x-ray fluences greater than about 10^3 J/cm^2 would be expected for a 1-keV device. This suggests an obvious—indeed, unavoidable—method of detection of true x-ray lasing action: One listens for a distinct report as a plasma plume jets out of the nearest surface irradiated by the laser's beam. The duration of the x-ray pulse has an upper-bound given by the length of the pumping wave pulse. For x-ray lasers pumped with presently available

high-power lasers, the pulse would not be longer than about 10^{-13} sec, the radiative lifetime of the upper state. The actual pulse duration might be much shorter than this, and could be as short as 10^{-16} sec, the reciprocal of the lasing linewidth.

The frequency coherence of the x-ray laser pulse will depend on the degree of line narrowing and other nonlinear effects in the medium, which might be significant if the gain is high enough. The spatial coherence of the laser beam will be determined by the number of transverse modes contributing to its output pulse. For an x-ray laser operating near threshold, the number of modes contributing will be quite high and the spatial coherence correspondingly poor. To produce spatially coherent beams, it will be necessary to develop x-ray lasers with very high gain and small angular divergence, so that only a few modes predominate. These lasers will therefore have to be relatively long ($l > 1$ cm), which will necessitate the development of means for sweeping a traveling-wave pumping pulse over such lengths.

Since the first x-ray laser will probably not produce very coherent beams, their first applications are likely to exploit only the exceedingly high brightness and short duration of the output pulses. Such extremely short x-ray pulses could be used for flash radiography where changes on subpicosecond time scales are important; for example, one could study aspects of the structure

of shock waves in solids in uniquely high space and time resolution. Examination of dense plasmas, such as those expected to be produced in laser-induced fusion experiments,[18] would also be greatly facilitated. The revolutionary advances in determining the structure of matter that we mentioned at the outset will follow as the pulse quality of x-ray lasers is refined by further work.

* * *

We wish to acknowledge many informative and illuminating discussions with our many colleagues in x-ray laser work, both at our own laboratory and elsewhere, as well as the joint support by the Materials Science Office of the Advanced Research Projects Agency and the Energy Research and Development Administration.

References

1. J. D. Watson, *The Double Helix*, Atheneum Press, New York (1968).
2. A. Guinier, *X-Ray Diffraction*, W. H. Freeman, San Francisco (1963).
3. M. A. Duguay, P. M. Rentzepis, Appl. Phys. Lett. **10**, 350 (1967).
4. T. C. Bristow, M. J. Lubin, J. M. Forsyth, E. B. Goldman, J. M. Soures, Optics Comm. **5**, 315 (1972).
5. B. Lax, A. H. Guenther, Appl. Phys. Lett. **21**, 361 (1972).
6. A. V. Vinogradov, I. I. Sobel'man, Soviet Phys—JETP **36**, 1115 (1973).
7. R. A. McCorkle, Phys. Rev. Lett. **29**, 982 (1972).
8. J. Kepros, E. Eyring, F. Cagle, Proc. Nat. Acad. Sciences **69**, 1744 (1972).
9. T. A. Boster, Applied Optics **12**, 433 (1973).
10. R. M. J. Cotterill, Appl. Phys. Lett. **12**, 403 (1968).
11. A. C. Selden, Phys. Lett. A **47**, 389 (1974).
12. L. Wood, G. Chapline, S. Slutz, G. Zimmerman, Preprint UCRL-75184, Univ. of Calif., Livermore (1973).
13. R. W. Schmieder, R. Marrus, Phys. Rev. Lett. **28**, 1233 (1972).
14. P. J. Mallozi, in *Proceedings of the Esfahan Symposium* (A. Javan and M. S. Feld, eds.) Wiley, New York (1973).
15. F. T. Arecchi, G. P. Banfi, A. M. Malvezzi, Optics Comm. **10**, 214 (1974).
16. D. Y. Cheng, Nuclear Fusion **13**, 458 (1973).
17. M. O. Scully, W. H. Louisell, W. B. McKnight, Optics Comm. **9**, 246 (1973).
18. J. Nuckolls, J. Emmett, L. Wood, PHYSICS TODAY, August 1973.
19. J. M. J. Maday, J. Appl. Phys. **42**, 1906 (1971).
20. R. H. Pantel, G. Soncini, H. E. Puthoff, IEEE J. Quantum Electron. **4**, 905 (1968).
21. G. Yonas *et al*, Appl. Phys. Lett. **30**, 164 (1974).
22. D. L. Morrow *et al*, Appl. Phys. Lett. **19**, 441 (1971).
23. P. Jaegle, G. Jamelot, A. Carillon, A. Sureau, P. Dhez, Phys. Rev. Lett. **33**, 1070 (1974).   □

Livermore group reports soft x-ray laser

Barbara G. Levi

PHYSICS TODAY / MARCH 1985

A group at Lawrence Livermore Laboratory has operated the first soft x-ray laser. As proof, they have extensive diagnostics from more than 100 test shots showing not only that radiation emitted at two wavelengths near 20 nm were up to 700 times more intense than those from spontaneous emission, but also that these lines behaved in other ways characteristic of a laser. Group leader Dennis Matthews, together with Mordecai Rosen, presented[1] Livermore's results at the Boston meeting of the APS Division of Plasma Physics last October. The announcement is a welcome milestone in the quest for an x-ray laser, which has been underway since the first theoretical suggestions of x-ray lasing schemes in the late sixties.[2] Other participants in that search, now more confident of their own eventual success, can be expected to join Livermore in the drive towards greater efficiency, higher power and ever-shorter wavelengths.

Admittedly, the very-short-wavelength region has intrinsic interest as a new frontier, but it has possible practical applications as well—such as x-ray holography of biological structures, x-ray photolithography for integrated circuits and more precise spectroscopy. Furthermore, proponents of strategic defense systems have high hopes for an x-ray laser weapon: According to one report,[3] Livermore has already tested an x-ray laser that is pumped by a nuclear explosion. However, Livermore's weapons program is formally separate from the project Matthews directs.

Also at the Boston meeting, Szymon Suckewer of the Princeton Plasma Physics Laboratory reported enhancements of soft x-ray light by a factor of 100. Since then, he and his colleagues have added an x-ray mirror to their experiment and feel that the results indicate stimulated emission. Both the Livermore and the Princeton experiments are outgrowths of fusion research.

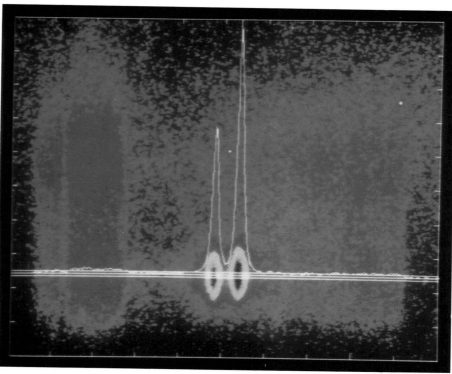

Evidence for soft-x-ray lasing from Livermore experiments with a selenium foil. The lasing lines appear in this computer-enhanced false-color photograph of data from a transmission grating spectrograph. In the plot, wavelength is on the horizontal axis and time on the vertical axis.

Livermore's experiment. One of the greatest challenges in producing an x-ray laser is to deliver a high enough power density to create a large population inversion within the plasma lifetime. Livermore pumped its soft x-ray laser with 450-psec pulses from the two-beam Novette 532-nm laser, in which each beam has a typical power density of 5×10^{13} W/cm^2. This light vaporized a target of high-Z material (most of their experiments used selenium), creating a plasma of ions with electronic structure similar to that of neon. Collisions with electrons within the plasma then created excited states of these neon-like ions, with a predicted population inversion between the 2p^{5}3p and the 2p^{5}3s levels. Transitions between the $J = 2$ and $J = 1$ states of these levels produced amplified radiation at 20.63 and 20.96 nm. Surprisingly, there was no amplification of the $J = 0$ to $J = 1$ line at 18.3 nm, where theoretical calculations had predicted the greatest gain. That one humbling aspect of the experiment is prompting some evaluation of the atomic models currently used in support of such laser work, as well as motivating further research in the lab.

In previous experiments, the Livermore group attributed the absence of any amplification to a steep density gradient by which the x rays were refracted out of the high-gain region. Rosen, Peter Hagelstein and others from Livermore collaborated with KMS Fusion Inc to model the hydrodynamics involved and, hence, to design an exploding foil target that would create a flat electron density. The optimum target design was a 75-nm layer of selenium vapor deposited on one side of a plastic substrate that is 150 nm thick. The foils were less than 0.1 cm wide by 1.1 cm long.

Laser diagnostics. The diagnostic instruments were critical to the success of the experiment, for several previous reports of significant amplification in the soft x-ray region have been either inconclusive or unsubstantiated. The Livermore experiment had two independent spectrographs viewing the axial direction and one measuring emissions at 77° from that axis. The time-resolved spectrum from one of these instruments showed that the two lines near 20 nm were both brighter and had shorter time duration than nearby spontaneous emissions. (See the figure on page 17.) The output pulse length grew shorter as the amplification increased. The off-axis spectrograph verified that the x-ray emissions were anisotropic, with the 20-nm lines not seen off axis at all.

But the most convincing evidence of laser action stemmed from the Livermore group's ability to vary the length of the target region up to 2.2 cm by adjusting the focal-spot size, by placing both laser beams side by side and by aligning two target foils end-to-end. In this way they determined that the energy appearing at the lasing wavelengths varied in a highly nonlinear way with the length of the foil. In the figure on this page, the data come from the grazing-incidence spectrograph, whose detector is gated in time to exclude background emissions occurring late in the experiment.) Theoretically, the intensity of a laser varies almost exponentially with the length of the gain medium; the constant in the exponent's argument is called the gain coefficient. Matthews and his colleagues concluded that the gain coefficient for each selenium line was (5.5 ± 1.0) cm^{-1}. Michael Kay (Rutherford Lab) told us he feels that for a "true" laser, the product of the gain coefficient and length should exceed 10 rather than the 5.5 measured by the Livermore group. "It should be possible," Key said, "to produce a proper laser device since very high gain is seen without any doubt."

The Livermore group also experimented with a higher-Z (and hence lower wavelength) material—yttrium—and measured amplification of

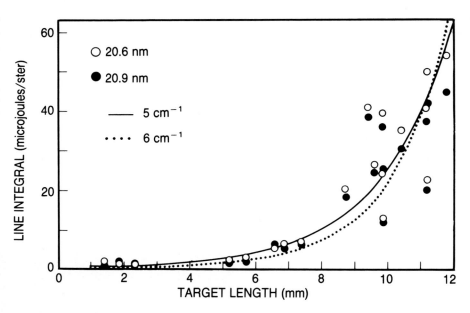

Nearly exponential growth of intensity of emitted radiation was a key indication of lasing in the Livermore experiment. Curves corresponding to gain coefficients of 5 and 6 cm are superposed on data points from lasing lines at both 20.6 and 20.9 nm. Additional measurements out to 2.4 cm indicated continuation of this exponential growth. (From reference 2.)

emissions at 15.49 and 15.72 nm.

Before publicly announcing their results, Livermore's director, Roger Batzel, invited a panel of laser specialists to review the data. Among those who examined the data were Charles Rhodes (University of Illinois, Chicago), William Silfvast (Bell Labs) and Ray Elton (Naval Research Laboratory), all of whom were favorably impressed. Silfvast commented to us that he found the crucial features of the demonstration to be both the instrumentation to detect the laser radiation and the creation of a long enough plasma region to give convincing growth.

The Novette laser is now shut down, to be replaced with the larger Nova facility, which should be operating this month. For six months starting around June, Matthews's group will have two of the laser beams from that facility devoted to the soft-x-ray experiments. The first objective in that work will be to try to achieve saturation in the laser. Matthews estimates that they might be able to increase the peak power by orders of magnitude with a new cylindrical lens designed to increase the length of the focal region up to 3 cm.

The Livermore team will also try to improve the efficiency of the scheme, partly by using x-ray mirrors developed by Troy Barbee (Stanford) to make a double pass through the lasing medium. These mirrors consist[4] of alternate layers of high-density and low-density material—molybdenum and silicon in this case. The thickness of the layers is tailored so that x rays reflected off the successive molybdenum layers will interfere constructively. Bar-

bee told us that measured overall reflectivities at normal incidence range from 50% to 60% for wavelengths from 16 nm to 22.8 nm. He is working on mirrors for x rays as short as 3 nm, although the reflectivity of these is so far barely 10%. Still shorter wavelengths will have to be reflected by crystals.

Matthews commented that his group will move towards shorter-wavelength lasers. The lowest they can anticipate with their collisional-excitation approach is about 10 nm, using molybdenum as a target. They are working on other laser schemes that might take them to 5 nm.

A published report indicates[3] that the nuclear-pumped x-ray laser operates at 1.4 nm. That report described one possible option for a weapon based on such a laser as consisting of a ring of about 50 dense metal lasing rods arranged around a low-yield nuclear warhead.

The Princeton group uses[5] a recombination lasing scheme. The technique is to use incoming laser light to vaporize a target of low atomic number, such as carbon, stripping the atoms completely of electrons in the process. While confined in a magnetic field, the resulting plasma is cooled radiatively as electrons repopulate the higher states. (Suckewer noted that the radiative cooling—as opposed to adiabatic cooling—helps to ensure a uniform plasma density and provides faster cooling.) The resulting hydrogen-like ions are expected to have a population inversion between the levels with principal quantum numbers $n = 3$ and $n = 2$. Transitions between these levels in carbon ions produce radiation whose wavelength is 18.2 nm. A similar method creates lithium-like ions.

In recombination schemes the electron temperatures and densities required for lasing are lower for the same wavelength region than those in the collisional-excitation method used at Livermore. These schemes are thus more adaptable to much smaller-scale facilities. Furthermore, the wavelengths at which lasing is predicted to occur decreases more rapidly as the atomic number of the target material increases.

In the Princeton experiment a 10–20-GW carbon dioxide laser is focused[5] onto a 200- to 400-micron-diameter spot size to produce energy densities on the order of 10^{13} W/cm^2. Suckewer and his colleagues measured the enhancement of the axial emission at 18.2 nm by comparing its intensity to that in the transverse direction. They also compared the ratio of axial to transverse intensities of this line with that of a nonlasing line. Using a solid carbon target, they obtained a product of gain coefficient and length equal to 6.5.

The geometry of the Princeton experiment prevents one from varying the length of the plasma to observe the growth in gain with length. Since the Boston meeting, the Princeton group has acquired from Barbee a spherical x-ray mirror with a radius of curvature of 2 m, to increase effectively the length of the target region. Although the Princeton workers feel they have not yet optimized the mirror arrangement, they observe that their enhancement is 2.0 times greater with the mirror in place, consistent with what they expect for stimulated emission with a mirror of the given reflectivity and small angle of acceptance. They plan further experiments with different targets, with an x-ray mirror of shorter radius, with lithium-like neon ions and with schemes to go to shorter wavelengths.

The x-ray laser schemes described so far have been pursued by others as well. In 1975 Elton suggested[6] a colli-

sional excitation scheme for vacuum ultraviolet light. Later, A. Zherikin, K. Koshelev and V. S. Letokhov (Institute for Spectroscopy, Moscow) described[7] methods of obtaining population inversion between the 3p and 3s levels in neon-like ions. About the same time, A. V. Vinogradov, I. Sobelman and E. Yukov of the Lebedev Physics Institute in Moscow were very active in investigating[8] several x-ray lasing schemes, including collisional excitation. More recently, Elton's colleagues at NRL, some in collaboration with Suckewer and Anand Bhattia (NASA Goddard), and, independently, Hagelstein from Livermore, have extended[9] this work with calculations of the shorter wavelengths expected from neon-like systems at higher atomic numbers. In 1977 a team headed by A. Ilyukhin claimed[10] to have achieved a laser cavity at 60 nm. According to Elton, their claim was never substantiated and the Russians have not published any results on this device since then.

Several experiments in Europe are based on the recombination scheme. Geoffrey Pert and his colleagues at the University of Hull (England) have a setup similar to that of the Princeton group, but with a thin carbon fiber as the target. They reported gain in 1980 and claimed[11] to have measured a gain-length product of approximately 5 on the 18.2-nm line. For the past two years, Pert and his colleagues have been collaborating with Michael Key and others at Rutherford. They are currently planning experiments on a larger scale similar to the Livermore work. At the University of Paris Sud, Orsay, a team led by Pierre Jaeglé is experimenting[12] with a recombination scheme involving lithium-like aluminum, with radiant emission at around 10.5 nm, from which they have reported small gains on the order of one.

There is certainly more than one way to pump a laser, and many other candidates are being studied. The recent success with a collisional excitation scheme does not necessarily indi-

cate the way for future experiments to go; this early in the game it simply speaks to the viability of the x-ray laser concept.

References

1. D. L. Matthews, P. L. Hagelstein, M. D. Rosen, M. J. Eckart, N. M. Ceglio, A. U. Hazi, H. Medecki, B. J. MacGowan, J. E. Trebes, B. L. Whitten, E. M. Campbell, C. W. Hatcher, A. M. Hawryluk, R. L. Kauffman, L. D. Pleasance, G. Rambach, J. H. Scofield, G. Stone, T. A. Weaver, Phys. Rev. Lett. **54**, 110 (1985).

2. L. I. Gudzenko, L. A. Shelepin, Sov. Phys. Doklady **10**, 147 (1965); M. A. Duguay, P. M. Rentzepis, Appl. Phys. Lett. **10**, 350 (1967).

3. C. Robinson, Aviation Week and Space Technology, 23 February 1981, page 25.

4. T. W. Barbee Jr, S. Mrowka, M. C. Hettrick, "Molybendum–silicon multilayer mirrors for the extreme ultraviolet (EUV)," Appl. Optics, to be published.

5. S. Suckewer, C. Keane, H. Milchberg, C. H. Skinner, D. Voorhees, in *Laser Techniques in the Extreme Ultraviolet*, S. E. Harris, T. B. Lucatorto, eds., AIP Conf. Proc. No. 119, Subseries No. 5 (1984), p. 55.

6. R. C. Elton, Appl. Optics **14**, 97 (1975).

7. P. Hagelstein, Plasma Phys. **25**, 1345 (1983); U. Feldman, A. Bhattia, S. Suckewer, J. Appl. Phys. **54**, 2188 (1983); U. Feldman, J. F. Seely, A. K. Bhattia, J. Appl. Phys. **56**, 2475 (1984); J. P. Apruzese, J. Davis, Phys. Rev., **A28**, 3686 (1983).

8. A. Zherikhin, K. Koshelev, V. Letokhov, Kvant. Elektron. (Moscow) **3**, 152 (1976) [Sov. J. Quantum Electron. **6**, 83 (1976)].

9. A. V. Vinogradov, I. Sobelman, E. Yukov, Kvant. Elektron. **4**, 63 (1977) [Sov. J. Quantum Electron. **7**, 32 (1977)].

10. A. Ilyukhin, G. Peregudov, E. Ragozin, I. Sobelman, V. Chirkov, Pis'ma Zh. Eksp. Teor. Fiz. **25**, 569 (1977) [JETP Lett. **25**, 536 (1977)].

11. D. Jacoby, G. J. Pert, L. D. Shorrock, G. J. Talents, J. Phys. **B15**, 3557 (1982).

12. G. Jamelot, P. Jaeglé, A. Carillon, A. Bideau, C. Moller, H. Guennou, A. Sureau, in *Proc. Int. Conf. Lasers 81*, New Orleans, La., December 1981, p. 178.

Prospects for a gamma-ray laser

Lasers generating wavelengths shorter than one angstrom unit would have many important applications. Ever since the discovery of the Mössbauer effect it has been realized that we would be able to stimulate nuclear transitions suitable for lasers in this range if we could prepare the population inversion on which lasers depend, and it has also long been known that some nuclear isomers are producible in a state of population inversion. Why then do we not yet have any operating gamma-ray lasers?

We propose here to consider this question and to show that, although these two phenomena have not yet been

G. C. Baldwin is a professor of nuclear engineering and science at Rensselaer Polytechnic Institute, Troy, N.Y.; R. V. Khokhlov is a professor of physics at Moscow State University as well as being Rector of the University.

achieved with the same transition, this may yet prove to be possible. Such a development would be an extremely difficult and challenging task, but we believe that it is possible and that it is worthwhile.

The trend toward short wavelengths has been an outstanding feature of the development of quantum electronic devices. Initially, only microwave radiation could be stimulated; today, lasers operate over a frequency range extending into the vacuum ultraviolet. (See figure 1.) Soft x-ray lasers, which a decade ago were generally conceded to be impossible, are now under active development. The appeal of new important applications continues to attract efforts to surmount the ever-greater challenge of higher frequencies, which is inherent in the λ^3-dependence of the ratio of Einstein's A and B coefficients.[1]

Although no absolute limit is predict-

ed by Einstein's ratio, it has occasionally been suggested that a practical upper limit may exist to the frequency that can be generated by stimulated emission. For example, an authoritative text on electrodynamics,[2] published in the middle 1950's, remarks in a footnote that ".... stimulated emission is unimportant at all temperatures less than [*the photon energy*] .... It may become important for radio waves." Several years later, in a paper that showed optical lasers to be possible,[3] this comment was made: ".... unless some radically new approach is found, they cannot be pushed to wavelengths much shorter than those in the ultraviolet region." However, a recent consideration of this question suggests that, if any fundamental limit indeed exists to laser action, it lies in the vicinity of the threshold for pair production.[4]

An obvious practical difficulty in ex-

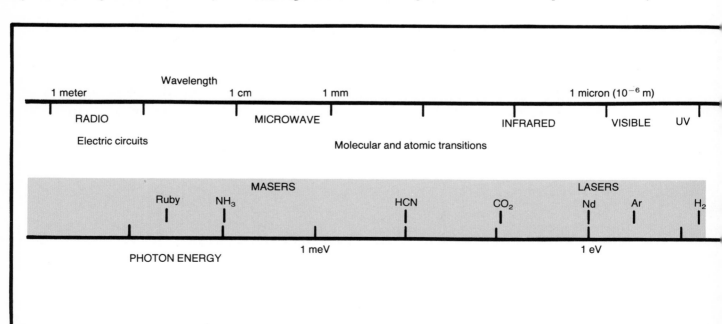

Development of a "graser" may be possible if ways can be found to achieve Mössbauer transitions and population inversion simultaneously in nuclear isomers.

G. C. Baldwin and R. V. Khokhlov

PHYSICS TODAY / FEBRUARY 1975

tending the laser principle to higher frequencies is that of providing high stored-energy density. Nuclear reactions offer a new approach to this problem; in fact, nuclear transitions offer the best hope for laser action at wavelengths shorter than 1 Å.

The possibility of a nuclear laser, often termed a "graser" (for "gamma-ray laser") or a "gaser," was recognized early in the 1960's,[5-9] and proposals were made to stimulate Mössbauer transitions in nuclides prepared by neutron bombardment, chemical isolation and crystallization. Obviously, long transition lifetimes would be required if these operations were to be completed before loss of population inversion by decay. It was soon realized that long-lived isomers involve greater difficulty because of the narrow natural linewidth.[5,3] After this difficulty became apparent and attempts to circumvent it

seemed unfruitful,[11] interest waned and the idea remained dormant until about two years ago,[12] when it became evident that the linewidth-versus-lifetime dilemma is not really fundamental, and can probably be overcome.

Some critics suggest that we await success in development of x-ray lasers, which is conceded to be extremely difficult, before attempting to develop gamma-ray lasers. However, the difficulties attending graser development are quite distinct from those of x-ray lasers; these difficulties, formidable as they may seem, may be overcome through research in certain specific topics that we shall identify during the course of this article.

Probable characteristics of grasers

Although the ultimate form and operating characteristics of grasers cannot be described at this stage, it is possible

to make some general assertions that will probably turn out to be correct. In particular, grasers will operate without mirrors, they will probably be single-pulse devices, they will be pumped by capture[7,9,14] of neutrons, they will use Mössbauer transitions, and their photon energy will be in the 5 to 200-keV range.[a]

Nuclear isomers. Because of extremely short lifetimes of the electric dipole transitions that emit characteristic x radiation, it appears unlikely that atomic states can be excited in sufficient density for lasing in this region of the spectrum; prohibitively high pump power density would be needed to establish population inversion. Nuclear transitions are available, however, with far longer lifetimes, and exoergic nuclear reactions can be employed to pump these transitions.

Figure 2 shows selected parts of the

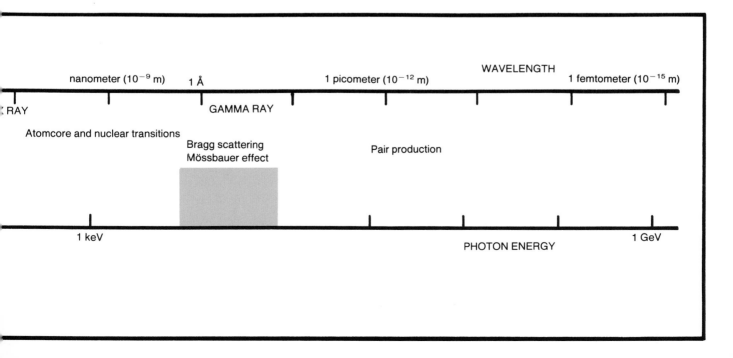

energy level scheme[13] for two typical examples of isomeric transitions, Ta[181] and In[114] (we do not mean to imply that these particular isomers are suitable for graser application).

In general, the formation of states of high angular momentum (which will be relatively long-lived) is not only possible, but probable, in the radiative decay of a highly excited compound nucleus, following its formation in a nuclear reaction. Isomer production ratios, unfortunately, are rarely known—although several cases are known (for example, In[114]) in which formation of the upper isomer is more probable than of the lower, so leading to population inversion. There are several isomeric nuclei (In[114], Zn [69]) in which the lower of two excited states has the shorter lifetime—another condition leading to population inversion. We shall show below that it is also possible to isolate nuclei in either of two isomeric states by physical methods.

Radiative transitions are possible between any pair of isomeric nuclear levels (except in the case of $I = 0 \rightarrow I = 0$) but, in many cases, internal conversion is more probable than gamma emission. Occasionally, beta decay from isomeric levels competes with gamma emission and internal conversion. Those properties of isomeric states most significant to the graser problem are lifetime τ, spin I, production cross sections, transition energy E, branching ratio β, and internal conversion coefficient α.

Beam formation. In the absence of a resonator, the geometry and transparency of the active medium together determine the ability of an amplifying medium to form a directed beam of stimulated radiation (figure 3). An elongated solid, such as the whisker crystals that have been proposed,[8] is required for this purpose.

If I_0 denotes the average in all directions of the gamma radiation from the graser body, the intensity I on its axis of greatest extension will be increased by the gain factor

$$G = I/I_0 = $$
$$[(\kappa - \mu)L]^{-1} \exp[(\kappa - \mu)L - 1] \quad (1)$$

in which

κ = coefficient of stimulation, cm^{-1}
μ = coefficient of nonresonant removal, cm^{-1}
L = total length

Ordinarily, κ is essentially zero, and the only observable effect is self-absorption of spontaneously emitted radiation. At any instant, few modes of the radiation field contain photons, and those that do are singly populated.

If κ exceeds μ sufficiently, there will be modes that contain many photons, populated to extents determined by the locations of the various initiating spontaneous emission events and the interplay of stimulation and absorption, governed by the coefficients κ and μ. Selection of the sense of beam propagation is possible by traveling wave methods[15] in which adjacent elements of the medium are made multiplying at successive instants of time.

Coefficient of removal. We first inquire concerning the coefficient of nonresonant removal, μ. (Resonant absorption is encompassed in our definition of κ.)

Elements of medium atomic number interact with gamma radiation of energy less than 200 keV principally by photoelectric absorption; in light elements the principal interaction is Compton scattering. The cross sections for these interactions are several orders of magnitude lower than nuclear resonance cross sections observed in Mössbauer experiments.[10,16]

Nonresonant removal does eventually exceed the maximum possible resonance cross section as the photon energy is increased;[4] but not in the energy range we are considering, where several orders of magnitude separate these cross sections.

It has been suggested that nonresonant photon losses may be reduced further in one of two ways: either by embedding the isomeric nuclei at low concentration in a crystal host of low atomic number (for example, Be)[14], or by the Borrmann effect.[8,17] In the latter, Bragg scattering gives rise to a spatially periodic modulation of the amplitude of an electromagnetic wave, such that wave modes exist with electric field nodes at the lattice planes; such modes have greatly diminished photoelectric absorption.[18] It might be objected that the interaction with nuclei would also be reduced, since they are located at the lattice nodes; however, it has been shown[19] that stimulated emission is not reduced[b] for nuclear isomeric transitions, which are of multipole order higher than electric dipole. Thus, when the direction of greatest extension of the solid coincides with a Bragg direction, the ratio μ/κ can be reduced by

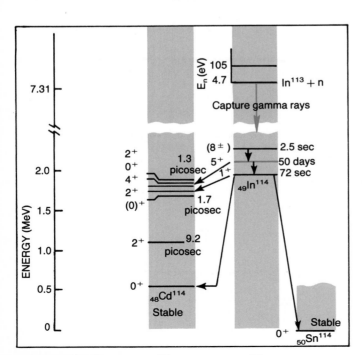
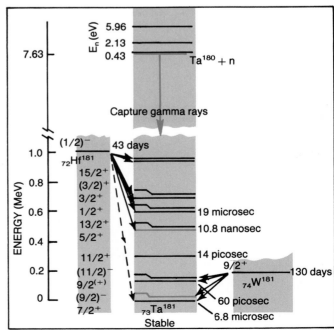

Energy-level diagrams for In[114] (left) and for Ta[181] (right). Only the relevant portions are shown here. Shortly after the compound nuclei have been formed by neutron capture, some intermediate states may be temporarily population-inverted. The two lowest states of In[114] are an extreme example of this phenomenon. The 6.8-microsecond state of Ta[181] is a Mössbauer isomer.   Figure 2

at least two orders of magnitude.[19] The Borrmann effect, by singling out certain modes, offers the prospect of eventually making a resonator for gamma radiation.

Coefficient of stimulation. The coefficient of stimulated emission for recoilless gamma emission is

$$\kappa = N^*\sigma \qquad (2)$$

where

$$N^* = N_2 - (g_1/g_2)N_1 \qquad (3)$$

is the net inversion density between an upper state 2 and a lower state 1, with statistical weights respectively g_2 and g_1, and

$$\sigma = \frac{\lambda^2}{2\pi} \frac{f}{\Gamma\tau} \frac{\beta}{1+\alpha} \qquad (4)$$

is the maximum value of the resonance cross section for a Mössbauer transition. Here

λ = wavelength of the radiation;
Γ = total linewidth;
τ = the effective lifetime of the transition, which if the two levels are both unstable, is the reciprocal of the sum of their respective reciprocal lifetimes;
β = branching ratio for the isomeric transition;
α = internal conversion coefficient; and
f = the "recoilless" Mössbauer fraction (Debye-Waller factor).

(Note that we employ the convention used in laser theory of applying the statistical weight factor to the population rather than to the cross section, instead of the convention common in Mössbauer work.)

Linewidth and transition lifetime

Figure 4 shows linewidth Γ as a function of transition lifetime τ. The solid, slanting line represents the natural width, τ^{-1}, of an unbroadened line. Homogeneous broadening through the factor $(1 + \alpha)\beta^{-1}$ causes the natural width of actual transitions to be somewhat greater than shown by this line; its effect appears in the cross section formula, equation 4.

The points in figure 4, experimental values from nuclear resonance absorption measurements on the indicated isotopes,[10] show that additional broadening is often present. For lifetimes longer than 10^{-5} seconds, the linewidth Γ appears to become independent of τ, so that $(\Gamma\tau)^{-1}$ decreases, reducing the resonance cross section at longer lifetimes (figure 5). For this reason, Mössbauer experiments are not performed with transitions of lifetime greater than 10 microseconds. For lifetimes shorter than 10^{-6} sec, the product $\Gamma\tau$ in equation 4 is essentially unity. Transitions in this lifetime range, showing a strong Mössbauer effect, would satisfy the condition for stimulated emission gain if inverted populations of the states could be generated. Under the usual condi-

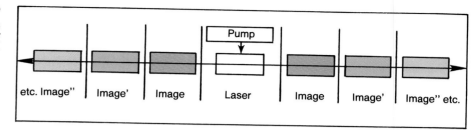

Mirror "resonators" extend the active medium in the majority of conventional lasers by an array of images, so that spontaneous emission into the axial direction receives greatest amplification. In the absence of mirrors, we extend the active medium itself—the so-called "amplified spontaneous emission lasers." Whisker crystals have been proposed for this purpose.   Figure 3

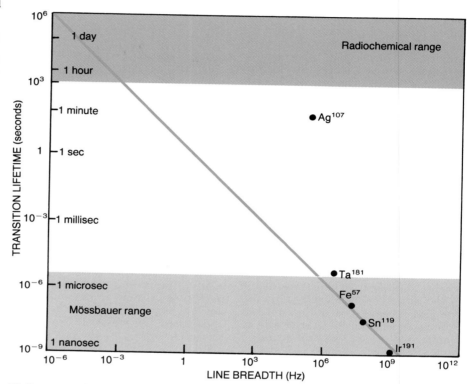

Lifetime versus bandwidth. The relationship of mean lifetime of a radiative transition to the frequency bandwidth of the radiation shown by the slanting colored line is for an unbroadened transition. The points show measured values, and indicate inhomogeneous broadening of long-lived isomeric transitions. The figure also indicates the range of isomer lifetimes in which line broadening does not interfere with observing Mössbauer effects, and also the range in which population inversion can be achieved by radiochemical procedures.   Figure 4

tions of Mössbauer experiments, however, only a very small number of excited nuclei are ever present simultaneously either in the source or in the absorber. It is the factor N^*, rather than σ, that requires attention if transitions in this lifetime range are to be used in grasers. This is indicated in the margin of the figure as the "Mössbauer range."

Figure 4 also indicates, along the margin, a range of lifetimes for which it is feasible to generate a condition of population inversion by neutron bombardment from conventional reactor sources, followed by conventional techniques of radiochemistry to isolate the transmuted material ("Szilard-Chalmers separation").[7]

A gap, nearly nine orders of magni-

tude, separates the "Mössbauer range" and the "radiochemical range" of nuclear transition lifetimes. To make a graser we must close this gap.

Proposed approaches

There are, consequently, two principal approaches to development of gamma ray lasers:

▶ Use long-lived isomers, and reduce or eliminate mechanisms that contribute to the total line breadth;[6,12]
▶ Use transitions in the Mössbauer range that can be inverted directly without chemical procedures by means of extremely intense pumping sources.[14,20]

Because neither approach may bridge the entire gap in figure 4, a combination of these approaches could be the most likely ultimate solution; for example, a combination of fast separation procedures.[20,21] intense pulsed neutron sources,[22,23] rapid assembly of the Mössbauer body.[21] and procedures for narrowing the Mössbauer line.[12,24–26]

Inhomogeneous line broadening

We first consider the excessive inhomogeneous linewidth of long-lived transitions.[5] Until about two years ago it was not evident that this problem could be resolved. Today it appears, not merely that the mechanisms that inhomogeneously perturb the Mössbauer line are known, but also that it is possible in principle to reduce the effect of each of them. The mechanisms include temperature broadening, hyperfine interactions between nuclei and locally generated fields, and gravitational broadening.

There are two contributions to temperature broadening, one of which, the second-order Doppler effect,[27] arises from gradients of temperature (more strictly, from gradients in the ratio of local temperature to Debye temperature). The second type of temperature broadening[28] is associated with local fluctuations in the value of the mean square lattice vibrational energy. Calculations[24] show that these two effects are reducible to insignificance at temperatures below 1 K.

Difficulty may be experienced with localized heating from radioactive decay of the isomer. A filamentary form for the graser body, such as the whisker crystals we mentioned earlier, assists in maintaining a low uniform temperature.[c]

Temperature broadening, therefore, is not an insurmountable problem.

Three types of hyperfine interaction also contribute to the inhomogeneous line breadth.[24] All are manifestations of the finite volume of the nucleus; they are associated, respectively, with its magnetic dipole moment, electric quadrupole moment, and radius. They contribute inhomogeneous line breadths in the range 10^4–10^6 sec^{-1}, and so are responsible for the trend evident in figure 2. Two of the hyperfine interactions are associated with nuclear spin.

The magnetic moments of isomeric states differ, and each may differ from that of other nuclei in the solid; therefore, even when the spins are aligned by a strong magnetic field at low temperature, inhomogeneous broadening occurs because of the random nature of that part of the local magnetic field which is created by nearby nuclei.

Recent nuclear magnetic resonance experiments show that the nuclear dipole–dipole interaction can be averaged to zero[29] by application of a cycle of radiofrequency $\pi/2$ pulses that reorient the nuclear moments. Reductions of linewidth of nearly four orders of magnitude have been achieved in nuclear magnetic resonance experiments with this technique.[30]

The same principle should be applicable to narrowing the gamma-ray line,[12,24] and therefore to extending the lifetime range accessible to the Mössbauer effect. We must note two significant facts, however:

First, two levels are involved in gamma-ray transitions, with unequal spins. Thus, to narrow the line, two different cycles of perturbing magnetic field pulses of unequal frequencies must be employed simultaneously.[24]

Second, the technique is limited to transitions of lifetime much greater than the period of the rf-pulse sequence, which, in turn, must be much greater than the Larmor precession period of either spin state. Thus, although rf perturbation will not be useful in conventional Mössbauer measurements with lifetimes near 10^{-5} sec, it should yield a significant effect in experiments of the Bizina[31] type, which used a long-lived transition in Ag[107].

The dipole–dipole interaction of nuclear spins that have been aligned by a strong magnetic field is governed by a Hamiltonian function of the form[29]

$$[\mathbf{I} \cdot \mathbf{I'} - 3I_z I_z'](r - r')^{-3}$$

summed over all interacting pairs of spins (lower-lower, lower-upper, and upper-upper) in the crystal. The interaction of nuclear quadrupole moments with electric field gradients in the crystal can be represented by a Hamiltonian in which the spin operators appear in

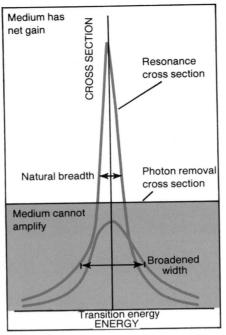

Broadening of the gamma-ray line reduces the maximum value of the effective resonance cross section; if it is thereby brought below the value needed to overcome losses, there can be no net gain.   Figure 5

similar form;[32] the Hamiltonian is

$$(\mathbf{I} \cdot \mathbf{I} - 3I_z I_z)$$

It follows that the same cycle of 90-deg rotations of nuclear spin that reduces the dipole–dipole interaction should also reduce the quadrupole breadth.[24] It may be noted further that the strong polarizing field can be internal in origin.

A third type of hyperfine interaction, inhomogeneous monopole shift[33] (also commonly called the "chemical shift" or the "isomer shift"), arises because the electron density at the nuclear site is not zero, and, furthermore, is modified by nearby chemical inhomogeneities or defects in the crystal lattice; in an isomeric transition, the nuclear volume changes, and with it, the interaction of nuclear charge with electrons lying within the nuclear volume.[33]

Estimates of the permissible concentration of lattice defects suggest that, with continued improvements in the art of growing perfect crystals, the monopole broadening contributed by lattice defects can be reduced to a tolerable amount.[24]

However, it may be possible to develop a technique with external radiofrequency fields that can compensate inhomogeneous chemical broadening as well. Owing to the so-called "Fermi contact interaction," there can be a net spin density at the nucleus. As long as minor changes of spin density and of the total s-electron density are proportional to each other, both can be affected by chemical inhomogeneities in a similar way. The hyperfine splitting produced by an internal magnetic field associated with electron spin density can be adjusted by varying an external radiofrequency field,[25] and this offers the possibility of compensating monopole shift by hyperfine splitting, for some single component of the hyperfine structure. That one component, then, would not be broadened by inhomogeneous chemical shifts. Although the cross section for stimulation of one component is lower than the theoretical maximum for an unsplit, unbroadened gamma-ray line, the increase in cross section from narrowing of a selected hyperfine component is a much greater effect.

The third contributor to inhomogeneous line broadening is the effect of gravity. Research on the Mössbauer effect has been closely associated with the gravitational red shift for many years,[10] and we know that the shift in transition frequency apparent to two nuclei with transition energy E, differing in gravitational potential by $\Delta\Phi$ is[5]

$$\Delta\omega = [\Delta\Phi/Mc^2][E/\hbar] \qquad (5)$$

For horizontal propagation in standard gravity this effect would limit the maximum vertical dimension of a graser body that could contribute to a growing wave, so that, for long-lived isomers, a diffraction limitation must be considered.

If we require that gravitational broadening not exceed the remaining part of the total linewidth, Γ', the vertical dimension of a graser body with its axis horizontal in a gravitational field of acceleration g cannot exceed

$$\delta z \; < \; \hbar c^2 \Gamma' / g E \; = \; \lambda c \Gamma' / g$$

where E is the transition energy and λ the corresponding wavelength. To reduce losses by diffraction, the transverse (vertical) dimension should be made many times the wavelength; in other words, $\delta z / \lambda$ must be a large number, but less than $\Gamma' c / g$. Even in standard gravity, it is apparent that gravitational broadening is not the principal problem, since Γ' is of the order of 10^5 sec^{-1} for the hyperfine interactions.

Moreover, elimination of gravity is a commonplace art today. The compensation of gravitational shift by temperature gradient[7] or even by deliberate chemical shift is also possible.

Thus, we see that all the interactions that inhomogeneously broaden the gamma-ray line are understood, and it appears likely that each can be eventually controlled, reduced, compensated or perhaps even eliminated.

It is difficult to estimate, without experimental evidence, the extent to which the linewidth can be reduced, and therefore, to decide whether this approach alone will suffice to close the lifetime gap and so make graser action feasible. We shall therefore next consider what advances are possible if we use the alternative approach, via short-lived transitions.

Transitions in the Mössbauer range

Many isomeric transitions of lifetime shorter than about 10^{-5} sec have resonance cross sections sufficient to be observable above a background of nonresonant absorption.[10] Producing population inversion, without destroying the conditions essential to the Mössbauer effect, is the problem with these transitions.

Although conventional Mössbauer transitions necessarily involve the ground state, for a graser there need be no such restriction, and, in fact, recoilless transitions between excited states would be preferable. Since Mössbauer cross sections of megabarns exist, the actual concentration of excited nuclei needed for observable gain can be as low as 10^{18} cm^{-3}. The total number also can be small, say 10^{13} or 10^{14}; however, the excitation process must be able to generate excited states in concentrations of this order of magnitude *within the mean lifetime of the transition;* reaction rates per unit volume of the order of 10^{23} cm^{-3} sec^{-1}, or higher, would be required of a *direct* excitation

process to satisfy this criterion.

Assuming parent isotopes in normal concentration and a 10-barn effective cross section for inversion, we would need a neutron flux of the order of 10^{24} cm^{-2} sec^{-1} to excite the nuclear inversion. This number usually suffices to discourage consideration of the short-lived graser possibility. Considerably higher neutron fluxes are indeed produced in nuclear explosions, but we prefer a less drastic method!

Nevertheless, it has been shown[14] that long filaments of beryllium containing approximately 100 parts per million of a parent isotope might be pumped by an explosive burst of neutrons to a lasing condition before their temperature would rise sufficiently (above 300 K) to destroy the Mössbauer effect.

Furthermore, current research on thermonuclear fusion demonstrates that intense laser irradiation of matter can compress it to extremely high densities[34] and, at the high temperatures and high electric fields in the resultant plasma, neutrons are generated. We also know that compression of fissile material in this way can lead to many orders of magnitude reduction in its critical mass,[22,23] so that nuclear fission explosions can be "miniaturized."

One suggestion for a neutron-pumped graser that uses transitions in the Mössbauer range[35] is to combine the neutron source—an annular region of fissile material, compressed by a cylindrical array of convergent laser beams—coaxial with a long filament of graser material—beryllium with a small admixture of a parent isotopic material.[14] Traveling-wave excitation[15] is achieved by so timing the

arrival of laser light at the fissile layer (figure 6), that the region of criticality advances at velocity c. Material damage from neutron recoil in the graser filament, essential to moderating the neutron energies, may be a serious problem in this scheme.

We would not need to go to this extreme if it were possible to furnish excitation at one site and transfer it rapidly to another site at increased concentration, as in the long-lived isomer approach. In fact, one proposal[36] suggests that we transfer excitation from bombarded material to a smaller volume at the lasing site, by means of resonant absorption at the second site. As this must be followed by a transition to an unoccupied upper sublevel of the ground state, the temperature at the second site must be extremely low.

Approaches involving material transfer

It may be unlikely that excitation, separation and reassembly of isomeric material can be carried out rapidly enough to allow use of transitions in the Mössbauer range. However, the restriction to lifetimes of hours or more, inherent in the early graser proposals[5-9] (which were based on radiochemical procedures) need not apply today.

If we grant the possibility of appreciable reduction of the inhomogeneous linewidth, perhaps by procedures suggested in earlier sections, but find that we still cannot bridge the gap of figure 4, there remains the possibility of combining line-narrowing methods with faster methods of chemical separation and, preferably, with more intense neutron sources.

Conceivably, recoil of nuclei trans-

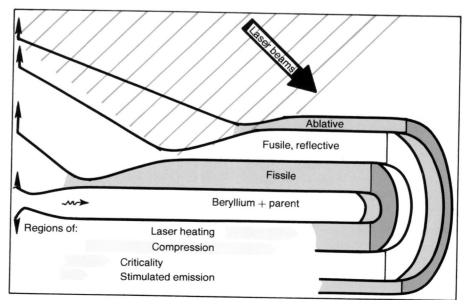

A conceptual graser employing transitions in the short (Mössbauer) lifetime range, pumped by a traveling wave of laser-driven fission. The reflecting material and the central beryllium core moderate neutrons generated in compressed fissile material; last to be compressed, the beryllium acts as a "flux trap" for neutrons, which excite isomeric states in target material. **Figure 6**

Labels in figure: Laser beams / Ablative / Fusile, reflective / Fissile / Beryllium + parent / Regions of: / Laser heating / Compression / Criticality / Stimulated emission

muted in the neutron burst might be used to transport them to a nearby surface, but quantitative estimates of the efficiency of the transplantation process have not been encouraging. Use of finely divided parent material in a scavenging gas stream has been proposed[20] for increasing the efficiency of the recoil separation process, with deposition of the active material upon a chemically active surface.

Recent progress in the separation of isotopes by photochemical reactions involving the use of tuned lasers offers the prospect of extremely rapid and specific isolation of nuclear isomeric material. We know that, because of their unequal spins, isomeric nuclear states have different optical spectra;[37] these differences are sufficient to provide a basis for quantitative separation. One obvious method is two-step photoionization,[38,39] in which the first step is absorption of sharply tuned laser light by one component of the hyperfine structure of the upper isomeric level, not present in the spectrum of the lower isomeric state. If this step is followed by rapid photoionization, by means of intense ultraviolet light, the upper-state isomer can be extracted and deposited by ion epitaxy on a nearby host surface[21] to form the laser material (figure 7). Many cases of *isotope* separation by laser radiation[40] have been reported; the principle is also directly applicable to *isomer* separation.

This proposal, suitable for transitions in the range from a few milliseconds to several seconds, involves many processes, hitherto untested; most of the data needed to evaluate it are unavailable. However, one attempt at an analysis concludes that this approach may be feasible.[21]

The most serious questions may turn out to concern, not the feasibility of the component steps, but their compatibility—that is, the possibility of nearly simultaneous excitation and separation, involving intense neutron bombardment, tuned lasers and an electrode system for implantation, with line narrowing, requiring cryogenic apparatus, intense magnetic fields, radiofrequency pulse generators and low gravity, establishing thereby conditions in the graser body conducive to a Mössbauer effect. Admittedly, it does seem to be a very difficult undertaking, if so many elements must be involved.

Research required

Realization of stimulated emission in nuclei evidently requires prior investigations into a variety of problems. We believe that this research should be undertaken with the stated objective of working toward development of gamma-ray lasers, rather than awaiting the normal course of developments in laser science and technology.

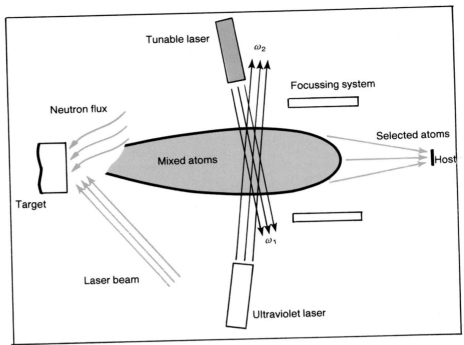

This graser system would employ transitions in an intermediate lifetime range, with laser vaporization of a neutron-bombarded target, photoexcitation and ionization of atoms containing the upper isomeric state, and electrical deposition by ion epitaxy onto a nearby host crystal to form the laser material. (Courtesy of V. S. Letokhov.)   Figure 7

For example, despite decades of research on nuclear isomerism, data needed for calculating the cross section for stimulated emission are adequate in only those few exceptional isomers, mainly of long lifetime, for which measurements of internal conversion coefficients, production ratios and branching schemes could be made conveniently.

Nuclear properties alone do not suffice for selecting appropriate isomers; chemical form and crystallographic properties are equally important. Ability to isolate and concentrate the isomer and to assemble it within a crystalline host can be developed only through experimentation with the graser objective in mind.

Although the theoretical basis of the possibility of narrowing the gamma-ray line needs further development, even now it should be possible to attempt to demonstrate line narrowing in experiments of the Bizina[31] type. Such experiments will be significant in connection with the hypothesis of a universal quantum of length.[41]

Those properties that are significant to the Mössbauer effect need to be studied in material prepared by implantation and in matter which has been subjected to intense neutron bombardment immediately prior to the measurements.

The development of intense neutron sources (for example, by laser compression techniques) needs to be encouraged. Neutron exposure within the isomer transition lifetime is the significant parameter that characterizes the neu-

tron source. Most of the development of pulsed neutron sources has concentrated on achieving pulse widths much shorter than the graser pump requires. Higher neutron intensity would greatly diminish demands on chemical separation yields.

We shall need to develop new methods for demonstrating stimulated emission at gamma wavelengths, probably based upon the statistical time structure of the gamma-ray counting rate in a photoelectric detector.[7]

Each of these ordinarily unrelated fields of investigation is closely interrelated to the others through the common requirement of graser development.

Nevertheless, each area of research promises dividends to science and technology; We might develop intense neutron sources, find new procedures for radioisotope production, or stimulate new applications of the Mössbauer effect.

Why develop a graser?

Early in this article, we asserted that development of a graser would be an extremely difficult and challenging task. The following sections amply confirmed that statement!

Is it worth the effort, and if so, is this the time to begin?

We believe it is worthwhile, because of the great potential for important application of radiation with enhanced coherence in this spectral region. In every region of the spectrum in which coherent sources exist, they have greatly multiplied the usefulness of radiation. Gamma rays have many important applications, because of their short wavelength, high penetrating power and ability to ionize. Controllable and directable sources of such radiation in which phase as well as intensity have significance, would greatly multiply its uses.

However, detailed speculation upon specific applications is clearly impossible at this stage, because the form, performance, output characteristics, cost—even the existence—of a graser are obviously impossible to describe at this stage of our knowledge, beyond the few characteristics enumerated above.

Moreover, it is well known that the most significant applications of new devices are rarely foreseen. No one, for example, in 1958 could have predicted the use of lasers in computers, surgery, nuclear fusion, isotope separation, and so on.

The most exciting possibility for coherent radiation in this short wavelength range is that of holography, which might permit direct observation of the structure of molecules, crystals, proteins, genes. Before this could be done, of course, one would need to solve the equally challenging problem of recording the hologram on a sufficiently fine scale.

Even coarse-grained gamma-ray holograms would be useful in radiographic inspection of thick material, however.

Precision frequency measurement, based on interferometric techniques made possible by coherent sources, now extends from very low frequency into the visible range; its extension into the nuclear region would greatly increase the precision with which fundamental nuclear constants are known.

Nonlinear optics has developed into a major field of current research, which has begun to make important contributions to technology, communication and metrology. Nonlinear optics exploits the high intensity of some laser sources and the exceptionally high coherence of others, so that very weak effects of nonlinearities become observable, either by higher-order intensity dependence or by coherently combined weak contributions from a medium of large extent. One can expect, by analogy, a new field of nonlinear nuclear physics to emerge if nonlinear (for example, multi-photon, stimulated Raman) processes are made possible by intense graser sources.

All of these, and the unforeseen applications that will ultimately prove the most rewarding, await resolution of the basic questions we have tried to identify here.

Admittedly, we have not given a recipe for constructing a graser, nor have we even proved that it is possible. We have demonstrated, however, that dogmatic assertions as to its impossibility are unwarranted.

The authors wish to express their appreciation to V. I. Goldanskii, Yu. Il'inskii, Yu. M. Kagan and V. S. Letokhov, who have read the manuscript and made helpful suggestions. One of us (Baldwin) wishes to acknowledge useful discussions with P. Casabella, H. M. Clark, D. Eccleshall, M. Maley, K. J. Miller, I. Preiss, R. Reeves, W. K. Rhim, R. Schnibman, R. Strong and J. Temperley.

This article is an adaptation of papers presented at the Spring Meeting of the Optical Society of America, in Washington D.C., April 1974 (Baldwin), and at the Eighth International Conference on Quantum Electronics, in San Francisco, June 1974 (Khokhlov).

References

1. A. Einstein, Phys. Zeitschr. **18**, 121, (1917); B. Lengyel, *Lasers*, Wiley, New York, (1962); section 3, pages 3–10; A. Yariv, *Quantum Electronics*, Wiley, New York (1967), pages 208–210.

2. L. Landau, E. Lifschitz, *Electrodynamics of Continuous Media*, English trans. by J. B. Sykes and J. S. Bell, Addison-Wesley, Reading, Mass., (1960); page 377.

3. A. L. Schawlow, C. H. Townes, Phys. Rev. **112**, 1949 (1958).

4. G. C. Baldwin, in *Laser Interaction and Related Plasma Phenomena*, Vol. 3, (H. Schwarz, H. Hora, eds.) Plenum, New York (1974), pages 875–888.

5. W. Vali, V. Vali, Proc. IEEE **51**, 182 (1963).

6. B. V. Chirikov, Sov. Phys.—JETP **17**, 1355 (1963).

7. G. C. Baldwin, J. P. Neissel, J. H. Terhune, L. Tonks, Trans. Am. Nucl. Soc. **6**, 176 (1963).

8. G. C. Baldwin, J. P. Neissel, J. H. Terhune, L. Tonks, Proc. IEEE **51**, 1247 (1963).

9. Unpublished reports on this subject include: L. Rivlin, Soviet Patent Disclosure Jan. 1961, cited by Chirikov in reference 6; G. C. Baldwin, J. P. Neissel, L. Tonks, Patent Disclosure October 1961 and Internal General Electric Reports 62GL22, 62GL178, February and December 1962; B. Podolsky, J. Mize, C. Carpenter, AVCO Internal Report 63AVCO-109, August 1963; R. Babcock, S. Ruby, L. Epstein, Westinghouse Report, December 1963. US Patents related to stimulated emission in nuclear states include: Baldwin et al, 3 234 099; Vali et al, 3 281,600; Eerckens, 3 430 046; Piekenbrock, 3 557 370.

10. *Mössbauer Effect*, (Selected reprints), American Association of Physics Teachers (1963).

11. J. H. Terhune, G. C. Baldwin, Phys. Rev. Lett. **14**, 589 (1965).

12. R. V. Khokhlov, Sov Phys—JETP Lett. **15**, 414 (1972); Paper A4, Proc. VII IQEC, Montreal, May 1972.

13. *Nuclear Data Sheets*, edited by the Nuclear Data Group, Oak Ridge National Laboratory (W. J. Horen, Director), published by Academic Press, New York; contains frequently revised compilations of isomer properties.

14. V. I. Goldanskii, Yu. Kagan, Sov. Phys.—JETP **37**, 49 (1973).

15. J. D. Shipman, Jr, Appl. Phys. Lett. **10**, 3 (1967).

16. G. W. Grodstein, "X-Ray Attenuation Coefficients from 10 keV to 100 MeV," NBS-583, Nat. Bur. Stds., Washington, 1957.

17. G. Borrmann, Zeits. Phys. **42**, 157 (1941).

18. P. P. Ewald, Rev. Mod. Phys. **37**, 46 (1965); B. Batterman, H. Cole, Rev. Mod. Phys. **36**, 681 (1964).

19. Yu. Kagan, Sov. Phys.—JETP Lett. **20**, 11 (1974).

20. V. I. Goldanskii, Yu. Kagan, Sov. Phys.—Uspekhi **16**, Jan.–Feb., 1974; Proc. Vth Int. Conf. on Mössbauer Spectroscopy, Bratislava, Sept. 1973.

21. V. S. Letokhov, Sov. Phys.—JETP **37**, 787 (1973); III Vavilov Conference on Nonlinear Optics, Novosibirsk, June 1973.

22. G. A. Askaryan, V. A. Namiot, M. S. Rabinovich, Sov. Phys.—JETP Lett. **17**, 424 (1973).

23. F. Winterberg, Proc. 3rd RPI Workshop Conference on Laser/Plasma Interaction, August 1973 (H. Schwarz, ed.) Plenum, New York (1974).

24. Yu. Il'inskii, R. V. Khokhlov, Sov. Phys.—Uspekhi **16**, Jan.–Feb. 1974.

25. V. I. Goldanskii, S. V. Karyagin, V. A. Namiot, Sov. Phys.—JETP Lett. **19**, 324 (1974).

26. Yu. Kagan, Sov. Phys.—JETP Lett. **19**, 373 (1974).

27. R. V. Pound, G. A. Rebka, Jr, Phys. Rev. Lett. **4**, 274 (1960); B. D. Josephson, Phys. Rev. Lett. **4**, 341 (1960).

28. Yu. Kagan, Sov. Phys.—JETP **47**, 366 (1964).

29. U. Haeberlen, J. S. Waugh, Phys. Rev. **175**, 453 (1968).

30. W. K. Rhim, D. D. Elleman, R. W. Vaughan, J. Chem. Phys. **59**, 3740 (1973).

31. G. Bizina, A. Beda, N. Burgov, A. Davydov, Sov. Phys.—JETP **18**, 973 (1964).

32. H. G. Dehmelt, Am. J. Phys. **22**, 110 (1954).

33. O. C. Kistner, A. W. Sunyar, Phys. Rev. Lett. **4**, 412 (1960).

34. J. Nuckolls, L. Wood, A. Thiessen, G. Zimmermann, Nature **239**, 139 (1972).

35. G. C. Baldwin, Laser Focus, March 1974, page 43.

36. V. I. Goldanskii, Yu. Kagan, V. A. Namiot, Sov. Phys.—JETP Lett. **18**, 34 (1973).

37. A. C. Melissinos, S. P. Davis, Phys. Rev. **115**, 130 (1958).

38. R. V. Ambartzumian, V. S. Letokhov, IEEE Jour. Quant. Elect. QE-7, 305 (1974); Appl. Optics **11**, 354 (1972).

39. V. S. Letokhov, Science **114**, 4 (1973).

40. PHYSICS TODAY, September 1974, page 17.

41. C. A. Mead, Phys. Rev. **143**, 990 (1966).

□

Notes added:
[a] Other reactions may be used to generate a long-lived isomer.
[b] Stimulated emission can be greatly *enhanced* for higher multipole orders.
[c] Decay during time required for line narrowing may eliminate these.

Ultrashort phenomena

Picosecond laser pulses are the key to several new techniques
for studying very rapid transient states that are being used by experimentalists
in biophysics, plasmas and condensed-state physics.

Robert R. Alfano and Stanley L. Shapiro

PHYSICS TODAY / JULY 1975

Physicists and chemists have gained an impressive amount of new information on rapid phenomena in materials by using sophisticated new techniques for handling picosecond light pulses. The reason for the great growth of interest is not solely the development of new techniques for probing this time region, but also its importance for understanding the most fundamental processes in materials physics and chemistry. Since particle-collision rates within materials are rapid, the picosecond time scale is the appropriate gauge for a larger number of energy-transfer processes in substances.

The attack by picosecond technology on such diverse areas as photosynthesis, vision, laser fusion and exciton decays thus basically stems from the simple fact that it is necessary for obtaining first-hand information on energy-transfer processes in materials. Any other method of attack, such as measuring the frequency spectrum, yields second-hand information and often will not substitute for direct measurement in the time domain. Figure 1 shows a dual-beam carbon-dioxide laser amplifier that is used for laser fusion at Los Alamos Scientific Laboratory with picosecond diagnostics to help interpret the results of experiments.

If there is an historical analogy to the present revolutionary time domain de-velopments in picosecond technology along with the new practical applications, we must look back fifty years to the discovery, in 1922, of the Raman effect, which led to the general method of identifying and understanding the structure of materials from spectral lines in the frequency domain.

This article briefly describes some recent rapid relaxation experiments in liquids, solids and plasmas, so that the reader can glimpse the type of work going on and get a feeling for the progress and direction in each field. It should become apparent that, although significant strides are being taken, the advancement is still in the beginning stages—indeed, in some areas the surface has hardly been scratched—and often the measurements themselves suggest new and exciting areas for further study.

The utilization of the mode-locked laser that emits light pulses of picosecond (10^{-12} sec) duration is the key to the study of rapid processes. In the techniques used to study the time development of rapid phenomena, the material is first disrupted from equilibrium by optical excitation with short intense light pulses, and then the rate of return of the nonequilibrium state to the normal state is studied by one of three basic techniques: the optical Kerr gate, a probe-beam technique and the streak camera. These techniques are described in the box on page 115. In this review article we will discuss the recent breakthroughs produced by measuring rapid phenomena with the new techniques in biophysics, plasma physics, solid- and liquid-state physics and chemical physics.

Biophysics

Two of the most important biophysical processes involve light—the photosynthetic processes responsible for life on Earth and, of course, the photovisual processes responsible for vision. Each process begins with light triggering the first step, and in both cases energy supplied by the light is absorbed and rapidly transferred via nonradiative processes in complicated molecules. Because of the importance of the photosynthetic and photovisual processes to us all, it is no surprise that scientists have already vigorously attacked the initial transfer and redistribution of light energy within chlorophyll and vision opsin molecules with picosecond technology, and the results appear to be exciting. We believe the primary energy-transfer steps in both processes may well be unravelled by the end of the decade.

It has long been established that the "vision" molecules undergo changes in molecular configuration as part of the visual process. Previous researchers had interpreted spectral changes observed when a vision molecule, rhodopsin, is optically excited in terms of the formation of intermediaries such as prelumirhodopsin, lumirhodopsin and metarhodopsin. Using picosecond pulses, George Busch, Meredith Applebury, Angelo Lamola and Peter Rentzepis found new evidence that the formation of prelumirhodopsin is extremely rapid. They excited rhodopsin obtained from

Robert R. Alfano is an associate professor of physics at the City College of the City University of New York, and Stanley L. Shapiro is a staff member of the laser division of Los Alamos Scientific Laboratory.

In this dual-beam carbon-dioxide laser amplifier being developed for experiments on laser fusion, two laser beams will enter the front windows to be amplified. Picosecond diagnostics help clarify the development of plasma interactions in the target.

Figure 1

the retinas of cattle eyes with a picosecond pulse at 5300 Å and immediately upon excitation they observed an increased absorption at 5600 Å. The increased absorption was interpreted as due to the creation of prelumirhodopsin with a time of formation shorter than 6 picosec, the pulsewidth of excitation. Since the formation of prelumirhodopsin is so rapid, one conclusion must be that no major structural change takes place between rhodopsin and prelumirhodopsin, for there is scarcely enough time for this to happen. A transformation involving only restricted changes in the geometry of the retinyl group and its local environment is more probable. The sequence of molecular photoproducts produced in the visual process is not known with certainty and picosecond pulse technology should be readily applicable for identifying the precursor molecules.

In photosynthesis a most important process is the energy transfer in the very early stages of the primary photophysical process from absorbed light quanta to chemical reactions. This energy transfer is so efficient that serious proposals have been made to convert solar energy into electricity or highly transportable high-energy fuels by photosynthetic processes. A second goal is the production of huge quantities of biological materials to feed humans and animals. An enhancement of solar-conversion efficiencies may be obtainable by thoroughly understanding the fundamental processes involved. In green plants chlorophyll a and other accessory pigment molecules absorb light and

sensitize two different photoreaction systems, PS I and PS II. Within a photosynthetic unit, resonance mechanisms transfer the optical excitation energy from a relatively large number of pigment molecules, including chlorophyll, to a few specialized molecules at the reaction center. This initiates a sequence of electron-transfer reactions that efficiently store the light energy. The rapid transfer within photosynthetic units can be monitored by the fluorescent emission and light absorption as a function of time from the chlorophyll and accessory pigment molecules. Several processes, such as fluorescence, nonradiative energy transfer and further chemical reactions, affect the migration of excitation energy in times shorter than 10^{-9} sec. Therefore it is extremely important to probe the subnanosecond region of photosynthesis *directly* to obtain direct experimental evidence for the fundamental processes in photosynthetic kinetics.

During the past fifteen years many groups have estimated the mean fluorescent lifetimes of photosynthetic units and have calculated the transfer time from antenna pigment molecules to the reaction center. However, the resolution is limited, since flash-photolysis pulse widths have been a few nanoseconds. Using the Kerr optical-gate technique with a resolution of 10 picosec, Michael Seibert, Alfano and Shapiro measured the time dependence of the fluorescent emission from chlorophyll. They found a fluorescence envelope with a 320 ± 50 psec decay time for *in vivo* chlorophyll in escarole chloro-

plasts. An interesting and surprising result is an nonexponential time-decay profile for the fluorescence. A theory attributing the nonexponential fluorescence to two independent photosystems has been recently proposed; this is in accord with a presently accepted two-photosystem model of photosynthesis. Supporting experiments on isolated PS I and PS II fractions obtained from spinach chloroplasts by William Yu, Philip Ho and Alfano yielded a fluorescent decay time of 60 psec for isolated PS I particles and 210 psec for PS II particles; the rise time was less than 10 psec for both systems. Further experiments by Anthony Campillo, Victor Kollman and Shapiro show that the nonexponential decays are observed in other systems and can be produced by nonradiative dipole–dipole transfer processes as predicted by Theodor Förster; see figure 2.

Additional picosecond work on photosynthesis has been done by Rentzepis, Tom Netzel and Jack Leigh, who used an absorption-probe technique to estimate an upper limit of 7 psec for the time to transfer energy from a particular pigment (bacteriophytin) to a reaction center in bacteriochlorophyll in a bacterium called *Rhodopseudomonas spheroides* strain R-26.

Plasma physics—practical applications

The most pressing problem facing technology today is the production of inexpensive energy. The controlled release of energy from nuclear reactions in a laser-produced plasma provides a possible source. How is picosecond

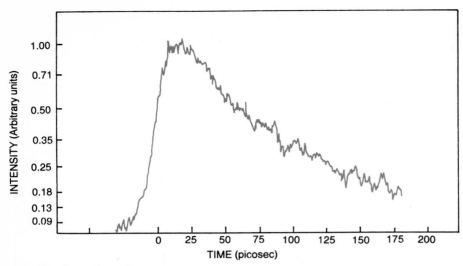

The decay of fluorescence from the alga *Anacystis nidulans* is nonexponential, as shown by this streak-camera photograph. An appropriate model for the first steps in photosynthesis, indicated by this result, involves the absorption of light by pigment molecules acting as antennas, followed by the migration of energy to a reaction center. Figure 2

technology used in this effort? In two ways: Picosecond pulses must be generated and amplified to obtain the energy and power levels necessary for testing the feasibility of laser fusion, and they are used to study the time-dependent interactions of the plasma with an intense light beam. These time-dependent interactions include formation of a hot plasma with a laser pulse, the compression of a small fusion pellet and the nonlinear interactions of the intense light pulse with the plasma. A detailed understanding of all the time-dependent processes is necessary for the proper design of the targets. To study these details it is necessary to probe on a shorter time scale than the pulse du-

ration, which is typically 30 to 1000 psec. Observations of the intensity of the scattered light and the spectral frequency shift as a function of time yield information on these time-dependent processes.

Recently, Barrett Ripin of the Naval Research Laboratories and his colleagues have applied picosecond diagnostic techniques to clarify some of the time-dependent processes originating in the targets. It had been noticed that a fraction of light was scattered back from the target, and stimulated Brillouin scattering (the scattering from an ion–acoustic wave) was proposed as the mechanism. It was important to clarify the nature of the interaction since light

scattered backwards is energy lost and unavailable for compressing the target. Ripin's group observed the backscattered light with a streak camera and was able to show that the envelope of the backscattered light does not follow the envelope of the incident pulse but that, after a certain interval, the backscattered light builds up suddenly, in less than 20 psec. The facts that this light process has a threshold and that it grows exponentially in intensity were consistent with the equations for stimulated Brillouin scattering. The picosecond diagnostics appeared to be absolutely necessary for clarifying this phenomenon.

The time variation of the spectrum is also very interesting. Initially the light scattered from the target is shifted toward the red, indicating perhaps that stimulated Brillouin scattering is the responsible mechanism, but at later times the spectrum shifts toward the blue and is numerically consistent with a Doppler blue shift due to motion of the plasma normal to the target surface.

Another important application is by Nicholas Bloembergen and his coworkers, who have effectively measured plasma-formation times in materials. According to a model proposed by Eli Yablonovitch and Bloembergen, avalanche ionization is often responsible for the breakdown of materials when laser pulses are propagated in materials. Within a small volume of material there are always a few initial free electrons present, and these may be built up exponentially by an avalanche process in the presence of a high-power laser beam. Once the avalanche begins, dense plasmas can be generated.

The avalanche-ionization model has been verified by measuring the field strengths at which optical damage was initiated in sodium chloride with laser-pulse durations of 15 and 300 psec. The field strength required for optical damage increases by almost one order of magnitude from 10^6 V/cm at 10^{-8} sec to more than 10^7 V/cm at 1.5×10^{-11} sec, in agreement with the electron-avalanche breakdown mechanism. Yablonovitch has also produced ultrashort pulses in gases that are breaking down rapidly.

Plasmas have also been optically generated in crystals with picosecond pulses by David Auston and Charles Shank of Bell Laboratories, who measured the time evolution of the plasmas with an ellipsometry technique. In this technique a weak pulse probes the plasma generated in germanium by an intense excitation pulse. This weak probe pulse passes through a polarizer and quarter-wavelength plates, reflects off the germanium crystal and then passes through an analyzer. When no intense excitation pulse is present, a tiny signal is detected because of the incomplete

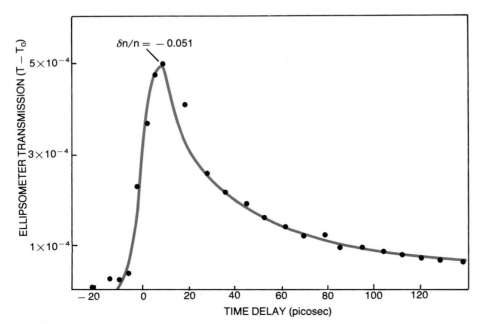

An intense picosecond light pulse in a germanium crystal generates a plasma that diffuses into the bulk medium. The change with time of the refractive index of the material is monitored by a picosecond probe, as described in the box. (From D. H. Auston and C. V. Shank). Figure 3

Three basic techniques for measuring picosecond events

Probe technique[1]

An intense pulse excites a sample. A weaker probe pulse derived from the initial pulse is reflected along a different and variable optical delay path to examine a parameter of the system such as light absorption or scattering as a function of delay time after excitation. Probe pulses can be obtained from harmonic generation, stimulated Raman scattering and self-phase modulation. The latter greatly expands, the bandwidth to the so-called "picosecond super continuum."

Optical-gate technique[2]

Light emitted from a sample can be examined with this technique. The emitted light is passed through an optical shutter composed of a carbon disulfide cell situated between a crossed polarizer and analyzer. No light is transmitted through the shutter unless an intense picosecond pulse is directed through the CS_2 cell. This intense pulse causes a short-lived birefringence due to orientation of the CS_2 molecules, and allows light from an event to be transmitted through the analyzer for a brief interval. By delaying the intense picosecond pulse through the CS_2 cell with respect to an event, different sections of the emitted light time history can be carved out for examination and an entire intensity profile of the event with time can be obtained.

Streak-camera technique[3,4]

Light produced from an ultrafast event enters a slit and is focussed onto a photocathode where electrons are released in proportion to the light intensity. These electrons are accelerated through an anode and are then deflected by a rapidly applied voltage ramp. The increasing voltage with time in the ramp streaks the electrons across a phosphorescent screen so that electrons released at different times strike the screen at different positions. The phosphorescent track is then photographed.

1. J. W. Shelton, J. A. Armstrong, IEEE J. Quantum Electron. **3**, 696 (1967).
2. M. A. Duguay, J. W. Hansen, Optics Comm. **1**, 254 (1969).
3. D. J. Bradley, B. Liddy, W. E. Sleat, Optics Comm. **2**, 391 (1971).
4. M. Ya. Schelev, M. C. Richardson, A. J. Alcock, Appl. Phys. Lett. **18**, 354 (1971).

extinction of the two polarizers. When the intense excitation pulse generates a plasma, a large signal is measured at a silicon detector; this is due to the change of refractive index caused by generation of a plasma that leads to a phase change, inducing an elliptical-polarization state.

The transmission as a function of delay for this method is shown in figure 3. The fall time is a measure of the change of refractive index as the carriers diffuse into the bulk medium. The peak change in refractive index, -0.05, was in excellent agreement with a simple Drude model of oscillators with a natural frequency of zero, for a free electron–hole gas. The decay time was consistent with a linear equation for ambipolar diffusion (electrons and holes diffusing together) provided a diffusion constant about 3.5 times that expected for a low-density plasma is chosen. At the high densities of plasma generated in their experiment, about $10^{20}/cm^3$, this higher diffusion constant could be predicted on the basis of a simple model that takes the degeneracy of the plasma into account.

Solid state physics

In our opinion picosecond pulses will continue to find application in plasma physics because theoretical calculations are often insufficient for predicting the highly complex behavior of plasmas.

Many electrical, optical and thermal properties of crystals are ultimately determined by the underlying physical phenomena produced by interactions among and between electrons, holes, phonons, excitons and other fundamental excitations. Most microscopic processes involving these excitations occur on a subnanosecond scale. Included among important lifetimes measured recently are those of the phonon, exciton and polariton.

The field of exciton interactions is an area in which the introduction of picosecond technology has proven particularly revolutionary. It is now possible to create high exciton densities in a time comparable with the exciton-relaxation time. This is a key element for, as has been pointed out in the pioneering work of Herbert Mahr and his associates, many-body effects become important. At high densities, the excitons strongly interact with one another. They collide with each other, form excitonic molecules and have many unusual properties.

An excitonic molecule is formed by the covalent coupling (like that of a hydrogen molecule) of two excitons. Mahr's group has discovered new emission bands attributable to the formation of excitonic molecules, to exciton collisions and to collisions of molecular excitons. Both they and Shigeo Shionoya and his coworkers have used a picosecond optical Kerr gate to study the time dependence of the luminescence from excitons. Their work has provided formation and decay times of the free exciton and the excitonic molecule, and collision times between two free excitons and between two excitonic molecules.

Perhaps the most striking use of picosecond pulses is the possible observation of Bose condensation of excitons. Excitons can be regarded as bosons since they are composed of an electron and hole bound together. It has been predicted that if excitons are produced rapidly at high enough concentrations, about $10^{17}/cm^3$, they may condense to a state of zero momentum at a suitably low temperature.

The luminescence spectrum of cadmium selenide at 1.8 K and 4.2 K exhibits an extremely sharp luminescence line under picosecond laser excitation on the lower-energy side of the spectral profile. This line was produced within a limited excitation range and the line shape was consistent with a theoretical Bose condensation model. Sharp lines can also be produced by impurities and stimulated emission processes—but the most intriguing interpretation is Bose condensation, in which the picosecond pulses would rapidly generate the required high density of excitons without heating the lattice.

Another example in which the optical phonon plays an important role is the exchange of energy in many solid-state processes. In a pure crystal, due to the anharmonic nature of the potential-energy curves of the lattice, phonons decay by scattering with other phonons in a crystal or by breaking up into multiple phonons, as depicted schematically in figure 4. The dominant decay mechanism for phonons at high temperature is collision with thermal phonons, while at low temperature a phonon spontaneously decays into two or more phonons. Using picosecond-pulse probe techniques, we found the lifetime of the 1086-cm^{-1} optical phonon of calcite to be 8.5 psec at room temperature and 19 psec at 100 K while Wolfgang

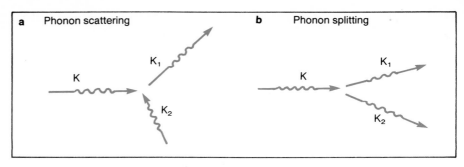

Phonon-decay processes due to lattice anharmonicity. At room temperature, phonons decay by colliding with the many other phonons present in the lattice (a), while at low temperatures the phonons decay spontaneously by splitting (b), with the coupling provided by the anharmonic crystal. Optical phonons have greater lifetimes at low temperatures.   Figure 4

Kaiser and his coworkers estimated the lifetime of optical phonons in diamond to be 2.9 psec at room temperature and 3.4 psec at 77 K. The lifetime lengthens at lower temperature because there are fewer thermal phonons present in the lattice to collide with the generated phonons.

Another recent experiment demonstrates interactions of particles not among themselves but rather with an impurity. Impurities are sometimes important in solid-state devices, often hindering their operation but in some cases playing an essential role. Dietrich von der Linde and Auston of Bell Labs have studied the dynamics of an optical rectification device with picosecond pulses. Optical rectification occurs when an optically excited impurity atom in a crystal decays nonradiatively and emits phonons that heat the lattice, thereby causing pyroelectric polarization in some crystals. The nonradiative decay is rapid and plays a key role in the development of the induced polarization.

To study this rapid transfer, a Cu^{2+} impurity ion in a lithium tantalate lattice was first excited with an intense picosecond pulse. The energy transfer to the lattice was then obtained by monitoring the decay of the populated state by measuring the absorption from the excited state to a higher charge-transfer state of a second probe pulse at 0.53 microns. The absorption decayed exponentially as a function of the delay between the excitation and the probe pulses. The lifetime ranged from 450 psec at 22 K to 10 psec at 423 K, which demonstrates the role played by phonons in this device.

Recently Auston has developed a silicon opto-electronic switch in which picosecond light pulses generate and manipulate electrical signals with picosecond precision. His invention opens up the possibility of a new high-speed electronic technology.

Liquids

Picosecond-probe techniques have also provided a great deal of insight on

how molecules interact in the liquid state. Molecular vibrations in a liquid have a depopulation time associated with transferring the energy internally to other vibrational, rotational and translational modes within the molecule or to other molecules. A second relaxation time, called a "dephasing time," is the duration for molecules that are vibrating together coherently to go out of phase relative to one another.

The equations governing the decay of both the population and the amplitude of molecular vibrations are similar to the Bloch equations that govern magnetization decay. The depopulation time is analogous to T_1 and the dephasing time to T_2 in the Bloch equations. Experimentally these two times can be conveniently separated. To measure the depopulation time of a vibration the probe is beamed on the vibration at an arbitrary angle of incidence and the light scattered from the vibrations is detected at an arbitrary angle. The amplitude of the scattered light depends only on the population. To measure a dephasing time for coherent vibrations one must observe the scattered probe beam at an angle such that conservation of momentum is preserved between the momentum vectors of the involved photons and the coherent molecular vibration. Albert Laubereau, von der Linde and Kaiser measured the decay of polyatomic molecules in the liquid phase. For ethanol they found that the depopulation time was 20 psec, with the dephasing time about 80 times shorter.

When the vibrations depopulate, new vibrations are formed. By examining the spectrum of the probe Raman light scattered from the vibrations, the dynamics of the formation and decay of these new daughter vibrations can be determined. Our first vibrational decay route measurement demonstrated that the methyl (CH_3) vibration at 2928 cm^{-1} in ethanol (CH_3CH_2OH) decays into daughter vibrations at 1464 cm^{-1}, where there are two vibrational modes. From the amplitude of the spontaneous scattering off these newly created vibrations we determined that this must be the main decay route in ethanol for the 2928-cm^{-1} vibration.

Another interesting measurement, by Laubereau, L. Kirschner and Kaiser showed that in a liquid mixture of trichloroethane (CH_3CCl_3) and deuterated methanol (CD_3OD), the 2938-cm^{-1} vibration of trichloroethane can de-excite by transferring its energy to the 2227-cm^{-1} methyl vibration in deuterated methanol. This mixture was chosen because of the possible three-quantum resonance energy transfer between the normal modes:

$$2938 \, cm^{-1}(CH_3) \longrightarrow$$
$$2227 \, cm^{-1}(CD_3) + 710 \, cm^{-1}(CCl)$$

The measured intensity of the daughter vibration is high, indicating that efficient transfer has occurred. Furthermore, by measuring the decay times as a function of concentration, two- and three-body collisional mechanisms were established for the exchange of vibrational energy by Laubereau and his associates and by Alfano.

Chemical physics

Picosecond pulses have already found numerous applications in chemical physics. They have been used to observe ultrafast chemical reactions, new short-lived transient states, "cage effects," rapid nonradiative decay processes, and vibrational and rotational decay constants of molecules in liquids. We will mention only a few of the more recent results.

Recently Tung Chuang and Kenneth Eisenthal of the IBM Research Laboratory (San Jose) have for the first time directly measured the rate of formation of a charge-transfer complex. These complexes arise from the transfer of an electron from a donor molecule to an excited-state receptor molecule to form a stable complex, as represented schematically by

$$A + D \xrightarrow{\hbar\omega} A^* + D \xrightarrow{k_{CT}} (A^- \!\!-\!\! D^+)$$

where k_{CT} is the rate constant for charge transfer. In Chuang and Eisenthal's studies the complex was formed by mixing anthracene (the acceptor molecule A) and N,N'-diethylaniline (the donor molecule D) in n-hexane solvent and exciting anthracene with a picosecond pulse at 3472 Å. Upon formation of the charge complex a new absorption is detected at 6943 Å because of the transition

$$(A^- \!\!-\!\! D^+) \longrightarrow (A^{-*} \!\!-\!\! D^+)$$

Since the absorption coefficient at 6943 Å is a measure of the number of charge-transfer complexes formed, the dynamics of charge-transfer complex formation may be obtained by monitoring the new absorption as a function of the delay time from excitation. Chuang and Eisenthal found that at the earliest and intermediate times transient terms such as those in the Smoluchowski diffusion model or the Noyes molecular-pair model are evident in their data. At longer times (≥ 300 psec), their results show that k_{CT} is close to the steady-state diffusion rate constant. Thus for the first time they were able to observe the complete transient behavior predicted by models of reaction kinetics.

Chuang and Eisenthal along with Geoffrey Hoffman have also made picosecond studies of a "cage effect" in liquids. Iodine atoms in a solvent are generated by dissociating I_2 by photoexcitation with an intense picosecond pulse at 5300 Å. In the liquid phase, unlike the gas phase, these iodine fragments have a high probability of recombining with their original partners, because the surrounding solvent molecules effectively form a cage and interfere with the escape of the iodine fragments. The number of iodine molecules present at any given time after excitation was monitored by measuring the absorbance of a weaker 5300-Å pulse, which was split off from the main pulse and variably delayed with respect to the photodissociating strong pulse.

The results show that the transmission of the probe pulse through a 10^{-2}-molar solution rises quickly, demonstrating immediate dissociation, and then decays in 140 psec, showing geminate recombination (recombination with original partners) of the iodine atoms. A residual transmission that becomes evident 800 psec later shows that some iodine atoms have escaped their original partners and slowly recombine via diffusion-controlled reactions with other iodine atoms. The data are in reasonable agreement with a theoretical model due to Noyes, which is based on the probability of encounter of fragments in a random walk. From these types of experiments useful information is obtained—not only on reaction kinetics of the species in liquids, but on the liquid state as well.

Impressive experiments have been recently carried out by Robin Hochstrasser's group on energy transfer between singlet and triplet states of dye molecules in various liquids. Measurements of the buildup and decay time of triplet–triplet absorption at 5300 Å of dyes have been made during and subsequent to the excitation of a singlet state of a dye by a 5-psec pulse at 3545 Å. One of the most interesting experiments of the series is for benzophenone dye dissolved in piperylene solvent. In this case there is a good overlap between the triplet states of benzophenone and piperylene. Energy transfer between the triplet states of both dye and solvent can readily occur. Hochstrasser and other members of his group found that triplet–triplet absorption of benzophenone builds up in 7 psec and then rapidly decays in 9 psec. This rise and decay indicates a very rapid singlet–triplet internal conversion in benzophenone dye and a rapid intermolecular triplet–triplet energy transfer between the dye and solvent.

Orientational relaxation times for molecules in liquids were first measured directly by Michel Duguay and John Hansen, and by Eisenthal and Karl Drexhage using picosecond techniques. Recently Shank and Eric Ippen of Bell Laboratories have used subpicosecond pulses (0.7 psec) generated by passive mode-locking of a CW dye laser together with an optical Kerr shutter to mea-

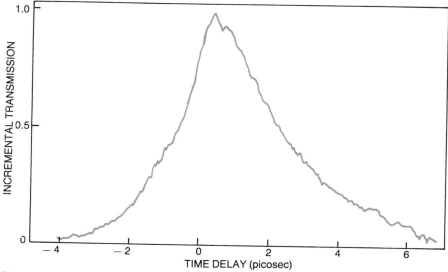

The orientational relaxation time of carbon disulfide molecules in the liquid phase is shown by this graph (after deconvolution) to be just 2.1 picoseconds. This result indicates that ultrashort lifetimes may be measured accurately with the optical-gate technique using CW mode-locked lasers, described in the box on page 33. (From E. P. Ippen and C. V. Shank.)   Figure 5

sure an orientational relaxation time (after deconvolution) of 2.1 psec in carbon disulfide, as shown in figure 5. This result demonstrates the capability of the subpicosecond laser for accurately measuring events of extremely short duration. The availability of a continuous source of subpicosecond pulses facilitates certain short-pulse experiments by allowing the use of averaging techniques required at low signal levels.

A streak camera with a resolution in the 10-psec range has been used to measure the time evolution of fluorescence from dye-vapor molecules by Shapiro, Campillo and Ronald Hyer of Los Alamos Scientific Laboratory. A pulse at 0.353 microns pumps samples of perylene and dimethylpopop molecules to an excited state, and the fluorescence generated by the pulse is projected into the streak camera. The fluorescence rises instantaneously (in less than 20 psec) indicating a rapid deactivation of the excited state of the molecule. This measurement is important because it shows that large molecules can relax in the absence of collisions. Their rapid deactivation is possible because they are so large (50 atoms) that their numerous vibrational modes allow them to relax internally by multi-vibrational processes.

Daniel Bradley, a coinventor of the streak camera, has now reported the development of a streak camera with a demonstrated time resolution of 0.5 psec. This very high resolution should be very important for uncovering phenomena in the subpicosecond domain.

Finally, we mention a new development due to Duguay and coworkers who have invented an ingenious picosecond sampling oscilloscope and have already used this device to measure picosecond x-ray pulses. The components of their device are very inexpensive and should be useful to many experimenters in the field.

We look forward to an increased use of laser technology with picosecond and subpicosecond times by experimenters whose interests have led them to study the fundamental processes of nature.

* * *

We gratefully acknowledge the support of the US Energy Research and Development Administration, the National Science Foundation, and AFOSR.

Bibliography

- P. M. Rentzepis, Science **169,** 17 (1970).
- A. De Maria, W. Glenn, M. Mack, PHYSICS TODAY, July 1971, page 19.
- R. R. Alfano, S. L. Shapiro, Scientific American, June 1973, page 42.
- A. Laubereau, W. Kaiser, Opto-electron. **6,** 1 (1974).
- N. Bloembergen, IEEE J. Quantum Electron. **10,** 375 (1974).
- K. B. Eisenthal, Acc. Chem. Res. **8,** 118 (1975). □

Sub-picosecond spectroscopy

Examples of ultrafast photoprocesses now being studied
include vibrational dynamics of molecules, primary photobiological mechanisms
and electronic processes in semiconductors.

Erich P. Ippen and Charles V. Shank

PHYSICS TODAY / MAY 1978

Ever since Man first became interested in movements and events that occurred too rapidly for his eye to follow, short flashes of light have been used to isolate moments of time. Each time shorter light pulses have become available, new areas of research have opened up, and new types of ultafast processes have become amenable to study. Spark photography can freeze the most rapid movement of macroscopic objects with flashes a tenth of a microsecond in duration. High-speed flashlamps and electronics play an important role in the study of fast photophysical and photochemical reactions with a resolution of a tenth of a nanosecond—10^{-10} sec. The last decade has seen dramatic advances in the development of short laser pulses and their application to the study of picosecond phenomena. Now, the extension of this technology into the sub-picosecond (10^{-13} sec) regime offers exciting possibilities for accurate studies of previously unresolved, ultrafast photoprocesses in physics, chemistry and biology. Investigations are being extended to such diverse and important topics as the vibrational dynamics of molecules and lattices, primary photobiological mechanisms and picosecond electronic processes in semiconductors.

Picosecond optical pulses became a reality about five years after the invention of the laser. In 1966, pioneering work at United Aircraft (now United Technologies) by Anthony DeMaria, D.A. Stetster

The authors are at Bell Laboratories, Holmdel, N.J., where Erich Ippen is a member of the physical-optics and electronics research department and Charles Shank is head of the coherent-wave physics department.

and H. Heynau[1] with a nonlinear absorber inside a flashlamp-pumped Nd:glass laser first led to the generation of pulses less than 10 picosec in duration—too short to measure by conventional electronic means. This work provided the stimulus for the rapid development of a variety of new methods for pulse measurement and diagnostics. In the decade that followed, these techniques were refined and new pulse sources were invented to provide even shorter pulses.[2] The dramatic progress made during this period is illustrated by the plot in figure 1 of available pulse duration versus year. The Nd:glass laser remained the workhorse of the picosecond field during this period. A well-designed system can provide pulses in a range about 4 picosec with a certain degree of reliability. This laser provides the high peak powers important to many applications; but it is inherently a very low repetition-rate system, and it is subject to statistical fluctuation from pulse to pulse.

The passively-modelocked, flashlamp-pumped dye laser, invented by W. Schmidt and Fritz Schäfer[3] at the Max Planck Institute at Göttingen and successfully developed by Dan Bradley and his co-workers at Queen's University, Belfast[2] and later at Imperial College, London, provides wavelength-tunable picosecond pulses with durations down to about 2 picosec. It is still a flashlamp system with fluctuation and low repetition rate, but it generates pulses in a way that is fundamentally different from that of the Nd:glass laser. It was this difference that made possible the next generation of picosecond pulse sources: passively modelocked, continuously operated

dye lasers. With the first operation of a continuous (cw) system[4] at Bell Labs, we reported pulses as short as 1.5 picosec. Subsequent improvements have led to systems that can produce pulses as short as 0.3 picosec.[5,6]

Perhaps even more important than the sub-picosecond pulse durations are the reproducibility from pulse to pulse and the very high repetition rates (greater than 10^5 pulses per second) obtained with cw systems. Powerful signal-averaging techniques can now be applied to picosecond studies. Measurements can be made with low-intensity pulses that do not distort the process under investigation. Recent experiments in our laboratory also indicate that a cw sub-picosecond pulse generator can provide the foundation for a high-power pulse source. Pulses from the generator can be amplified to high intensity in a series of laser-pumped dye amplifiers without significant change in pulse duration. It will become increasingly difficult for conventional flashlamp lasers to compete with such a system. In this paper we will therefore concentrate on the characteristics of the cw sub-picosecond laser, techniques being developed to take advantage of it, and some applications.

Sub-picosecond pulse generation

The process of sub-picosecond pulse generation in a dye laser is an interesting phenomenon in itself. With a proper adjustment of parameters it can occur without the presence of any picosecond recovery time in the dye media. Geoffrey New[7] first described the conditions under which dramatic pulse shortening occurs, and Hermann Haus[8] has obtained analytical solutions for steady-state pulse shape and stability in such systems.

Pulse narrowing in a single pass through the laser may be discussed qualitatively with the help of figure 2. Two dyes are involved in the process. A slowly recovering saturable absorber acts to steepen only the leading edge of the pulse. Saturation of the gain dye in combination with linear loss discriminates against the trailing edge. One condition for short pulse formation is simply that the absorber saturate more easily than the gain; this allows net gain at the peak of the pulse and loss on either side. The process

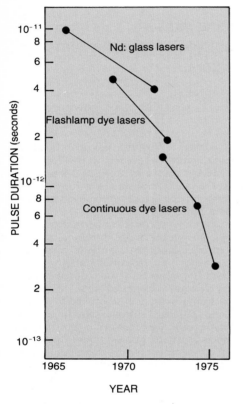

is repeated as the pulse makes multiple passes through the laser. Both dyes substantially recover during the several nanosec between passes. Shortening continues until dispersive effects prevent the pulse from becoming narrower. This situation contrasts with the transient build-up in the Nd:glass laser, in which the gain cannot recover in a pulse round-trip time and a fast recovering saturable absorber is required to select the pulse from noise. Operation of the dye laser is continuous, with the result that the pulses always reach a reproducible, steady-state shape.

Figure 3 is a schematic illustration of a sub-picosecond laser system that was used in the experiments described here. The system is pumped by several watts of continuous power from a commercial argon-ion laser. The dye gain medium, Rhodamine 6G in a free-flowing thin stream of ethylene glycol, is located at a focal point approximately in the center of the resonator. Near one end of the resonator is a second, free-flowing ethylene glycol stream containing the saturable absorber dyes. The laser mode is focussed more tightly in the absorber stream than in the gain stream.

Sub-picosecond pulses can be obtained with the saturable dye known as DODCI alone in the absorber stream. DODCI has a measured recovery time of 1.2 nanosec, so that pulse formation must occur in the manner described above. Experiments show that the addition of a second dye, malachite green, improves the overall stability and allows operation well above laser threshold. Malachite green alone will not induce modelocking, but it does provide supplementary absorption recovery on a picosecond time scale. With this combination of absorbers the laser in figure 3 is a reliable tool for sub-picosecond spectroscopy.

An acousto-optic deflector positioned near the other end of the resonator is used to "dump" single pulses from the laser. This scheme is preferable to using the pulse output from a partially transmitting mirror—the individual pulse energy is

greater and the pulse repetition rate is adjustable. Single pulses can be dumped at rates greater than 10^5 pps to maximize the extent of signal averaging. Alternatively, the repetition rate can be reduced if the system being studied needs more time between pulses for complete recovery.

The sub-picosecond pulses produced by this laser each have an energy of approximately 5×10^{-9} joule. Beam quality is such that the pulses can easily be focussed to energy densities of several millijoules/cm^2 or photon densities greater than 10^{16}/cm^2.

Pulse measurement

New experimental techniques are necessary before these sub-picosecond laser pulses can be used to study ultrafast processes. The fastest photodetectors and oscilloscopes are simply not capable of recording events with sub-picosecond resolution. The only sensors with this resolution are the pulses themselves; so measurement techniques have been developed that use one pulse to detect another, or one pulse to probe the response of a material to excitation by another pulse. Many of the techniques that were used in conjunction with flashlamp picosecond pulse sources can now be modified and improved for high resolution studies with continuous sub-picosecond sources. Reference 2 gives a chronology and summary of these techniques.

The first step in any sub-picosecond study is a careful characterization of the pulses themselves. We show an experimental pulse-measuring arrangement particularly suited for use with high repetition rate systems in figure 4. The pulse train from the laser is divided into two beams by a partially reflecting beamsplitter. The two beams then follow different paths in a modified interferometer and emerge parallel but not quite colinear. A relative delay between the two beams is obtained by changing the length of one path with a stepping-motor controlled translation stage. The stepping-motor drive also controls a multi-channel signal averager. Data from the measurements are then stored and averaged as a function of delay between the two beams. Notice that no high speed is required of the photo-electronic detection system; all the temporal resolution in these experiments is obtained by short optical pulses and accurate mechanical control of delay. The delay time τ is related to a stage translation Δz by $\tau = 2\Delta z/c$ where c is the speed of light. Delay reproducibility can be better than 10^{-14} sec.

In our pulse measurement, we focus the two beams by a simple lens through the same region of a thin crystal of potassium dihydrogen phosphate (KDP). This nonlinear crystal is oriented in such a way that phase-matched second harmonic generation occurs at an angle bisecting the two beams only when pulses from each

Decrease in laser-pulse duration during the last decade. In this article we consider primarily the production of sub-picosecond pulses with continuous lasers.   Figure 1

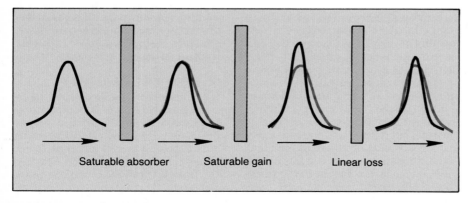

Formation of sub-picosecond pulses in a dye laser. The saturable absorber steepens the leading edge of the original pulse, and the gain dye and the linear loss combine to discriminate against the trailing edge. The process is repeated during multiple passes through the laser.   Figure 2

beam are present simultaneously in the crystal. There is no background signal from each beam individually, and accurate measurement of overlap in the pulse wings is possible. In mathematical terms, one measures the autocorrelation function $G(\tau)$ of the pulse intensity $I(t)$:

$$G(\tau) \propto \int_{-\infty}^{\infty} I(t)I(t + \tau)\, dt$$

Although the pulse shape $I(t)$ cannot be deduced uniquely from $G(\tau)$, very good estimates are possible. Furthermore, analysis of time-resolved spectroscopic measurements generally requires accurate knowledge of only the correlation function and not of $I(t)$.

The efficiency of this second-harmonic generation scheme for pulse measurement is high. An accurate trace can usually be obtained by taking the average result of very few scans. Nevertheless, an important feature, for many applications, of the experimental arrangement of figure 4 is that it allows averaging over many rapid scans to improve the ratio of signal to noise. This guards against possible distortion due to system drift during a single, slow scan.

The pulses produced by the cw generator of figure 3 have by now been accurately characterized.[5] The shortest pulses are produced at wavelengths near 615 nanometers. When they are dumped from the laser they have a duration just under a picosecond; they are asymmetric in temporal shape and are frequency swept ("chirped") in a way consistent with normal dispersion. For experimental purposes they may be compressed in time and filtered to produce Fourier transform-limited pulses with a duration of about 0.3 picosec (as shown in figure 5).

Pump–probe experiments

With slight modification, the system shown in figure 4 can be applied to a great variety of sub-picosecond studies. The sample under investigation replaces the KDP crystal, and the detector is moved to monitor the transmission of one of the beams (the probe beam) through the sample. The probe beam is made considerably weaker than the other (pump) beam. At negative delay the probe pulses precede the pump pulses through the sample and are unaffected by any action of the pump. At positive delay, the probe pulses experience increased or decreased transmission due to the excitation. The change in transmission describes the dynamical behavior of the sample. With high-repetition-rate pulses and automated delay scanning, one has in effect a sub-picosecond optical sampling oscilloscope.

Pump-probe measurements can monitor saturation, depletion and recovery of the original species. They can be applied to studies of induced absorption due to excited states, energy acceptors or photochemical products. With the addition

of polarization analysis, one can evaluate the dynamic contributions of induced dichroism or birefringence. The experimental arrangement is also easily extended to measure amplitude and polarization changes in the surface reflectivity of solids.

A measurement made over several minutes with a cw system allows averaging over more than 10^7 pulses. Changes in sample transmission as small as 10^{-4} are easily detected. With such small perturbations one avoids the distorting effects of stimulated emission and other nonlinear processes that often accompany high-power laser measurements.

Accurate interpretation of pump-probe measurements near zero delay also requires that one analyze the effects of pulse coherence. Consider the response expected from an ideal, instantaneously responding system. The transmission change $\Delta T(t)$ produced by the pump pulse $I(t)$ alone is given by the proportionality

$$\Delta T(t) \propto \int_{-\infty}^{t} I(t')\, dt'$$

if we assume that the system does not recover on the time scale of interest. After convolution with a probe pulse that has the same shape and is coherently related

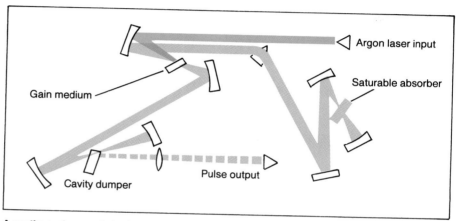

A continuously operated sub-picosecond pulse generator. Pumping (top right) is by a commercial argon-ion laser. The two dyes are Rhodamine 6G in ethylene glycol (labeled "gain medium") and the saturable absorber, which contains a mixture of DODCI and malachite green. The "cavity dumper" (lower left) is an acousto-optic deflector positioned to dump single pulses out of the laser at high repetition rate.

Figure 3

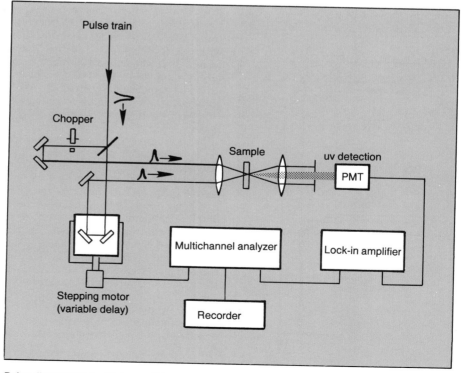

Pulse diagnostics for high-repetition-rate, sub-picosecond pulses. The stepping motor moves a translation stage to provide variable delay between two beams. Pulse measurements are made by nonlinear correlation in a sample of potassium dihydrogen phosphate, KDP.

Figure 4

to the pump, the measured response is

$$M(\tau) \propto \int_{-\infty}^{\tau} G(\tau')\, d\tau' + E(\tau)$$

The first term is the expected response if

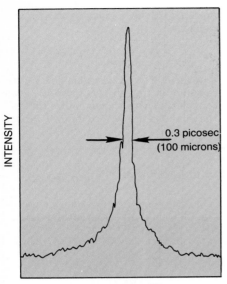

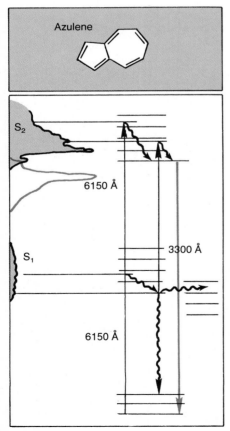

Autocorrelation measurement of a sub-picosecond pulse produced by the generator shown in figure 3 and measured by the arrangement of figure 4. **Figure 5**

Azulene, $C_{10}H_8$, has unusual fluorescence properties that can be studied in the experimental arrangement of figure 4. **Figure 6**

the pulses were not coherently related. It can be obtained simply by integration of the measured pulse autocorrelation function $G(\tau)$. The second term is a coherence artifact inherent in nonlinear spectroscopy. It is significant only when the pump and probe pulses overlap in time and interfere, so that $E(\infty) = 0$. One can also show that

$$E(0) = \int_{-\infty}^{0} G(\tau')\, d\tau' = \frac{1}{2} M(\infty)$$

Thus, $M(\tau)$ has already reached its eventual level at $\tau = 0$ and does not suffer the delay that one might expect from the finite pulsewidth. In a more general case the coherence contribution near $\tau = 0$ is affected by the dynamic response of the system, but its contribution can still be calculated for comparison with experiment.[2]

Optical gating

The concept of gating adds an additional dimension to ultrafast spectroscopic studies, allowing time resolution of emission due to fluorescence or light scattering. The most widely used picosecond gate is the optical Kerr shutter developed by Michel Duguay and his co-workers at Bell Labs.[7] In this device a picosecond light pulse induces a transient birefringence in a Kerr active material, situated between two polarizers set initially for extinction. The induced birefringence permits momentary passage of the light signal under investigation. Relative timing of the gate and the signal is controlled in the manner outlined above for pump-probe experiments.

We have operated a carbon-disulfide Kerr shutter with pulses from the cw sub-picosecond generator. High repetition rate allows a new mode of operation with polarizations biased for maximum incremental transmission, but gate resolution is limited by the 2 picosec (orientational) response time of CS_2.[8] With amplified pulses, efficient Kerr gating will be possible with faster (electronic) Kerr materials.

An alternative to the Kerr shutter for gating is frequency up-conversion. The optical signal to be sampled is mixed in a nonlinear crystal with the picosecond gating pulse. Conversion to the sum frequency occurs only in the presence of both optical fields. The advantage over the Kerr shutter is that there is less background signal in the absence of the gating pulse. Gating efficiencies can be comparable. Herbert Mahr and Mitchell Hirsch of Cornell recently employed up-conversion successfully in conjunction with a cw picosecond system.[9] More widespread application to time-resolved spectroscopy can be expected in the near future.

Frequency conversion

To expand the potential applicability of sub-picosecond spectroscopy even

further, one would like to generate pulses at a variety of wavelengths—not only to have new excitation frequencies, but also to do simultaneous probing at different wavelengths. Advances in this area rely primarily on nonlinear optical frequency conversion.

Sub-picosecond pulses from the cw dye laser can be converted into sub-picosecond ultraviolet pulses near 307 nm by phase-matched second-harmonic generation in lithium iodate. Because of the short pulse duration one must guard against pulse spreading by group-velocity mismatch between the two pulse frequencies. In $LiIO_3$ a transit-time mismatch of 1 picosec/mm between the red and the uv requires the use of a thin crystal. With a crystal thickness of about 0.2 mm, a conversion efficiency of 15% is still possible. The sub-picosecond uv pulses produced can be used in conjunction with visible probe pulses to study relaxation from highly excited states to lower intermediate levels.[10]

With high power pulses a greater number of nonlinear optical techniques[2] become possible for generating frequencies from the vacuum ultraviolet to the infrared region of the spectrum. Particularly useful is the technique known as continuum generation.[11] Recently, in our laboratory, multiple-stage amplification of sub-picosecond pulses from the stable cw oscillator has resulted in peak powers approaching a gigawatt at a repetition rate of 10 pps. Focussing these pulses into water produces efficient conversion to a continuum of light extending from the near infrared to the near ultraviolet. Measurements at selected wavelengths indicate that these continuum pulses, like the pulses producing them, have durations less than a picosecond.

Recent applications

It is worth reiterating that sub-picosecond pulse generation is reproducible with the system we have described. This reproducibility, combined with the high repetition rate, greatly extends the potential of time-resolved spectroscopy. To illustrate the use of some of the different measurement techniques, we describe now some recent experimental applications. The particular examples, by necessity limited in number, have been chosen to emphasize the interdisciplinary nature of work in this area.

Nonradiative relaxation in molecules. The kinetics of energy redistribution and excited-state relaxation play an important role in the chemistry and photochemistry of molecules. Nonradiative processes can be especially fast, and they are not always amenable to study by conventional techniques. This is particularly true of the molecule azulene, which theoretical chemists find interesting because of its unusual fluorescence properties. The molecular structure and a schematic energy diagram of azulene is given in figure

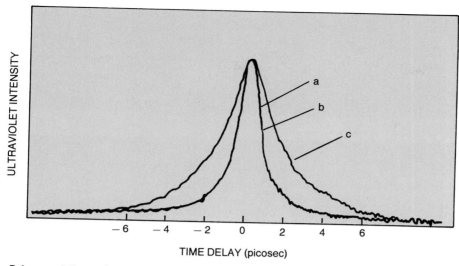

Pulse correlations. Curve **a** is by second-harmonic generation, curve **b** by two-photon fluorescence and curve **c** by two-step induced fluorescence in azulene. Curve **c** is broadened by the picosecond lifetime of the S_1 state.

Figure 7

ond-harmonic generation autocorrelation and the other by two-photon fluorescence, are indistinguishable on a 0.1-picosec scale. The third curve is clearly broadened by the S_1 lifetime. Proper analysis and deconvolution yield a simple exponential behavior with a time constant of 1.9 ± 0.2 picosec. That we obtained the same result in a variety of solvents indicates, not surprisingly, that the surrounding solvent molecules have little influence on this time scale. More recently, we found that deuteration of the azulene has no dramatic effect. Ground-state recovery measurements made with amplified sub-picosecond pulses have also yielded a time of 1.9 ± 0.5 picosec and point to the conclusion that the lifetime of state S_1 is limited by direct coupling to the ground state. These experiments and others like them provide a basis for comparing theoretical models and will, we hope, lead to a better understanding of nonradiative relaxation.

Carrier dynamics in semiconductors. Short optical pulses provide a unique way to observe, directly, the dynamics of hot-carrier distributions in semiconductors. The experimental method is to generate a distribution of hot carriers at the surface of a semiconductor by band-to-band absorption of an optical pulse of frequency much greater than the bandgap. A second, less intense, optical pulse then probes the incremental reflectivity change at a delayed time relative to the generation. While the carriers relax by phonon emission the reflectivity is altered (due to the continuously changing occupation probabilities of different electronic states) until the carrier distribution is in thermal equilibrium with the lattice. Details of the relationship between the optical reflectivity and the carrier distribution depend on the frequency of the probing pulse and the band structure of the particular semiconductor.

The case of gallium arsenide excited with a short pulse at 4 eV and probed with a 2-eV pulse is shown in figure 8. In this example the sign of the reflectivity changes goes from negative to positive as the carriers relax through the probing pulse energy. The energy loss rate for electrons and holes can be estimated from the zero crossing to be 0.4 eV/picosec.

Biological processes. It is becoming apparent that time-resolved spectroscopy can make important contributions to the better understanding of biological molecules.[2] Although conventional measurement techniques can determine that a biological reaction has taken place, the complexity of these molecules often masks the actual atomic or molecular rearrangement that has occurred. Recent experiments indicate that many of these processes do occur with picosecond speeds. Dynamic measurements may be used to distinguish between the different possible reaction pathways; for example, we have used sub-picosecond pulses to

6. Of particular interest is the lifetime of the lowest excited single state, S_1.[12] A very low quantum efficiency of fluorescence indicates that this level is rapidly depopulated by nonradiative decay. In this case, the fact that fluorescence can be observed following excitation to high states suggests a method for monitoring the population in S_1.

The experimental arrangement of figure 4 lends itself well to this purpose. With the KDP in place, ultraviolet light is detected only when the two pulses arrive at the same time. If the KDP is replaced by a thin cell containing a solution of azulene, the situation is different. A single pulse can excite molecules to S_1 where they remain for some short time τ.

If the second (delayed) pulse arrives within this time, some molecules are elevated further to a state from which fluorescence can occur. Each individual pulse can also induce two-step excitations, but this process only gives a constant background independent of delay. By measuring the average fluorescence as a function of delay, one can deduce τ. A reference curve (one for which $\tau = 0$) is obtained by substituting a dye ($\alpha - NPO$) that has no intermediate state; the experiment then becomes simply an autocorrelation measurement of the pulses by two-photon fluorescence.

Typical experimental curves we obtained in this study appear in figure 7. Two reference curves ($\tau = 0$) one by sec-

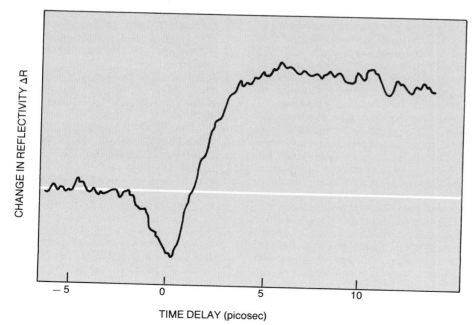

Reflectivity dynamics in gallium arsenide, excited with a short pulse at 4 eV and probed with a 2-eV pulse. The reflectivity change goes from negative to positive as the hot carriers at the semiconductor surface relax through the probing pulse energy.

Figure 8

investigate the rate of photodissociation of hemoglobin compounds,[13] and, most recently, the ultrafast dynamics of bacteriorhodopsin's photochemistry.

Bacteriorhodopsin is the object of intense research interest because of its similarities to visual pigment rhodopsins and because of its biological role as an energy converter. The absorption of light is known to induce very rapid formation of a more energetic species with a redshifted absorption maximum.[14] We have applied the pump-probe technique discussed above to resolve the dynamics of this primary step: bacteriorhodopsin→bathobacteriorhodopsin. The pump excitation is expected to result in an increase in absorption of a probe at 615 nm. The colored curve in figure 9 shows the experimental result, which is consistent with the expectation. The black curve is the calculated (and experimentally verified) response of the measurement system to an instantaneous rise. It includes the effects of coherence near zero delay. The dotted line shows an ideal 1.0-picosec response. The actual curve deviates somewhat from this simple curve, but nevertheless approaches its final value with a time constant of 1.0 ± 0.5 picosec.

This result argues against any major change in molecular configuration. It does, however, support a proposal of Aaron Lewis[15] in which the instantaneously excited electron distribution in the chromophore induces a conformational transition in the neighboring protein. Further sub-picosecond measurements are now planned to distinguish specific conformational changes.

Future directions

Sub-picosecond pulse lasers similar to that shown in figure 3 are now being put into operation in a number of different university and industrial laboratories. As they begin to be used in increasingly varied applications, existing techniques will certainly be improved and new sub-picosecond techniques will be developed. Lasers with different dye combinations will be found to generate sub-picosecond pulses in other regions of the spectrum, and we expect that pulses even shorter than 0.3 picosec will eventually be produced.

Several other sub-picosecond pulse sources based on the cw dye laser are in the early stages of development. Workers at the University of California at Berkeley[16] have obtained pulses as short as 0.6 picosec from a doubly modelocked laser. In this mode of operation the saturable absorber dye is allowed to lase in synchronism with the gain medium; short pulses are produced at two wavelengths simultaneously. Workers at Bell Labs have recently obtained independently tunable pulse outputs at two wavelengths by synchronously locking two dye lasers

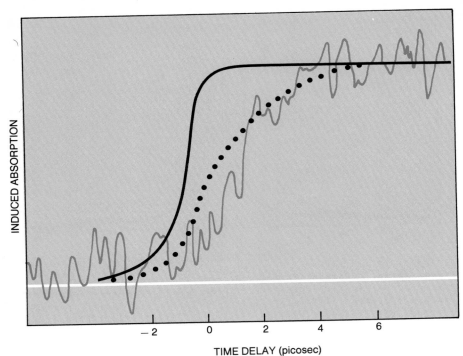

Picosecond time-delayed absorption increase in bacteriorhodopsin. Excitation by a "pump" pulse increases the absorption of a "probe" pulse at 615 nanometers. The colored curve is experimental data, resolving the delay in the photochemical reaction. The black curve is the calculated response to an instantaneous rise, and the dotted line shows a 1.0-picosecond response. Figure 9

to the same modelocked dye laser.[17] The same group produced tunable, sub-picosecond pulses by operating two synchronously modelocked lasers in tandem.[18]

Amplification of sub-picosecond pulses to high peak powers will be important for many applications. We have already demonstrated that amplified pulses can be converted efficiently to new wavelengths and that they can be used to generate a sub-picosecond continuum for studies encompassing a broad spectral range. High-power pulses can now be applied to the study of low density materials and low cross-section interactions. Even more important may be new investigations of weak nonlinear processes and highly excited systems. Because the threshold for material breakdown is higher for shorter pulses, amplified sub-picosecond pulses can be focussed to optical intensities of record magnitude.

The importance of picosecond processes and devices to both fundamental and applied research is becoming increasingly apparent. This month the first international Topical Conference on Picosecond Phenomena is being held in Hilton Head, South Carolina. Topics to be discussed include photosynthesis, vision processes, electron solvation, nonradiative molecular dynamics, hot electron effects in semiconductors, short wavelength generation, laser-induced fusion, and picosecond electronics. Important advances will certainly be made and new areas of importance uncovered now that these investigations can be pursued on a time scale of 10^{-13} sec.

References

1. A. J. DeMaria, D. A. Stetster, H. Heynau, Appl. Phys. Lett. **8**, 22 (1966).

2. See contributions to *Ultrashort Light Pulses* (S. L. Shapiro, ed.), Springer-Verlag, 1977.

3. W. Schmidt, F. P. Schäfer, Phys. Lett. **26A**, 558 (1968).

4. E. P. Ippen, C. V. Shank, A. Dienes, Appl. Phys. Lett. **21**, 348 (1972).

5. E. P. Ippen, C. V. Shank, Appl. Phys. Lett., **27**, 488 (1975).

6. I. S. Ruddock, D. J. Bradley, Appl. Phys. Lett. **29**, 296 (1976).

7. M. A. Duguay, in *Progress in Optics XIV*, (E. Wolf ed.), North Holland (1976); page 163.

8. E. P. Ippen, C. V. Shank, Appl. Phys. Lett. **26**, 92 (1975).

9. H. Mahr, M. D. Hirsch, Optic Commun. **13**, 96 (1975).

10. C. V. Shank, E. P. Ippen, O. Teschke, Chem. Phys. Lett. **45**, 291 (1977).

11. R. R. Alfano, S. L. Shapiro, Chem. Phys. Lett., **3**, 407 (1971).

12. P. M. Rentzepis, Chem. Phys. Lett. **2**, 117 (1968).

13. C. V. Shank, E. P. Ippen, R. Bersohn, Science, **193**, 50 (1976).

14. K. J. Kaufmann, P. M. Rentzepis, W. Stoeckenius, A. Lewis, Biochem. Biophys. Res. Comm. **68**, 1109 (1976).

15. A. Lewis, Proc. Natl. Acad. Sci. USA **75**, 2 (1978).

16. Z. A. Yasa, A. Dienes, J. R. Whinnery, Appl. Phys. Lett. **30**, 24 (1977).

17. R. K. Jain, J. P. Heritage, Appl. Phys. Lett. **32**, 41 (1978).

18. J. P. Heritage, R. K. Jain, Appl. Phys. Lett. **32**, 101 (1978). □

30-femtosecond light pulses at Bell Labs

Bertram M. Schwarzschild

PHYSICS TODAY / DECEMBER 1982

The development of continuously pumped, mode-locked dye lasers in the early 1970s made possible the first production of sub-picosecond optical pulses in 1974. For the remainder of the decade, attempts to generate significantly shorter pulses of light were not particularly successful. Two new developments reported early last year have now been brought together by a group at Bell Labs to produce optical pulses only 30 femtoseconds long—3×10^{-14} seconds, or 14 cycles of 6200-Å red light.

Early last year Charles Shank and his Bell Labs colleagues reported[1] the first production of optical pulses briefer than 100 fsec with a newly developed colliding-pulse ring dye laser. At a Gordon Conference on nonlinear optics that summer, Shank heard Daniel Grischkowsky of IBM report on a new technique[2] for the compression of picosecond light pulses in a single-mode optical fiber. Adapting the IBM group's method to the femtosecond regime, the Bell Labs group has now succeeded[3] in compressing the 90-fsec output of their colliding-pulse laser by a factor of three.

The development of subpicosecond pulses in the 1970s made possible the detailed investigation of previously inaccessible ultrafast processes in condensed matter (PHYSICS TODAY, May 1978, page 41)—non-radiative relaxations, configurational changes in molecules, carrier dynamics in semiconductors, fast photochemical processes in biological systems, fluorescence and the like. With pulses on the order of ten femtoseconds, one should now be able to probe an entirely new class of processes. This is the time scale of typical molecular vibrational periods in liquids. Shank expresses the hope that one will now be able to study the evolution of excitation processes in liquids and solids before they have been brought to statistical equilibrium by collisions and phonon interactions—while the system still has "phase memory."

The time resolution of such studies is of course determined by the width of the available optical pulses. One initiates an excitation with one pulse and probes it with another. It is *not* necessary to rely upon the fast repetition of successive pulses from a laser. Rather, one splits a single pulse into two components and delays one relative to the other by making it traverse a longer path to the system under study. To measure time evolution on a femtosecond scale one would need to vary the difference between optical paths with something like a stepper motor accurate on a scale of microns.

Pulse compression. The IBM group has made use of a pulse-compression scheme whose basic idea goes back more than thirty years to radar technology. To circumvent the power limitations of radar pulse-generating tubes, radar engineers sought to raise peak power by compressing pulses in time. One begins by "chirping" the initial radar pulse—imposing a linear frequency sweep on the carrier frequency. The chirped pulse is then sent down a dispersive delay line, in which the group velocity is strongly frequency dependent. If the length of the delay line is properly chosen, the trailing edge of the pulse—which had been chirped to higher frequency—will catch up with the lower-frequency leading edge, producing a time-compressed output pulse at the end of the line.

One can look at this idea from the point of view of a Fourier transform. The minimum time width of any pulse is essentially the inverse of its frequency bandwidth. Broadening the bandwidth by chirping the carrier frequency makes it possible to compress the pulse in the dispersive delay line.

For optical pulses on the order of picoseconds, one cannot chirp the carrier frequency electronically, as one does with radar. More than ten years ago F. Gires and P. Tournois in France and Joseph Giordmaine and his colleagues at Bell Labs suggested that one might compress fast optical pulses by chirping them in optically nonlinear liquids. In a strongly nonlinear liquid

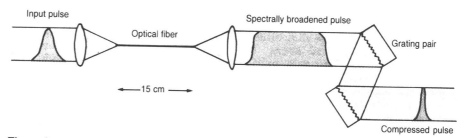

The pulse-compression technique developed at IBM has been adapted at Bell Labs to compress 90 fsec light pulses from a colliding-pulse ring laser (see cover photo) to 30 fsec. The input pulse is spectrally (and temporally) broadened by chirping in a 15-cm-long optical fiber. Subsequently, the blue-shifted trailing edge of the pulse catches up with the red-shifted leading edge in a pair of diffraction gratings that serve as a delay line, compressing the original pulse by a factor of three.

such as carbon disulfide, the index of refraction has a significant nonlinear term proportional to the light intensity. When the light passing through the liquid is sufficiently intense, its electric field becomes comparable to the intramolecular electric fields, thus inducing polarization and increasing the index of refraction. The medium is nonlinear in the sense that refraction of light depends on the intensity of the light itself. It turns out that this effect produces a time-dependent shift of the pulse carrier frequency proportional to the time derivative of the pulse-intensity envelope.

A decade of attempts to exploit this "self phase modulation" effect in nonlinear liquids did not succeed in producing reproducible, stable light-pulse compression—though some compression was demonstrated. The liquid medium posed several intractable problems, Grischkowsky explained to us: The self phase modulation tends to produce "catastrophic" self-focusing of the light pulses. Furthermore, because the beam profile in liquids is not uniform, the self phase modulation and hence the chirp will vary across the wave front, preventing optimal compression in the subsequent dispersive delay system.

Optical fibers. The IBM group believes it has found the ideal nonlinear medium to get around these difficulties—single-mode glass optical fibers. Because such fibers propagate only a single transverse electromagnetic mode, the optical wave front is quite uniform. Self-focusing is eliminated and the chirp is radially uniform. Furthermore, because the width of the fibers is so small—only a few microns—one gets adequate light intensity for self phase modulation with very modest laser power inputs.

A "happy accident," Grischkowsky told us, makes the single-mode fibers even more attractive. The dispersive effect of traversing many meters of fiber produces a long linear chirp of the proper sign (frequency increasing with time), with only small leading and trailing edges having the wrong chirp sign. Without adequate dispersion, the self-phase-modulation effect alone, being proportional to the derivative of the pulse envelope, produces large "wings" of the wrong sign in liquids, which cannot subsequently be compressed.

In their most recent demonstration[4] of this technique, Grischkowsky and Bernhard Nikolaus (also at IBM) have succeeded in compressing 5.4-psec pulses from a 1-kW mode-locked dye laser down to 450 fsec—a factor of twelve, with a threshold increase of peak power. To compress the pulse after chirping it in 30 meters of optical fiber, they pass the pulse through a dispersive diffraction-grating system

that permits the later, high-frequency end of the pulse to catch up with the lower-frequency leading edge. This is the analog of the delay line used at microwave (radar) frequencies. This diffraction-grating delay line also cancels the broadening of the pulse due to dispersion in the optical fiber.

Earlier this year Grischkowsky, and his IBM colleague Anne Balant had performed a theoretical analysis[5] predicting that such a system should produce a tenfold pulse compression, and pointing out that the clipping of the wrong-sign chrip wings by the dispersive properties of the fiber itself should leave more than 90% of the initial pulse energy in the final compressed pulse. In a nonlinear liquid, about 40% of the pulse energy would be lost to those uncompressed wings. "We think it particularly useful," Grischkowsky told us, "that we were able to demonstrate an order-of-magnitude pulse compression with a conventional, off-the-shelf, low-power, 5-psec pulsed dye laser."

Colliding-pulse laser. Passive mode locking by means of a saturable absorber is by now a well established technique for the production of picosecond pulses with dye lasers. A thin jet of saturable dye in the laser cavity serves as a highly nonlinear absorber. Because the dye is quickly bleached (saturated) by the passage of light through it, the fractional absorption is nonlinear—it decreases with increasing light intensity. Thus, as a light pulse passes through the absorber, its low-intensity wings are preferentially absorbed and the central peak is preferentially amplified, resulting in a much compressed pulse. Viewed in another way, the different resonant frequency modes in the laser cavity are locked into a definite phase relation by the thin layer of saturable absorber, thus adding coherently to produce short, intense pulses separated in time by $2L/c$, twice the transit time of a light pulse through a laser cavity of length L.

Shank and his Bell Labs colleagues Richard Fork and Benjamin Greene have developed a variant of this standard mode-locked dye laser that has by now produced light pulses as short as 60 fsec. This new device is a colliding-pulse ring dye laser that enhances the mode-locking effect of the saturable absorber by having countercirculating light pulses pass through a 10-micron-wide jet of absorber dye (and each other) simultaneously. In place of the two mirrors at the ends of a conventional laser, the ring-laser cavity contains three mirrors that send two streams of pulses around the cavity in countercirculating triangular paths. The coherent colliding pulses set up a standing-wave pattern in the ultrathin absorber jet. Because the fractional absorption

is least where the colliding pulses interfere constructively, the pulses tend to narrow and synchronize so that they both fit simultaneously into the saturable absorber jet, thus producing a stable train of pulses as short as 60 fsec.

The Bell Labs colliding-pulse ring laser is pumped by a cw argon laser, and the active gain dye medium is physically separated from the saturable absorber in the laser cavity. Gerard Mourou and Theodore Sizer at the University of Rochester have recently reported[6] the production of light pulses shorter than 70 fsec with a somewhat different mode-locked dye laser. Instead of the passive mode locking employed in the cw-pumped Bell Labs ring laser, they achieve mode locking by synchronous pumping of the dye gain medium with a pulsed YAG laser. Furthermore, the saturable absorber and gain dyes are combined in a single jet in the Rochester dye laser.

Combining the techniques. Early this year, Shank, Fork, Richard Yen, Rogers Stolen and William Tomlinson reported[3] that they had successfully adapted the IBM fiber-optic pulse-compression technique to compress 90-fs pulses from the Bell Labs ring laser down to 30fs. "This," their *Applied Physics Letter* tells us, "is the shortest optical laser pulse ever generated." The output of the ring laser is focused into a polarization-preserving, single-mode optical fiber 15-cm long, which chirps its bandwidth from an initial 60-Å spread to a final 180 Å, thus permitting a threefold pulse compression in time. The chirped fiber output is then diffracted off a pair of parallel gratings that provide a shorter optical path for the blue-shifted trailing edge of the pulse to catch up with the red-shifted leading edge.

How does one measure so short a pulse—only 14 cycles of 6200-Å red light? One uses an autocorrelation technique "that makes the pulse measure itself," as Shank describes it. The compressed pulse is split into two components, which are then directed onto a 0.2-mm-wide crystal of potassium dihydrogen phosphide by different optical paths of variable length. In the crystal, pairs of 6200-Å photons combine to generate 3100-Å ultraviolet light. Only when both pulses are simultaneously present in the narrow, frequency-doubling crystal is there sufficient photon intensity to produce an observable level of this second-harmonic generation. The ultraviolet output intensity is thus a measure of the overlap of the two pulses in the crystal. By varying the path length difference in submicron steps (light travels a third of a micron in a femtosecond) one produces an autocorrelation function of ultraviolet output intensity that yields the pulse width. A time measurement has thus been converted into a measurement of distance. One might imagine that with a pulse width of only 30 fs, dispersion in the lenses and other optical components would seriously distort the measurement. Happily, Shank explained us, the grating pair can be adjusted to cancel precisely all these unwanted dispersions.

With two stages of fiber-optic compression, Shank told us, the Bell Labs group hopes ultimately to get down to "just a few cycles of light—perhaps 5 fsec." With uncompressed 70-fsec pulses from the ring laser, the group has already performed a study[7] of the picosecond dynamics of photoexcited gap states in polyacetylene, a polymer that behaves like a one-dimensional semiconductor. Greene has investigated time-resolved induced birefringence with a time resolution of about 100 fsec.

Mourou and his colleagues have also recently employed[8] a ring laser similar to the Bell Labs instrument to probe ultrafast electrical transients in GaAs photoconductive detectors with a temporal resolution better than 2 psec, using 100-MHz train of 120-fsec pulses. At Cornell, Jean-Marc Halbout and C. L. Tang have recently used 70-fsec pulses from a similar colliding-pulse ring laser to do time-resolved observation of the orientational relaxation of molecules in liquids. Tang told us that in 1979 his group had been able to achieve a significant reduction[10] in the width of pulses from a synchronously-pumped, mode-locked, cw dye laser by replacing the usual linear laser cavity with a ring cavity.

Starting with a modest kilowatt of power and using a multistage laser amplifier, the Bell Labs group has been able to deliver *gigawatts* of peak power to various semiconductor materials with these extraordinarily brief light pulses. "It's the most intense way we have of interacting with condensed matter without destroying it," Shank told us.

References

1. R. L. Fork, B. I. Greene, C. V. Shank, Appl. Phys. Lett. **38**, 671 (1981).
2. H. Nakatsuka, D. Grischkowsky, A. C. Balant, Phys. Rev. Lett. **47**, 910 (1981).
3. C. V. Shank, R. L. Fork, R. Yen, R. H. Stolen, W. J. Tomlinson, Appl. Phys. Lett. **40**, 761 (1981).
4. B. Nikolaus, D. Grischkowsky, Appl. Phys. Lett. (Jan. 1983), to be published.
5. D. Grischkowsky, A. C. Balant, Appl. Phys. Let. **41**, 1 (1982).
6. G. A. Mourou, T. Sizer II, Optics Comm. **41**, 47 (1982).
7. C. V. Shank, R. Yen, R. L. Fork, J. Orenstein, G. L. Baker, Bell Labs (Holmdel, N. J.) preprint (1982).
8. J. A. Valdmanis, G. Mourou, C. W. Gabel, J. Quant. Electronics, to be published (1982).
9. J. M. Halbout, C. L. Tang, Appl. Phys. Lett. **40**, 765 (1982).
10. S. Blit, C. L. Tang, Appl. Phys. Lett. **36**, 16 (1980).

CHAPTER 3

LASER SPECTROSCOPY

It had been known since Newton that a light beam could be dispersed into a band of colors—a "spectrum"—by means of a prism; and around 1865 both Bunsen and Kirchhoff had shown that different chemical elements emit characteristic spectral patterns when heated to incandescence. As spectrometers were improved in the late 1800s the science of spectroscopy flourished as a separate branch of optics. In 1868 a French astronomer measured the spectrum of sunlight during a total eclipse and was unable to identify one bright yellow line. The British astronomer Norman Lockyer compared the position of this yellow line with those of various elements and decided that this new line was produced by an element in the sun that was not present—or at least had not yet been discovered—on Earth. He named this element helium, from the Greek word for the sun. In 1895 Lockyer and William Ramsey analyzed the spectra of some gas trapped in a uranium mineral and found helium on Earth. This was a widely heralded triumph for spectroscopy. Spectrochemical analysis quickly became an established tool for analytical chemists; for the physicists the challenge remained of understanding why different elements had characteristic spectra. Niels Bohr was able in 1913 to postulate a model for the spectral absorption and emission of light by the hydrogen atom, and by the middle 1920s the science of quantum mechanics was born. With each improvement in the resolution of spectrometers or the sensitivity of radiation detectors new advances in spectroscopy became possible. But now, with the availability of the laser, a new renaissance is occurring in spectroscopy: the laser itself can be tuned to sweep through a frequency interval, bringing to optical spectroscopy the high resolving power and sensitivity not formerly available at optical frequencies. Spectroscopy has now become the new field of laser spectroscopy. The articles in this chapter show the variety of applications of lasers to this exciting new field.

John N. Howard

CONTENTS

Nobel Physics Prize to Bloembergen, Schawlow and Siegbahn

Gloria B. Lubkin

PHYSICS TODAY / DECEMBER 1981

The 1981 Nobel prize in physics has been awarded to Nicolaas Bloembergen of Harvard University, Arthur L. Schawlow of Stanford University and Kai M. Siegbahn of Uppsala University. Bloembergen and Schawlow will receive half the prize "for their contribution to the development of laser spectroscopy." The other half will go to Siegbahn "for his contribution to the development of high-resolution electron spectroscopy." The prize, to be awarded in Stockholm on 10 December, this year amounts to one million Swedish kroner (worth about $180 000).

Masers. In the early 1950s, workers in the US and the Soviet Union were trying to make use of stimulated emission in molecular systems to amplify weak microwave signals and to design oscillators based on such systems. This work led to the maser, first demonstrated by Charles Townes and his collaborators in the US; at the same time the maser had been suggested by Nikolai Basov and Alexander Prokhorov in the Soviet Union.

Schawlow told us that he and Townes (his future brother-in-law) shared a hotel room at the APS Washington meeting in Spring, 1951. To avoid disturbing Schawlow's sleep one morning, Townes sat outside on a park bench and worked out the idea for the ammonia maser. Three years later, Townes, James Gordon and Herbert J. Zeiger reported the operation of such a maser.

In 1948, Bloembergen recalled as we sat in his Harvard office, he, Edward M. Purcell and Robert V. Pound published a paper on relaxation effects in nuclear magnetic resonance. In this work, they "burned a hole" in the nmr spectrum of protons in water in an inhomogeneous magnetic field and obtained a narrow dip. Later, in 1963, a closely related spectral hole-burning effect in lasers (based on an inhomogeneous velocity distribution rather than an inhomoge-

Arthur Schawlow and Nicolaas Bloembergen on the day their Nobel prize was announced. Schawlow (top photo, right) receives standing ovation from Walter Meyerhof's physics class at Stanford. Bloembergen (bottom photo, center), at Harvard press conference, shares champagne with his wife Deli and Paul Martin.

neous spatial distribution) was discovered independently by William Bennett Jr, Willis Lamb and R. A. McFarlane and independently by Abraham Szoke and Ali Javan. This so-called Lamb dip is the basis of laser saturation spectroscopy because it allows one to get rid of Doppler broadening and to determine the center of a spectral line with great accuracy. These line-narrowing phenomena in inhomogeneous systems also allow one to take advantage of magnetic resonance in organic materials and in liquids.

Work on magnetic resonance led Purcell and Pound to introduce in 1950–51 the ideas of inverted populations and negative temperatures. In 1953 Albert Overhauser found the effect that now bears his name, an effect in which you saturate the electron spin transitions in a metal and get a very large change in population distributions between the nuclear spin levels. Inspired by these discoveries, Bloembergen in 1956 developed the concept of a three-level solid-state maser. To make a steady-state inverted population, one takes advantage of the combined action of a "hot" pump and relaxation to a "cold" reservoir. Bloembergen argued that masers can be considered as a type of heat engine operated between two temperatures. The hot and cold temperatures occur in the same volume element in space, unlike that of the usual thermodynamic engines, in which there is a hot part in which fuel is burned and another part where the system is cooled.

Schawlow told us the first really useful maser was Bloembergen's proposal. Although the ammonia maser was useful for atomic clocks, it was not yet a broadband, tunable low-noise amplifier. Following Bloembergen's proposal of the three-level maser, the first such solid-state maser was demonstrated by Derrick Scovil, George Feher and Harold Seidel at Bell Labs using lanthanum ethyl sulfate doped with gadolinium and cerium. Bloembergen built a solid-state maser using $CoK_3(CN)_6$ doped with chromium. A ruby maser was built by three groups, one of them headed by Townes. Bloembergen reminisced about the occasion in 1959 when he and Townes received the Morris Leibmann award for contributions to the maser art. Townes had presented his wife, Frances, with a medallion made from the ruby he used for his maser. When Bloembergen's wife, Deli, admired the medallion and asked for a similar memento of *his* maser, he said, "Well, dear, my maser works with cyanide."

The solid-state maser had low noise and could be tuned with a magnetic field because of its unpaired spins. The multi-level pumping scheme led to the development of low-noise microwave receivers. One such maser was used by Arno Penzias and Robert W. Wilson in 1965 to discover the 3-K blackbody radiation from the Big Bang. Subsequently all lasers have used similar schemes in a combination of multilevel pumping and relaxation.

Laser spectroscopy. The Swedish Academy of Sciences notes that the idea of extending the maser principle to the infrared or optical region "arose in different quarters in the late 1950s. The wholly decisive contribution in the realization of this idea was made in 1958 by Schawlow and Townes, who then published a work analyzing the preconditions necessary for such a design, theoretical as well as practical. Prokhorov around the same time proposed a similar design for the generation of longer waves. Other suggestions based on the same idea were also presented at that time. However, it was primarily the work by Schawlow and Townes which initiated the whole dynamic field which we now associate with the concept of 'laser.'" The Academy also noted that although the 1964 Nobel prize in physics was awarded to Townes, Prokhorov and Basov for "fundamental work in the field of quantum electronics, which has led to the construction of oscillators and amplifiers based on the maser–laser principle," subsequent developments, particularly in lasers, "have made this field increasingly deserving of additional rewards."

Starting in mid-1957, Schawlow recalls, he and Townes analyzed the problems of extending the maser to shorter wavelengths. By December 1958 they had shown the possibility of making an optical maser with a structure that had mirrors on either side of the cavity. The race to build the first one was won by Theodore Maiman in Spring 1960; he used light ruby, taking advantage of isolated chromium ions. During the race, Schawlow told us, he outsmarted himself, believing that because no one had yet made a laser, it was difficult to make one work. So he tried using pairs of chromium ions, a more complex approach than Maiman used. By the end of 1960, five different lasers were operating, including Schawlow's.

The Swedish Academy notes that Schawlow and his coworkers at Stanford have emphasized the kind of nonlinear spectroscopy based on saturation phenomena, which occur in the absorption of laser light because of its high intensity. Schawlow told us that even in his 1958 paper with Townes they had mentioned that lasers could be used for spectroscopy provided one could get some degree of tunability, but narrow lines were needed for high gain in the laser. During the 1960s, he said, some spectroscopy was done to investigate the details of the same lines as those producing the laser action, such as the neon line in the He–Ne laser.

A particularly sensitive and simple method of saturated absorption spectroscopy now known as the Hänsch–Bordé method was used in 1970 by Christian Bordé in Paris and independently in Stanford by Theodor W. Hänsch, Marc D. Levenson, Schawlow and Peter Smith, following earlier experiments at the University of Heidelberg by Hänsch and Peter Toschek. The absorbing gas sample is kept in a cell outside the laser resonator. The beam is split into a strong saturating beam and a weaker probe beam, which are sent by mirrors in nearly opposite directions through the absorber. When the saturating beam is on, it bleaches a path through the cell, allowing the probe signal to be received more strongly at the detector. The probe signal will be modulated when the laser is tuned near the center of the Doppler-broadened absorption line, so that both beams interact with the same atoms. This approach eliminates Doppler broadening due to motion of the atoms.

Schawlow recalls that initially he had failed to see the advantages of using the high intensity of one laser to produce lasing action in another medium. In 1966, at IBM, Peter Sorokin made a laser from an organic dye; similar work was done by Fritz Schäfer at the University of Marburg. The organic dye laser made broadband tunability possible; the high light intensity needed to pump the dyes is most easily obtained from another laser. In 1971 Hänsch, Issa Shahin and Schawlow used a pulsed dye laser pumped by a nitrogen laser; they introduced a telescope to expand the beam and an etalon for very fine tuning. The result was linewidths as narrow as 300 MHz (5 parts in 10^7).

In 1972 the same three Stanford collaborators used a pulsed laser and the external saturation method to observe for the first time Doppler-free optical spectra of hydrogen. They obtained optical resolution of the Lamb shift, allowing a precision measurement of the absolute wavelength of the hydrogen lines. Thus they were able to determine the Rydberg constant with much higher precision than previously possible (initially a factor of ten better). Every couple of years since, the determination of the Rydberg constant has improved by factors of two or three.

In 1974 Bernard Cagnac in France, Bloembergen and Levenson at Harvard and Hänsch, G. Maisel, Ken C. Harvey and Schawlow all showed the feasibility of Doppler-free two-photon spectroscopy. (The idea had earlier been suggested by Veniamin Chebotayev.) In this form of high-resolution nonlinear laser spectroscopy, gas atoms in a standing-wave field are excited by absorbing two photons from opposite directions. Their first-order Doppler shifts are equal and opposite; so the sum frequency is unchanged.

The use of lasers to detect trace elements was pioneered by William Fairbank Jr, Hänsch and Schawlow. In 1975 they were able to detect resonance fluorescence from sodium atoms when as few as 100 atoms/cm³ were present (or as little as one atom at a time in the beam).

In 1976 Schawlow and his collaborators started using lasers to simplify complicated spectra. The technique is now known as lower-level labeling or population labeling; it has been extended to include polarization labeling.

Asked to compare and contrast his work with that of Bloembergen over the past two decades, Schawlow said the Stanford workers have been trying to simplify spectra, eliminate Doppler broadening and do it very sensitively with a small number of atoms and molecules. They have concentrated on changing populations 'of stationary states. On the other hand, he said, Bloembergen and his collaborators have been concentrating on changing the susceptibility of a medium by nonlinear optics. The Harvard workers have emphasized bulk properties (which are of course determined by atoms) whereas the Stanford workers have emphasized individual atomic and molecular levels.

Bloembergen told us that since lasers were first built in 1960 he has been exploiting them to study the properties of matter at high light intensities, particularly after 1961, when Peter Franken and his collaborators demonstrated second harmonic generation of light, in which red light from a ruby had its wavelength halved to the ultraviolet. This discovery opened the field of nonlinear optics to experimenters. Bloembergen noted that lots of nonlinear effects were already known from the work done shortly after World War II with microwave spectroscopy, nuclear magnetic resonance and electron paramagnetic resonance.

Two papers in 1962 by Bloembergen and his collaborators—John A. Armstrong, Jacques Ducuing and Peter S. Pershan—developed a general framework to describe a large number of nonlinear optical phenomena applicable to liquids, semiconductors, metals and so on. The electric polarization can be expanded in a power series in **E**, the electric-field amplitudes.

$$\mathbf{P} = \chi^{(1)}\mathbf{E} + \chi^{(2)}\mathbf{EE} + \chi^{(3)}\mathbf{EEE} + \cdots$$

$\chi^{(1)}$ is the linear susceptibility and $\chi^{(2)}$ is the lowest-order nonlinear susceptibility, a third-rank tensor.

In 1964 Bloembergen and Yuen-Ron Shen worked out analogies to the Kramers–Heisenberg relation for non-

linear behavior, including damping for nonlinear susceptibilities and showing the properties of the imaginary parts of the equation. Then, Bloembergen told us, he and his collaborators did experiments to back up the theory. For example, in 1969 Bloembergen and Hansen Shih predicted the nonlinear analog to conical refraction (whose theory had been developed by William R. Hamilton in 1833); in 1977 Anita Schell and Bloembergen observed nonlinear conical refraction.

The Swedish Academy announcement singled out the contributions of Bloembergen and his collaborators to four-wave mixing, in which three coherent light waves act together to produce a fourth light wave in a new direction. Although first demonstrated experimentally by Robert W. Terhune and his collaborators at Ford Research Labs, Bloembergen and his collaborators have studied four-wave mixing systematically as a function of the three frequencies in many different materials and obtained the dispersive characteristics. By varying the angle between beams, one can also vary the wave vectors and alter the polarization of each beam. "Clearly, nonlinear properties are a much richer field than linear spectroscopy," Bloembergen remarked.

One example of four-wave mixing is phase conjugation. Bloembergen and his collaborators recently used four light beams of almost equal frequency to show the effect of collision-induced coherence, a paradoxical behavior because collisions usually destroy coherence. Another example of four-wave mixing is Coherent Anti-Stokes Raman Scattering or "CARS," a technique developed by Bloembergen and various Harvard collaborators and done in many other labs.

Electron spectroscopy. The Nobel prize in physics to Kai Siegbahn is in a sense the first to be given for surface physics since Clinton Davisson was honored in 1937. Kai Siegbahn's father, Manne Siegbahn, received the Nobel prize in physics in 1925 for the development of high-resolution x-ray spectroscopy. The son is honored this year for the development of high-resolution electron spectroscopy.

As early as 1913, H. Robinson in Ernest Rutherford's lab in Manchester had used photoelectron spectroscopy to obtain information on the electron structure of a given sample. Despite two decades of work, Robinson, because he did not have a high-resolution spectrometer, could not distinguish an elastic from an inelastic peak in a plot of intensity *vs.* electron kinetic energy. Because of the relatively low resolution, photoelectron spectroscopy for materials research was considered less useful than x-ray spectroscopy until

SIEGBAHN

the 1950s, the Swedish Academy said.

In the mid-1950s, Kai Siegbahn and his collaborators at Uppsala, Carl Nordling and Evelyn Sokolowski, began analyzing photoelectrons with the aid of a high-resolution double-focusing spectrometer, originally designed for nuclear beta-ray spectroscopy. Because the spectrometer was iron-free, its magnetic field was directly proportional to the coil current. The spectrometer had high accuracy and stability (so spectral lines did not drift during a measurement) and a resolution a factor of ten better than previously available.

Until the work of Siegbahn and his collaborators the binding energies of core electrons of atoms were not well known. Previously, these binding energies had been found through x-ray absorption, in which one measures the position of an absorption edge rather than a peak. Finding the edge had limited the accuracy of the x-ray absorption method as well as earlier photoelectron spectroscopy until the Uppsala work. The group in Uppsala irradiated a sample with x rays having a characteristic line spectrum. (One popular line is the copper Kα line at about 8 keV.) These x rays expel electrons from atomic levels in the target, and these electrons undergo energy losses through inelastic scattering on their way out of the target. At low electron-energy resolution such as used by Robinson, the inelastically scattered electrons could not be distinguished from the elastically scattered electrons. The resulting spectrum had an edge shape similar to an absorption spectrum.

The Uppsala group discovered that at very high resolution the inelastically scattered electrons are well separated from the elastic peak because the energy-loss mechanism is quantized in energy. The resulting electron spectrum therefore shows very well-defined peaks (electron lines) corresponding to the various atomic levels of the target. The widths of these peaks are determined by the natural width of the atomic levels. In this way the kinetic energy of the photoelectrons could be measured very precisely and the atomic binding energies could be deduced with high accuracy.

During the late 1950s Siegbahn, Nordling and Sokolowski made a systematic study of electron binding energies in different elements, a study which is still a major source of information on inner atomic levels.

In 1958 Stig Hagstrom (now at Xerox Research Center and Stanford) joined the group, which started to work on developing photoelectron spectroscopy for quantitative elemental analysis of light elements.

In a compound such as Na_2SO_4 one will observe a peak from sodium, a peak from sulfur and a peak from oxygen. By measuring relative intensities and knowing the composition, one could calibrate the spectrometer, allowing one to study unknown materials. In 1964, while they were studying Na_2SO_4, partly by mistake and partly for convenience, Hagstrom recalled to us, he, Nordling and Siegbahn looked at $Na_2S_2O_3$. In this compound the two sulfur atoms are in two different chemical positions, one in the $+6$ and the other in the -2 oxidation state. From the sulfur K level in fact they saw two peaks, and they realized that one came from the $+6$ state and the other from -2. Chemical shifts had previously been seen in x-ray absorption and also by Nordling, Siegbahn and Sokolowski in 1957. At that time the group thought the shift was from changes in valence electrons. But through the double peak from the sodium thiosulfate compound it became obvious that the core level was shifted, not the valence.

These chemical shifts in core levels were caused by a difference in the way the atom was bound in the molecule or crystal, that is, the different electron densities in the vicinity of the atoms. Similar shifts had been seen in x-ray spectra but were much more subtle and

difficult to interpret. The Swedish Academy notes that "in the development of electron spectroscopy, a practically useful analytical method had been obtained [by the Uppsala group] with which it was possible to study not only which atoms are included in a sample but also in what chemical environment these atoms exist. At this time the concept of 'ESCA' (Electron Spectroscopy for Chemical Analysis) was created. . . " Siegbahn and his collaborators immediately realized the great potentiality of ESCA. Not only would it be useful for quantitative elemental analysis (how much of a given element is present), it would also be sensitive to various oxidation states of an element.

The early Uppsala work was done on various elements and compounds. After finding the chemical-shift effect, they made a systematic study of noble gases and simple gases such as CO and CO_2. Later they went on to study solids. About 1973–74, in a *tour de force*, Siegbahn and his collaborators did ESCA on liquids, employing a continuous wire (attached to a pulley), which dipped into the liquid bath. Above the liquid surface was an x-ray tube. The group observed photoelectrons from the wet wire.

X-ray photoelectron spectroscopy (also called XPS) allows one to study the outermost several atomic layers (between two and ten layers or 5–20 Å). After the Siegbahn group's original work on ESCA it was realized that one can optimize the surface sensitivity by using grazing take-off angles. To study the composition of deeper layers, one can remove surface atoms. By sputtering with argon ions, for example, one can remove surface atoms and obtain the composition as a function of depth. Such probing can also be done by mechanical or chemical means, depending on how deep one wishes to probe. However, all these removal techniques can leave undesirable artifacts, for example, decomposing a chemical compound. One advantage of ESCA is that the incident x rays are not very damaging compared to the electron beams used with Auger-electron spectroscopy of surfaces.

After the discovery of the chemical

shift by Siegbahn and his collaborators, a number of other labs became involved. One such group was established in 1964 by Manfred Krause and Thomas Carlson at Oak Ridge. Another was the group at Berkeley started in 1965 by Hagstrom, Charles Fadley, David Shirley and Jack Hollander.

By the late 1960s, several commercial companies started making instruments for ESCA. By now photoelectron spectroscopy is being applied in various forms at hundreds of labs. It has been used for studies of many surface-chemistry processes such as catalysis and corrosion.

Meanwhile synchrotron radiation sources have become increasingly available, allowing one to tune the x-ray energy. Thus one can improve surface sensitivity by choosing the photoelectron energy to correspond to the minimum mean free path for a given material.

Vital statistics. Bloembergen received a BA in 1941 and an MA in physics in 1943, both from the University of Utrecht. In 1946 he went to Harvard, where he did his thesis with Purcell. His PhD was awarded in 1948 from the University of Leiden. From 1949 to 1951 he was a junior fellow in the Society of Fellows at Harvard, where he has been ever since, except for visiting professorships. Since 1957 he has been Gordon McKay Professor of Applied Physics and since 1980 Gerhard Gade University Professor.

Schawlow earned all three degrees at the University of Toronto, BA in 1941, MA in 1942 and PhD in 1949. At that time he became a research associate at Columbia University, staying until 1951, when he joined Bell Labs. He remained there until 1961. He then joined Stanford where he served as department chairman 1966–70. Since 1978 he has been J. G. Jackson–C.J. Wood Professor of Physics at Stanford. He was president of the Optical Society of America in 1975 and is president this year of The American Physical Society.

Siegbahn earned his PhD at Stockholm University in 1944. Since 1954 he has been professor of physics at Uppsala University.

High-resolution spectroscopy of atoms and molecules

New laser techniques, pulsed and continuous, which make it possible
to see optical spectra without Doppler broadening, to label energy levels,
and to enhance sensitivity, are now opening new applications.

Theodor W. Hänsch

PHYSICS TODAY / MAY 1977

Lasers are rejuvenating, even revolutionizing, the field of spectroscopy of atoms and molecules. Compared with the light of conventional sources, laser light is more—sometimes dramatically more—powerful, directional, spectrally pure and coherent. Laser light can be generated in extremely short pulses. Furthermore, tunable lasers can operate at wavelengths at which intense conventional sources such as spectral lamps simply have not been available. Lasers thus can enormously enhance the sensitivity and application range of classical spectroscopic methods, such as absorption spectroscopy, fluorescence spectroscopy and Raman spectroscopy.

At the same time, lasers have made it possible for us to observe new nonlinear spectroscopic phenomena. Intense laser light can change level populations and bleach absorbing transitions. The polarizabilities and dielectric susceptibilities of atoms or molecules themselves become functions of the light's field strength, and we encounter such effects as two-photon and multiphoton transitions, the generation of new light at the sum or difference frequency of the incident waves and stimulated light scattering. Laser light can prepare atoms effectively in coherent superpositions of quantum states, and it can give rise to fascinating coherent transient phenomena.

These effects are not only interesting in their own right and deepening our understanding of the nature of light and its interactions with matter, but some of them can be harnessed in nonlinear spectroscopic techniques to provide us with powerful new experimental tools. They allow us to probe the structure of atoms and molecules with an unprecedented depth of scrutiny.

The various ways in which lasers have already been used for atomic and molecular spectroscopy are much too numerous even to list here. Although several good reviews have been written[1,2] new and sometimes surprising results are being reported at an almost breathtaking pace. In this article I shall only attempt to look at some recent progress in one class of nonlinear spectroscopic techniques in which lasers have been particularly successful: high-resolution spectroscopy without the *Doppler broadening* that so often blurs important details in the spectra of free atoms and molecules.

Although some methods of Doppler-free laser spectroscopy have been known and used for more than a decade, numerous new approaches have been developed or suggested in the past few years. These techniques not only permit us to study fine spectral structures in unprecedented detail; they are also providing important tools for precision metrology and for fundamental physics research. Some of the same methods can also be used also to simplify and unravel complex absorption spectra. To illustrate what can be done I should like to describe only a few examples, mostly chosen from the research carried out by my collaborators and me at Stanford University.

Spectra without Doppler broadening

Atoms or molecules in gases are relatively free and undisturbed, but their spectral lines appear spread out by the Doppler effect over a range of wavelengths (typically about 1 part in 10^5) because they are moving in all directions with high thermal velocities. Those atoms moving towards an observer appear to emit or absorb light at larger frequencies than those at rest; those receding emit or absorb at lower frequencies. The resulting Doppler broadening often masks spectral fine and hyperfine structure, although each individual atom still retains its typically much narrower natural linewidth.

In the oldest approach to eliminate Doppler broadening, a well collimated atomic beam is used to select just a group of atoms that move nearly perpendicularly to the observer's line of sight. Excitation of the atoms with a tunable laser can be used with great advantage in this method because such lasers can easily be monochromatic to one part in 10^8 or better, and rather spectacular results have recently been obtained in this way.[1,2] Nevertheless, atomic-beam spectroscopy has its limitations; it can, for instance, not be applied easily to rare or expensive substances or to atoms in short-lived excited states.

Doppler broadening has also been reduced by the cooling of a gas sample. Because the velocity spread diminishes only with the square root of the temperature and because of condensation, this approach tends to be much less effective. But it has recently been pointed out[3] that the radiation-pressure force exerted by intense resonant laser light could be used to reduce the thermal velocities of gas atoms very rapidly—within microseconds—to very low values, corresponding to temperatures of less than one kelvin.

Vladilen Letokhov has suggested[4] that very slow atoms be trapped and suspended in the nodes of a strong standing-wave light field. These stationary atoms could then be observed over extended periods of time, and their spectrum would be completely free of Doppler broadening. Radiation cooling and trapping could be combined for this purpose, although light shifts in the strong trapping field are likely to limit the resolution in this as yet untried scheme. (See article by Letokhov on p. 23 of May 1977 issue of PT.)

The high intensity of laser light fortunately makes it possible to eliminate Doppler broadening in gas samples without any need for cooling. Two entirely different approaches have so far been used very successfully:

▶ In saturation spectroscopy and polarization spectroscopy, a group of atoms in

Theodor W. Hänsch is a professor of physics at Stanford University, Stanford, California.

a narrow interval of axial velocities are "labelled" by their nonlinear interaction with a monochromatic travelling laser wave and a Doppler-free spectrum of these selected atoms is then observed, generally with a second laser beam as a probe.

▶ Two-photon excitation with two counterpropagating laser beams, so that the first-order Doppler shifts cancel, permits high-resolution spectroscopy of gases despite high thermal velocities.

Saturation spectroscopy

The first step towards saturation spectroscopy was the realization by Willis Lamb[5] that the two waves travelling in opposite directions inside a laser could work together to saturate the emission of those atoms that happen to have a zero component of velocity along the laser axis. Thus the power output would decrease when the laser length was adjusted to produce the light wavelength that would interact with those atoms. This "Lamb dip" was soon observed by R. A. Macfarlane, William Bennett Jr and Lamb[6] and independently by Abraham Szöke and Ali Javan.[7] "Inverse Lamb dips" have later been observed with an absorbing gas sample inside a laser resonator.[8]

This method was soon applied in several laboratories to high-resolution spectroscopy. Although it was limited for the time to studying the laser transitions themselves or those few molecular lines that happened to coincide with gas-laser wavelengths, it produced spectacularly

narrow resonance lines. Through continuous refinements, John Hall and Christian Bordé[9] have recently been able to push the resolution of Lamb dips in methane, observed with a helium–neon laser near 3.39 microns, to about 2 parts in 10^{11}. They not only observed close molecular hyperfine splittings but also a line splitting due to the radiative recoil of the molecules that absorb or emit infrared photons.

Lamb-dip-stabilized gas lasers have become invaluable tools for precision metrology. They not only have established new *de facto* standards of length, but they have opened the way for new precision measurements of fundamental constants, and for stringent new tests of special relativity and of quantum electrodynamics as well.

A particularly sensitive and simple method of saturated absorption spectroscopy is one that was first used in 1970 by Bordé in Paris[10] and independently at Stanford,[11] following earlier related experiments at Heidelberg.[12] The absorbing gas sample is contained in a cell outside the laser resonator. A beam splitter divides the laser output into a strong saturating beam and a weaker probe beam, which are sent by mirrors in nearly opposite directions through the absorber. When the saturating beam is on, it bleaches a path through the cell, and the probe signal is received more strongly at the detector. As the saturating beam is alternately stopped and transmitted by a chopper, the probe signal is modulated. That, however, happens only when the laser is tuned near the center of the Doppler-broadened absorption line, where both beams are interacting with the same atoms, those that are standing still or at most moving transversely to the laser-beam direction.

Other techniques of Doppler-free saturation spectroscopy have also been reported.[1,2] Particularly noteworthy is the method of observing the combined satu-

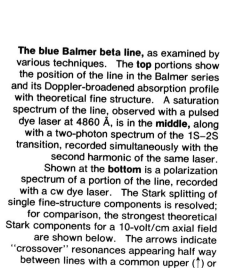

The blue Balmer beta line, as examined by various techniques. The **top** portions show the position of the line in the Balmer series and its Doppler-broadened absorption profile with theoretical fine structure. A saturation spectrum of the line, observed with a pulsed dye laser at 4860 Å, is in the **middle,** along with a two-photon spectrum of the 1S–2S transition, recorded simultaneously with the second harmonic of the same laser. Shown at the **bottom** is a polarization spectrum of a portion of the line, recorded with a cw dye laser. The Stark splitting of single fine-structure components is resolved; for comparison, the strongest theoretical Stark components for a 10-volt/cm axial field are shown below. The arrows indicate "crossover" resonances appearing half way between lines with a common upper (↑) or lower level (↓). **Figure 1**

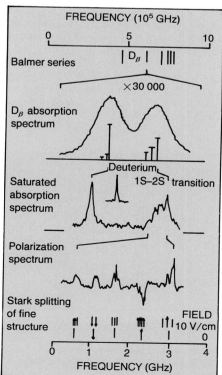

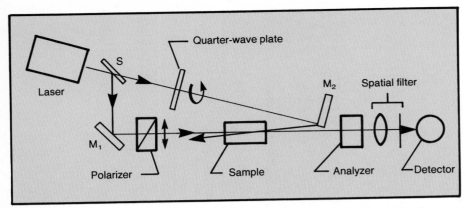

Apparatus for Doppler-free polarization spectroscopy. A circularly polarized beam from the pump laser orients the rotation axes of atoms in a narrow range of velocities. These atoms can be detected with high sensitivity by a counterpropagating probe beam because they change its polarization, causing light to pass through the analyzer into the detector.   Figure 2

ration by two counterpropagating laser beams in the fluorescent sidelight rather than in absorption, as first demonstrated by Charles Freed and Javan.[13] Very dilute samples of negligible absorption can be studied in this way, in particular if the sensitivity is increased by chopping the laser beams at two different frequencies and by monitoring the modulation of the fluorescence at the sum or difference frequency, as reported by Arthur Schawlow and Michael Sorem.[14] The narrow-band spectrum of a selected group of molecules can also be observed by probing the dispersive refractive index rather than the absorption, as Bordé demonstrated with a gas sample inside a ring interferometer.[15]

All these methods were first used with gas lasers of very limited tuning range. Although broadly tunable visible lasers, especially dye lasers, began to become available about the same time, they generally did not produce sufficiently narrow spectral lines. But in late 1970 it was demonstrated at Stanford that linewidths as narrow as 300 MHz (5 parts in 10^7) can be produced by a pulsed dye laser when a beam-expanding telescope and an etalon were introduced inside the resonator.[16] Further narrowing could be obtained as needed by a confocal interferometer used as a passive filter. Continuous-wave dye lasers with linewidths much less than 1 MHz have become available as well.[17] But pulsed dye lasers continue to be useful for high-resolution spectroscopy, because they span a much wider conveniently accessible wavelength range and because their high peak power makes it possible to produce tunable ultraviolet or infrared radiation efficiently by nonlinear frequency mixing in crystals and gases.

Such a pulsed laser and the external saturation method made it possible to observe, for the first time, Doppler-free optical spectra of the simplest of all stable atoms, hydrogen.[18] As is well known, hydrogen has been intensively studied by

spectroscopists for almost a century, because its simplicity permits detailed and accurate comparison with theoretical models.

The red Balmer alpha line ($n = 2$ to 3) of hydrogen and deuterium was studied in the initial experiment.[18,19] Subsequently, Sui Au Lee and her co-workers[20] observed the second member of the Balmer series, the blue Balmer beta line ($n = 2$ to 4), by the same technique; the results obtained for deuterium are illustrated in figure 1. The atoms were excited to the absorbing $n = 2$ stage in the low-pressure glow discharge and a saturated absorption spectrum was recorded with oppositely directed saturating and probe beams from a pulsed dye laser. Doppler broadening is particularly troublesome because the atoms are so light. Even with the heavier isotope deuterium, a conventional absorption spectrum at room temperature shows only two peaks partly resolved (see figure 1, top) although theory and rf studies agree on the underlying fine structure indicated. The saturation spectrum shown in the center of the figure gives a much clearer indication of the fine structure, but it still falls short of being completely resolved.

The earlier saturation spectra of the red

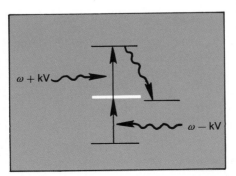

In two-photon spectroscopy without Doppler broadening, gas atoms in a standing-wave field are excited by absorbing two photons from opposite directions. Their first-order Doppler shifts are equal and opposite.   Figure 3

Balmer alpha line were actually better resolved, and revealed—for the first time—single fine-structure components. The $n = 2$ Lamb shift could be clearly observed in the optical spectrum. More interesting than the splittings, measured previously by rf techniques, are the absolute wavelengths of the isolated components. An accurate measurement of the Balmer alpha wavelengths yielded the new value of the Rydberg constant with about ten times higher accuracy than had previously been possible.[19]

The blue Balmer beta line is more difficult to observe because the fine structure splitting of the upper level is smaller. In addition, this line is more susceptible to splitting by the electric field in the discharge tube. It is also weaker and so requires stronger discharge current and a more intense saturating laser beam for a good signal, resulting in additional line broadening.

The spectrum was nonetheless good enough to serve as reference line for the first measurement of the hydrogen 1S ground-state Lamb shift. As discussed below this was done by comparing its wavelength with that of the 1S–2S two-photon transition, which was simultaneously observed with the frequency-doubled dye laser output. But it clearly seemed worthwhile to try to improve the resolution of the Balmer beta spectrum.

Polarization spectroscopy

This goal led to the development of "polarization spectroscopy," a sensitive new method of Doppler-free laser spectroscopy.[21] It is related to the saturated-absorption method, but the nonlinear interaction of the two counterpropagating laser beams is monitored *via* changes in light polarization instead of intensity. As shown in the scheme of figure 2, the saturating beam is circularly polarized. This beam labels atoms of selected axial velocity by orienting them through optical pumping. Normally, atoms in a gas have their rotation axes distributed at random in all directions. Because the absorption cross section for circularly polarized light generally depends on the atomic orientation, the saturating beam will preferentially deplete atoms with particular orientations, leaving the remaining ones polarized. Those atoms can then be detected with high sensitivity because they can change the polarization of a probe beam.

The linearly polarized probe light can be thought of as a combination of right-handed and left-handed circularly polarized waves of equal intensity. If the laser is tuned to the center of the Doppler-broadened line, where the probe beam sees the atoms polarized by the pump beam, more of the left-handed wave, say, is then absorbed than of the right-handed one, and the light passing through the sample becomes elliptically polarized. In addition, the polarization axis is rotated, because the two waves also see different refractive indices. The probe beam thus acquires a component that can pass through a crossed polarizer, which otherwise blocks the beam from the detector.

Polarization spectroscopy has an important advantage in its signal-to-noise ratio. There is almost no transmission of the probe beam until its polarization is changed by the atoms pumped by the saturating beam. With essentially no background, the signal is not easily obscured by noise or intensity fluctuations of the laser. This method can thus easily be used with fewer atoms and lower laser power.

For very small signals it is actually advantageous to uncross the polarizers slightly. The detector will then register some finite background, but the signal caused by saturated dispersion can now interfere with this background. Dependent on the sign of the dispersive polarization rotation, the intensity at the detector can either increase or decrease. Such a dispersive resonance line can be very useful for locking the laser frequency to an absorption line. It can be electronically differentiated to give a resonance peak of width less than half the natural linewidth.

The improved resolution obtained in this way for the hydrogen Balmer beta line can be seen in the lower part of figure 1. The shown portion of the spectrum corresponds to the cluster of lines at the right-hand side of the saturated absorption spectrum in this figure. The high sensitivity of polarization spectroscopy has made it possible to work with a highly monochromatic cw dye laser of low power. It has also been possible to operate with a very mild glow discharge with only few hydrogen atoms (1–2% absorption). Not only are the fine-structure lines and crossover lines much better resolved, but even the Stark splittings are clear enough to permit measurement of the electric field on the axis of the discharge tube. (As in saturated absorption spectroscopy, additional "crossover lines" are observed between any two resonant lines that share a common upper or lower level.)

The line components originating in the short-lived 2P state and in the metastable 2S state clearly have different widths. The narrowest observed linewidth so far has been about 25 MHz, but it may be possible to reduce the linewidth of the quasi-forbidden 2S–4S component (third line from the right) to as low as 1 MHz in the future.

Even the resolution of about 4 parts in 10^8, which has already been achieved, sets a new record for optical spectroscopy of this simplest of the stable atoms. It would certainly be possible to improve the accuracy of the Rydberg constant by another order of magnitude by carefully measuring the absolute wavelength of one of the resolved line components. If the

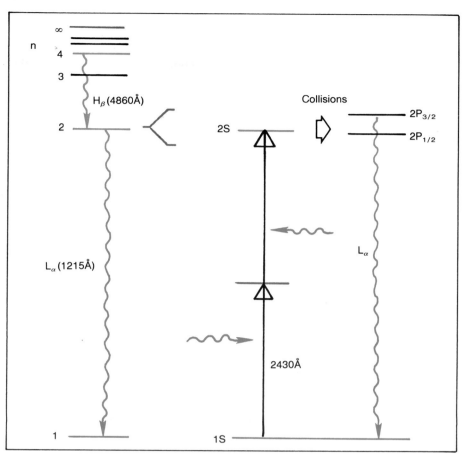

A simplified level scheme for atomic hydrogen. Two-photon excitation of the metastable 2S state from the 1S ground state promises to yield extremely high resolution—ultimately higher than 1 part in 10^{15}. The excitation can be monitored by observing the collision-induced 2P-1S Lyman alpha fluorescence in the vacuum-ultraviolet spectral region. **Figure 4**

isotope shift for hydrogen and deuterium (about 167 000 MHz) could be measured to better than 0.1 MHz, it could be used to confirm or improve the present value of the ratio of electron mass to proton mass. But it is expected that these quantities will soon be determined better by two-photon spectroscopy of the hydrogen 1S–2S transition. The Balmer beta spectrum remains nonetheless valuable as a reference for a measurement of the hydrogen ground-state Lamb shift.

Two-photon spectroscopy

About three years ago, researchers in three different laboratories—at the University of Paris, at Harvard and at Stanford—demonstrated the feasibility of Doppler-free two-photon spectroscopy,[2] a particularly simple and elegant technique of high-resolution nonlinear laser spectroscopy. It has already been used widely for novel studies of atoms and molecules.

It has long been known that atoms can jump from the ground state to some excited level of the same parity by absorption of two photons that together provide the required energy, but experiments in the optical region have become possible only since the advent of strong laser sources. Two-photon spectroscopy has already become a valuable complement to single-photon spectroscopy because of its different selection rules, and because it permits one to reach high-lying states with longer wavelengths, which are often more readily generated and measured.

A particularly interesting situation arises if a gas sample is irradiated simultaneously with two monochromatic laser waves travelling in opposite directions. The atoms can then be excited by absorbing two counterpropagating photons, one from each beam, as diagrammed in figure 3.

As L. S. Vasilenko and his collaborators pointed out,[22] there is no net first-order Doppler effect in this case because, as seen from a moving atom, the two waves have equal and opposite Doppler shifts so that the sum frequency is constant, independent of the atomic velocity. (More generally, the net Doppler effect in multiphoton excitation vanishes if the momentum vectors of the participating photons add to zero.[23]) The two-photon excitation can often be observed with high sensitivity, for instance by monitoring the fluorescence light emitted by the excited atoms. All the atoms contribute to the

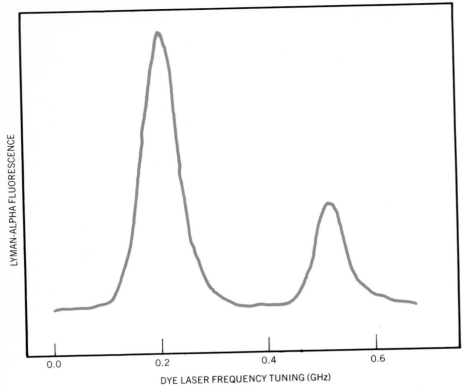

LYMAN-ALPHA FLUORESCENCE

DYE LASER FREQUENCY TUNING (GHz)

0.0   0.2   0.4   0.6

Doppler-free two-photon spectrum of the hydrogen 1S–2S transition, as recorded with a frequency-doubled cw dye-laser oscillator and a pulsed dye amplifier.

Figure 5

resonant Doppler-free two-photon signal, not just a few with selected velocity, and a narrow resonance line is observed superimposed on the low, wide background produced by each beam separately.

After the new technique was first suggested, numerous authors have pointed out that a particularly interesting subject would be the transition from the 1S ground state of atomic hydrogen to the metastable 2S state; see figure 4. If there are no external perturbations, the 2S state decays by spontaneous two-photon emission with a lifetime of about $\frac{1}{7}$ sec, and the natural linewidth should be on the order of only 1 Hertz. In principle, it should then be possible to achieve a resolution of better than 1 part in 10^{15} and to measure the line center to perhaps 1 part in 10^{17}.

In practice it will of course be extremely difficult to approach this ultimate resolution, because there are numerous possible causes of line broadening, including the second-order relativistic Doppler effect, light shifts in the intense laser field, transit time broadening and collision effects. But the main obstacle at present is the lack of a suitable strong, highly monochromatic laser at the required ultraviolet wavelength of 2430 Å. Photons at this wavelength can be generated readily by frequency doubling an intense pulsed dye laser beam at 4860 Å, but the linewidth can then not be narrower than the Fourier-transform limit (about 50 megahertz for the typical 10-nanosecond pulse length of a dye laser that is

pumped by means of a nitrogen laser).

In initial experiments at Stanford it was possible to detect the Doppler-free two-photon resonance in atomic hydrogen in this way.[24] The ground-state hydrogen atoms were produced in a low-pressure gas discharge but observed outside the discharge plasma. After two-photon excitation, the metastable 2S atoms are induced by collisions to decay *via* fluorescence at the Lyman alpha line, which can be detected with high sensitivity. Figure 5 shows the 1S–2S two-photon spectrum for hydrogen, recorded very recently by Carl Wieman at Stanford with a somewhat improved laser. It combines a single-frequency cw dye-layer oscillator with a three-stage pulsed dye-laser amplifier, which permits the generation of several kilowatts peak power at the second-harmonic frequency. The resolution is close to the theoretical transform limit, and the $F = 1$–1 and 0–0 hyperfine components for hydrogen are fully resolved. For deuterium the $F = \frac{3}{2}$–$\frac{3}{2}$ and $\frac{1}{2}$–$\frac{1}{2}$ components are partially resolved.

Using earlier, somewhat less well resolved spectra, Lee and her associates were able to measure[20] the hydrogen–deuterium isotope shift to about 1 part in 10^4, improving earlier experimental values for the Lyman-alpha isotope shift by several orders of magnitude.

The same experiment yielded also, for the first time, an experimental value for the Lamb shift of the hydrogen 1S ground state, by comparing the 1S–2S two-photon spectrum with a saturated absorption

spectrum of the Balmer beta line, which was simultaneously recorded with the fundamental dye-laser output, as discussed earlier. If Bohr's formula were correct, the latter $n = 2$–4 interval would be exactly equal to one-fourth the Lyman-alpha interval, and we would find the two resonances at exactly the same dye-laser frequency. The actual displacement is entirely caused by relativistic and quantum-electrodynamic corrections plus some small nuclear-structure effects. A precise measurement is a very sensitive test of the ground-state Lamb shift, which can not be observed by rf spectroscopy because there is no nearby P reference level. The accuracy of this first measurement was limited to about 1.5% by the relatively poor resolution of the Balmer beta line. New measurements, substituting the much superior polarization spectrum of the Balmer beta line, promise at least a tenfold improvement, and should also give first experimental evidence for an additional 23.4-MHz "Dirac shift" of the 1S state, which is caused by relativistic effects in the nuclear recoil. This effect can not be observed by rf spectroscopy, even for the excited levels, because it shifts all fine-structure levels of a given n by the same amount.

The resolution of even the best present hydrogen 1S–2S two-photon spectrum clearly falls far short of the ultimate limit and dramatic future improvements can be expected, which are likely to send the theorists back to their computers. Particularly promising in this context are some recently suggested or demonstrated new techniques of high-resolution laser spectroscopy that make use of multiple light fields.

Optical Ramsey fringes

The most obvious improvement for two-photon spectroscopy of hydrogen 1S–2S, which would overcome the Fourier-transform limit of pulsed laser spectroscopy, would be the use of a cw laser. Highly monochromatic cw radiation at 2430 Å has recently been generated at Stanford by summing the frequency of a blue krypton-ion laser and a yellow dye laser in a cooled crystal of ammonium dihydrogen phosphate. The power of about 0.1 mW may be sufficient for Doppler-free two-photon excitation of hydrogen 1S–2S if the beam is tightly focussed, but the short transit time of the atoms moving through the narrow beam waist would certainly cause considerable transit-time broadening.

A rather ingenious solution to this transit-time problem has recently been suggested by Ye. V. Baklanov and his collaborators.[25] An atomic beam could be sent through two consecutive transverse standing-wave light fields separated in space. After passing through the first field the atoms will be in a coherent superposition of states and will oscillate at the two-photon resonance frequency. The effect of the second light field depends on the phase of the radiation relative to the atomic oscillations, so that the atoms passing through this field will either be further excited or returned to the ground state by stimulated two-photon emission. When monitoring the net excitation by the two fields one should observe the optical analog of the well known Ramsey fringes, which are routinely utilized in rf spectroscopy of atomic beams.[26]

The spectral resolution is then limited by the travel time between the two fields rather than by the transit time through each waist. Doppler-free two-photon excitation with standing waves ensures that the fringe structure is not smeared out by dephasing due to the unavoidable small spread of transverse atomic velocities. Attempts to use these narrow spectral fringes for very-high-resolution spectroscopy of atomic beams are presently under preparation in several laboratories. (Similar optical Ramsey fringes have recently been observed by James Bergquist and his associates in saturated-absorption spectroscopy of a beam of fast neon atoms with spatially separated laser fields.[27])

To an atom traversing the separated field regions, the laser field appears as a succession of light pulses. This suggests that optical Ramsey fringes in two-photon excitation should also be observable with a pulsed laser source, without the need for an atomic beam, if a gas cell is irradiated by two standing-wave light pulses separated by a delay time T.

Although a single pulse produces a broad spectrum, as shown at the top of figure 6, two sequential pulses give rise to sinusoidal fringes (center) if the pulse delay T is scanned together with the light wavelength. The spacing between neighboring fringes is equal to the reciprocal of the pulse separation. Such interference fringes have recently indeed been observed by Michael Salour and Claude Cohen-Tannoudji,[28] who excited the sodium 3s–4d transition with two standing-wave dye-laser pulses produced with an optical delay line. The observation of spectral fringes of a width well below the Fourier-transform limit of an individual laser pulse opens new prospects for high-resolution spectroscopy, although the sinusoidal fringe structure would make it rather difficult to resolve spectral line components that are closely spaced.

These two-field excitation experiments can be regarded as spectral analogs to wave diffraction at a double slit. This analogy immediately suggests an important next step. If two-photon transitions are excited with a whole train of phase-coherent light pulses, as shown at the bottom of figure 6, the spectral fringes should condense into narrow lines, which, in the analogy, can be compared to the

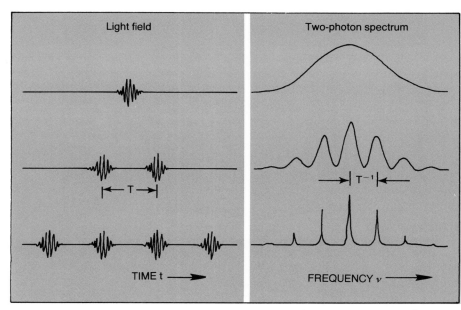

High spectral resolution is possible in two-photon spectroscopy with multiple short standing-wave light pulses. A single pulse produces a relatively broad, transform-limited spectrum (top); excitation with two sucessive light pulses can give rise to sinusoidal "optical Ramsey fringes" (center), and a train of pulses can produce narrow spectral lines (bottom). Figure 6

diffraction orders of a multislit aperture or grating.

Such narrow multipulse resonance lines have recently been observed by Richard Teets and his co-workers for the 3s–5s transition of atomic sodium with a very simple experimental setup.[29] The pulse train is produced by injecting a single short dye-laser pulse into an optical cavity formed by two mirrors. The gas sample is placed near one end-mirror so that the atoms see a pulsed standing-wave field once during each round trip, when the pulse is being reflected by the mirror. For spectral fine tuning the resonator length is scanned with a piezotranslator. The observed sharp multipulse interference fringes can be interpreted in terms of the modes of the optical resonator. The resolution is limited only by the natural atomic linewidth and by the losses of the resonator, not by the laser bandwidth.

The same scheme can provide a dramatic signal increase if the laser pulse is injected into the resonator without loss, for instance with some acousto-optic light switch. As long as atomic relaxation can be neglected, the probability of two-photon excitation for small intensities is proportional to the square of the number of pulse roundtrips; if the pulse recurs a hundred times, the two-photon signal will be 10^4 times stronger than in a single-pulse experiment. This enhancement can make it possible to use larger, less intense beams, reducing light shifts and transit-time broadening.

Two-photon excitation with multiple light pulses promises to extend the range of Doppler-free laser spectroscopy to new wavelength regions, in particular the ultraviolet and vacuum ultraviolet, where only short-pulse laser sources are available at present.

Polarization labelling of spectra

We have illustrated how nonlinear laser spectroscopy can provide a "microscope" to study the finest spectral details in high resolution, free of Doppler broadening. This ability has opened numerous new possibilities for interesting and fundamental research, even if we restrict our attention to the simplest atom. Naturally, a much richer field lies ahead if we include more complex atoms and molecules—studies of more complex aggregations of matter have in the past revealed surprising and important new effects that were not at all obvious from the basic laws of quantum mechanics. But for some of the more complex atoms and molecules even the coarse features of their absorption spectra are far from understood, and some means of classifying and unravelling these features can often be more important than the ability to see fine details.

It is quite remarkable that some of the same methods of nonlinear high-resolution laser spectroscopy can also be used to simplify atomic and molecular spectra and to unravel their complexities. Polarization spectroscopy can be a particularly powerful tool for this purpose, as demonstrated in recent experiments by Teets and others at Stanford.[30]

One of the difficulties in analyzing a complicated spectrum is that spectral lines from unrelated levels can occur at nearly the same wavelength. Even the absorption spectrum of a simple diatomic molecule such as Na_2 is greatly complicated by these accidental coincidences and near-coincidences, as the upper part of figure 7 indicates.

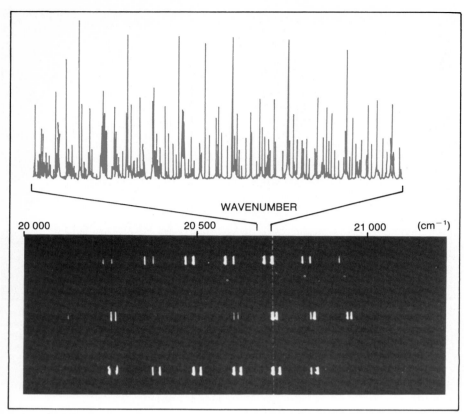

A small portion of the visible absorption spectrum of diatomic sodium molecules, shown at the top, is compared with spectra simplified by polarization labelling, bottom.　　Figure 7

These spectra resemble those from laser-excited fluorescence, but they reveal all absorption lines with a common lower state rather than emission lines with a common upper state. They can provide direct information about the spectroscopic constants and quantum numbers of the excited state. The vibrational quantum numbers can, for instance, immediately be inferred from figure 7 by simply counting down until the doublets end at $v' = 0$. Any irregularity in the series of doublets can reveal perturbations by neighboring triplet states, which would be very difficult to discover in the forest of lines of the ordinary absorption spectrum.

Collision processes in the presence of buffer gas produce a series of satellite lines around each doublet. These originate from ground-state levels close to the pumped level, and differ from it by $2, 4, 6, \ldots$ units of the rotational quantum number. They indicate that collisions change the magnitude of the angular momentum without destroying the orientation of the molecules.

It is also possible to pump a line of one band while observing some other band. For example, one could pump a transition in the infrared, where the spectroscopic constants are often well known, and probe a transition in the ultraviolet.

Once the spectral lines are identified by polarization labelling, their wavelengths can obviously be measured with very high precision. If highly monochromatic tunable dye lasers are used for pumping and probing, the polarization spectrum is free of Doppler broadening, as we have seen before. With accurate, fringe-counting digital wavelength meters it is thus easily possible to determine much-improved spectroscopic parameters, even for molecules that were studied as thoroughly in the past as diatomic sodium.

New territory

The few examples illustrate how nonlinear laser spectroscopy of atoms and molecules makes it now possible to explore new territory and to gather information that would be difficult or impossible to determine by classical spectroscopic methods.

I have not mentioned many other interesting recent applications of laser spectroscopy, such as studies of atoms in high Rydberg states or in highly excited autoionizing states. Nor have we been able to discuss other interesting laser techniques, such as methods of ultrasensitive spectroscopy, which are providing analytic tools of unprecedented power. (Trace absorptions can be detected sensitively by placing the absorber inside a dye laser cavity, or by monitoring the sound generated by absorption of modulated laser light. Fluorescence spectroscopy with lasers has been carried to densities of less than one atom, on the average, in the beam, and single atoms or

At ordinary temperatures, numerous vibrational and rotational energy levels are populated in the electronic ground state. Each such level can absorb light at many different wavelengths, forming a series of doublets or triplets of spectral lines. The individual lines in such a group are those in which the rotational quantum number, J, of the upper, electronically excited state, differs from that of the lower state by $-1, 0$ or 1, according to the selection rules for dipole transitions.

The absorption spectrum would obviously be dramatically simplified if only a single lower state were populated. For the lowest levels such a situation can sometimes be approximated by the cooling of a gas through rapid expansion during flow through a supersonic nozzle. But this approach does not help one in identifying and studying the absorption lines that originate in the higher levels. Mark Kaminsky and his co-workers[31] have demonstrated recently that laser light can be used to "label" any selected lower level and to identify its absorption lines. The population of the chosen level is depleted by saturating a selected absorption line with a tunable laser, and a second probe laser beam identifies all lines with the same lower state by their weakened absorption. Much higher sensitivity has been achieved subsequently by combining this method of lower-level labelling with the technique

of polarization spectroscopy so as to provide a simplified spectrum that can be seen or photographed.

The apparatus for polarization labelling is shown in figure 8. For convenience, nitrogen-pumped dye lasers were used both for labelling of a chosen level and for probing. The pumping laser is monochromatic and can be circularly or linearly polarized. The probe is a broad-band laser with a continuous output spectrum 300 Å wide. Crossed polarizers are placed into the probe beam, in front of and behind the absorption cell. The polarized pump-laser beam then selectively removes molecules that happen to be oriented so as to absorb it, leaving the remaining ones in that level with a complementary alignment. Thus the gas (Na_2 in these preliminary experiments) can change the polarization of the probe at the wavelengths of all absorption lines from the polarized level. Those wavelengths can then pass through the crossed analyzer and appear as bright lines on a dark background in a photograph of the spectrum.

In the spectrum in the lower part of figure 7, the probe-laser wavelengths extend on either side of a pump at about 4830 Å and cover transitions to the low vibrational states of the B band $B^1\Pi_u \leftarrow X^1\Sigma_g^+$ of the diatomic sodium molecule. As the pump wavelength is shifted very slightly, different lower levels are polarized and the various spectra are observed.

molecules in selected quantum states can be detected among 10^{19} background molecules through resonant multi-step photoionization.)

The high sensitivity and resolution of laser techniques has stimulated interesting new approaches to challenging problems in fundamental physics, such as the detection of parity-violating neutral-current effects in atoms or molecules, or the search for quarks and other elusive species, or new sensitive tests to reveal "ether drifts" that might cause a directional anisotropy of the velocity of light.

We do not have to be overly optimistic to expect a rich harvest of new results—and perhaps surprising discoveries—from the new techniques in the future, and physics, chemistry and biology will surely be among the sciences to benefit.

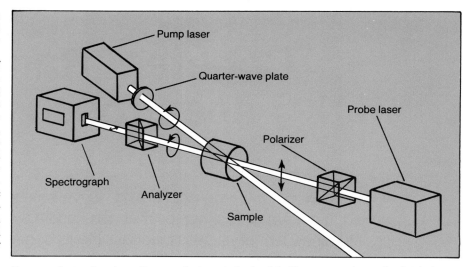

To unravel complex absorption spectra by polarization labelling, a monochromatic circularly polarized laser beam is used in this apparatus to orient molecules in one selected lower level. The gas then changes the polarization of a broad-band beam from a probe laser at all wavelengths that correspond to absorption lines from this level.   Figure 8

References

1. *Laser Spectroscopy of Atoms and Molecules* (H. Walther, ed.), Springer, New York (1976).

2. *High Resolution Laser Spectroscopy* (H. Shimoda, ed.), Springer, New York (1976).

3. T. W. Hänsch, A. L. Schawlow, Opt. Commun. **13**, 68 (1975).

4. V. S. Letokhov, B. D. Pavlik, Appl. Phys. **9**, 229 (1976).

5. W. E. Lamb Jr, Phys. Rev. A **134**, 1429 (1964).

6. R. A. Macfarlane, W. R. Bennett Jr, W. E. Lamb Jr, Appl. Phys. Lett. **2**, 189 (1963).

7. A. Szöke, A. Javan, Phys. Rev. Lett. **10**, 521 (1963).

8. P. H. Lee, M. L. Skolnick, Appl. Phys. Lett. **10**, 3641 (1967).

9. J. L. Hall, C. J. Bordé, K. Uehara, Phys. Rev. Lett. **37**, 1339 (1976).

10. C. Bordé, C. R. Acad. Sci. Paris **271**, 371 (1970).

11. T. W. Hänsch, M. D. Levenson, A. L. Schawlow, Phys. Rev. Lett. **27**, 707 (1971).

12. T. W. Hänsch, P. Toschek, IEEE J. Quant. Electr. **QE-4**, 467 (1968).

13. C. Freed, A. Javan, Appl. Phys. Lett. **17**, 53 (1970).

14. M. S. Sorem, A. L. Schawlow, Opt. Commun. **5**, 148 (1972).

15. C. Bordé, G. Camy, B. Decomps, L. Pottier, Colloques Internationaux du C.N.R.S. No. 217, Paris (1974), page 231.

16. T. W. Hänsch, Appl. Optics **11**, 895 (1972).

17. B. B. Snavely, in *Dye Lasers* (F. P. Schäfer, ed.), Springer, New York (1973), page 91.

18. T. W. Hänsch, I. S. Shahin, A. L. Schawlow, Nature **235**, 63 (1972).

19. T. W. Hänsch, M. H. Nayfeh, S. A. Lee, S. M. Curry, I. S. Shahin, Phys. Rev. Lett. **32**, 1396 (1974).

20. S. A. Lee, R. Wallenstein, T. W. Hänsch, Phys. Rev. Lett. **35**, 1262 (1975).

21. C. Wieman, T. W. Hänsch, Phys. Rev. Lett. **36**, 1170 (1976).

22. L. S. Vasilenko, V. P. Chebotaev, A. V. Shishaev, JETP Letters **12**, 113 (1970).

23. G. Grynberg, F. Biraben, M. Massini, B. Cagnac, Phys. Rev. Letters **37**, 283 (1976).

24. T. W. Hänsch, S. A. Lee, R. Wallenstein, C. Wieman, Phys. Rev. Lett. **34**, 307 (1975).

25. Ye. V. Baklanov, V. P. Chebotaev, B. Ta. Dubetsky, Appl. Phys. **11**, 201 (1976).

26. N. F. Ramsey, *Molecular Beams*, Oxford UP, London (1956), page 124.

27. J. C. Bergquist, S. A. Lee, J. L. Hall, Phys. Rev. Lett. **38**, 159 (1977).

28. M. Salour, C. Cohen-Tannoudji, Phys. Rev. Lett. **38**, 757 (1977).

29. R. Teets, J. Eckstein, T. W. Hänsch, Phys. Rev. Lett. **38**, 760 (1977).

30. R. Teets, R. Feinberg, T. W. Hänsch, A. L. Schawlow, Phys. Rev. Lett. **37**, 683 (1976).

31. M. E. Kaminsky, R. T. Hawkins, F. V. Kowalski, A. L. Schawlow, Phys. Rev. Lett. **36**, 671 (1976).   □

Coherent Raman spectroscopy

Once exotic and time-consuming, wave-mixing spectroscopy has burgeoned into a set of techniques that can handle systems—flames, plasmas, luminescent crystals—inaccessible to conventional methods.

Marc D. Levenson

PHYSICS TODAY / MAY 1977

In 1928 Chandrasekhara Raman reported a process in which a material would simultaneously absorb one photon and emit another. The energies of the two photons differed by an amount corresponding to the energy difference between two quantum-mechanical levels of the medium. Raman scattering, as the phenomenon came to be known, provided a tool for the spectroscopic investigation of energy levels not accessible by the usual absorption and emission techniques. For the first thirty-five years Raman scattering was a laborious and exotic technique, important more for the quantum-mechanical principles it illustrated than for its practical applications.[1] The development of gas lasers completely revolutionized the practice of Raman spectroscopy. Gone were the discharge lamps and hours-long photographic exposures; they were replaced by the cw laser, tandem monochromator and cooled photomultiplier. What had been a difficult and exotic technique became a routine analytical procedure for studying vibrational and other elementary excitations of materials.[1] The development of powerful tunable lasers now promises a second revolution. Rather than randomly scattering photons as in present techniques, the Raman modes of a medium studied by coherent Raman techniques are made to emit a beam of coherent radiation containing the details of the spectrum. Samples in which the spontaneous Raman scattering is intrinsically weak, or masked by fluorescence and black-body radiation, can now be studied.

The advantages of the coherent Raman techniques result from the fact that the laser fields at two different frequencies can force a particular Raman mode of a medium to produce an oscillating dielectric constant which then interacts with one of the fields to produce a coherent output beam. The power in this beam can be many orders of magnitude larger than that in the spontaneously scattered radiation, and spatial filtering can be used to separate the output beam from unwanted radiation.

The essence of the process can be derived from a simple model. If we describe a Raman mode by a normal coordinate Q, then the first-order dependence of the polarizability α_{ij} of the medium upon Q is

$$\alpha_{ij} = \alpha^0_{ij} + \alpha'_{ij}Q \qquad (1)$$

where i and j are two Cartesian coordinates and α' is just the conventional Raman susceptibility tensor. (In solids, the coordinate Q corresponds to a phonon with wavevector equal to the difference in wavevectors of the incident beams. For a polariton—a mode in a crystal lacking a center of inversion symmetry—the treatment is similar, but slightly more complex.) Because the electromagnetic free energy of such a medium contains a term proportional to QE^2, there will be generalized force on the coordinate Q that is bilinear in the applied optical fields. If the equation of motion for Q is that of a damped harmonic oscillator with frequency ω_R, one can use Newton's second law to calculate the response to this force

$$\ddot{Q} + 2\Gamma\dot{Q} + \omega_R^2 Q$$
$$= \frac{1}{2}(N_1 - N_2)\sum_{ij}\alpha'_{ij}E_iE_j \qquad (2)$$

As a concession to quantum mechanics I have included on the right a factor of $N_1 - N_2$, giving the difference in the population of the two levels separated by the Raman circular frequency ω_R.

If the electric field has frequency components at ω_1 and ω_2, the force will have Fourier components at $\pm(\omega_1 - \omega_2)$, which can drive the Raman mode resonantly when $|\omega_1 - \omega_2| \approx \omega_R$. The oscillating coordinate then gives a modulated polarizability according to equation 1. An electric polarization $\mathcal{P}$ at the frequency $\omega_4 = \omega_3 \pm (\omega_1 - \omega_2)$ results from the product of this oscillating polarizability and a Fourier component of the field at ω_3:

$$\mathcal{P}_i(\omega_4) \propto$$
$$\sum_{jkl}\left\{\frac{(N_1-N_2)\alpha'_{ij}\alpha'_{kl}}{\omega_R{}^2 - (\omega_1-\omega_2)^2 + 2i\Gamma(\omega_1-\omega_2)}\right\}$$
$$\times E_j(\omega_3)E_k(\omega_1)E_l{}^*(\omega_2) \qquad (3)$$

This polarization, which is cubic in the incident electric field amplitudes, acts as a source term in Maxwell's equation to produce[2] the output beam at ω_4. The quantity in braces is sometimes termed the Raman contribution to the third-order nonlinear susceptibility, $\chi^R{}_{ijkl}(-\omega_4, \omega_3, \omega_1, -\omega_2)$. If $N_2 = 0$, the overall four-photon parametric mixing process can be described by the level diagrams in figure 1.

There are other processes involving molecular reorientation and real or virtual electronic transitions that also contribute to the radiated signal. These nonresonant background signals are pretty much independent of $\omega_1 - \omega_2$ and parametrized by another term in $\chi^{(3)}$ denoted $\chi^{nr}{}_{ijkl}$. In some experiments the background level is interesting, in others merely a nuisance. The resonant Raman term interferes constructively and destructively with the background, producing a line-shape function with maxima and minima.

All the processes leading to output at ω_4 can be described by a third-order nonlinear susceptibility tensor $\chi^{(3)}{}_{ijkl}(-\omega_4, \omega_3, \omega_1, \omega_2)$. This is a fourth-rank tensor; of its four frequency arguments only three

Marc D. Levenson is an assistant professor of physics and electrical engineering at the University of Southern California, Los Angeles.

are independent because $\omega_1 + \omega_2 + \omega_3 - \omega_4 = 0$. By convention the frequency arguments and polarization subscripts are paired and can be permuted as long as their pairing is respected.[3] Symmetry considerations fortunately reduce the number of independent nonzero elements and permit classification of the symmetries of modes observed by coherent Raman techniques in different polarization conditions. Unlike the second-order nonlinear susceptibility responsible for optical second-harmonic generation, $\chi^{(3)}$ does not vanish identically for any symmetry group.[2] Thus the techniques of coherent Raman spectroscopy are generally applicable.

One major advantage of these techniques is the large signal produced at ω_4. The formalism of nonlinear optics can be used to estimate the power (in watts) radiated at this frequency as

$$P_i(\omega_4) = 2 \times 10^{-26} \frac{\omega_4{}^2}{n^4} \frac{l^2_{\text{eff}}}{A^2} \left| \sum_{jkl} \chi^{(3)}{}_{ijkl} \right|^2$$
$$\times P_j(\omega_3) P_k(\omega_1) P_l(\omega_2) \quad (4)$$

Factors of order 4 resulting from frequency degeneracy have been suppressed in equation 4, and $P_j(\omega_1)$, $P_k(\omega_2)$, and $P_l(\omega_3)$ are the incident laser powers in watts at circular frequencies ω_1, ω_2 and ω_3 polarized in the j, k and l directions, A is the area of the beams where they interact (in cm^2), the linear index of refraction of the medium is n, and the nonlinear susceptibility $\chi^{(3)}$ is in electrostatic units. The effective interaction length (l_{eff} in equation 4) is always less than the distance over which the beams coincide. It depends in a rather complex way upon the wavevector mismatch of the interacting beams $\Delta k = |\mathbf{k}_1 - \mathbf{k}_2 + \mathbf{k}_3 - \mathbf{k}_4|$ and upon the details of the interaction geometry. Generally the best performance is obtained when Δk is as small as possible; if it is large, the interaction length scales as $1/\Delta k$. Negligible values of Δk occur automatically in several techniques of coherent Raman spectroscopy. In others, the propagation directions of the incident beams must be adjusted to approach optimum wavevector matching.

The radiated power depends upon the absolute square of the Raman susceptibility and upon the square of the interaction length. For typical liquids and solids, these parameters have values of about 10^{-13} esu and 0.1 cm, while at standard temperature and pressure, typical gases have nonlinear susceptibilities 100 times smaller but permit interaction lengths 10 times longer. In a coherent Raman experiment in which 10-kW lasers are focussed into an area of 10^{-5} cm^2, equation 4 gives an output of 0.055 W for $\chi^{(3)} = 10^{-13}$ esu and $l_{\text{eff}} = 0.1$ cm, and 0.002 W for $\chi^{(3)} = 10^{-15}$ esu and $l_{\text{eff}} = 2$ cm, for output at 5000 Å.

The data-collection rate, however, is proportional to the rms output power.

For lasers pulses 6 nanosec long and a repetition rate of 15 pulses per sec, the rms output is 16.5 microwatts for a "typical" solid, and 0.6 microwatt for a "typical" gas. In a spontaneous scattering experiment in which 10^7 photons at 5000 Å are collected per second, the rms signal power is 4×10^{-12} W.

According to equation 4, the rms output power is proportional to the time average of a cubic product of laser powers. Taken literally, this implies that large, slowly firing pulsed lasers will give the best performance. In practice, modest pulsed lasers that have high repetition rates are more convenient. Excellent results have been obtained with cw lasers, especially when the sample can be inserted into the laser cavity.

The several advantages of observing Raman spectra by means of nonlinear optical mixing have been realized in a variety of coherent Raman spectroscopic techniques. Acronyms and neologisms such as CARS, RIKES, HORSES, CSRS, "Submarine," "Helicopter," "Asterisk," and so on have been invented to designate each variation of the basic four-photon parametric mixing process.[4-7] As a spectroscopic tool each technique has its own set of advantages and disadvantages, and it is important to match the technique properly to the investigation. The main alternatives are individually reviewed in the following sections.

Coherent anti-Stokes Raman spectra

In the most widely practiced technique of coherent Raman spectroscopy, two in-

cident laser frequencies are employed, and $\omega_3 = \omega_1$ in the level diagram shown in part **a** of figure 1. The output frequency is then $\omega_4 = 2\omega_1 - \omega_2$. If ω_1 corresponds to the laser frequency in a spontaneous-scattering experiment and ω_2 to the Stokes-scattered photon, the output occurs at the corresponding anti-Stokes frequency. (If ω_1 is less than ω_2, the analogous technique is called Coherent Stokes Raman Spectroscopy or CSRS—pronounced "scissors.") Paul Maker and Robert Terhune, using discrete frequencies, initially demonstrated this technique in 1965.[3] It did not become a practical spectroscopic tool until 1972, when several groups began to employ repetitively pulsed dye lasers for continuous scanning of the Raman spectrum.[11,12] An excellent review article recounts many of the recent results obtained using this technique, so there is no need for more than an overview here.[8]

Figure 2 depicts a typical CARS experimental setup, along with the outputs used in other coherent Raman techniques. Two lasers are focussed into a sample with an angle between the beams that best fulfills the wavevector-matching condition for the overall three-wave mixing process: $\Delta k = 0$. In gases this angle is essentially zero, but in condensed phases it depends upon $\omega_1 - \omega_2$ and the dispersion of the index of refraction.[9] The beam that emerges from the sample at ω_4 is selected by means of filters or a simple monochromator, and its intensity is detected photoelectrically. To scan, the difference frequency $\omega_1 - \omega_2$ is varied by

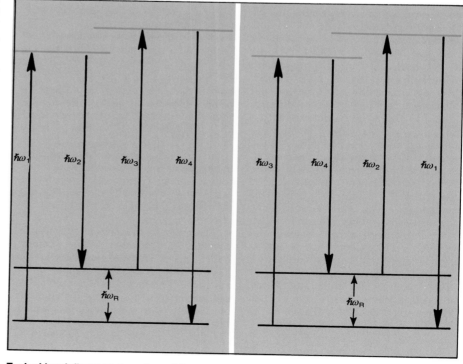

Typical level diagrams for coherent Raman processes. Physical optical fields must have both positive- and negative-frequency components, but the detected frequency must correspond to some sum of the inputs. The diagram on the left shows four-photon parametric mixing of the "CARS" type, and that on the left, a process of the "RIKES" type.

Figure 1

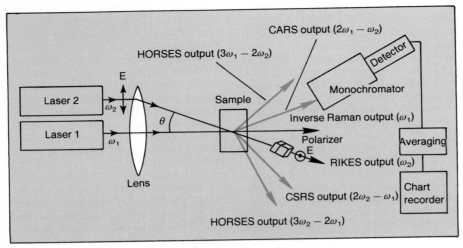

The layout of a typical CARS experiment, also showing the directions of the output beams for some other coherent Raman techniques: RIKES, CSRS, HORSES and the inverse Raman effect. In four-wave mixing an additional beam would be added, collinear with that from laser 1. The RIKES output has its electric-field vector polarized normal to the beam.   Figure 2

tuning one or both lasers. Plotted as a function of $\omega_1 - \omega_2$, the output reflects the Raman spectrum of the sample.

The tremendous power of this technique for taking Raman spectra of gases can be illustrated by a few examples. With laser powers of 2 MW at ω_1 and 0.2 MW at ω_2, and $\omega_1 - \omega_2$ at the Q branch of the breathing-mode ($\nu_1 = 0 \rightarrow 1$) transition in methane, Albert Harvey and Joseph Nibler found that the CARS output beam from a 0.1-Torr sample was strong enough to be seen by eye. Similar lasers 1000 times stronger are already being used in studies related to isotope separation. Conservatively, an improvement in sensitivity of three orders of magnitude can be expected when these more powerful lasers are used for CARS.

In an ingenious experiment with cw lasers, A. Hirth and K. Vollrath obtained the Q-branch spectrum of atmospheric pressure nitrogen that appears in figure 3—in ten milliseconds! Conventional scattering techniques would take nearly an hour to give a comparable resolution and signal-to-noise ratio. Even more rapid data collection is possible when the intensity is measured simultaneously in many separate frequency channels. That can be done conveniently by making laser 2 in figure 2 oscillate over a band of frequencies. The various frequency components of the output beam can then be separated with a spectrometer and recorded simultaneously on a multichannel detector such as a photographic plate, optical multichannel analyzer or vidicon camera. Won Roh, Paul Schreiber and Jean-Pierre Taran employed such a system to resolve the Q branch of the vibration quantum number v = 0 → 1 transition in H_2 with a single 20-nsec laser pulse. Rapidly evolving systems such as explosions and shocks can obviously be studied by this technique if the lasers are properly synchronized.

Wolfgang Kaiser and A. Laubareau

have used synchronized picosecond lasers in CARS-related experiments to measure the dephasing and decay times of phonons and of molecular vibrations. In these experiments, a delayed pulse at ω_1 samples the amplitude of the oscillation previously excited by simultaneous pulses at ω_1 and ω_2. The decrease in coherent anti-Stokes intensity with increasing delay time gives the dephasing, while the decrease in spontaneously scattered anti-Stokes radiation parametrizes the decay. Relaxation times of one or two picoseconds can be measured with good accuracy, and some questions related to the mechanisms for broadening of Raman lines can be answered.[10]

Some other virtues of the CARS technique deserve mention.

▶ The CARS technique has considerable potential for precision spectroscopy. The resolution of these experiments is limited by the laser linewidth, which can be readily reduced to 15 MHz and perhaps below. Inhomogeneous broadening due to the Doppler effect contributes a width of $\omega_R[(kT/2mc^2)\ln 2]^{1/2}$ where m is the mass of the molecule and all beams are collinear. That means that precision measurements of rotational and vibrational Raman frequencies can now be made precise enough to elucidate the structure of molecules lacking microwave absorption bands.

▶ The intensity of coherent anti-Stokes radiation emitted from a region of an inhomogeneous sample is related to the local concentration of the Raman active material. This fact has been exploited to produce images showing the distribution of substances in a flame or jet, and it might be applicable to living cells.[11]

Figure 4 illustrates one difficulty in applying CARS technology to solids, liquids and solutions. In that figure, the colored curve is the CARS spectrum of calcium fluoride in the region of the relatively weak 320 cm^{-1} mode. That mode

appears as a slight variation in the level of a signal dominated by a nonresonant background due to virtual electronic transitions.[12] For comparison, the black curve shows the spectrum obtained with up-to-date spontaneous scattering technology![13] An analysis of the line shape in the CARS spectrum reveals the ratio of the peak Raman contribution to the nonresonant background term. The nonresonant third-order susceptibility is of some interest in nonlinear optics, and because the Raman term can be related to well determined spontaneous-scattering cross sections, this technique gives accurate measurements of this quantity. However, to see weak spectroscopically interesting modes by coherent techniques it is necessary to suppress the nonresonant background. Without such suppression, the coherent Raman techniques are actually less effective than spontaneous scattering in detecting Raman spectra of condensed phases.

Raman-induced Kerr effect (RIKES)

The simplest method promising background suppression employs the Raman-induced Kerr effect proposed originally by Robert Hellwarth.[5] This phenomenon results from a source polarization of the form given in equation 4, but with $\omega_3 = -\omega_1$; its level diagram is shown in part **b** of figure 1. The wave that results from the coherently driven vibration is at the same frequency as one of the inputs (that is, $\omega_4 = -\omega_2$) but polarization selection rules are employed to ensure that the radiated field is in a different state of polarization than the input laser. Essentially, the driven vibration produces an intensity- and frequency-dependent birefringence, which alters the polarization condition of a wave probing the sample. The wavevector-matching condition is automatically fulfilled, and if the ω_1 beam is circularly polarized, the nonresonant background is eliminated.

The early RIKES experiments employed a linearly polarized probe beam at ω_2 and a circularly polarized pump at ω_1. After the sample, a crossed polarization

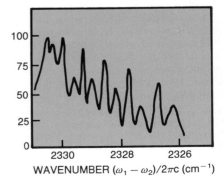

The Q branch of nitrogen at atmospheric pressure as it appears in a CARS trace. The ordinate is the square root of the intensity at $2\omega_1 - \omega_2$, which is roughly linear in the Raman scattering cross section.   Figure 3

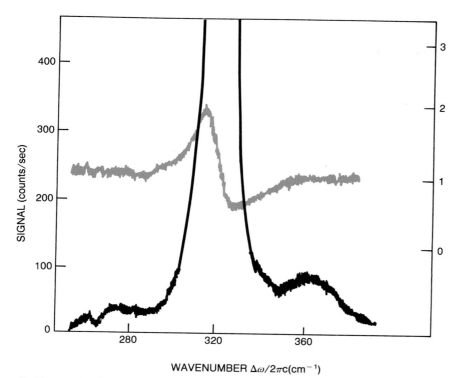

Two traces of a phonon mode in calcium fluoride. The colored line is a CARS trace; the black line is the same 320-cm^{-1} mode, as resolved with spontaneous scattering and pulsed lasers. The vertical scales have been adjusted to give about equal noise levels.   Figure 4

analyzer blocked the probe intensity except when $|\omega_1 - \omega_2| = \omega_R$, in which case a small transmission was detected. The overall experimental scheme was otherwise quite similar to figure 2. Slight birefringence due to strains in the sample or optics led to a small background transmission of the probe frequency even when the pump beam was blocked.

This background, however, could be used to enhance the sensitivity of the technique. The field radiated as the result of the coherent Raman process interferes constructively or destructively with the background, producing an enhanced change in transmitted intensity. If the pump laser is then modulated, electronic techniques can detect the quite small modulations in the transmitted intensity caused by the coherent Raman process. Background-free Raman spectra are obtained with a sensitivity limited only by the fluctuations in the probe-laser intensity. This technique also can be used to separately determine the real and imaginary parts of the complex nonlinear polarization in equation 3 as well as the real and imaginary parts of the total nonlinear susceptibility tensor. The differential cross section due to spontaneous Raman scattering is proportional to the imaginary part of this susceptibility, and so spectra taken with this technique can have a familiar appearance. The signal also scales linearly with density and cross section, facilitating the identification of modes by their relative intensities.

Spectra similar to figure 4 appear when

a pump is employed linearly polarized at an angle to the polarization of the probe. Analysis of the resulting line shape relates the Raman cross section, optical Kerr constant and other terms in $\chi^{(3)}$ to one another. RIKES also has advantages for studying low-frequency modes. When $\omega_1 - \omega_2$ approaches zero, the CARS wave-vector-matching condition requires collinear propagation, making separation of the signal beam difficult. No such geometrical restriction applies to RIKES, and polarization selection and spatial filtering easily extract the signal.

Four-wave mixing techniques

Providing a third input frequency adds valuable extra degrees of freedom to CARS and RIKES experiments. If the detected frequency is at $\omega_4 = \omega_1 + \omega_3 - \omega_2$, Raman resonances occur when $|\omega_1 - \omega_2| = \omega_R$ and when $|\omega_3 - \omega_2| = \omega_R$. The line shapes observed when the two different frequencies excite different Raman modes determines the ratio of the Raman cross sections. Alternatively, $\omega_3 - \omega_2$ can be set to a frequency at which the Raman contribution from one mode nearly cancels the nonresonant background. The sensitivity with which Raman modes are detected near $\omega_1 - \omega_2$ is markedly increased.[14]

Nonresonant background signals can be completely eliminated with CARS and four-wave mixing in certain polarization configurations. The most flexible of these is the "Asterisk" configuration shown in figure 5, which works for the four-wave analogs of CARS ($\omega_4 = \omega_1 + \omega_3 - \omega_2$) and of

RIKES ($\omega_4 = \omega_2 - \omega_1 + \omega_3$).[7] The planes of polarization of the input waves at ω_1, ω_2 and ω_3 are at 45° to one another, while a polarization analyzer selects the component of the output at the angle ϕ. For some particular value of ϕ—generally near 45°—the background level will vanish. With the background gone, photons appear at frequency ω_4 only when a Raman resonance condition exists. Weak Raman modes can be detected with a sensitivity limited only by the quantum nature of light. The difference between CARS and Asterisk spectra is demonstrated in figure 6.

The strength of the observed intensity depends quadratically upon the Raman cross section and the concentration, and on the product of the intensities of the three lasers. The 1977 practical limit in liquid solution corresponded to the detection of a 0.03-molar solution of benzene in a Raman-inactive solvent. That may be enough sensitivity to study biologically interesting materials in dilute solution if the Raman tensor is enhanced by preresonance phenomena. With stronger lasers lower concentrations and weaker modes can be investigated.

Other techniques and applications

Gain and loss due to the stimulated Raman effect can also be used for spectroscopic purposes.[15] These phenomena are described by a polarization of the form in equation 3 with $\omega_3 = -\omega_1$ and $\omega_4 = -\omega_2$ as in RIKES, but no polarization selection is necessary. An input laser at ω_2 will be amplified by the Raman interaction if $\omega_1 - \omega_2 \approx \omega_R$ in the stimulated Raman effect, and attenuated if $\omega_2 - \omega_1 \approx \omega_R$ by the so-called "inverse Raman effect." With enough intensity at ω_1, large signals at the Stokes frequency can be built up from noise and large anti-Stokes intensities can be completely absorbed. These techniques enjoyed considerable interest at one time, but enthusiasm waned when it

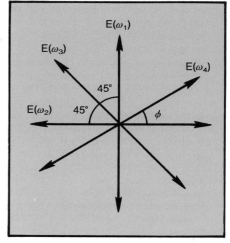

The "asterisk" polarization condition. All the input waves are plane polarized as shown; an analyzer selects the desired component of the output wave.   Figure 5

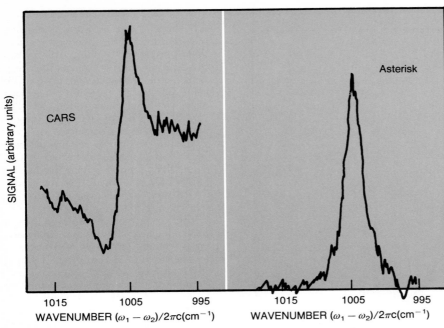

CARS and asterisk spectra of the 1005-cm^{-1} mode of sodium benzoate in a $\frac{1}{3}$-molar water solution. The nonresonant background due to the solvent has been suppressed in the asterisk spectrum by a factor of more than a thousand, enhancing the detection sensitivity. **Figure 6**

became apparent that only the strongest Raman modes produced measurable differences in the intensity at ω_2.

Improved technology—especially the development of stable cw dye lasers and the application of interferometric techniques to nonlinear spectroscopy—has revived interest in these methods. Using a Jamin interferometer, Adelbert Owyoung has measured the Raman amplification and intensity-dependent dispersion due to 992-cm^{-1} mode of benzene to determine the absolute scattering cross section. Changes in laser intensity of one part in 10^5 due to the stimulated Raman effect are detectable. Improved laser stabilization and detection strategies have made the quasi-cw gain techniques the most sensitive methods in coherent Raman spectroscopy of gases.

When the output beam of an optical mixing experiment becomes strong enough, it too can mix with one of the incident fields to drive a vibration and create a new output frequency. In this way coherent second Stokes and second anti-Stokes beams have been generated in the imaginatively named Higher Order Raman Spectral Excitation Studies (HORSES), which oddly enough were demonstrated after CARS.[6] These extra frequency components may prove useful in doing Raman spectra of absorbing samples.

The technology of coherent Raman spectroscopy can also be used to study other kinds of transitions. One- and two-photon absorption processes lead to resonances in the "nonresonant" contribution to the third-order nonlinear susceptibility and result in observable variations in the output-wave intensity.[2]

Steven Kramer and Nicolaas Bloembergen have used such a two-photon resonance to probe the Z_3 exciton in CuCl.[16] The homogeneous linewidth of an inhomogeneously broadened one-photon transition can be estimated from the linewidth of the CARS "Rayleigh resonance" observed when $\omega_1 - \omega_2 \to 0$ in an absorbing material. The analogous RIKES experiment is a variation of the "polarization spectroscopy" technique reviewed on page 34 in this issue of PHYSICS TODAY by Theodor Hänsch. Tatsuo Yajima has pointed out that a detailed analysis of the line shape of the Rayleigh resonance permits estimation of both the longitudinal and transverse relaxation times for an inhomogeneously broadened two-level system.[17]

When to use these techniques?

Coherent Raman spectroscopy today provides a valuable supplement to the conventional technique of spontaneous scattering. But the techniques are all rather cumbersome, relying on newly developed technology, while the apparatus necessary for scattering has benefited from fifteen years and more of engineering development. When is the extra effort—and extra expense—worthwhile?

At present, if an investigation *can* be performed by using spontaneous scattering with modest lasers, it *should* be done in that way. Certain systems—low-pressure gases, flames, plasmas, luminescent molecules and crystals—can not be easily studied by scattering. Again, certain parameters—nonlinear susceptibility tensor elements, dephasing and decay times, rotational constants—can not be measured with sufficient precision, if at all. These must be investigated by coherent techniques.

As time goes on, coherent Raman spectroscopy will become more convenient and less expensive, and the difficult tasks reserved for coherent Raman spectroscopy will become more routine. Continued technological innovation may ultimately make the coherent Raman techniques as accessible as spontaneous scattering is today.

* * *

This work has been supported by the National Science Foundation and the Alfred P. Sloan Foundation.

References

1. M. C. Tobin, *Laser Raman Spectroscopy*, Wiley, New York (1971); T. R. Gilson, P. J. Hedra, *Laser Raman Spectroscopy*, Wiley, New York (1970).

2. C. Flytzanis, in *Quantum Electronics*, volume 1A (H. Rabin, C. Tang, eds.), Academic, New York (1975); C. Flytzanis, N. Bloembergen, in *Progress in Quantum Electronics* (J. H. Sanders, S. Stenholm, eds.), volume 4, part 3, Pergamon, New York (1976).

3. P. D. Maker, R. W. Terhune, Phys. Rev. A **137**, 801 (1965).

4. R. F. Begley, A. B. Harvey, R. L. Byer, Appl. Phys. Lett. **25**, 387 (1975).

5. D. Heiman, R. W. Hellwarth, M. D. Levenson, G. Martin, Phys. Rev. Lett. **36**, 189 (1976).

6. I. Chabay, G. K. Klauminger, B. S. Hudson, Appl. Phys. Lett. **28**, 27 (1975).

7. J. J. Song, G. L. Eesley, M. D. Levenson, Appl. Phys. Lett. **29**, 567 (1976).

8. W. M. Tolles, J. W. Nibler, J. R. McDonald, A. G. Harvey, Appl. Spectroscopy **31**, 96 (1977).

9. A. B. Harvey, J. R. McDonald, W. M. Tolles, in *Progress in Analytical Chemistry*, Plenum, New York (1977).

10. W. Kaiser, A. Laubareau, in *Tunable Lasers and Applications* (A. Mooradian, I. T. Jaeger, P. Stoketh, eds.), volume 3, Springer, New York (1976), page 207.

11. P. R. Regnier, in *Laser Raman Gas Diagnostics*, (M. Lapp, C. M. Penney, eds.) Plenum, New York (1974).

12. M. D. Levenson, N. Bloembergen, Phys. Rev. B **10**, 4447 (1974).

13. P. Yaney, J. Raman Spectroscopy **5** (to be published).

14. H. Lotem, R. T. Lynch Jr, N. Bloembergen, Phys. Rev. A **14**, 1748 (1976).

15. W. T. Jones, B. P. Stoicheff, Phys. Rev. Lett. **13**, 657 (1964).

16. S. D. Kramer, N. Bloembergen, Phys. Rev. B **14**, 4654 (1976).

17. T. Yajima, Opt. Comm. **14**, 378 (1975). □

Coherent optical transients

A new branch of optical spectroscopy that deals with the optical
analogs of spin transients such as NMR is providing unique ways to explore dynamic
interactions in optically excited atoms, molecules and solids.

Richard G. Brewer

PHYSICS TODAY / MAY 1977

The dynamics of nuclear spins in atomic and molecular environments have been examined in exquisite detail over the past 27 years by the methods of pulsed nuclear magnetic resonance.[1] The early discovery of the spin echo by Erwin Hahn[2] and the transient nutation effect by Henry Torrey[3] initiated the field by introducing coherent transient phenomena to radio-frequency spectroscopy. An entire class of coherent spin transients could be observed by a simple variation of the rf pulse sequence, and from the decay characteristics the various spin-dephasing mechanisms could be examined separately. In short, these methods allowed decomposition of the spin-transition linewidth into its broadening components, and also offered new and versatile ways for performing high-resolution rf spectroscopy.

Optical coherence is also a mature subject. Well known examples predating the laser are Young's two-slit interference effect, which demonstrates spatial coherence; the Michelson interferometer, which demonstrates temporal interference, and the Brown–Twiss experiment, which demonstrates an intensity-correlation effect. With laser light, however, a new class of optical coherence phenomena is now at hand, consisting of effects that are the *optical analogs* of spin transients. In this article I review this new branch of optical spectroscopy, which is providing unique ways for exploring dynamic interactions in optically excited atoms, molecules and solids.

The possibility of bringing the methods of pulsed NMR to the optical region was not immediately obvious. Because coherent radiation is required, optical studies were excluded prior to the laser, that is, before 1960. Furthermore, it was

Richard G. Brewer is an IBM Fellow at the IBM Research Laboratory in San Jose, California.

not clear whether optical electric-dipole transitions behaved in the same way as the rf magnetic-dipole transitions of spin systems. This situation was clarified in a well known paper[4] of Robert Dicke, who showed that the two are equivalent. In either case, a collection of two-level quantum systems can be prepared coherently in superposition states, and these constitute a phased array of dipoles (electric or magnetic), which can emit coherent radiation as prescribed by Maxwell's equations.

The photo on the cover of May 1977 issue of PHYSICS TODAY shows one example of a coherent transient in the optical region; it is observed by a laser frequency-switching technique described below.

Maxwell–Bloch equations

The equivalence of magnetic- and electric-dipole transients was further demonstrated by Richard Feynman, Frank Vernon and Robert Hellwarth,[5] who transformed the Schrödinger equation into the three-dimensional vector equation

$$\frac{d\boldsymbol{\beta}}{dt} = \boldsymbol{\omega} \times \boldsymbol{\beta} \qquad (1)$$

This has the form of the classical torque equation of motion for a spin precessing in a magnetic field and was used originally by Felix Bloch[6] to describe NMR. Equation 1 is commonly referred to as the Bloch equation. It is a geometric representation of a two-level quantum system interacting resonantly with a radiation field, acting either through an electric- or a magnetic-dipole interaction. The precessional motion of the Bloch vector $\boldsymbol{\beta}$ about an effective field $\boldsymbol{\omega}$ (in frequency units) therefore applies to either situation, where we understand that the coordinate system in equation 1 rotates with

the angular frequency of the radiation field about the z axis.

For the spin-½ case, the Bloch vector $\boldsymbol{\beta}$ is a magnetic dipole moment with three projections in real space, and the effective field $\boldsymbol{\omega}$ is the vector sum of a static magnetic field (along the z axis) and an rf magnetic field (along the x axis) in the rotating frame, all multiplied by the gyromagnetic ratio.

For a two-level quantum system interacting with an optical wave, the Bloch vector

$$\boldsymbol{\beta} = \mathbf{i}u + \mathbf{j}v + \mathbf{k}w \qquad (2)$$

does not lie in physical space, but rather in a mathematical space with components

$$u = \tilde{\rho}_{12} + \tilde{\rho}_{21}$$
$$v = i(\tilde{\rho}_{21} - \tilde{\rho}_{12}) \qquad (3)$$
$$w = \rho_{22} - \rho_{11}$$

where ρ is the density matrix and the rapidly oscillating factor is removed with the substitution $\rho_{12} = \tilde{\rho}_{12} \exp[i(\Omega t - kz)]$. This assumes an electric-dipole interaction $V = -\boldsymbol{\mu}\cdot\mathbf{E}$ for an optical field $E = E_0 \cos(\Omega t - kz)$, which resonantly excites a transition $1 \rightarrow 2$ having a dipole matrix element μ_{12} and a transition frequency ω_{21}. Here, u and v represent the in-phase and out-of-phase components of the optically induced dipole

$$p = \text{Tr}(\mu\rho) \qquad (4)$$

while the population difference of the transition levels is given by Nw, N being the atomic number density.

The effective field now takes the form

$$\boldsymbol{\omega} = -\mathbf{i}\chi + \mathbf{k}\Delta \qquad (5)$$

where χ is the Rabi flopping frequency $\mu_{12}E_0/\hbar$. For the case of a moving atom with a velocity component v_z along the laser beam, the resonant tuning parameter, Δ, equals $\Omega + kv_z + \omega_{21}$ and the Doppler shift is kv_z.

The optically prepared dipoles given by equation 4 generate a coherent signal field

$$E_s(z,t) = \bar{E}_s(z,t)e^{i(\Omega t - kz)}$$
$$+ \text{ complex conjugate} \quad (6)$$

which obeys Maxwell's wave equation in the slowly-varying-envelope approximation

$$\frac{\partial E_s}{\partial z} = -2\pi i k N \langle \bar{p} \rangle \quad (7)$$

when the sample is optically thin, the dipole amplitude being $\bar{p} = \mu_{12}\bar{\rho}_{12}$. Because atomic and molecular environments are always inhomogeneous, the dipoles radiate with a distribution of frequencies and exhibit temporal interference. The angular bracket in equation 7 therefore sums over the *inhomogeneous* line broadening, which is due to Doppler broadening for a gas and to crystalline strain broadening for a solid.

Decay phenomena

With the inclusion of damping terms due to *homogeneous* line broadening, the coupled Maxwell–Bloch equations are sufficiently general to describe many spin and optical coherent transient phenomena. Decay due to atomic collisions, spontaneous radiative emission or other causes can depopulate the lower and upper transition levels, labelled 1 and 2 respectively, with phenomenological decay rates γ_1 and γ_2.

In the density-matrix formulation of the Schrödinger equation,[7] these rates correspond to decay of the diagonal elements ρ_{11} and ρ_{22}. As a consequence, the dipole and the off-diagonal element ρ_{12} dephase at the rate

$$\gamma = \tfrac{1}{2}(\gamma_1 + \gamma_2) \quad (8)$$

There also exist processes that disrupt the phase of the dipole without depopulating either level so that equation 8 becomes

$$\gamma = \tfrac{1}{2}(\gamma_1 + \gamma_2) + \gamma_\phi \quad (9)$$

where γ_ϕ expresses the rate of phase interruptions.

When the radiative frequency changes with time because of spectral diffusion within the inhomogeneous line shape, the equations of motion must also be modified to allow for a decay that is no longer a simple exponential (see equation 11, below). In the language of NMR, the dipole terms u and v decay at the rate $1/T_2$ and the population term w decays at the rate $1/T_1$. In the above terms $\gamma = 1/T_2$, but T_1 is defined only when w is characterized by the single decay rate $1/T_1$, for example, when $\gamma_2 = \gamma_1$ or $\gamma_1 = 0$. The description in terms of T_1 and T_2 appears to be valid for transitions of molecular vibrational states in the infrared,

as in NMR, but in the visible and ultraviolet regions the spontaneous emission rate can be large so that γ_1 and γ_2 are required,[8] as well as equations 8 and 9.

Echoes, free induction decay, nutation

The spin-echo concept, extended to the optical region in 1964 by Norman Kurnit, Isaac Abella and Sven Hartmann, was the first example of this class of optical coherence effects.[9] The group irradiated a ruby crystal with *two* short pulses of coherent light from a ruby laser and observed *three* equally spaced pulses in transmission, as shown in figure 1. The third pulse was a delayed spontaneous burst of coherent light, which they called a *photon echo* by analogy with the spin echo.

These studies revealed that the electron spin of the optically excited chromium ion of ruby dephased due to the presence of the surrounding aluminum nuclei through the mutual flipping of the electronic and nuclear spins. The photon-echo method has been applied also in other ways and to other systems, for example, by Hartmann and co-workers to photon-echo nuclear double resonance in ruby and most recently to low-temperature organic solids.[10]

An example of the photon-echo effect in a molecular gas[11] is shown in figure 2. The photon echo is an interference effect involving a coherent set of oscillating dipoles that dephase in the first pulse interval because of a spread in their frequencies (destructive interference) and rephase in the second pulse interval (constructive interference). The Bloch vector model shows this symmetric time behavior as a sequence of four precessional motions in figure 2 where the solutions of equation 1 are obtained by inspection.

Nutation For an initial laser pulse sufficiently long and intense, an atom will be driven first to the upper state (stimulated absorption) and then back to the lower state (stimulated emission), the cycle repeating thereafter until the end of the pulse. Since the laser beam is alternately absorbed and emitted by the sample, the intensity of the transmitted beam will display an oscillation as shown[11] in figure 3. This is the optical analog of the spin–nutation transient; it was first seen by G. B. Hocker and Chung Tang,[12] using a pulsed CO_2 laser, in an infrared transition of SF_6.

In terms of equation 1, the Bloch vector $\boldsymbol{\beta}$ points in the $-z$ direction for an atom initially in the lower state. With the application of a pulse, $\boldsymbol{\beta}$ precesses about the effective field $\boldsymbol{\omega} = -\mathbf{i}\chi + \mathbf{k}\Delta$, causing the level-population difference (the projection of $\boldsymbol{\beta}$ on the z axis) and the dipole (the projection of $\boldsymbol{\beta}$ on the x–y plane) to oscillate at $\omega = (\chi^2 + \Delta^2)^{1/2}$, the precession frequency. For atoms exactly on resonance ($\Delta = 0$), the precession frequency is the Rabi frequency χ, and the higher the light intensity the faster the oscillation.

If the pulse width t_1 is now reduced so that the precession angle χt_1 equals $\pi/2$, the Bloch vector of the resonant group will be totally in the x–y plane at the end of the pulse and the induced dipole will be a maximum, as in part **a** of figure 2. From equation 4, the off-diagonal element ρ_{12} will be a maximum also, corresponding to an equal admixture of the upper- and lower-state wave functions.

Free induction decay Immediately after the pulse, the dipoles are in phase and, according to equations 6 and 7, emit an intense coherent beam of light. This emission, shown in figure 4, was demonstrated initially by Richard Shoemaker

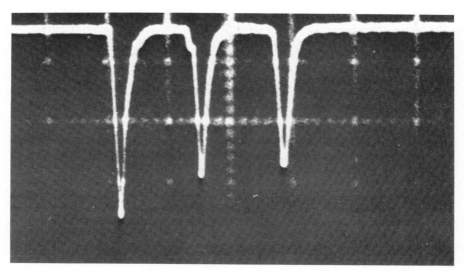

A photon echo from a ruby sample at 4.2 K appears as a delayed third pulse after two successive ruby-laser pulses excite the crystal coherently. Time increases to the right at 100 nanosec/division. The photon echo was the first of a useful new class of optical coherence effects. Photo from N. A. Kurnit, I. D. Abella and S. R. Hartmann, reference 9. **Figure 1**

and me[13]; it is the optical analog of the free induction decay first seen by Hahn in NMR. Notice that because the fields of the N dipoles add in phase, the emission intensity will vary as N^2 and so will far exceed spontaneous emission, which varies as N. Because of momentum (or **k**) conservation, coherent emission can only occur in the forward direction.

In short, the emission due to free induction decay resembles the laser light that produced it, but with one difference: Because of inhomogeneous line broadening the dipoles radiate with a distribution of frequencies and thus interfere as time evolves, causing the emission to decay. The dephasing behavior of these frequency "packets" is seen in figure 2**b**, which shows the Bloch vector processing about $\omega = \mathbf{k}\Delta$ (since $\chi = 0$) for different frequencies Δ. We will now see that the echo is a free induction decay that has rephased.

Photon echo The second laser pulse tips the Bloch vector so that the sign of the dipole phase is reversed. This is shown in figure 2**c** for the resonance case, where the pulse duration t_2 gives a precession angle $\chi t_2 = \pi$. Other tipping angles and Δ's give echoes with smaller amplitudes. Following the second pulse, each Bloch vector in figure 2**c** precesses about the effective field $\omega = \mathbf{k}\Delta$ in the same sense as during the first pulse interval, but because the slow packets (labelled "s" in the figure) lead the fast ones, "f," it is clear that all packets will come into phase precisely at time 2τ, where τ is the pulse delay time. At this point the sample has a macroscopic dipole moment and will emit a burst of coherent light—the photon echo. By this dephasing–rephasing mechanism, the echo amplitude

is unaffected by the large inhomogeneous linewidth, whereas the irreversible dephasing effects of homogeneous broadening remain.

Stark switched molecules

Optical studies have progressed rather slowly compared to the field of spin transients, because pulsed lasers are more difficult to control than rf pulse generators. A new method for observing coherent optical transients, particularly in the infrared, was introduced[11] by Shoemaker and me in 1971, and possesses several distinct advantages. In this case, the *exciting laser radiation* is *cw* and the *molecular level splitting* is *pulsed*. The apparatus is diagrammed in figure 5.

The underlying idea rests on the fact that molecules at a pressure of a few millitorr can exhibit exceedingly sharp optical resonances. Laser light, which is essentially monochromatic, will be strongly absorbed only when its frequency closely matches, to one part in 10^8–10^9, the molecular resonance frequency. If by some method the molecular transition frequency is shifted suddenly outside its homogeneous linewidth, the absorption can be switched on or off nonadiabatically. The switching mechanism used is the Stark effect, in which the application of a pulsed dc electric field allows molecules to be tuned to or away from the laser frequency. Instead of using pulsed laser light, which is difficult to control, the molecule's energy-level spacing is pulsed while the laser's frequency and intensity remain fixed in time.

The transients illustrated in figures 2 and 3 have been obtained in this way for a particular $C^{13}H_3F$ vibration–rotation transition in the 10-micron infrared re-

gion, which will be referred to throughout this article. The beam from a stable cw CO_2 laser passes through the gas sample, which is contained in a Stark cell, before striking a photodetector that monitors the absorption or emission transients.

Imagine in figure 6 that, prior to a Stark pulse, a laser with frequency Ω selects from a molecule's Doppler line shape a particular velocity group v_z, which it coherently prepares under steady-state conditions. When a Stark pulse appears, this group is switched out of resonance and freely radiates an intense, coherent beam of light at frequency Ω' in the forward direction—the free-induction-decay transient. At the same time, a second velocity group v_z' will be switched into resonance and will alternately absorb and emit laser radiation—the nutation effect. If two Stark pulses are applied, the group v_z' emits a third pulse—the photon echo.

The important advantages inherent in the Stark switching technique now become evident:

▶ The transient observed is the desired coherent transient itself, and the difficulty of separating a small transient signal from a time-coincident laser pulse of large amplitude is avoided.

▶ Heterodyne detection occurs because the emission transient is displaced in frequency from the laser light by the Stark shift and together with the laser beam strikes a photodetector. This increases the free-induction-decay and echo signals by about three orders of magnitude and facilitates measuring their decay.

▶ The entire class of coherent optical transient effects can be monitored, because the electronic pulse sequence and pulse shape can be tailored conveniently to the particular experiment of interest. To date, some ten different coherent optical transient effects have been observed in this way.[14]

Molecular collisions

The Stark switching method is well suited to the study of molecular collision phenomena.[15] Elastic and inelastic scattering can be examined easily with two-pulse echoes and without the complication of Doppler broadening. This work complements the more elaborate molecular-beam technique and contrasts with traditional steady-state linewidth measurements, in which the broadening mechanisms are rarely disentangled. The two-pulse sequence of figure 2 provides two independent measurements as the pulse delay time is advanced:

▶ The echo amplitude decreases because collisions dephase the coherently prepared dipoles.

▶ The second or delayed nutation signal grows, due to collisions that partially restore the equilibrium population from the imbalance created by the first pulse.

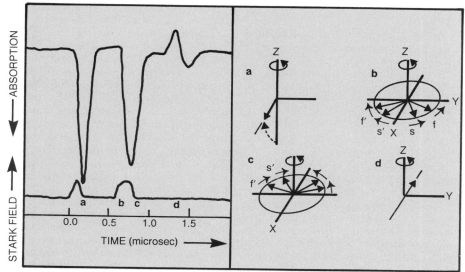

The third pulse in the upper trace at the left is an infrared photon echo for a vibration–rotation transition in $C^{13}H_3F$. The 60 V-cm Stark pulses shown in the lower trace switched the gas sample into resonance twice with a cw carbon-dioxide laser at 9.7 microns. The four diagrams at the right indicate four stages in the precessional motion of the Bloch vector, corresponding to the times marked **a**, **b**, **c** and **d** on the pulse sequence. From reference 11.   Figure 2

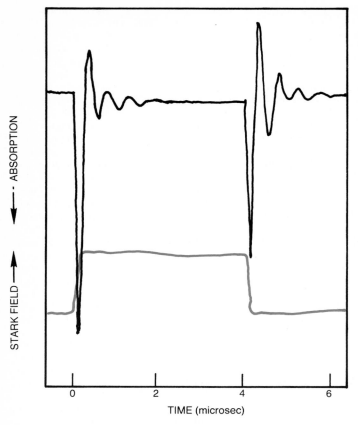

The optical nutation effect in methyl fluoride, $C^{13}H_3F$, irradiated by a carbon-dioxide laser at 9.7 microns. The apparatus for this experiment as well as that of figure 2 is shown schematically in figure 5. The Rabi oscillations appear here because the 35-volt/cm Stark pulse (colored line) is longer than those in figure 2. From R. G. Brewer and R. L. Shoemaker, reference 11.   Figure 3

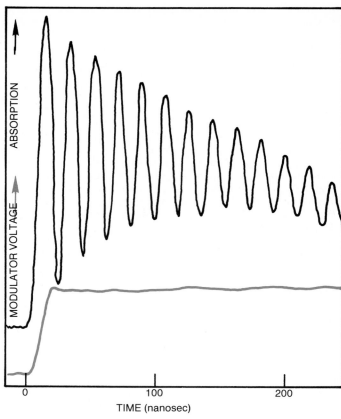

Optical free induction decay in I_2 vapor produces a heterodyne beat signal with frequency-shifted laser light. The sample is prepared with 5896-Å light from a cw dye laser; it radiates coherently when the frequency is abruptly switched 54 MHz by a 100-volt pulse. The slowly varying background is a nutation signal from a second velocity group. Figure 9 shows the experimental arrangement. (Ref. 18).   Figure 4

The delayed nutation and echo amplitudes for $C^{13}H_3F$ are shown in figure 7 as a function of pulse delay time. The nutation signal (upper curve) displays the decay time (T_1) for inelastic collisions, the cross section being about 500 Å^2. Elastic collisions, on the other hand, alter the molecular velocity, causing the echo signal to deviate from the upper curve. Because a collision-induced velocity change Δv_z translates as a Doppler shift $k \Delta v_z$ and a corresponding phase shift $k \Delta v_z t$ in time t, it is clear that even if Δv_z is small the dipoles will eventually get out of phase.

Writing the echo electric field in the form

$$E \propto \langle \exp(ik \Delta v_z t) \rangle$$

where $\langle \ \rangle$ denotes a collision average, we can use a simple Brownian-motion argument[15] to derive the echo time dependence due to elastic collisions. Let Δu be $\sqrt{2}$ times the rms change in velocity per collision and τ the pulse delay time. When $k \Delta u \tau \gg 1$, any collision produces destructive phase interference so that the only contribution to the field average E that survives will be the one where no collision occurs up to the time of echo formation $t = 2\tau$. Since the associated probability is $e^{-\Gamma t}$, where Γ is the rate of elastic collisions, one finds for long times

$$E(t = 2\tau) \propto \exp(-\Gamma t)$$
$$\text{when} \quad k \Delta u \tau \gg 1 \quad (10)$$

If, on the other hand, $k \Delta u \tau$ is much less than 1, each collision produces a small phase change such that $E \propto 1 - \frac{1}{2} k^2 \langle (\Delta v)^2 \rangle \tau^2$, where $\langle \Delta v \rangle = 0$. With the collision average $\langle (\Delta v)^2 \rangle = \Gamma t \Delta u^2 / 2$, we get for short times

$$E(t = 2\tau) \propto \exp[-\Gamma t^3 (k \Delta u)^2 / 16]$$
$$\text{when} \quad k \Delta u \tau \ll 1 \quad (11)$$

It agrees essentially with a more rigorous theory based on the Boltzman transport equation.[15]

These calculations also agree with the experiments of figure 7 and constitute primary evidence that elastic molecular collisions involving small changes in velocity *play a crucial role in infrared photon echoes.* The cubic decay law (equation 11) is reminiscent of the dephasing that occurs in spin echoes when spin diffusion in liquids takes place in a magnetic-field gradient.[2] In these optical experiments diffusion occurs in velocity space due to low-angle elastic scattering, for which the characteristic velocity jump per collision in $C^{13}H_3F$ is only $\Delta u = 85$ cm/sec and the corresponding total cross section is about 430 Å^2.

Carr–Purcell echoes A variation on this theme is to replace the two-pulse echo by an n-pulse echo sequence, which is easily done by applying $n + 1$ Stark pulses.[15] If the pulse interval τ is short enough, velocity or spectral diffusion can be made negligible, and the decay time of the echo envelope function is the residual dipole-dephasing time, simply T_2. This is demonstrated in figure 7, where we see, from the coincidence of the Carr–Purcell and nutation-decay curves, that $T_2 = T_1$. The n-pulse photon echo is an optical version of the NMR method of Herman Carr and Edward Purcell.[16]

The basic concept can be explained by noticing that for an n-pulse echo train with short pulse intervals τ, the amplitude of the last echo at time $t = 2n\tau$ is given by the nth power of the field E in equation 11; that is, $[\exp(-8K\tau^3)]^n = \exp(-Kt^3/n^2)$. In comparison to that given by equation 11 for a two-pulse echo over the same time interval t, it is evident that the Carr–Purcell decay time is n^2 times longer.

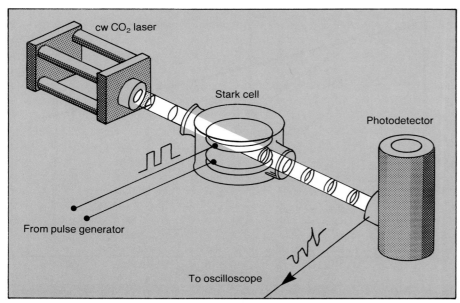

Schematic of the Stark switching apparatus that is used for observing optical transients such as those illustrated in figures 2 and 3. From reference 11.   Figure 5

Fourier-transform spectroscopy

The method of Fourier transforming transient phenomena from the time to the frequency domain has proven to be an extremely versatile technique in pulsed nuclear magnetic resonance. With it, ultrahigh-resolution NMR spectroscopy can be performed quickly and with high sensitivity in a set of densely spaced lines. Furthermore, because the NMR signals display coherent transient behavior, dynamic information about nuclear spin interactions can be derived in a selective manner for each transition as well.

An application of this technique in the optical region[17] is shown in figure 8. The spectrum is obtained from several $C^{13}H_3F$ transitions that produce two-pulse photon echoes simultaneously. By advancing the pulse delay time, the echo decay for each transition can be monitored also. This yields a three-dimensional diagram of echo amplitude versus frequency and elapsed time. Measurements are performed with the Stark switching apparatus, which is now interfaced with a computer for rapid Fourier analysis. The four lines are actually heterodyne beat signals of the same Stark-split $C^{13}H_3F$ vibration–rotation transition, the ν_3 mode $(J,K,M) = (4,3,M) \rightarrow (5,3,M)$ mentioned earlier, each beat frequency being due to two transitions with $M \rightarrow M$ and $-M \rightarrow -M$. The lines are 170 kHz wide, are spaced at 0.83-MHz intervals and clearly display a Doppler-free behavior, as the Doppler width is 66 MHz. From echo and free-induction-decay measurements of this kind, elastic and inelastic scattering can be studied *independently* again, but since the lines are now resolved, the cross sections can be obtained as a function of *quantum state* and *molecular ve-*

locity. This degree of flexibility offers stringent tests of existing scattering theories and is being used to study the force laws that dominate molecular collisions.

Laser frequency switching

Azriel Genack and I recently reported a general method for observing coherent transients in the visible–ultraviolet region[18] (Search and Discovery, PHYSICS TODAY, October 1976, page 17). The

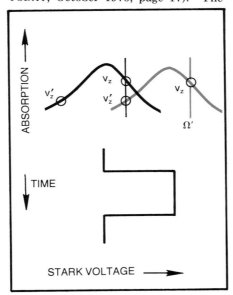

Stark switching principle for a Doppler-broadened transition. The laser, of fixed frequency Ω, initially excites molecules of velocity v_z. A Stark pulse that abruptly shifts the molecular transition frequency from the black to the colored curve causes the velocity group v_z to emit the free induction decay signal at frequency Ω', while the group with velocity v_z' exhibits nutation. With two pulses, the group v_z' emits an echo.   Figure 6

technique, diagrammed in figure 9, requires switching the laser frequency into or out of resonance with the sample; it is equivalent to the Stark switching method. However, when the source is a stable tunable cw dye laser, this approach is more universal in its application, particularly because the need for Stark tunable molecules is removed. Laser frequency switching is exceedingly simple and can be achieved merely by driving an electro-optic crystal (ammonium dihydrogen phosphate), which is inside the dye laser cavity, with the desired sequence of low-voltage pulses.

The time-dependent variations in the refractive index of the electro-optic element produce corresponding changes in the optical cavity length, and hence in the laser frequency. Dynamically the effect is equivalent to the Doppler shift the laser beam would experience on being reflected by a moving end mirror of the cavity. The frequency shift therefore occurs almost instantaneously. The switching time is limited by the transit time of light through the ADP crystal, about 50 picosec, and not by the rise time of the voltage pulse or by the laser-cavity ringing time, about 25 nanosec. In addition, the frequency-shifted light is stored in the optical cavity and amplified in the same way that the unshifted light was.

Clearly, a resonant sample exposed to frequency-switched laser light will experience coherent transient behavior. Because the experiment is controlled electronically, all of the advantages inherent in the Stark switching technique are preserved here as well. Moreover, with the large tuning range of the dye laser, measurements are equally feasible in atoms, molecules and solids.

In initial studies in molecular iodine vapor, effects such as free-induction-decay (figure 4), nutation and echoes are easily monitored in the visible. Because the signal-to-noise ratio is high, about 10^3, it has been possible to test the theory of these transient phenomena quantitatively and in complete detail. From two-pulse nutation and echo studies of I_2 it is found that elastic collisions are not of the velocity-changing type, as in the infrared, but rather are due to collision-induced frequency shifts. In this case, the echo decays exponentially with a decay rate given by equation 9, in contrast to equation 11. Apparently, excited electronic states are more sensitive to perturber-induced frequency shifts than vibrationally excited states are. This appears to be the first optical-coherence measurement of phase-interrupting collisions.

In a solid, laser frequency switching has been used at IBM to measure the dephasing time of coherently prepared impurity ions of praseodymium in a low-temperature LaF_3 host crystal.[19] Free-induction decay is monitored and possesses certain advantages over the echo technique. As can be seen in figure 4, the entire decay is obtained in a single burst following a step-function switching pulse. Because of the sensitivity offered by heterodyne detection, weak transitions can be studied at laser powers as low as 50 microwatts. At laser power levels, where $\chi^2 \ll 1/(T_1 T_2)$, the free-induction-decay time is simply $\frac{1}{2}T_2$ where the $\frac{1}{2}$ factor arises because T_2 determines the inhomogeneous bandwidth during steady-state preparation and adds an additional $1/T_2$ decay rate. The rather long Pr^{3+} dephasing time of 0.38 microsec corresponds to an extremely narrow linewidth of 830 kHz (resolution: 6×10^8) and is unaffected by phonons below 4 K, power broadening, Pr^{3+}–Pr^{3+} interactions or the large inhomogeneous width of 4 GHz arising from crystal strains. On the other hand, it far exceeds the 640-Hz limit due to spontaneous radiative decay of the upper level.

Magnetic hyperfine interactions constitute a possible dephasing mechanism currently under study by several groups including Hartmann (Columbia University), Lynden Erickson (NRC, Ottawa) and Raymond Orbach (UCLA). Thus these measurements begin to invade a time scale ranging from microseconds to milliseconds and will impose significant demands on laser frequency stability.

At the other extreme, the fastest dephasing time measured by laser frequency switching is about 8 nanosec, as seen in the free induction decay of the sodium D lines. It will be challenging to see whether this method can be extended in the future to even shorter times, perhaps to the 50-picosec range.

Picosecond pulse generation

A related development, intense optical pulses of 30 picosec duration generated by free induction decay in a high-density gas, has attracted interest in laser fusion.[20] In the preparative step, a CO_2 laser pulse 200 nanosec long and 1 MW in power passes through a pair of lenses before resonantly exciting a CO_2 gas sample at a pressure of about 250 Torr. At a certain time during the pulse the excitation ends abruptly, due to dielectric breakdown of the gas in the focal region between the lenses; at this point free induction decay commences. Because nearly all of the laser light was absorbed during preparation, the emission will be intense initially and thereafter will decay in the time $T_2/(\alpha L)$ when $\alpha L > 1$; here α is the absorption coefficient and L the sample length. Hence, in this high-density regime, the decay can be considerably faster than the dipole-dephasing time T_2.

Two-photon transients

The one-photon processes considered up to this point are for the most part optical analogs of NMR transient phenomena. Coherent two-photon transients have been observed in the optical region

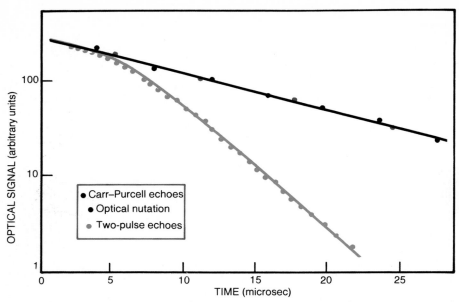

Decay curves for three coherent infrared transient effects, observed in $C^{13}H_3F$ by Stark switching. The Carr–Purcell and nutation curves coincide. From reference 15.   Figure 7

also and provide additional information about dephasing processes.

Picosecond studies Foremost among the two-photon processes are the Raman transients, which were exploited in simple liquids by Wolfgang Kaiser and his coworkers,[21] who used picosecond laser pulses, as from a mode-locked neodymium glass laser of wavelength 1.06 microns. (See also the accompanying article by Marc Levenson on steady-state aspects of Raman scattering in this issue of PHYSICS TODAY.) Vibrational decay times on a picosecond time scale have been obtained by first generating a coherent vibrational excitation and Stokes light through the stimulated Raman process. A weak delayed second pulse then probes the remaining vibrational excitation by generating phase-matched anti-Stokes radiation. The vibrational dephasing time is monitored through the dependence of the anti-Stokes intensity on pulse delay time. In liquid carbon tetrachloride, for example, the totally symmetric vibrational mode shows a dephasing time T_2 of 3.6 picosec. Moreover, the decay exhibits beats at a frequency of about 10^{11} Hz, due to the different isotopic species Cl^{35} and Cl^{37}.

Raman echoes Stimulated spin-flip Raman scattering has been used to induce coherent spin states[22] of bound donors in n-type CdS at 1.6 K, in the same way that vibrational modes are prepared coherently in molecular liquids. This is accomplished by exciting the sample with a pulsed dye laser at 4905 Å with two modes having a frequency difference $\omega_L - \omega_R = 32$ GHz, which is just resonant with a pair of Zeeman-split spin states. After pulse preparation the spins dephase, as in free induction decay, and if a second optical pulse is applied the spins will rephase, as in an echo experiment. The spin dephasing–rephasing behavior can be monitored optically by suitably delaying

a weak probe laser pulse at 4880 Å to produce Stokes scattering. The probe Raman scattering is enhanced at the time the spins have rephased and is called the Raman echo. Echoes of this type, which were predicted by Hartmann, have been observed[22] with delay times to 162 nanosec in dilute samples of CdS (of density 8 $\times 10^{15}$ cm^{-3}).

Coherent Raman beats A different manifestation[23] of a two-photon transient can be found in a molecular gas with the Stark switching technique when three molecular levels are prepared initially in superposition by means of continuous laser radiation. We imagine that two of the levels (1 and 2) are degenerate and connect optically with a third (3) during a resonant steady-state preparative phase. With the sudden application of a dc Stark field, the molecular level degeneracy is broken and the laser beam is no longer in resonance with the one-photon transitions, $1 \leftrightarrow 3$ and $2 \leftrightarrow 3$. However, the laser beam can now scatter off the coherently prepared sample in a two-photon process $1 \leftrightarrow 2$; that is, coherent forward Raman scattering can occur in the presence of the same laser field during the nonresonant condition. The two beams—the laser beam with frequency Ω and the Raman light with frequency $\Omega \pm \omega_{21}$—strike a photodetector, where they produce a coherent beat at frequency ω_{21}. This is just the level splitting between initial and final states in the two-photon process.

From energy conservation, we see that the first-order Doppler shift kv_z in the first half of the two-photon process ($1 \leftrightarrow 3$) essentially cancels that of the second half ($3 \leftrightarrow 2$), making the overall process $1 \leftrightarrow 2$ independent of Doppler shift. As a result, inhomogeneous dephasing and elastic collisions that change the molecular velocity do not alter the decay rate of the Raman beat signal. That the decay

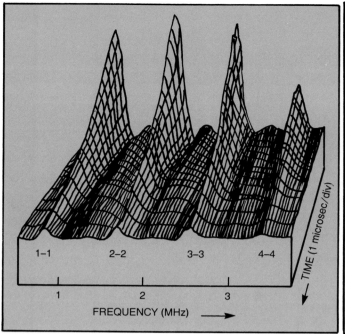

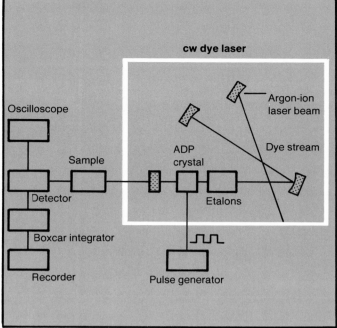

A Fourier-transform spectrum of $C^{13}H_3F$, derived from Stark-switched two-pulse photon echoes, shows beats in this three-dimensional diagram of echo amplitude as a function of frequency and pulse-delay time. The vibration–rotation transitions 1–1, 2–2, . . . , exhibit decay with increasing pulse delay time. From reference 17. **Figure 8**

Laser frequency switching apparatus for observing coherent optical transients. In the box is a commercial cw dye laser with an electro-optical crystal of ammonium dihydrogen phosphate. When the crystal is driven by low-voltage dc pulses its refractive index changes, abruptly changing the laser frequency. From reference 18. **Figure 9**

will be determined only by inelastic collisions, in contrast to the two-pulse echo results, is corroborated by theoretical arguments and experiments.[15,23]

Two-photon nutation A two-photon transient can be induced also by the sudden application of two light pulses of frequency Ω_1 and Ω_2, such that the transition $1 \to 2$ satisfies the resonance condition $\Omega_1 + \Omega_2 = \omega_{21}$. In this case the energy of the intermediate level 3 lies roughly half way between 1 and 2, and the two waves must be counterpropagating so that the Doppler shifts nearly cancel, as in the Raman beat effect. Exact pulse or steady-state density-matrix solutions[23] for the three-level problem show in fact that the Raman beat effect with a single beam and two-photon absorption of oppositely directed laser beams are basically the same phenomenon, the difference being one of level configuration. The pulse solutions also reduce in an appropriate limit to a simple Bloch vector model,[24] resembling equation 1, as it involves only the initial and final states in the overall transition $1 \to 2$. The intermediate level does not enter explicitly when the pulses are applied slowly (adiabatically) with respect to the frequency offset of the intermediate level, that is, when the intermediate transitions are nonresonant.

Michael Loy[25] has observed an infrared two-photon vibrational transient of this type in ammonia, which is induced by two counterpropagating CO_2 laser pulses. The transient oscillates because of the nutation effect, in quantitative agreement with the two-photon vector model. It is interesting that two-quantum transients in NMR are now also being explored.

Milliseconds to picoseconds

New ways for observing a class of coherent optical transient phenomena— measurements that resemble pulsed NMR transients—are now possible with the availability of laser radiation. Other novel effects[26] such as superradiance, quantum beats, adiabatic rapid passage and self-induced transparency, which were not discussed here, could have been included as well. Many of these coherence effects are just beginning to be useful in examining energy transfer and dephasing processes in optically excited atoms, molecules and solids, with time scales spanning the range from milliseconds to picoseconds.

* * *

The author gratefully acknowledges the thoughtful comments and criticisms of A. Z. Genack and A. Szöke. This work was supported in part by the US Office of Naval Research.

References

1. A. Abragam, *The Principles of Nuclear Magnetism*, Oxford, London (1961).
2. E. L. Hahn, Phys. Rev. **80**, 580 (1950); PHYSICS TODAY, Nov. 1953, page 4.
3. H. C. Torrey, Phys. Rev. **76**, 1059 (1949).
4. R. H. Dicke, Phys. Rev. **93**, 99 (1954).
5. R. P. Feynman, F. L. Vernon, R. W. Hellwarth, J. Appl. Phys. **28**, 49 (1957).
6. F. Bloch, Phys. Rev. **70**, 460 (1946).
7. M. Sargent III, M. O. Scully, W. E. Lamb Jr, *Laser Physics*, Addison-Wesley, Reading, Mass. (1974), page 84.
8. A. Schenzle, R. G. Brewer, Phys. Rev. A **14**, 1756 (1976).
9. N. A. Kurnit, I. D. Abella, S. R. Hartmann, Phys. Rev. Lett. **13**, 567 (1964); Phys. Rev. **141**, 391 (1966); S. R. Hartmann, Scientific American, April 1968, page 32.
10. T. J. Aartsma, D. A. Wiersma, Phys. Rev. Lett. **36**, 1360 (1976); A. H. Zewail, T. E. Orlowski, D. R. Dawson, Chem. Phys. Lett. **44**, 379 (1976).
11. R. G. Brewer, R. L. Shoemaker, Phys. Rev. Lett. **27**, 631 (1971).
12. G. B. Hocker, C. L. Tang, Phys. Rev. Lett. **21**, 591 (1968).
13. R. G. Brewer, R. L. Shoemaker, Phys. Rev. A **6**, 2001 (1972).
14. R. G. Brewer, in *Proceedings of the Les Houches Summer School Lectures, Volume I, Applications of Lasers to Atomic and Molecular Physics*, Les Houches, France, 1975 (R. Balian, S. Haroche, S. Liberman, eds.), North-Holland (in press).
15. P. R. Berman, J. M. Levy, R. G. Brewer, Phys. Rev. A **11**, 1668 (1975).
16. H. Y. Carr, E. M. Purcell, Phys. Rev. **94**, 630 (1954).
17. S. B. Grossman, A. Schenzle, R. G. Brewer, Phys. Rev. Lett. **38**, 275, 1977.
18. R. G. Brewer, A. Z. Genack, Phys. Rev. Lett. **36**, 959 (1976).
19. A. Z. Genack, R. M. Macfarlane, R. G. Brewer, Phys. Rev. Lett. **37**, 1078 (1976).
20. E. Yablonovitch, J. Goldhar, Appl. Phys. Lett. **25**, 580 (1974); **30**, 158 (1977).
21. A. Laubereau, G. Wochner, W. Kaiser, Phys. Rev. A **13**, 2212 (1976).
22. P. Hu, S. Geschwind, T. M. Jedju, Phys. Rev. Lett. **37**, 1357 (1976).
23. R. L. Shoemaker, R. G. Brewer, Phys. Rev. Lett. **28**, 1430 (1972); R. G. Brewer, E. L. Hahn, Phys. Rev. A **11**, 1641 (1975).
24. D. Grischkowsky, M. M. T. Loy, P. F. Liao, Phys. Rev. A **12**, 2514 (1975).
25. M. M. T. Loy, Phys. Rev. Lett. **36**, 1454 (1976).
26. L. Allen, J. H. Eberly, *Optical Resonance and Two Level Atoms*, Wiley, New York (1975).

□

Laser linewidth

Detailed theoretical and experimental study of the quantum effects that limit the spectral purity of lasers has led to semiconductor lasers whose output is highly monochromatic and coherent.

Aram Mooradian

PHYSICS TODAY / MAY 1985

One of the most important properties of laser light is its spectral purity and coherence. This unique quality has been important for the study of many new physical phenomena using laser sources that operate from the vacuum ultraviolet to the far infrared. An understanding of the mechanisms responsible for the broadening of the linewidth is necessary for the development of laser sources with sufficient spectral purity for various applications.

The broadening of the linewidth in lasers is caused by two mechanisms: the so-called "technical" noise due to mechanical vibrations of the laser cavity and other sources of external noise, and the "fundamental" line broadening due to quantum fluctuations. In this article I will discuss the fundamental limits of spectral purity, with particular emphasis on the semiconductor laser. I will also show that one can reduce the fundamental linewidth of a semiconductor laser by coupling it to an external cavity (see figure 1).

The need for spectral purity

A large number of research and technical applications require lasers with a high degree of spectral purity. The linewidth and phase stability of the light source are very important for fiber-optical sensors. The sensors detect changes in the signal phase that are caused by external influences—such as temperature, pressure, or magnetic fields—in a laser signal passing through an optical fiber.

The introduction of tunable lasers with narrow linewidths revolutionized optical spectroscopy. High-resolution spectroscopy using tunable lead-salt diode lasers emitting in the infrared has already proven to be useful. Linear-absorption spectroscopy of Doppler-broadened gases involves linewidths of several tens of megahertz in the infrared to about one gigahertz in the visible region. Nonlinear-absorption spectroscopy requires linewidths that are 100–1000 times narrower. Recent experiments aimed at detecting gravity waves require ultra-narrow-linewidth lasers having widths and stabilities in the subhertz region.[1] Small perturbations of massive Earth-based interferometers could in principle detect the emission of gravitons emitted by large intergalactic objects such as black holes or binary stars. Optical-fiber and line-of-sight laser communication using heterodyne detection require that the spectral width of the laser source usually be less than 1% of the transmitted data rate to achieve a low enough error rate in the information signal. For example, a 1-gigabit data-rate system would require a laser spectral width of less than 10 MHz.

Fundamental linewidth

Two processes contribute to the generation of photons in a laser: stimulated emission, the process responsible for laser amplification in which the radiation emitted by an atom is exactly in phase with the radiation that surrounds the atom; and spontaneous emission due to the transition between upper and lower laser levels caused by the finite lifetime of the upper level.

Because the field from a laser is usually well polarized and nearly monochromatic, we can represent the electric field as the real part of the complex field

$$\mathbf{E}(\mathbf{r}, t) = E(t)\hat{n}e^{i\mathbf{k} \cdot \mathbf{r}}$$

where $\hat{n}$ is the unit polarization vector and $\mathbf{k}$ and ω are the wavevector and frequency of the emitted light. The time variation of the complex amplitude $E(t)$ due to spontaneous emission photons gives rise to the finite linewidth of the radiation. If we plot the values $E(t)$ takes on over a period of time, we get a graph like that shown in figure 2a for a laser. For incoherent light, values of $E(t)$ in the complex plane will take on all the values shown in figure 2b.

In addition to the stimulated emission photons, the spontaneous emission photons that radiate into the laser mode add as a small vector to the laser field vector (see figure 2a) to produce a new laser field vector with field amplitude $E + \Delta E$ and phase angle $\phi + \Delta\phi$. Because spontaneous emission is incoherent, its phase is random; the result-

Aram Mooradian is leader of the quantum electronics group of the Massachusetts Institute of Technology Lincoln Laboratory.

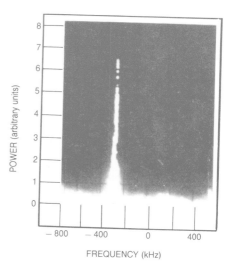

a

Diode lasers: The upper photo **(a)** shows a compact external-cavity diode laser; the adjoining trace shows a heterodyne beat note between two external-cavity diode lasers of a different construction, indicating a linewidth of less than 15 kHz. The lower photo **(b)** shows a semiconductor laser without an external cavity. The adjoining trace shows that the linewidth is much broader.

Figure 1

b

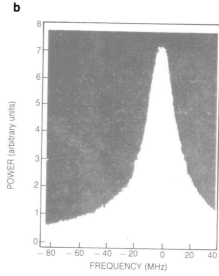

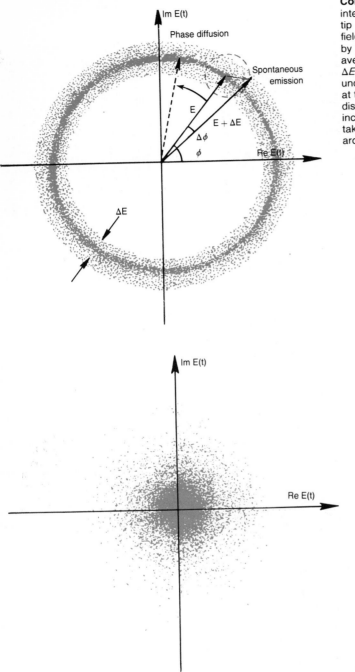

Complex representation of the field intensity. Each dot represents the random tip of the field vector. In **a** we show the field intensity of laser light as it is modified by spontaneous emission photons. The average spread of radial field intensity is ΔE, while the laser-field phase angle undergoes a diffusion from its initial value at time $t = 0$. For comparison, the distribution of the complex amplitude of incoherent light is shown in **b**. There $E(t)$ takes on a set of random values distributed around the origin.   Figure 2

large number of photons. A convenient way of describing[2] the phase diffusion, which is equivalent to a linear Brownian motion, is by the correlation of the field at time t with the field at time zero: $\langle E^*(t)E(0)\rangle$. It can be shown that the changes in ϕ have a Gaussian distribution. In this case,

$$\langle E^*(t)E(0)\rangle \approx E(0)^2 e^{i\omega t - \langle\Delta\phi^2\rangle/2} \quad (2)$$

From equations 1 and 2 we see that the correlation function decays exponentially

$$\langle E^*(t)E(0)\rangle = E(0)^2 e^{i\omega t - t/\tau_c} \quad (3)$$

where $\tau_c = 4I/R$ is the laser coherence time (a measure of the average time it takes for the laser phase to become uncorrelated from its initial value), ω is the laser frequency and the brackets denote a statistical average. The Fourier transform of equation 2 leads to a Lorentzian line shape with a spectral linewidth given by:

$$\Delta\nu = \frac{R}{4\pi I} \quad (4)$$

An alternative way of expressing equation 4 in terms of experimentally measurable parameters is

$$\Delta\nu = \frac{\pi h\nu\Gamma^2}{P} \quad (5)$$

where P is the power in the laser mode and Γ is the "cold cavity" Q of the laser, that is, the linewidth of the passive cavity resonator. In their first paper on the laser, Arthur L. Schawlow and Charles H. Townes derived[3] a value for the linewidth that is twice that given in equation 5. Later, Lax showed that the Schawlow–Townes analysis only applies to lasers operating below threshold. Above threshold, the lasing action stabilizes the field-amplitude fluctuations, reducing the predicted linewidth by a factor of two and leading to the field distribution shown in figure 2. The change in linewidth in going from below threshold to above threshold was calculated[4,5] and verified[6] experimentally in a herculean experiment. Equation 5 is the well-known modified Schawlow–Townes linewidth[2] that had been accepted for many years as describing the laser linewidth. The treat-

ing field is anywhere on the small circle shown in figure 2a. The magnitude of the field intensity does not remain at the new value but is restored to the average value by the coupling of the radiation field to the population inversion in the laser medium. The differential equations of motion that describe this coupling show that the perturbed field amplitude returns to its equilibrium value by undergoing damped oscillations, as shown in figure 3.

The phase angle, however, does not have a force that restores it to the original phase. The repeated occurrence of spontaneous emission events thus causes the phase angle to diffuse

from its initial value in a random walk, in which the mean square phase change $\langle\Delta\phi^2\rangle$ increases linearly in time. Melvin Lax showed[2] that

$$\langle\Delta\phi^2\rangle = \frac{R}{2I}t \quad (1)$$

where R is the total spontaneous emission rate (photons/sec) and I is the total number of photons—or the optical intensity—in the mode. As shown in figure 2a, the laser field at time t has an uncertainty in its phase compared with the value at $t = 0$. Each dot near the circumference of the circle represents the possible position of the field vector after the spontaneous emission of a

ment I have given here is similar in many respects to that developed, even before the laser, for electronic oscillators driven by electrical noise. In a laser oscillator above threshold, the population inversion reaches a stable level, and any increase in the excitation rate shows up as increased laser power. Because the amount of spontaneous emission is proportional to the population in the upper level, the spontaneous emission rate also becomes stabilized above threshold. Figure 2a shows that the magnitude of phase-angle fluctuation $\Delta\phi$ decreases linearly with field amplitude, providing a physical argument for the inverse power dependence of $\langle\Delta\phi^2\rangle$ and the laser linewidth.

Most of the early work in the study of linewidth was on gas lasers. The parameter values for, say, a typical helium–neon gas laser will, according to equation 5, lead to linewidths of about a hundredth of a hertz. In practice, such linewidths are difficult to measure and even more difficult to achieve. For such lasers, the observed linewidth is dominated by technical noise. Most laser systems require exceptional efforts to stabilize and reduce the effects of technical noise to approach more closely the fundamental linewidth. At present, the best results for laser-frequency control have been obtained with a dye laser locked to an acoustically isolated solid-quartz Fabry–Perot etalon. Two such independent lasers, when heterodyned together, showed a beat frequency stable to a few hertz.[7]

Semiconductor laser linewidth

The modified Schawlow–Townes description of the fundamental laser linewidth remained the established view for many years. For gas lasers under most operating conditions, this was indeed an adequate description. The first measurements[8] of the fundamental linewidth of semiconductor lasers were performed on a lead-salt diode laser operating near a wavelength of 10 microns. By heterodyning the output of the diode laser with a stable carbon-dioxide laser whose linewidth had already been demon-

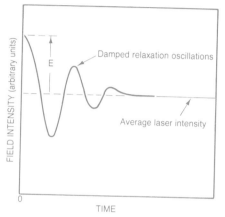

Laser-field intensity fluctuations following an impulse of spontaneous emission that produces an impulse change of field intensity of ΔE. The field-intensity scale is exaggerated.   Figure 3

strated to be less than a few hundred hertz, the lineshape of the diode laser was shown to be Lorentzian with a width predicted by equation 5. The demonstration of the first continuously operating diode lasers at room temperature, using crystals of the alloy GaAlAs, resulted in widespread interest in the practical use of such devices and accelerated the development of high-quality, low-cost lasers. The use of precision microelectronic fabrication techniques produced lasers that operated reliably in a single frequency, allowing careful measurements of the fundamental linewidth. Our measurements[9] showed the line-broadening mechanisms for GaAlAs and later GaInAsP diode lasers to be more complex than predicted by equation 5.

The first precision measurements[9] of the linewidth of a GaAlAs diode laser at room temperature showed that the width increased linearly with increasing reciprocal output power (see figure 4a) as expected, but with a magnitude some 50 times greater than that predicted by equation 5.

An additional broadening mechanism that modifies equation 5 comes from the incomplete population inver-

sion between the laser energy levels. While in most lasers the lower energy level is usually empty, this is not the case for semiconductor lasers, especially at room temperature. Laser photons can be absorbed by exciting electrons from the valence band (or an impurity level) to the conduction band, from which they may reradiate either stimulated or spontaneous photons. The additional spontaneous photons thus created increase the fluctuations in field intensity. The linewidth given by equation 5 is increased by a factor n_s, the spontaneous emission factor, which is the ratio of the spontaneous emission rate per mode to the stimulated emission rate per laser photon. This factor is about 2.5 at room temperature and becomes unity by 77 K for GaAlAs lasers. The remaining discrepancy factor of 20 was explained[7] by Charles Henry; it is due to spontaneous emission events that alter the field amplitude as shown in figure 2a. The discrepancy had been considered[2] earlier for gas lasers in which the cavity mode is detuned from the atomic transitions, but it had not been observed experimentally.

In semiconductors, laser action takes place at energies below the strong interband absorption edge, and this discrepancy effect is dominant. When the field intensity changes, it returns to the steady-state average value by undergoing damped relaxation oscillations, as shown in figure 3. During this time (a few nanoseconds), both the real and imaginary part of the refractive index change as the electron population density also changes in response to the field. The imaginary index change provides changes in gain that restore the steady-state amplitude. The change in the real part of the refractive index produces a phase change that adds an additional broadening to the laser linewidth, which can be significantly larger than the broadening produced by the instantaneous phase changes. The expression for the power-dependent linewidth is given by

$$\Delta\nu = \Delta\nu_{\mathrm{ST}}(1 + \alpha^2)n_s \qquad (6)$$

where $\Delta\nu_{\mathrm{ST}}$ is the modified Schawlow–

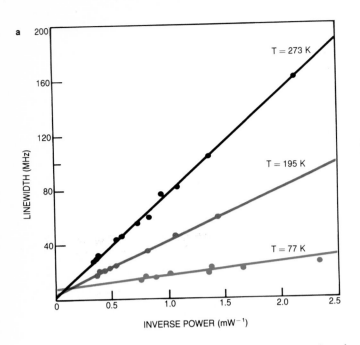

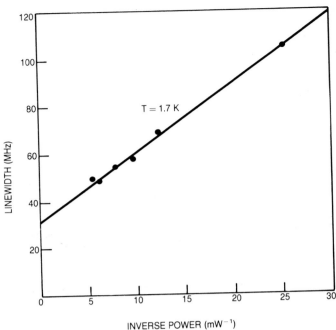

Linewidth of a single-frequency GaAlAs diode laser as a function of reciprocal output power at temperatures of 273, 195, 77 (**a**) and 1.7 K (**b**). The straight lines are least-square fits to the data.   Figure 4

Townes linewidth given by equation 4 or 5 and α is the ratio of the change of the real part to the change in the imaginary part of the refractive index. We verified[11,12] experimentally that equation 6 accounts for the observed power-dependent linewidths to within 10–15% at four temperatures between 273 and 1.7 K for GaAlAs diode lasers. The experimental results are shown in figure 4.

Additional phenomena contribute to the laser lineshape. The Fabry–Perot scan of the output from a GaAlAs diode laser (see figure 5) shows the presence of sidebands separated by about two gigahertz from the main laser line. These sidebands come from the damped relaxation oscillations that occur every time the laser-field intensity is perturbed by the spontaneous emission noise.[13,14] The relaxation oscillations cause $\langle \Delta\phi^2(t) \rangle$ to undergo damped sinusoidal oscillations at short times.[15,16] The temporal behavior of these intensity fluctuations was already shown in figure 3. The integrated intensity of these sidebands relative to the integrated intensity of the main peak increases linearly with increasing reciprocal output power. For typical GaAlAs diode lasers, this fractional value is a few tenths of a percent at output power levels of about ten milliwatts. The presence of these sidebands can interfere with many applications requiring a high degree of spectral purity.

An additional source of broadening is evident in figure 4. The linewidth intercept is nonzero and the magnitude of the intercept becomes significantly larger at low temperature. This is equivalent to the presence of a non-power-dependent line broadening,[17] which adds to the power-dependent effects already discussed above. The nonzero intercept may be explained by the very small actual size of these semiconductor lasers. The region in which population inversion occurs is on the order of $0.1 \times 2 \times 200$ microns or 4×10^{-11} cm^3. The absolute number N of conduction electrons is also relatively small, ranging from about 10^8 at room temperature down to about 10^6 at 1.7 K for typical lasers. If one assumes that the number of these conduction electrons fluctuates statistically so that the root-mean-square fluctuation in the electron number is $\sqrt{N}$ we can construct a phenomenological model for the fluctuation of the laser frequency that involves the fluctuation of the Fabry–Perot resonant frequency of the laser cavity. The cavity-mode frequency fluctuations are related to changes in the refractive index n via the phenomenological relation

$$\delta v = (v/n)(\mathrm{d}n/\mathrm{d}N)\sqrt{N} \qquad (7)$$

where $\mathrm{d}n/\mathrm{d}N$ is evaluated at the laser frequency v. The parameters in this equation, evaluated[17,12] at four temperatures, lead to a remarkably close agreement between the observed and calculated power-independent linewidths.

At first sight it is not entirely clear why the conduction-electron density should fluctuate at all according to the equations of motion, except at the relaxation frequency, and this only leads to the relaxation frequency sidebands discussed above. A number of experimenters, however, have observed[18,19] that both the amplitude and frequency of the laser fluctuate with a power spectrum that depends inversely on the frequency. The presence of $1/f$ noise in semiconductor electronic devices is well known and can stem from carrier traps that exist in the material. Carrier trapping,[20] or any other mechanism that could influence the electron number density or temperature[21] without being affected by the laser field, could in principle contribute to the power-independent linewidth. More work needs to be done to understand this phenomenon completely.

All of the line-broadening mechanisms I have described above are for monolithic diode lasers where the natural reflectivity of the cleaved ends of the diode provides the mirrors that define the cavity. The fundamental spectral width is substantially larger than most other types of lasers and inadequate for a number of applications requiring greater spectral purity. One of the ways to overcome all of the fundamental broadening is by coupling the laser to an external resonator to increase the cavity Q. By increasing the total length of the laser so that most of the cavity consists of air or vacuum, the fractional contribution to

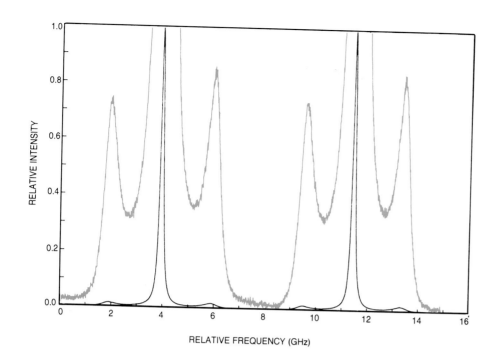

RELATIVE INTENSITY

RELATIVE FREQUENCY (GHz)

Fabry–Perot spectral scan of the output of a single-frequency GaAlAs diode laser at room temperature over a range of two free spectral scans. Sidebands separated by about 2 GHz are due to relaxation oscillations that are driven by the spontaneous emission noise in the device. An amplification of the trace by a factor 50 is shown in red.

Figure 5

the fundamental linewidth can be reduced by several orders of magnitude. This follows immediately from equations 4 and 5. Increasing the cavity size without increasing the amount of active material increases the total optical intensity I (the total number of photons in the mode) while leaving the spontaneous emission rate R unchanged. Figure 1a shows an external-cavity laser structure together with a heterodyne-beat signal between two external-cavity GaAlAs diode lasers of a somewhat different design. For comparison, the output of a monolithic diode laser without such a cavity is shown in figure 1b. Linewidths for external-cavity diode lasers have been shown to be less than 15 kHz and governed by only the technical noise. These devices can be made very stable by the incorporation of active control techniques such as locking to stable passive cavities or to microwave standards[22] to reach sub-hertz stabilities. The cost of such lasers could be much lower than other laser sources, and experiments requiring multiple lasers could be carried out more readily. A semiconductor laser such as GaAlAs of a given alloy composition can be continuously tuned over a broad emission bandwidth of about 10 nm. Diode lasers of different materials and alloy compositions have operated in the range from 0.6 to 35 microns and are a useful source of high-resolution tunable radiation.

Ultra-stable diode lasers locked to narrow atomic transitions or radio-frequency standards could be useful sources as secondary frequency standards. A number of efforts are currently underway to improve microwave frequency standards by using GaAlAs diode lasers for optical pumping of cesium.[23]

Because of their small size, high efficiency and low cost, as well as the high spectral purity and stability they achieve, these semiconductor diode laser devices will find many applications in industry and research.

References

1. See for example: R. W. P. Drever *et al.*, in *Proc. Int. Conf. Laser Spectroscopy*, A. R. W. McKellar, T. Oka, B. P. Stoicheff, eds., Springer-Verlag, New York (1981), p. 25.
2. M. Lax, Phys. Rev. **160**, 290 (1967).
3. A. L. Schawlow, C. H. Townes, Phys. Rev. **112**, 1940 (1958).
4. R. D. Hempstead, M. Lax, Phys. Rev. **161**, 350 (1967).
5. H. Richen, H. Vollmer, Z. Physik **191**, 301 (1967).
6. H. Gerhardt, H. Welling, A. Guttner, Z. Physik **253**, 301 (1972).
7. J. Hall, private communication.
8. E. D. Hinkley, C. Freed, Phys. Rev. Lett. **23**, 277 (1969).
9. M. W. Fleming, A. Mooradian, Appl. Phys. Lett. **38**, 511 (1981).
10. C. H. Henry, IEEE J. Quantum Electron. **18**, 259 (1982).
11. D. Welford, A. Mooradian, Appl. Phys. Lett. **40**, 865 (1982).
12. J. Harrison, A. Mooradian, Appl. Phys. Lett. **45**, 318 (1984).
13. B. Daino, P. Spano, M. Tamburrini, S. Piazzola, IEEE J. Quantum Electron. **QE-19**, 266 (1983).
14. K. Vahala, C. Harder, A. Yariv, Appl. Phys. Lett. **42**, 211 (1983).
15. C. H. Henry, IEEE J. Quantum Electron. **QE-19**, 1391 (1983).
16. E. Eichen, P. Melman, paper K-2, 9th IEEE International Conference on Semiconductor Lasers, Rio de Janeiro (1984).
17. D. Welford, A. Mooradian, Appl. Phys. Lett. **40**, 560 (1982).
18. Y. Yamamoto, S. Saito, T. Mukai, IEEE J. Quantum Electron. **QE-19**, 47 (1983); Y. Yamamoto, IEEE J. Quantum Electron. **QE-19**, 34 (1983); A. Dandridge, H. F. Taylor, IEEE J. Quantum Electron. **QE-18**, 1738 (1982).
19. M. J. O'Mahony, I. D. Henning, Electron. Lett. **19**, 1000 (1983); K. Kikuchi, T. Okoshi, Electron. Lett. **19**, 812 (1983); G. Tenchio, Electron. Lett. **12**, 562 (1976); G. Tenchio, Electron. Lett. **13**, 614 (1977); M. Omtso, S. Kotajima, Jpn. J. Appl. Phys. **23**, 760 (1984); K. Kikuchi, T. Okoshi, R. Arata, Electron. Lett. **20**, 535 (1984); R. Schimpe, W. Harth, Electron. Lett. **19**, 136 (1983).
20. H. J. Zeigler, private communication.
21. K. Vahala, A. Yariv, Appl. Phys. Lett. **43**, 140 (1983); R. Lang, K. Vahala, A. Yariv, to be published in IEEE J. Quantum Electron.
22. R. DeVoe, R. G. Brewer, Phys. Rev. A **30**, 2827 (1984).
23. M. Arditi, Metrologia **18**, 59 (1982). □

Tunable coherent infrared techniques show progress

Gloria B. Lubkin

PHYSICS TODAY / JUNE 1973

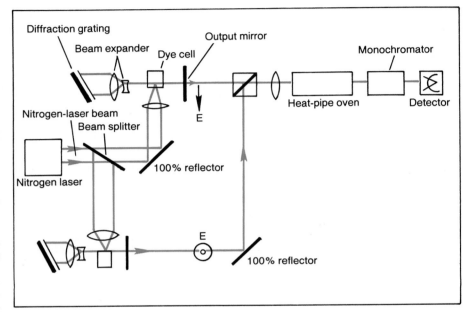

Tunable coherent infrared source has two dye lasers pumped by a nitrogen laser. The beams interact within a chamber of alkali vapor. The dye lasers are tunable by using different dyes and diffraction gratings. The source is tunable from 2–24 microns.

Recently several groups have produced tunable coherent infrared sources; some of these sources are tunable over a broad range of wavelengths. They vary in their ability to be continuously tuned. The most novel of these sources was reported by James J. Wynne in an invited paper at the Washington meeting of the American Physical Society on 23 April. Wynne, Peter Sorokin and John Lankard, all of IBM Research Center, reported[1] that they had produced a continuously tunable source over the range 2–24 microns with no apparent fundamental limitation to achieving longer-wavelength generation. Groups at MIT National Magnet Laboratory, Bell Laboratories, Rice University and the University of California at Berkeley have all recently reported on tunable coherent infrared sources.

These sources have many applications. They can be used to study elementary excitations in solids and complex biological molecules. A number of air pollutants show strong absorption in the infrared. It has been suggested that because molecules of different isotopes of the same element have different vibrational frequencies, one could achieve efficient and economical separation of these isotopes. For astronomy the sources could be used as a tunable local oscillator in a heterodyne system.

In the IBM experiment beams from two separate dye lasers, which are pumped simultaneously by a nitrogen laser, interact within a chamber of alkali vapor, the vapor being maintained in a heat-pipe oven. Each dye laser can be tuned from about 0.36 to 0.7 microns by using different dyes and diffraction gratings. The experimenters used a "four-wave parametric" conversion technique (see figure on page 20). Most of the work was done with potassium. The first dye laser is chosen so that its output wavelength is tunable in the neighborhood of the 4s–5p resonance lines of potassium. The beam does three things: It provides one of the waves needed in the four-wave mixing process; it generates a second wave needed for the four-wave process by inducing stimulated electronic Raman emission and producing a Stokes wave at a frequency $\nu_S = \nu_L - E(5s)/h$ it allows the four-wave mixing process to become phase matched through fine adjustments of its frequency.

A third wave, ν_P, is provided by the second dye laser. The waves ν_L, ν_S and ν_P beat together to create a polarization at

$$\nu_R = E(5s)h - \nu_P = \nu_L - \nu_S - \nu_P$$

This polarization can radiate, thus providing a tunable infrared output, because the frequency of the second laser is tunable.

As originally reported,[2] with the use of one vapor (potassium) alone, tunable infrared output was limited to the range about 2–4 microns. This range was determined by the fact that the first dye laser had to be simultaneously tuned to generate high-intensity Raman light and yet satisfy the phase-matching condition for the four-wave mixing process. However, by the addition of a second non-resonant alkali-metal vapor (sodium) to provide linear dispersion, the phase-matching condition can always be satisfied with the first dye laser tuned for maximum Raman–Stokes generation. In this way, by "biasing" the vapor for long-wavelength output, the IBM group generated[1] coherent infrared over the range 2–24 microns.

The two-laser method extends the tuning range of the dye laser, which was previously tunable only from the near-ultraviolet, across the visible spectrum and into the near-infrared to about 1.2 microns.

The technique produces peak powers between 0.1 and 100 milliwatts, using a 100-kW nitrogen laser pump. The infrared output is proportional to the cube of the pump power; thus a 1-MW nitrogen laser is expected to increase the infrared output by 1000.

In a related experiment, Sorokin, Wynne and Lankard have also generated laser infrared at a number of discrete wavelengths in the range 12–220 microns, forming a "picket-fence" spectrum. This is done by tuning the second harmonic of one dye laser to various resonance lines in potassium.

The IBM group told us that among the advantages of their scheme are the continuously tunable nature of the output (that is, it does not display mode hopping) and an absence of the need for cryogenic temperatures.

At Bell Labs Kumar Patel and his collaborators have been using two carbon-dioxide lasers to obtain step-tunable far-infrared radiation, work first discussed by them[3] in 1969. Now T. J. Bridges and Van Tran Nguyen of Bell reported at the March meeting of the American Physical Society that they have used one carbon-dioxide laser to pump a spin-flip Raman laser, producing a source that is tunable from 80 to

120 microns.

In a spin-flip Raman laser, a magnetic field is applied to an indium-antimonide sample. Raman scattering occurs when the conduction electrons in the crystal flip their spins in the magnetic field, B. The frequency ω_S of the shifted light varies as $\omega_S = \omega_0 - g\mu_B B$; ω_0 is the pump frequency, μ_B is the Bohr magneton, and g is the effective gyromagnetic ratio of the electrons in indium antimonide.

When the carbon-dioxide laser pumps the spin-flip Raman laser, two frequencies come out—ω_0 and $\omega_0 - g\mu_B B$. The radiation enters a second indium-antimonide crystal and difference-frequency mixing occurs, resulting in an output of frequency $g\mu_B B$. The output is tunable because the magnetic field can be varied.

The Bell experimenters have produced output powers of about 1 microwatt with a pulse repetition frequency of about 400 times per second, an average power of 10^{-8}–10^{-9} watts. Patel notes that this power is more than sufficient to do any kind of spectroscopy desired. The group has measured the absorption of carbon monoxide with a signal-to-noise ratio of about 10 000. There is not enough power to do further nonlinear optics, however. To do that they will need to increase the input power, since in a difference-frequency mixing experiment, the output is proportional to the product of the two input powers.

In the spectroscopic measurements the Bell group's narrowest lines are about 0.1 cm^{-1}. The line width arises because the spin-flip Raman laser is not continuously tunable; it has its own modes, which are separated by about 0.1 cm^{-1}. Patel notes that this is very close frequency spacing, at least compared to doing difference-frequency mixing between two carbon-dioxide lasers. In the present spin-flip Raman laser the crystal faces are parallel so that light bounces back and fourth, making the crystal act like the modes

of a cavity. On the other hand, tilting one of the ends of the crystal eliminates the cavity effect, and a superradiant spin-flip Raman laser, which is essentially a single-pass device, is produced. Nguyen is trying this arrangement now, using a pumping laser with enough intensity to produce enough single-path gain through the sample to build up intensity of the Raman-scattered light from spontaneous emission so that the device can be continuously tuned without mode hopping.

In principle the Bell technique should be applicable over a much wider range. Using two fixed-frequency carbon-dioxide lasers Patel and his collaborators had earlier shown that one could go from 2 cm^{-1} to 140 cm^{-1}. So Patel believes that with the new technique they should be able to tune in the range from 1 mm to 70 microns.

At MIT's National Magnet Laboratory R. L. Aggarwal, Benjamin Lax and their collaborators have demonstrated an infrared tunable source in the 5-micron region in which they combined a high-power spin-flip Raman laser with a 10.6-micron carbon-dioxide laser to produce the sum frequency. To bridge the far-infrared-microwave gap, they extended this general technique to produce the difference frequency. This was accomplished by a noncollinear technique of phase matching in gallium arsenide. The group reported their results in the 1 April issue of *Applied Physics Letters*.[5]

In more recent results reported at the March APS meeting in San Diego, the group produced step tuning using two carbon-dioxide lasers; they obtained 60 lines from 70 microns to 2 mm. Lax told us that it is evident that if one uses gratings for both carbon-dioxide lasers, it is possible to obtain 3000 lines between 70 microns and 1 cm.

In the present experiment one laser has 30 kW of peak power and the other 100 kW of peak power. The output observed was about 1 milliwatt at 100 microns; the MIT group had calculated that they would obtain 700 milliwatts. In work now in progress, the group is making a number of obvious engineering improvements that they expect will enhance the peak-power capability by two orders of magnitude and the repetition rate from three per second to at least ten per second. Using these improvements and a 1-kW spin-flip Raman laser that they had developed earlier, the group expects to obtain continuous tuning for the difference frequency.

At the March meeting the group also reported on a scheme to produce cw operation by employing a collinear phase-matching technique that takes advantage of a periodic variation of the refractive index.

Other groups. In the same issue of *Applied Physics Letters* that carried the MIT report was an article[6] by C. D.

Decker and F. K. Tittel of Rice University on difference-frequency mixing between the output from a tunable narrow-linewidth ruby-pumped infrared dye laser and a Q-switched ruby laser in a proustite crystal. The Rice experimenters obtained peak infrared powers in the kilowatt range; they told us that their source was tunable from 3.20 to 6.47 microns. They believe that by using proustite crystals cut at a different angle to the optic axis, tunability from 6.47 to 13 microns should be possible.

In a more recent experiment, Decker and Tittel have used two independent dye lasers, mixing them in a proustite crystal and obtained tunability from 5.82 to 7.25 microns with peak powers in the 20–100-watt range. Provided appropriate nonlinear mixing crystals can be found, they say that this dual dye-laser frequency-mixing arrangement could generate radiation from 5 microns to greater than 100 microns.

At the March APS meeting, Patrick Yang, James Morris, Paul L. Richards and Yuen-Ron Shen of the University of California at Berkeley reported generation of continuously tunable far-infrared radiation over a factor of 20 in wavelength from 52 to 2000 microns. In these experiments the laser source was a one-dye-cell dual-beam dye laser. The laser cavity contains two diffraction gratings to tune, individually, two orthogonal polarizations of laser light that are spatially separated by a Glan-Thomson prism. This arrangement minimizes alignment and synchronization problems. Using a ruby-laser pump, the dye-laser output is 300 kW in each beam. Collinearly phase-matched far-infrared generation in room-temperature lithium niobate and zinc oxide yields about 5 milliwatts of far-infrared power from 52 to 800 microns. The power falls below 1 milliwatt at 2000 microns. These power levels are more than adequate for spectroscopy in the energy-starved far-infrared region of the electromagnetic spectrum, Richards said.

Tuning is achieved by rotating one of the laser gratings and simultaneously changing the angle of the nonlinear crystal. The tuning range can be extended to shorter wavelengths by using other nonlinear crystals.

References

1. J. J. Wynne, Bull. Am. Phys. Soc. **18**, 534 (1973).
2. P. P. Sorokin, J. J. Wynne, J. R. Lankard, Appl. Phys. Lett. **22**, 342 (1973).
3. C. K. N. Patel, N. Van Tran, Appl. Phys. Lett. **15**, 189 (1969).
4. C. P. Pigeon, B. Lax, R. L. Aggarwal, C. E. Chase, F. Brown, Appl. Phys. Lett. **19**, 333 (1971).
5. R. L. Aggarwall, B. Lax, G. Favrot, Appl. Phys. Lett. **22**, 329 (1973).
6. C. D. Decker, F. K. Tittel, Appl. Phys. Lett. **22**, 411 (1973).

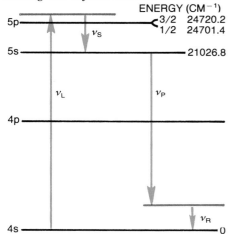

ENERGY (CM^{-1})

5p — 3/2 24720.2 / 1/2 24701.4
ν_S
5s — 21026.8
ν_L ν_P
4p
ν_R
4s — 0

Four-wave parametric mixing in potassium. The waves ν_L, ν_S and ν_P beat together to create a polarization at $\nu_R = \nu_L - \nu_S - \nu_R$.

Lasers and physics: a pretty good hint

**The monochromaticity, directionality and intensity of laser light
make possible spectroscopic investigations of previously unimagined precision
whose results give hints of more new physics to come.**

Arthur L. Schawlow

PHYSICS TODAY / DECEMBER 1982

In the early years of lasers, it was common for popular accounts to refer to the laser as "a solution looking for a problem." That description was sometimes even attributed to me, although I have tried to point out clearly that I did not believe it. Indeed, any comic-strip artist could think of lots of "real" uses, such as drilling tunnels through mountains, or shooting down Klingons a million miles away. But for such important applications, the capabilities of real lasers were—and still are—pathetically inadequate. For these and many other problems lasers did not provide solutions, but they did give a new and promising place to look for attainable devices that could give answers to worthwhile problems.

As is now well known, the light produced by lasers is much more powerful, directional, coherent and monochromatic than that of ordinary light sources. This is true even for small lasers, but different lasers vary enormously in the extent to which they possess these properties. Moreover, we can make lasers for a very wide range of wavelengths, ranging from microwaves though infrared and visible to the vacuum-ultraviolet region of the spectrum. There are lasers having continuous, narrow-band tunability in some, but not yet all regions.

The monochromaticity, directionality and intensity of laser light are making possible a wide range of scientific investigations that would have been unimaginable without them. With lasers we can measure spectral lines with great precision and resolve previously hidden details. We can detect and

Arthur Schawlow, who won the Nobel Prize for Physics for his work on lasers, is professor of physics at Stanford University and was president of the American Physical Society in 1981.

study single atoms and molecules. Figure 1, for example, shows an apparatus for making high-precision measurements of the excitation spectra of molecules before and after they are scattered from a surface. Even so, we are still sometimes limited by the properties of available lasers, and have to try to extend laser technology. Perhaps someday we will have tunable gamma-ray lasers that we can use to excite coherent superpositions of nuclear energy levels and to alter radioactive decay of nuclei. The existence of lasers has not provided solutions to all our problems. But it has given us a pretty good hint as to where some interesting solutions might be found.

Sharpening the lines

You can see that the light generated by an ordinary helium–neon laser is quite monochromatic just by viewing it or photographing it through a diffraction grating. Whereas the light emitted spontaneously by the helium–neon gas discharge produces many colors, the laser beam shows just one. But on closer examination with the aid of a Fabry–Perot interferometer, one sees that the output of the laser consists of several nearby wavelengths—these are just the axial modes of the laser resonator that lie within the wavelength band in which the laser amplification exceeds the losses. To select just one of them usually requires one or more additional tuning elements, such as a Fabry–Perot etalon, inside the laser resonator. These resonances due to standing waves between the laser's end mirrors are the chief factors determining the exact wavelengths, within the amplifying band, at which the laser oscillates. If the output frequency of, say, an argon-ion laser is to be stable to

within one part in 10^8, its length must be held constant with that precision. It must be maintained, not only against slow drifts but against all fluctuations at audio and even higher frequencies. Stiff spacers of low-expansion materials, such as quartz or invar, may be used to provide a fixed spacing between the end mirrors. Vibrations—even from sound waves—must be minimized, and the temperature of the laser structure must be controlled. This is a continuing challenge,[1] but there are strong motivations to meet it, not only from high-resolution spectroscopy but also from other areas of physics—the interferometers planned for detecting gravity waves, for example, will require finely tuned and ultrastable lasers. Currently perhaps the most stable laser—with a stability approaching one part in 10^{14}—is one built by John L. Hall and his coworkers at the Joint Institute for Laboratory Astrophysics in Boulder.[1]

For most spectroscopy, however, it is necessary not just to have a stable laser but to tune it across some band of wavelengths. During most of the first decade of lasers, until the late 1960s no laser was very broadly tunable. While one could generate many wavelengths in the infrared, visible and even ultraviolet, there were large gaps between them. Any one laser line could only be tuned by a small fraction of its wavelength: for a gas laser, typically across the Doppler width of the gas, or about one part in 10^5. Some solids, such as ruby, could be tuned by as much as one part in 10^3 by changing the temperature of the laser material. With these, some molecular lines that happened to coincide with the laser wavelength, or the lines of the laser medium itself, could be investigated in fine detail.

Apparatus used to measure excitation spectra of molecules, before and after scattering from a surface. In one such experiment, performed by Richard Zare and collaborators, a beam of NO molecules was cooled by expansion to supersonic speeds; laser-excited fluorescence showed that the rotational states of the molecules had a temperature of about 11 K. After the molecules were scattered by a silver crystal at 650 K, the fluorescence spectrum showed a much richer set of rotational lines, indicating a temperature of 375 K. Figure 1

Lasers based on organic dyes with very broad lines made continuous coverage of the visible spectrum possible in the late 1960s. By pumping various dyes with uv from nitrogen lasers, one can generate short (5 nsec) pulses of laser light at any visible wavelength. Some dyes provide continuous tuning over a range of several hundred angstrom units. With such broadly tunable lasers, the task of defining an accurate, narrow-band output was even more difficult than for the older gas and solid-state lasers. In the dye lasers one generally makes a rough wavelength selection by using a diffraction grating for one of the end mirrors of the laser. Theodor W. Hänsch added[2] a telescope to the laser, placing it between the dye cell and the tuning grating, thereby both improving the

collimation of the light at the tuning grating and further increasing the resolution by illuminating more rulings on the grating. Such a laser typically gave a linewidth of about 0.03 Å, which one could further refine to 0.004 Å by placing an etalon inside the laser resonator. A passive resonator to filter the output can reduce the linewidth even further, but at the expense of increasing the pulse-to-pulse amplitude fluctuations. Thus, in the 1970s we had widely tunable visible-light lasers, with linewidths much smaller than the Doppler width, even though the narrowness of the line was limited by the shortness of the pulse.

Doppler-free spectra

The narrowness of the lines from the tunable dye lasers together with the

other properties of the laser light, directionality and high intensity, provided just what was needed to display spectral lines without Doppler broadening. Hänsch at Stanford and, independently, Christian Bordé in Paris developed the method shown schematically in figure 2, which is now known as saturation spectroscopy. The light from the laser is divided into two parts that traverse the gas sample in opposite directions. One of these beams is chopped at an audio frequency. When it is on, it bleaches a path through the sample, so that less of the probe beam is absorbed; thus, as the saturating beam is chopped on and off, the probe beam is modulated. However, this happens only when the laser is tuned so that both beams interact with the same atoms or molecules, those with a zero

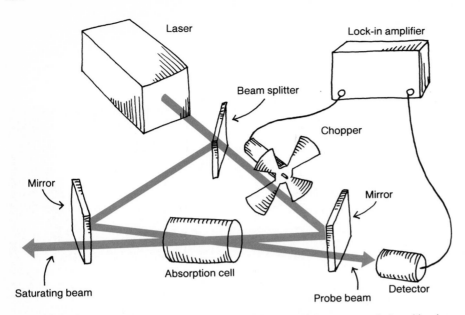

Saturation spectrometer. An intense beam from a laser saturates an atomic transition in a sample, so that a probe beam at the same frequency (obtained from the same laser by a beam splitter) suffers less absorption. Because the laser line is much narrower than the Doppler width of the absorption line, only those atoms having the same velocity with respect to both beams will affect the saturation—that is, atoms with zero velocity. The lock-in amplifier picks up only that part of the probe signal that is modulated by the saturation effects.   Figure 2

velocity component along the direction of the two oppositely directed beams. Moving molecules, on the other hand, see the saturating and probe beams as Doppler shifted, one up in frequency and the other down, so they cannot be simultaneously resonant to both beams. Consequently, the probe signal displays the spectrum of the stationary molecules without Doppler broadening. This method is an extension of the earlier work of Willis Lamb, Ali Javan and Paul H. Lee and M. Skolnick on saturated absorption or stimulated emission of atoms inside the laser resonator.

With the saturation method and the broadly tunable laser, Hänsch and Issa Shahin[3] were able to study fine structure of the well-known red line of atomic hydrogen, Hα, which over the last century has revealed so much of fundamental importance for physics.[4] Figure 3 shows, with increasing resolution, the Balmer spectrum of hydrogen, the Hα line as it would be seen at room temperature with a perfect conventional spectrograph and the Doppler-free spectrum of the same region. The light hydrogen atoms have an exceptionally large Doppler broadening so that most of the fine structure is obscured in the conventional spectrum; it is, however, known from theory and from radiofrequency spectroscopy, and is indicated on the figure. The Doppler-free laser technique reveals details, including the Lamb shift, which had not previously been resolved optically in hydrogen. In 1976 Hänsch and Carl Wieman introduced[5] a polarization technique, which further increased the sensitivity of

Doppler-free laser spectroscopy. In this method, a polarized saturating beam preferentially excites atoms with a particular orientation, leaving the remainder with a complementary orientation; the gas is then optically anisotropic at the wavelengths that can be absorbed by the selected atoms. Thus, a plane-polarized probe beam, tuned to interact with these atoms, is depolarized and can pass through an analyzer whose axis is perpendicular to that of the polarizer. Consequently, the probe signal appears with very little background, rather than as a small fractional change of a large signal, such as one obtains from the saturation method. The polarization method yields good signals with lower atomic densities and lower laser powers than other methods, so that line widths can be further decreased by reducing the gas density and the laser power, thereby avoiding pressure and power broadening. With the individual components of the hydrogen lines accessible as narrow, well-resolved lines, Hänsch and his associates were able to make improved measurements of their absolute wavelengths and thereby improve the precision with which the Rydberg constant is known[6] by a factor of 30. Shaikh Amin, Denise Caldwell and William Lichten at Yale have recently been able to improve[7] on this precision by using a beam of hydrogen atoms for their measurements instead of a cell, thereby avoiding the need to make corrections for pressure broadening, discharge effects, and other shifts.

A different kind of investigation of

hydrogen involves a precise comparison of the wavelength of the transition between the 1s and 2s levels with the wavelength of the 2s–4s transition. According to the Bohr theory, the frequency of the transition from $n = 1$ to $n = 2$ is exactly 4 times that of the transition $n = 2$ to $n = 4$. Nowadays, we do not expect this exact ratio, because the s levels are displaced by the relativistic corrections and the Lamb shift, which are greatest for the 1s level. However, while the 2p level provides a convenient nearby reference level for the 2s state, there is no such convenient reference for the 1s state. Thus, for many years the Lamb shift for the 1s level remained unmeasured, even though it is expected theoretically to be about eight times larger than for the 2s level.

Wieman and Hänsch have recently found a clever way around this problem by using two-photon absorption. One can excite the 1s–2s transition (the Lyman α line at 1215 Å) with two photons at 2430 Å. The photons come from two oppositely directed beams and have opposite circular polarizations, thus allowing energy, momentum and angular momentum to be conserved—and simultaneously avoiding Doppler broadening. The frequency of these photons is very near that of the Balmer line, at 4860 Å, corresponding to the transition $n = 2$ to 4. In the experiment a dye laser is scanned across the 4860-Å region to observe the fine structure of Hα line with saturation or polarization spectroscopy. Part of the output of this laser is amplified and doubled in frequency, to obtain the 2430-Å radiation needed for the two-photon transition. Thus the frequencies of these transitions are derived from the same laser, and can be compared with enough accuracy to measure the Lamb shift of the hydrogen 1s ground state.

This two-photon transition is likely to be of interest for some years, perhaps decades, because it is in principle extraordinarily narrow. The upper state, the 2s, is truly metastable, decaying only by two-photon emission. Its lifetime is $1/7$ second, so that the linewidth due to its lifetime must be about one hertz, for a fractional width of about 10^{-15}. It is usually possible to measure the absolute frequency or wavelength to about one percent of the linewidth, which in this case would be about one part in 10^{17}. But there are no standards of that precision, nor techniques for measuring anything with that precision. We will have to overcome many difficulties before we can make use of that narrow linewidth. The first of these is the large linewidth (about one part in 10^7) of the available lasers. Furthermore, the 2430-Å wavelength is short enough that the commonly used frequency-doubling crystals are very quickly damaged; after considerable efforts, Hänsch and his associates have quite recently (in September) succeeded in obtaining a continuous-wave beam of sufficient power at that wavelength.

Recently, Steven Chu and Allen Mills have observed[9] the corresponding two-photon transition in an even simpler atom, positronium. The great difficulty in observing any part of the positronium spectrum is that the atoms annihilate less than a seventh of a microsecond after they are formed. Positron sources of reasonable strength, therefore never yield many positronium atoms—on the average there is much less than one atom present at any one time. The positrons come out of their source at random times, and they are not concentrated in a small volume to which a laser could be focused. To be sure of exciting every positronium atom and then ionizing it for detection, one would, according to some estimates, need a continuous laser at 4860-Å with a average power around 100 kW. This is at least a thousand times greater than one could currently obtain. Chu and Mills have been able to do the experiment with a pulsed laser of low duty factor, by storing positrons in a magnetic trap from which they are released to form positronium just as the laser is flashed.

Resonant scattering

In 1970, Otis Peterson, Sam Tuccio and Benjamin Snavely obtained[10] continuous-wave action from a dye laser. They used a beam of green light from an argon-ion laser to excite a fluorescent organic dye, with the exciting beam focused into the dye along nearly the same direction as the laser beam. Because continuous-wave dye lasers tend to have only very little optical amplification, any tuning elements with more than very low optical losses could prevent laser oscillation. For that reason, making cw dye lasers simultaneously monochromatic and broadly tunable appeared to be much more difficult than for pulsed lasers.

In our laboratory at Stanford, William M. Fairbank Jr built a cw dye laser with the wavelength-selecting prism outside the main laser resonator, only weakly coupled to it through one of the end mirrors, to avoid excessive losses. This laser had several independent adjustments, and was rather difficult to tune. Although we could tune it to any desired wavelength in its range, we could not scan it smoothly over any substantial wavelength range.

However, we were able to tune this primitive dye laser to one or the other of the well-known yellow D lines of atomic sodium. We found we could measure the amount of the laser light scattered by even a very small number of sodium atoms by using a glass cell containing sodium metal and carefully constructed to minimize light from the walls. For laser intensities low enough to avoid saturation, the scattered light intensity was proportional to the number of atoms in the beam, so we could measure the variation of the sodium-vapor density with temperature. So sensitive was the method, that Fairbank was able to measure[11] the light scattering down to $-30\,°C$. At that temperature, the density of sodium atoms in the cell was around 100 per cubic centimeter, and there were only one or two atoms in the beam at any time.

At this point we realized that the light-scattering method is inherently more sensitive than, say, radioactive techniques. We can scatter many light quanta from an atom without destroying it, whereas each radioactive atom can decay only once. Fairbank considered ways in which the laser method might be used to search for very rare atoms, such as those with a fractionally charged quark in the nucleus. However, that search had to wait for more widely tunable lasers.

Subsequently other, even more sensitive, methods—such as resonance ionization—have been discovered,[12] which can indeed detect a single atom of a chosen species in the presence of a large background of other gases. Ingenious radiochemical methods have been suggested[13] that might make it possible to detect the very small number of atoms produced by interactions of low-energy solar neutrinos.

We have already seen that resonant scattering of laser light itself is sensitive enough to detect an average number of one or two atoms. If the atoms would just stay there a few seconds, we could count them one by one. Hänsch and I have shown[14] that you could slow down the rapidly moving atoms, that is, cool them, by using intense laser light of a frequency slightly below the resonance frequency for a stationary atom. Atoms moving toward the laser would see it as Doppler-shifted up into resonance, and would scatter photons, thereby losing momentum. Hans Dehmelt and his group at the University of Washington, however, have shown[15] that a more effective way to keep the atom stationary is to ionize it and hold it in a trap of suitably configured fields, a so-called Penning trap. Figure 4 shows photographs of their trap containing three, two and finally just one

FREQUENCY ν (Thz)

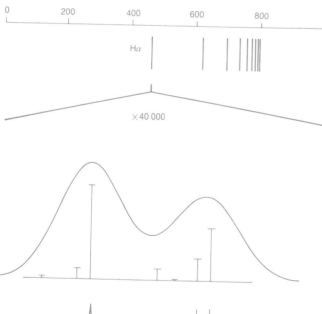

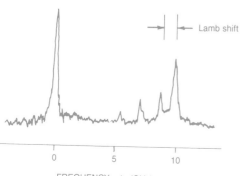

FREQUENCY $\Delta \nu$ (GHz)

The Balmer α line. The top graph shows a sketch of the Balmer spectrum of hydrogen as it might be seen in an ordinary student spectroscope. Below it is the Hα line, expanded by a factor of 40 000, as it would be seen at room temperature by an ideal spectrograph; the calculated intensities of the fine-structure components are indicated by the bars. At bottom, to the same scale, is the Doppler-free spectrum obtained by Theodor Hänsch and Issa Shahin, which resolves the fine structure, displaying the Lamb shift directly in the visible. Figure 3

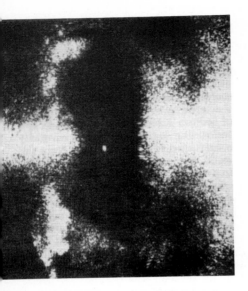

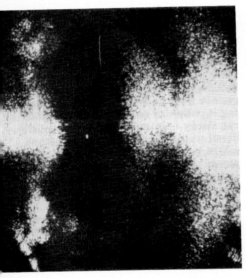

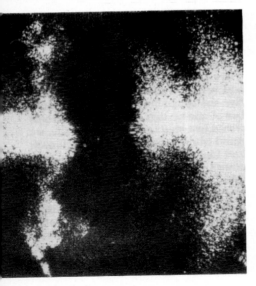

Barium ions in a Penning trap. The central spots show light emitted from three atoms (at top), two atoms and, finally, one atom held by electric and magnetic fields. The grainy areas surrounding the bright spots (which indicate the positions of the atoms) are out-of-focus images of the trap electrodes.   Figure 4

ion. Trapped ions can give extremely sharp spectra, for the interaction time between the light and the atom is not limited by the transit time, and thermal broadening is eliminated as well.

Simplifying complex spectra

The very high sensitivity of laser-induced fluorescence makes it possible to study the distribution of molecules among the various possible rotational states, and thus to measure the rotational temperature of even a small number of molecules. Richard Zare and his associates at Stanford have, for example, used an ultraviolet laser to measure the change in effective temperature of NO molecules in a beam when they are scattered at a metal surface.[16] (See figure 1.) In their experiment, the beam is cooled—by expansion to supersonic velocity—so that only a few of the lowest rotational states of the molecule are populated. Light from a uv laser tuned to excite molecules from a particular rotational state gives rise to fluorescence whose intensity is proportional to the number of molecules available to be excited, that is, to the population in that state. The wavelengths near 214.4 nm needed for this excitation were obtained from a pulsed dye laser (pumped by a frequency-doubled YAG laser), whose output was frequency-doubled and finally shifted in wavelength by stimulated Raman scattering in liquid hydrogen. The cold NO beam exhibits only a small number (about a dozen) of rotational lines in a compact range extending roughly from $44\,375$ cm^{-1} to $44\,220$ cm^{-1}, which indicates a rotational temperature of only 11 K. After the molecules had been scattered by the 111 surface of a clean silver crystal, the effective rotational temperature was raised to 375 K, with many more rotational levels populated and a spectrum extending more than 100 cm^{-1} on each side of the central peak at 44 200 cm^{-1}. However, the rotational states did not attain full thermal equilibrium with the silver crystal, whose temperature was 650 K.

The sensitivity of laser-induced fluorescence can also simplify complicated spectra. Even the simplest diatomic molecules have many vibrational and rotational levels for each electronic state. For example, while the sodium atom has only two absorption lines in the visible region of the spectrum, the diatomic sodium molecule has thousands. Molecular spectroscopists have been able to use the regularities in these spectra to analyze many of these levels, but only six excited electronic levels were known for Na$_2$. In our lab we were able to make use of the intensity and monochromaticity of a tunable laser to excite just one line, and thereby populate a single excited level.

The resultant fluorescence is limited by the selection rules to a small number of lines. For a diatomic molecule, for example, the rotational quantum number is allowed to change only by plus or minus one unit, or sometimes by zero. Thus, from the excited level there would be only two or three lines for each vibrational level of the lower state. Because the spectrum has only a manageable number of vibrational lines, the vibrational and rotational spacing of the ground electronic state become apparent.

Selectively excited fluorescence is indeed a very useful tool for learning about the low-lying states of complex molecules such as NO$_2$. However, the information may be already familiar, for if anything at all is known about a molecule, it is likely to be the constants of the ground state, which can also be studied by microwave, infrared, or Raman spectroscopy.

We can explore the excited states by inverting the process, and using the pumping laser to alter the population of a level.[17] For instance, when the laser depletes the population of a chosen lower level, all absorption lines from that state are weakened. As the pumping laser beam is chopped alternately off and on, the absorption of all the lines from the chosen level is modulated, and they can easily be identified, even though they may be displaced by perturbations from their expected positions.

Alternatively, one can use[18] a polarized pumping beam to orient the molecules in the chosen level; the initially isotropic medium then becomes doubly refracting at the wavelengths of the absorption lines from the oriented level, and the absorption lines from the labeled level can be recorded as bright lines through crossed polarizers. (See figure 5.)

In our laboratory at Stanford Nils Carlson, Kevin Jones, Frank Kowalski, and Antoinette Taylor have used[19] a variation of this polarization-labeling technique to identify excited levels in the Na$_2$ molecule. The light from a pulsed dye laser both raised some of the molecules to a chosen level in the first excited electronic state and oriented them. The absorption spectrum of the excited level was probed by a broadband beam from another laser. Carlson and his coworkers succeeded in finding and analyzing 23 new electronic levels to add to the six previously known. Many of the new levels belong to one of two series of Rydberg states. In these, the two sodium ions are bound by one inner 3s electron, while the other electron is in a state much farther from the core, with a high principal quantum number. It is thus possible to extrapolate each of these series to obtain reasonably good values

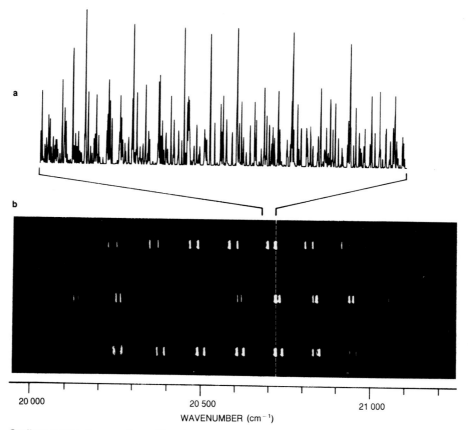

Sodium molecular spectrum. The photo shows a conventional high-resolution spectrogram of Na_2 molecules in the range 4750–5000 Å. The graph shows a small region of that spectrum, as recorded by polarization-labeling spectroscopy. The technique is similar to that illustrated in figure 2, but the pump beam is polarized, making the gas optically active. The probe beam is sent through crossed polarizers, so the signal is observed directly instead of as the decrease from a large value, thus improving the signal-to-noise ratio greatly.   Figure 5

for the constants of the ground state of the Na_2^+ molecular ion.

These studies, and others carried on at laboratories all over the world, have clearly not produced rock drills or death rays for Klingons. But the early lasers gave us some good hints for building powerful scientific tools to investigate some long-standing problems in atomic physics, and the current investigations still give us pretty good hints as to where more solutions can be found.

* * *

This article is adapted from a talk given at the APS Annual Meeting in January 1982.

References

1. A. R. W. McKellar, T. Oka, B. P. Stoicheff, eds. *Laser Spectroscopy V*, Springer-Verlag, New York (1982).
2. T. W. Hänsch, Appl. Opt. **11**, 895 (1972).
3. T. W. Hänsch, I. Shahin,, A. L. Shawlow, Nature **235**, 63 (1972).
4. T. W. Hänsch, A. L. Schawlow, G. W. Series, Sci. Am. **240**, March 1979, page 72.
5. C. Wieman, T. W. Hänsch, Phys. Rev. Lett. **36**, 1170 (1976).
6. J. E. M. Goldsmith, E. W. Weber, T. W. Hänsch, Phys. Rev. Lett. **41**, 1525 (1978).
7. S. R. Amin, C. D. Caldwell, W. Lichten, Phys. Rev. Lett. **47**, 1234 (1981).
8. C. Wieman, T. W. Hänsch, Phys. Rev. **A22**, 1 (1980).
9. S. Chu, A. P. Mills Jr, Phys. Rev. Lett. **48**, 1333 (1982).
10. O. G. Peterson, S. A. Tuccio, B. B. Snavely, Appl. Phys. Lett. **17**, 245 (1970).
11. W. M. Fairbank Jr, T. W. Hänsch, A. L. Schawlow, J. Opt. Soc. Am. **65**, 199 (1975).
12. G. S. Hurst, M. G. Payne, S. D. Kramer, J. P. Young, Rev. Mod. Phys. **51**, 767 (1979).
13. V. S. Letokhov, V. I. Mishin, M. Eshkobilov, A. T. Tursunov, Opt. Comm. **41**, 331 (1982).
14. T. W. Hänsch, A. L. Schawlow, Opt. Comm. **13**, 68 (1975).
15. W. Neuhauser, M. Hohenstatt, P. E. Toschek, H. G. Dehmelt, Phys. Rev. **A22**, 1137 (1980); and in *Spectral Line Shapes*, B. Wende, ed., de Gruyter, New York (1981).
16. G. M. McClelland, G. D. Kubiak, H. G. Rennagel, R. N. Zare, Phys. Rev. Lett. **46**, 831 (1981).
17. M. E. Kaminsky, R. T. Hawkins, F. V. Kowalski, A. L. Schawlow, Phys. Rev. Lett. **34**, 683 (1976).
18. R. Teets, R. Feinbert, T. W. Hänsch, A. L. Schawlow, Phys. Rev. Lett. **37**, 683 (1976).
19. N. W. Carlson, A. J. Taylor, K. M. Jones, F. V. Kowalski and A. L. Schawlow, Phys. Rev. **A24**, 822 (1981).   □

CHAPTER 4

HOLOGRAPHY AND PHASE CONJUGATION

Holography—3D photography—was widely touted as the first practical application of lasers. (Then skeptics asked what is a practical application of holography?) Light from a laser is split into two beams, one of which strikes an object. Then the light scattered from the object and the unscattered reference beam are focused together again at a photographic plate. The interference between the two beams leaves a corresponding intensity pattern in the photographic emulsion. Then when the photograph is illuminated by the reference beam alone, the image of the object is reconstructed with three-dimensional realism, as the reference beam supplies the necessary phase information for viewing the object at various angles.

Holography now has many practical applications besides clever optical illusions: the reconstruction of charged-particle tracks in bubble chambers, the monitoring of crystal growth, and even as memory storage for computers.

Phase conjugation is somewhat similar to holography, and is one of the many interesting phenomena that results when laser light interacts with matter. The intense electric fields of the incident laser light can cause electrostriction (a density increase) in the material, which generates a backscattered optical wave (a conjugate beam) traveling in the opposite direction (and with opposite phase) to the incident beam. When ordinary light passes through a distorting medium and then is reflected back through the medium again the resulting beam will be doubly distorted. If, however, laser light passes through a distorting medium, different parts of the beam change phase due to the distorting material. If then, the beam is incident on a reflector of material that scatters a conjugate beam, that beam then traverses the medium with opposite phase and the phase distortions of the second pass cancel those from the first pass, and the emerging beam appears to be undistorted. Phase conjugation may have applications in communications, laser fusion, and many other areas where signal aberrations are a problem.

Phillip F. Schewe

CONTENTS

Nobel prizes: physics to Gabor, chemistry to Herzberg

Gloria B. Lubkin

PHYSICS TODAY / DECEMBER 1971

The 1971 Nobel prize in physics has been awarded to Dennis Gabor for his invention and development of holography. The 1971 Nobel prize in chemistry has been awarded to Gerhard Herzberg "for his contributions to the knowledge of the electronic structure and geometry of molecules, particularly free radicals." Each prize is worth more than $87 000 this year.

Gabor. Commenting on the physics prize, the Swedish Academy noted that Alfred Nobel said in his will it should be given to "the person who has made the most important discovery or invention within the field of physics." The Academy said that "Gabor's achievement can be referred to both as discovery and invention."

In a paper published in 1948 and in subsequent papers[1,2] Gabor showed that if you allow highly coherent light to be diffracted by an object and the light from a highly coherent reference wave to interfere, the resulting interference pattern would contain all the information needed to reconstruct the object completely. The intensity distribution could then be photographed. Although only sources of very poor coherency were available at the time, Gabor was able to demonstrate the validity of the concept of holography. He published a photograph showing a somewhat blurred reconstruction of a hologram.

Gabor originally proposed that holography be used to sharpen electron micrographs that were blurred by the spherical aberration of electron objectives. The main problem was that two images were formed when the hologram was illuminated. (Image deblurring of electron micrographs was subsequently achieved with a method suggested by George Stroke in 1957.)

For many years only modest effort was applied to holography because of the lack of adequate coherent-light

GABOR

HERZBERG

sources. Then in 1956 G. L. Rogers[3] described the formation of holograms from radio waves reflected from inhomogeneities in the ionosphere.

As the Academy noted, the development of holography became rapid after Emmett Leith and Juris Upatnieks at the University of Michigan "published the first of their important contributions in 1962." They exploited the offset procedure of coherent radar, and using the off-axis reference-beam technique they eliminated the overlap of the two images.[4] Then using light from a gas laser they made the first laser hologram.[5]

Over the years holography has become an enormously wide-ranging field, to which Gabor has continued to make many contributions. In a recent article in *Science,* some of the applications

cited by Gabor, Stroke and Winston Kock[6] were: photographic storage of several hundred pictures in one emulsion, pattern and character recognition, new microscopic methods, production of spectroscopy gratings, looking through turbulent media, synthesis of images of nonexisting objects, three-dimensional cinematographic image projection, microwave and acoustical holography and ultrasonic "sonoradiography."

The Academy also said that Gabor "has made fundamental analyses of the information content of wave fields." In addition, Gabor, who holds more than 100 patents, has done research and development on a high-speed cathode-ray oscillograph, worked on a shrouded magnetic lens, and on communication theory, information theory, physical optics, predicting machines, plasmas, gas discharges and television.

Gabor was born in Budapest in 1900. He received a Dr Ing at the Technische Hochschule in Charlottenburg, Germany and then was a research engineer at Siemens & Halske in Berlin until 1933. From there he went to British-Thomson-Houston in Rugby, England, where he stayed until 1948. At that time he joined the faculty of Imperial College, London where he became professor of applied electron physics. He now divides his time between Imperial College and CBS Laboratories in Stamford, Conn., where he is a staff scientist. He has written *The Electron Microscope, Inventing the Future* and recently, *Innovations.*

Herzberg and his group at the National Research Council in Ottawa have for many years been investigating the structure and electronic spectra of molecules. The Academy said that Herzberg's investigations provided "extremely precise information on molecular energies, rotation, vibrations and electronic structures which, in turn, yield data on molecular geometries, that is, the distances between atoms of a molecule. From such investigations many results of fundamental importance for chemical physics and quantum theory were obtained. The work on the hydrogen molecule is especially outstanding." Herzberg's original measurements on the hydrogen molecule were extremely precise, and his data on the vibrational–rotational levels are still quoted in all the handbooks.

Through his determinations of the Lamb shifts in the ground state of deuterium and the ground and first excited states of helium and Li$^+$, Herzberg helped confirm predictions of quantum electrodynamics.

In 1931 Herzberg found what are now called the "Herzberg bands" of O_2, which are important for the absorption spectrum of air and the understanding of the upper atmosphere. Later he developed multiple-reflection equipment to produce a long path for absorption spectra. He subsequently used this equipment to analyze other forbidden transitions in O_2, and to study the infrared spectra of H_2 and HD.

In the 1950's Herzberg did a series of experiments on free radicals, which proved to be valuable to chemistry in general, the Academy said. Free radicals had been very difficult to analyze spectroscopically, because they react violently and it is very difficult to get an atmosphere of them sufficiently dense to obtain a spectrum. To do his free-radical studies Herzberg used and further developed the flash-photolysis method originally pioneered by Ronald Norris and George Porter (who won the Nobel chemistry prize in 1967).

The Academy said that in the 1960's Herzberg obtained results of fundamental importance for many free radicals of particular interest to organic chemistry. It said that Herzberg's "ideas and discoveries have stimulated the whole modern development, from chemical kinetics—dealing with forces that influence the motion of bodies and with gases used as refrigerants—to cosmic chemistry."

Herzberg has had a continuing interest in astrophysics, successfully applying his spectroscopic studies to identification of certain molecules in planetary atmospheres, comets and interstellar space.

For many years Herzberg's books have been veritable bibles to spectroscopists. They are *Atomic Spectra and Atomic Structure (1937)* and *Molecular Spectra and Molecular Structure,* Volumes I (1939), II (1945) and III (1966). His latest is *The Spectra and Structures of Simple Free Radicals.*

Born in Hamburg in 1904, Herzberg earned a Dr Ing at the Technical University in Darmstadt. He taught there until 1935, when he went to the University of Saskatchewan, remaining there until 1945. For three years he was at the Yerkes Observatory of the University of Chicago, when he returned to Canada to join the National Research Council. Herzberg's wife, Luise, was a scientific collaborator of his until her recent death. He served as president of the Canadian Association of Physicists (which named a medal after him), as president of the Royal Society of Canada and as vice-president of the International Union of Pure and Applied Physics.

References

1. D. Gabor, Nature **161,** 777 (1948).
2. D. Gabor, Proc. Roy Soc. Ser A **197,** 464 (1949); Proc. Phys. Soc. London, Sect. B **64,** 221 (1951).
3. G. L. Rogers, Nature **177,** 613 (1956).
4. E. N. Leith, J. Upatnieks, J. Opt. Soc. Amer. **52,** 1123 (1962).
5. E. Leith, J. Upatnieks, J. Opt. Soc. Amer. **53,** 1377 (1963).
6. D. Gabor, W. E. Kock, G. W. Stroke, Science **173,** 11 (1971).

Progress in holography

As this method of imaging becomes better understood, it can offer the best solution for problems in fields as diverse as architecture, medicine and mechanical engineering.

Emmett N. Leith and Juris Upatnieks

PHYSICS TODAY / MARCH 1972

With the award of the 1971 Nobel prize in physics to Dennis Gabor, holography has reached a new pinnacle of prestige. Gabor won his prize for the invention of holography, a form of wavefront reconstruction in which a coherent reference wave appears to unlock a three-dimensional replica of an object from a two-dimensional standing-wave pattern.

The science of holography has taken some curious turns in its relatively short 25-year history. The time divides itself naturally into three periods: The first could be considered a precursor stage, when the aim was to record wavefronts diffracted from crystals that had been irradiated with x radiation or electron waves. If reconstruction with visible light were successful, a highly magnified image of the crystal lattice would result. Recording the phase of the radiation was, however, a difficult problem that was soluble only for rather special cases. (For a description of this early work, carried out principally by Sir Lawrence Bragg and Hans Boersch, see reference 1.)

In 1948, Gabor developed a way to introduce a coherent background or reference wave, and the second period began.

Emmett Leith and Juris Upatnieks are both at the University of Michigan, Ann Arbor; Leith is Professor in the Electrical Engineering Department, and Upatnieks is a Research Engineer at the Institute of Science and Technology.

The third stage began in the early 1960's, when high-quality holographic imagery was demonstrated; the beginnings of this stage coincided in time with the development of the first lasers. Our attention here shall center on the latter two stages, particularly on the more recent work, for the history of holography gives us indications of what it can be reasonably expected to do in the future.

Early holography

Gabor's method, simple and elegant, solved in quite a general way the basic problem of recording the phase, as well as the amplitude, of a wave. In three principal papers between 1948 and 1951,[2] he developed the theory in considerable depth and offered convincing experimental results. Gabor's original purpose, to record electron waves and regenerate them at optical wavelengths, thereby compensating with optical techniques for the uncorrectable aberrations of electron lenses, is an historical point that today receives at best a passing notice. However, the holographic process has been revealed to have far more potential than one could, at that time, have imagined.

The excitement that must have attended the first accomplishment of the method is known, perhaps, only to Gabor and his associates. Our own excitement in 1960, when we duplicated for our own curiosity the experiments he described, was considerable, even

though the expected results were known beforehand. With an eyepiece, we followed the optical paths of the reconstructed wavefronts downstream to their focal position and observed a sharp, well defined image that had the startling property of having no apparent antecedent; nowhere in the optical system was there an object with which to associate the image (see figure 1). The mathematics, simple and easily understood, fully predicted the result; yet, the physical realization seemed altogether mysterious. The first time must have produced a powerful impact indeed!

Interest in holography continued strong for several years afterward, and produced some notable pioneers. Gordon Rogers explored the new technique in many ways, uncovering new ramifications and new insights; one of his best known contributions[3] is his extensive development of holographic image-forming principles in terms of Fresnel zone-plate theory, by which one can grasp in a highly intuitive way the first-order imaging properties of holograms. M. E. Haine, James Dyson and T. Mulvey applied[4] holography to electron microscopy. In the US, Ralph Kirkpatrick and his two students, Albert Baez and Hussein El-Sum, became interested in holography, particularly in its application to x-ray imagery. El-Sum produced the first doctoral thesis in holography.[5] In Germany, Adolf Lohmann first applied communi-

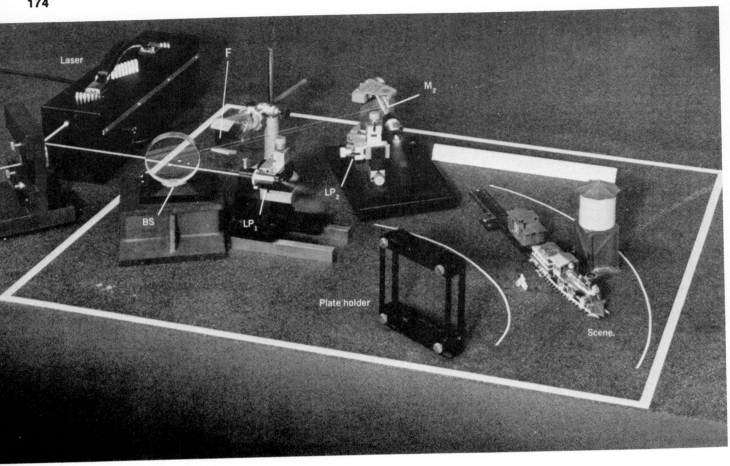

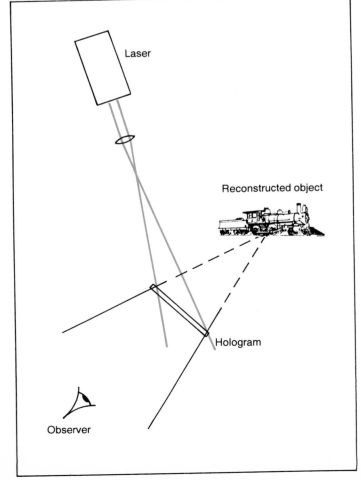

Typical arrangement for holography of a three-dimensional, opaque object (photo, top). The light emerging from the laser is deflected by mirror M_1 to beam splitter BS. The transmitted beam is diverged by lens LP_1, so that it illuminates the scene at right. The reflected beam, after passing through the variable attenuator F, is reflected by mirror M_2 to lens LP_2. The now diverging reflected beam illuminates the plate (held in the plate holder) as the reference beam. When the fringes produced by interference between object and reference beams are recorded on the plate, the result is a hologram. Under illumination by the reference beam alone (diagram, bottom) the developed plate, or hologram, regenerates the object wave, and an observer who looks through the hologram sees what appears to be the original object.
Figure 1

Principle of the coherent background.
When the object wave o is attended by a strong coherent background (or reference) wave r, the resultant wave (color) has a phase that varies only slightly from that of the background wave, and the loss of phase of the total field is relatively unimportant. Both the phase and the amplitude of the object wave are preserved to a high degree through its interference with the background wave.
Figure 2

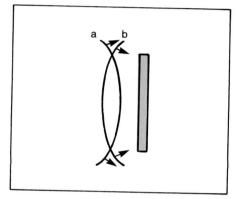

The twin image occurs because the two wavefronts a and b, which have equal but opposite curvature, produce the same fringe pattern when "interfered" with a coherent background wave. On reconstruction, the hologram produces both waves, regardless of which had been present during the recording.
Figure 3

cation-theory techniques to holography and, in consequence, suggested the single-sideband method[6] for removing the "twin image," one of the residual defects of the process. Despite the initial impetus, however, interest in holography waned in the middle 1950's, although activity never completely ceased.

The principal reason for the loss of interest was the relatively poor imagery, due mainly to the previously noted twin image, which occurs because the recording process is sensitive only to the intensity of the incident radiation. As a consequence, the reconstruction process not only recreates the original wave, it also creates a conjugate wave that, under collimated illumination, forms an

image in mirror symmetry to the "true" image with respect to the plane of the hologram. Whichever image one elects to use, he must view it against the out-of-focus background of the other, and the result is a noisy image.

The origin of the conjugate image can be described in various ways. Basically, the introduction of the coherent background wave renders the inevitable loss of phase of the total wave (signal plus coherent background) relatively unimportant; the phase of the total wave can vary only slightly from the background wave if the latter is strong (see figure 2), and the phase of the signal is manifested by its interference with the background. Nevertheless, the recording process has a fundamental ambiguity: It cannot distinguish between signals with equal and opposite phase, relative to background (figure 3), because both produce the same intensity. Alternatively, we may say that the surface that records the hologram cannot distinguish an object wave from the left from an object wave in a position of mirror symmetry on the right; the resulting interference pattern is the same. In either view, the reconstruction process resolves the dilemma by generating waves corresponding to both situations.

High-quality imagery

In the early 1960's, several papers appeared that proved to be forerunners of the great explosion of activity that ushered in the next stage of holography. We announced at the October 1961 Optical Society of America meeting a number of new concepts, including the off-axis or spatial-carrier frequency method of holography, which removed the twin-image problem in a simple and practical way.[7] In this method, the reference and object waves are brought together at an angle, to form a rather fine fringe pattern. The resulting hologram, behaving like a diffraction grating, produces several nonoverlapping diffracted orders. The zero-order wave produces the usual inseparable twin images which, in combination with other defects of in-line holography, result in poor imagery; but each first-order diffracted wave produces an image of high quality.

When, however, we extended the process to continuous-tone object transparencies instead of the black and white transparencies used in holography until then, another defect became prominent. The difficulty was the well known "artifact" problem of coherent light—each extraneous scatterer (for example, a dust particle) produces a wake of diffraction patterns that contaminate the resultant image. We surmounted this problem with *diffused* coherent illumination, which properly used, smooths the field produced by these scatterers.

During the course of this work, the laser became available; we used it, as well as the mercury arc source, although it was not at all necessary. Its brightness reduced the exposure time and its great coherence made careful equalization of the object and reference beam paths unnecessary. It offered, however, a thousand times the needed coherence, violating the basic rule that, in a coherent system, one should not be more coherent than necessary. As a consequence, the artifact problem was aggravated. On balance, with transparencies, either the laser or the mercury arc source should do quite well. Indeed, with proper optical system design, the coherence requirements for off-axis holography reduce exactly to those for in-line holography.[8]

Finally, we exploited the great potential of the laser by using three-dimensional, reflecting objects. For such objects, the coherence length should be of the order of the scene depth, a requirement that generally cannot be met with the mercury source. The laser also permits enormous quantitative advances: larger holograms and large objects.

About the same time, Yu. N. Denisyuk of the USSR introduced a new concept into holography, the "volume hologram," which combines holography with the Lippman color process.[9] Object and reference beams are introduced from opposite sides of the recording plate, and the resulting fringes are embedded within the emulsion as surfaces running nearly parallel to the emulsion surface, with half a wavelength spacing between them. Typically, the number of fringes in a cross section is about 50, although for very thick (a few mm) recording materials, there may be thousands. The twin image is eliminated by the thickness effect and, in addition, the holograms, because of their wavelength selectivity, can be viewed in white light derived from a point source. Denisyuk's work is a cornerstone of modern holography. Related work was reported by Pietr J. van Heerden,[10] whose concern was principally with holographic memories. During this same period, Brian Thompson and George Parrent used Gabor's in-line method, in combination with the pulsed laser, for their particle-sizing work, an application ideally suited for this configuration because the objects are extremely simple and a noise-free image was not needed.[11]

The great surge

By late 1964, holography had become probably the most active field of research in optics, engaging hundreds of groups throughout the world. Discovery and invention dominated the next three years; this period produced several techniques of color holography,

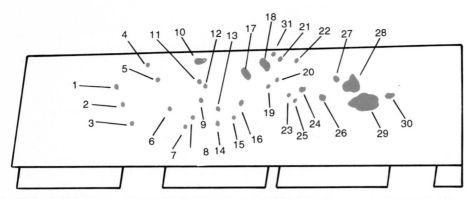

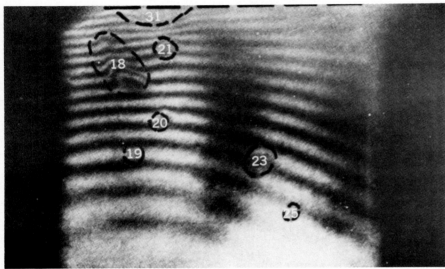

Defects in a section of an airplane trim tab are examined with double-exposure hologram interferometry. The trim tab was holographed a section at a time; each segment was recorded in two exposures, with thermal stress applied for the second exposure. In this way, the various defects were mapped (top). In the holographic image of one section of the wing (bottom), the defects are seen as defects in the fringe patterns that overlay the image. The wing was also examined with conventional, ultrasonic testing methods; the two methods gave substantially similar results. Courtesy C. Vest and D. Sweeney.

Figure 4

hologram interferometry in its various forms, techniques for holographic imagery through scattering and aberrating media, and many other basic concepts. The vast potential that had been inherent in holography now emerged with astonishing force.

By 1967, however, the discoveries began to diminish and the inventions took on an increasingly restrictive aspect, becoming more concerned with the design of specific configurations. The holographic world appeared to be consolidating its gains. Serious analysis of the holographic process in its various aspects became more evident, and, undoubtedly due to pressure by management and in response to the prevailing economics (in the US), applications received greater attention. Despite the shift in emphasis, however, the activity in holography remained at a high level.

Holography was found capable of an astonishingly wide variety of tasks that were normally done in other ways. Holographic methods could be useful in optical metrology and offered some interesting possibilities for spectroscopy. Optical memories using holographic techniques seemed destined to make significant inroads into the huge computer-memory field. Optical reading and feature-recognition machines that used holography were visualized. Microscopy, at least in the visible wavelength range, seemed promising. With hologram interferometry, a wide variety of nondestructive testing techniques became available, ranging from early detection of fatigue failure, to the detection of "debonds" in multilayered materials such as tires and honeycomb panels (figure 4), to the determination of heat flow in transparent materials, to the study of bending moments and to the dynamic operation of audio speakers and other sound-transducing equipment. A most ingenious and unlikely application of hologram interferometry is the determination of the complex mode structure of a laser.[12] Merely to list the applications of hologram interferometry in reasonable completeness would fill a page. Even in conventional interferometry, it was found that one could apply to the hologram essentially all the techniques—such as schlieren, dark field, and phase contrast—normally done with the actual object.[13]

Holography, we have noted, can be used to detect and examine aerosol particles, such as atmospheric pollutants. Not only was holography unique for visual displays of many kinds, including portraiture, but it could also be used in instrumentation for conventional stereo imagery. Holographically produced optical elements showed promise for improving the performance of optical elements. The versatility of holography seemed limitless. (See figures 5 and 6 for examples.)

Practical uses

Serious efforts to commercialize holography became visible in the late 1960's. Many small companies dedicated to a particular aspect of holography sprang up, and in larger companies, the basic-research programs were reoriented toward product development. Equipment was designed for bringing holographic measuring and testing techniques into the factory, and prototype holographic memories began to develop.

By the end of the 1960's, some defects in the holography bandwagon showed themselves. The hoped-for commercial products either failed to materialize or were not competitive with conventional products. The tremendous versatility of holography proved insufficient to ensure its success; holography would have to be in some way an improvement over the established methods. Often, however, the old ways were better, at least within present state-of-the-art limitations. Thus, holography programs were curtailed or eliminated,

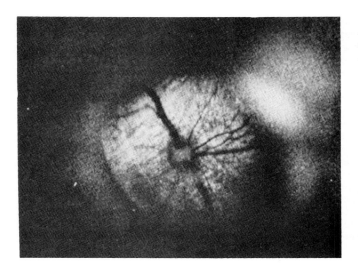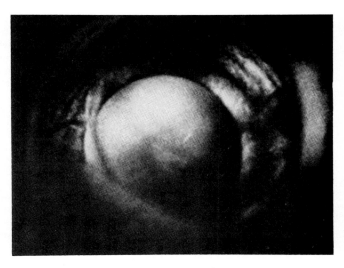

Medical application for holographic three-dimensional imagery. The eye of an anesthetized cat was holographed. From this single hologram it was possible to image any plane throughout the eye: Photo at the left shows the retina, with the blood-vessel structure and optic disk clearly visible, and photo at the right shows the iris. Courtesy J. Caulkins and C. Leonard.

Figure 5

and some small companies disappeared. In many quarters an atmosphere of gloom was observed, but whether or not the level of activity has diminished is problematical: As some groups abandoned holography, others appeared. The number of papers published has decreased slightly but remains high.

In retrospect, this course appears to have been predictable. The attempt to apply holography indiscriminately to all conceivable situations tended to force it into unsuitable molds. In addition, when the game is played for the multi-million dollar markets, the long-shot strategy enters the picture, and to date none of the longshot, large dollar volume areas has paid off.

Increasingly, we find evidences of an alternative approach. As holography has become more widely understood, persons with specific problems and goals have examined it and, occasionally, have discovered that holography appears to offer the best available solution. Additionally, holographers have become more sophisticated in their judgments of holography's capabilities

and have searched more deeply for applications. The result is an increase in the number of applications that withstand critical evaluation; such applications are typically those that are highly specific or those in which some fairly simple holographic principle is incorporated harmoniously into a basically nonholographic system. These approaches stand in contrast with the bolder, albeit less successful, ones that pit holography in a broad way against a well developed and well established technology.

A recent example is the work of Ralph Wuerker and Lee O. Heflinger, who used[14] pulsed-laser hologram interferometry to examine the thermally produced deformations in an antenna to be carried on a satellite. The antenna, placed in a space-simulation vacuum chamber, was heated with a solar-radiation simulator. A hologram was made by a double-exposure process, so that the resulting image formed by the hologram was a coherent superposition of the image formed during each exposure. The deformation occurring in the interval between exposures was thus manifested as a fringe pattern overlaying the image; the greater the deformation, the finer the fringe pattern. Here the task is one to which holography is highly suited, and to which other techniques are ill suited.

Anthony Vander Lugt used[15] a holographically produced spatial filter in an optical processing system to analyze cloud motion. The input was a succession of pictures of cloud cover taken by a satellite. Certain frames are converted into Fourier-transform holograms and used as spatial matched filters for the next few successive frames. Portions of the cloud pattern that remain completely unaltered produced a bright cross-correlation spot in the output. Portions that are unaltered except for linear displacement produce displaced cross-correlation peaks. Picture segments can be identified with the cor-

responding output peaks by covering portions of the input data by an adjustable aperture. If such operations are performed on successive frames, the velocities of the various cloud segments can be found. Rotation of cloud masses can be measured by adjustment of the filter orientation. Alteration of the cloud structure can be determined by measurement of the correlation peaks. As the cloud structure changes with time, later frames can be introduced as spatial filters, so that the filters can be continuously updated. The spatial matched filter thus fits centrally into the system, performing in a simple way precisely the tasks required.

Most promising areas

Presently, the areas of holography that appear most promising are probably hologram interferometry, holographic memories and hologram optical elements.

Shortly after it was announced in 1965, hologram interferometry became the major type of application, a position it retains to the present day and will likely retain at least into the near future. Recently other techniques involving coherent light have developed.[16] For example, interferometers using the well known "speckle" phenomenon of coherent light can do many of the jobs to which holography is suited, such as detection of vibration and measurement of deformations. This method is easier to carry out than hologram interferometry but generally lacks its precision and sensitivity. Thus the method is complementary to, rather than competitive with, holography.

Hologram interferometry has proved quite valuable in certain specific situations. In general situations (such as nondestructive testing techniques for detection of cracks and debonds), it has shown itself to be competitive with such standard techniques as x-ray and ultrasonic analysis (figure 4). However,

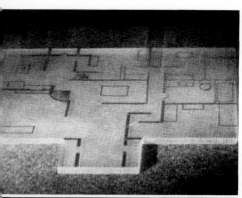

Architectural structures are visualized through holography. A model of a building, in various stages of construction, was formed by multiple storage of images; each image can be read out separately by using the Bragg diffraction effect of thick emulsions. A 5.5-deg rotation of the plate produces the next successive image, and intermediate orientations produce adjacent pairs of images in superposition. As the observer rotates the plate, he can see the structure at any desired stage of completion, and because the images are fully three-dimensional, he can perceive the various structural relations vividly. Courtesy L. Fader and C. Leonard.
Figure 6

hologram interferometry has as yet made few if any inroads into the standard nondestructive testing methods.

Holographic memories show promise but face an uncertain future. The absence of an erasable, reusable and sufficiently sensitive recording material has generally limited holography to read-only memories, which seem promising when viewed in terms of broad design concepts, such as cost per bit and memory size versus access time. John La Macchia,[17] however, has indicated that serious engineering problems become evident at the prototype stage: variation of diffraction efficiency from one bit to another; inability to achieve satisfactory high diffraction efficiencies and high signal-to-noise ratios simultaneously; detectors that lack uniform sensitivity; expensive and bulky light deflectors; and lasers with inadequate power, insufficient lifetimes and high operating costs.

These are not fundamental problems, but merely problems of engineering design and, quite possibly, they are solvable either by pushing present technology to its limit or by further advances in the appropriate technologies. We can expect that these problems will be surmounted within a time period of five to ten years. Holographic-memory development is apparently being actively pursued by various groups in the US, Germany and Japan.

The outlook for holographic optical elements is presently quite good. There are two basic reasons for this optimism. First, the aberrations of holograms have by now been explored in considerable depth. Second, experimental techniques for producing high-quality, high diffraction-efficiency, low-noise holograms have been developed. John Latta[18] has pioneered in the use of the computer to investigate these aberrations in great detail, and has explored the tandem combination of optical elements in arrangements that greatly reduce the aberrations (figure 7).

Diffraction gratings can be produced by holographic methods.[19] Indeed, because this process involves merely the interference of unmodulated light

beams, it should perhaps be viewed as conventional interferometry rather than holography and, in such a context, was explored by James Burch many years ago. The recent advances in holographic recording technology, however, are generally applicable here also.

Major problems to be solved

The intensive research of the past seven years has resolved many difficulties, but severe problems remain that block many of the hoped-for attainments in holography.

The lack of recording materials is one of the foremost problems. At present, high-resolution photographic film is the most commonly used material. Its deficiencies are severe and include lack of optical flatness, variation from plate to plate and among batches of emulsion, nonlinearity, a messy chemical-development process that swells the emulsion and generally leaves some permanent distortions, insensitivity to light in comparison with conventional emulsions, and lack of erasability and reusability. These deficiencies have promoted a search for alternative materials, but the results have generally been disappointing; photographic film remains, in general, the best material.

Bleaching techniques, in which the silver deposits in the developed negative are changed into a transparent material with an index of refraction different from the surrounding emulsion, has led to dramatic increases in diffraction efficiency of holograms. A hologram formed from a diffusely scattering object generally has a diffraction efficiency such that only about 0.5 to 1.0% of the incident light is converted into the reconstructed wave. Bleaching under controlled conditions can raise this conversion to 15 to 20% while preserving good image quality. Even higher diffraction efficiencies can be achieved, but at the expense of increased noise.

Alternative materials include photochromic glasses, which, by virtue of their molecular grain size, offer extremely good resolution and low noise but are several orders of magnitude less sensitive than even the very slowest photographic emulsions. They are erasable but generally have extremely low diffraction efficiency. Holograms of excellent quality have been produced on photopolymers, dichromated gelatin and lithium niobate. Magneto-optic materials, such as MnBi, have produced holograms that can be rapidly formed and erased. But none of these alternative materials is likely to displace photographic film as the most commonly used recording medium.

Another severe problem area is laser speckle, a defect that never escapes the notice of those viewing holograms. Yet, the problem lies not with holography, but with the coherent light that holo-

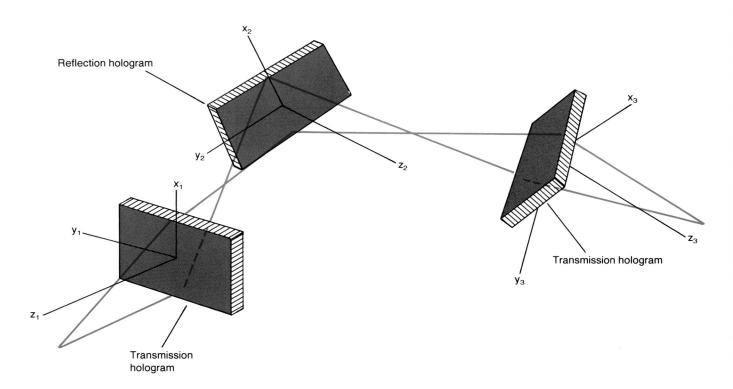

Reflection hologram

Transmission hologram

Transmission hologram

Computer analysis of holographic optical elements in tandem.   In this model, developed by John Latta, the geometry is extremely general; it allows for any orientation and spacing of the various elements, as well as for both transmissive and reflective elements.   Computer programs that optimize the design of such holographic systems are available.
Figure 7

graphy requires.   The problem is generally solvable only with the introduction of massive redundancy into the viewing system.   For the case where the final image is a recording of the holographically formed image, the redundancy takes the form of a multiplicity of superimposed images (ten or more), each the same except for uncorrelated speckle patterns.   Alternatively, the redundancy can take the form of an imaging system with the f-number far smaller than necessary for the desired resolution.   When the hologram is to be viewed directly, these methods do not apply, except in the awkward case of projecting a multiplicity of pictures at a frame rate higher than the retention time of the eye.   In general, the proposed solutions to the speckle problem come with a high price tag and are no panacea.

The light source is a problem that has been yielding to advances in technology. We obtain 1000 times more power, at 1/100 the price per milliwatt, than was available when lasers were first applied to holography.   Coherence is better than ever; even pulsed lasers now have coherence properties suitable for holography.   Holographic portraiture is one quite dramatic result of pulsed-laser holography.   There is, however, need for further improvement in producing powerful, highly coherent, inexpensive and compact lasers.

The future prospects for holography evidently depend on the development of further viable applications, without which holography would become only a laboratory curiosity.   Much work remains to be done in the aforementioned problem areas; unfortunately, the reduction of government and industrial research funding for nonmission-oriented and long-term research has adversely affected prospects here.   In any event, whatever its usefulness may be, holography will remain a subject of fascination to all who have the opportunity to make holograms and, indeed, even to those who must content themselves merely with viewing them.

References

1. H. Kiemle, D. Roess, *Einführung in die Technique der Holographie*, Akademische Verlagsgesellschaft, Frankfort, 1969 (English translation to be published this year).

2. D. Gabor, Nature **161**, 771 (1948); Proc. Roy. Soc. (London) **A197**, 454 (1949); Proc. Phys. Soc. (London) **B64**, 449 (1951).

3. G. L. Rogers, Nature **166**, 237 (1950); Proc. Roy Soc. (Edinburgh) **A63**, 193 (1950–1951).

4. M. E. Haine, J. Dyson, Nature **166**, 315 (1950); M. E. Haine, T. Mulvey, J. Opt. Soc. Am. **42**, 763 (1952).

5. H. M. A. El-Sum, *Reconstructed Wavefront Microscopy*, PhD thesis, Stanford University (1952).

6. A. Lohmann, Opt. Acta **3**, 97 (1956).

7. E. Leith, J. Upatnieks, J. Opt. Soc. Am. **52**, 1123 (1962); **53**, 1377 (1963); **54**, 1295 (1964).

8. Yu. N. Denisyuk, Sov. Phys.-Dokl. **7**, 543 (1962); Opt. Spectrosc. **15**, 279 (1963).

9. P. J. van Heerden, Appl. Opt. **2**, 393 (1963).

10. E. N. Leith, J. Upatnieks J. Opt. Soc. Am. **57**, 975 (1967).

11. G. B. Parrent, B. J. Thompson, Opt. Acta **11**, 183 (1964).

12. C. Aleksoff, Appl. Opt. **10**, 1329 (1971).

13. M. Horman, Appl. Opt. **4**, 333 (1965).

14. R. F. Wuerker, *Proceedings* of the Society of Photo-optical Instrumentation Engineers Seminar on Developments in Holography, Boston, April 14, 15, 1971; R. F. Wuerker, L. O. Heflinger, Pulsed Laser Holography II, Technical Report No. AFAL-TR-71-323, December 1971.

15. A. Vander Lugt, *Proceedings* of the Society of Photo-optical Instrumentation Engineers Seminar on Developments in Holography, Boston, April 14, 15, 1971.

16. J. M. Burch, *Proceedings* of the Society of Photo-optical Instrumentation Engineers Seminar on Developments in Holography, Boston, April 14, 15, 1971.

17. J. La Macchia, *Proceedings* of the Society of Photo-optical Instrumentation Engineers Seminar on Developments in Holography, Boston, April 14, 15, 1971.

18. J. Latta, Appl. Opt. **10**, 2698 (1971).

19. A. Labeyrie, J. Flamard, Opt. Comm. **1**, 5 (1969).

Optics as scattering

The art of deriving information about an object from the radiation it scatters, once limited to visible light, now includes much of modern physical research.

Giuliano Toraldo di Francia

PHYSICS TODAY / FEBRUARY 1973

One of the most efficient ways for a physicist to collect information about the outside world is through scattering processes: We aim a suitable beam of particles at the target or object to be investigated and observe the recoil particles or, more generally, the end products of the process as in figure 1. Classically, we use the results of this observation to derive, by theory and computation, some properties of the target that are assumed to be more fundamental than the mere scattering data. Alternatively we could take a very cautious attitude and assume that the scattering matrix, without further elaboration, fully describes our target, so far as that kind of primary particle is concerned. How do we get the most useful scattering matrix? And why, for so many years, did the only beam of "particles" exploited to any extent remain electromagnetic radiation within a certain limited frequency band—why did optics, the physics of visible light, develop first?

The negentropy principle

For us to collect information from a scattering process two conditions must be fulfilled:

Giuliano Toraldo di Francia is professor of physics at the Istituto di Fisica Superiore, Università di Firenze (Italy).

▶ A beam of primary particles must be available with sufficient intensity for the end products to be distinguishable from the background as well as from the particles spontaneously emitted by the target.

▶ The observer must be provided with a detector of suitable sensitivity and resolving power.

Thermodynamically, the first condition implies a nonequilibrium situation and is related to the entropy balance. At thermodynamic equilibrium between beam and background no information can be acquired about the details of the surroundings. To collect information we must have a source of negative entropy or *negentropy*.

The negentropy principle of information was developed many years ago by Léon Brillouin.[1] The negentropy of an isolated system in a given state may be defined as the difference between the maximum admissible (or equilibrium) entropy of the system and the entropy in that state. Negentropy can be either dissipated through irreversible processes or be converted into information. Information, in turn, can be transformed back into negentropy. The second law of thermodynamics may be written in the form

$$\Delta(N + I) \leq 0$$

where N is negentropy and I informa-

tion, measured in the same units. One information bit is equivalent to $k \log 2$ thermodynamical units, k being the Boltzmann constant. In other words, one bit of information must be paid for with at least $k \log 2$ units of negentropy.

The only important source of negentropy for us is the Sun. That is, the Sun–Earth system has a lot of negentropy to be spent. Negentropy from the Sun is carried by radiation of three main types: neutrinos, electromagnetic radiation and solar wind.

Neutrinos reach the ground freely but cannot be used for acquiring information about terrestrial objects: They are hardly ever scattered by material objects and, at least at present, we have no appropriate detector. As far as the other two types of radiations are concerned, we find ourselves in a very peculiar condition. We are living inside a black box whose walls are represented by the atmosphere. The atmosphere is virtually impenetrable to electromagnetic and particle radiation.

Luckily enough, the builder of the box, through a small oversight, has inadvertently left a tiny *crack* in the wall, which lets in visible light, plus a bigger *hole* for microwaves. Nature has been conscientious in exploiting the possibilities offered by the radiation coming through the tiny crack.

Classic scattering experiment. Here the "beam of particles" is a single heavy rock and the end products observed are splashes of water and waves.   Figure 1

An overwhelming majority of living creatures is provided with detectors for visible light, and very often such devices are extremely refined and sophisticated.

Why only visible light?

This wonderful display of efficiency makes us ask the following question: Why did nature disregard the microwave radiation coming through the bigger hole? There are indeed very good reasons. An obvious reason is seen in figure 2.

But there is something else. The average number n of photons per degree of freedom contained in electromagnetic radiation at temperature T is represented by

$$n = \frac{1}{\exp(h\nu / kT) - 1}$$

In the case of visible light, $n \simeq 0.05$ for solar radiation and $n \simeq 10^{-26}$ for terrestrial radiation. The scattered solar photons, then, can be perfectly distinguished from the very few photons emitted by the target, even when solar light does not arrive directly at the target but is previously scattered one or more times by other objects (atmosphere, moon, walls of a room). Incidentally, because background noise is absent, a detector of the highest sensitivity, of the order of one photon, is

useful. The human eye has such a sensitivity.

The situation is different for microwaves. In this case we have $h\nu \ll kT$ both for solar and for terrestrial radiation, so that $n \simeq kT/h\nu \gg 1$. Since the Sun's surface is about 20 times as hot as the Earth's surface, the value of n for solar radiation is only 20 times greater than the value for terrestrial radiation. Even for direct illumination and very small absorption, a target that scatters within an angle of more than 2.5 deg sends out more noise than scattered radiation. Nature has apparently decided not to take the trouble to provide animals with such an inefficient system for collecting information.

This is probably the reason why, of all possible scattering processes that can be used to obtain information about the physical world, light scattering has for so many thousand years been virtually the only one known to human beings. Consequently *optics*, as is revealed by its Greek etymology, has been thought to be necessarily related to the eye and vision.

Now we can ask ourselves: Is there any sensible reason to go on and preserve as a separate science the science that deals with the collection of information by means of light scattering? The limitation to a very small band of

electromagnetic radiation appears to be due rather to an accidental condition of the Sun–Earth system than to fundamental or conceptual reasons.

Of course, the hardware used to deal with visible light is different from that used for other radiation. Consequently, the technology of optical instruments may still represent a separate body of knowledge. And we must not forget that very peculiar and intriguing optical instrument—the human eye. This system is still the object of profound investigation.

However, the emphasis of present-day research is placed rather on a number of methods and devices that do not belong specifically or necessarily in the visible range of electromagnetic radiation. The main problem and *leitmotiv* of the research is to derive information about an object from the radiation scattered by it or sent out spontaneously. On this account one may even be tempted to say that optics encompasses a great part of modern physics.

Man realized very early that when solar radiation was not available (at night, or inside a cave) he could easily produce more or less the same radiation by means of fire. From that discovery he proceeded to the construction of a host of artificial sources such as oil and gas lamps, candles, torches and so on. He did not know that some animals, such as bats, were provided by nature with sources and detectors of a different type of radiation, namely ultrasonic radiation, and that the system could be very efficient for gathering information, especially for measuring distances. No wonder that when an extremely serviceable physical agent, electricity, was discovered, it was first applied to the production of conventional light.

Meanwhile the art of gathering information from light had been highly refined, on the basis of geometrical optics. Conventional optical instruments were brought to near perfection, with wave optics considered only as a sort of unfortunate disturbance, that set a limit to resolving power, due to the appearance of the Airy disc of diffraction.

Information from radiation

The first important steps toward a conceptual revolution were due to Max von Laue and Ernest Rutherford. Working at about the same time, they showed that other kinds of radiation, different from visible light, could be applied to gather essential information about crystals and atoms respectively. The *inverse scattering problem* was making its decisive appearance in physics. Since then, the method has been applied with all conceivable sorts of radiation and with enormous success.

It took some time for optics researchers to realize that what they had long been doing was nothing but a particular case of what a greater and greater number of physicists were now doing with different kinds of radiation. The eye, like the other optical instruments of classical type, is an analog computer that elaborates the information carried by scattered light and presents it in a convenient form to the mind of the observer. However, information can also be displayed in different and sometimes more convenient ways, and much of the information carried by the scattered light is missing in the image formed by a conventional instrument.

Two main factors are responsible for the missing information:

▶ Some *real* and all evanescent waves scattered by the object are not collected by the instrument; consequently their information is lost.

▶ Classical instruments do not measure the *phase* of scattered light, and the corresponding information is therefore also lost.

As is well known, the latter shortcoming was removed between the first and second world wars by Frits Zernike[2] with the invention of the phase-contrast method, and in general with the method of the coherent background. The idea is to introduce a proper

Why only visible light? If we were equipped instead with, say, microwave detectors, a device with suitable resolving power would be too big for convenience.   Figure 2

phase shift in the light scattered by a small object and to make it interfere with the light of a coherent background. Mere phase differences are thus converted into amplitude differences.

Today, this idea appears so simple and natural that one may miss its historical significance and wonder why it had such an influence on all later developments. For the first time after the pioneering efforts of Ernst Abbe,[3] *coherent light* entered the scene as an important tool for investigating the structure of the visual world. It turned out to be so useful that soon after World War II many refined methods and devices were developed to take full advantage of the possibilities it offered. Note that this happened long before coherent light could be produced with substantial efficiency by means of lasers.

Zernike's phase contrast is very efficient but applicable only to particular microscopic objects. The large-scale utilization of interference with a coherent background came with the introduction of holography. The first ideas about holography occurred to Dennis Gabor in 1948,[4] when he was thinking about electron microscopy, particularly the difficulty of correcting the spherical aberration of electron lenses, and he thought it possible to compensate for spherical aberration by wavefront reconstruction. However his method has never found a practical application in electron microscopy but has instead

come into prominence for the use with laser light.

Let S represent the signal, the complex amplitude of a beam of coherent light scattered by the object, and R the coherent background, or reference beam as in figure 3a. When they impinge simultaneously on a photographic plate, the exposure E is proportional to the square modulus of their sum:

$$E \propto SS^* + RR^* + RS^* + R^*S$$

By controlling the developing time, a positive can be made from this negative plate such that the amplitude transmission is proportional to E (in photographic terms, with $\gamma = -2$). Therefore, if we illuminate with the reference beam, the transmitted amplitude A will be

$$A = (SS^* + RR^*)R + R^2S^* + RR^*S$$

Under the usual conditions of holography, the first term represents the reference beam almost unmodified, the last term represents the signal multiplied by the constant (or nearly constant) factor RR^*, and the second term represents the complex conjugate of the signal, or "twin image."

Holography with laser light

In his original work, Gabor had to use the "in-line" method, where the signal and reference beams were approximately in one line. This was because of the limited coherence of the light used (a high-pressure mercury lamp), which did not allow the production of interference at large angles. As a result, the image, the twin image and the reference beam were very inconveniently superimposed. This difficulty was overcome by Emmett Leith and Juris Upatnieks in 1962, when lasers were available, by taking a skew reference beam (see figure 3b).[5] Thus in the reconstruction the three beams were angularly separated. The assumption that $\gamma = -2$ is useful for an elementary discussion, but can easily be dropped. If γ does not equal minus two, we get not only the ordinary and twin images but also a series of "harmonic terms" or of angularly separated images. Black-and-white holograms, bleached holograms and phase holograms have become possible.

Black-and-white holograms can be computed and synthesized. By taking advantage of the thickness of the emulsion, one can combine the principle of Lippmann color plates with holography, as suggested by Yu. N. Denisyuk,[6] and get color holograms.

One feature that makes holography very useful is that information about a given wavefront remains "frozen in" in the emulsion and can be added linearly to the information of a different wavefront. This property is used in holographic interferometry, which has

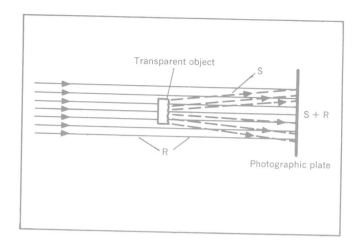

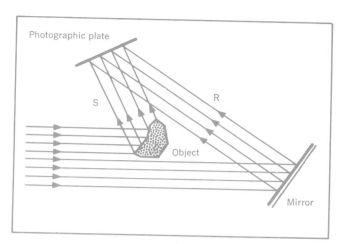

Hologram recording systems. In the early "on-line" system (left) a beam of coherent light is scattered by the object. This beam (S) and the coherent background beam (R) impinge simultaneously on a photographic plate and are recorded. To view the hologram, the plate is illuminated with the reference beam. Holography with a skew reference beam (right) was developed in 1962, when lasers were available. This method overcame the problems of superposition of image, twin image and reference beam (see reference 5). Figure 3

many industrial applications. Two holograms of a given object, taken at different times, are recorded on the same plate. The two wavefronts obtained in the reconstruction can interfere, and from the interference pattern one can tell whether the object has moved or been deformed between the recordings (see figure 4).

In 1971, looking back to his intended electron-optics application, Gabor recalled: "Why should one bother in light optics, with such a complicated two-stage process, [with] coherent light so weak and uncomfortable to use, when we had such perfect lenses, even achromatic ones? Little did I think at that time that after 24 years the application of holography in electron

microscopy would still be in a primitive state, while the simple optical experiments, which I considered only as model or feasibility experiments, would give rise to a new branch of optics, with some 2000 papers and a dozen books!"[7]

Image formation with coherent light can be described for a typical case of a transparent object. A coherent plane wave illuminates the plane of the object O and is scattered by "inverse interference" into many plane waves with different directions, plus a set of evanescent waves. Each plane wave is brought to a focus by lens L_1 at a point of the focal plane F, conventionally termed the "pupil plane." Lens L_2 transforms the wave back into a plane

wave. All the output plane waves are brought together to direct interference at the plane of the image (see figure 5).

Each one of the scattered plane waves represents a spatial frequency or a component of the Fourier spectrum of the object. This Fourier analysis shows that there is an analogy between an optical signal as a function of space coordinates and an electrical signal as a function of time, so that the ordinary techniques of frequency manipulation can be applied. The only difference is that in optics we have two dimensions instead of one.

Let us first consider the question of resolving power. If the instrument could collect all the scattered waves and bring them back to interference onto plane I, the image would be similar to the object. However, any instrument has a finite aperture, and some real waves as well as all the evanescent waves do not enter the system and are missing in the image. There is a cutoff in the spatial frequencies of the image, and consequently the finest details of the object are lost. This is tantamount to saying that an infinite number of different objects should have one and the same image: The image is ambiguous.

The question of how many and what objects have one and the same image is closely related to the problem of finding the number of degrees of freedom of the image. This question dates back at least to Laue's discussion of the degrees of freedom of electromagnetic radiation. In more recent times it has been argued that, due to the finite width of the spatial frequency band of the image, one can apply the sampling theorem: The result is that the number S of degrees of freedom is proportional to the object area times

Holographic interferometry is a technique with many industrial applications. Two holograms of an object are taken at different times and recorded. The interference pattern indicates whether the object has moved or been deformed between recordings. Figure 4

the entrance solid angle of the instrument, divided by the wavelength squared (this is the "Shannon number").[8]

The validity of using the sampling theorem has been repeatedly refuted for the reason that when the object has finite extension, the frequency distribution over the pupil plane is an analytic function and as such can be completely known when we know its behavior even over the limited domain of the pupil aperture region. This remark is true, but has no practical value. The question has been completely clarified by means of the prolate spheroidal functions, analyzed by D. Slepian and H. O. Pollack.[9] In this approach, an object distribution represented by a (properly scaled) spheroidal function ψ_n has an image similar to the object; that is, an image represented by the same distribution multiplied by a factor λ_n. The ψ_n form a complete set of orthogonal functions for the object, which can therefore be expanded in series of the ψ_n functions. Then each coefficient of the series corresponds to a degree of freedom of the object. It turns out that λ_n equals about one up to values of n approximately equal to S, whereas for n greater than S the factor λ_n drops practically to zero. The corresponding degree of freedom is lost for the image, no matter what physically conceivable type of detector we use. Therefore S represents the "physical" number of degrees of freedom.

The problem of noise

Of course, the evaluation of S represents only a preliminary stage in the application of information theory to optics. A complete theory can be developed only by taking noise into account. Noise in optics can arise from different sources, but one source that can never be eliminated is represented by the photon nature of light.

The following argument, due to Gabor,[10] is very instructive: A photographic plate collects energy $RR^* \equiv E$ from the reference beam, $SS^* \equiv e$ from the signal, and from their interference, $RS^* + R^*S \equiv 2(Ee)^{1/2}\cos\phi$. The last term, which represents the interference fringes, carries the information and can be made as large as we want by increasing E, no matter how small the signal energy e.

In the limit one could (seemingly) transmit information without energy. Of course this is absurd. The signal cannot be recognized once it is small enough that it is drowned out in the fluctuations of E. Now the mean-square amplitude of the fringes is $2\,Ee$. We postulate that the signal becomes unrecognizable at some minimum value of e, called ϵ, at which the mean square fluctuation of the reference ener-

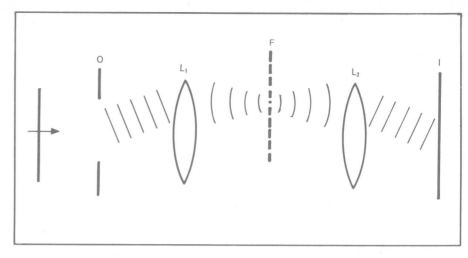

Image formation for a transparent object. A coherent plane wave illuminates the object O and is scattered into many plane waves with varying directions (only one is shown above), plus a set of evanescent waves. Each plane wave is focussed by lens L_1 at a point along the focal plane F (the "pupil plane") and is transformed back into a plane wave by lens L_2. All the output plane waves are again brought together, by direct interference, at the image plane I. Fourier analysis of this process shows an analogy between the optical signal in space and an electrical signal in time, so that the ordinary techniques of frequency manipulation are applicable to optical as well as to electrical waves. Figure 5

gy E exceeds this by a factor k. Thus

$$\langle\Delta E^2\rangle = 2kE\epsilon$$

or, if $n \equiv E/\epsilon$

$$\langle\Delta n^2\rangle = 2kn$$

By putting $k = 1/2$ we get Poisson's law

$$\langle\Delta N^2\rangle = n$$

which indicates that the fluctuation of the energy is of the nature of shot noise. We recognize that the result is correct so long as the photons are distributed over a very great number of cells of phase space. Otherwise there is a correction term, because photons are not classical particles. The problem, we see, is similar to one of fluctuations of blackbody radiation in an enclosure.

Whatever the source of noise, we can code and process the information so as to reduce its effects. For instance we can filter out unwanted frequencies from the plane of the pupil, a procedure pioneered by André Marechal and Paul Croce.[11] To mention a few interesting cases:

The object can be periodic, giving rise to a series of dots on the image plane. By blocking these frequencies with black dots, we can remove the image of the "perfect" object and reveal only its faults. Conversely, if one lets through only the periodic frequencies by means of a diaphragm with holes, one can remove the faults and restore a perfect image of the object.

In general the restoration of an image damaged by blurring or other causes requires filters for both amplitude and phase in the plane of the pupil. This was a formidable requirement until holography was applied to it by Anthony van der Lugt and George

Stroke.[12] Holography allows us to build a plate that gives to the impinging wavefront any wanted modulation of amplitude and phase.

One of the first things made possible by this filtering technique is the restoration of images blurred by defocusing, movement or other causes. Another application of great interest is in pattern recognition. If the pattern to be recognised is p, its Fourier spectrum P is formed in the focal plane and holographically recorded. The hologram will contain both P and P*. The latter defines the filter to be used. When P* is multiplied with the Fourier spectrum of the object, one obtains in the image plane the correlation function of the object and the pattern. Bright spots appear in the places where the object contains the pattern to be searched. This procedure suggests the possibility of building a reading machine; however, no practical device appears so far to have been built on this principle. One of the serious problems in applying this technique is that the pattern may have different size and orientation from that used in the recording.

Nature of coherent light

All these developments appear to have been made possible by coherent light. What is coherent light? For a long time people were content with the vague notion of coherent light being highly monochromatic and coming from a very small source. A quantitative analysis was pioneered by Zernike[13] and developed extensively in the last decades. Let us denote by $V(t)$ the complex wavefunction (or *analytic signal*) whose square modulus $V(t)V^*(t)$

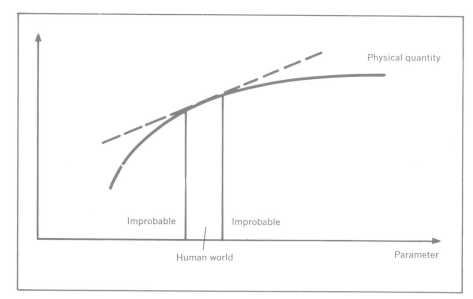

Nature is linear. A modern rephrasing of *natura non facit saltus* (nature does not jump) might be that any curve representing a physical phenomenon can, within a limited interval, be replaced by a straight line. The important philosophical question here is why the interval of physical interest is usually limited to this straight line portion of the curve.　　Figure 6

$= I(t)$ represents the light intensity. For two points P_1, P_2 we define the *mutual coherence function*

$$\Gamma_{12}(\tau) = \langle V_1(t + \tau)V_2{}^*(t)\rangle$$

as a time average taken over a sufficiently long period. The normalized function

$$\gamma_{12}(\tau) = \frac{\Gamma_{12}(\tau)}{\sqrt{\langle I_1\rangle\langle I_2\rangle}}$$

is called the *complex degree of coherence*. It can be shown that the maximum value of $|\gamma_{12}|$ is a measure of the visibility of the fringes formed by light coming from P_1 and P_2 in a Young interferometer.

There is a beautiful theorem, due to P. H. van Cittert[14] and Zernike,[13] on the value of γ_{12} for two points illuminated by a planar quasimonochromatic source S: γ_{12} is the normalized complex amplitude that would be produced at P_2 by a spherical wave centered at P_1 and diffracted through an aperture equal to S. This theorem easily explains why Albert Michelson could measure star diameters with his stellar interferometer, despite its shortcomings. Scintillation and the difficulty of maintaining the interferometer arm length constant to within a quarter wavelength make it impossible for the arm length to exceed a few meters.

However, if light can be considered as a random Gaussian process, the fluctuations of *intensity* obey the equation

$$\langle\Delta I_1(t + \tau)\Delta I_2(t)\rangle = I_1 I_2|\gamma_{12}(\tau)|^2$$

Intensity fluctuations are much slower than amplitude oscillations and the measurement is much easier; this is the basis of the intensity interferomet-

er that R. Hanbury-Brown and R. Q. Twiss[15] have constructed both for visible light and microwaves. Radiation is collected by two mirrors, at P_1 and P_2, and sent to two phototubes that reveal intensity. Signals from both phototubes are sent to a multiplier and correlated. In this way one can measure $|\gamma_{12}|$ for two points even hundreds of meters apart. By recording the signals, one can also compare fluctuations at places thousands of kilometers apart.

Phototubes can be sensitive to individual photons, and one may wonder if classical wave theory is adequate to treat these phenomena. However, semiclassical theory shows that, because photons are bosons, fluctuations of photocounts obey substantially the same laws as fluctuations of intensity.

A more important question is whether and when the assumption that light is a Gaussian random process is justified. It is certainly justified when light is generated by many independently excited and spontaneously emitting atoms. But what about lasers?

The question can be clarified only by an appropriate quantum treatment. Pioneering work in this direction has been done by Roy J. Glauber.[16] As is well known, the electric field E can be represented by an expansion:

$$E = \Sigma[a_k u_k(r)e^{-i\omega Kt} + a_k{}^+ u_k{}^*(n)e^{i\omega kt}]$$

where a_k, $a_k{}^+$ represent the annihilation and creation operators. The annihilation and creation parts of E will be represented by E_+ and E_- respectively.

If by $x_1\ldots x_{2N}$ we denote $2N$ points of space–time and by ρ the density matrix of the states of the field, we can

introduce the correlation function of Nth degree by

$$G^{(N)}(x_1\ldots x_{2N}) =$$
$$\Sigma[\rho E_-(x_1)\ldots E_-(x_N)E_+(x_{N+1})\ldots E_+(x_{2N})]$$

It turns out that $G^{(1)}(x_1\, x_1)$ is proportional to the probability P_{x1} of having a photoelectron produced at x_1; that is, to the intensity at x_1. Similarly $G^{(N)}$ $(x_1\ldots x_N, x_1\ldots x_N)$ is proportional to the probability $P_{x1}\ldots x_N$ of revealing one photoelectron at each point $x_1\ldots x_N$ (N-fold coincidence). Let us introduce a normalized correlation function

$$g^{(1)}(x_1 x_2) = \frac{G^{(1)}(x_1, x_2)}{\sqrt{G^{(1)}(x_1, x_1)G^{(1)}(x_2, x_2)}}$$

and similar expressions $g^{(N)}(x_1\ldots x_{2N})$ for the higher-order correlation functions. A field will be said to be coherent in the first order if $g^{(1)} = 1$, in the Nth order if $g^{(N)} = 1$ everywhere.

If $g^{(N)} = 1$, then the probability $P_{x_1 x_2}\ldots x_N$ of a N-fold coincidence turns out to be equal to the product

$$P_{x_1 x_2}\ldots x_N = P_{x_1} P_{x_2}\ldots P_{x_N}$$

of the probabilities of revealing one photon at each point $x_1\ldots x_N$. Therefore such probabilities are independent and there is no correlation in the photocounts. Coherent light obtained with ordinary sources (that is, filtered Gaussian light) is coherent only to the first order and therefore shows twofold correlations, whereas the field radiated by a classical antenna is coherent to any order. Laser light lies in between. It is coherent to a high, but not infinite, order.

The invention of the laser is just one more step in a process that has been going on since the beginning of mankind: At first, Man played only a role analogous to that of Maxwell's demon, selecting from the random noise of all natural objects and phenomena those improbable cases that were advantageous for his survival. Then, little by little, he learned to build the improbable and useful things right away, without having recourse to noise. Thus coherent light from a natural or conventional source is only filtered noise. Making a laser is much more clever.

Noise, however, can be used in a very subtle way. We are accustomed to thinking that random fluctuations mar the signal sent out by an object and diminish obtainable information. Nevertheless random fluctuations carry some valuable and quantitative information about the object. This was first shown by Einstein with his theory of Brownian motion. Another good example is Rayleigh scattering of light by the sky. A more recent and striking case is correlation interferometry.[15]

Nonlinear optics

Departing from filtered noise and natural phenomena occurring around us very often brings us to discovering nonlinearities. When a young student first encounters the mathematical expressions of elementary physical laws, he may be tempted to conclude that *nature is linear.* Is this conclusion right?

A more correct statement appears today to be that nature is analytic, which in turn is a modern rephrasing of the old statement *natura non facit saltus,* or nature does not make jumps. Any curve representing a physical phenomenon has a continuous tangent and within a limited interval can be replaced by a straight line (figure 6). The important philosophical question to answer is why the interval of interest is in most cases limited to the straight portion of the curve. Why is the usual departure of the parameters from their equilibrium values so small? Is this an essential feature of the human world? We live in a degenerate world, very close to absolute zero, the world of molecular forces. But even if we include the Sun in our system, we remain in the realm of modest energies as compared with the high energies that we know to be physically possible. Near equilibrium only the lower states tend to be filled, and the more we want to depart from equilibrium, the more ingenuity and effort we must spend in order to make a very improbable case to occur. Laser light can easily be made a million times hotter than the Sun's surface. No wonder therefore that the straight-line limits are exceeded and optics may become nonlinear.

Classical optics is based on the circumstance that in an ordinary material the electric polarization is proportional to the electric field; waves scattered by the atoms have the same frequency as the incident wave. However, when E becomes sufficiently large, we discover that P is a nonlinear function of E:

$$P = a_1E + a_2E^2 + a_3E^3 + \dots$$

The second term of the expansion gives rise to scattered waves with double frequency; the third term gives rise to a third harmonic, and so on. In this way one can multiply the frequency of optical radiation, as was first shown by Peter A. Franken.[17] Because two or more photons of the incident radiation give rise to one photon of the harmonic, there is of course a condition of momentum conservation: $\hbar K_1 + \hbar K_1 = \hbar K_2$. Since K defined as 2π is proportional to the refractive index, this is equivalent to an index-matching condition $n_1 = n_2$. This condition can be met in an anisotropic medium in a particular direction.

Nonlinear optics has opened a huge field of prospective applications in communications systems, where many well known radiofrequency techniques such as mixing, heterodyning, modulating and so on can be transferred to the domain of optical frequencies.

More light

Light, the first gift of God to Man, one of the first physical phenomena to be investigated, still continues to supply a wealth of problems for our mind, a wealth of applications for our welfare. This is a gratifying realization. However, no discussion of optics and its latest applications would be complete if we ignored the fact that with all our light we are not yet able to illuminate human brains and to defeat human stupidity. Many people working with lasers were simply horrified when they realized that the most effective application of lasers is today represented by the guidance of deadly missiles. What could have been a monument to man's intelligence has turned into a shame for all mankind.

Mehr Licht! More light! Like Goethe on his deathbed, we desperately need more light. More light for the human mind.

* * *

This article was adapted from a talk given at the September 1972 General Assembly of the International Union of Pure and Applied Physics in Washington, D.C. The original talk will appear in the Proceedings, to be published by the US National Academy of Sciences.

References

1. L. Brillouin, J. Appl. Phys. **24**, 1152 (1953).

2. F. Zernike, Phys.Z. **36**, 848 (1935).

3. E. Abbe, *Gesammelte Abhandlungen,* vol. 1, G. Fischer, Jena (1904), page 45.

4. D. Gabor, Nature **161**, 777 (1948).

5. E. N. Leith, J. Upatnieks, J. Opt. Soc. Am. **52**, 1123 (1962).

6. Yu. N. Denisyuk, Opt. Spectrosc. **15**, 279 (1963).

7. D. Gabor in *Optical and Acoustical Holography* (E. Camatini, ed.), Plenum, New York (1972), page 10.

8. G. Toraldo di Francia, J. Opt. Soc. Am. **59**, 799 (1969).

9. D. Slepian, H. O. Pollack, Bell System Tech. J. **40**, 43 (1961).

10. D. Gabor, reference 7, page 32.

11. A. Maréchal, P. Croce, Compt. Rend. **237**, 706 (1953).

12. A. B. van der Lugt, IEEE Trans. Inform. Theory **IT-10**, 2 (1964); G. W. Stroke, Optica Acta **16**, 401 (1969).

13. F. Zernike, Physica **5**, 785 (1938).

14. P. H. van Cittert, Physica **1**, 201 (1934).

15. R. Hanbury-Brown, R. Q. Twiss, Phil. Mag. **(7) 45**, 663 (1954).

16. R. J. Glauber, Phys. Rev. **130**, 2529 (1963).

17. P. A. Franken, A. E. Hill, C. W. Peters, G. Weinreich, Phys. Rev. Lett. **7**, 118 (1961).   □

Applications of optical phase conjugation

Light waves that are, in effect, time-reversed images of their original can serve to restore severely aberrated waves to their original state.

Concetto R. Giuliano

PHYSICS TODAY / APRIL 1981

The recently discovered generation of conjugate waves in the course of observing several nonlinear optical phenomena has attracted a lot of attention in the international technical community. Much of the interest arises from a fascination with the idea of "time reversal" and with impressive, almost magical demonstrations in which severely distorted optical beams can be restored to their original, unaberrated state.

These intriguing properties suggest a number of potential applications for optical phase conjugation, which I will explore here. Let me emphasize, however, that this is not intended to be a review paper. Amnon Yariv has published[1] an excellent review on the field as it existed in 1978. (Actually, so many new results have appeared since that publication, that perhaps another such paper is in order.) I will attempt to discuss only those findings that are most relevant to the understanding of how phase conjugation comes about and that can be connected to a specific application.

To accomplish these goals, I start with an explanation of optical phase conjugation. This is followed by a discussion of how we make conjugators, that is, what are the nonlinear optical processes that give rise to conjugate-wave generation and the workings of the two most promising types. Next comes a discussion of several potential applications, one of the most promising being adaptive optics. Finally, I describe the wide range of wavelengths

and experimental conditions under which optical phase conjugation has been observed.

What is phase conjugation?

To understand the properties of conjugate fields, consider an optical wave of frequency ω moving in the $+z$ direction,

$$E = \text{Re}[\psi(x, y, z)e^{i\omega t}]$$

where

$$\psi(x, y, z) = A(x, y)e^{i(-kz + \varphi(x, y))}$$

and A is real. The phase conjugate of wave E is defined as

$$E_{\text{PC}} = \text{Re}[\psi^*(x, y, z)e^{i\omega t}]$$
$$= \text{Re}[A(x, y)e^{i(kz - \varphi(x, y))}e^{i\omega t}]$$

That is, the phase conjugate of wave E contains the complex conjugate of only the spatial part, leaving the temporal part unchanged. The conjugate wave corresponds to a wave moving in the $-z$ direction, with the phase $\varphi(x, y)$ reversed relative to the incident wave. We can also think of the process as a type of reflection combined with phase reversal. It is equivalent to leaving the spatial part of E unchanged and reversing the sign of t; in this sense, phase conjugation is equivalent to "time reversal."

One can obtain an intuitive feeling for phase conjugation by comparing reflections from an ordinary mirror and from a "conjugate mirror." As figure 1 shows, a diverging spherical wave striking an ordinary mirror at an angle θ leaves it at an angle $-\theta$ and continues to diverge. In contrast, the same wave striking the conjugate mirror is converted to a *converging* wave that retraces the path of the incident wave.

To elaborate a bit further, suppose the wave incident on the conjugator were aberrated. An example is a uniform plane wave passing through a glass plate with a hole, a distorting medium such as a piece of bottle glass, a turbulent atmosphere, or a severely strained optical element (figure 2). Such a wave incident on a conjugator results in an output wave that is as severely aberrated as the input wave (with phase reversed). When the output passes back through the aberrator, however, it will emerge completely free of distortion. Compare this with the use of an *ordinary* mirror, which would *double* the distortion.

It is just this type of demonstration that was the subject of the first published observations in 1971 and 1972 of optical phase conjugation by researchers in the Soviet Union. B. Ya. Zel'dovich, O. Yu. Nosach and their coworkers[2] observed wave-front conjugation resulting from the nonlinear optical phenomenon, stimulated Brillouin scatttering (SBS). In his laboratory B. I. Stepanov observed[3] conjugation while experimenting with real-time holography. This phenomenon is similar to degenerate four-wave mixing, (DFWM), a nonlinear phenomenon that Robert Hellwarth recognized is a conjugation process;[4] D. M. Bloom and Gary Bjorklund subsequently demonstrated it experimentally.[5] Other nonlinear optical processes that give rise to conjugate-wave generation are backward stimulated Raman scattering and three-wave down conversion.

Stimulated Brillouin scattering

Stimulated Brillouin scattering involves the generation of a coherent acoustic wave when an intense optical

Concetto Giuliano is the manager of the quantum electronics program of the optical physics department at Hughes Research Laboratories in Malibu, California.

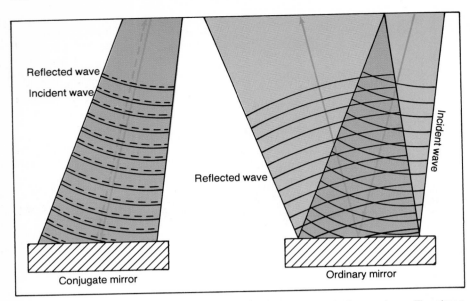

Ordinary and conjugate mirrors. The ordinary mirror reflects light in the usual way. The phase conjugate mirror produces a time-reverse image of the incident beam.　　　　Figure 1.

wave interacts with a nonlinear medium. The mechanism generally acts through electrostriction—that is, the tendency of the medium's density to increase in proportion the electric field intensity. Above some threshold intensity a light wave allowed to propagate in a nonlinear medium produces an intense back-scattered optical wave, and its frequency is down-shifted by an amount equal to the acoustic frequency.

The collinear process, which involves counterpropagating incident and scattered light waves, has the highest gain of all the possible scattering processes and is the only one observed in stimulated Brillouin scattering. The acoustic wave propagates in the same direction as the incident wave; it can be thought of as a moving mirror or stack of dielectric plates, from which the incident wave reflects to generate a Doppler-shifted scattered wave. In fact, the wavelength of the acoustic wave is half that of the optical wave in the medium; thus the acoustic wave serves as a moving half-wave dielectric stack.

When thinking of stimulated Brillouin scattering in terms of electrostriction, note that the two counterpropagating optical waves form a moving interference pattern. The speed of the pattern, given by $V = (\omega_1 - \omega_2)/(k_1 + k_2)$, is equal to the speed of sound in the medium. In fact, it is this condition that allows for the buildup of both the acoustic wave and the scattered optical wave, at the expense of the incident optical wave.

Analysis indicates that, under certain conditions, the process for which the Brillouin gain is greatest is the one in which the scattered wave is the conjugated of the incident wave.[2] Consequently, an aberrated input wave generates an equally aberrated acoustic wave through stimulated Brillouin scattering, with a phase surface that matches it exactly. One can think of the nonlinear process as creating (in the medium) a deformable mirror whose surface is just right to reverse the phase of the reflected wave from that of the incident wave. Thus, when the reflected wave retraces the incident path, the medium removes from it whatever phase errors were introduced in the first pass. If it were possible to take a moving picture of the incident wave the complete behavior of the conjugate wave would be portrayed by running the film backwards.

Stimulated Brillouin scattering can be made to occur in a highly controlled manner and with efficiencies approaching unity. This is especially true under the conditions where it works best as a conjugator, in multimode optical-waveguide configurations. It has been observed over a wide range of wavelengths and under both continuous and pulsed conditions.

Degenerate four-wave mixing.

Four-wave mixing is a nonlinear process in which three input waves mix to yield a fourth (output) wave. The three input waves consist of two planar counterpropagating pump waves, labeled E_f and E_b (f for forward and b for backward), and a probe wave E_p, entering at an arbitrary angle to the pump waves. All three couple through the third-order susceptibility, $\chi^{(3)}$, to yield a fourth wave, E_s, which is proportional to the spatial complex conjugate of E_p.

The third-order polarization—which yields the conjugate wave, $P_{nl} = \chi^{(3)}E_f E_b E_p{}^*$—is proportional to the product of the amplitudes of the three input waves. More specifically, the

nonlinear polarization that yields the conjugate of E_p can be shown to arise (for isotropic media) from the contributions of three separate terms:

$$\mathbf{P}_{nl} = A(\theta)(\mathbf{E}_f \cdot \mathbf{E}_p^*)\mathbf{E}_b$$
$$+ A(\pi - \theta)(\mathbf{E}_b \cdot \mathbf{E}_p^*)\mathbf{E}_f$$
$$+ B(\mathbf{E}_f \cdot \mathbf{E}_b)\mathbf{E}_p{}^*$$

(For anistropic media the situation is more complex; I will not discuss it here). The first two terms are responsible for the analogy between the degenerate mixing and holography. Each contains a scalar product corresponding to the interference between one of the pump waves and the probe wave; the product is multiplied by the field of the other pump wave. Thus, each term corresponds to the creation of a hologram from one of the pumps fields and the probe while simultaneously reading it out with the other pump. This is illustrated in a simple way in figure 3, which shows the holographic (or dual-grating) picture of degenerate four-wave mixing. The formation and readout processes are shown separately here, although they actually take place at the same time. The formation process is shown as the generation of two overlapping grating structures (fringe patterns), also shown separately for simplicity. Each one consisting of a series of planes, with normals in the directions $\mathbf{k}_f - \mathbf{k}_p$ and $\mathbf{k}_b - \mathbf{k}_p$; the separation of the planes is given by

$$D = \lambda/(2 \sin\theta/2)$$

One refers to the pattern arising from the interference between forward pump and probe as the large-spaced grating; the one between the backward pump and probe is the small-spaced grating.

The readout or playback process occurs when the backward pump scatters from the large-spaced grating and the forward pump scatters from the small-spaced grating, yielding the conjugate wave. Thus, the phenomenon described by the first two terms contributing to P_{nl} can be viewed as one in which the refractive index of the nonlinear material is spatially modulated as a result of the interference between pump and probe. This is followed by scattering by the other pump.

The term $B(\mathbf{E}_f \cdot \mathbf{E}_b)\mathbf{E}_p$ has no holographic analog. The scalar product of $\mathbf{E}_f$ and $\mathbf{E}_b$ corresponds to a nonlinear index, which has no spatial modulation but which oscillates at a frequency 2ω. The probe wave interacting with this driven coherent excitation at 2ω creates a polarization that results in the generation of a conjugate wave.

The relative magnitudes of the coefficients A and B depend strongly on the properties of the nonlinear medium chosen for the four-wave interaction. In particular, if the nonlinear medium has an optical resonance for a single quantum transition at a frequency near ω (the wave frequency used in the experiment), large enhancements of the four-wave mixing signals[6] arising from the first two terms are possible compared with that obtainable from a nonresonant system.

For example, consider a nonlinear medium consisting of an ensemble of two-level atoms. The near-resonant contribution to the nonlinear index will manifest itself as a spatial modulation of the populations of the lower state relative to that of the upper state. The gratings formed as a result of interference between the pumps and the probe would be "population gratings," that is, if one were to walk in the direction $\mathbf{k}_f - \mathbf{k}_p$ in the nonlinear medium, one would notice that the population of atoms in the excited state relative to the ground state varies sinusoidally with a period $D = \lambda/(2 \sin\theta/2)$. Alternatively, if the medium possesses a pair of energy levels of the same parity, allowing them to couple coherently through an interaction involving two quanta, then the third term in P_{nl} may be dominant in contributing to the four-wave signal.

Unlike stimulated Brillouin scattering where the conjugate-wave intensity cannot exceed the input intensity, degenerate four-wave mixing allows for conjugate reflectivities in excess of unity. That is,

$$I_{signal}/I_{probe} > 1.$$

Many such examples have been observed experimentally (see below). This fact has practical implications for four-wave mixing.

Another characteristic of the mixing process that can have practical implications involves its behavior for waves of different polarization. In fact, experiments in which the polarization of pump and probe waves have been manipulated to achieve the desired result have enhanced our understanding of the four-wave mixing.

Note that each of the terms in the equation for P_{nl} involves the scalar product of two fields multiplied by a third field. Thus, a given term will contribute to the nonlinear polarization only if the scalar-product term is nonzero—that is, only if the fields in the scalar product have polarization components along a common direction. This fact can be exploited to explore the fundamental properties of four-wave mixing. For example by performing an experiment in which $\mathbf{E}_f$ and $\mathbf{E}_p$ are linearly copolarized while $\mathbf{E}_b$ is cross polarized, one is examining the contribution of only the first of the three terms in P_{nl}, that is, the large-spaced grating. Hence, by appropriate selection of co- and cross-polarized combinations, the contributions of the dif-

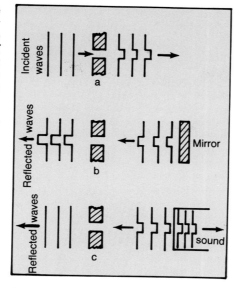

Stimulated Brillouin scattering to produce phase-conjugate reflection: (a) a distorted wave produced by a medium; (b) the effect of the medium on a reflected wave, (c) reflection from a medium with stimulated Brillouin scattering: The sound in the crystal compensates for the aberation of the incident wave.   Figure 2

ferent terms in P_{nl} can be examined for various nonlinear materials.

Let's carry this one step further by examining what might be expected in the following situation: We have a gaseous, nonlinear medium that consists of atoms or molecules having a two-level, optically-allowed transition in the vicinity of the frequency of our four-wave-mixing experiment. Now, because of the resonant enhancement, we would expect the contribution from the first two terms—the "population grating" terms—in P_{nl} to be orders of magnitude greater than that from the third term.

The interference patterns between pump and probe waves are fixed in space; if the atoms were stationary, the population gratings would also be fixed in space. The depth of modulation of the population gratings (and hence the four-wave reflectivity) would depend both on the light intensity—which determines how fast atoms are promoted to the upper level—and on the lifetime of the upper level prior to decay back to the ground state. Now, if the atoms can move an appreciable distance—say, an appreciable fraction of a grating period (or fringe spacing) during an excited-state lifetime—we have another mechanism by which the four-wave signal can be degraded: a grating "wash-out" effect.

How long does it take for an average atom to move the length of a grating period? That depends on how big the grating period is. In fact, it suggests a nice experiment. Remember, the grating spacing, $D = \lambda/(2 \sin\theta/2)$, can vary all the way from $\lambda/2$ to infinity, de-

pending on θ. This means that we would expect the large-spaced grating to suffer less from degradation due to atomic motion than the small-spaced grating. By using polarization tricks, we can choose to look at the effects of one grating at a time, comparing the results with what we would expect from a theory that takes the effects of atomic motion into account. This, in fact, has been done: the agreement between theory and experiment is excellent.[7]

A contrasting feature of our expression for P_{nl} is that the third term—the 2ω coherent term—is not expected to degrade in any way due to motion effects, a characteristic which has practical implications.

Fidelity of aberration correction

One of the questions arising in connection with the conjugation process is, how good is it? The answer is, perfect—within the limits of measurement employed so far. Evidence of the ability of a conjugator to correct for optical distortion is depicted in figure 4. These far-field photographs of a laser beam show the unperturbed beam (a), the beam after passing through a random aberrator, (an etched glass plate) (b), and the profile after the aberrated beam was conjugated and passed back through the etched glass plate (c). When the detailed profile in (a) is compared with that in (c) using a multiple-exposure photographic technique, they are seen to be identical within experimental measurement error.[8]

These measurements extend from the center of the beam out into the wings, where the intensity is as low as 10^{-5} of the axial peak intensity. The most severe random aberration used in this work degraded the unperturbed, diffraction-limited beam to about 35 times the diffraction limit. This degradation corresponds to reduction of the axial intensity to about 1/1000 of the unperturbed condition. Such effects have been demonstrated for both stimulated Raman scattering and degenerate four-wave mixing.[8]

The only apparent limitation to the fidelity of the conjugation process is the effective aperture of the conjugator itself. This is equivalent to saying that the laws of diffraction still hold. In other words, if a conjugator having finite aperture a is located a distance L away from a random-phase plate, the smallest transverse scale of aberrations that will be compensated will be of the order L/a, Smaller-scale transverse bumps (aberrations) in the phase fronts emanating from the phase plate will not be compensated when the conjugate wave passes back through.

Another way of saying this is that light coming from small-scale regions (smaller than L/a) will diffract by the

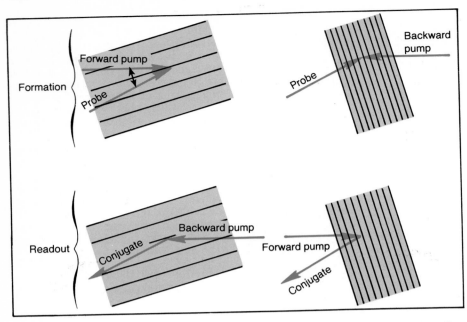

Degenerate four-wave mixing. The upper sketch shows the generation of the two "gratings"—actually periodic modulations in the polarization—produced by interference of the probe with the pump beams (they are shown separately for clarity). The lower sketch shows the production of the conjugate wave as the pumps are scattered by the "gratings."   Figure 3

time it gets to the conjugator so that an appreciable fraction will spill over the collecting aperture; and if you do not collect all the light, you cannot conjugate it completely.

Adaptive Optics

It wasn't until several years after the first observations of phase conjugation that a potential application to adaptive optics was first described.[9] The ability of phase conjugators to restore severely aberrated waves to their original state on passing through the distorting medi-um twice suggests a potential application to adaptive optics. The concept is illustrated in figure 5. The goal is to deliver energy from a laser system to a target or receiver, while minimizing the effects of distortions that tend to spread the beam and reduce the energy density at the target. A practical situation where this might be applied is laser fusion, where energy is to be deposited onto a target pellet.

The first step is the generation of a reference. This is done, for example, by illuminating the target with a pulsed laser source having a wavelength within the gain bandwidth of the system's amplifiers. Some of the light reflected from the pellet is captured in the aperture of the focusing element (shown in the figure as a lens) and enters the optical system. Note that in a real system containing many optical elements and possibly a propagation medium the reference wave accumulates phase distortions; these can cause its wavefront to deviate substantially from what it would be if it propagated in free space and encountered perfect optical components.

The second step is amplification of the distorted reference wave; the third is conjugation. After conjugation, the wave undergoes a second amplification and propagates back through the optical train; this time, because of the conjugation process, the phase distortions accumulated on the first pass are eliminated in a reverse sequence, as the wave makes the second pass through the system. The result is delivery to the target of an intense pulse of light that is virtually diffraction limited. More precisely, the pulse wavefront incident on the target is a replica of the reference wavefront that radiated from the target in the first place.

The beauty of the scheme is that once the reference is created everything else follows automatically. It has particular value in laser-fusion systems that irradiate the target from several directions. Here, the problem of beam alignment, pointing, and focusing is extremely complex in the conventional technology, demanding that the target be precisely located within a narrowly defined field of view and requiring sophisticated sensor/servo systems. This is to say nothing of the large number of optical elements, turning mirrors and so forth within the optical train, each of which is a source of optical distortion.

Phase conjugators in such multiple-arm systems have the potential for eliminating this complexity. A single reference pulse illuminating the target to initiate the process can result in precise delivery of energy from a multiple-arm system. Pointing and focusing are provided automatically, no matter where the pellet is located in the target chamber (within reason). Simultaneous of arrival if the intense pulses can be ensured by making all arms the same length to fairly loose tolerances—no more than a small fraction of the pulse length (1 nsec or 30 cm, approximately).

Similar considerations apply to the delivery of laser energy to a remote receiver or target through a turbulent atmospheric path. Quite often, the energy must be delivered in a beam that has minimal distortion. In "conventional" adaptive optics, one mea-

Observations of phase conjugation

Wavelength	Laser/ operation	Interaction	Medium (length)		Reflectivity	Power density	Comments	Ref.
10 000nm	CO$_2$/pulsed	DFWM	SF$_6$	2cm	37%	200 kW/cm^2		18
		DFWM	HgCdTe	0.5mm	10%	100 kW/cm^2		19
		DFWM	Ge	15cm	25%	10 MW/cm^2		20
3800nm	DF/pulsed	DFWM	Ge					21
1060nm	Nd:YAG /pulsed	DFWM	Si	1mm	180%	6 MW/cm^2		22
		SBS	CS$_2$,CH$_4$		10-90%			16
		TWM	Li Formate				poor correction	15
		DFWM	BDN dye		600%			23
694nm	Ruby/pulsed	SBS	CS$_2$		10-90%			2
		SBS	CH$_4$					4
		DFWM	Cryptocyanine		30%		preliminary	14
		DFWM	CdSCdSe glass		5%			14
		DFWM	CS$_2$	40 cm	100%			12
589nm	Dye/cw	DFWM	Na		17%		narrow band	14
	Dye/pulsed	DFWM	Na		10^4%		narrow band	24
532 nm	Nd:YAG/ pulsed	DFWM	CdSCdSe glasses		30%			14
								14
		DFWM	I_2 vapor		0.1%		preliminary	14
		DFWM	Rhodamine 6G	1mm	100%			14
		DFWM	Rhodamine B		10%		preliminary	14
510nm	Dye/pulsed	DFWM	CdSCdSe glass		1%		preliminary	14
480nm,	Ar$^+$/cw	DFWM	BiSiO$_4$		1%		slow response	25
514nm		DFWM	BaTiO$_3$		100%		slow response	12

 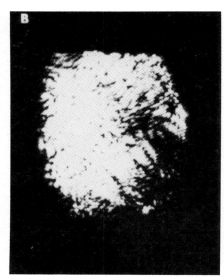 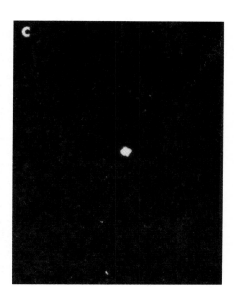

Correction of aberation. The photo at left shows an unperturbed laser beam; at center is the same beam after it has passed through an etched glass plate. The photo at right shows the beam after a phase-conjugate reflection and a second pass through the plate.   Figure 4

sures in some way the atmospheric turbulence along the propagation path; one then uses deformable mirrors to predistort the outgoing laser beam in a way that compensates exactly for the atmospheric distortions to be encountered along the propagation path to the target. The atmospheric path errors can be sensed by measuring the phasefront of a reference wave that originates from a bright glint or a beacon located at the target. The reference wave from the target glint is received at the aperture of the laser transmitter. This aperture is divided into a number of subapertures; a local wavefront tilt for each of these is measured—with, for example, Hartmann sensors or shearing interferometers. The size of each subaperture is chosen to be consistent with the scale of atmospheric turbulence expected at the particular operating conditions. The wavefront-tilt information is used to drive actuators on a deformable mirror, from which the outgoing laser beam reflects on its way out to the target. Through this sequence of wavefront measurement, error signal generation, and mirror deformation, the outgoing wave is transmitted as the phase conjugate of the incoming reference wave.

Here we see the essential contrasting features between conventional adaptive optics and nonlinear adaptive optics. In the conventional case, the reference wave is measured and discarded, and the results of the measurement used to obtain the necessary settings on the deformable-mirror actuators. This is followed by transmission of a beam from a different source, the laser (presumably, one relatively free of distortion), through the optical system. In the nonlinear case, the reference wave is not discarded, nor is it measured in the usual sense of the word. It is amplified, conjugated, and retransmitted.

One distinct advantage of the nonlinear over the conventional approach to wavefront correction of laser beams is that compensation will still occur even if the reference wave has substantial amplitude variation over the wavefront, as will be the case in the event of severe turbulence; the conventional approach does not compensate for amplitude variations—only phase variations.

In the conventional, adaptive-optics approach the reference wave is measured and then discarded and it therefore does not need to have the same wavelength as the wave that is ultimately transmitted. It can be derived from a completely different source that is in no way related to the laser. Of course, if the reference wave and transmitted wave have vastly different wavelengths, the compensation process may suffer because of dispersive effects. This fact, nevertheless represents a potential advantage over the nonlinear approach in which the reference wave length must be compatible with the amplification process and the conjugation process. If the conjugator is based on degenerate four-wave mixing then the reference wave must satisfy specific coherence requirements relative to the pump waves, consistent with the response time of the nonlinear medium—namely that the difference between the pump and reference frequencies be on the order of the reciprocal of the response time.

Optical resonators

An interesting and potentially valuable application for phase conjugate optics is the use of a phase conjugator as an element in an optical resonator. The essential features of a phase-conjugate resonator is an optical resonator in which one (or both) of the conventional mirrors is replaced by a conjugate mirror. Several papers[10,11] have predicted the properties of these devices, and a few of these properties have been demonstrated.[12]

There are several unique properties that phase-conjugate resonators are expected to exhibit. One is that such a resonator will not possess longitudinal modes that depend on the cavity length. An ordinary optical resonator possesses longitudinal modes separated in frequency by $c/2L$, where L is the cavity length. This results from the boundary condition that after one round trip (two reflections) the wave that corresponds to a resonant mode must constructively interfere with itself. This requires that the net accumulated phase after one round trip must be an integral multiple of 2π, in other words, the only waves that "fit" in the resonator are those for which $n\lambda = 2L$ (n an integer).

In a phase conjugate resonator on the other hand, the phase that is *accumulated* as the wave propagates from the ordinary mirror to the conjugate mirror is *subtracted* by the same amount on the way back to the ordinary mirror; in one round trip, the net accumulated phase is thus always zero. Consequently, a phase conjugate resonator of length L can support any wavelength consistent with the bandwidth of the gain medium and the conjugator itself. Moreover, a conjugate resonator oscillating at a particular wavelength will continue to oscillate at that wavelength, independent of variations of the cavity length. This is in contrast to an ordinary resonator, whose spectral output will exhibit "mode hopping" and frequency drift if the cavity length is allowed to vary.

Another property of a phase-conjugate resonator makes it highly attractive for application to high-power oscillators: It can compensate for intracavity optical distortion. One can show that, when light is extracted from the "ordinary-mirror" end of the resonator, the transverse phase of the wave only depends on the output mirror's figure (its detailed surface shape); the phase will not depend on any other sources of distortion within the body of the resonator.[10] This feature has been demonstrated qualitatively in laboratory experiments.[12]

Power and image transmission

One intriguing potential application of phase conjugation is in the transmission of high power electromagnetic radiation from a space-borne power generating station to a terrestrial site. Many concepts have appeared over the years that involve conversion of solar energy to electrical energy in a space station, followed by beaming of the energy to earth via coherent optical or microwave radiation. NASA is also interested in the possibility of direct conversion of solar energy for pumping high power space-borne lasers. An intriguing possibility arises from the concern for safety, in requiring a provision for accurate and reliable pointing of the multi-megawatt beams from the space station to the earth station.

Phase conjugation suggests a unique approach for solving this potential problem. The space-based power-generating station consists of a power source (solar radiation), a coherent oscillator (high-energy laser), and four-wave mixing medium. The space-borne high-energy laser provides the counterpropagating pumps for the four-wave mixing medium; in one configuration, the medium could be contained within the laser's resonant cavity. Ideally, the pump power and the nonlinear medium would be chosen to provide coupling from the laser resonator through the medium, via a pilot beam originating at the terrestrial receiving station. The conjugation process returns the earth-directed output beam along the same path on which the input beam is received. This ensures that the only energy delivered to earth is along the pilot-beam path. Moreover, the energy circulating inside the space-borne laser will be coupled out only when the pilot beam is present.

Optical phase conjugation can be useful for a number of applications requiring delivery of a "special" field distribution from one point to another. An example of a special field distribution is an image. One of the first such applications suggested has to do with the transmission of images along optical fibers.[1] A problem in transmitting an image along an optical fiber and recovering it at the other end arises from

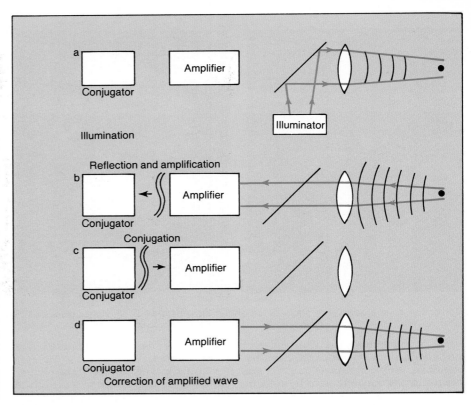

Adaptive optics, as applied, for example, to laser fusion. The sequence of events is (a) illumination of the pellet with a probe beam; (b) reflection and amplification, resulting in a distorted wave; (c) conjugation; and (d) a second pass through the distorting medium, producing coherent illumination of the pellet.
Figure 5

modal dispersion in the fiber; this causes a scrambling of spatial information, which can seriously degrade the image. By sending the image through a length of fiber, into a conjugator and finally through another fiber identical to the first, with the conjugator at the midpoint of the path, one could offset the effects of modal dispersion and recover the original image. The real question here is whether or not it would be possible to find lengths of fiber with "identical" modal-dispersion characteristics, an issue yet to be addressed experimentally.

Another area involving imaging and optical phase conjugation is in photolithography. Projection of complex patterns onto photo-resist layers is of great technological importance in the microelectronics industry. Projection systems using conventional optical techniques are very complex, due to the needs for near-diffraction-limited, low *f*-number performance.

One approach to a lensless 1:1 projection system employing optical phase conjugation is shown in figure 6. The object, in this case a mask or transparency, is illuminated from the back with a laser. The image is formed on the photoresist surface of a substrate, after being reflected from the conjugator by the way of the beam splitter. The advantage of this scheme is that a diffraction-limited performance can be achieved without using expensive optical components. (The only element requiring high optical quality is the beam-splitter). This approach achieves the same goal as contact photolithography, without placing the mask in direct contact with the sample (a step that may be highly undesirable). Projection of high-quality images using such a technique has recently been demonstrated.[13]

A variation on this application is one where a special field distribution is to be delivered to a target plane at intensities that are high enough to damage a mask or transparency. This scheme is also shown in figure 6, but this time including the amplifier. In this case, the mask is illuminated with a low-power beam; this is amplified, conjugated, reamplified, and delivered to the target as a high-intensity beam having a special field distribution. The scheme could be applied, for example, to situations involving laser annealing in cases where only specific samples areas are to be irradiated and no others. Another possibility is laser fusion, where it may be necessary to illuminate the target pellet with other than nearly uniform illumination. Still another potential application is irradiation of a wire with a high-power laser to create an intense, linear x-ray source.

Pump Manipulation

Four-wave mixing is more complicated experimentally than stimulated

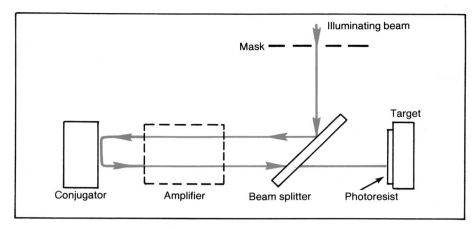

Application to photolithography. A light beam is used to form an image of the mask onto a target. The ability of a phase-conjugate "mirror" to provide an undistorted image is particularly useful if the mask could be damaged by the high intensities used to form the image; the conjugate mirror restores the distortions introduced by the amplifier.   Figure 6

Brillouin scattering because it requires the use of auxiliary pump waves. This very fact, however, gives it greater flexibility. The reason is that the pump parameters can be modified to yield an output wave which, in addition to being the spatial conjugate of the input wave, has some other desirable property that can be exploited for a specific application. For example:

▶ Pointing control by pump "misalignment." Suppose the target in the laser-fusion example in figure 5 were moving transverse to the direction of the beam. If its speed is sufficiently large, it will have moved an appreciable fraction of a beam diameter in the time it takes light to travel from the target to the conjugator and back to the target again. In this case, delivery of the laser energy to the target requires that it be possible to override the tendency of a good conjugator to produce a true time-reversed beam. We can deviate from the precise "retro" behavior by intentionally misaligning the pumps from the counterpropagating geometry. The degree to which the

direction of the conjugate wave deviates from the reference wave, as the pumps are misaligned, can be calculated in a straightforward manner from the phase-matching conditions; it is a function of the pump-misalignment angle and the pump-probe angle. The conjugate-wave reflectivity drops off as the pumps are misaligned—but only a few percent for misalignments of the order of a few milliradians.[14]

Thus by controlling the precise propagation direction of the pump waves one can control the direction of the conjugate wave. It is important to remember, of course, that compensation for optical-path errors will be less perfect if the conjugate wave is not allowed to retrace exactly the path taken by the reference wave. The extent to which the compensation is degraded depends upon the detailed spatial structure of the propagation path errors, and will ultimately limit the amount of point-ahead angle (or lead angle) that can be tolerated in an adaptive optical system.

▶ Pump polarization manipulation. As

discussed earlier, it is possible to generate a conjugate wave whose polarization is orthogonal to that of the reference wave. This fact allows the physical separation of the two waves through the use of polarization splitters and opens the door to a variety of applications. One is repointing the conjugate wave away from the backward direction, another is refocusing to a plane other than that from which the reference originates. Figure 7 shows schematically how such a polarization-manipulation scheme might work.

▶ Pump temporal modulation. Another, similar, application of four-wave mixing is in covert optical communications that require information to be conveyed to one or more remotely located, mobile receiver sites from an air- or space-born platform, without broadcasting over a wide area. Here again, the four-wave mixer is situated in the transmitter and the receivers are equipped with interrogating lasers. At prearranged times, the interrogators illuminate the remotely located transmitter with lasers tuned to a predetermined operating frequencey. For this example, the interrogating beam and one of the pump beams operate continuously and the other pump is pulse-modulated in an appropriate fashion. The information is transmitted back to the interrogation site as a modulated conjugate wave. Only those sites possessing proper interrogating capability can obtain the information. An added advantage is that the total power required to operate the transmitter can be many orders of magnitude smaller than that required for a broadcasting system.

Several other pump parameters can be varied, with potentially interesting results. The imposition of phase variations on the pump waves result in a transfer to the signal wave, and has the potential for imposing focus on other phase information on the conjugate wave. This could be important in certain applications of adaptive optics.

Correlation and convolution

The fact that the four-wave signal is proportional to the product of the three input fields suggests an application that, though having little to do with the conjugation property of four-wave mixing, is sufficiently interesting to mention in passing. One can readily obtain the spatial Fourier transform of a field through the appropriate use of lenses. By multiplying the Fourier transforms of probe and pump fields having transversely varying phase or amplitude, one can obtain a resultant signal field that has the properties of a correlation or convolution between the input fields.

Thus, to obtain correlation between two of the input fields, the probe wave and one of the pump waves are chosen

Modulation of polarization to produce a conjugate wave whose direction or focus can be changed. Because the conjugate and probe beams are polarized orthogonally, the conjugate beam can be separated and thus refocussed or realigned.   Figure 7

to have transverse information and the other pump wave to be a point source.[15] To perform a convolution, information is placed on the pump waves and the probe wave is made a point source. Autocorrelation and autoconvolution as well as other more complex operations are also possible with this approach. Considering the potential for high-speed, real-time information processing for spatially complex optical fields, such four-wave-mixing applications have exciting possibilities.

Accomplishments to date

After relatively few years in this young field, most of the fundamental physics of optical phase conjugation is well understood. Many predictions of the details of degenerate four-wave mixing have been experimentally confirmed, with good agreement; deviations from theoretical predictions are at least qualitatively understood. Many demonstrations have been made under both pulsed and cw conditions, over a wide range of wavelengths, and for many different materials. These are summarized in the table, which includes pertinent comments (percent reflectivities, power densities, conjugator longitudinal dimensions) along with references. Note that a number of impressively high reflectivities have been observed at modest power densities, in samples of modest dimensions.

This is only the beginning; much remains to be done, especially in finding new nonlinear materials tailored for specific applications. The only demonstrations carried out so far have been on a laboratory scale using well behaved, low-average-power lasers. None have yet been done at high average powers with conjugators larger than about a centimeter in cross section. In addition, we have yet to experiment by taking a weak input wave, amplifying it and conjugating it, with sufficient overall gain to produce an output wave with adequate energy for practical applications. And there are several other issues to be considered:
▶ the fidelity of conjugation via four-wave mixing under very high reflectivity conditions
▶ the extent to which other nonlinear phenomena compete with the desired conjugation process
▶ how a conjugator behaves under extremely weak signal conditions
▶ how the four-wave-mixing reflectivity degrades as the probe–pump ratio approaches unity.

* * *

This paper was inspired by the intensive activity in optical-phase conjugation at Hughes Research Laboratories. This activity has been highly interactive, involving the participation and key contributions of several highly competent scientists and engineers. I wish to acknowledge the influence of R. L. Abrams, W. P. Brown, D. T. Hon, R. K. Jain, M. B. Klein, J. F. Lam, R. C. Lind, R. A. McFarlane, T. R. O'Meara, D. G. Steel, S. M. Wandzura, and V. Wang.

References

1. A. Yariv, IEEE J. Quantum Electronics **QE14**, 650 (1978).
2. B. Ya. Zel'dovich, V. I. Popovichev, V. V. Ragul'skii, F. S. Faizullov, Sov. Phys. JETP **15**, 109 (1972); O. Yu. Nosach. V. I. Popovichev, V. V. Ragul'skii, F. S. Faizullov, Sov. Phys. JETP **16**, 435 (1972).
3. B. I. Stepanov, E. V. Ivakin, A. S. Rubanov, Sov. Phys. Doklady **16**, 46 (1971).
4. R. W. Hellwarth, J. Opt. Soc. Amer. **67**, 1 (1977).
5. D. M. Bloom, G. C. Bjorklund, Appl. Phys. Lett. **31**, 592 (1977).
6. R. L. Abrams, R. C. Lind, Opt. Lett. **2**, 94 (1978); Opt. Lett. **3**, 205 (1978).
7. D. G. Steel, R. C. Lind, J. F. Lam, C. R. Giuliano, Appl. Phys. Lett. **35**, 376 (1976); S. M. Wandzura, Opt. Lett. **4**, 208 (1979).
8. R. C. Lind, C. R. Giuliano, Conf. on Laser Engineering and Applications, Washington, D. C., June 1979 (unpublished).
9. V. Wang, C. R. Giuliano, Opt. Lett. **2**, 4 (1978).
10. J. F. Lam, W. P. Brown, Opt. Lett. **5**, 61 (1980).
11. J. M. Bel'dyugin, M. G. Galushkin, E. M. Zemskov, Sov. J. Quantum Electron. **9**, 20 (1979); J. AuYeung, D. Fakete, D. M. Pepper, A. Yariv, IEEE J. Quantum Electron. **QE-15**, 1180 (1979); P. A. Belanger, A. Hardy, A. E. Siegman, Appl. Opt. **19**, 602 (1980).
12. D. M. Pepper, D. Fakete, A. Yariv, Appl. Phys. Lett. **33**, 41 (1978); J. Feinberg and R. W. Hellwarth, unpublished; R. C. Lind and D. G. Steel, unpublished.
13. M. D. Levenson, Opt. Lett. **5**, 182 (1980).
14. R. C. Lind, T. R. O'Meara, R. K. Jain, D. G. Steel, R. A. McFarlane, Hughes Research Laboratories preprints.
15. P. Avizonis, F. A. Hopf, W. D. Bomberger, S. F. Jacobs, A. Tomita, K. H. Womack, Appl. Phys. Lett. **31**, 435 (1977).
16. D. T. Hon, J. Opt. Soc. Amer. **70**, 635 (1980).
17. D. M. Pepper, J. AuYeung, D. Fekete, A. Yariv, Opt. Lett. **3**, 7 (1978).
18. R. C. Lind, D. G. Steel, M. B. Klein, R. L. Abrams, C. R. Giuliano, R. K. Jain, Appl. Phys. Lett. **34**, 457 (1979).
19. R. K. Jain, D. G. Steel, Appl. Phys. Lett. **37**, 1 (1980).
20. E. E. Bergmann, I. J. Bigio, B. J. Feldman, R. A. Fisher, Opt. Lett. **3**, 82 (1978).
21. D. DePatie, D. Haneisen, Opt. Lett. **5**, 252 (1980).
22. R. K. Jain, M. B. Klein, R. C. Lind, Opt. Lett. **4**, 328 (1979).
23. E. I. Moses, F. Y. Wu, Opt. Lett. **5**, 64 (1980).
24. D. M. Bloom, P. G. Liao, N. P. Economu, Opt. Lett. **2**, 58 (1978).
25. J. P. Huignard, J. P. Herriau, P. Aubourg, E. Spitz, Opt. Lett. **4**, 21 (1979). □

CHAPTER 5 —

LIGHTWAVE COMMUNICATIONS, FIBER AND INTEGRATED OPTICS

Communication—sending signals to a distant receiver—has always been an important aspect of human endeavor: torches and beacons to guide ships back to harbor, drums and trumpets to alert military units, and so on. In the early 1800s the French devised an elaborate semiphore system with signalers on hills relaying dispatches from Paris all the way to Marseilles. Samuel F. B. Morse invented his electric telegraph just before the Civil War, and the armies of both sides were immediately followed by specialists stringing wires to convey dispatches back to commanders. William Thomson (later Lord Kelvin) was knighted for his part in helping to lay the first transatlantic cable from Scotland to Newfoundland. Then the telephone and finally radio completed communication links all around the world. But even with all this advanced technology in place, when the laser was invented the communication engineer recognized immediately that a laser beam—a source of coherent radiation operating at optical frequencies—would be capable of carrying information at rates that were hundreds or even thousands times the capacity of earlier systems. So one of the earliest goals of optical engineers has been to perfect techniques of optical communication. The papers in this chapter focus on this still-emerging technology.

John N. Howard

CONTENTS

Lightwave communications: An overview

The dazzling information-carrying capacity of light guided within fine, light-weight glass fibers is opening new vistas in signal processing, information display and communication.

Solomon J. Buchsbaum

PHYSICS TODAY / MAY 1976

One hundred years ago Alexander Graham Bell invented the telephone. Shortly thereafter Bell also invented a photophone—a contraption that transmitted human speech on a beam of light. Bell, who had a deep interest in the problems of the deaf, sought ways to make "visible speech." He had to resort to heroic measures to demonstrate the photophone, using the Sun as the source of light and a Rube-Goldberg-like scheme as a modulator. Although the photophone worked, Bell wisely gave it up in favor of copper wires and electrical transmission.

The modern impetus for lightwave communications came a decade and a half ago with the invention of the laser, a source of coherent radiation scarcely dreamed of by earlier scientists and communication engineers. Back-of-the-envelope calculations showed the dazzling information-carrying capacity of coherent light; and the race was on to exploit this new source for communications, just as microwave sources were exploited after World War II. Very quickly it became apparent, however, that a source of light alone does not a communications system make—especially if it is as crude, bulky and inefficient as the early lasers were. There had to be a panoply of other devices: modulators, detectors, filters, amplifiers, regenerators and such, all smoothly fitting and working together—and, most importantly, there had to be a reliable transmission medium. The atmosphere will do as a transmission medium for light for some applications, but the vagaries of the weather make the atmosphere unsuitable for those applications, such as telephony, in which reliability is a must. (Fog is impenetrable over long distances—even by laser light.)

The stage to which industry has advanced in the development of lightwave communications is suggested by the photo on page 25, taken at a test of a prototype system at Bell Laboratories' Atlanta facility.

In the articles in this special issue of PHYSICS TODAY the various elements of a lightwave communication system are discussed including the transmission medium of choice today: The optical fiber, or as we now prefer to call it, the *fiber guide*. (See the article by Alan Chynoweth on page 218 of this book.) Before launching into these topics, however, let us dwell briefly on the question: "Why lightwave communications?" and on some general features of such a system.

In figure 1 is shown the evolution of the carrier network in the Bell System over the past three decades. (By "carrier network" we mean that part of the network, primarily interconnecting switching machines in central offices, in which, for efficiency and economy, telephone calls and other information are bundled together and transmitted on an electromagnetic carrier such as microwave radio.) Other communication networks in the US and elsewhere in the world have experienced similar growth.

Yet, in spite of the tremendous growth that has occurred, the communication explosion still is in its infancy. While the present communication technologies—microwave radio, coaxial cable and paired-wire cable—can be harnessed to satisfy any future need, these systems have many limitations, not the least of which is their cost. Hence one hears of the new communication technologies—satellites, millimeter waveguides and lightwave communication systems. It should be kept in mind that any new communication technology must be able to compete economically with the present workhorses—radio, cable and wire—to be adopted widely under free-market forces. Nevertheless the opportunities and the promise are enormous and that is why nearly every R&D laboratory in the world that is even remotely related to communications, spurred by the invention and development of semiconductor-junction light sources in the 1960's and by the prospect of low-loss lightguides, is busily engaged in research on some aspect of lightwave communications.

Errors and margins

The essential elements of any communication system are a transmitter, a transmission medium and a receiver. For fiber-guide transmission the transmitting source is either an injection laser or a high-radiance light-emitting diode (LED). The heart of the receiver is the photodetector. The operating capabilities of a lightwave communication system are determined by
▶ the average amount of light power the source can inject into the fiber,
▶ the loss and dispersion characteristics of the fiber, and
▶ the sensitivity of the photodetector.
Many modulation and coding techniques

Solomon J. Buchsbaum is vice president, network planning and customer services, Bell Telephone Laboratories, Holmdel, N. J.

Why "lightwave"?

We have adopted the term "lightwave" communication to describe the use of light as a carrier of information. The term is consistent with existing communications terminology such as microwave and millimeter wave. We have found from experience, especially in discussing the subject of these articles with nonspecialist audiences, that the term "optical" communication often needs further definition because its colloquial usage also encompasses phenomena and functions related to vision. "Lightwave" has the advantage of being accurate, descriptive and, we hope, more understandable.   —SJB

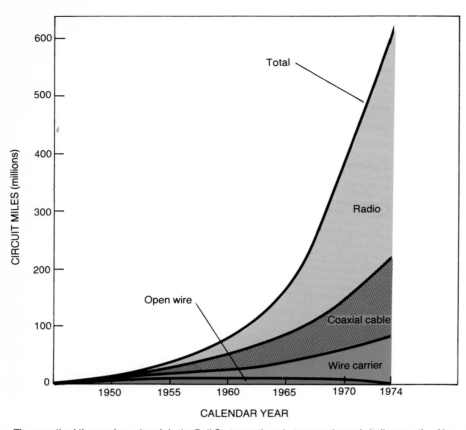

The growth of the carrier network in the Bell System; others have experienced similar growth. New technologies, such as lightwave communications, must compete economically with the facilities now in use: microwave radio, and coaxial and paired-wire cable. Figure 1

can be used, in principle, to enable the light to carry the desired information. In practice binary on–off modulation of the light source and square-law detection are quite suitable for lightguide systems. Such modulation requires, of course, that

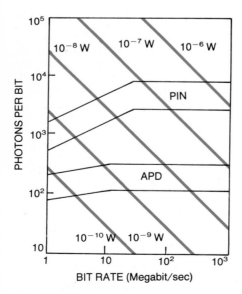

The sensitivity of optical receivers is indicated by the colored areas for receivers with (APD) and without (PIN) avalanche gain. The diagonal lines show the threshold power per pulse at 0.82 microns. (From Tingye Li, Bell Lab. Record, Sept. 1975) Figure 2

any analog information, such as voice or video, be digitally encoded.

The information content is therefore carried along the fiber as a stream of light pulses and the detector is asked to detect these pulses with little or no error. If the detector were 100% efficient, inherent fluctuations (quantum noise) would set a limit on the average number of photons (n) per pulse (and therefore per bit) that are needed to achieve a given error probability (P_e). For example, $n = 10.5$ for $P_e = 10^{-9}$. Practical photoreceivers are presently one to two orders of magnitude less sensitive than this, as indicated by figure 2. The sensitivity of the receiver, of course, sets the lower limit on the level of optical power that must reach it and be detected.

For the optical transmitter, both LED's and injection lasers can be made with emitting areas that match the dimensions of the fiber and with power outputs of several milliwatts. (See the article by Henry Kressel and his co-workers on page 200 of this book.) With the laser, because of the much greater coherence of its light output, most of this power can be injected into the fiber rather easily. Not so with the LED, which spews its energy more or less uniformly in all directions so that typically less than about 0.05 mW of optical power is coupled into the fiber. The price of the greater efficiency of the laser is the greater complexity of the device and

of the associated circuitry required for stability of laser operation.

Given the input source power cited above and the receiver sensitivities shown in figure 2, we can easily obtain the maximum available margin for the loss between receiver and transmitter, or between repeaters. (A repeater is a receiver and a transmitter put back to back with amplification—really, regeneration—between them.) We see that this margin depends on the desired information rate (the bit rate). For very high information rates (hundreds of Mbit/sec) and with simple PIN detectors and LED's, the margin is so small as to be of no broad practical interest except for very short distances. For low bit rates (a few Mbit/sec), injection lasers and avalanche photodetectors, the margin is huge (about 10^7, equivalent to 70 dB). Other combinations are, of course, possible—and are, in fact, more practical than the two limiting ones. Practical margins generally are in the range between 40 and 60 dB.

Given this margin, we can immediately obtain the maximum allowable separation between receiver and transmitter by dividing the margin by the fiber loss (measured in dB/km), making sure also to allow for the losses incurred in cabling the fibers, in splicing the the cable and in other interconnections. As an example, for a margin of 50 dB and fiber-guide cable with a loss of 10 dB/km, the maximum distance between repeaters is 5 km.

At low bit rates, the total loss in the fiber-guide cable is the sole determinant of the separation. At high bit rates, dispersion rears its ugly head and can further limit the information-carrying capacity of the system. For multimode fibers with a step in the refractive index (typically of about 1%), in which light is guided by total internal reflection, modal dispersion limits the information rate to about 20 Mbit·km/sec (that is, we can reach 5 km at 4 Mbit/sec, for example).

In a multimode, graded-index fiber, where light guidance is by refraction, the figure of merit can be as high as several Gbit·km/sec, depending on how accurately the grading has been achieved. In any case, at very high bit rates dispersion rather than loss determines repeater spacing. In single-mode fiber, which is dominated by material rather than model dispersion, the figure of merit is about 50 Gbit·km/sec. In multimode fibers the figure of merit can be increased somewhat by inducing mode coupling in the fiber, in

In a field test of lightwave communication, Michael Buckler of Bell's Atlanta lab is using an optical-fiber patch cord to connect to a fiber from an underground cable. Each of the fibers from the cable ends in an optical coupler, the covers of which are the yellow "buttons" visible on the front panel. The rack-mounted equipment includes regenerators, which amplify and retransmit the digital light signals.

which case dispersion increases with the square root of the distance for distances greater than the coupling length.

The payoff

The designer of a lightwave communication system thus has the task of juggling these (and other) parameters to obtain an optimum match between loss, dispersion, available source power and receiver sensitivity. Needless to say, economic and reliability considerations weigh heavily in any such design.

What are some of the attractive features of lightwave communication systems? Fibers are small and so are fiber-guide cables, a potential boon in metro-politan areas where duct space is crowded and expensive. Fibers are light compared with copper wires, an important potential advantage where weight is at a premium, as in aircraft, missiles and satellites. There is no metal in fibers; they are therefore essentially immune from electrical interference. The very weak dependence of optical loss on frequency and temperature in fibers simplifies the design of receiver electronics—no equalization is required. And last, but certainly not least, is the hope—nay, the expectation—that this technology will turn out to be less costly than present technologies based on wire, cable or radio.

Research to date, as discussed in the other articles in this issue, has already shown the technical feasibility of fiber-guide communication. Its economic feasibility remains to be demonstrated. Still, optical electronics is well on its way, and it is difficult to overstate the vistas it opens to signal processing (see the article by Esther Conwell on page 240.) and information display, as well as to lightwave communications. It is claimed by some that optical electronics will have an impact similar to those of the transistor in the 1950's and of integrated circuits in the 1960's. If even a fraction of these promises materialize, the results will amply repay the current investment in this new technology. □

Light sources

Light-emitting diodes are simpler to construct, while heterojunction lasers produce lower dispersion and couple to fibers more efficiently; both types have now been operating for well over one year.

Henry Kressel, Ivan Ladany, Michael Ettenberg and Harry Lockwood

PHYSICS TODAY / MAY 1976

The selection of a system for optical communications is largely governed by the availability of the necessary components. Although through-the-atmosphere systems have some applicability, they are generally restricted to small installations where their inherent limitations are tolerable. The optical transmission medium of greater importance for the future is a glass or plastic fiber, and therefore the properties of these fibers—their absorption, dispersion, mechanical strength, and so on—are all-important.

The other determining factor is the requirement that the sources, fibers and detectors used in the system be compatible with one another. Fortunately this compatibility exists—and it is for this reason that optical communications is such a rapidly expanding field. It is generally true that commercially available optical fibers have moderate loss and small dispersion (both modal and material) in the wavelength range 0.8–1.1 microns. In this article we will therefore discuss the current status of semiconductor light sources spanning this wavelength region, and because there are applications for both coherent and incoherent sources we will treat both lasers and LED's (light-emitting diodes).

We will limit our discussion in this article to discrete components. Semiconductor laser components can be expected in the future to move into the area of integrated functions, along the lines described by Esther Conwell in her article on page 240 of this book.

Heterojunction structures

It is only since the advent of the first practical aluminum gallium arsenide–gallium arsenide heterojunction laser structures[1-5] that the potential of laser diodes for optical communications and other applications has been realized. The GaAs homojunction lasers available earlier had such high threshold and operating currents as to make them impractical for room-temperature operation. In some heterojunction devices we will describe, threshold current densities have been reduced two orders of magnitude to 500 A/cm^2 from the homojunction values of about 50 000 A/cm^2. This progress has made practical a compact continuous-wave laser with high radiance and power output in the range of 10 mW, which can be directly modulated to hundreds of MHz. Figure 1 vividly illustrates the small size and brightness of a cw laser similar in construction to those used in optical communications.

Since optical fibers with losses under 20 dB/km are now commercially available, multi-kilometer, high-data-rate transmission systems (even without repeaters) are expected to become competitive with the existing coaxial systems of electrical communication.

There are also many applications for fiber-optic links over short distances as well as at low data rates (for example, in local distribution systems). The light-emitting diode satisfies many of the requirements of such systems, and because it is not a threshold device its output power is less sensitive than the laser to small changes in operating current or ambient temperature. Much of the technology of LED's is comparable to that of lasers, so that the two devices have tended to be developed together. In fact, state-of-the-art LED's deliver several milliwatts of output power at modulation rates of 100–200 megahertz.

The most highly engineered and tested lasers and LED's for optical communications are those derived from the ternary alloy system $Al_x Ga_{1-x}As$. Thin layers (0.05–1 micron) are sequentially grown with a high degree of perfection by liquid-phase epitaxy on substrates of GaAs. By varying the alloy composition of the recombination region of the direct-band-gap material,[6] the emission wavelength can be varied over a useful range of 0.8–0.9 microns. Emission, and indeed lasing, can be achieved at still shorter wavelengths but attenuation and dispersion in the fiber begin to become appreciable at shorter wavelengths. The purest (OH-free) fibers available have attenuation and dispersion that decrease to negligible values in the wavelength region of 1.0–1.1 microns, so that there is a real interest in developing new alloy systems and a heterojunction technology for emission in this range also.

In the $Al_x Ga_{1-x}As$ system the lattice parameter varies by only 0.14% as x goes from zero to unity. Heterojunctions in this system consequently have negligible strain-induced defects and demonstrate long-term reliability. In most other ternary alloys, the band gap and lattice parameter vary significantly between the extremes of the binary alloys from which they are derived; $In_x Ga_{1-x}As$, $In_x Ga_{1-x}P$ and $GaAs_x P_{1-x}$ are examples of such alloys. As a result heterojunction structures in these materials contain many defects due to the inevitable strain at the heterojunction interfaces. However, lattice-matching heterojunction structures of indium gallium arsenide–indium gallium phosphide, for example, can be defect free.

An alternative approach to obtaining efficient device performance at 1.0–1.1 microns is to fabricate heterojunction structures in quaternary alloys such as $In_x Ga_{1-x}As_y P_{1-y}$ and $Ga_x Al_{1-x}As_y$-

---

Like a Lilliputian beacon, this double-heterojunction cw laser is dwarfed by an ordinary sewing needle. The aluminum gallium arsenide source shown here, emitting at about 7500 Å, is similar in construction to devices for optical communications, which are designed to emit at about 8200 Å (a wavelength that records poorly on color film). The emitting region of this laser is about 15 microns wide by about 0.5 micron thick. (Photograph courtesy of RCA Laboratories.)

Figure 1

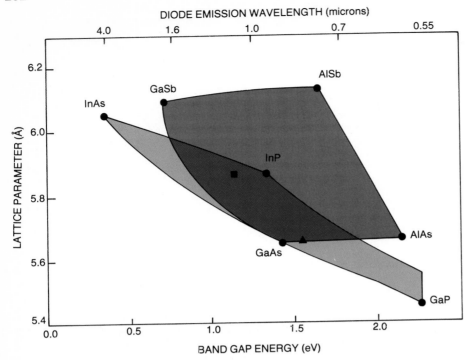

Extent of band gaps and lattice parameters covered by the quaternary alloys $Al_xGa_{1-x}As_ySb_{1-y}$ (gray area) and $In_xGa_{1-x}As_yP_{1-y}$ (colored area). The boundaries of the areas represent ternary, and the vertices binary, alloys. Two particularly useful alloys are indicated: The square (■) locates $In_{0.80}Ga_{0.20}As_{0.35}P_{0.65}$, the lattice parameter of which matches InP substrates and would make diodes that emit at 1.1 microns; the triangle (▲) shows the position of $Al_{0.1}Ga_{0.9}As$, which matches the GaAs lattice and emits at about 0.82 microns.   Figure 2

Sb_{1-y}. In these alloys, which cover the desired emission range, the band gap and lattice parameter can be adjusted independently, as figure 2 shows. At some cost in simplicity, this extra degree of freedom permits the fabrication of strain-free heterojunction devices.

As evidenced by a growing literature on the subject, progress towards useful lasers and LED's at about 1.1 microns is being made. Much work, however, remains to ensure their reliability, which is handicapped by lattice defects. Whether the complexity of fabrication will be justified by the moderate improvements in attenuation and dispersion will ultimately be determined by cost and long-term reliability.

The problem of lattice-parameter match at heterojunctions can be discussed in terms of mismatch-dislocation networks. For example, for a 1% lattice mismatch at an interface, a dislocation will be generated at approximately every 100 atom planes. Since the dislocation core consists of nonradiative recombination centers, such a high dislocation density will depress the internal quantum efficiency of the device. Furthermore, dislocations are not always confined to the mismatched interface but can propagate through multilayer structures.

The effect of dislocations on radiative efficiency is dramatically illustrated in figure 3, where we compare a transmission electron photomicrograph of a dislocation array in a mismatched heterojunction structure[7] with a cathodoluminescence scan of a similar structure. The dark lines and spots in the cathodoluminescence micrograph are areas of low radiative efficiency, which correspond to dislocations lying parallel and perpendicular to the plane of the surface viewed. Therefore, in designing heterojunction structures for LED's and lasers, it is of extreme importance either to choose a totally lattice-matched system or else to remove dislocations from the active region by compositional grading.

Continuous-operation lasers

Although there are numerous potential configurations for cw laser diodes, the symmetric double heterojunction with a stripe contact has been widely adopted as the simplest laser geometry and that most suitable for optical communications. The schematic and actual cross section of a typical laser, figure 4, shows the optical cavity (recombination region) defined by its higher-band-gap dielectric walls as well as the outer n and p type GaAs regions to which ohmic contact is made.

The efficient operation of a laser diode requires effective minority-carrier and radiation confinement to the optical cavity, which is also the recombination region of the device. Both functions are provided by heterojunctions. Radiation confinement is provided by the dielectric discontinuity, while carrier confinement results from a potential barrier created by the difference in band gap between the materials that form the heterojunction. The average carrier-pair density injected into the recombination region of a double-heterojunction device for a current density J is

$$\Delta N \approx J\tau/ed$$

where e is the electron charge, τ is the carrier lifetime and d is the width of the recombination region (that is, the heterojunction spacing). In a typical GaAs-laser diode, ΔN at lasing threshold is about 2×10^{18} cm^{-3} at room temperature.[8] To minimize the threshold current density, we restrict the width of the recombination region by placing the heterojunction that forms the potential barrier for minority carriers closer to the injecting heterojunction than the diffusion length. It is, however, essential that the heterojunction interfaces be relatively defect free in order to prevent excessive nonradiative recombination of the injected carriers.

The nonradiative loss of carriers at an interface is characterized by the recombination velocity S of that interface. Under the usual laser operating conditions we can express the effective recombination rate due to the presence of the interface as

$$\frac{1}{\tau_{eff}} \approx \frac{1}{\tau_0} + \frac{2S}{d}$$

where $1/\tau_0$ is the rate in the absence of an interface. The internal quantum efficiency is given by

$$\eta_i \approx \tau_{eff}/\tau_0$$

so that

$$\eta_i \approx (1 + 2S\tau_0/d)^{-1}$$

In typical cw laser diodes $d = 0.3$ microns, and $\tau_0 = 10^{-9}$ sec. Therefore, for an internal quantum efficiency of 50% (a reasonable lower limit), we would require $S \leq 2 \times 10^4$ cm/sec.

The single most important contribution to S is from nonradiative-recombination states introduced at the heterojunction due to the lattice-parameter mismatch between the two materials. If this mismatch is kept below 0.1%, S will be under 2×10^4 cm/sec and cw operation can be anticipated. Experimental data concerning S in GaAs–$Al_xGa_{1-x}As$ heterojunctions indicate that $S \approx 5 \times 10^3$ cm/sec in practical laser structures (where $\Delta a_0/a_0 \lesssim 0.07\%$), a value that is fully satisfactory for narrow-recombination-region devices.

To ensure wave propagation along the plane of the junction, as well as low threshold current density and high efficiency, means must be provided for confinement of stimulated radiation to the region of inverted population (or its close

Henry Kressel is the Head of Semiconductor Device Research, and Ivan Ladany, Michael Ettenberg and Harry Lockwood are members of the technical staff, at the David Sarnoff Research Center of RCA Laboratories.

proximity). Two heterojunctions, as indicated in figure 4, provide a controlled degree of radiation confinement due to the higher refractive index in the lower-band-gap recombination region, with the fraction of the radiation confined dependent on the heterojunction spacing d and the refractive-index steps Δn at the lasing wavelength;[9] in $Al_xGa_{1-x}As$–GaAs structures, $\Delta n \approx 0.62\,x$ at $\lambda \approx 0.9$ micron. In general it is desirable to equalize the refractive-index steps Δn at all the heterojunctions to prevent the loss of waveguiding that can occur in thin asymmetrical waveguides. However, even with the symmetric double-heterojunction laser, wave confinement within a heterojunction is gradually reduced as its spacing becomes small.

The fraction of the radiation confined to the recombination region of the double-heterojunction laser affects the radiation pattern and threshold current density. The radiation pattern (far-field intensity distribution) is affected because it is determined by the effective source size (near-field intensity distribution); the threshold current density J_{th} is determined primarily by the optical gain at threshold—only that portion of the optical flux within the recombination region is amplified. Figure 5 shows calculated and experimental values of J_{th} as functions of d for various values of Δn corresponding to varying heterojunction barrier heights. The lowest J_{th} value[10] of 475 A/cm^2 is obtained with $d = 0.1$ micron and $\Delta n = 0.4$ (data corresponding to an $Al_{0.6}Ga_{0.4}As$–GaAs–$Al_{0.6}Ga_{0.4}As$ structure). It should be noted that the barrier height also affects the high-temperature performance of lasers.[11]

The maximum desirable threshold current density for cw operation is below 2000 A/cm^2, and double-heterojunction laser diodes have routinely operated steadily at room temperature with 0–12% AlAs concentration in the active recombination region. Because of the corresponding band-gap variation this corresponds to an emission wavelength range of 0.9 to 0.8 microns. Room-temperature cw operation of lasers at a wavelength of 1.0 micron has recently been reported for the $Ga_xAl_{1-x}As_ySb_{1-y}$ system,[12] where $J_{th} \approx 2000$ A/cm^2 was achieved; for the InGaAsP–InP heterojunction system another recent paper reported room-temperature pulsed operation at low thresholds.[12]

Stripe geometry

Laser diodes are prepared by cleaving two parallel facets to form the Fabry–Perot cavity. The cw laser diode utilizes a strip geometry to define the lateral dimension, as shown in figure 6. The advantages of this design feature are:

▶ The radiation is emitted from a small region, which simplifies coupling of the radiation into fibers with low numerical aperture.

▶ The operating current can be minimized because it is relatively simple to form a small active area with photolithographic contacting procedures.

▶ The thermal dissipation of the diode is improved because the heat-generating active region is imbedded in an inactive semiconductor medium.

▶ The small active diode area makes it simpler to obtain a reasonably defect-free area.

▶ The active region is isolated from an open surface along its two major dimensions, a factor believed to be important for reliable long-term operation.

In the simplest stripe-contact[13] structure, the active area is defined by opening a stripe in a deposited SiO_2 film. The surface of the diode is then metallized, with the ohmic contact formed only in the open area of the surface. Other methods that have been used to define the stripe contact include the implantation of deep ionic centers to increase the lateral resistance, and etched moats. Various aspects of the technology were discussed in papers in a special issue on semiconductor lasers of the *Journal of Quantum Electronics*.[14]

Diodes for cw operation are generally designed with the thin p side mounted on copper heat sinks to minimize the thermal resistance of the structure and with a soft solder such as indium to minimize strains in the devices.

The lateral width of the emitting region can be adjusted for a desired operating level by adjusting the stripe width. For example, 100 mW of cw power (from one facet) is obtainable for a stripe width of 100 microns. However, for the typical power levels needed in optical communications (5–10 mW), stripe widths of about 13 microns are used, a dimension that represents a suitable compromise between low operating currents and an appropriate power-emission level. A typical curve of power output as a function of diode current is shown in figure 7. The junction temperature for such devices is only a few degrees above the heat-sink temperature. For example, a temperature differential of 7 K is calculated with a typical power input of 0.5 W at a diode current of 0.3 A, and a thermal resistance of 14 K/W.

The electromagnetic modes of the laser-diode cavity are separable into two independent sets, with transverse-electric and transverse-magnetic polarization. The mode numbers m, s and q define the number of sinusoidal half-wave variations along the three axes of the cavity, transverse, lateral and longitudinal, respectively.

The allowed *longitudinal* modes are determined from the average index of refraction and the dispersion seen by the propagating wave. The Fabry–Perot mode spacing is several angstrom units in typical laser diodes. The *lateral* modes depend on the method used to define the two edges of the diode. In stripe-geometry lasers generally only low-order modes

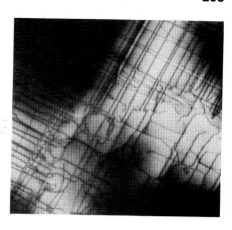

Misfit-dislocation arrays in a compositionally graded $In_xGa_{1-x}P$ vapor-grown epitaxial layer on a GaP substrate. The top photo is a transmission electron micrograph (ref. 7). Below is shown a cathodoluminescence scan, in which dislocations near the surface appear as nonradiative regions. **Figure 3**

are excited; their mode spacings are 0.1–0.2 Å and they appear as satellites to each longitudinal mode. The *transverse* modes depend on the dielectric variation perpendicular to the junction plane. In the devices discussed here, only the fundamental mode is excited, a condition achieved by restricting the width of the waveguiding region (the heterojunction spacing) to values well under one micron. Therefore, the far-field radiation pattern consists of a single lobe in the direction perpendicular to the junction. (Higher-order transverse modes would give rise to "rabbit-ear" lobes, undesirable for fiber coupling.)

For a laser operating in the fundamental transverse mode, the full angular beam width at half power *perpendicular* to the junction plane is a function of the near-field radiation distribution. The narrower the emitting region in the direction perpendicular to the junction plane, the larger the beam width. In practical cw laser diodes the beam width is about 30°–50°. The beam width in the plane of the junction (lateral direction) is

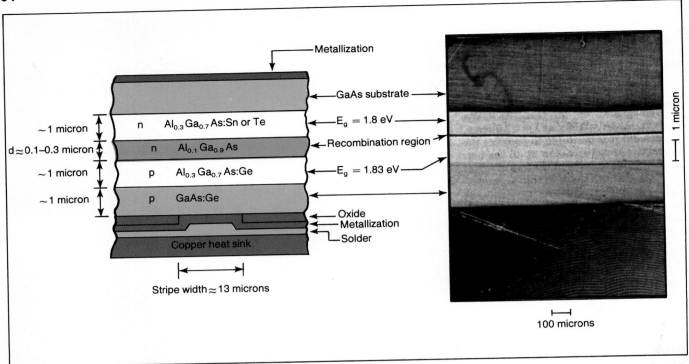

The cross section of a typical laser for optical communications is depicted in a schematic illustration (left, not to scale) and in a photomicrograph of a sample that has been polished at a shallow angle to produce high magnification in the transverse direction (right). The "terracing" effect evident on the lower portion of the micrograph, a growth artifact, causes some interface roughness.   Figure 4

typically 5°–10° and varies only slightly with the diode topology and internal geometry. At least one half of the power emitted from one facet can be coupled

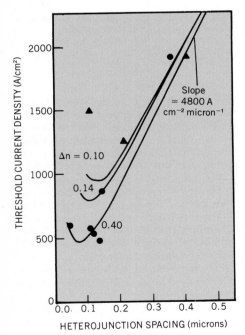

Threshold current density as a function of the heterojunction spacing for $Al_xGa_{1-x}As$ double-heterojunction lasers. The experimental data points are for aluminum-concentration steps Δx of 0.20 (triangles) and 0.65 (circles). The theoretical curves are for the discontinuities in the refractive index Δn shown, where the relation $\Delta n = 0.62 \Delta x$ has been assumed to apply.   Figure 5

into a multimode step-index fiber with a numerical aperture of 0.14 and a core diameter of 80 microns.

While operation in the fundamental transverse mode is easily achieved, most narrow-stripe laser diodes operate with several longitudinal modes, and therefore emit over a 10–30 Å spectral width, although some units can emit several milliwatts in a single mode. Figure 8 shows the emission spectrum from such a device operating in the fundamental lateral and transverse mode and a single longitudinal mode. The line width is 0.15 Å and the power emitted is 3 mW.

Methods of modulating the laser output vary widely depending upon the application. For fast-pulse modulation, the diode is biased with a current near the threshold current and then pulsed to an appropriate level above threshold. If no bias is applied, there is a lasing delay (related to the spontaneous carrier lifetime) before the carrier population becomes fully inverted and the device turns on. This delay, of several nanoseconds in a typical situation, vanishes if the laser is biased to threshold.

Light-emitting diodes

The spectral bandwidth of the LED is typically $1–2 \ kT$ (300–600 Å) at room temperature, hence one to two orders of magnitude broader than that of the typical laser diode. Because of the spectral dispersion in fibers, this may limit the long-distance applications of fiber communications with LED's. Furthermore, the coupling efficiency of LED's into low-numerical-aperture fibers is much lower than that of laser diodes. However, the LED has the advantage of a simpler construction and a smaller temperature dependence of the emitted power. For example, the spontaneous output from an LED may decrease by a factor of only 1.5–2 as the diode temperature increases from room temperature to 100 deg C (at constant current), while the output of a laser diode would be typically decreased by more than a factor of 3.

The topology of light-emitting diodes is designed to minimize internal reabsorption of the radiation, allow high-current-density operation and maximize the coupling efficiency into fibers. While the structures used are applicable to all materials, most of the work on devices suitable for communications reported so far has been on $Al_xGa_{1-x}As$.

Two basic diode configurations for optical communications have been reported: surface emitters[15] and edge emitters.[16,17] In the *surface emitter*, the recombination region is placed close to a heat sink, as shown in figure 9. A well is etched through the GaAs substrate to accommodate a fiber. The emission from such a diode is essentially isotropic. The *edge-emitting* heterojunction structures, similar to the laser geometry of figure 6, uses the partial internal waveguiding of the spontaneous radiation due to the heterojunctions to obtain improved directionality of the emitted power in the direction perpendicular to the junction plane. The lateral width of the emitting region is adjusted for the fiber dimension, but is typically 50–100 microns.

Surface-emitting and edge-emitting structures provide several milliwatts of power output in the 0.8–0.9 micron spectral range operated at drive currents of 100–200 mA (2000–4000 A/cm²). The coupling loss into step-index fibers with a numerical aperture of 0.14 is about 17–20 dB for surface emitters and 12–16 dB for edge-emitting diodes, compared to about 3 dB for an injection laser. Since the coupling loss decreases as the inverse square of the numerical aperture, much more power can, of course, be coupled into larger-numerical-aperture fibers. But with these coupling losses, $Al_{0.1}Ga_{0.9}As$ double-heterojunction LED's can provide about 0.1 mW into a 0.14 NA, 80 micron diameter fiber at drive currents of about 200 mA with an applied voltage of 1.7 V.

Let us turn now to the modulation problem. The relation between the optical power output of an LED (with constant peak current) and the modulation frequency is given by

$$\frac{P(\omega)}{P_0} = \frac{1}{[1 + (\omega\tau)^2]^{1/2}}$$

where τ is the lifetime of the injected carriers in the recombination region and P_0 is the dc power emission. (However, parasitic circuit elements can reduce the modulated power range below this value.)

It is evident that a high-speed diode requires the lowest possible value of τ, but without sacrifice in the internal quantum efficiency. Low values of τ are obtained at high doping levels; but in GaAs and related compounds, a high density of nonradiative centers is formed when the dopant concentration approaches the solubility limit at the growth temperature. Of the devices reported so far, germanium-doped double heterojunction LED's have exhibited modulation capability (at

the 3-dB point) to about 200 MHz.[16] The use of germanium is advantageous because it can be incorporated into GaAs to concentrations of about 2×10^{19} cm⁻³, thus providing lifetimes on the order of one nanosecond without unduly reducing the internal quantum efficiency.

With regard to LED's for 1.0–1.1 micron emission, surface-emission outputs of about 1 mW have been achieved with InGaAs structures[18] and further progress is expected, particularly with lattice-matched InGaAs–InGaP heterojunction structures.[19]

Device reliability

In any practical optical communications system, component reliability is of great concern. It has been a major research goal to identify and correct the myriad of failure mechanisms that plagued early electroluminescent devices.[20] The failure modes have since been identified as either facet- or bulk-related; they can be of a gradual or a catastrophic nature. Facet degradation is specifically a laser problem because the facets are the mirrors that define the Fabry–Perot resonator. Bulk degradation, on the other hand, can occur in both LED's and lasers. Most failure mechanisms have been eliminated as the overall technology (crystal growth, device fabrication) has matured, but active concern still exists for those remaining few that may limit ultimate operating life. Insufficient data exist for mean time to failure, which, for telephone applications, is in the 100 000-hr range. A comprehensive model for laser-facet failure does not exist but considerable phenomenological data have been accumulated.

Facet failure due to intense light fields is a well known phenomenon in solid-state lasers, and has been found to occur in

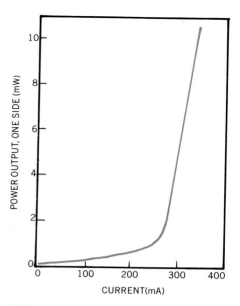

Optical power output from one facet of a typical aluminum gallium arsenide cw laser as a function of the drive current, measured at room temperature.   Figure 7

semiconductor lasers of all types under varying conditions when the optical power density in the recombination region reaches the order of 10⁶ W/cm². The appearance of the damaged laser facets suggests local dissociation of the material as well as "cracking" in some cases.

The critical damage level is also a function of the pulse length t, decreasing as $t^{-1/2}$, over the range of 20 to 2000 nsec. It is not surprising, therefore, that facet damage can occur in room-temperature laser diodes operating cw at their maximum emission levels (even with relatively low total power). Because of the nonuniform radiation distribution in the plane of the junction in stripe-contact lasers, it is difficult to establish damage-level criteria in terms of the total power output. However, the damage threshold for 100 nsec pulses is about ten times higher than for cw operation of diodes selected from the same group.[21]

In addition to the dependence on optical flux density and pulse length, it has been found that ambient conditions, specifically moisture and surface flaws (scratches, dirt particles), can lead to premature facet failure. To remove these limitations, lasers must be operated at a specific maximum power level and the facets must be passivated with protective dielectric coatings for isolation from their surroundings. For cw lasers the operating range of linear power density is about 1 mW per micron of stripe width. Facet failure, at least in its early stages, commonly manifests itself as a decrease in differential quantum efficiency without an increase in threshold current density.

Bulk degradation, the other failure mode, is accompanied by an increase in threshold current density and a decrease

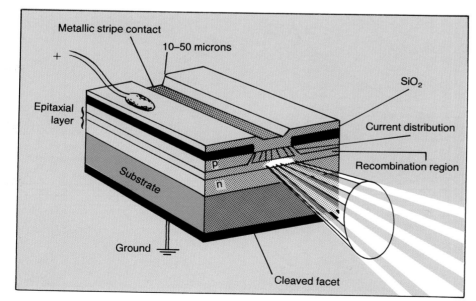

Schematic of a typical cw heterojunction laser, drawn upside down relative to figure 4 to show the stripe contact.   Diffraction causes the vertical spreading of the beam.   Figure 6

in differential quantum efficiency due to a reduction in *internal* quantum efficiency and an increase in the absorption coefficient. The reduction in output power may be small but need not be—if the device is a laser operated near threshold it may turn off completely. In either case a small adjustment in current will restore the output to its initial value. With such a feedback system, a definition of operating life becomes somewhat arbitrary and system-dependent.

The available evidence suggests that the gradual degradation process results from an increase in the concentration of nonradiative centers in the recombination region. These defects are initiated by the growth of flaws initially present in the recombination region of the diode. Point defects may also diffuse into the active region from adjacent flawed regions. Detrimental flaws include dislocations and impurity precipitates. One prominent effect in some degraded lasers is the formation of "dark lines" in regions where the luminescence is gradually extinguished.[22] These regions have been identified as large dislocation networks that, having started as smaller pre-existing dislocation networks, grow by the immigration of vacancies or interstitials.[23] In addition, more dispersed nonradiative centers such as native point defects apparently contribute to the degradation process.

It has been suggested that a multiphonon emission process resulting from nonradiative electron–hole recombination gives rise to an intense vibration of the center, which reduces its displacement energy.[24] (Whether any point defects are actually *formed* within the recombination region remains unclear.) Hence, if nonradiative electron–hole recombination occurs, say at the damaged surface of a diode (as in the case of the sawed-edge diode experiment described in reference 25), it accelerates the motion of point defects into the active region of the device.

The stoichiometry of the material in the active region or in close proximity to it is believed to affect the reliability of the device, as does the nature of the dopant.[26] A modification of initial stoichiometry may account for the improved degradation resistance of diodes fabricated with $Al_{0.1}Ga_{0.9}As$ rather than GaAs in the recombination region. Finally, regions where nonradiative recombination occurs will tend to grow in size, leading to the strongly nonuniform degradation process that is commonly observed.

Enormous progress has been made since 1970 in eliminating many degradation mechanisms in lasers and LED's by careful attention to the liquid-phase epitaxial growth and device processing. Facet passivation with dielectric coatings has virtually eliminated facet damage as an important failure mode in cw lasers. The ultimate operating lifetime of state-of-the-art devices remains undetermined,

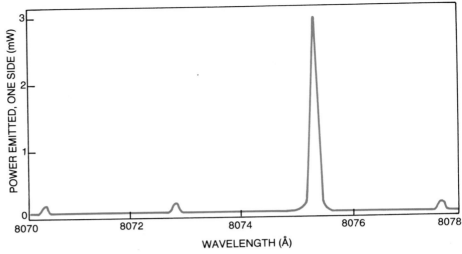

The spectral output from one facet of an AlGaAs cw laser. It emits 3 mW when operating in a single longitudinal mode and in the fundamental transverse and lateral mode.   Figure 8

but accumulated data exists on devices in operation *at constant current* for times well in excess of 14 000 hr without a serious drop in output power. Two examples of lifetime data taken over many hours of operation on our more recently fabricated diodes are:

▶ In a lot of AlGaAs LED's emitting about 1 mW at 0.8 microns, the emitted power remained constant within 5% in 14 000 hours of operation;

▶ In AlGaAs cw lasers emitting between 5 and 10 mW from one facet at a wavelength of 0.82 microns, the maximum deviation was less than 20% in 10 000 hours.

Laser diodes with unpassivated facets have operated for more than 15 000 hours but with more significant reduction in power output at fixed current. These data contrast sharply with those of a few years back, when the best operating life

was a few hundred hours under similar conditions.

Future devices

With regard to future work, heterojunction structures of AlGaAs emitting in the 0.80–0.85 micron spectral range have progressed to the point where practical systems applications of lasers and LED's are becoming possible. This has been the result of extensive research on the properties of materials and methods of synthesis as well as the identification of major factors affecting the operating lifetime of these devices. Research is now under way on devices emitting in the 1.0–1.1 micron range, which offers some potential advantages in reduced fiber absorption and dispersion. These devices involve more complex material problems than the AlGaAs devices because of the need to use dissimilar alloys to match the lattices.

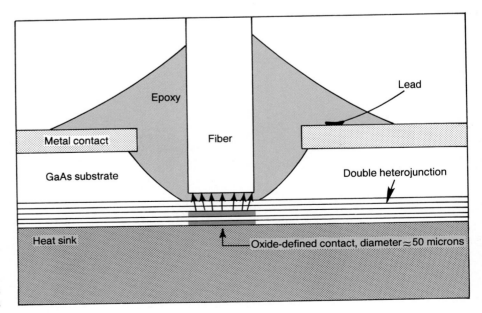

An etched-well surface-emitting LED designed for fiber optics, in a schematic cross section. The emission from this type of diode is essentially isotropic. (Reference 15)   Figure 9

References

1. H. Kressel, H. Nelson, RCA Review **30,** 106 (1969).
2. I. Hayashi, M. B. Panish, P. W. Foy, IEEE J. Quantum Electron. **5,** 211 (1969).
3. Zh. I. Alferov, V. M. Andreev, E. L. Portnoi, M. K. Trukan, Sov. Phys. Semiconductors **3,** 1107 (1970).
4. I. Hayashi, M. B. Panish, P. W. Foy, S. Sumski, Appl. Phys. Letters **17,** 109 (1970).
5. H. Kressel, F. Z. Hawrylo, Appl. Phys. Lett. **17,** 169 (1970).
6. C. B. Duke, N. Holonyak Jr, PHYSICS TODAY, December 1973, page 23.
7. G. H. Olsen, J. Crystal Growth **31,** 223 (1975).
8. F. Stern, IEEE J. Quantum Electron. **9,** 290 (1973); A. Yariv, *Quantum Electronics,* 2nd edition, Wiley, New York (1975); J. I. Pankove, *Optical Processes in Semiconductors,* Prentice-Hall, Englewood Cliffs (1971); S. E. Miller, Tingye Li, E. A. J. Marcatili, Proc. IEEE **61,** 1726 (1973).

9. J. K. Butler, H. Kressel, I. Ladany, IEEE J. Quantum Electron. **11,** 402 (1975); W. P. Dumke, *ibid,* page 400; P. R. Selway, A. R. Goodwin, J. Phys. D **5,** 9041 (1972); H. C. Casey Jr, M. B. Panish, J. L. Merz, J. Appl. Phys. **44,** 5470 (1973).
10. M. Ettenberg, Appl. Phys. Lett. **27,** 652 (1976).
11. A. R. Goodwin, J. R. Peters, M. Pion, G. H. B. Thompson, J. E. A. Whiteway, J. Appl. Phys. **46,** 3126 (1975).
12. R. E. Nahory, M. A. Pollak, E. D. Beeke, J. C. DeWinter, R. W. Dixon, Appl. Phys. Lett. **28,** 19 (1976); J. J. Hsieh, Appl. Phys. Lett. **28,** 283 (1976).
13. J. C. Dyment, Appl. Phys. Lett. **10,** 84 (1966).
14. IEEE J. Quantum Electron. **11,** 381–562 (July 1975).
15. C. A. Burrus, B. I. Miller, Opt. Commun. **41,** 307 (1971).
16. M. Ettenberg, H. F. Lockwood, J. Wittke, H. Kressel, *Technical Digest* of the 1973 International Electron Devices Meeting, Washington, D.C., page 317.

17. H. Kressel, M. Ettenberg, Proc. IEEE **63,** 1360 (1975).
18. C. J. Nuese, R. E. Enstrom, IEEE Trans. Electron. Devices **19,** 1067 (1972).
19. C. J. Nuese, G. H. Olsen, Appl. Phys. Lett. **26,** 528 (1975).
20. H. Kressel, H. F. Lockwood, J. de Physique C **3,** Suppl. 35, 223 (1974).
21. H. Kressel, I. Ladany, RCA Review **36,** 230 (1975).
22. B. C. DeLoach, B. W. Hakki, R. L. Hartman, L. A. D'Asaro, Proc. IEEE **61,** 1042 (1973); R. Itoh, H. Nakashima, S. Kishino, O. Nakada, IEEE J. Quantum Electron. **11,** 551 (1975).
23. P. Petroff, R. L. Hartman, Appl. Phys. Lett. **23,** 469 (1973).
24. R. D. Gold, L. R. Weisberg, Solid-State Electron. **7,** 811 (1964).
25. I. Ladany, H. Kressel, Appl. Phys. Lett. **25,** 708 (1974).
26. M. Ettenberg, H. Kressel, Appl. Phys. Lett. **26,** 478 (1975); P. G. McMullin, J. Blum, K. K. Shih, A. W. Smith, G. R. Woolhouse, Appl. Phys. Lett. **24,** 595 (1974).   □

ADDENDUM

Since 1976, a great deal of progress has been made in the purification of glass fibers leading to a greatly reduced attenuation in the 1.3 micrometer range. As a result, devices of InGaAsP emitting in that region are currently of major commercial importance.

Advances in semiconductor lasers

Diode lasers are providing coherent monochromatic light for an enormous variety of applications, including fiber-optic communications, audio and video recording systems and aligning structures and tunnels.

Yasuharu Suematsu

PHYSICS TODAY / MAY 1985

In the past decade the semiconductor laser—also known as the diode or junction laser—has become a key device in optical electronics because of its pure output spectrum and high quantum efficiency. Not coincidentally, its output can be modulated at very high speeds; this property and its compact size make it useful in a vast range of applications, from fiber-optic communications to optical radar.

Because they are compact and highly compatible with semiconductor circuits, laser diodes are becoming an important part of semiconductor technology and research. Laser diodes were invented soon after the light-emitting diode was devised in the early 1960s. The earliest diodes consisted[1] of p–n junctions made from gallium arsenide. Because of their large lasing threshold, such lasers can only operate in the pulsed mode. Subsequent developments reduced the threshold, making cw GaAs lasers possible. More recently, it has become possible to make junction lasers from a wide range of materials, with chemical and physical properties tailored to specific applications.

Yasuharu Suematsu is professor of physical electronics at the Tokyo Institute of Technology.

As the capabilities of laser diodes have grown, so has the range of applications contemplated for them. Laser–disk audio systems and optical-fiber communications systems are by now in wide use, and there are exciting prospects for further integrating optical and electronic components.

With these prospects in mind, I will review in this article current ideas about useful laser structures, materials technology for laser diodes, models for their behavior and, finally, developments in laser applications.

Device structures

The light emitted by a semiconductor diode comes from the recombination of electrons and holes at a p–n junction as a current flows through the diode in the forward direction: An electron from the n-layer conduction band recombines with a hole from the p-layer, emitting a photon. If the current is large enough and the recombination layer is small enough, this radiative process dominates; nonradiative processes (such as phonon emission) always provide a loss mechanism. The electrons that arrive at the junction to recombine can come from a forward current driven through the diode; such diodes are said to be injection pumped. Other pumping schemes include optical pumping and electron-beam pump-

ing, but these are less well suited to the cw operation of laser diodes as elements in photonic circuits.

To fabricate a junction laser, one must combine the optical gain provided by a laser diode with the optical feedback provided by a cavity. Usually this cavity is made by polishing the ends of the diode at right angles to the p–n junction channel. Reflections from the polished ends provide the build-up of intensity required for laser action in one of the cavity modes. The technical problem is to reduce the number of lasing cavity modes and their linewidths so that the output is as collimated and monochromatic as the particular application requires. Aram Mooradian discusses the limits on laser linewidths in his article on page 42.

The original laser diodes were homojunctions, that is, they were made from one material, usually GaAs, selectively doped to form a p–n junction; it is difficult to make the boundaries between the doped layers sharp enough to make good lasers from homojunctions. Junctions made from different materials—called heterojunctions—can be manufactured with much more precise control. (With modern fabrication techniques, material can be grown epitaxially on the substrate, so that the crystal planes are continuous throughout the diode, as in the homojunction.)

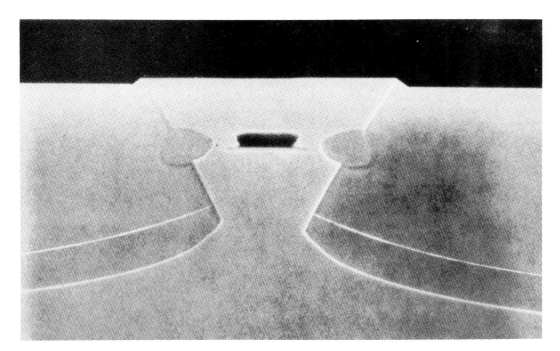

Buried heterostructure laser, seen end on. The active layer and waveguide are the small, black, rectangular area in the center of the photo; it is about 1 micron wide, 0.1 micron thick and 300 microns long. The active material consists of InGaAsP grown epitaxially on p-type InP (the hourglass-shaped area), and is covered with an epitaxial layer of n-type InP. To keep the current flowing only through the laser junction, it is surrounded by "blocking" layers of p- and n-doped InGaAsP. The sketch shows details of the composition. To delineate the various layers, the end of the diode was polished and selectively etched before it was imaged in a scanning electron micrograph. (Courtesy of T. Ikegami, NTT.)   Figure 1

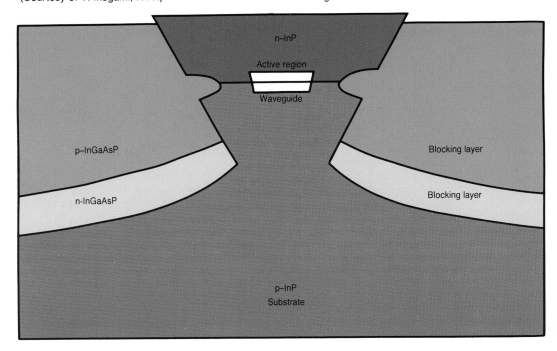

The development,[2] in the early 1970s, of the GaAs/AlGaAs double-heterostructure laser (lattice-matched across the boundaries), operating at room temperature with a cw output, provided a trigger for many subsequent developments. The new heterojunction lasers had a substantially lower laser threshold than the earlier homojunction devices, and they operated at somewhat shorter wavelengths—in the 0.75–0.9-micron range. Concurrent developments[3] greatly improved the response of lasers to variations in the injection current, allowing for high-speed direct modulation of the laser output.

To control the collimation and spatial purity of the emitted laser light, it is necessary to control[4] the transverse modes of oscillation of the radiation in the laser cavity. The currently preferred technique is to confine the p–n junction to a narrow channel or stripe, only 1 to 3 microns wide. An early example of such a structure is the buried heterojunction.[5]

Because the resonating lightwave inside the laser interacts strongly with the electrons and holes in the small active region of the laser, the modes of oscillation of the cavity are distorted by nonlinear effects and spurious modes are introduced. For most advanced applications, one requires light of considerable spectral purity, so it is essential to control the axial mode of oscillation of the laser. Among the many different approaches used to stabilize[6] the axial oscillations of laser diodes are: distributed feedback, distributed Bragg reflection, and the cleaved-coupled-cavity laser. Several groups have developed lasers that can operate in a single mode, even under rapid modulation.

Lasers diodes are, of course, developed with applications in mind—applications whose technological requirements drive the direction of development. One important goal[7] is to provide laser sources for high-capacity optical-fiber communication, which Tingye Li discusses in his article on page 24. Because the glass fibers have the lowest losses around 1.3 microns, there is a great incentive to develop efficient lasers at that relatively long wavelength. An early candidate was the GaInAsP/InP quarternary laser, developed[8] in the mid-1970s, whose output is in the region of 1.2–1.6 microns.

Other goals for semiconductor laser development include narrow spectral linewidth, injection locking, high output power, and ultrashort pulses. Thus far, the wavelengths produced by semiconductor lasers have generally been longer than 700 nm; shorter wavelengths would be useful for many applications, such as laser disks, and such lasers are being developed. Bistable lasers, which would be useful as elements of purely optical switching devices, have been studied.

Heterostructure diodes

As I have mentioned, the preferred type of diode for lasers consists of an epitaxially grown heterostructure. Most commonly, the structure consists of a ternary or quaternary compound on a binary substrate: AlGaAs on GaAs or GaInAsP on InP, for example. Particularly in the case of quaternary compounds, the lattices can be well matched across the junction. Lattice matching is important because lattice defects provide sites for nonradiative electron–hole recombination and because the defects can migrate through the structure, eventually aggregating and causing the diode to fail.

Figure 1 shows the cross section of a single-mode diode laser. The index of refraction and the band structure change relatively abruptly at the boundary of the active region in such a buried double-heterostructure diode. These discontinuities serve both to confine the carriers to a precisely defined active region and to provide a waveguide for the emitted radiation. By making the active stripe thin and long enough—in the illustrated diode it is only 1 micron wide—one can suppress most of the transverse modes of oscillations to provide a well-collimated output.

In a conventional laser diode, the ends of the crystal are polished to provide reflecting surfaces. Effectively, the active stripe becomes a Fabry–Perot resonator. In principle, this confines the allowed frequencies of the laser to a discrete number of longitudinal modes. However, the intrinsic emission from a diode has a broad spectral linewidth (electrons and holes of a relatively broad range of energies can recombine), so that the laser can excite many longitudinal modes simultaneously.

One way to reduce the number of axial, or longitudinal, modes is to provide some periodic structure within the active region that couples dynamically to the radiation field: One can construct the waveguide that forms the active region with a periodic pattern of grooves on its surface. Such grooves form a distributed reflector that turns the waveguide itself into a wavelength-sensitive resonator; we have called this technique "distributed feedback." Another scheme for dynamic control of the axial mode of oscillation is distributed Bragg reflection, in which the distributed reflector is formed on a waveguide region separate from but coupled to the active region. Figure 2 illustrates the progress in controlling the output of semiconductor heterojunctions—from light-emitting diodes to dynamic-single-mode laser diodes.

Unlike the conventional Fabry–Perot cavity, which changes its frequency by hopping from mode to mode, the dynamic-single-mode laser remains in the same mode of oscillation as, for example, the temperature changes. Figure 3 shows the differences in behavior. Changing the temperature of the active region alters the effective band gap because the average energy of the carriers changes. As a result, the index of refraction of the active region also changes. This variation in turn leads to a change in the wavelength of the dynamic-single-mode laser of about 1 Å per centigrade degree. The same variation is seen in the Fabry–Perot laser, but in addition, the oscillations hop from mode to mode, leading to a further variation of the laser wavelength amounting to 3–5 Å per degree.

Single-mode laser diodes can thus be tuned continuously—electrically as well as thermally—over a wavelength range of several angstroms. The electrical tunability arises because the index of refraction of a semiconductor depends on the carrier density, which in turn depends on the injection current. Again, for the Fabry–Perot diodes the tuning takes place by mode hopping, while in dynamic-single-mode lasers the mode of oscillation remains stable as its frequency shifts slightly.

One elegant narrow-band but broadly tunable system is the cleaved-coupled-cavity laser (PHYSICS TODAY, October 1983, page 20). This laser con-

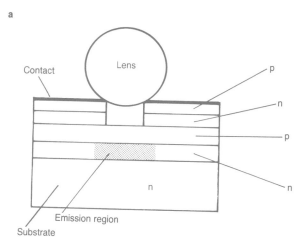

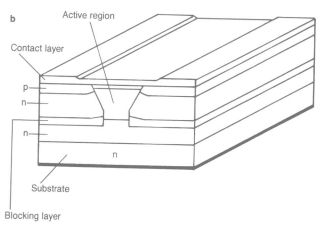

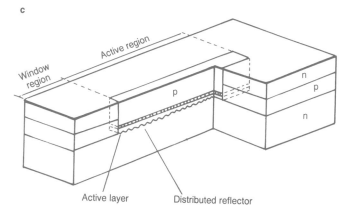

sists of two Fabry–Perot cavities coupled by a small air gap. Only those wavelengths that are common to both cavities are allowed, so that the output is highly monochromatic. By adjusting the currents in the two lasers, one can shift the frequency from mode to mode, tuning the laser—in steps—over more than 100 Å.

A typical buried-heterostructure diode, such as the one shown in figure 1, has an active region about 300 microns long, 1 or 2 microns wide and 0.1 microns thick. The table on page 36 shows the range of operating parameters of typical laser diodes. The threshold currents of such devices vary from 2.5 to 100 milliamps—a typical value is 20–50 milliamp. The cw output power ranges from a few to a few hundred milliwatts; pulsed power levels reach 800 mW. The output radiation is polarized horizontally and collimated to roughly 30° horizontally and 70° vertically. The ohmic resistance of these devices tends to be several ohms, and the thermal impedance is 30–50 °C per watt. The table also shows the temperature dependence of threshold currents of some typical diode lasers.

In a buried heterostructure laser, the current is confined to the active region by blocking layers that surround the active region with an electric potential barrier to carrier penetration. Because the blocking layer itself is a semiconductor, the whole structure behaves as a pnpn thyristor, which can act to limit the bias current. Various structures, such as the double-channel planar buried heterostructure, have been devised to reduce the leakage currents.

The output optical power is limited by the ability of the active region to dissipate the heat generated by nonradiative processes as well as its internal resistance. It is also limited by the maximum light power density that the resonator can support. Experimentally one finds that the maximum output power is typically 10 mW per micron of stripe width. For very-high-power lasers one can increase the transverse cross section of the resonators by adopting a coupled-multistripe structure—that is, by making a coherently coupled array of junction lasers. Such lasers have achieved 2.6 W of output power.

Materials

Figure 4 shows some of the materials

Evolution of the semiconductor laser.
The drawings show a typical light-emitting diode; a conventional buried-heterostructure laser diode, or junction laser; and a dynamic-single-mode laser diode. The buried-heterostructure laser controls emission into transverse modes of radiation by being thin, long and narrow; the dynamic-single-mode laser diode restricts emission into a single axial mode by means of the grooves in the waveguide surface, which form an active distributed reflector.   Figure 2

Output characteristics of diode lasers

Type	Wavelength (microns)	Typical power (mW)	Maximum power (mW)	Quantum efficiency %	Device efficiency %	T_0 (°C)	T_{max} (°C)	I_0 (mA)
GaAlAs/GaAs								
LED	0.85	10	200*		45	120–150	150 (cw), 276 (pulsed)‡	4.5*
LD	0.85	5–10	200†	50–70	40	200		2.5‡
LD (quantum well‡)	0.85			80				
LD (phased array)	0.85		2600§	60				
GaInAsP/InP								
LED	1.3	1	140#	50–70	43	50–70	142 (cw)#	5.5*
LD	1.3	5–10	40#	40–60		40–60	115 (cw)#	12#
LD	1.55	5–10						
LD (single mode)	1.3	5–10	60#,@	30 (per facet)				
LD (single mode)	1.55	5–10	20#,¶	20 (per facet)				

* Obtained at Hitachi   † Obtained at Matsushita
‡ Bell Labs   § Obtained at Xerox
# Obtained at NEC   @ Obtained at NTT
¶ Obtained at Fujikura   • Obtained at MIT

that have been used for semiconductor heterojunction lasers, together with the range of wavelengths emitted by each, as well as other characteristics. The emitted wavelength is determined by the band gap E_g of the region in which the recombination takes place: The wavelength λ is just hc/E_g. The active materials listed in the figure are all alloys—$Al_x Ga_{1-x} As$, for example—and for these the band gap varies with composition; the range of variation of the possible emission wavelength is shown in the figure. For short wavelengths, in the range 0.75–0.9 microns, $Al_x Ga_{1-x} As/GaAs$ is the preferred system; it is the earliest and probably the most studied of the diode-laser materials. For longer wavelengths, in the range 1.2–1.65 microns, $Ga_x In_{1-x} As_y P_{1-y}/InP$ is often used. Most of the compounds used involve elements from groups III and V of the periodic table.

One active area of experimental research is a search for materials that have a shorter lasing wavelength. The GaInAsP/GaAs system has the potential to reach wavelengths as low as 0.65 microns—at the red edge of the visible spectrum. Compounds of materials from groups II and VI, as in the ZnSSe/GaAs system shown in the figure, are being studied for making blue-emitting lasers. Experimental efforts are underway to give these materials better electrical properties and make them efficient emitters of radiation. Diode lasers emitting at wavelengths above 2 microns have also been developed.[9]

Liquid-phase epitaxy is the most frequently used technology for growing laser-diode structures. This process produces very reliable diodes, but the ability to control crystal area and size is rather limited. To overcome these limitations various new technologies have been developed, such as metal-organic vapor-phase epitaxy (chemical vapor deposition), or halide vapor-phase epitaxy. These techniques show great promise.[10] The ultimate tool in control of crystal growth—down to the level of individual atomic layers—is molecular-beam epitaxy.[11]

Laser diodes degrade in use, in part because of the high current densities at which they operate—on the order of 1 kA/cm^2. At present the operating life of a long-wavelength laser diode appears to be limited not by defects in the crystal itself but by the mounting and other structural characteristics. The operating lifetime of a GaInAsP/InP laser, for example, is more than 100 000 hours. Future laser diodes are likely to be even more reliable. Short-wavelength lasers, such as AlGaAs/GaAs lasers, generally have lifetimes shorter than this. Many groups have worked[12] on increasing the lifetime of junction lasers, and present-day laser diodes have a satisfactorily long life.

Models

There are several models for calculating the gain of semiconductor lasers. These models must take into account not only the observed spectral spread in the vicinity of the band edge but also the suppression of gain in modes other than the lasing mode. An early model for the linear laser gain, called the band-tail model, did account for the spectral spread,[13] but it had some difficulty not only in explaining the gain suppression but also in making direct connection with conventional semiconductor parameters. To overcome these difficulties, a model akin to that used for gas lasers has been developed.[14] The model is based on a density-operator analysis with electronic relaxation. The spectral spread near the band edge is, according to this theory, an effect of electronic relaxation; the gain suppression of neighboring modes appears as a result of mode competition. The gain suppression is closely related to the distribution of the intensity in the lasing modes. Because the electronic relaxation time in most III-V compounds is on the order of 0.1 picosec, the gain suppression in these materials extends a few nanometers from the lasing wavelength. At longer wavelengths the effect is larger, and thus there is a stronger tendency toward cw single-mode operation at longer wavelengths.

While the lasers I have mentioned thus far consist of junctions that are large on the atomic scale, several groups have been able to grow "superlattices," that is, crystals whose chemical composition varies over a few atomic layers.[15] Such multiple-quantum-well structures have many interesting properties and may lead to lasers with a large gain and low threshold as well as broad tunability, an ability to operate at high temperatures and other useful properties (PHYSICS TODAY, April 1979, page 20). We have extended the theory of laser gain to the case of quantum-well lasers.

The efficiency of diode lasers is affected by many factors, including absorption losses in the active region, nonradiative recombination processes, and leakage currents (through the blocking layer) that bypass the active region. Quater-

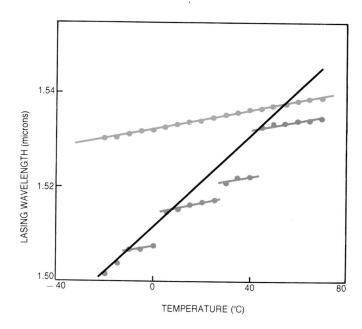

Lasing wavelength as a function of temperature of a conventional laser diode with a Fabry–Perot cavity (black) and a dynamic-single-mode laser diode with distributed feedback (color). Note that the Fabry–Perot cavity adjusts its frequency by hopping from mode to mode as well as by an intrinsic frequency drift.   Figure 3

nary materials, for example, with a relatively small energy gap and relatively large recombination rates due to the Auger effect, have relatively large losses. Because the nonradiative recombination rates increase with temperature, the laser threshold current also increases with temperature, especially for long-wavelength lasers.[16] To reduce the leakage currents, one must increase the barrier height of the blocking layer. In short-wavelength lasers the barrier should be at least 0.2 eV.

The table on the opposite page shows that typical device efficiencies (total radiated power out divided by electrical power in) are on the order of 40–45%. The differential quantum efficiency (incremental output light power divided by incremental input power) does not include effects associated with the laser threshold and is thus always larger than the overall efficiency. For GaAs lasers the differential quantum efficiency is 50–80%; for 1.3-micron quaternary lasers it is 50–70%, and for 1.5-micron lasers it is 40–70%.

The threshold current for laser operation depends exponentially on temperature, according to the empirical formula

$$I = I_0 e^{T/T_0}$$

The table gives some characteristic values for I_0 at room temperature and T_0. Note that for typical GaAs short-wavelength lasers T_0 is in the range 120–140 °C, while for longer-wavelength lasers the values of T_0 are considerably smaller. Like the quantum efficiency, the threshold current—and its temperature dependence—is a function of the internal loss mechanisms, and it is therefore not surprising that systems

with large quantum efficiencies, such as the GaAs short-wavelength lasers, also have large values of T_0.

Short-wavelength diode lasers have operated at temperatures as high as 150 °C in cw operation and at 276 °C in pulsed operation. Long-wavelength lasers have not been stable at such high temperatures—for 1.3-micron diodes, for example, the hottest operating temperatures have been around 140 °C.

For single-mode lasers, the spectral width is broadened by spontaneous emission into the lasing mode. The measured widths—ranging from tenths to several megahertz—are in agreement with theoretical predictions based on the Schawlow–Townes relation. M. Ohtu has recently verified that these theoretical limitations on laser linewidths could be decreased significantly by electronic feedback to the bias current of the intensity fluctuations of the light output.

Modulation

One of the most attractive properties of laser diodes is their capacity for rapid direct modulation of the output by varying the injection current. The modulation frequency can clearly be no larger than the frequency of the relaxation oscillations of the laser field. The relaxation frequency f_r depends upon both the carrier lifetime and the photon lifetime. The carrier lifetime in turn decreases as the current is increased above the threshold current. The theoretical relationship is

$$f_r = \frac{1}{2\pi} \sqrt{\frac{I/I_{th} - 1}{\tau_s \tau_p}}$$

where τ_s is the carrier lifetime at

threshold and τ_p is the photon lifetime. In a typical direct-gap semiconductor, τ_s is on the order of a nanosecond; for indirect-gap materials it is some three or four orders of magnitude larger. In a conventional laser diode some 300 microns long, τ_p is a few picoseconds. Thus, when the injection current is about twice the threshold current, the maximum modulation frequency is on the order of a few gigahertz.

The modulation rate is, of course, also limited by the frequency response of the electronic circuit of which the diode is a part. Thus, for example, the laser diode must have a small capacitance if it is to be capable of high-speed modulation. Specially designed laser diodes have reached modulation rates of 10 GHz.

For low modulation frequencies, the modulation sensitivity of a laser diode is independent of the modulation frequency. As the frequency approaches f_r, however, the modulation sensitivity shows a sharp resonance. This resonance increases the spectral noise in the output radiation and leads to relaxation oscillations in the output power for pulsed operation.[17]

Spectral broadening due to multimode operation, mode hopping, dynamic line shift ("wavelength chirping") and an increase in noise are all a direct result of carrier oscillations, which are enhanced by relaxation oscillations. One can, to a great extent, suppress the relaxation oscillations by making the width of the active region nearly equal to the diffusion length. In that case the carrier diffusion will compensate for effects due to carrier oscillations. Because the diffusion length is on the order of microns, the stripe width

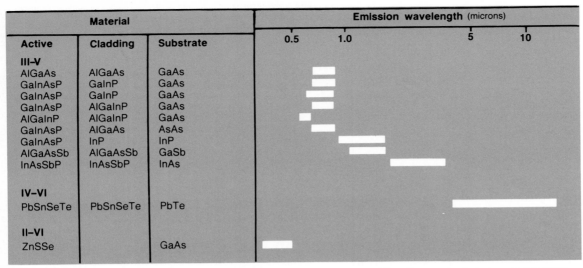

Material			Emission wavelength (microns)
Active	**Cladding**	**Substrate**	
III–V			
AlGaAs	AlGaAs	GaAs	
GaInAsP	GaInP	GaAs	
GaInAsP	GaInP	GaAs	
GaInAsP	AlGaInP	GaAs	
AlGaInP	AlGaInP	GaAs	
GaInAsP	AlGaAs	AsAs	
GaInAsP	InP	InP	
AlGaAsSb	AlGaAsSb	GaSb	
InAsSbP	InAsSbP	InAs	
IV–VI			
PbSnSeTe	PbSnSeTe	PbTe	
II–VI			
ZnSSe		GaAs	

Materials for semiconductor light sources. The chart above shows some of the alloys and substrates used in laser diodes, as well as the wavelength ranges at which they operate.   Figure 4

required to compensate for oscillations is about the same as the width required to suppress the transverse modes of oscillation. An additional advantage of narrow stripes is that the threshold current is small.

Reflection of emitted light back into the laser cavity is an additional source of noise. The reflected light can decrease the wavelength stability and increase any tendency toward mode hopping. Because the mirror reflectivity of the polished ends of semiconductor lasers is fairly small (about 30–40%), the effect is particularly large for these devices. Optical isolators can largely eliminate this unwanted effect.

In steady cw operation, a typical index-guided laser—that is, a laser in which radiation is confined by changes in the index of refraction—operates in a single mode. Neighboring modes are suppressed by the competition from the lasing mode. However, when the injection current is modulated rapidly, the mode suppression—which is based on slight differences in gain between the lasing mode and neighboring modes— is less effective, and the laser oscillation hops from mode to mode. There is then little difference between the lasing mode and neighboring modes. Figure 5 shows the transition from single-mode to multimode operation for a typical Fabry–Perot laser diode as the modulation frequency is increased to 1.5 GHz.

To suppress this multimode operation, one must impose additional losses on the neighboring modes. The dynamic-single-mode, or stabilized single-frequency, laser was developed for just this purpose. As described above, it consists of a tightly mode-controlled resonator for the transverse modes that contains an additional wavelength selection mechanism for controlling the longitudinal mode. Figure 5 also shows

the behavior of such a dynamic-single-mode laser at various modulation frequencies. At conditions of modulation that lead to oscillations of many secondary modes in the Fabry–Perot laser, the dynamic-single-mode laser exhibits only a slight line broadening.

Another technique to stabilize the laser frequency that has recently been developed involves injection-locking a slave laser diode to a stabilized laser diode. Such a system has also been tested for its response to modulation.

Superlattice structures appear to provide a novel method to control the recombination lifetime and thus promise to allow very-high-speed modulation.[18] Mode-locked diode lasers whose output is controlled by an external mirror have produced pulses as short as a picosecond.

Frequency modulation of the laser output is a very interesting possibility. Bistable lasers with two stable output frequencies have been demonstrated— an example is the cleaved-coupled-cavity laser.

Applications

Laser diodes are applied in many areas of optics and electronics. One of the most exciting areas of research and development—and recently application—is their use in optical communication. The stability, quantum efficiency and ease of modulation of a laser diode make it an ideal light source for communications.[19] Coupling it to an extremely fine, low-loss silica fiber produces a communication system that has many desirable properties: low loss, wide bandwidth, immunity to induction noise and interference, low cost, and ease of installation (because the cables are thin and light). In many places, fiber-optics and laser diodes are already providing high-density, high-capacity channels of communication.

Because the laser substrate materials—GaAs and InP—have high electron mobilities, they may also make useful substrates for high-speed electronic devices. Much recent interest has therefore focused on the possibility[20] of integrating laser diodes directly into monolithic integrated electronic circuits; the circuits that control the laser—such as field-effect transistors— and the laser itself thus are part of the same chip.

At the other end of the laser, one has the possibility of integrated optical devices in which the laser is integrated on the same semiconductor substrate with the output waveguide, filter, and any optical switches and modulators.[21]

While some research has been done on such opto-electronic integration, semiconductor integrated optics is still in an experimental stage. The chief effort in optical and electronic research on laser diodes is still to improve device properties. For example, laser arrays that emit light from a two-dimensional surface are under development.

Another application that makes use of many of the properties of laser diodes is in optical memories or laser–disk memory systems. The density of information stored on a disk is limited by the size of the spot that can be focused on it. For laser light the spot size is diffraction limited, so shortening the wavelength is a crucial problem: The memory density is inversely proportional to the square of the wavelength. The laser–disk systems are finding wide application in video and audio systems; they are also being developed for high-capacity computer memories.

Lasers with oxide-stripe structures, in which the laser mode is gain guided, generally exhibit multimode operation. On the other hand, buried-heterojunction lasers, in which the radiation is confined by the index of refraction,

Laser spectra under direct modulation: **(a)** for a conventional buried heterostructure with a Fabry–Perot cavity, and **(b)** for a dynamic-single-mode laser diode. Both lasers are operating at a current 1.2 times their threshold, and the modulation depth is 100% for both. Note that the dynamic-single-mode laser diode continues to emit an extremely narrow spectrum at modulation frequencies that cause the conventional laser to emit over a very broad range of modes. Figure 5

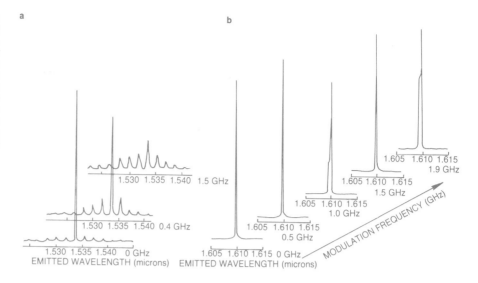

normally operate in a single mode. For gain-guided structures the focal point of the emitted light is unstable, varying with the current distribution, whereas for index-guided structures the focal point is stable. Optical recording and pickup technology requires a stable focal point for the light source, so only index-guided structures are suitable for these applications.

Laser diodes are replacing other varieties of lasers in many different optical applications. One well-known example of an application of laser beams is in optical ranging systems for measuring distances, finding locations and aligning structures and tunnels. Semiconductor ir lasers are used to measure velocities and flows in situations as diverse as burglar alarms, building security systems and clinical monitoring systems. In metrology, lasers are used for length and shape measurements. The recent invention of the optical-fiber gyroscope makes it possible to use lasers to measure rotations and accelerations (PHYSICS TODAY, October 1981, page 20). Finally I should mention the use—important in medicine and technology—of lasers and bundles of optical fibers to illuminate and observe inaccessible areas.

References

1. N. G. Basov, B. M. Vul, Y. M. Popov, Zh. Eksperim. Teor. Fiz. **35**, 587 (1959) [Sov. Phys. JETP **8**, 406 (1959)]; M. I. Nathan, W. P. Dumke, G. Burns, F. H. Dill Jr, G. Lasher, Appl. Phys. Lett. **1**, 62 (1962); M. Quist, R. H. Rediker, R. J. Keyes, W. E. Krag, B. Lax, A. L. McWhorter, H. J. Zeiger, Appl. Phys. Lett. **1**, 91, (1962); R. N. Hall, G. H. Fenner, J. D. Kingsley, T. J. Soltys, R. D. Carlson, Phys. Rev. Lett. **9**, 366 (1962).

2. H. Kroemer, Proc. IEEE **51**, 1782 (1963); Zh. I. Alferov, V. M. Andreev, E. L. Portnoi, M. K. Trukan, Fiz. Tekh. Poluprov. **3**, 1328, (1969) [Sov. Phys. Semicond. **3**, 1107 (1970)]; I. Hayashi, M. B. Panish, P. W. Foy and A. Sumski, Appl. Phys. Lett. **17**, 109 (1970).

3. B. S. Goldstein, R. M. Weigand, Proc. IEEE **53**, 195 (1965); J. Takamiya, F. Kitasawa, J. I. Nishizawa, Proc. IEEE **56**, 135 (1968); T. Ikegami, Y. Suematsu, IEEE J. Quantum Electron. **QE-4**, 148 (1968).

4. J. C. Dyment, Appl. Phys. Lett. **10**, 84 (1967).

5. T. Tsukada, J. Appl. Phys. **45**, 4899 (1974); W. Susaki, H. Namizaki, H. Kan, A. Ito, J. Appl. Phys. **44**, 2983 (1973).

6. H. Kogelnik, C. V. Shank, Appl. Phys. Lett. **18**, 152 (1971); M. Nakamura, A. Yariv, H. W. Yen, S. Somekh, H. L. Garvin, Appl. Phys. Lett. **22**, 515 (1973); W. Tsang, S. Wang, Proc. 9th Int. Quantum Electron. Conf. June 1976, p. 38. A. Yariv, M. Nakamura, IEEE J. Quantum Electron. **QE-13**, 233 (1977); Y. Suematsu, S. Arai, K. Kishino, IEEE J. Light Technol. **LT1**, 161 (1983); Y. Sakakibara, K. Furuya, K. Utaka, Y. Suematsu, Electron. Lett. **16**, 456 (1980); K. Utaka, S. Akiba, K. Sakai, Y. Matsushima, Electron Lett. **17**, 961 (1981); T. Matsuoka, H. Nagai, Y. Itaya, Y. Noguchi, U. Susuki, T. Ikegami, Electron Lett. **18**, 27 (1982); W. T. Tsang, N. A. Olsson, R. A. Linke, R. A. Logan, Electron. Lett. **19**, 415 (1983).

7. Y. Itaya, Japan J. Appl. Phys. **18**, 1795 (1979).

8. A. P. Bogatov, L. M. Dolginov, P. G. Eliseev, M. G. Milvidskii, B. N. Sverdlov, E. G. Shevchenko, Sov. Phys. Semicond. **9**, 1282 (1975); J. J. Hsieh, J. A. Rossi, J. P. Donnelly, Appl. Phys. Lett. **28**, 709 (1976).

9. I. Melngailis, IEEE Trans. Geosci. Electron. **GE-1**, 7 (1972).

10. G. H. Olsen, C. J. Nuese, M. Ettenberg, Appl. Phys. Lett. **34**, 262 (1979); R. D. Dupuis, P. D. Dapkus, Appl. Phys. Lett. **33**, 68 (1978); J. P. Hirtz, J. P. Duchemin, P. Hirtz, B. DeCremoux, T. Pearsall, M. Bonnet, Electron. Lett. **16**, 275 (1980).

11. A. Y. Cho, R. W. Dixon, H. C. Casey Jr, R. L. Hartman, Appl. Phys. Lett. **28**, 501 (1976); W. T. Tsang, F. K. Reinhart, R. C. Miller, F. Capasso, J. A. Ditzenberger, paper presented at the 2nd Int. Symp. Molecular Beam Epitaxy and Related Clean Surface Techniques, Tokyo, Japan, August 1982.

12. H. Kressel, H. Nelson, RCA Rev. **30**, 106 (1969); G. H. B. Thompson, P. A. Kirkby, Electron Lett. **9**, 295 (1973); H. Yonezu, I. Sakuma, T. Kamejima, M. Ueno, K. Nishida, Y. Nannichi, I. Hayashi, Appl. Phys. Lett. **24**, 18 (1974); B. C. DeLoach Jr, B. W. Hakki, R. L. Hartman, L. A. D'Asaro, Proc. IEEE (Lett.) **61**, 1042 (1973); P. Petroff, R. L. Hartman, J. Appl. Phys. **45**, 3899 (1974); T. Kobayashi, T. Kawakami, Y. Furukawa, Jap. J. Appl. Phys. **14**, 508 (1975).

13. G. Lasher, F. Stern, Phys. Rev. **A133**, 553 (1964).

14. Y. Nishimura, K. Kobayashi, T. Ikegami, Y. Suematsu, Monthly Meet. Tech. Group, Quantum Electron. **QE71-22**, IECE Japan, September 1971; M. Yamada, Y. Suematsu, IEEE J. Quantum Electron. **QE-15**, 743 (1979).

15. N. Holonyak Jr, R. M. Kolbas, R. D. Dupuis, P. D. Dapkus, IEEE J. Quantum Electron. **QE-16**, 170 (1980); W. T. Tsang, Appl. Phys. Lett. **40**, 217 (1982).

16. M. H. Pilkuhn, H. Rupprecht, Solid-State Electron. **7**, 905 (1964); R. E. Nahory, M. A. Pollack, J. C. DeWinter, Electron. Lett. **15**, 659 (1979); A. R. Adams, M. A. Asada, Y. Suematsu, S. Arai, Jap. J. Appl. Phys. **19**, L621 (1981); G. H. B. Thompson, Proc. Inst. Electr. Eng. **128**, 37 (1981).

17. M. Yamanishi, I. Suemune, Jap. J. Appl. Phys. **22**, L22 (1983).

18. J. E. Ripper and T. L. Paoli, Appl. Phys. Lett. **18**, 466, (1971).

19. D. Boetz, J. Hershkowitz, Proc IEEE **68**, 689 (1980).

20. I. Ury, S. Margalit, M. Yust, A. Yariv, Appl. Phys. Lett. **34**, 430 (1979).

21. Y. Suematsu, M. Yamada, K. Hayashi, Proc. IEEE **63**, 208 (1975); F. K. Reinhart, R. A. Logan, C. V. Shank, Appl. Phys. Lett. **27**, 45 (1975); C. E. Hurwitz, J. A. Ross, J. J. Hsieh, C. M. Wolf, Appl. Phys. Lett. **27**, 241 (1975).

Novel laser for fiber-optic communication

Bertram M. Schwarzschild   PHYSICS TODAY / OCTOBER 1983

A new kind of semiconductor laser, developed by Won-Tien Tsang and his colleagues at Bell Labs, has demonstrated a number of remarkable capabilities that promise to be of considerable importance for fiber-optic communication. This cleaved-coupled-cavity semiconductor laser, nicknamed the C^3 laser, is able to generate essentially monochromatic pulses of infrared light at repetition rates in excess of a gigahertz. Furthermore, one can electronically switch this pulsed monochromatic output from one wavelength to another on a nanosecond time scale.

In the infrared wavelength regime suitable for long-distance fiber-optic communication (1.3 to 1.6 microns), the pulsed output of conventional semiconductor diode lasers has a wavelength spread of about fifty or a hundred angstroms. This spectral width presents dispersion problems that severely limit the rate of data transmission and the distance over which one can transmit information without reamplification. One is faced with a trade-off. Single-mode silica optical fibers are most transparent at 1.55 microns. But at this wavelength the fibers suffer from considerable dispersion. Different frequency components of the pulse travel at different speeds, eventually washing out the signal structure and thus limiting the rate at which one can send information over long distances. If one chooses, on the other hand, to transmit information at a wavelength of 1.3 microns, dispersion is minimal. But at 1.3 microns the fiber is significantly less transparent, making it necessary to reamplify the signal along its way more often than one would like.

The new C^3 laser, developed by Tsang, Anders Olsson and Ralph Logan, uses[1] a coupled-cavity resonance technique to get rid of all but one of the half dozen or so Fabry–Perot modes that are responsible for the spectral width of a conventional single-cavity diode laser. The result is a pulsed output with a spectral width of less than an angstrom at gigahertz rates of amplitude modulation. With this extraordinarily monochromatic pulsed output one can forget about dispersion and transmit data at a wavelength of

1.55 microns, where the optical fibers are at their clearest. Thus Bell Labs has recently demonstrated[2] the transmission of digital information at a rate of 10^9 bits/second over 104 kilometers of optical fiber, with an error rate of less than 1 in 10^9 and no reamplification along the way. At this rate, Tsang likes to point out, one can transmit the entire text of the *Encyclopedia Britannica* in less than half a second.

This impressive long-distance, high-rate demonstration was done at a fixed infrared frequency, with pulse amplitude modulation providing the binary information. But the C^3 laser offers an alternative method of encoding digital data that can enhance the information transfer rate still further. At a gigahertz rate one can switch the output wavelength among as many as a dozen modes spaced about 20 Å apart. Thus, instead of single-frequency transmission with low- and high-power pulses representing binary 0 and 1, respectively, one could simply switch between two frequencies. Switching among four (or eight) frequencies would yield pulses with twice (or three times) the information content of a binary bit.

The cleaved-coupled-cavity laser is made by cleaving a conventional semiconductor diode laser along a crystal plane parallel to its end faces to produce two shorter diode lasers of slightly different length. Starting, for example, with a GaInAsP laser 250 microns long, one might end up with two lasers with lengths of 130 and 120 microns. It turns out, contrary to the pessimistic expectations of many, that the precise difference between the two lengths doesn't matter. One can therefore mass-produce the C^3 laser by conventional fabrication techniques without having to worry about the exact point of cleavage.

Each of these half-length diodes in isolation would behave like a conventional semiconductor laser. Its two end faces would serve as the mirrors of a Fabry–Perot interferometer cavity, supporting only those lasing modes for which the diode's length is an integral multiple of half the wavelength. About half a dozen such Fabry–Perot modes, spaced roughly 20 Å apart, would

generate a laser output with a spectral width of more than a hundred angstroms.

The trick of the C^3 laser is to suppress all but one of these modes by coupling the two half-length diodes together optically. The cleaving is done with the original diode fixed on a substrate film, so that the active lasing strips of the resulting short diodes will be precisely aligned, with an air gap of about 5 microns between them. Thus the laser light is reflected back and forth between their active stripes.

Because their lengths are slightly different, the Fabry–Perot mode spacings of the two coupled laser cavities will differ; their lasing modes will not, in general, overlap. If, for example, the lengths of the two diodes differ by 10%, only every tenth Fabry–Perot mode can be the same in both cavities. (Think of a Vernier scale.) The coupled laser system will support only these *common* resonant modes. A lasing mode in one diode that does not correspond to a Fabry–Perot mode of the other will not be coherently reinforced. With com-

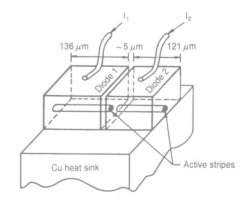

The cleaved-coupled-cavity semiconductor diode laser with which the Bells Labs group demonstrated gigahertz frequency switching in the 1.3-micron wavelength regime. Laser light is reflected back and forth across the 5-micron air gap between the GaInAsP active stripes of the two diodes, so that only the resonant mode common to both cavity lengths is reinforced, producing monochromatic output pulses at gigahertz repetition rates. Fine tuning and frequency switching are accomplished by independently varying the injector currents, I_1 and I_2, in the two diodes.

mon resonant modes thus spaced an order of magnitude further apart than the modes of a single diode laser, the stimulated radiative recombination of electrons and holes in the active lasing medium will generate only a single, monochromatic Fabry–Perot mode in the coupled system.

The idea of coupling laser cavities for single-frequency selection is not new. Herwig Kogelnik and Kumar Patel applied it with a three-mirror system of gas lasers at Bell Labs more than twenty years ago. In 1978, Louis Allen and his colleagues at McDonnell–Douglas achieved single-mode lasing with a cleaved diode laser configuration similar to Tsang's. This single-frequency lasing, however, was reported only for continuous (cw) laser operation, as distinguished from the high-rate pulsed operation necessary for fiber-optic communication. Allen has told us that he had, in fact, succeeded in achieving pulsed single-frequency operation at high modulation rates. But, he explained, this work could not be reported in the open literature at the time, for reasons of commercial and military secrecy.

Electrical tuning. The crucial feature that permits the Bell Labs C^3 laser to generate single-mode output in high-rate pulsed operation is the *independent* control of injection current in each of the coupled diodes. The effective index of refraction of a semiconductor diode laser depends on the carrier density, which increases with the imposed injection current. Increasing the current in one of the diodes decreases the internal wavelength of the stimulated emission, thus shifting the Fabry–Perot modes of the cavity.

If one does not have independent electrical control of the refractive indices of the two coupled diodes, the overlap of their Fabry–Perot patterns is fixed by the precise position and orientation of the cleavage, and by the width of the air gap it opens. No practical cleaving technique, Tsang told us, can control these geometric parameters with sufficient precision to assure adequate single-mode coincidence between the two Fabry–Perot patterns. Unless one lasing mode in the first cavity coincides quite precisely with a mode in the second, single-frequency operation will break down in pulsed operation.

The C^3 laser has therefore been designed as a three-terminal electro-optical device, permitting one to control separately the injection current in each diode. Thus electrical fine tuning can easily adjust the effective optical lengths of the coupled cavities to optimize common-mode resonant overlap, compensating for the vagaries of mass production. "We didn't know about the earlier work," Tsang told us. "If we

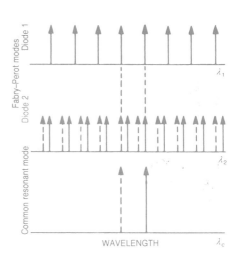

Frequency switching in the C^3 laser. The solid arrows in the top two rows represent the wavelengths of the Fabry–Perot lasing modes in each diode. Because their lengths are slightly different, there can be only one mode in the high-gain region common to both diodes. This common mode (bottom row) is the only one reinforced in the coupled-cavity configuration. A small change in the injection current of diode 2 shifts its Fabry–Perot pattern (dashed arrows), producing a much larger shift of the common output mode.

had, we probably wouldn't have tried." The prevailing opinion, he explained, was that the fabrication precision required to make a successful C^3 laser would be prohibitive.

Allen's 1978 coupled-cavity laser had been a two-terminal device, permitting only a common injection current to flow through both diodes. But, he told us, the temperature changes generated by this current did afford some degree of fine-tuning capability.

In 1981, a Bell Labs group led by Larry Coldren developed a three-terminal, coupled-cavity diode laser, using chemical etching rather than cleavage. Last year the group reported[3] that they had achieved single-mode, 1.3-micron pulses of sub-nanosecond duration. Coldren cautions, however, that his device is a high-current-threshold, gain-guided laser—a laboratory device with dimmer prospects for immediate practical application than the index-guided laser of Tsang. Furthermore, he adds, the chemical etching technique produces a less reliable yield of good reflecting surfaces than one gets with cleavage.

Frequency switching. Separate control of the two injection currents has another crucial consequence. It lets one switch the C^3 laser from one single-mode frequency to another in about a nanosecond. To achieve this gigahertz frequency modulation, Tsang, Olsson and Logan maintain the injection current in one of the coupled diodes above lasing threshold while its coupled partner is kept below threshold to serve as a Fabry–Perot etalon, suppressing all but the one frequency common to both diodes. If one now increases the injection current in the etalon slightly, its Fabry–Perot pattern will shift slightly, yielding a much larger jump in the common resonant mode. If, for example, the Fabry–Perot modes are about 20 Å apart and the two diodes differ in effective length by 10%, one need shift the mode pattern by only 2 Å to get a full 20-Å shift in the frequency

of the common mode. (Again it is helpful to think of a Vernier scale.)

By this technique, which they call cavity-mode-enhancement frequency modulation, the Bell Labs group has recently been able to tune a 1.55-micron C^3 laser over fifteen Fabry–Perot modes—a range of 300 Å. The frequency shifting responds very sensitively to the etalon injection current; it requires little more than a milliamp to shift the common mode by 20 Å. Spectral analysis of the output pulses indicates that they remain quite monochromatic and stable even at frequency-modulation rates in excess of a gigahertz.

Replacing the usual binary amplitude modulation of digital fiber-optics communication by four-frequency modulation would double the amount of information carried per pulse. Using only two frequencies, Tsang points out, a C^3 laser can serve as an optoelectronic logic gate, performing the full set of logic operations (and, or, exclusive or, inversion).

Another scheme for exploiting the frequency tuning capabilities of the C^3 laser to enhance fiber-optic digital communication capacity is referred to as wavelength-division multiplexing. One would couple several C^3 lasers to the same fiber, each tuned to a different frequency. They could then transmit several independent messages simultaneously along the single optical fiber. Such a scheme requires stable, monochromatic pulses of a quality that only the C^3 laser has thus far demonstrated.

References

1. W. T. Tsang, N. A. Olsson, R. A. Logan, Appl. Phys. Lett. **42**, 650 (1983).
2. R. Linke, B. Kasper, J. Ko, I. Kaminow, R. Vodhanel, Proc. 4th. Int. Conf. on Integrated Optics and Opt. Fiber Commun., Tokyo (1983), Pub. by Jap. Inst. of Electronic and Commun. Engineering.
3. K. Ebeling, L. Coldren, B. Miller, J. Retchler, Electron. Lett. **18**, 901 (1982). □

The fiber lightguide

The central component of a lightwave communication system is a fiber no thicker than a human hair, accurately formed of high-purity glass, with low dispersion and losses as low as 1 dB per kilometer.

Alan G. Chynoweth

PHYSICS TODAY / MAY 1976

The realization that the open atmosphere would be a very unreliable medium for the transmission of light led, soon after the invention of the laser, to consideration of the use of conduits, possibly evacuated pipes, for sending the light beams from one place to another, thereby providing a controlled atmosphere. If necessary, such conduits could be fitted with lens and mirror systems to provide beam path correction and beam path redirection. Servomechanisms could be devised to adjust the mirrors and lenses to compensate for such changes as might arise from thermal expansion or other distortions. Such systems could be made to work[1] but undoubtedly they would be very cumbersome and if practical at all, would be so only for very heavy communications traffic routes.

A variation on the conduit approach that was explored in the mid-1960's was to form what is called a gas lens. The scheme was to fill the conduit with gas but to provide a radial temperature gradient by means of a suitable arrangement of heater elements. This radial temperature gradient in turn gave rise to a radial gas-density gradient and hence a radial gradient in the refractive index. Such a radially graded refractive index could provide a waveguiding action for a light beam travelling more or less along the axis of the gas lens. But while experiments established the technical correctness of this approach, it was again likely to be an extremely cumbersome one in practice,[2] with uncertain reliability.

A way that had been known and practiced for a long time to transmit light over relatively short distances was to use light pipes in which the light is guided principally by internal reflection along internally silvered, hollow tubes or along plastic rods. Such schemes, while useful in many equipment applications, were not regarded as practical for long-distance communication because of the relatively high attenuations they exhibited. Charles Kao and G. A. Hockman of the Standard Telecommunications Laboratories in England were the first persons to suggest publicly (in 1966) that light pipes or glass fiber guides could be made with sufficiently low loss to be useful as transmission media for relatively long distances.[3] At that time some of the best optical glasses had attenuations of the order of several thousand dB/km. What was needed to make a practical communications system was thought (somewhat naively) to be attenuations of 20 dB/km or less. These figures give some indication of the enormous improvement in glass technology that it was realized would be necessary, but Kao felt it might be possible. He was encouraged in his belief by measurements on pure silica that exhibited losses as low as a few tens of dB/km.

The photograph in figure 1 shows a present-day form of a glass fiber used as a medium for light transmission.

Basic types of optical waveguides

From the outset it was recognized in a number of laboratories that the basic guiding structure would be one consisting of a core of a given refractive index surrounded by a cladding layer with a slightly lower refractive index. Such structures could carry propagating optical waveguide modes, the modes being internally reflected or refracted at the core–cladding interface. Extremely pure, low-optical-loss glass would have to be used both for the core and the cladding, but especially for the core as it carries most of the optical energy.

Two basic types of waveguide were considered; these are shown in figure 2. The first is the multimode fiber with a core of relatively large diameter. This fiber has a relatively large numerical aperture, making it suitable for collecting the light from an incoherent light source such as a light-emitting diode. However the simple multimode fiber has a disadvantage: Because light rays or modes tracing relatively coarse zigzag paths down the guide would take longer to reach the other end of the guide than light travelling along the axis, a multimode guide exhibits more dispersion. Thus, a narrow pulse of light fed in at one end becomes broadened as it travels down the guide. This sets an upper limit on the bandwidth that can be transmitted down a guide of given length.

A way around this difficulty is to make a single-mode fiber, one in which only the modes that travel at very small angles to the axis will propagate. This can be done either by making the core with an extremely small diameter or by making the refractive index difference between the core and the cladding very small. Both approaches set higher demands on the precision of the waveguide-fabrication processes.

What in many ways is a more attractive and more elegant solution to this problem is the graded-index fiber. This is really a solid-state analog to the gas lens mentioned earlier. In this fiber, the refractive index is graded radially, being highest at the axis of the core of the guide and dropping off in some predetermined way towards the circumference of the core. To a first approximation, a parabolic profile for the refractive index is the appropriate one. A mode or ray travelling at a relatively large angle to the axis is swept into a region of progressively lower refractive index. It is bent back by this gradient but for much of the time it is travelling in material with a lower refractive index and therefore has a higher velocity that the ray that travels along the axis in the region of highest refractive index. By choosing the right profile, we can keep these various rays in phase. Thus the multimode fiber with a graded index and sufficiently low optical loss became the prime objective in this search

Alan G. Chynoweth is the executive director of the Electronic Device, Process and Materials Division, Bell Laboratories, Murray Hill, New Jersey.

for an optical-fiber waveguide—a fiber-guide, as it is now called.

Initial uncertainties about glass

From the point of view of the physicist, it is rather sobering to realize how few quantitative answers he could provide to questions asked by the scientists embarking upon the search for a suitable glass technology. Sufficiently detailed, comprehensive information about such basic quantities as the refractive index and the thermal-expansion coefficient of glass as a function of its composition was lacking ten years ago. This made it very much an empirical matter of seeking the optimum combinations for core and cladding materials. It was known, of course, that unprecedented levels of purity had to be achieved in the glass to obtain the required low optical attenuation and it was also known that various transition-metal ions and water were likely to be the most troublesome impurities. But again, the amount of information in the literature on the actual absorption spectra of ions in various glassy hosts was inadequate. Often, it was not known with any great confidence what concentration levels of impurities could be tolerated in the glasses, or even the valence states in which they occurred. Furthermore, there were few analytical approaches available for measuring trace amounts of impurities in the raw materials and the glasses made from them. One of the spin-offs from this glass program has been an increase in the sophistication of related analytical chemical techniques. Neutron activation analysis has proved particularly valuable in analyzing for trace impurities and has enabled the detection of levels down to a few parts per billion.

Another question was, what is the fundamental lower limit to attenuation in a pure glass? Is this due to band-edge tail effects, such as those being postulated for amorphous semiconductors, for example, or to scattering effects from compositional and structural inhomogeneities? As it turned out, no evidence has been found in these oxide glasses that band-edge tail effects do provide such a limit of practical significance.

By combining absorption spectra obtained on bulk glasses in spectral regions where the absorption is relatively high with spectra obtained on fibers where the optical absorption is extremely low, an absorption spectrum for glass has been achieved[4] that is probably more complete than for any other material, ranging over ten orders of magnitude in absorption coefficient, and from 0.05 to 200 microns in wavelength. The absorption spectrum of a soda-lime silicate glass is shown in figure 3. But in all glasses examined, at the edge of the transmission band the actual absorption level could always be attributed to residual impurities. Further purification always resulted in a further lowering of the absorption coefficient.

Density and structure fluctuations give

BELL LABORATORIES

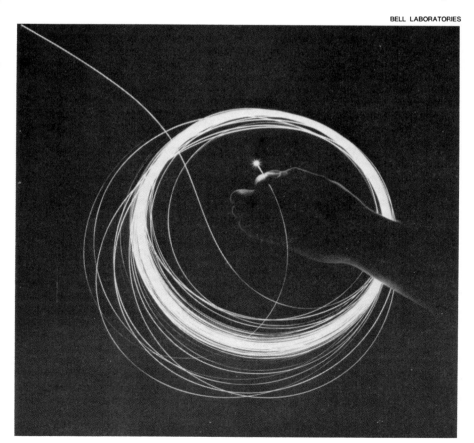

Conversation piece? Optical waveguides such as this glass fiber may carry your voice, digitally encoded, in the not-too-distant future. The fibers are strong, flexible and can be cabled together; light leakage at gentle bends is not a serious problem.

Figure 1

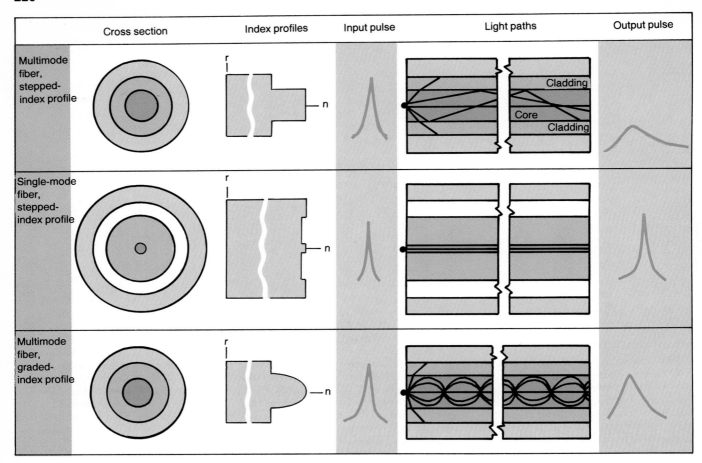

| | Cross section | Index profiles | Input pulse | Light paths | Output pulse |

Schematic representation of the basic types of optical waveguides, showing the cross sections of the glass fibers, their refractive-index profiles, the paths of typical light rays and the way light pulses spread in traversing the lengths of the fibers.   Figure 2

rise to scattering losses—again there was little prior knowledge of what the actual scattering loss levels would be in practical glasses with various compositions. Other sources of light-scattering losses could be anticipated, such as from bubbles, precipitates, cracks, separated phases and irregularities along the core–cladding interface. Then there were other questions that, it was known, would have to be answered eventually, such as: What is the mechanical strength of glass fibers that can be assured over long times and in various types of environment? In the face of all these unknowns, much of the progress had to be empirical.

Choice of glass systems

There are basically two families of glass that can be considered for fabricating optical fibers. There are the multicomponent glasses, which contain a number of oxides, and the high-silica glasses, which are basically fused silica, pure or doped. Because they melt at much lower temperatures and because they can be prepared in a wide variety of compositions, the compound glasses proved attractive to various laboratories, principally because of the belief that they offered more versatility and fewer processing problems than the high-temperature silica glasses. Major problems, however, have been the preparation of

sufficiently pure raw materials from which to make the glasses and that of keeping the glass pure during manufacture. High-purity glass manufacture has traditionally been done in platinum crucibles, and contamination from the crucible and the furnace has proved a major problem to many. Nevertheless, impressive progress with multicomponent glasses has been made, particularly by Japanese scientists with the development of their Selfoc glass fibers,[5] and more recently by scientists at the British Post Office.[6]

The original Selfoc fibers were a particularly interesting early development in the optical-fiber field in that a homogeneous glass rod was converted to the core–cladding configuration by an ion-exchange process performed by placing the glass rod into a molten salt bath. Ion exchange between the rod and the bath gave rise to a graded-composition profile and hence the radially-graded refractive index that produces the self-focussing action of multimode fibers.[6]

More recently the Japanese scientists, like those at the British Post Office, have used a double-crucible technique for forming fibers directly from the melt. Some grading of the index occurs as a result of interdiffusion during the drawing process.[7]

The first really major breakthrough

from an optical-communications viewpoint was made by scientists at Corning Glass Works when they developed fibers based on high silicas. In 1970 they announced[8] fibers with attenuations as low as 20 dB/km; subsequently they were able to improve this figure to 4 and then 2 dB/km.[9] They achieved these low-loss fibers by a flame-hydrolysis method in which silicon tetrachloride with, if desired, a suitable dopant halide, would react with oxygen to form a white silica soot deposited on a silica mandrel. Rotating the mandrel and translating the flame back and forth causes a buildup of the white soot deposit. By varying the dopant concentration during this buildup, the Corning scientists were also able to achieve a gradient in the composition. Subsequently the soot is consolidated into a clear glass by an appropriate heat treatment, after which the mandrel is removed. From this "preform," as it is called, a fiber can be drawn, those layers deposited first becoming the core and the later layers the cladding.

Meanwhile it was recognized at Bell Labs also that silica has many attractions to offer: its chemical simplicity and the low losses already known to be achievable, even in commercial types of silica. Ideally it was thought that pure fused silica should be used for the core but the problem was to find a material with a lower refractive index that could then be used as cladding. The breakthrough occurred with the discovery that boron-doped silica, namely borosilicate glass, could be prepared with a refractive index lower than that of silica alone.[10] Immediately various configurations exploiting this combination of a silica core and a borosilicate cladding were conceived. It was realized that these could be fabricated by variations of chemical-vapor deposition processes of the sort very familiar to those in semiconductor technology. The basic scheme is to start with an ordinary, commercially available quartz tube and to feed silicon tetrachloride and appropriate doping gases into the tube where they can react, at an appropriate temperature, to deposit a glass layer on the inside of the tube. Again, by varying the gas composition during a run, a radial distribution in the glass composition, and therefore its refractive index, can be achieved.

The usual chemical-vapor deposition process[11] in which an oven is placed around the quartz tube is a relatively slow process. A major step forward occurred with the development of a modified chemical-vapor deposition process[12] in which the reaction occurs in the tube where heat is applied by an external oxyhydrogen flame; this process is shown in figure 4. In this way the deposition process has been considerably speeded up and made practical—furthermore, excellent control can be exercised over the radial distribution of the glass composition.

The modified chemical-vapor deposition process has provided a considerable degree of flexibility in the design of waveguide configurations. Silica, the basic material, can be doped with boron

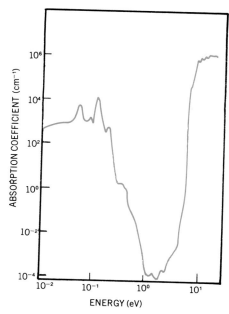

Absorption spectrum of a soda-lime silicate glass, obtained by combining measurements made on bulk glass in the high-attenuation regions with data from fibers in the low-loss regions. (From B. G. Bagley *et al*, reference 4.)
Figure 3

to *reduce* the refractive index, as mentioned already, or with germania or other compounds to *increase* the refractive index. Thus, if needs be, the inside of the tube may first be coated with borosilicate and then gradually graded towards germania-doped silicate so as to obtain a large numerical aperture. Other variations of this theme can be readily imagined.

After one of these preforms is prepared

by the modified chemical-vapor deposition method, the reactive gas supply is turned off, the temperature of the flame increased to take the tube up to the softening temperature and the tube is collapsed into a solid rod. In this way the last layers deposited become the core of the waveguide and the first layers deposited, the cladding. The original tube acts as a supporting and protecting element. It turns out that, with care, Nature is very cooperative and allows the collapse of the tube to occur very symmetrically so that the concentric geometry that is necessary for the optical fiber waveguide is retained.

Fiber pulling

After the solid rod preform has been prepared a fiber is pulled from the heated tip of the preform. A fiber several kilometers long can be pulled from most preforms; it is then wound on a drum. With a steadily and smoothly running machine and with suitable control of the various elements, the diameter of the fiber can be kept within very close tolerances. Again Nature has turned out to be surprisingly cooperative—the cross section of the fiber is simply a scaled-down version of the cross section of the very much larger preform. In a typical situation a 1-cm diameter preform is drawn down into a fiber with an outside diameter of about 100 microns.

An alternative approach to fiber pulling that has many attractive control and feedback features is the use of a high-power CO_2 gas laser as the heat source,[13] as depicted in figure 5. Silica is opaque to the 10.6-micron radiation of the laser, which thus provides a very clean and controllable heat source.

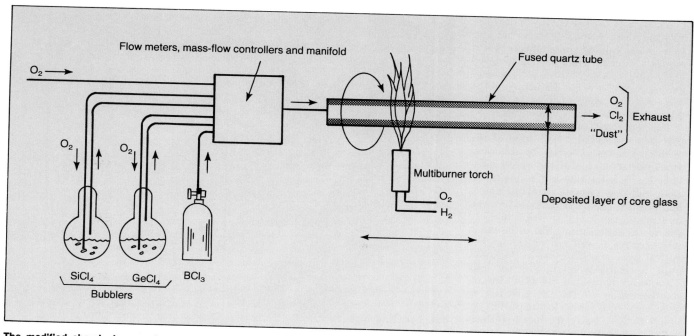

The modified chemical-vapor deposition process for preparing the preforms from which fibers are drawn. In this process, developed by J. B. MacChesney and P. B. O'Connor, the first layers deposited become the cladding and the last layers, the core.
Figure 4

One of the little understood properties of glass fibers is that their mechanical strength rapidly deteriorates if they are left unprotected. Pristine fibers, immediately after leaving the drawing head, may have tensile strengths approaching a million pounds per square inch. To preserve this high strength it is necessary to provide a protective coating, usually a polymer of some sort. This is applied, typically, by passing the fiber through a coating cup on its way to the takeup drum.

In addition, a protective plastic jacket, either loosely or tightly fitted, can be placed around the fiber by passing it through an extrusion nozzle through which the molten polymer flows.

Optical characteristics of fibers

A typical absorption curve for a fiber produced by the modified chemical-vapor deposition process is shown in figure 6.[14] It basically represents a background provided by the scattering limit on which is superimposed a small absorption peak due to the OH group. One of the attractions of the processes used for making preforms is that the raw materials are already available in ultrahigh purity for the semiconductor industry, so that other impurity-absorption peaks are usually absent. On the shorter-wavelength side of the OH band, attenuations run typically in the range 3–5 dB/km while on the longer-wavelength side attenuations as low as 1 dB/km at 1.06 microns have been obtained, and they generally fall below 2 dB/km. It is attenuations as low as these, which can now be routinely and reproducibly obtained, that have been the real breakthrough and have aroused the interest of communications-systems engineers in optical communication.

Tight control over the cross-sectional dimensions of the fibers is necessary not only to reduce waveguide losses but also to facilitate cabling and connector techniques. Much progress is being achieved in the stabilization of fiber-pulling techniques; control over the diameter and the core–cladding concentricity to within ±1% is now relatively commonplace.

The importance of being able to control precisely the radial distribution of the refractive index in order to minimize mode dispersion has already been mentioned. To eliminate mode dispersion alone calls for a nearly parabolic distribution of the refractive index. However, in general, material dispersion (akin to chromatic abberration) has to be taken into account as well. Even though the optical wavelength spread in the input pulse is relatively small, over the long distances represented by fibers material-dispersion effects can give rise to significant broadening of the pulse. To minimize these material dispersion effects calls for an optimum profile slightly different from the parabolic. Figure 7 shows a typical profile of this type. (It is worth

noting that for most glasses the material dispersion decreases at longer wavelengths and in fact may become zero between 1.3 and 1.4 microns.)

By carefully preparing a series of fibers with slightly different refractive-index profiles, the variation of the pulse dispersion has been measured as a function of the profile in a borosilicate fiber system.[15] Near the optimum profile (for the fiber compositions under study) the pulse broadening is as little as 0.17 nanosec per kilometer. This figure is a factor of about 70 times better than would be achieved by a step-profile fiber with core and cladding refractive indices corresponding to those along the axis of the graded index fiber and the periphery, respectively.

One of the big attractions of optical fiber waveguides—compared, for example, with microwave waveguides—is their flexibility. The flexibility of glass is such that the fibers can be led around equip-

ment and relatively sharp corners quite readily without breaking. However, the tendency of the optical modes to leak further into the cladding at bends introduces additional optical losses. Long constant-radius bends can be lossy but random small changes in curvature, called "microbends," are more serious in practice. These bending losses can be minimized by a judicious choice of the refractive-index difference between core and cladding, and by choosing overall fiber dimensions and supporting coatings that impart a suitable stiffness to the fiber.

Other requirements for optical fibers

From the above discussion, it is correct to draw the conclusion that as far as the optical characteristics of glass fibers are concerned, the desired transmission objectives have been largely met. But there are a number of other requirements that have to be met before fiberguides can be

A fiber is being drawn from the tip of a silica-based preform that has been heated by the mirror-focussed beam of a high-power carbon-dioxide laser. (Bell Labs photo.)   Figure 5

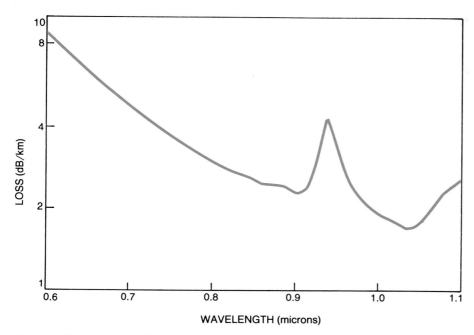

An absorption spectrum typical of glass fibers drawn from preforms that have been prepared by the modified chemical-vapor deposition process; the peak is due to the OH group.   Figure 6

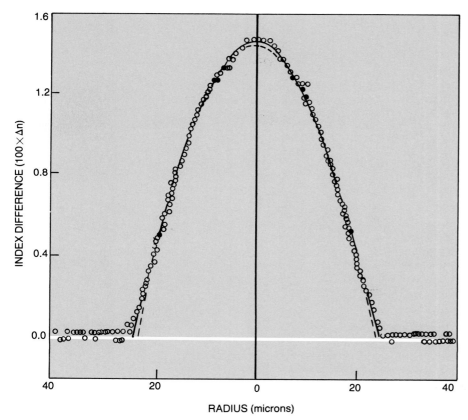

The refractive-index profile of a fiber, measured with an interference-fringe analysis technique. The nearly parabolic radial distribution minimizes mode dispersion.   Figure 7

but in practice many bare samples show strengths only a tenth of this or even much less, down to a hundredth in some cases. The high strengths generally exhibited by freshly made fibers drop off very rapidly with time unless the surface of the fiber is protected in some way. The interplay between microcracks and the surface and chemical attack by water molecules, for example, is very poorly understood at the present time. In the absence of adequate understanding of such stress–corrosion phenomena resort has yet again to be made to the empirical approach; namely, by providing protective coatings to the fibers as soon as possible after drawing. With such methods considerable improvements in the measured strengths of fibers, even after several weeks, have been obtained recently.

Given fibers that meet the desired optical and mechanical characteristics, the next step is to incorporate them into cables made up of a number of fibers. Here again new technologies are being developed. Some of the more important and largely novel features that such cables must possess are designs that:

▶ avoid subjecting the individual fibers to intolerable levels of bending;

▶ maintain all the individual fibers in some predetermined spatial registry with respect to each other so as to facilitate making splices and connections;

▶ result in small overall size so as to retain the basic advantage that optical fibers, each no thicker than a human hair, have over electrical conductor systems, and

▶ facilitate splicing sections together, when necessary, in the field.

Progress is being made, however. In some cases, splices with losses averaging about 0.1 dB can be quite routinely produced in development-laboratory facilities. Thus, with fiber optical losses of the order of 4 or 5 dB/km and splice losses of the order of 0.1 dB—bearing in mind that the total loss that can be tolerated between source and receiver for typical optical systems under development can be about 55 dB—it is clear that optical waveguides will be able to run for several kilometers before needing repeaters.

Basic questions

From the foregoing account, it is evident that dramatic progress has been made in the technology of optical waveguides. Within a few years glass technology, responding to the demands made on it by the prospect of optical-communications systems, has lowered the loss exhibited by glass from several thousand dB/km (in typical optical glass) to about 1 dB/km. At the outset of the program these objectives might have appeared almost unobtainable, but one by one the technical problems have been solved.

Much of this progress has been empirical rather than based on rigorous quantitative prediction. This contrasts, for example, with the situation now in semi-

regarded as really *practical* transmission media. Perhaps the most important of these other requirements is that of mechanical strength and stability. If cables made up of numbers of fibers are to be routinely handled and pulled through ducts, for example, they must be able to withstand the large tensile forces to which

they would be subjected in practice.

As mentioned earlier, one of the peculiarities of glass fibers, as yet little understood, is the considerable variation in the tensile strength that is observed from sample to sample. Upper limits to the strength of glass fibers appear to approach a million pounds per square inch,

conductor crystals. There the state of sophistication of our knowledge has reached such levels that the chemical compositions and composition structures required for a device to perform a desired function can often be entered into the design from first principles. This is by no means the state of affairs with glass—in many ways the current state of our fundamental knowledge of glass and its properties compares with what we knew of silicon in the early to mid 1950's.

There are very many basic questions that require answers. For example, can glasses with even better optical absorption and scattering properties be discovered? Can ways be found to reduce pulse dispersion further? Can better processes be invented for forming the glasses out of which the fibers are made? Can a higher degree of concentricity and diameter control be achieved in fiber-drawing processes? And what are the practical long-term strengths of optical fibers under various environmental conditions? What are the mechanisms of stress–corrosion cracking of glass? What materials are most suitable for protecting glass from chemical attack? How do such materials adhere to the glass surface? Finally, what are the effects of long-term exposure to low-level ionizing nuclear and cosmic radiation?

References

1. R. Kompfner, Applied Optics **11**, 2412 (1972).

2. D. Gloge, Proc. IEEE **58**, 1513 (1970).

3. K. C. Kao, G. A. Hockman, Proc. IEE (London) **113**, 1151 (1966).

4. B. G. Bagley, E. M. Vogel, W. G. French, G. A. Pasteur, J. N. Gan, J. Tauc, J. Non-Cryst. Solids, to be published.

5. K. Koizumi, Y. Ikeda, I. Kitano, M. Furukawa, T. Sumimoto, Appl. Opt. **13**, 255 (1974).

6. K. J. Beales, W. J. Duncan, G. R. Newns, in *Optical Fibre Communications,* IEE Conference Publication No. 132, page 27 (1975).

7. K. Koizumi, I. Ikeda, in *Optical Fibre Communications,* IEE Conference Publication No. 132, page 24 (1975).

8. F. P. Kapron, D. B. Keck, R. D. Maurer, Appl. Phys. Lett. **17**, 423 (1970).

9. R. D. Maurer, in *Proceedings* of the 10th International Congress on Glass, 6 (1974).

10. L. G. VanUitert, D. A. Pinnow, J. C. Williams, T. C. Rich, R. E. Jaeger, W. H. Grodkiewicz, Mater. Res. Bull. **8**, 469 (1973).

11. W. G. French, A. D. Pearson, G. W. Tasker, J. B. MacChesney, Appl. Phys. Lett. **23**, 338 (1973).

12. J. B. MacChesney, P. B. O'Connor, H. M. Presby, Proc. IEEE **62**, 1278 (1974).

13. R. E. Jaeger, Am. Ceram. Soc. Bull. **52**, 704 (1973).

14. G. W. Tasker, W. G. French, Proc. IEEE **62**, 1282 (1974).

15. L. G. Cohen, G. W. Tasker, W. G. French, J. R. Simpson, Appl. Phys. Lett., **28**, 391 (1976).   □

Lightwave telecommunication

High-speed semiconductor lasers transmit billions of bits of data per second to sensitive solid-state detectors through ultra-low-loss glass fibers more than a hundred kilometers long.

Tingye Li

PHYSICS TODAY / MAY 1985

In less than 20 years, the transmission of information by optical fibers has advanced from a mere theoretical proposal to a pervasive commercial reality. The rapid progress that has led to important applications (figure 1) has been marked by innovative breakthroughs as well as steady advances in all areas of lightwave technology.[1] The demonstration of the laser in 1960 sparked intense interest in light as a medium for transmitting information. At that time, scientists began extensive research on optical devices, techniques and subsystems for processing signals, and on a variety of transmission media such as line-of-sight atmospheric paths and beam waveguides with periodic focusing elements.[2] During the early 1960s, work on optical fibers as waveguides was mainly theoretical, as the available glass fibers exhibited transmission losses around 1000 decibels per kilometer—about two orders of magnitude too large for use in telecommunications.

In 1966, Charles Kao and George Hockham at Standard Telecommunication Laboratories in England proposed using clad glass fiber as a medium for telecommunications. Within just a few years, scientists obtained bulk silica samples with optical attenuations as low as 5 dB/km at a wavelength of 850 nm. Then came two key breakthroughs, both in 1970: the attainment at Corning of a 20-dB/km single-mode fiber and the achievement at Bell Labs of a continuous-wave semiconductor laser that operates at room tempera-

ture near 850 nm. These two developments had been earnestly sought by the lightwave research community, and served as signals to accelerate work already in progress. Soon the attenuation of 850-nm light in multimode silica fibers doped with germanium dioxide was reduced to 4 dB/km, and the operating life of the AlGaAs semiconductor laser was increased to hundreds of thousands of hours. These accomplishments, together with advances in silicon avalanche photodiodes, low-noise receivers, fiber cables, splices and connectors, led workers at Bell Labs in 1976 to test an integrated 45-megabit/sec optical fiber transmission system under simulated field conditions. Encouraged by the results of tests, communications engineers installed trial systems for interoffice trunking the following year, marking the first time that an optical-fiber transmission system carried mass commercial traffic. Many such 850-nm multimode fibers systems are now in service in telephone plants throughout the world.

Just as the 850-nm systems entered the commercial market in 1976, there were two significant research advances that had a pivotal impact on the course of lightwave technology: at Nippon Telegraph and Telephone in Japan, the reduction of fiber loss to 0.5 dB/km at 1300 nm and 1500 nm; and at Lincoln Laboratories at MIT, the demonstration of a continuous-wave room-temperature InGaAsP semiconductor laser capable of emitting in the 1000-nm to 1700-nm wavelength region. Subsequent work has brought the loss of single-mode fibers close to the limit imposed by Rayleigh scattering—below 0.2 dB/km at 1550 nm. Much of today's research on lasers, passive components

and signal-processing techniques is aimed at developing single-mode fiber systems that operate near this wavelength of lowest loss.

In this article I will look at all of the elements of lightwave telecommunication, from light sources to detectors, beginning with a discussion of the glass fibers that carry the information.

Multimode fibers

The transmission bandwidth of a multimode fiber is determined by modal dispersion, which arises because different modes have different group velocities, and material dispersion, which arises from the variation of refractive index with wavelength. A near-parabolic refractive-index profile in the fiber cross section minimizes modal dispersion, the exact profile shape depending on the operating wavelength. Material dispersion is minimum near 1300 nm for doped-silica materials. Fortuitously, a local loss minimum of about 0.4 dB/km occurs near this wavelength. Pulse spreading due to material dispersion is proportional to the spectral width of the source; thus systems based on light-emitting diodes, which have large spectral widths, operate best at this wavelength.

One measure of the quality of a fiber is the product of its bandwidth and the distance it can carry a signal before pulse spreading leads to an unacceptable error rate, which for long-distance transmission is usually taken to be 10^{-9}. With today's advanced fiber manufacturing techniques,[3] one can produce multimode fibers whose bandwidth–distance products, as limited by modal dispersion, exceed 2 GHz km. For light from an LED with a spectral width of 120 nm, material dispersion

Tingye Li is head of the lightwave systems research department at AT&T Bell Laboratories, in Holmdel, New Jersey.

Cable ship retrieving an optical-fiber line after a deep-water sea trial in 1982. An engineer is seen lovingly stroking a repeater unit. A transatlantic optical cable containing six single-mode fibers is scheduled for service in 1988. The repeaters will be powered from the shore ends and will be spaced more than 50 km apart. They will operate at 296 megabits/sec with InGaAsP lasers that emit at 1300 nm, with p–i–n InGaAs photodetectors and with state-of-the-art silicon integrated circuits. The $335-million cable will be able to carry up to 40 000 telephone conversations. (Photograph courtesy of Gordon Reinold and Peter Runge.)   Figure 1

puts an upper limit of 2.4 GHz km on the bandwidth–distance product. Edge emitters with narrower spectral widths perform better. Hence a 100-megabit/sec, or 0.1-GHz, LED-based multimode system can send signals between repeaters 20 km apart if fiber loss is of no consequence; this performance is more than adequate for most local-distribution applications. Indeed, to connect to local areas, telephone companies today are deploying large numbers of LED-based multimode fiber systems that operate from a few megabits/sec to 100 megabits/sec. Light-emitting diodes

are favored over lasers because they afford superior reliability, lower cost and much less sensitivity to temperature variations.

Single-mode fibers

Single-mode fibers have lower transmission loss than multimode fibers because their cores have less dopant and therefore less Rayleigh scattering. Figure 2 shows the loss spectrum of a typical state-of-the-art single-mode fiber.[3] Rayleigh scattering dominates the losses in the short-wavelength region, whereas infrared absorption due

to multiphonon processes presents a sharp edge above 1700 nm. The tails of these two loss processes combine to produce the minimum in the wavelength region of 1500 nm to 1600 nm. The absorption peak at 1390 nm is the first overtone of the OH stretch vibration; at this wavelength an OH concentration of about 25 parts per billion produces an excess loss of 1 dB/km. The lowest loss in a single-mode fiber reported to date is 0.16 dB/km near 1570 nm.

As the colored curve in figure 2 indicates, the conventional single-mode

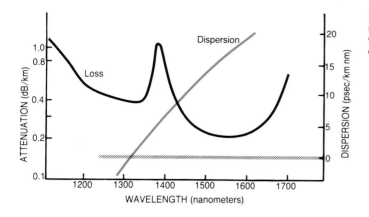

Loss and dispersion (the left and right scales, respectively) as a function of wavelength for a typical single-mode doped-silica fiber.   Figure 2

fiber with a step refractive-index profile exhibits a chromatic dispersion minimum at a wavelength λ_0 slightly above 1300 nm. Thus the transmission bandwidth of a fiber operating at the wavelength λ_0 can be very large, especially with a source having a narrow spectral width. This is illustrated in figure 3, which gives the maximum transmission bandwidth, or bit rate, of a conventional single-mode fiber, 100-km long, as a function of wavelength for sources of various spectral widths $\Delta\lambda$. The maximum transmission bandwidth of a single-mode fiber with a refractive-index profile more complicated than a conventional single-step can be quite different.

The calculations[4] that figure 3 represents take into account higher-order waveguide and material effects as well as the degradation of performance due to interference between bits, as caused by dispersion. The case labeled $\Delta\lambda = 0$ is akin to radio transmission in that the spectal width of the optical carrier is much smaller than the signal bandwidth, or bit rate B. When this condition prevails, the maximum transmission bandwidth B_{max} over a distance L satisfies the relation $(B_{max})^3 L =$ constant for wavelengths near the dispersion minimum λ_0; away from λ_0 the relation is $(B_{max})^2 L =$ constant. On the other hand, when the spectral width of the optical carrier is much greater than the bandwidth of the signal, the maximum transmission bandwidth is related to the transmission distance by $(B_{max})^2 L =$ constant near λ_0 and by $B_{max} L =$ constant away from λ_0. With these relations and figure 3, one can

scale maximum transmission distances for various bandwidths.

It is clear from figure 3 that at 1550 nm, where fiber attenuation is minimal, a source with a wide spectral width would severely limit the transmission bandwidth of a conventional single-mode fiber. The chromatic dispersion at 1550 nm is about 18 psec/km nm. Therefore, to maximize repeater spacing, it is necessary to use lasers with small linewidths or to shift the wavelength of minimum dispersion of the fiber to 1550 nm. Yasuharu Suematsu, in his article on page 32 of this issue, discusses single-frequency lasers with small linewidths.

One can modify the dispersion characteristics of a fiber by changing its refractive-index profile, and this approach has met with varying degrees of success. One quadruply clad fiber showed a very low dispersion of ± 2 psec/km nm over a broad wavelength range of 1300–1700 nm, but had higher losses than conventional single-mode fibers. Most recently, a fiber whose index of refraction decreases linearly in the radial direction in the core and has a brief increase in the cladding has its wavelength of minimum dispersion, λ_0, shifted to 1550 nm, while maintaining losses as low as those of conventional fibers.[5]

The 1450–1650 nm region of minimum loss corresponds to an immense bandwidth of 25 000 GHz. The challenge to fiber designers and fabricators is to minimize fiber dispersion in this region while maintaining low loss, and the challenge to systems and components workers is to devise practical

techniques for using this bandwidth efficiently.

Lasers and LEDs

If low-loss optical fibers opened the door to lightwave communications, then reliable, miniature semiconductor lasers and light-emitting diodes led the way to commercial applications of lightwave systems. The technology of AlGaAs lasers and LEDs has now reached maturity; tens of thousands of these devices operating in first-generation systems have accumulated an aggregate device lifetime exceeding a billion hours, proving their predicted excellent reliability.

Progress in the development of InGaAsP lasers that operate at 1300 nm to 1500 nm has been swift because much of the technology for the AlGaAs systems also applies. Device fabricators seeking InGaAsP lasers that perform well and are easy to manufacture have built and studied many structures. Figure 4 shows six of these structures. The stripe-geometry structure, used to build the first semiconductor laser that operated continuously at room temperature, is easiest to fabricate but has a high threshold current and poor modal properties. The buried-heterostructure lasers have an active region that has a wider band gap and higher refractive index, thereby giving good carrier confinement and close light guidance, leading to low threshold currents of about 20 mA. The ridge-waveguide and inverted-rib structures offer weak index guiding with correspondingly larger mode volumes and,

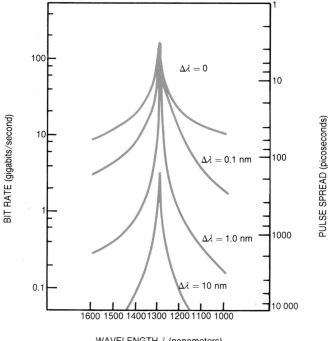

Maximum bit rates. These theoretical curves show the maximum transmission bandwidth of a conventional single-mode fiber 100-km long. The four curves are for sources of different spectral widths Δλ. The fiber's dispersion minimum is at 1300 nm.   Figure 3

hence, the potential for higher output. These structures are also easier to fabricate and can be modulated at higher speeds. Nevertheless, the etched-mesa buried-heterostructure laser is more fully developed than others, and engineers are using it in single-mode systems that are under development and deployed in the field. The double-channel planar buried-heterostructure laser has a low threshold current of about 20 mA and a high power of more than 100 mW. Its novel and effective current-confinement scheme allows it to operate continuously at temperatures above 100 °C. Because of these properties, this laser is now being used in many studies.

Much research today focuses on fundamental laser properties that influence the performance of fiber systems. Critically important questions include modulation bandwidth, single-frequency operation, chirp (the rapid variation of wavelength with time), linewidth and noise. Some examples: The amenability of a laser to high-speed direct modulation depends on its structure and the interaction of carriers and photons in its cavity; one can achieve single-frequency operation by using coupled-cavity or distributed-feedback structures; chirp, arising from variations in the refractive index of the laser medium during current modulation, can severely limit the distance between repeaters in a multigigabit/sec, single-mode fiber system; phase noise, which gives rise to spectral broadening, can degrade severely the performance of a system based on coherent modulation and detection. The articles on semicon-

ductor lasers and laser linewidth in this issue treat these and other questions of current interest in more detail.

Whereas the wavelength of interest for lasers lie in the region of 1500 nm to 1600 nm, where fiber transmission losses are least, the obvious choice for LEDs is 1300 nm, where the chromatic dispersion of multimode fibers is minimum. A surface-emitting InGaAsP LED with a monolithically formed lens for efficient coupling can launch 50–100 microwatts of light into a multimode fiber, depending on its numerical aperture and core diameter. Although LEDs are used mostly for low-speed to moderate-speed applications, their modulation bandwidths can be quite high. A group at Nippon Electric Company in Japan recently fabricated[6] a small-diameter InGaAsP LED with an active layer highly doped to decrease carrier lifetime; they used it as a directly modulated source in a 2-gigabit/sec experimental system.

Some recent calculations and experiments demonstrate the feasibility of using LEDs in applications where single-mode fibers are being installed in anticipation of future growth. Lasers at present are expensive and do not perform well in the harsh environment of a local telephone system. LEDs are much more robust, but they do not couple well to single-mode fibers because the light they emit is spatially incoherent. A surface-emitting LED ten microns in diameter radiates in about a thousand spatial modes. Assuming that a one-milliwatt LED's power is uniformly distributed among all modes, then there should be a 30-dB

coupling loss to a single-mode fiber, making about one microwatt of power available in the fiber for transmission. This power is quite sufficient for many short-distance applications. Edge-emitting LEDs have fewer modes and better coupling efficiencies. A recent experiment[7] used an edge-emitting LED to transmit a 140-megabit/sec signal over a single-mode fiber 22.5 km long.

Detectors and receivers

One needs photodetectors to convert optical signals into electrical currents for amplification and processing. The essential performance requirements are fast response, meaning large bandwidth, and good sensitivity, meaning low noise and high quantum efficiency. Two types of photodiodes are available: simple p–i–n diodes, in which photons produce carriers, or electron–hole pairs, on a one-to-one basis, and avalanche photodiodes, which feature internal gain, or carrier multiplication. Silicon is an ideal photodiode material for operation at 850 nm because its bandgap is about 900 nm, thereby allowing high quantum efficiency and fast response, with low dark current. Furthermore, its widely differing impact ionization rates for electrons and holes permit large avalanche gains with low excess noise. Designers have optimized both types of silicon photodiodes, and these devices are employed in the first-generation fiber systems. (See Hans Melchior's article on detectors for lightwave communication, PHYSICS TODAY, November 1977, page 32.)

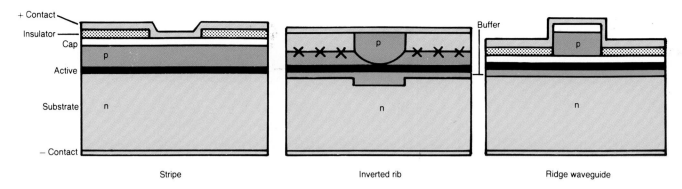

+ Contact
Insulator
Cap
p
Active
Substrate
n
— Contact

Stripe

p
×××
×××
n

Inverted rib

Buffer
p
n

Ridge waveguide

Structures of InGaAsP lasers grown on indium phosphide substrates. Various layers are labeled. The crosses indicate reverse-biased junctions, which serve to confine the current. (After a diagram by Ivan Kaminow.) Figure 4

Because silicon is transparent for wavelengths longer than 1000 nm, one must use materials with narrower bandgaps to detect light in the longer-wavelength region of 1200–1600 nm. Germanium is a good candidate, but the typical 0.1-microamp dark current in germanium devices is three orders of magnitude larger than that in silicon devices, and because carrier ionization rates in germanium are nearly equal, germanium avalanche photodiodes are noisy. In spite of these shortcomings, device designers have developed practical germanium avalanche photodiodes with sufficient sensitivity to be useful, and these detectors are now employed in high-speed fiber systems experiments and applications.

Because one can vary the bandgap of InGaAsP by varying its composition, the InGaAsP/InP system offers more flexibility for photodetector design than does germanium. Besides, the materials technology for InGaAsP is well developed for making lasers and LEDs. High-performance InGaAs p–i–n photodiodes with dark currents below 10 nanoamps are part of the second-generation fiber systems now being deployed in the field. Device researchers have fabricated p–i–n photodiodes with response times as fast as 20 picoseconds for experimental use. The realization of a useful InGaAsP avalanche photodiode, however, proved to be much more difficult. Two problems had to be solved: the high dark current due to tunneling, and the degradation of frequency response due to the accumulation of charge at the heterojunction interface. Employing a structure consisting of a narrow-bandgap InGaAs layer for absorption and a wide-bandgap InP layer for the multiplication of carriers solved the problem of dark current. Adding a graded layer of InGaAsP at the heterojunction interface solved the problem of charge

accumulation. The resulting device, known as a separate absorption grading multiplication avalanche photodiode (shown in figure 5), has worked well in experimental high-speed systems.[8] This device is currently the best-performing photodetector for multigigabit/sec experiments at 1500 nm.

High-speed, low-noise preamplifiers are as crucial as high-performance photodetectors in attaining high sensitivity in optical receivers. An accepted definition of the sensitivity of a receiver in a digital repeater is the minimum input power required to achieve an error probability of 10^{-9}. The sensitivity of an optical receiver is ultimately limited by the quantum noise, or shot noise, of the primary photogenerated signal current in the detector. This noise is due to random variations in the number of electron–hole pairs generated by identical light pulses. In practice, thermal noise from the input circuit, shot noise due to leakage currents, excess noise from the photodetector and device noise from the input transistor amplifier all contribute in varying degrees to degrade the performance of the receivers.

Figure 6 summarizes the performance of state-of-the-art optical receivers designed to operate in the spectral region of 1300–1500 nm. The ordinate is the average number of primary signal photoelectrons required in a time interval equal to the reciprocal of the bit rate to achieve an error probability of 10^{-9}, assuming an equal distribution of ones and zeros. This number is proportional to the average energy per bit, or pulse. The abscissa is the bit rate. The black curves are the result of calculations based on parameters of currently available devices, such as InGaAs p–i–n photodiodes and GaAs field-effect transistors. The colored diagonal grid lines convert the

energy per bit into power, in decibels referenced to one milliwatt, by assuming a wavelength of 1500 nm and a photodiode quantum efficiency of 0.7. The horizontal line at the bottom of the chart represents the quantum limit of 10 photoelectrons. The two solid lines with slopes proportional to the square root of the bit rate represent the theoretical performance for a receiver consisting of a p–i–n photodiode and a gallium arsenide field-effect-transistor preamplifier with no leakage current, a 50 milliamp/volt transconductance and a total capacitance of either 1 or 2 pF. The transistor's transconductance is the change in its source-to-drain current per unit change in the potential on its gate. The determining factor for the preamp's sensitivity is the field-effect-transistor's channel conductance noise, which is proportional to the square of the total capacitance divided by the transconductance. The circular dots represent the best experimental values for p–i–n receivers reported to date in the literature, while the square dots represent those for avalanche photodiode receivers.

As the figure indicates, the best sensitivity achieved is about 17 dB from the quantum limit; this is to be compared with about 13 dB for the best silicon avalanche photodiode receivers, which have lower dark current and less excess noise. Closing this performance gap remains a challenge to device researchers.

Optical repeaters. As an optical signal travels along a fiber, it is attenuated and distorted by absorption, scattering and dispersion. Hence, communication systems must use optical repeaters to restore weakened and corrupted signals to their original state. Figure 7 is a block diagram of a typical repeater for an optical fiber system. Aside from the photoreceiver and the laser transmitter, the repeater is very similar to

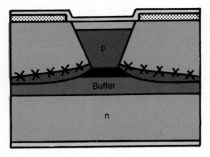

Etched mesa
buried het

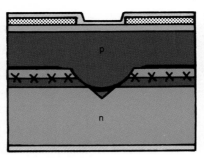

V-substrate
buried het

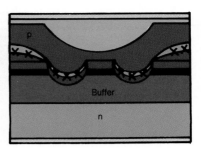

Double channel
planar buried het

those used in digital coaxial-cable systems.

To demonstrate the feasibility of transmission systems and to explore the limits of technology, research workers do experiments to see how individual components affect the performance of other components and the entire system. Recent experiments on high-speed systems found the laser to be the major factor limiting the transmission distance and speed.[9] Operating at a single frequency and reducing the laser's modulation-induced mode hopping and chirp pushed the performance limits of other components such as the receiver and the fiber.

In the 12 years since the construction of the first repeater designed specifically for optical fiber communication, engineers have tested many systems at many bit rates. A pair of recent experiments[10] using 1550-nm light achieved 4-gigabit/sec transmission over a conventional single-mode fiber more than 100-km long. One of the experiments used a directly modulated distributed-feedback laser, while the other featured a coupled-cavity laser with an external Ti-diffused LiNbO$_3$ waveguide modulator. Researchers pushing the available state-of-the-art devices and technologies to perform their best have set transmission distance records at various bit rates. Examples are 233 km at 34 megabits/sec and 117 km at 4 gigabits/sec.

Coherent transmission

One area of photonics that has aroused considerable interest recently is coherent transmission and detection,[11] in which information is coded as phase, frequency or amplitude shifts in coherent light. Optical coherent transmission was investigated in the late 1960s in connection with beam waveguides. Recent interest is motivated by the potential for increased receiver sensitivity, efficient demultiplexing of densely packed channels that are multiplexed by wavelength, and the use of simple but extremely wideband optical (laser) amplifiers to boost wavelength-multiplexed signals.

The theoretical sensitivity of a coherent-detection system can approach the quantum limit of direct detection. For example, the sensitivity limit of an ideal receiver in a coherent digital system employing phase modulation and homodyne detection is 9 photoelectrons per bit, as compared with 10 for direct detection, while that for heterodyne detection is 18 photoelectrons per bit. For a system employing frequency modulation or amplitude modulation (intensity modulation), with heterodyne detection, the sensitivity limit is 36 photoelectrons per bit.

We must solve many problems before we can approach in practice these ideal sensitivities. Foremost among the problems is laser spectral purity; the linewidths of conventional and distributed-feedback semiconductor lasers are

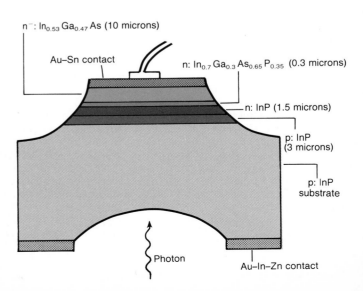

Avalanche photodiode structure. The solid-state light detector represented here is known as a "SAGM" device, because it has separate regions that absorb light, grade the heterojunction interface and multiply the photocarriers. Photons of 1200-nm to 1600-nm wavelength enter the detector through a number of transparent layers: the InP substrate, the p and n epitaxial InP multiplication layers and the InGaAsP grading layer. The relatively thick InGaAs layer absorbs the photons, and the photo-generated holes drift into the high-field region near the p–n junction, where they undergo multiplication.[8]   Figure 5

n$^-$: In$_{0.53}$Ga$_{0.47}$As (10 microns)

Au–Sn contact

n: In$_{0.7}$Ga$_{0.3}$As$_{0.65}$P$_{0.35}$ (0.3 microns)

n: InP (1.5 microns)

p: InP (3 microns)

p: InP substrate

Au–In–Zn contact

Photon

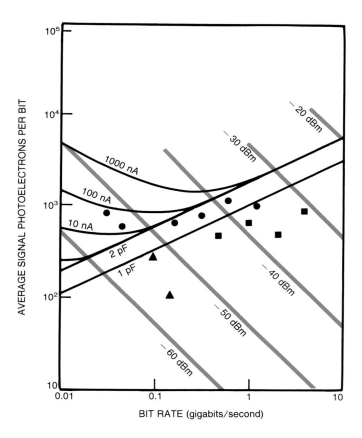

Receiver sensitivities of state-of-the-art optical repeaters, plotted as functions of bit rate. The sensitivities are expressed as the average number of signal photoelectrons per bit required to achieve an error probability of 10^{-9}. The curves are labeled with the total leakage current from the photodiode and preamplifier that follows it. The circles and squares represent tests of p–i–n and avalanche photodiode receivers, respectively. The triangles represent the results of experiments based on coherent modulation and detection.   Figure 6

very large, lying between a few MHz and a few hundred MHz. (See Aram Mooradian's article on laser linewidths, page 42.) These large spectral widths—large compared to those of gas lasers, for example—are caused by phase noise arising from fluctuations in the refractive index and from relatively high levels of spontaneous emission in the laser cavity. One can narrow the linewidth to tens of kHz by coupling a relatively long external cavity to the laser. In practice, laser linewidths need to be two to three orders of magnitude smaller than the bit rate to achieve an error probability of 10^{-9}.

Researchers interested in coherent transmission are now addressing the important issues of laser spectral purity and control, as well as temperature stabilization and wavelength tunability. Other studies include direct frequency modulation of lasers, noise in heterodyne receivers, optical phase-locked loops for homodyne detection, nonlinear effects such as Brillouin scattering, and the polarization properties of fibers and components. Polarization effects are important because coherent detection requires that the polarization state of the optical field be known at the receiver. Because the output polarization of conventional single-mode fibers tends to fluctuate randomly, one must find a way to maintain polarization in the fiber or to compensate at the receiver for changes in the polarization. Recent research indicates that one can track the polarization at the receiver.

Experimental coherent systems using helium–neon gas lasers as transmitters and semiconductor lasers as local oscillators have demonstrated receiver sensitivities within several decibels of the sensitivities that theory predicts. Figure 6 shows the results of two recent experiments using semiconductor lasers for both transmitters and local oscillators: The upper triangle represents the result of an amplitude modulation heterodyne experiment involving two distributed-feedback lasers,[12] and the lower triangle, that of a frequency modulation heterodyne experiment involving two external-cavity lasers.[13] The latter experiment achieved a repeaterless transmission distance of 200 km in a single-mode fiber.

Commercial systems

The first commercial optical-fiber transmission system went into service in 1977, operating at 45 megabits/sec with multimode fibers, AlGaAs lasers and silicon detectors. Many such systems are now carrying traffic between telephone switching offices in metropolitan areas.

Rapid advances in technology soon made these first-generation systems obsolete, as research reduced the losses in multimode fibers to below 1 dB/km at 1300-nm wavelength. Second-generation fiber systems, operating at 1300 nm with InGaAsP lasers, LEDs and detectors and low-loss multimode fiber cables, began service between central switching offices in 1981. To withstand the harsh local "subscriber loop" environment, engineers designed systems based on LEDs, which are more robust than lasers, to carry traffic between central switching offices and remote branching points for distribution to subscribers. Telephone companies are now deploying these 12- and 45-megabit/sec systems in large numbers.

The second-generation fiber systems will soon be superseded by 1300-nm, single-mode long-haul systems now under development for intercity and intercontinental applications.[14] These third-generation systems operate at several hundred megabits/sec using transmitters based on conventional multifrequency lasers and receivers built around p–i–n photodiodes and field-effect transistors or germanium avalanche photodiodes. It should be possible to upgrade these systems to speeds greater than one gigabit/sec.

Future "fourth-generation" systems will most likely be super-high-speed single-mode systems operating at the minimum-loss wavelength of 1550 nm.

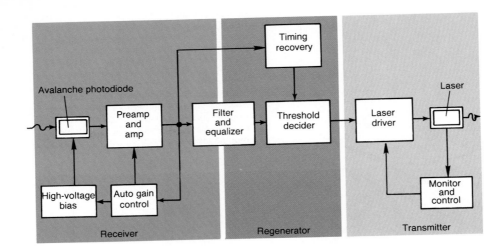

Looking ahead, optical fibers will be deployed widely in local areas for the distribution of a wide variety of broadband services; major population centers will be connected by long-haul fiber systems, and so will continents, by undersea fiber cables such as the one shown in figure 1. Multiplexing will help relieve congestion as traffic grows; this would involve increasing the bit rate or making full use of the fiber's low-loss spectral window. Within a few years, optical fibers will be ubiquitous in society's network of communication. Lightwave systems are sure to play an increasingly important role in society as the economy continues its shift toward information-based industries.

References

1. For a general review, see T. Li, IEEE J. Sel. Areas Comm. **SAC-1**, 356 (1983).

2. S. E. Miller, Science **170**, 685 (1970).

3. T. Li, ed., *Optical Fiber Communications, Vol. 1, Fiber Fabrication*, Academic, New York (1985).

4. D. Marcuse, C. Lin, IEEE J. Quantum Electron. **QE-17**, 869 (1981).

5. T. D. Croft, J. E. Ritter, H. Bhagavatula, in *Tech. Digest Conf. Optical Fiber Comm.*, OSA, Washington, D.C. (1985), paper WD2.

6. A. Suzuki, K. Minemura, H. Nomura, in *Conf. Proc. (Post Deadline Papers) Tenth Europ. Conf. Optical Comm.*, VDE-Verlag, Berlin (1984), paper PD5.

7. P. W. Shumate, J. L. Gimlett, M. Stern, M. B. Romeiser, N. K. Cheung, in *Tech. Digest (Post Deadline Papers), Conf. Optical Fiber Comm.*, OSA, Washington, D.C. (1985), paper PDO3.

8. J. C. Campbell, A. G. Dentai, W. S. Holden, B. L. Kasper, in *Tech. Digest Int. Elect. Dev. Meeting*, IEEE, New York (1983), paper 18.2.

9. R. A. Linke, in. *Tech. Digest Conf. Optical Fiber Comm.*, OSA, Washington, D.C. (1985), paper WB3.

10. A. H. Gnauck, B. L. Kasper, R. A. Linke, R. W. Dawson, T. L. Koch, T. J. Bridges, E. G. Burkhardt, R. T. Yen, D. P. Wilt, J. C. Campbell, K. C. Nelson, L. G. Cohen, in *Tech. Digest (Post Deadline Papers) Conf. Optical Fiber Comm.*, OSA, Washington, D.C. (1985), paper PD02; S. L. Korotky, G. Eisenstein, A. H. Gnauck, B. L. Kasper, J. J. Veselka, R. C. Alferness, L. L. Buhl, C. A. Burrus, T. C. D. Huo, L. W. Stulz, K. C. Nelson, L. G. Cohen, R. W. Dawson, J. C. Campbell, paper PD01.

11. T. G. Hodgkinson, D. W. Smith, R. Wyatt, D. J. Malyon, in *Tech. Digest Conf. Optical Fiber Comm.*, OSA, Washington, D.C. (1985), paper MH1.

12. M. Shikada, K. Emura, S. Fujita, M. Kitamura, M. Arai, M. Kondo, K. Minemura, Electron. Lett. **20**, 164 (1984).

13. R. Wyatt, D. W. Smith, T. G. Hodgkinson, R. A. Harmon, W. J. Devlin, Electron. Lett. **20**, 912 (1984).

14. See the special issue on undersea lightwave communications, IEEE/OSA J. Lightwave Tech. **LT-2** (December 1984).

Detectors for lightwave communication

Silicon photodiodes, particularly those of the avalanche type,
offer speed, efficiency, sensitivity and reliability—both for the systems
now in development and future ones at longer wavelengths.

Hans Melchior

PHYSICS TODAY / NOVEMBER 1977

Telecommunication with light waves today stands at the threshold of widespread use.

The successful development of optical communication started in earnest when Arthur Schawlow, Charles Townes, A. M. Prokhorov and Robert Dicke invented the laser in 1958. The initial attempts at exploitation of the immense information-carrying capacities promised by the laser were not too successful—in addition to the vagaries of the early-day lasers, the atmosphere imposed severe limitations, due to scattering and attenuation. Progress only came through significant advances in all the major components that make up a communication link: the *light sources* in the transmitter, the *transmission medium* and the *photodetectors* in the receiver. In this article we shall be concerned primarily with photodetectors, in which a signal carried by light waves is converted into an electrical signal. However, the requirements the photodetectors must meet depend strongly on the wavelength of the source and the properties of the medium.

A more suitable medium was found in the form of thin glass fibers. These have been developed into light-guiding structures with low loss and small dispersion. Semiconductor light sources for them became practical with the achievement of continuous operation of gallium aluminum arsenide lasers at room temperature over extended periods of time. The emission wavelengths of these GaAlAs sources between 0.8 and 0.9 microns closely match those spectral regions of the glass fibers in which their losses are as low as a few decibels per kilometer. Such

Hans Melchior is a professor of electronics at the Institutes of Electronics and Technical Physics at the Swiss Federal Institute of Technology, Zurich, Switzerland.

semiconductor lasers and the concurrently developed light-emitting diodes both proved well suited for lightwave communication over glass fibers. These miniature light sources can be current-modulated directly, generating optical pulses sufficiently fast to accommodate not only voice and data, but video signals as well. However, only moderate powers can be launched into the fibers—at most a few milliwatts in the case of the laser and a few tens of microwatts in the case of the light-emitting diode.

Despite these shortcomings, fiber-optical links with spans of up to several kilometers in length have been built with these sources and well designed optical receivers, and successfully tested. To these important developments in lightwave communication, PHYSICS TODAY devoted its entire May 1976 issue. Although fiber lightguides hold the spotlight, other activities in lightwave communication, particularly short atmospheric links and space applications, are being pursued actively.

Requirements

Lightwave communication links all need photodetectors and optical receivers to demodulate the optical signals and to convert them into electrical outputs. (So far the alternative of direct amplification of weak light signals in an optical repeater with an active medium is not practical.) The performance criteria for practical applications of lightwave communications put stringent requirements on both the photodetectors and the receivers. The photodetectors must have a high response at the operating wavelength, with sufficient bandwidth and response speed to accommodate the information rate. Leakage currents and shunt conductances should be low, and the optical receiver should be designed so as to minimize the

influence of disturbances on the weak signals. Even at the end of a long optical link, where the light signals are quite weak, the light detectors and the receivers should be able to detect the presence or absence of a signal with only a small margin of error.

Since its beginnings and throughout its advancements, lightwave communications has provided, through its requirements, strong incentives and guidance for the improvement and the development of photodetectors. Early interest in high-bit-rate communication at visible and infrared wavelengths, including the 1.06-micron neodymium yttrium-aluminum-garnet emission line, greatly stimulated the development of fast photodetectors[1,2] and resulted in high-speed photomultipliers[3,4] with new types of photocathodes.[5] New semiconducting photodiodes made of silicon or germanium provided the needed high speed of response[6] throughout the visible and the infrared to 1.6 microns. A novel type of solid-state photodiode, with internal gain for the photocurrent, evolved: the *avalanche photodiode*.[7]

As an illustration of the short response times of the best of these Si and Ge[8] photodiodes of both the simple junction and avalanche types, figure 1 depicts the output from a Si photodiode[6,9] connected to a 50-ohm load and excited with pulses from a dye laser. These high-speed photodiodes[9] have found numerous applications in their own right. Not of least importance has been their availability to the physicist for the investigation of fast optical phenomena such as pulse spreading and dispersion in optical fibers.

The breakthrough of communication with glass lightguides at the GaAlAs emission wavelengths of 0.8–0.9 microns brought the development of silicon photodetectors specifically adapted and optimized for this purpose.[10-12] To live up to its full potential of minimal losses in long links at the highest data rates fiber-optical communication will need a shift to longer wavelengths, between 1.1 and 1.4 microns. This awaits the development of new light sources and detectors—on which an encouraging start has already been made.

Basic types

For the detection of light we can draw upon a large number of different types of photodetectors, from a simple thermal detector, which responds to the energy of the incident radiation, to a sophisticated photon detector, which responds to the arrival rate of the photons. After an initial search and trial period, three types of photon detectors were found to be most suitable for lightwave communication at visible and near-infrared wavelengths in a room-temperature environment. They are:

▶ photomultipliers,
▶ solid-state photodiodes and
▶ avalanche photodiodes.

These types best combine good response and speed with low dark currents, negligible shunt conductance, long term stability and high sensitivity to weak light signals. The longer wavelengths in the infrared require cooled detectors as described by Henry Levinstein in PT, Nov. 1977, p. 23, or they could utilize heterodyne detectors, which even at these longer wavelengths pose difficult alignment problems between the signal and the optical local oscillator.

As their basic principle of operation, photon detectors absorb incident photons and liberate current carriers. The absorbed photons excite electrons over energy barriers to higher states where they can be separated from the holes in the lower-energy states by an electric field, giving rise to a photocurrent signal. The photocurrent I_{ph} in a photon detector is proportional to the arrival rate of the incident photons $P_{opt}/h\nu$:

$$I_{ph} = \eta \frac{e\,P_{opt}}{h\nu} \qquad (1)$$

Here e is the unit electronic charge, P_{opt} is the incident optical power, ν is the frequency and $h\nu$ the energy of the photons, and h is Planck's constant. The factor η denotes the carrier-collection (quantum) efficiency.

Over the quantum efficiency the responsivity, I_{ph}/P_{opt}, of a photon detector depends strongly on the material and on the construction of the detector. In addition, it changes considerably with the wavelength, in contrast to thermal detectors with a uniformly black surface. The responsivity of the latter is independent of wavelength over an extended range, a property that renders thermal

detectors useful for calibration and for monitoring.

It is a consequence of the wavelength dependence of the optical absorption that the responsivity of a photon detector is high only over a narrow range of wavelengths. Careful consideration therefore has to be given to the choice of detector type as well as to the detector material and its construction to achieve optimal responsivity, sufficient speed of response and good sensitivity to weak light signals at a particular wavelength of operation. Best responsivity and high quantum efficiency demands a choice of photon-detector material with a band gap (barrier energy) that is slightly lower than the energy of the photons at the longest wavelength of operation. Under these conditions the photon-excited carriers encounter the least competition from thermally activated carriers, a situation that helps achieve low dark currents and good sensitivity to weak light signals.

To meet the demands of lightwave communication, among other applications, each of the three types of photon detector has been subjected to intensive development—first towards high speed of response in the visible, and then towards high sensitivity for room-temperature operation at wavelengths in the near infrared.

Speed and efficiency

The classical photomultiplier, with its photocathode that absorbs photons and emits electrons and a dynode chain that multiplies the emitted photoelectrons through secondary emission into a much larger output current, has undergone a complete overhaul. New photocathodes with higher responsivity were introduced. These include the so-called "negative-electron-affinity cathodes,"[5] made of cesiated III–V compounds as well as similarly constructed dynodes with larger secondary-emission gain. To achieve a

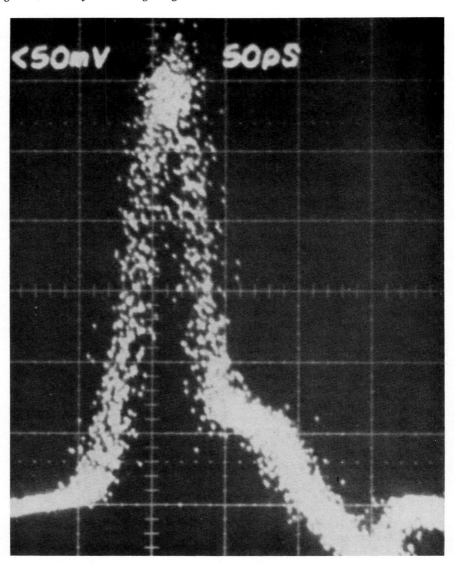

Fast response times such as this are vital to rapid information transfer in optical communications. Shown is the output of a high-speed silicon photodiode with a 50-ohm load; it is being excited by pulses from a 6150-Å dye laser (refs. 1, 9). (Courtesy of E. Ippen, Bell Labs)   Figure 1

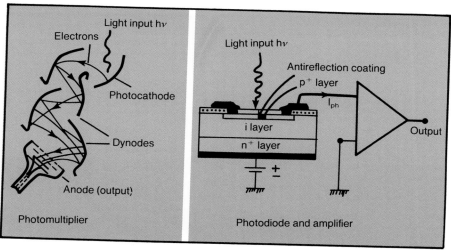

Two types of optical receivers. The photomultiplier **(left)**, with a negative-electron-affinity cathode, electrostatically focussed dynode chain and anode with coaxial 50-ohm output, has the high gain necessary for optical communications in the atmosphere and in space. The solid-state photodiode receiver **(right)**, consisting of an antireflection-coated silicon p-i-n diode connected to an amplifier, is suitable for detection of light signals from glass fibers.   Figure 2

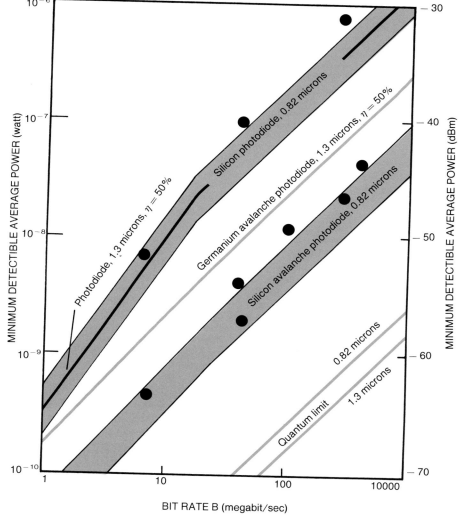

Receiver sensitivity as a function of the rate at which information arrives. The left-hand scale shows the minimum optical power needed on average to detect pulses with errors less than 10^{-9}; that on the right is in decibels above a 1-mW reference. The bands show the sensitivity of silicon photodiodes with gain (avalanche type) and without gain, at 0.82 microns and 50% duty cycle. Dots mark achieved values. Shown as well are the sensitivities expected at 1.3 microns, both for photodiodes with 50% quantum efficiency and for germanium avalanche photodiodes. Ideal receivers would reach the quantum limit, shown in color.   Figure 3

high speed of response in the subnanosecond region the time of flight of the electrons between the dynodes have had to be better controlled and equalized. In the classical structure, on the left in figure 2, only a small number of high-gain dynodes is utilized, together with electrostatic focussing to minimize time-of-flight differences,[3] resulting in short response times. In the more exotic types of photomultipliers, the dynamic and the static cross-field photomultipliers,[4] high response speeds are achieved through a tight bundling of the electrons in crossed electric and magnetic fields. Among a great number of applications these photomultipliers—with quantum efficiencies of 20% at 0.53 microns and a reported efficiency of 9% at 1.06 microns[13] (a dramatic improvement over the old S-1 photocathodes)—are of interest for short atmospheric links as well as for optical communication in space at gigabit rates.

Although photomultipliers, with their large gains of 10^4 to 10^6, are convenient and therefore utilized wherever possible in laboratory and field applications, the semiconductor photodiodes are more suitable for lightwave communication, especially in combination with optical fibers. This is because the diodes are less bulky, usually do not need high bias voltages, and are potentially much cheaper.

The solid-state photodiode[1,2,6,8] usually takes the form of a simple reverse-biased p-n junction with an open center area that is antireflection coated to give access to the incident light, as indicated schematically on the right-hand side of figure 2. Additional types of construction that are often used to advantage include p-i-n diodes with fully depleted high-resistivity i-type regions, metal–semiconductor junctions and heterojunction structures. In all these photodiodes the absorbed photons excite electrons from the valence band into the conduction band. The photon-excited holes and electrons that reach the space-charge layer are then separated by the electric field to induce a photocurrent across the junction. To achieve both high carrier collection efficiencies and short response times, these diodes are constructed so as to absorb the incident photons within or close to the space-charge layer and operated in the reverse-bias range to keep the carrier transit times and the diode capacitances low. (Carrier diffusion from outside the space-charge region is undesirable for high speeds of response.)

Optical signals generate only small photocurrents in photodiodes even when the quantum efficiencies are high. An optical signal of 1 microwatt typically generates a photocurrent of only 0.5 microamps. These small photocurrents must be amplified, either in a low-noise preamplifier or by a gain mechanism within the detector itself, before they can be handled easily by regular electronic circuitry. The solid-state equivalent of a photomultiplier therefore consists not only of a simple photodiode but of a photodiode combined with an amplifier, as indicated in figure 2 (right).

Receiver sensitivity

The performance of an optical receiver for the detection of weak light signals is ultimately limited by fluctuations at the input, the noise of the dark current or the statistical fluctuations inherent in the photodetection process. (Background noise is usually of no consequence in lightwave communication links because either it is not present or it can be blocked out easily by virtue of the spectral narrowness of the light sources.) In photodiode receivers the major sensitivity limitation is set by the amplifier noise. In the photomultiplier the situation is more advantageous. The photocurrents are multiplied within it, overcoming any disturbing influences of the output circuit, even with a 50-ohm load. As a result the sensitivity of a photomultiplier is basically limited by the Poisson statistics of the carrier-generation process and only slightly degraded by the small gain fluctuation within the dynode chain and by the dark current of the photocathode.

Compared with the simple junction photodiodes, avalanche photodiodes allow a considerable increase in the sensitivity of optical receivers. Through their internal current gain they amplify the photocurrents before they reach the amplifier. These avalanche photodiodes[1,2,7] are especially constructed to operate at high reverse-bias voltages close to the avalanche breakdown. There the photon–excited carriers gain sufficient energy to release new electron–hole pairs by ionization; these new carriers provide gain for the photocurrent.

The gain mechanism of these avalanche photodiodes is not, however, free of statistical fluctuations.[1,2,14] The mean square of the noise fluctuations of the photocurrent I_{ph} of a photodiode over bandwidth df is

$$\langle i_{ph}^2 \rangle = 2 e \langle I_{ph} \rangle \, df \qquad (2)$$

However, through the process of carrier multiplication by a factor M, it increases in avalanche photodiodes to a much higher value,

$$\langle i^2 \rangle = 2 e \langle I_{ph}M^2 \rangle \, df$$
$$= 2e \langle I_{ph} \rangle \langle M \rangle^2 F(M) df \qquad (3)$$

The noise factor $F(M) = \langle M^2 \rangle / \langle M \rangle^2$, a measure of the degradation as compared to an ideal noiseless multiplier, increases considerably[14] with the average gain $\langle M \rangle$. It depends strongly on the detector material, the device construction and the wavelength of operation.

Despite the drawback of a noise factor that increases with gain, avalanche photodiodes are used to great advantage in

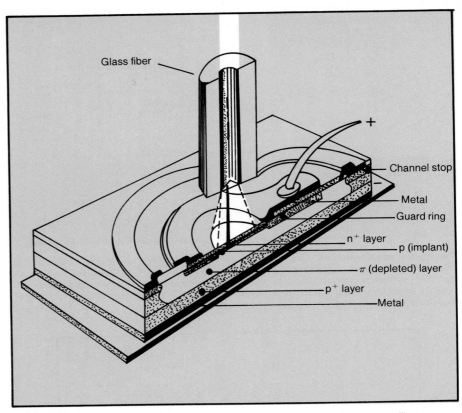

A "reach-through structure," the Si n^+-p-π-p^+ avalanche photodiode is suited for fiber communication at 0.8–0.9 microns. The electric field reaches from n^+ to p^+ layers.   Figure 4

optical receivers to overcome at least part of the noise of the amplifier. While the receiver sensitivity increases initially with current gain as long as the amplifier noise $\langle i_{ampl}^2 \rangle$ dominates, a gain value exists that optimizes performance[8] before the noise fluctuations of the avalanche diode degrade the sensitivity of the receiver again. The gain values of these avalanche photodiodes for optimal performance range typically from 10 to 200, depending on amplifier, detector and bit rates.

The sensitivity that can be reached by different types of solid-state photodiodes and avalanche photodiodes with state-of-the-art preamplifiers is illustrated in figure 3. This graph also shows the minimum average optical power that is needed to detect the presence or absence of pulses as a function of bit rate B, according to the expression

$$P_{opt,min} = \frac{h\nu}{\eta} \left\{ [\text{erfc}^{-1}(2P_E)]^2 F(M)B \right.$$
$$\left. + \frac{1}{e\langle M \rangle} \text{erfc}^{-1}(2P_E) \langle i_{ampl}^2 \rangle^{1/2} \right\} \quad (4)$$

Here the inverse complementary error function is, as usual, denoted by erfc^{-1}. The data in the figure are for a 50% duty cycle and an error margin, P_E, of 10^{-9}.

Photomultipliers reach high sensitivity throughout the visible and into the infrared, to about 0.9 microns, and in the case of cesium-oxide-activated indium, gallium, arsenic and phosphorus photo-

cathodes[13] even to 1.06 microns. Depending on the on–off ratios (duration of optical signal to that of dark current) of the photocathodes as little as a few hundred photons per bit of information suffice to keep the error margin of detection below 10^{-9}, which translates into optical powers of magnitude similar to the ones shown in figure 3 for silicon avalanche photodetectors.

For the simple high-speed silicon p-n junction photodiodes and for similarly constructed avalanche photodiodes with response times of the order of 100 picosec, the responsivity peaks in the visible and does not extend much into the infrared. The sensitivities of the infrared detectors presented in figure 3 refer to silicon p-i-n and avalanche photodiodes that are especially optimized for the 0.8–0.9-micron range and to germanium, one of the detector materials under consideration for the 1.1–1.4-micron range.

Photon detectors for 0.8–0.9 microns

Gallium-arsenide photodiodes would, if their problems with surface passivation could be solved, be well suited for the wavelength range between 0.8 and 0.9 microns, where interest in fiber-optical communication now concentrates.

Nevertheless, silicon is still the preferred detector material, not only because of its highly developed technology, but also because its avalanche carrier multi-

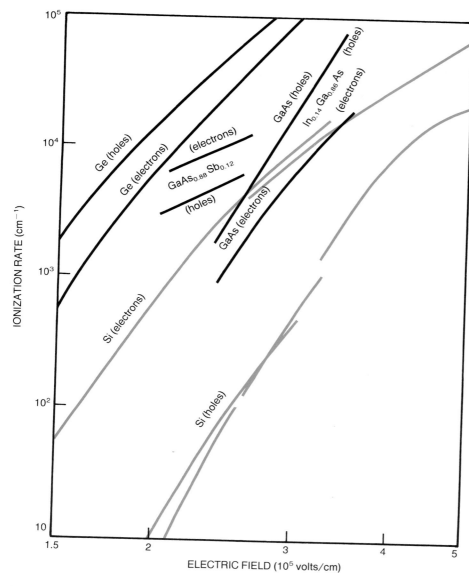

Ionization rates at 300 K for electrons and holes as a function of electric field, for silicon, germanium, GaAs, InGaAs and GaAsSb (see references 15–19 respectively).　Figure 5

tiplication is initiated. A large difference between the ionization coefficients is beneficial for low noise, provided the avalanche is initiated by the carrier type, electron or hole, with the higher ionization coefficient. Ionization coefficients, which give the probability per unit length that a carrier undergoes an ionizing (pair-producing) collision, have been experimentally found to be different for each detector material. Furthermore, they increase with the electric field, as depicted in figure 5. Silicon, the most suitable material for avalanche photodiodes, exhibits the largest difference between the ionization coefficients of electrons and holes, especially at low fields.[5]

Increased sensitivity

The n^+-p-π-p^+ reach-through structures take full advantage of the possibilities offered by silicon for almost fluctuation-free carrier multiplication. Because light is predominantly absorbed in the π region, a relatively pure electron current is injected into the high electric field of the n^+-p zone, where the carrier multiplication takes place. In the avalanche photodiodes with the lowest noise[10] the noise is only four times the shot-noise limit at an average gain of 100; in these the light enters from the back and passes through the p^+ contact into the carrier-collection region.

With front illumination through the n^+ contact, as illustrated in figure 4, a somewhat poorer noise performance results, due to the mixed injection of electrons and holes into the avalanche region. The noise penalty can, however, be kept as low as $F(M) = 5$, with $M = 100$ at 0.82 microns, through the tailoring of the gain region by means of an essentially triangular field profile. With their epitaxial layers the front-illuminated diode structures have the advantage of easier fabrication in an integrated-circuit facility.

The way the current gain increases with bias voltage in these avalanche photodiodes is demonstrated in figure 6. As the response times indicate, the useful operating range of these reach-through structures extends from the knee of the gain–voltage curve (70–100 volts) to over 300 volts. Although these operating voltages might be reduced somewhat in future devices, they basically constitute the penalty we have to pay for high efficiency, short response times and carrier multiplication with low noise fluctuations at the light wavelengths to which silicon is relatively transparent. An additional drawback of these reach-through devices is that the gain-voltage characteristics change considerably with temperature; a 1–3-volt increase in bias is needed per deg C to keep the gain constant. In most optical receivers this is of no practical consequence, because the overall amplification is controlled by using feedback to keep the signal amplitudes constant at the threshold of the regeneration circuit.

plication exhibits low noise, allowing high receiver sensitivity. At these infrared wavelengths light penetrates deep into the silicon before it is fully absorbed. It is for this reason that silicon p-i-n photodiodes (consisting of a p-doped, an intrinsic and an n-doped layer) and avalanche photodiodes with high quantum efficiencies at 0.8–0.9 microns need photon-absorption and carrier-collection regions as wide as 20–50 microns. Despite such widths, these detectors still achieve response times as short as one nanosecond, which makes them useful for bit rates in the range, up to several hundred megahertz, currently of interest in fiberguide communication.

A good number of silicon p-i-n photodiodes, offering a variety of combinations of responsivity and speed of response are available commercially; surface-passivated devices with extremely low dark currents and small areas well adapted to

the diameters of glass fibers also have been reported.[11]

Silicon avalanche photodiodes for the infrared take the form of n^+-p-π-p^+ structures,[10,11,12] often with guard rings, channel stops and field plates, as figure 4 shows schematically. (The π layer is of a basically intrinsic material that inadvertently, because of imperfect purification, retains some p doping.) These so-called "reach-through structures"—in operation the electric field reaches through from the n^+ to the p^+ region—combine optimally the desiderata of high quantum efficiency and high speed with a carrier multiplication that is relatively free of excess noise.

In an avalanche photodiode the noise factor $F(M)$ of the carrier-multiplication process depends, Robert McIntyre[14] has pointed out, both on the ratio between the ionization coefficients for electrons and for holes and on the way the carrier mul-

Both the silicon p-i-n photodiodes and the n^+-p-π-p$^+$ avalanche photodiodes with their highly optimized performance in the 0.8–0.9-micron range of the optical spectrum are now in practical use in various fiberguide communication links. The excellent sensitivity achieved by these devices at various bit rates is indicated in figure 3. As compared with simple optical receivers, the silicon receivers with silicon avalanche photodiodes are considerably more sensitive (by about 15 decibels), an improvement that allows a significant lengthening of the distance between transmitter and receiver in a fiberguide link.

Future devices

In the future fiberguide communication may well switch to longer wavelengths. As can be seen from part **a** of figure 7, both the attenuation and the dispersion of optical signals in silica fibers decrease with increasing wavelength, reaching minimal values in the spectral range between 1.1 and 1.4 microns. Communication at these wavelengths thus promises considerably longer links than the present 3–10 km and higher information rates as well.

The development of light sources for this wavelength range is already well underway; various III–V compound light-emitting diodes and semiconductor lasers, as well as miniature yttrium-aluminum-garnet solid-state lasers, have been reported to operate continuously for thousands of hours.

The detector picture is somewhat more complicated—it is not clear yet which detector material is best suited for these longer wavelengths.

The spectral response of silicon p-i-n photodiodes and of n^+-p-π-p$^+$ avalanche photodiodes[10] with sufficiently wide carrier-collection regions extends to the yttrium-aluminum-garnet wavelength of 1.06 microns. Limits set to the quantum efficiency and to the speed of response by the poor light absorption in these front- or back-illuminated structures could be relieved considerably by bringing the light in from the side. The response of silicon, however, does not extend much beyond 1.1 microns, being basically limited by the 1.12-eV energy of the bandgap.

Detectors for longer wavelengths in the range between 1.1 and 1.4 microns have to use materials with narrower bandgaps. Germanium is a good candidate.[8,21] Mixed crystals, such as indium gallium arsenide,[18,20] gallium arsenide antimonide[19] and mercury cadmium telluride,[22] are also under active investigation for this purpose.

Germanium responds, as can be seen in figure 7b, from the visible out to 1.6 microns in the infrared. A variety of small-area mesa-etched[8] and planar[21] p-n and p-i-n[6] junction photodiodes and avalanche photodiodes[8,21] have been reported, some with response times in the subnanosecond region.[9] With careful

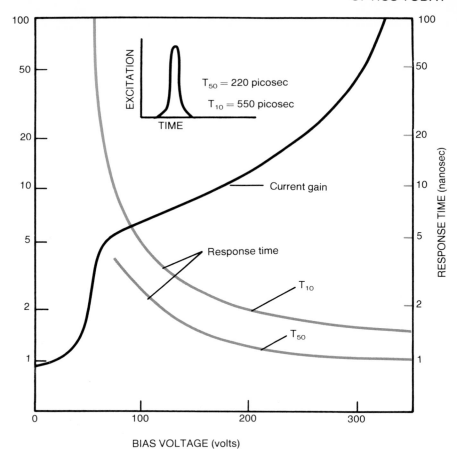

EXCITATION

T_{50} = 220 picosec

T_{10} = 550 picosec

TIME

Current gain

Response time

T_{10}

T_{50}

BIAS VOLTAGE (volts)

RESPONSE TIME (nanosec)

The gain characteristics (black) and pulse widths of the response (color) versus bias voltage for a silicon n^+-p-π-p$^+$ avalanche photodiode at a wavelength of 0.838 microns. The inset schematically indicates the widths of the light pulse at 10% and 50% of the pulse height; the colored curves show the corresponding widths for the electrical output. Figure 6

surface passivation,[21] the dark currents of these germanium devices can be as low as[2,8] 10^{-8} A. The detection of even weak optical signals therefore appears assured in a room-temperature environment. Receiver sensitivities that are possible with these germanium photodiodes and avalanche photodiodes at 1.3 microns are depicted in figure 3.

Germanium photodiodes appear to be well suited for the wavelength range of 1.1 to 1.4 microns, especially if their quantum efficiencies can be further improved through optimal antireflection coatings and shallower p$^+$ or n$^+$ junction contacts. Germanium avalanche photodiodes[8]—with their almost equal electron and hole ionization coefficients[16]—suffer from the excess noise of a less-than-ideal carrier multiplication. Nonetheless germanium avalanche photodiodes offer significant increases in sensitivity (by about 8 dB compared with photodiodes, as depicted in figure 3) to render them interesting for fiberguide communication—especially at high bit rates, where the amplifier noise tends to become more of a limiting factor.

A basic reason for the investigation of additional detector materials is in the hope of finding an avalanche photodiode for this wavelength range that has a gain process lower in noise than germanium.

An ideal detector material would be one in which only one type of carrier—either the holes or the electrons—undergo ionizing collisions. Finding such a material is a tedious job; because of the lack of theoretical guidance each material with a suitable bandgap has to be investigated experimentally. In each case it is only after the formidable materials and device-fabrication problems are overcome —to allow a spatially uniform carrier multiplication relatively free of edge breakdown and microplasma inhomogeneities and without excessive leakage currents—that meaningful measurements can be made of the ionization rates and of the noise fluctuations of the avalanche-carrier multiplication process.

These difficulties notwithstanding, appreciable avalanche gains have been observed in $In_xGa_{1-x}As$[18,20], $GaAs_{1-x}Sb_x$[19] and in $Hg_{1-x}Cd_xTe$.[22] Diodes of GaAsSb have even been utilized in experimental gigabit communication links at 1.06 microns. To the extent that ionization coefficients have been reported for these materials, they are included in figure 5. Results for $In_{0.14}Ga_{0.86}As$ and $GaAs_{1-x}Sb_x$ to date indicate that, just as in the case of GaAs,[17] the difference between the ionization coefficients for holes and electrons is somewhat larger than for germanium. Hopes for dramatic im-

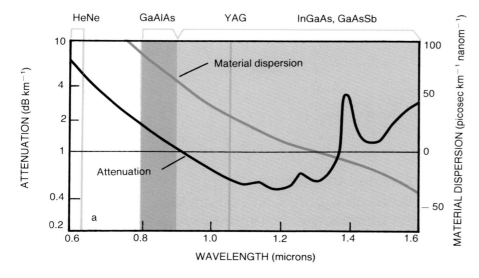

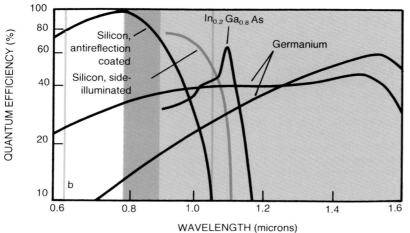

Matching the components. The wavelength dependences of (**a**) the loss spectrum and material dispersion of a low-loss fiber with a phosphosilicate core, and (**b**) the response of photodetectors are compared; the emission wavelengths of suitable light sources are shown at the top. While silicon photodiodes are well optimized for the 0.8–0.9-micron emissions of GaAlAs light sources, glass fibers have both low loss and low dispersion around 1.3 microns. This indicates a need for new sources (InGaAs, GaAsSb) as well as new detectors chosen from such materials as germanium, InGaAs, GaAsSb and HgCdTe. The fiber spectra in **a** are from M. Horiguchi, Electron. Lett. **12**, 311 (1976) and F. P. Kapron, Electron. Lett. **13**, 97 (1977). Figure 7

provements are held out by Soviet findings[23] for InAs and indium-rich InGaAs, which at 77 K show an ionization rate for holes that is much larger than for electrons (the ratio between the two approaches the values for silicon at room temperature). Avalanche photodiodes of InAs thus should exhibit low noise in their carrier multiplication at 77 K, a temperature at which the current with no illumination is low.

To help progress towards low-noise avalanche photodiodes that operate at room temperature in the 1.0–1.4-micron range it would be helpful if a theorist were to find an explanation as to why the ionization rates for electrons are so much larger than those for holes in silicon, slightly larger for holes in germanium, GaAs and $In_{0.14}Ga_{0.86}As$, apparently changing[19] with composition in GaAsSb, and so much larger for holes in indium arsenide.

Prospects look good for photodiodes in the infrared to 1.4 microns, while im-

proved avalanche photodiodes await further developments. In addition to the work on germanium, efficient devices are forthcoming from compound semiconductors the composition of which can be attuned to optimal response at a particular wavelength.[20]

Because of their low losses and small dispersion, fiber-optical communication links in the 1.1–1.4-micron range would outperform, even with existing photodiodes, the present links by factors of two or more, both in length and in information capacity, provided the cabling losses of the fibers were not excessive.

To hint at future developments, let me suggest that photodetectors will undoubtedly become closely integrated into optical-communication links: On the electrical side, the detectors will be combined with preamplifiers into densely packaged hybrids or even fully integrated circuits; on the optical side, a tighter coupling from the fibers might lead to integrated optical structures.

References

1. L. K. Anderson, B. J. Mc Murtry, Proc. IEEE **54**, 1355 (1966).

2. H. Melchior, M. B. Fisher, F. Arams, Proc. IEEE **58**, 1466 (1970); H. Melchior, J. Luminescence **7**, 390 (1973); H. Melchior, in *Laser Handbook 1*, North Holland (1972), page 725.

3. D. E. Persyk, Laser Journal **1**, 21 (1969).

4. O. L. Gaddy, D. F. Holshauser, Proc. IRE **50**, 207 (1962); **52**, 616 (1964); R. C. Miller, N. C. Wittwer, IEEE J. Quantum Electron. **1**, 49 (1965).

5. J. J. Scheer, J. Van Laar, Solid State Commun. **3**, 189 (1965); R. L. Bell, W. E. Spicer, Proc. IEEE **58**, 1738 (1970).

6. R. P. Riesz, Rev. Sci. Instr. **33**, 994 (1962).

7. R. L. Batdorf, A. G. Chynoweth, G. C. Dacey, P. W. Foy, J. Appl. Phys. **31**, 1153 (1960); K. M. Johnson, IEEE Trans. Electron Dev. **12**, 55 (1965); L. K. Anderson, P. G. McMullin, L. A. D'Asaro, A. Goetzberger, Appl. Phys. Lett. **6**, 62 (1965).

8. H. Melchior, W. T. Lynch, IEEE Trans. Electron Dev. **13**, 829 (1966).

9. H. Melchior, Verh. Deutsche Phys. Ges. **6–2**, 57 (1967).

10. H. Ruegg, IEEE Trans. Electron Dev. **14**, 239 (1967); P. P. Webb, R. J. Mc Intyre, J. Conradi, RCA Rev. **35**, 234 (1974); J. Conradi, P. P. Webb, IEEE Trans. Electron Dev. **22**, 1062 (1975).

11. H. Melchior, A. R. Hartman, in *Technical Digest* of the International Meeting on Electron Devices (Washington, DC), IEEE, New York (1976), page 412.

12. T. Kaneda, H. Matsumuoto, T. Sakurai, T. Yamaoka, J. Appl. Phys. **99**, 1151 (1976); H. Kanbe, T. Kimura, Y. Mizushima, K. Kajiyama, IEEE Trans. Electron Dev. **25**, 1337 (1976); J. Muller, A. Ataman, in volume cited in reference 11, page 416.

13. J. S. Escher, G. A. Antypas, J. Edgecumbe, Appl. Phys. Lett. **29**, 153 (1976).

14. R. J. McIntyre, IEEE Trans. Electron Dev. **13**, 164 (1966); **19**, 703 (1972).

15. C. A. Lee, R. A. Logan, J. J. Kleimack, W. Wiegman, Phys. Rev. A **134**, 716 (1964).

16. S. M. Miller, Phys. Rev. **99**, 1234 (1955).

17. G. E. Stillman, C. M. Wolfe, J. A. Rossi, A. G. Foyt, Appl. Phys. Lett. **24**, 471 (1974).

18. T. Pearsall, R. Nahory, M. Pollack, Appl. Phys. Lett. **27**, 330 (1975); T. P. Lee, C. A. Burrus, M. A. Pollack, R. E. Nahory, IEEE Trans. Electron Dev. **22**, 1062 (1975).

19. R. C. Eden, Proc. IEEE **63**, 32 (1975); F. W. Scholl, K. Nakano, R. C. Eden, in volume cited in reference 11, page 424; T. P. Pearsall, R. E. Nahory, M. A. Pollack, Appl. Phys. Lett. **28**, 403 (1976).

20. G. E. Stillman, C. M. Wolfe, A. G. Foyt, W. T. Lindley, Appl. Phys. Lett. **24**, 8 (1974).

21. J. Conradi, Appl. Optics **14**, 1948 (1975); T. Shibata, Y. Igarashi, K. Yano, E. C. L. Tech. J. NTT **22**, 1069 (1974).

22. D. A. Soderman, W. H. Pinkston, Appl. Optics **11**, 2162 (1972); T. Koehler, A. M. Chiang, Proc. Soc. Photo-Opt. Instr. Eng. **62**, 26 (1975).

23. M. P. Mikhailova, N. N. Smirnova, S. V. Slobodchikov, Sov. Phys. Semicond. **10**, 509 (1976). □

Integrated optics

Improved stability and efficiency, obtained by miniaturizing optical components and mounting them on a common substrate, will benefit the communications field and other areas of technology.

Esther M. Conwell

PHYSICS TODAY / MAY 1976

The appearance of "Integrated Optics: An Introduction" by Stewart E. Miller in the September 1969 *Bell System Technical Journal* signalled the birth of an activity that now occupies thousands of researchers. Bell Labs, among others, had been concerned with optical communications for some time, encouraged in this activity by the advent of the laser. Miller pointed out that the then typical optical telephone repeater, involving a laser, modulator, detector, lenses, and so on, spread out on an optical bench, was a form of extremely short-range radio communication and as such suffered from a number of difficulties. The apparatus was sensitive to ambient temperature gradients, to temperature changes, to mechanical vibrations of the separately mounted parts. The elegant solution to these problems proposed by Miller was to combine the separate components on the same substrate or chip, connecting them by miniature transmission lines or waveguides. Because the size of the components need only be of the order of the wavelength of light in one, and possibly two, dimensions, the substrate could be quite small—centimeters or less. He proposed calling such an assembly of components an "integrated optical circuit," in view of its analogy to the assembly of electrical components on an integrated circuit chip.

Although Miller's emphasis and that of most of the ensuing work has been on communications, it does not take much vision to realize that this type of treatment might benefit many optical systems other than those used in communications. In addition, as will be detailed subsequently, the fallout of this work should include important applications to spectroscopy and other studies of thin films.

As the term is currently used, "integrated optics" encompasses many areas. It includes guided-wave optics, or, if you like, the use of microwave techniques at optical frequencies to perform such functions as transmission, modulation, switching, mixing and upconversion. All sorts of optical and electro-optical techniques are brought together to achieve these functions. Integrated optics involves miniaturization of components such as lasers, modulators, detectors, and so on. This is a necessary step before integration can take place, although, of course, in some cases achieving the miniaturized component may be a useful end in itself. Finally, it does involve integration of different optical functions, and ultimately perhaps also different electrical functions, on the same substrate.

In long-distance transmission it is clearly desirable to use low-loss glass-fiber guides of the type discussed in the article in this issue of PHYSICS TODAY by Alan Chynoweth. For short distances, however, such as are required to connect one component of an integrated optical circuit to another, it is preferable to use guides with planar or rectangular geometry. These are easier to fabricate and have simpler field patterns. As will be seen, such guides also form the basis for most components in integrated optics. We therefore begin our discussion with a description of these waveguides. The source of the guided waves, at this evolutionary stage, is usually an external laser. In the case of a multimode glass fiber it is quite satisfactory to simply butt the laser or LED (light-emitting diode) up against the fiber, or to focus its light on the end of the fiber. It is much more difficult to get the laser beam efficiently into a planar or rectangular guide by these techniques, because the beam width is generally much larger than the confining dimension (or dimensions) of the guide. We devote a section to the methods designed for efficient coupling of a laser beam into a waveguide. Sections on modulators and

switches, devices based on periodic waveguides, detectors and miscellaneous components follow. Lasers and LED sources are for the most part covered in the article by Henry Kressel and his colleagues, although we shall mention some developments with peculiarly "integrated optics" flavor. We conclude with a discussion of the present status of integration and a view of the future.

The theory of the basic components—waveguides, couplers, switches, and so on—has numerous analogs in many fields of physics. We shall point out many of these as we go along. To supplement our necessarily brief discussion we include a bibliography of recent review articles. Where possible, we have left it to the bibliography also to provide references to (and credit for) the original work because it is not possible to do this thoroughly in these few pages.

Waveguides

Waveguides may be constructed so as to confine the light in one dimension or in two. The former case will be considered first, since it is much easier to analyze.

The simplest guide of the one-dimensional confinement type has a sandwich structure, with the center of the sandwich a thin film with height of the order of the light wavelength. The outer layers of the sandwich are conveniently referred to individually as "substrate" and "cover"

A guided wave in an eight-micron-thick film of Lexan (polycarbonate). At the left, where the wave is coupled in, through a prism, from a helium–neon laser, the film is attached to a glass substrate. The bend in the guided light occurs where the film has been pulled away from the substrate, thus showing the guided light following the film. The end of the light track coincides with the end of the Lexan film. (After J. Maher, Xerox Webster Research Center, Webster, N.Y. Figure is printed by permission of Xerox.) **Figure 1**

(or "superstrate") and collectively as the "cladding." All three layers should have low absorption for the propagating light and, for reasons that will become clear subsequently, the dielectric constant of the film must be positive and larger than those of the cladding materials. Air is the most usual cover. The thin films have been made of a large variety of materials and by a variety of fabrication techniques. They include sputtered glasses, sputtered Ta_2O_5, sputtered and epitaxial ZnO, epitaxial GaAs, epitaxial garnets, sputtered and epitaxial $LiNbO_3$, nitrobenzene liquid, nematic liquid crystals and a number of other organics and polymers.[1,2,3] An example is shown in figure 1. Planar guides have also been formed by ion bombardment, which creates a thin layer of higher index at the top of a bulk sample, the remainder then acting as substrate. This technique has been particularly useful for forming guides in GaAs where ion bombardment raises the dielectric constant to a depth of the ion range by providing traps that remove free carriers.[3]

To obtain a detailed description of the guided waves we turn to the wave equations for the electric and magnetic fields. Taking the magnetic permeability equal to unity, we write for the former[4]

$$\nabla^2 \mathbf{E} = \frac{\epsilon}{c^2} \frac{\partial^2 \mathbf{E}}{\partial t^2} \qquad (1)$$

For specificity we choose the x direction of our coordinate system as the direction in which the waves travel and the z direction as the one in which they are confined. We may then consider the sandwich infinite in the y direction, neglect the y dependence of the fields and take our solutions as plane waves, which may be written

$$\mathbf{E} = \mathbf{e}F(z) \exp[i(k_x x - \omega t)] \qquad (2)$$

where $F(z)$ represents the amplitude of the field and $\mathbf{e}$ its direction, k_x the propagation vector and ω the angular frequency. A similar equation may be written for $\mathbf{H}$. For an isotropic medium or one with cubic symmetry, it turns out that electric and magnetic field components separate into two independent groups[4]

$$\begin{array}{ll} E_y, H_x, H_z & \text{TE} \\ H_y, E_x, E_z & \text{TM} \end{array} \qquad (3)$$

As indicated, solutions involving the first set of three are called TE modes, those involving the second set of three, TM modes. Separation into TE and TM modes may also be made for waves traveling in appropriate directions in lower symmetry media, such as uniaxial crystals ($LiNbO_3$, for example). The properties of TE and TM modes are overall quite similar and in what follows we shall generally speak of TE modes for specificity.

Esther Conwell is a principal scientist at Xerox Webster Research Center, Rochester, N.Y.

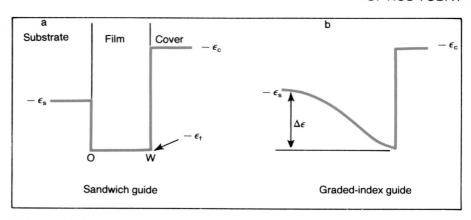

Variation of dielectric constant, ϵ, across the width of a sandwich guide (a) and a graded-index guide (b). The sandwich guide is made up of layers, the center of the sandwich being a thin film with thickness of the order of the light wavelength. The graded-index guide can be constructed by creating a thin layer of higher refractive index at the surface of a suitable single crystal or amorphous material by diffusion or by ion exchange. **Figure 2**

For TE modes, using equation 2, we may write the wave equation

$$\frac{d^2F}{dz^2} + k_0^2 \left(-\frac{k_x^2}{k_0^2} + \epsilon(z) \right) F = 0,$$
$$k_0 \equiv \omega/c \qquad (4)$$

Compare this with the one-dimensional Schrödinger equation[5]

$$\frac{d^2\psi}{dz^2} + \frac{2m}{\hbar^2} \left(E - V(z) \right) \psi = 0 \qquad (5)$$

It is clear that equation 5 is identical in form with equation 4 if $-\epsilon(z)$ is taken as "the potential" and $-k_x^2/k_0^2$ the eigenvalue. A plot of $-\epsilon(z)$ versus z for a typical sandwich guide is shown in figure 2. Because the requirement of continuity of tangential $\mathbf{E}$ and $\mathbf{H}$ at the boundaries $z = 0$ and W is, for TE modes, equivalent to the requirement of continuity of F and dF/dz at these boundaries, the problem of determining the TE modes in a sandwich guide is precisely that of determining the eigenfunctions for a particle in a square well. (For TM modes there is a slight difference because the boundary conditions are continuity of F and of $(1/\epsilon)dF/dz$.) The only difference from the usual quantum-mechanical problem is that the two sides of the well are generally of different heights. The guided modes correspond to the eigenfunctions of the particle bound in the well, decaying exponentially for $z > W$ and $z < 0$. We know then that the sandwich guide may have a number of modes, each characterized by its own value of k_x, the number being larger the larger W and the larger the difference between ϵ_f and ϵ_s ("well depth"). The modes may be numbered $m = 0, 1, 2, \ldots$, according to the number of zeroes they possess, the $m = 0$ mode having the largest k_x, $k_x^{(0)}$, $m = 1$ the next largest, $k_x^{(1)}$, and so on. Some typical modes are shown in figure 3a. A guide that will support only the $m = 0$ mode, called a "single-mode" guide, is particularly useful for some applications.

It is easily seen that there are limits to the possible values of k_x for guided modes. If equation 4 is to have an oscillatory solution inside the film, where $\epsilon(z) = \epsilon_f$, then $\epsilon_f > k_x^2/k_0^2$. If the solution is to decay exponentially in the cladding, $(k_x^2/k_0^2) > \epsilon_s, \epsilon_c$. Combining these two inequalities and using the fact that the index of refraction, n, is the square root of ϵ, we may write

$$n_f\omega/c > k_x > n_s\omega/c \qquad (6)$$

where we have assumed, as is generally the case, $n_s > n_c$. Solutions to equation 4 exist also for $k_x < n_s\omega/c$. They are oscillatory in the substrate, or both in the substrate and cover, and are called "radiation" modes.[1,6]

For many purposes it is useful to look at the guided modes from a different point of view, involving ray optics. In figure 3 the modes are shown as standing waves in the z direction. Inside the film their variation may be represented by $A_f\cos(k_z z - \phi_0)$, where, to satisfy the wave equation,

$$k_z = (k_0^2\epsilon_f - k_x^2)^{1/2} \qquad (7)$$

and A_f and ϕ_0 are determined from the boundary conditions. The standing wave could just as well be decomposed into two progressing waves, with wave vectors $k_x \pm k_z$, respectively, as indicated in figure 4. F in the film would then be given by

$$F_f = A_f\{e^{-i\phi_0}e^{ik_z z} + e^{i\phi_0}e^{-ik_z z}\}$$
$$W \geq z \geq 0 \qquad (8)$$

To represent the guided modes in the cladding we take exponentially decaying F's

$$F_c = A_c e^{-p_c(z-W)} \qquad z > W \qquad (9)$$

and

$$F_s = A_s e^{p_s z} \qquad z < 0 \qquad (10)$$

where to satisfy equation 4

$$p_c = (k_x^2 - k_0^2\epsilon_c)^{1/2}$$
$$p_s = (k_x^2 - k_0^2\epsilon_s)^{1/2} \qquad (11)$$

and the A's are to be determined from the boundary conditions. From a ray-optic point of view, the progressing waves hit the walls of the guide repeatedly, undergoing total internal reflection each time. Upon each reflection the amplitude does not change but there is a change in phase, which we may take as $-2\phi_0$ for reflection at $z = 0$, $-2\phi_W$ for reflection at $z = W$. The magnitude of ϕ_0 may be determined from the continuity requirements on F and dF/dz (that is, on, E_y and H_x) at $z = 0$. This leads to

$$\tan\phi_0 = p_s/k_z \qquad (12)$$

Similarly, from the continuity requirements at $z = W$ it must be true that

$$\tan\phi_W = p_c/k_z \qquad (13)$$

However, if we treat the situation at $z = W$ formally, using equations 8 and 9 and writing down the conditions for equality of F and dF/dz at $z = W$, we get

$$\tan(k_z W - \phi_0) = p_c/k_z \qquad (14)$$

Clearly, compatibility of equations 13 and 14 requires that

$$k_z W - \phi_0 - \phi_W = m\pi, \quad m = 0, 1, 2, \ldots \qquad (15)$$

Equation 15, the dispersion relation for the guided modes, is called the "waveguiding condition." It is readily found that m must be identified as the mode number.

If we multiply equation 15 through by the factor 2, we can see that it has a simple physical interpretation. Following the exposition of P. K. Tien,[6] we consider an observer who moves synchronously with the wave in the x direction. What he sees is the pair of plane waves $e^{\pm ik_z z}$ folding back and forth. For them to represent a guide mode, they must interfere constructively; in other words the total phase change in one round trip must be an integer, m, times 2π. Because $-2\phi_0$ and $-2\phi_W$ represent the phase changes on reflection, and $-2k_z W$ the phase change due to the path length in one round trip, equation 15 expresses precisely the condition for constructive interference.

To complete the description of the modes from the ray point of view, each mode is characterized by a bounce angle θ_m, indicated in figure 4. This angle is largest for the fundamental and gets successively smaller with increasing mode number.

Most of the materials with desirable electro-optic or nonlinear properties for active waveguides or waveguide devices are single crystals, and it is difficult to obtain high-quality thin films of such materials. Notable examples are lithium niobate and lithium tantalate. A clever way to obtain waveguiding in such a case is to create a thin layer of higher refractive index at the top surface of a suitable single crystal by diffusion or ion exchange. With lithium niobate, for example, satisfactory guides may be made by heating in vacuum to diffuse out Li_2O, or by diffusing in various metallic impurities, such as titanium. A plot of $-\epsilon$ versus z for the resulting guide looks like figure 2b. Exact solutions of equation 4 for the modes can only be obtained for the cases where ϵ varies linearly or exponentially.[1] Since ϵ usually varies slowly (that is, it changes little in the distance of one wavelength), the WKB approximation[5] may be used to obtain approximate solutions for the modes and propagation vectors. In figure 3b we show some typical modes for the case in which ϵ decreases monotonically with depth. As we see in the figure, the modes in such graded index guides differ from those in the sandwich guide in that the amplitude and spacing of the oscillations increases with depth. Also, the penetration of the modes into the substrate increases with increasing mode number.

The guides we have been discussing so far confine light in one dimension only, the z dimension. Two-dimensional confinement can be obtained with various types of channel or strip waveguides. The additional confinement is desired for a number of purposes, such as steering the guided wave around curves or obtaining greater energy density from a given light source for integrated-optics devices such as lasers or second-harmonic generators. A raised strip guide, for example, may be fabricated by starting with a planar guide, masking the desired strip region and removing the surrounding film by sputtering or etching techniques.[2] Diffusion through a mask or ion implantation may be used to produce an embedded strip guide. Obviously many variations of these two principal types of strip waveguide are possible.

Mathematical analysis of guides with two-dimensional confinement is much more complex than that for planar guides, and no exact solutions for the modes are

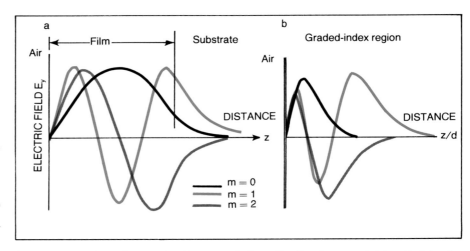

Guided modes, shown here as standing waves in the z direction, for a sandwich guide (a) and a graded-index guide (b). Electric-field component E_y is plotted as a function of distance z for the three lowest TE guided modes. In the sandwich guide ($\epsilon_f - \epsilon_c$) is much greater than ($\epsilon_f - \epsilon_s$), the typical condition for an epitaxial semiconductor film in air on a semiconductor substrate. (From H. F. Taylor, J. Vac. Sci. Technol. **11**, 150, 1974.) In the graded-index guide (b), we have $\epsilon(z) = \epsilon_s + \Delta\epsilon e^{-z/d}$; $\Delta\epsilon$ is much less than ϵ_s and ($\epsilon_s - \epsilon_c$) is much greater than $\Delta\epsilon$, as is typical for a diffused lithium-niobate guide in air. **Figure 3**

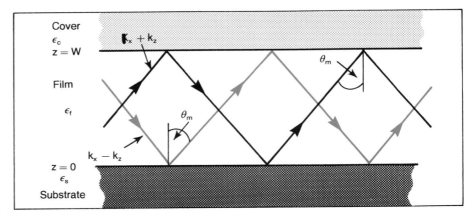

Ray picture of guided modes, shown in this figure as two progressing waves with wave vectors $k_x + k_z$ and $k_x - k_z$. The angle θ_m is given by $\tan\theta_m = k_x^{(m)}/k_z^{(m)}$. **Figure 4**

available. It is clear that when one dimension of the guide is much greater than the other the modes will differ little from the TE and TM modes of the planar guide. In any case, most phenomena are affected only quantitatively in going from a planar to a strip or channel waveguide. We refer the interested reader to the review by Herwig Kogelnik[1] for further discussion and references to the original calculations for these guides.

Couplers

The devices that convert the energy of an external light beam into a mode (or modes) propagating in a waveguide are called "beam couplers." The beam couplers in common use are prisms, gratings and tapered thin films, and we shall discuss them in that order.

As indicated earlier, it is possible to couple a laser beam into a guided mode by focussing it on the end face of the guide, but to do this with reasonable efficiency requires a well-polished, clean face and delicate adjustments. It is readily seen that it is not possible to launch a guided mode by having the laser beam incident on the top surface of the guide. Assume that the wave is traveling in air, in a direction that makes an angle θ_c with the normal to the film surface. The x component of the propagation vector in air is then $(\omega/c)n_c \sin \theta_c$. For this wave to feed the mth guide mode, with propagation vector $k_x^{(m)}$, continuity of tangential $\mathbf{E}$ and $\mathbf{H}$ at the film-air interface requires that

$$k_x^{(m)} = (\omega/c)n_f \sin\theta_m = (\omega/c)n_c \sin\theta_c \tag{16}$$

where θ_m is the angle shown in figure 4. (Equation 16 may also be thought of as the requirement of conservation of momentum.) We have already shown, however, that θ_m is larger than the critical angle. Thus $n_f \sin\theta_m > n_c$, and equation 16 cannot be satisfied, proving our statement that it is not possible to couple into a mode by simply having the laser beam incident on the top of the film.

Consider now what happens when the laser beam is incident on a prism with index of refraction $n_p > n_f$, as shown in figure 5. If the prism is not very close to the film surface the beam will be totally internally reflected at the interface between the prism and air. If, however, the prism is brought very close to the surface (say, within a wavelength of light) and the condition

$$(\omega/c)n_p \sin\theta_p = k_x^{(m)} \tag{17}$$

is satisfied, there will be a leakage across the air gap into the mth mode, as indicated in the figure. (This can be seen in figure 1.) In optical language this is frustrated total internal reflection. In quantum-mechanical language it is tunneling. As seen in the "potential diagram" of figure 5, tunneling takes place from the medium with $-\epsilon_p$ through the

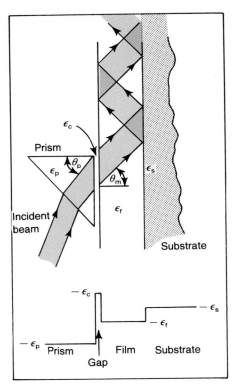

Prism coupler (above) and the corresponding "potential barrier" (below). Leakage of light across the gap can be expressed in quantum-mechanical terms as tunneling through the potential barrier. **Figure 5**

square barrier of "height" $-\epsilon_c$. The transmissivity, the ratio of the transmitted power per unit area of the boundary to the incident power per unit area, has been calculated in a number of different ways, one of the earliest involving ray optics.[6] Using the point of view stated above, we find that for TE modes it is precisely the quantum-mechanical transmissivity of a square barrier, provided allowance is made for oblique incidence of the light and the different heights of the two sides of the barrier. Thus the transmissivity depends strongly on the width of the gap, on the decay rate of the evanescent field within it (given by p_c, equation 11) and on the component of $\mathbf{k}$ in the prism normal to the "barrier" (given by equation 7 with ϵ_f replaced by ϵ_p).

Provided $n_p > n_f$, any mode of the guide can be coupled in by a prism simply by adjusting the angle θ_p so that equation 17 is satisfied for that mode. It must be noted, however, that an *input* coupler also serves to couple light *out*. Thus, in order to maximize the energy that stays in the guide, it is customary to use a rectangular prism and arrange it so that, as shown in the figure, the right edge of the laser beam is as close as possible to the rectangular corner of the prism. Nevertheless, the maximum coupling efficiency of the prism coupler is only about 80%, provided the gap is uniform and the intensity in the incoming laser beam is uniformly distributed[6] or Gaussian.[7] Better efficiency

can be achieved, for example, by decreasing the gap at the edge where the light starts coming in and increasing it where the light energy is substantial to make the "coupling-out" process relatively less important. Finally, it is clear that the prism coupler can be made 100% efficient when it is used as an output coupler, simply by making it long enough. By the reciprocity of linear optics, therefore, it should be possible to make a 100% efficient input coupler by tailoring the intensity distribution across the input beam to match the perfect output-coupler case.[6]

The prism coupler is useful as a laboratory tool, but clearly many of the potential advantages of integrated optics would be lost if it were used in integrated optical circuits. Fortunately the other two couplers mentioned are compatible with the realization of these advantages. The grating coupler is a phase grating usually made of photoresist that has been exposed to the interference pattern obtained by splitting and recombining a suitable laser beam, and then developed.[2] It is shown schematically in figure 6. Actually, the profile need not be rounded and symmetrical as shown, but may be angular and asymmetric, depending on the photoresist and the developing procedure. It is usually preferable to make the grating as thin as possible to perturb the guide modes the least. As with the prism coupler, the operation may be analyzed in many different ways.[7] One way of describing the action of a diffraction grating is to say it exchanges momentum with the light incident on it, the momentum being in units of $2\pi/d$ where d is the grating spacing. Thus the grating makes it possible to feed light into a guide mode with propagation vector $k_x^{(m)}$ provided

$$\frac{\omega}{c}\sin\theta_c + \frac{2\pi l}{d} = k_x^{(m)}, \qquad l = \pm 1, \pm 2, \ldots \tag{18}$$

By varying θ_c one should be able to couple into all of the guide modes. The grating also acts as an output coupler. Thus, just as in the case of the prism, an incident beam of uniform intensity couples in most efficiently if its boundary just intersects the end of the grating, as shown in figure 6. Also, as for the prism, the maximum coupling efficiency for an incident uniform beam is approximately 80%. The main disadvantage of the grating coupler is that a substantial portion of the incident energy may be transmitted through the film, as shown in figure 6, and lost in the substrate. Thus, grating couplers are not easily made with efficiencies as high as prism couplers. However, theory predicts that an efficiency close to 100% can be attained if the grating thickness and profile are suitably chosen.[7]

The tapered film coupler consists of a thin film that tapers down onto the substrate. A mode, guided in the film, that reaches the taper undergoes bounces with

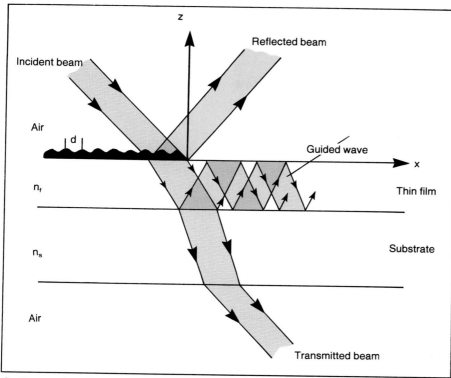

Grating coupler. The grating is usually made of photoresist that has been exposed to an optical interference pattern and then developed; it need not be as rounded and symmetrical as shown here. It is not easy to obtain an efficiency as high as with prism couplers, although theory predicts efficiencies approaching 100% with appropriate grating design. **Figure 6**

progressively smaller angle θ_m, until θ_m becomes smaller than the critical angle. Beyond that point energy is refracted into the substrate so that ultimately most of the energy is transformed into an outgoing beam. By reversing the process, one can also couple a light beam into the film by focussing it on the tapered edge through the substrate.

Periodic waveguides

Waveguides with periodic variations in the properties of the thin film, usually in width or in index of refraction, form the basis for a number of integrated optical devices. As will be seen, these devices require very small spatial periods, not greater than one micron. Width variations, or corrugations, with such periods have been made by coating a uniform waveguide with a thin layer of photoresist, exposing this to the interference pattern of an ultraviolet (He–Cd) laser, developing and then ion milling.[2,3] This process can yield a periodicity of about 0.3 micron. To obtain a smaller period, about 0.1 micron, the photoresist is exposed through a high-index prism and a layer of index-matching immersion oil.[3] Periodic width variations may also be provided by an acoustic wave propagating along the guide. Periodic index variation may be achieved, for example, by exposing to an ultraviolet interference pattern a material, such as PMMA (polymethyl methacrylate) or gelatin, whose index is permanently changed by ultraviolet ir-

radiation.

Mathematically the problem of finding the modes of the periodic waveguide is similar to that of finding the electron energy levels in a one-dimensional lattice. The guide modes may be taken as sums of Bloch–Floquet waves. The dispersion, ω versus k_x, shows bands of frequencies that are allowed to propagate in the structure separated by bands of frequencies that may not propagate, corresponding to the allowed and forbidden energy bands for electrons in a periodic potential.[8] As in the one-dimensional lattice, the gaps occur at the value of k_x given by the Bragg relation

$$k_x = \pm\pi/d \qquad (19)$$

where d is the spatial period. Waves with this value of k_x cannot propagate because they are strongly reflected, the reflections in each period or "cell" adding in phase. The size of the gap, or the discontinuity in allowed frequencies, that occurs at this k_x depends on how strongly the uniform guide is perturbed by the periodicity. For the typical situation in integrated optics, the perturbation is small. In any case, the existence of the gaps in transmission suggests the use of a periodic waveguide as a filter, and such filters have been made by corrugating a sputtered glass waveguide with the technique described earlier. They have shown reflectivities better than 75% and bandwidths less than 2Å for visible light.[1]

The fact that there is strong reflection

of a wave with k_x given by equation 19 leads to a much more important application than filters, the distributed feedback laser. It was realized early that it would be difficult to fabricate a laser suitable for an integrated optical circuit with conventional end mirrors. Kogelnik and Charles Shank were able to demonstrate a successful dye laser with feedback provided by a periodic structure.[9] Since then, distributed feedback has been successfully demonstrated in GaAs injection lasers. To satisfy the Bragg condition in this case requires $d \approx 0.1$ micron, because the laser wavelength in air is approximately 0.8 micron and the index of refraction of GaAs at this wavelength is 3.6 (we take $\sin \theta_m \simeq 1$). This periodicity can be obtained with the high-index-prism and immersion-oil technique described earlier. Distributed-feedback lasers have been found to be very stable and frequency selective, frequently oscillating in a single longitudinal mode. (For further discussion of the properties of GaAs lasers the reader is referred to the article in this issue by Kressel.) More recently, the high reflectivity of periodic structures has been found useful for making end mirrors for integrated-optics lasers.

As noted earlier, in typical cases the periodic variation may be considered a small perturbation on a uniform guide. This makes possible another type of analytical approach to such devices, the "coupled mode" formalism.[10] Because the modes of the uniform guide constitute a complete orthonormal set, the modes of the periodic or perturbed guide may be expanded in terms of them. It will be found, however, that two modes interact particularly strongly. They are the modes i and j related by

$$k_x^{(i)} - k_x^{(j)} = \frac{2\pi l}{d} \qquad l = \pm 1, \pm 2, \ldots$$

$$(20)$$

(In solid-state physics terminology equation 20 is the statement that the perturbation of the periodic lattice strongly mixes states differing by a reciprocal lattice vector.[8]) The interaction is strongest for $l = \pm 1$. Note that equation 19 is a special case of equation 20 with $l = 1$ and $k_x^{(j)} = -k_x^{(i)}$, as is appropriate for reflection. The effect of the perturbation is to transfer energy from one mode to the other (corresponding to scattering from one state to the other in the solid-state case). In the coupled-mode formalism this exchange is described by a pair of coupled first-order differential equations derived from Maxwell's equations with the aid of the assumption that the energy transfer takes place slowly. As has been pointed out by Amnon Yariv,[10] the coupled-mode approach may be used to describe many other types of phenomena—electro-optic, acousto-optic and magneto-optic modulation, second-harmonic generation, wave coupling in gratings, directional couplers, and so on.

Equation 20 suggests that a periodic deformation may also be used to convert from one mode to another within a waveguide, and such a device has been demonstrated.[7] In another type of mode converter, the periodic deformation provided by an acoustic wave was used to couple a mode out of the guide into a substrate radiation mode, that is, a mode oscillatory in the substrate.[10]

Modulation

Optical modulation is the process by which information is put on a light wave. Usually this is done by a modulator external to the light source, although in the case of semiconductor lasers (as well as LED's) direct modulation of the amplitude may be obtained simply by varying the diode current. By prebiasing the laser just below threshold, direct modulation at rates as high as one gigabit per second has been obtained. However, even for applications where this frequency is adequate, separate modulation may be preferable because the laser has a tendency to oscillate, making the modulation hard to control.[3] An external modulator operates by altering some detectable property—amplitude, phase, polarization or frequency—of the light through a change in the index of refraction. The index change may be caused by an applied electric or magnetic field or an acoustic wave, the time-variation of which represents the information to be conveyed. Switching is the process by which the spatial location of a coherent light wave is changed in response to an applied field or acoustic wave. A device may act as both a modulator and a switch. In what follows we shall take up modulators first, describing a few of the more important types. For further discussion the reader is referred to some recent review articles.[10,11,12]

A number of the modulators to be described are similar in principle to bulk modulators, the main difference being that the light is confined within a waveguide. Since for maximum efficiency the modulating fields must be applied over the volume of the light beam, the power needed for a given depth of modulation decreases as the cross-sectional area of the light beam decreases. Thus a power savings of more than one order of magnitude should be possible for one-dimensional confinement, more than two orders of magnitude for two-dimensional confinement, and such power savings have actually been realized. By a similar argument, in the case of a modulation process that depends on electric field, voltage savings of more than an order of magnitude are achievable by applying the voltage across the small dimension of the guide.

A number of the outstanding modulators for integrated optics are based on the

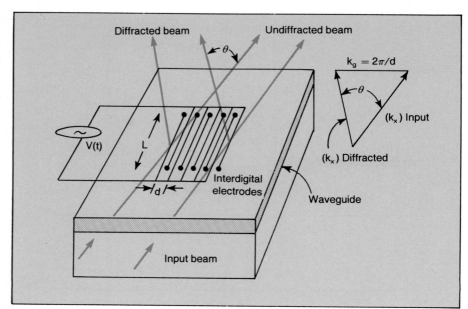

Electro-optic diffraction modulation in a thin-film waveguide. Voltage applied to the interdigital electrodes causes a periodic variation in refractive index, which deflects part of the guided beam through an angle θ. (Adapted from reference 10.) **Figure 7**

Pockels or "linear electro-optic" effect, so called because the change in index is linear in electric field. This effect occurs in crystals without a center of symmetry. Lithium niobate and GaAs are compounds with relatively strong electro-optic effects. In a typical electro-optic waveguide modulator, a linearly polarized light beam is coupled into a channel waveguide and the linearly polarized beam is decomposed into two orthogonal polarization modes, one TE and one TM. An electric field is applied across the guide through electrodes alongside it. This changes the index of refraction, and thus the phase velocity, differently for the two modes, resulting in a phase difference $\Delta\phi$ between them when they leave the guide. A second polarizer, oriented the same as the first, turns this into an amplitude-modulated output with intensity given by $I_m \cos^2(\Delta\phi/2)$, I_m being the maximum intensity. (Note that if the TE and TM modes have different phase velocity in the absence of the field, this effect also makes a contribution to $\Delta\phi$, but one that can be cancelled out by a compensator.) In a recent modulator of this type, with the channel guide made by Ti-diffusion in lithium niobate, for a phase shift of 1 radian at 0.63 micron the modulating voltage required was 0.3 volts and the power 1.7 microwatts per megahertz of bandwidth.[13]

In another promising type of modulator, voltage applied between two sets of interleaved electrodes on an electrooptic material (usually lithium niobate or tantalate) generates a periodic index-of-refraction variation or phase grating. As shown in figure 7, the top of the crystal is a planar guide, made by a suitable diffusion. Modulation occurs due to diffraction of the guided wave by the grating;

details of the diffraction process differ depending on the ratio of L to the grating spacing d.[10,11,12] For a large ratio, or a thick grating, which is the situation shown in figure 7, the situation is similar to that of x-ray (Bragg) diffraction by a crystal lattice. Conservation of momentum, expressed by the vector diagram at the right of figure 7, leads to the Bragg condition for the angle θ through which the beam is deflected

$$\sin\left(\frac{\theta}{2}\right) = \frac{\lambda_w}{2d}, \quad \lambda_w = \frac{2\pi}{|(k_x)_{\text{inc}}|} \quad (21)$$

This type of modulator, which is also a switch, has the advantage of giving amplitude modulation without the need for an additional polarizer. The intensities in the undiffracted and diffracted beams, respectively, are $I_m(\cos^2\phi/2)$ and $I_m(\sin^2\phi/2)$, ϕ being the electro-optically induced phase shift.

The Bragg grating that produces the modulation we have just discussed may also be produced by an acoustic wave, through the photoelastic effect. The modulator that results is called an acousto-optic modulator. In the version of acousto-optic modulator used in integrated optics the acoustic waves are usually surface waves rather than bulk waves, conveniently produced by interdigital transducers set on top of the waveguide. This requires that the waveguide (or possibly the substrate) be piezoelectric. To produce the grating shown in figure 7 the transducers would be out at the left- and right-hand sides. Such modulators have been realized on lithium niobate with the waveguiding layer produced by diffusion.[11]

The two mode converters mentioned in the last paragraph of the section on periodic guides also function as modulators if the amplitude of the periodic deformation, or of the acoustic wave, is controlled with an electrical signal by the techniques we have just been discussing.[10]

An important group of integrated-optics switch modulators, that (unlike most of those discussed so far) have no bulk analog, are based on the principle of directional coupling. This provides another example of coupled modes, in this case modes with the same k_x propagating in two parallel channels coupled by their evanescent fields, thus having a separation not greater than the light wavelength. As in the case of coupled pendulums, if energy is launched in one guide it will transfer gradually to the other, the transfer being complete in a distance L_0 called the "coupling length." This distance depends strongly on the separation between the guides, increasing rapidly as the separation increases. The energy will return to the first guide in the next L_0 and thereafter continue to oscillate between the guides with this characteristic period. If the k_x's of the guides differ somewhat, some energy will still be exchanged between them periodically, but the fraction exchanged decreases rapidly to zero as the difference between the k_x's increases. Assume now that we have a pair of coupled guides, covered with electrodes, which are matched so that, with no voltage applied, all the energy put into guide 1 initially will emerge from guide 2. If the material is electro-optic, a voltage applied between the two electrodes will cause the two k_x's to move apart and decrease the transfer. With enough voltage the coupling will be spoiled and energy put into guide 1 will all emerge from guide 1. For a configuration of this type, with guides formed by Ti diffusion into lithium niobate, 6 volts was required to spoil the coupling.[14] The guide widths were 2 microns, the separation between them 3 microns and the length of the interaction region 3 mm ($= 3L_0$). It is seen that the technology required to make such a switch is quite demanding. A similar switch has also been made with channel guides in GaAs defined by metal strips along them.[15] It is interesting to note that optical switching has been accomplished with 1 pico-joule, which is comparable to the energy required for semiconductor switching.[13]

Some applications to physics

Many of the techniques and phenomena described above are useful for physics applications. We take this opportunity to point out a few, and the reader will no doubt see others. One useful application is to the accurate determination of index of refraction and thickness of thin films, in particular films that can support two or more guided modes. By measuring the angles θ_p for which energy couples into (or, more conveniently, out of) a prism, with a knowledge of the prism index one can determine the $k_x^{(m)}$ with equation 17.

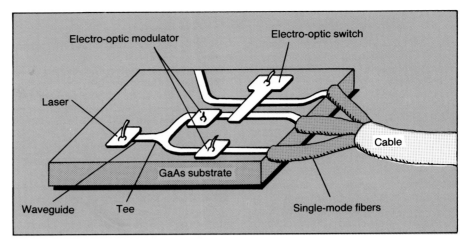

Monolithic GaAs integrated-optics transmitter. In this artist's impression, the output is shown coupled to single-mode optical fibers. (From ref. 15.)   **Figure 8**

With that information one can determine n_f and W by using the waveguiding condition (equation 15) and the various auxiliary relationships. With care, an accuracy of 0.0001 in the index can be obtained.[16] If the film or a surface of a solid has an index that varies with depth, it may be possible to determine the profile by determining the $k_x^{(m)}$ of a group of modes guided by the varying index region with a prism and using analytical techniques based on the WKB approximation.[17]

Attenuation of guided waves may be determined by monitoring the energy coupled out as a function of distance from the input coupler; therefore one may also use waveguiding to do spectroscopy in thin films. Clearly this affords a much greater path length than the thickness. In some studies of this kind attenuation has been found to be sensitive to the presence of a monolayer of organic molecules.[18] Spectroscopy of guided waves should also provide an ideal method for studying some of the excitations or modes characteristic of thin films or surfaces.

Prisms and gratings have, of course, been used by many investigators to couple surface polaritons into solids. These studies have been limited to relatively small values of k_x in the grating case because of the coarseness of the gratings. With gratings made by holographic techniques developed for integrated optics it should be possible to study these phenomena at much larger k_x values.

Integration—the future

It should be clear, even from the relatively brief presentation above, that a wide variety of miniaturized components suitable for integration, some quite sophisticated, has been developed. There have also been some cases of integration of a few components. For example, a taper-coupled semiconductor laser has been integrated with a linear electro-optic modulator.[19] The functions of deflection (produced by an acousto-optic Bragg modulator), waveguiding and detection of the guided light (by p–n junction photodetectors) have been integrated on a silicon chip.[20] The latter is an example of "hybrid" thin-film technology, so called because the different components are made of different materials, chosen presumably to optimize each function. Considerable thought has been given to monolithic integrated optics, in which all optical functions—generation, guiding, coupling, modulation/switching, detection—are obtained in a single material. The only material in which all these functions have been obtained to date is GaAs, usually alloyed with other elements of columns III and V, notably Al. Indeed, because it was clear early that GaAs is the most likely material for monolithic integrated optics, more effort has been put on it than any other material.

Since most of the work in this general area has been done with optical communications in mind, a likely early candidate for an integrated system would be an optical transmitter or receiver. A program for the development of a GaAs integrated-optical circuit transmitter, such as shown in figure 8, is being carried out at the research labs of Texas Instruments. In figure 8 the laser output is shown coupled into channel waveguides. Electro-optical modulation and switching by means of a directional coupler are performed in the different channels and controlled separately. The outputs are shown coupled to single-mode optical fibers that join to form a transmission cable. As we have seen above, the individual devices have all been made; the task now is to integrate them and couple the circuit outputs to fiber-optic cables, for which coupling, incidentally, satisfactory technology is not yet available.

Despite the availability of components, integration and the introduction of fiber cables may not proceed very rapidly. A possible exception to this is in installations for ships and planes, where the smaller weight and freedom from electromagnetic interference and radiation are important considerations. It may also be that application of integrated-optics techniques to "solid-stating" of some existing optical systems, such as laser scanners, will move ahead quickly. However, the most significant advantage this technology offers—the greater available bandwidth—is for the most part not really needed yet. When industry does get around to exploiting this bandwidth, many new services, such as TV access to stores, educational resource centers, libraries, medical centers, and so on, can be made available to office and home.

References

1. H. Kogelnik, "Theory of Dielectric Waveguides" in *Integrated Optics,* (T. Tamir ed.) Springer-Verlag, New York (1975).
2. F. Zernike, "Fabrication and Measurement of Passive Components" in *Integrated Optics* (see ref. 1).
3. E. Garmire, "Semiconductor Components for Monolithic Applications" in *Integrated Optics* (see ref. 1).
4. See, for example, M. Born, E. Wolf, *Principles of Optics,* Pergamon, New York (1970).
5. See, for example, L. I. Schiff, *Quantum Mechanics,* McGraw-Hill, New York (1949).
6. P. K. Tien, Appl. Optics **10,** 2395 (1971).
7. T. Tamir, "Beam and Waveguide Couplers" in *Integrated Optics* (see ref. 1).
8. See, for example, C. Kittel, *Quantum Theory of Solids,* Wiley, New York (1963).
9. H. Kogelnik, C. V. Shank, Jour. App. Phys. **43,** 2327 (1972).
10. H. F. Taylor, A. Yariv, Proc. IEEE **62,** 1044 (1974).
11. I. P. Kaminow, IEEE Trans. Microwave Theory and Techniques, **MTT-23,** 57 (1975).
12. J. M. Hammer, "Modulation and Switching of Light in Dielectric Waveguides" in *Integrated Optics* (see ref. 1).
13. I. P. Kaminow, L. W. Stulz, E. H. Turner, Appl. Phys. Lett. **27,** 555 (1975).
14. M. Papuchon *et al,* Appl. Phys. Lett. **27,** 289 (1975).
15. J. C. Campbell, F. A. Blum, D. W. Shaw, K. L. Lawley, Appl. Phys. Lett. **27,** 202 (1975).
16. R. Ulrich, R. Torge, Appl. Optics **12,** 2901 (1973).
17. G. B. Hocker, W. K. Burns, J. Quantum Electronics, QE-11, 270 (1975).
18. J. D. Swalen, J. Fischer, R. Santo, M. Tucke, *Technical Digest* of Integrated Optics Meeting, Salt Lake City, 1976.
19. F. K. Reinhart, R. A. Logan, Appl. Phys. Lett. **27,** 532 (1975).
20. G. B. Brandt, M. Gottlieb, G. E. Marx, *Technical Digest* of Integrated Optics Meeting, Salt Lake City, 1976. □

Observation of optical bistability confirms prediction

H. Richard Leuchtag

PHYSICS TODAY / DECEMBER 1977

An effect predicted in 1969 has been observed experimentally: An intense beam of light can cause an optical medium to pass from one state to another, opaque to transparent. In a complete cycle (opaque–transparent–opaque), the medium exhibits hysteresis.

The discovery at Bell Labs of the effect, known as optical bistability, promises to contribute significantly to the growing technology of lightwave communications (PHYSICS TODAY, May 1976 issue).

The possibility of making bistable optical devices based on saturable absorbers was proposed in 1969 by Harold Seidel of Bell Labs (who was awarded a patent on it in 1971) and by Abraham Szöke (Lawrence Livermore), V. Daneu, Julius Goldhar and Norman Kurnit (Los Alamos).

Sodium vapor was the nonlinear medium in the first successful demonstration of optical bistability, first announced in 1975, by Hyatt Gibbs, Samuel McCall and T.N.C. Venkatesan of Bell Labs, Murray Hill. In this and later experiments, the medium was placed in the resonant cavity of a Fabry–Perot interferometer, with its length adjusted so that one of its resonances is close to one of the absorption lines of the medium. For the sodium experiment this was one of the D lines, to which a single-mode dye laser was also tuned. The gas-filled resonator transmitted very little optical power until the incident beam attained a power density of about 500 mW/cm^2; then the output power jumped dramatically. Above this value, transmitted power increased linearly with beam power. On decreasing the beam power, the experimenters found that the transmitted power decreased linearly until the input beam dropped to about 400 mW/cm^2, then dropped sharply as the sodium vapor became virtually opaque again; thus the cyclic response forms a hysteresis curve.

The experiment disclosed a surprise: The apparatus functioned best when slightly detuned, demonstrating that the dominant effect was nonlinear *dispersion* (refractive index) rather than nonlinear *absorption*, as some of the theoretical predictions had assumed. In their paper,[1] Gibbs, McCall and Venkatesan developed a model of the effect encompassing both nonlinear absorption and dispersion.

The device can be operated in a binary mode, or it can be biased at a steep region of the input–output power curve to produce ac gain. The experimenters expect to develop "three-port" devices analogous to triodes, which may be driven in cascade.

Seeking to demonstrate the same effect in a solid, Venkatesan and McCall constructed a Fabry–Perot device with a ruby element as the nonlinear medium.[2] They drove the resonator with a liquid-nitrogen-cooled cw ruby laser at 6934 Å. The experimenters expected to find bistability at liquid-nitrogen but not room temperature because of the wavelength shifts of the relevant levels. They nevertheless observed bistable operation from 85 to 296 K; they explained the unexpected results as a laser-induced population redistribution between ground and excited states.

Both sodium and ruby experiments showed differential-gain, clipper and discriminator characteristics as well as bistability, "demonstrating the physical principles from which practical miniaturized optical memories . . . may be developed"[1] The authors foresee "a practical active component for integrated optical circuits."[1]

Another approach to obtaining nonlinearity is that of Peter Smith and Edward Turner, of Bell Labs's Holmdel facility. While the experiments at Murray Hill involve media with a nonlinear refractive index, this experiment is a hybrid

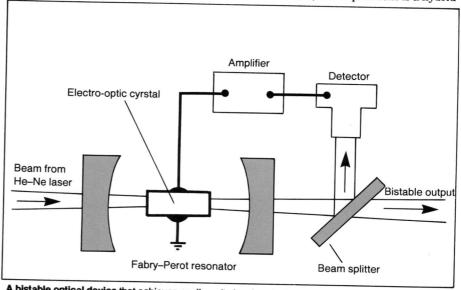

A bistable optical device that achieves nonlinearity by electrical feedback on an electro-optic crystal; the crystal is in a cavity near resonance with the laser beam. Earlier work showed bistability in purely optical systems with media of nonlinear refractive index. (From ref. 3)

approach using an electrical feedback loop. Smith and Turner placed an electro-optic crystal (of KDP, potassium dihydrogen phosphate) into the cavity of their Fabry–Perot resonator. A beam splitter sends a portion of the output beam to a detector, the amplified output of which is applied to electrodes on two faces of the crystal. The field, which varies with light intensity, modulates the refractive index of the crystal. Together with Ivan Kaminow they built an integrated version of this device using a lithium niobate crystal, which responds to incident light levels as low as 10^{-12} joules. Such a device will operate over a broad wavelength range; its switching times are limited by the response of the detector, not of the nonlinear media.

The advantages of thus creating an "artificial" nonlinearity are that
▶ the device can be made to operate at very low optical power, limited only by the sensitivity of the detector;
▶ it can be switched electrically as well as optically, and
▶ known techniques can be used to integrate the device.

In recent work with P. J. Maloney, Smith and Turner have also demonstrated "optical triode" operation as well as the possibility of multilevel optical logic and switching operations.

Many of the experimental findings were predicted in advance by theoretical models, which gave accurate values of operating parameters. In a study of electromagnetic waves in plasmas, John Marburger (University of Southern California) and Frank Felber (now at Gulf–General Atomic) realized, they told us, that a non-dissipative (dispersive) nonlinearity would be effective in producing a bistable device at about the same time the results of the Gibbs group became available.[4]

Marburger also told us of his current work with Elsa Garmire (Center for Laser Studies, USC) which resulted in their discovery of a new class of bistable devices, based on Smith's electro-optic approach, that do not require a Fabry–Perot cavity, can use multimode media and do not require precise temperature control. One concept, worked out in collaboration with Marc Levenson of USC, uses second-harmonic generation in a birefringent electro-optic material, with the phase mismatch between first and second harmonics tuned electro-optically. In a simpler, more general scheme, optical switches such as directional couplers are made into bistable devices.

A theoretical approach oriented towards incorporating bistability theory into a more quantum-mechanical picture has been pursued by Rodolfo Bonifacio and Luigi Lugiato, at the University of Milan.[6] Their steady-state analysis, made in the mean-field approximation, yields a number of striking results:
▶ The absorption of light in nonlinear media exhibits strong cooperative behavior.
▶ Even though the system is far from equilibrium, its behavior is analogous to a first-order phase transition.
▶ Critical slowing down is predicted at the phase-transition threshold.

This approach was extended to account for quantum fluctuations by Lorenzo Narducci and his co-workers Robert Gilmore and Da Hsuan Feng of Drexel University, together with Girish Agarwal (University of Hyderabad) in two papers presented at the June Conference on Coherence and Quantum Optics at Rochester. They showed that at the onset of this transition the relaxation time of one mode diverges (critical slowing down). They were also able to extend the analysis so as to relate it to catastrophe theory, demonstrating that the system behaves as a "cusp" catastrophe. The approaches of both the Milan and Drexel groups are general enough to include the more standard theory of lasers.

References

1. H. M. Gibbs, S. L. McCall, T. N. C. Venkatesan, Phys. Rev. Lett. **36**, 1135 (1976).
2. T. N. C. Venkatesan, S. L. McCall, Appl. Phys. Lett. **30**, 282 (1977).
3. P. W. Smith, E. H. Turner, Appl. Phys. Lett. **30**, 280 (1977).
4. F. S. Felber, J. H. Marburger, Appl. Phys. Lett. **28**, 731 (1976).
5. R. Bonifacio, L. A. Lugiato, Optics Commun. **19**, 172 (1976).

Sensitive fiber-optic gyroscopes

Bertram M. Schwarzschild

PHYSICS TODAY / OCTOBER 1981

Special relativity teaches us that no contrivance, mechanical or electromagnetic, confined in a closed vehicle, can detect the uniform rectilinear motion of that vehicle. But anyone who suffers from motion sickness can attest to the fact that nonuniform motion (acceleration or rotation) is a different story. Thus it is possible to direct ships, aircraft or missiles by inertial navigation systems, once an initial direction is established, without reference to magnetic compasses or radio beams.

Although Georges Sagnac demonstrated in 1913 that the rotation of a closed optical path about an axis normal to its plane changes the interference pattern of two light beams traversing the loop in opposite directions, navigational rotation sensors have until now been exclusively mechanical—spinning gyroscopes. But optical rotation sensors making use of the "Sagnac effect" are about to encroach upon this monopoly. Ring-laser gyroscopes developed by Honeywell will soon make their commercial debut aboard the new Boeing 767 and 757 airliners.

Two papers appearing in this month's *Optics Letters*[1,2] appear to promise that a second generation of optical rotation sensors—fiber-optics gyroscopes—is well on its way to achieving the sensitivity required for navigation. John Shaw and his colleagues at Stanford and an MIT group headed by Shaoul Ezekiel have constructed single-mode fiber-optic Sagnac gyroscopes with noise levels sufficiently low to permit detection of rotation rates about a hundredth of the Earth's rotation. Although this is still an order of magnitude less sensitive than what's needed for navigation (and what has already been achieved by ring-laser gyroscopes), both groups express confidence that no serious obstacles lie in the way of attaining a sensitivity of 0.01°/hour, a thousandth of "Earth rate." If this optimism is vindicated, fiber-optic gyroscopes would have a number of important practical advantages over ring-laser rotation sensors and mechanical gyros.

The Sagnac effect can be derived by a simple-minded argument that gives the Sagnac phase shift correctly to first order in v/c, the tangential velocity of the rotating optical loop divided by the speed of light. If two coherent light beams are made to travel in opposite directions around a stationary circular ring of radius R from a common source fixed on the ring, they will still be in phase when they return to the source. If we now rotate the ring and its light source with a tangential velocity v, the wavelength λ, this would result in a Sagnac phase difference $\phi_s = 8\pi^2 Rv/\lambda c$ between the two beams after a single traversal of the loop, and hence an observable rotation-sensitive interference fringe pattern. For a single loop enclosing an area A and rotating with angular velocity Ω, the phase difference is given by $8\pi A\Omega/\lambda c$, independent of the shape of the loop.

The problem is that this Sagnac phase shift is exceedingly small for modest rotation rates. Shortly after Sagnac's original demonstration on a rapidly rotating table, Albert Michelson constructed a Sagnac interferometer, using about eight kilometers of evacuated sewer pipes, to detect the Earth's rotation. Even with an interferometer of such outlandish size, the Earth's rotation produces a phase shift at only about a tenth of a fringe, the smallest shift that could be detected with the instruments then available.

Ring-laser gyroscopes. With the development of the helium–neon laser (the first continuous-wave laser) in the early 1960's, a Sagnac rotation sensor became a practical possibility. W. M. Macek and D. T. M. Davis demonstrated the first such device at Sperry in 1963. In a ring-laser gyroscope, the He–Ne laser discharge tube is an integral part of the closed optical path of the interferometer. The usual end mirrors of the gas laser are replaced by the three or four mirrors that send laser light in opposite directions around the triangular or square optical path of the cavity. At rest, this laser system resonates in two degenerate modes—a clockwise and a counterclockwise traversal of the loop at a single frequency. When the system is rotated, the Sagnac effect breaks the degeneracy, producing a small frequency difference between the clockwise and counterclockwise modes.

This frequency difference is a direct measure of the rotation rate. The ring-laser gyroscope actually measures integrated rotation rather than rotation rate, by counting interference beats beam rotating with the ring will have an optical path longer than the counter-rotating beam by a distance $4\pi Rv/c$. For monochromatic light of

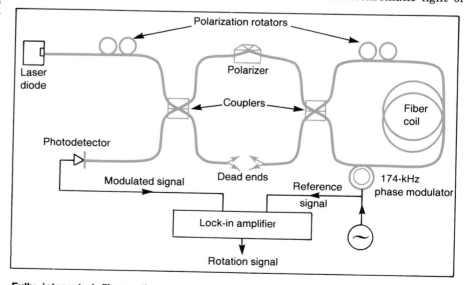

Fully integrated fiber-optic gyroscope developed at Stanford is shown schematically. Infrared light from the GaAs laser diode is split by an integrated coupler into two components, which then traverse 580 meters of single-mode optical fiber (wound around a 14-inch-diameter spool) in opposite senses. Rotation of this fiber-optic loop causes a proportional "Sagnac" phase shift between the counter-rotating beams. The resulting rotation-sensitive interference pattern is sensed by the photodiode detector. With integrated polarizer, polarization rotators and piezoelectric phase modulator, the splice-free integrated fiber-optical path avoids the reflecting interfaces that contribute to noise, achieving a rotation sensitivity of 0.1°/hour.

between the counter-rotating modes. To achieve the sensitivity necessary for navigation (10^{-3} of the Earth's rotation rate), the ring laser gyroscope must be able to detect a frequency difference of less than one part in 10^{17}—one of the most delicate measurements in all of physics, Ezekiel told us.

A problem that has plagued ring-laser gyros is "lock-in" at slow rotation rates. At such rates, where the Sagnac frequency shift is exceedingly small, the laser tends to lock the two counter-rotating modes into a single intermediate frequency. Honeywell has solved this problem by "dithering" the gyroscope, subjecting it to a continuous low-amplitude sinusoidal vibration so that it's never rotating at low rate long enough for lock-in to take hold. But dithering introduces its own problems. The mechanical vibration can generate cross-coupling of the three orthogonal gyroscope axes and other sources of error.

Fiber-optic gyroscopes. In 1976, when low-loss optical fibers had become available, Victor Valli and Richard Shorthill at the University of Utah constructed the first fiber-optic gyroscope. They were able to detect Sagnac phase shifts on a rapidly rotating table, but not until 1979 did groups at Telefunken (Ulm, Germany) and Thomson CSF (Orsay, France) succeed in getting the noise level in fiber-optic gyros down sufficiently to detect Earth rate.

If one can reduce the noise level enough to measure 10^{-3} of Earth rate, fiber-optic rotation sensors would in several respects be more attractive than mechanical or ring-laser gyroscopes. Mechanical gyroscopes suffer from acceleration sensitivity, to which the "massless" optical sensors are immune. Furthermore, mechanical gyros require a long start-up time as one waits for precession to die out, and they tend to drift as the spinning top winds down. They are also not very robust; the delicately suspended mechanical gyroscopes do not like to be turned through large angles.

Because the light source in a fiber-optic interferometer is external to the Sagnac loop, it does not suffer from lock-in, and hence requires no dithering. Fiber-optic gyroscopes are also likely to be more compact, rugged and inexpensive than ring-laser gyros. The light source of an integrated fiber-optic gyroscope would be a solid-state laser diode, or even a simple LED. Gas lasers are of course much larger and more expensive, and one has to worry about spurious phase shifts (Fizeau drag) resulting from the flow of the gas discharge. An integrated fiber-optic system does not require the careful alignment of mirrors and laser beams necessary for a ring-laser device. A fiber-optic navigational gyroscope

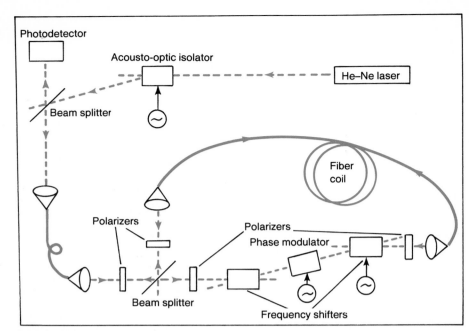

MIT fiber-optic gyroscope has achieved the same rotation sensitivity as the Stanford device, despite retaining bulk optical components such as a gas laser, half-silvered-mirror beam splitters, polarizers, electro-optic phase modulator and acousto-optic isolator. Acousto-optic Bragg cells shift the frequencies of the counter-rotating beams to restore a null interference signal when the gyroscope is rotating. The net frequency shift becomes the calibration-free measure of rotation. Colored solid lines indicate the 200 meters of fiber-optical path.

could probably be fitted into a container the size and shape of a circular box of chocolates, Shaw told us.

Whereas in a ring-laser gyroscope the counterrotating beams pass repeatedly around the Sagnac loop that serves as the laser resonator, the light beams traverse a fiber-optic gyroscope only once before being made to interfere at a detector. But this fiber-optic Sagnac path can be made thousands of meters long by winding many turns of fiber around a circular spool less than a foot in diameter. Much enthusiasm was generated at a 1978 San Diego conference on optical rotation sensors, organized by Ezekiel and George Knausenberg (Air Force Office of Scientific Research), when a number of photon-noise-limit calculations were presented, indicating that fiber-optic sensors could in principle be made at least as rotation sensitive as ring-laser gyros.

In the afterglow of the San Diego conference, fiber-optic gyroscope experiments were undertaken in a number of laboratories in the United States and Europe. But the results were disappointing. Observed noise levels turned out to be much higher than the optimistic predictions presented at San Diego. The most sensitive devices constructed in the next two years were produced by Hervé Arditty and his colleagues at Thomson CSF and by a collaborative effort of the University of Hamburg and Telefunken, led by Reinhold Ulrich. They succeeded in reducing noise levels by two to three orders of magnitude, but their devices were still

too noisy to measure rotation rates much slower than Earth rate ($15°$/hour). The primary noise sources appeared to be scattering at mirrors and other interfaces in the system, and coherent Rayleigh backscattering in the optical fibers.

Progress at Stanford and MIT. To deal with these problems, Shaw and his Stanford colleagues, Ralph Bergh and Hervé Lefevre, have now produced a fully integrated single-mode fiber-optic gyroscope. In place of the half-silvered-mirror beam splitters, polarizers and other bulk optical elements that produce undesirable interface scattering, they have developed integrated fiber-optic couplers, polarizers and polarization rotators. In fact, the entire 580-meter fiber-optical path of the Stanford gyroscope (most of it wound on a spool of radius 7 cm) has only a single interface—a splice joining the GaAs laser-diode light source to the rest of the system.

The infrared diode-laser light is split by an integrated coupler into two counter-rotating components that then circle the spool more than a thousand times in opposite directions before returning to the coupler to be reunited and diverted to a silicon photodiode interference detector. Before and after circling the spool, the light passes through an integrated polarizer to insure that one is observing the interference of only a single polarization mode. (This is required to guarantee the identity of the clockwise and counterclockwise optical paths.) A segment of the fiber is bonded to a piezoelectric cylinder that imposes a 174-kHz modulation on the light signal. One determines the rotation-induced Sagnac phase angle by measuring the amplitude of the first harmonic of this modulation signal in the interferometer's output. This gets around the problem that small phase shifts only slightly displaced from an interference maximum are difficult to measure precisely.

Most fibers used in optical communication systems are "multi-mode fibers." That is to say their cores—with diameters on the order of a hundred microns—permit the transit of many spatial modes at a given optical frequency. When one wants to cut off the propagation of higher modes in a microwave guide, one reduces the cross-sectional area of the wave guide. Similarly in fiber optics, if one makes the core small enough, it will transmit only the lowest electromagnetic mode at a particular frequency. With a core diameter of 8 microns, the optical fiber used in the Stanford gyroscope is a "single-mode fiber," transmitting only the lowest mode at the infrared frequency of the diode laser.

It is important to use single-mode fibers in a gyroscope because different modes have effectively different path lengths and propagation velocities. A fiber that supports multiple modes tends to degrade the coherence necessary for a clear Sagnac rotation signal.

Coherent Rayleigh backscattering light off the disordered arrangement of atoms in the fiber is another serious contributor of noise. To reduce this noise, one wants to *weaken* the coherence between backscattered light and the light source. This can be accomplished by reducing the temporal coherence of the source. A solid-state laser diode, with its broader natural bandwidth, is intrinsically less coherent than a gas laser. Thus Shaw and his colleagues found earlier this year that the noise level in their fiber-optic gyroscope fell dramatically when they replaced the original He–Ne laser source with a GaAs diode laser.

With these innovations, the Stanford group reports that they have now achieved a reduction of the rms noise level on the Sagnac phase angle to

1×10^{-6} radians, with measurements averaged over 30 seconds. This gives them a rotation sensitivity of 0.1°/hour—about 10^{-2} of Earth rate. For a given phase noise level, rotation sensitivity increases as the product of the total fiber length and the spool diameter. In the absence of unforseen noise increases coming with increased fiber length, Shaw expects soon to be able to increase the sensitivity of his instrument to the 10^{-3} of Earth rate required for navigation, by winding more turns of optical fiber around the spool and reducing the noise level on the phase angle still further.

Ezekiel and James Davis at MIT, on the other hand, have managed to construct a fiber-optic gyroscope of comparable noise level and rotation sensitivity without for the moment going to the full integration of the Stanford device. The nonintegrated bulk elements of their system include a He–Ne laser, polarizers, mirrors, acousto-optic frequency shifters and an electro-optic phase modulator.

The temporal coherence of the gas laser is artifically degraded by vibrating its mirrors. Rayleigh scattering noise is further reduced by the MIT technique for measuring the Sagnac phase shift. In addition to the phase modulation used at Stanford, the MIT instrument shifts the frequencies of the counter-rotating beams with Bragg-cell acousto-optic frequency shifters until the interference returns to its zero-rotation value. The frequency shift required to get this null reading is then the measure of the rotation rate.

This frequency shifting is done primarily to obtain a calibration-free rotation measurement and to render the Sagnac phase-angle measurement immune to the random intensity fluctuations to which measurements of the first modulation harmonic are sensitive. But it also reduces the noise effect of Rayleigh scattering; the counter-rotating beams are now at slightly different frequencies, raising the frequency of the Rayleigh interference noise to higher frequencies that are largely filtered out. This effect, plus the vibrational broadening of the laser output, also helps get rid of interference noise from scattering at the numerous

optical interfaces of the MIT system.

With 200 meters of single-mode fiber wound around a spool of 9.5-cm radius, Davis and Ezekiel report an rms rotation noise level close to 0.1°/hr, also for measurements averaged over 30 seconds. They conclude that this noise level is within a factor of two of their photon noise limit. For measurement integration times, t, up to 30 seconds, their noise does indeed fall like $t^{-1/2}$, as one expects for quantum-limited noise. Ezekiel does not expect, therefore, that full fiber-optic integration will bring his phase-angle noise down much further; but it will provide a more compact instrument. Like the Stanford group, he plans to increase rotation sensitivity by winding more turns of fiber around the spool. One can of course reduce the fractional error introduced by quantum noise by increasing the number of photons—that is to say raising the intensity of the light source.

Work on fiber-optic gyroscopes is also continuing at Thomson CSF, Telefunken and about a dozen other industrial, academic and military laboratories in the US and Europe. Some military applications, for example submarine navigation, will require very high rotation sensitivity and stability, while others, such as "smart bombs," will need less sensitive sensors that are rugged, compact and inexpensive. Ezekiel and Arditty are organizing an "International Conference on Fiber-Optic Rotation Sensors and Related Technologies," to be held next month at MIT.

Sagnac's original work was inspired by his desire to disprove Einstein's special relativity. Now, seven decades later, Ezekiel is planning an experiment to measure the Earth's rotation with high precision by means of the Sagnac effect—essentially to test whether the "fixed stars" provide a good inertial reference frame.

References

1. R. A. Bergh, H. C. Lefevre, H. J. Shaw, Optics Lett. **6**, 502 (1981).
2. J. L. Davis, S. Ezekiel, Optics Lett. **6**, 505 (1981).

CHAPTER 6
LASERS IN CHEMISTRY AND INDUSTRY

Isotope Separation

The monochromatic nature of laser light makes possible selective excitation of the isotopes of an element. For example, a natural element consisting of several isotopes can be vaporized in an atomic beam and irradiated by an intense laser beam that excites one specific isotope but not the others. Then the excited ions can be separated by electric or magnetic fields. Such laser separation techniques promise to be more efficient than traditional methods, such as selective diffusion, centrifugation, or distillation.

Laser Fusion

A major goal of those studying high-energy lasers is the effort to obtain controlled nuclear fusion. At present the controlled nuclear power generators in operation in the world utilize the energy generated by fusion of heavy radioactive elements. The process contaminates the furnace structure and also produces radioactive waste. A much cleaner process is nuclear fusion, in which two light nuclei, such as deuterium and tritium, fuse together into a heavier nucleus, helium, and in so doing release a surplus of energy. However, before such reactions can begin, the electrostatic repulsions of the nuclei must be overcome, requiring temperature and density conditions approaching those inside the sun.

Fusion reactor designers have given much attention to the tokamak, in which the plasma is contained in a doughnut-shaped vessel. The plasma particles are "magnetically contained" by large fields produced by magnet coils wrapped around the reaction vessel. An attractive alternative to the bulky tokamak is laser fusion, using inertial confinement. Here a tiny (0.1 mm) deuterium–tritium fuel pellet is struck simultaneously on all sides by several high-power laser beams. This causes the pellet to implode, bringing about high plasma density and high temperatures. The momentary fusion reaction that follows produces energetic neutrons that are absorbed in a blanket of lithium surrounding the fusion vessel. The heat generated in this way may then be used to generate electricity. A new fuel pellet is then dropped in place and the process begins again.

Phillip F. Schewe

CONTENTS

Laser selective chemistry —is it possible?

With sufficiently brief and intense radiation, properly tuned to specific resonances, we may be able to fulfill a chemist's dream, to break particular selected bonds in large molecules.

Ahmed H. Zewail

PHYSICS TODAY / NOVEMBER 1980

NOTE: Recent experimental advances with ultrashort laser pulses have made it possible to obtain the dynamics of vibration energy redistribution in molecules—see the article by N. Bloembergen and A. Zewail entitled "Energy Redistribution in Isolated Molecules and the Question of Mode-Selective Laser Chemistry Revisited" [J. Phys. Chem. **88**, 5459 (1984)], and on chemical reactivity—see the article by Scherer, Duany, Zewail, and Perry entitled "Direct Pico-second Time-Resolution of Unimolecular Reactions Initiated by Local Mode Excitation" [J. Chem. Phys. **84**, 1932 (1986)].

One of the main goals of chemists is to understand the "alchemy" that leads to the building and breaking of molecules. There are many different ways of approaching this goal. One of these is photochemistry, the cracking of molecules by adding energy in the form of light to break bonds in the molecules. The resulting bond breakage is in most cases limited by statistical thermodynamic laws. With sufficiently brief and intense laser radiation properly tuned to specific resonances, we hope to bypass the statistical laws and break molecules precisely where we want to break them. Intellectually this is a challenging problem; if we succeed, laser selective chemistry may also have application in various areas of pure and applied chemistry and, perhaps, in medicine.

Laser chemistry involves two basic questions: How can we break molecules selectively with lasers, and what happens when molecules are subjected to heavy doses of laser radiation? These questions on selectivity open the door to many other fundamental questions pertaining to absorption and emission light by molecules, intramolecular vibrational relaxation, dephasing and coherent pumping by nonlinear optical processes. This article and the article by Vladimir Letokhov will explore what happens during and after the interaction between molecules and the laser field, covering single-photon and multiphoton absorption events. The other two articles in this issue by Richard N. Zare and Richard B. Bernstein, and by Yuan T. Lee and Ron Y. Shen are more concerned with state-to-state chemistry using lasers as effectors or detectors.

Ahmed Zewail is professor of Chemical physics at the California Institute of Technology. He has received an Alfred P. Sloan Fellowship and the Camille and Henry Dreyfus Foundation Teacher–Scholar Award.

In large molecules (that is, those with more than several atoms) the bonds are weak or strong depending on the atomic constituents and on the shape of the molecule. When the atoms take on energy, the bonds vibrate according to well-known rules of physics. In addition to vibrations, the molecules can convert the input energy to translational or rotational motion. It takes different amounts of energy to excite these different degrees of freedom; for example, the quanta of vibrational energy are larger than those for rotation.

When molecules are heated indiscriminately, the energy will be distributed statistically among all degrees of freedom; as the internal energy increases the weakest bonds will break, so that one can induce only a limited set of reactions in this way. However, because lasers can produce much more power in very short times and over much narrower spectral ranges than conventional light sources, they can, in principle, supply energy only to certain bonds in a molecule, leaving the other bonds cold. Thus, chemists could direct chemical reactions by breaking specific bonds that are not necessarily the weakest ones. But how do we go about such selective heating? To answer this question we must first know what goes on inside these large molecules—how the bonds "communicate" with each other, how fast the heat (or energy) spreads among the bonds or the different vibrational states. We must also understand why certain lasers can do the job while others cannot. In other words, we must still resolve some problems standing in the way of a happy marriage between lasers and molecules—laser chemistry.

Chemical selectivity

When the degrees of freedom of a large molecule are efficiently coupled, energy supplied to one of the bond vibrations is soon distributed among them. Even a light source well tuned to one vibrational mode heats the molecule uniformly, and all vibrations reach equilibrium in accordance with the laws of statistical thermodynamics. The reactivity of the molecule under these conditions has been described by a well-known theory advanced by Oscar K. Rice, Herman C. Ramsberger, Louis S. Kassel, and Rudolph A. Marcus (the RRKM theory), in which the intramolecular vibrational relaxation is very rapid.[1] In some sense, by using selective bond chemistry we hope to deviate from the RRKM scheme. There are several situations in which this statistical theory may not apply. For example, the reaction may depend on the energy present in *each* vibrational mode. Noel B. Slater has used a nonstatistical harmonic-oscillator theory to describe some isomerization reactions such as

$$ \underset{C-C}{\overset{C}{\diagdown}} \rightarrow C=C-C $$

that ought to follow such a scheme.

Another way in which the statistical theory can break down is that energy is supplied to one degree of freedom much faster than the energy can be distributed to the others— because the intramolecular vibrational relaxation is slow on the time scale of the experiment, the incoming power is so large and mode-specific that one bond breaks before the vibrational modes can communicate.

Intramolecular dephasing

A large molecule containing N atoms has $3N - 6$ vibrational modes. The potential energy of these vibrations becomes anharmonic for large oscillations, as the graph in figure 1 shows, reflecting the finite energy needed to break the molecular bonds. As a result of the anharmonicity, the spacing between vibrational levels decreases with

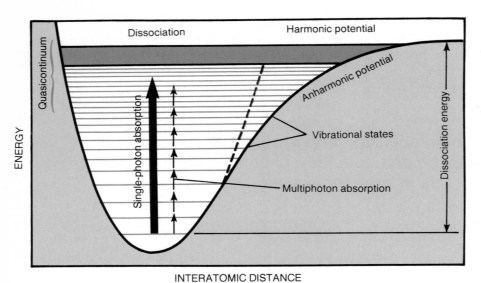

Vibration of a polyatomic molecule. The graph shows a plot of the potential energy of large molecule as a function of intratomic distances. At low energies the potential is nearly harmonic, so the vibrational quantum states are evenly spaced. For large bond-stretches the vibrations become anharmonic and the levels are more closely spaced, producing a quasicontinuum below the true continuum of dissociated states. The arrows show two possible modes of excitation, with single or multiple photons. Figure 1

increasing energy, producing a "quasi-continuum" of levels below the dissociation energy. For example, in benzene at about two electron volts of excitation energy there are about 10^8 states per reciprocal cm (about 10^{-4} eV—see the table on page 26). The variation of energy-level spacing caused by the anharmonic potential also permits different vibrational modes to exchange energy at excitations where the spacings in different modes match up. This happens, of course, particularly at high energies, so that many modes are coupled in the quasicontinuum.

In molecules with many degrees of freedom it is useful, for many purposes, to divide modes into two groups, a "relevant" part and an "unimportant" part. In other words, we may say that the molecule has some modes that we excite ("R-type" modes) and many others ("U-type" modes) that are not excited directly but interact with the excited R-modes (see figure 2). Accordingly, the "relevant" R-modes form a subsystem in the molecule while the remaining modes, the unimportant U-modes, make a large system that functions as a "heat bath" or energy sink for the R-modes. Unlike the heat baths encountered in other fields, such as magnetic resonance spectroscopy, the bath here is entirely intramolecular:

here we are dealing only with interactions within isolated molecules. In other words, the large number of vibrational degrees of freedom in a large molecule is sufficient for the molecule to be its own energy sink. A similar formalism describes another type of transition, "electronic radiationless" transitions, in large molecules.[2] The vibrational problem we are considering is similar in many ways to the electronic radiationless transition problem.

Consider, for example, benzene (figure 3). With 12 atoms it has thirty vibrational modes; there are oscillations of the carbon–hydrogen bonds, out-of-plane vibrations and many others. The carbon–hydrogen bond resonates at a wave number of approximately 3000 cm^{-1}; all other modes have lower resonant frequencies. We may consider the C–H stretch to be the relevant part and all the other normal modes to form the bath. The interaction between them is due to the off-diagonal anharmonicity terms of the Hamiltonian.

The theoretical formalism for this system-bath picture of a molecule may be borrowed from the literature. We can partition the Hamiltonian for the molecule into parts dealing with system, the bath, and their interaction:

$$H_{tot} = H_{sys} + H_{bath} + H_{int}(sys–rad) + H_{int}(sys–bath) \qquad (1)$$

Equation 1 includes a term describing the interaction of the system with an external radiation field; for dipole transitions it is of the form $\mu\mathcal{E}$, where μ is the dipole-moment operator for the system and $\mathcal{E}$ is the laser-field amplitude. (We may regard $\mathcal{E}$ as a given parameter rather than a dynamical variable, so we need not include a term H_{rad} in the Hamiltonian. Only if we consider spontaneous emission do we need H_{rad}.)

The equation of motion for the density matrix of the system is

$$i\hbar \frac{d\rho}{dt} = [H, \rho] \qquad (2)$$
$$= [H_{sys}, \rho] + [\mu\mathcal{E}, \rho] + (relaxation)_1 + (relaxation)_2$$

The two relaxation terms arise from the interaction of the system with the bath:

▶ The first describes the loss of energy from the system to the bath; this is an irreversible process. We shall denote the characteristic time with which it takes place as T_1.

▶ The second describes the "elastic" part of the interaction which leads only to phase changes in the system; it also contributes to the broadening of the system transitions. The time constant for the phase relaxation is T_2.

To make the separation in equation 2 of the energy-relaxation and phase-relaxation terms one must assume that both these processes are slow compared to the rate at which the bath reaches

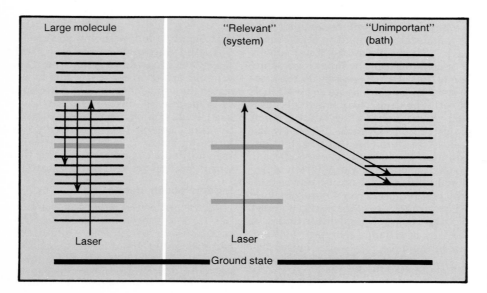

Energy levels in a large molecule. Only some of the levels interact directly with the laser field; the remaining levels form a heat bath for this "relevant" part of the molecule. Figure 2

internal equilibrium. That is, one assumes

$$T_1, T_2 \gg \tau_c \qquad (3)$$

where τ_c is the correlation time of the bath; it reflects the random fluctuations in the bath and is typically on the order of the vibrational period of the bath. If this criterion is satisfied, the resonance of the system will be homogeneous. On the other hand, in cases where τ_c is long, we expect complete or partial inhomogeneous broadening, depending on the strength of nonradiative coupling and relaxation in isolated systems. To understand non-RRKM chemistry we must understand the origin of T_1 and T_2. We shall come to this point later.

Bond localization

To break a specific bond we must be able to supply energy to that bond rather than to some broadly spread-out, nonlocal vibrational mode. In large molecules what are the criteria for bond locality? How does locality relate to T_1 and T_2? By "locality" here we mean spatial or temporal localization of vibrational excitation energy. Let us turn to some relevant experimental results.

Classical chemical activation experiments have provided vibrational relaxation times for some molecules. Seymour Rabinovitch's group at the University of Washington has measured[3] product distributions from reactants activated by different chemical means; they concluded that the vibrational relaxation time is of the order of 10^{-12} sec. This is a rather short time, and implies, unfortunately, that ultrashort laser pulses may be required for laser-induced selective chemistry. The theoretical group at the University of Southern California[4] have argued, however, that the measured quantity for chemical activation and also for beam experiments is not very sensitive to departures from RRKM behavior.

Unlike chemical activation methods, lasers can, in principle, produce monochromatic excitation of certain vibrations. One can excite the molecule with single a photon or with multiple photons. In the former case the molecule is provided with the required energy in a single shot. In multiphoton excitation the photon energy is much smaller than the necessary dissociation energy, but the molecule successively absorbs many of the low-energy photons among its many vibrational levels until it has accumulated sufficient energy to dissociate. The quasicontinuum helps the "climbing up" process because it contains many states whose separation can match the energy of the infrared photons. Nicolaas Bloembergen and Eli Yablonovitch discussed this mechanism earlier in PHYSICS TODAY.[5]

With single photons one may, in principle, excite either normal or local

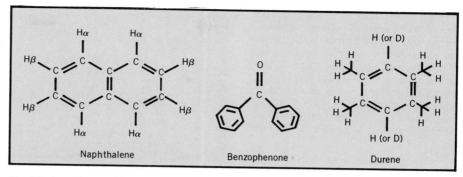

Naphthalene, benzophenone and durene. The "relevant" parts of these molecules are the virational modes of the C–H bonds in naphthalene, the C=O bond in benzophenone, and the C–H bonds of the methyl groups in durene.
Figure 3

vibrational modes of the molecule. Vibrational spectroscopists, a long time ago, told us that when the molecule is excited to the fundamental region (that is, a first excited vibrational state) the excited states are normal modes whose wavefunctions have a symmetry determined by the full point group of the molecule. Recently, however, some experimental and theoretical results have given support for the "local-mode" picture. This picture, is, of course, what we want for laser selective chemistry.

In molecule such as benzene one expects the C–H-stretch modes to be different in character from the rest of the modes simply because of the frequency mismatch: the C–H modes have the highest vibrational frequency. Thus, in a very simple picture, one would expect the C–H modes to be local and to have a large diagonal anharmonicity and small off-diagonal anharmonic couplings to other modes.

Several research groups have examined the local-mode character of large molecules, both experimentally and theoretically. Here we shall discuss our own work,[6] which is concerned with large molecules at low temperatures, (around 1.7 Kelvin); as we shall see, this is important for obtaining high-resolution information on intramolecular dynamics. The other groups have focussed their attention on liquid and gas spectra.[7,8] The experimental work of Andy Albrecht and Robert Swofford in the US and of Brian Henry in Canada has provided the foundations for understanding the energetics of local modes. The gas-phase experiments of Robert Bray, Michael Berry, and Don Heller have raised important questions regarding the broadening of overtones as a function of energy.

We have investigated bond locality in naphthalene and benzophenone (figure 3). We excited the molecules to C–H or C=O vibrational states with single photons, over a range which covers the low- and high-energy limits. The low temperatures allowed us to avoid rotational congestion and also kept the quasicontinuum out of the picture, because no

molecules could be in those states at the start of the experiment—at 1.7 K all molecules are essentially in the ground state.

Cooling of large molecules is very important and may be done by:

▶ isolating the molecules in a cold host matrix

▶ cooling a crystal of these molecules, for example, to liquid helium temperatures

▶ expanding the molecules with an inert gas through a pinhole nozzle.

Although in all cases the spectra will be sharpened considerably because of the removal of thermal congestion, in the matrix-isolation and the cold-crystal methods there is some perturbation due to guest–host interactions and the crystal field. However, if these interactions are small compared to the intramolecular interactions, then these two methods are ideal for studying vibrational and electronic dynamics. Our group at Caltech has used the second method of cooling to study the spectroscopy of high-energy vibrational states and the third method of cooling, supersonic expansion of a molecular beam, to examine intramolecular relaxation processes. In the latter case, the molecules are rotationally very cool (0.3 K) and vibrationally cool (around 40 K) after they leave the nozzle of the apparatus; they are then excited by a laser beam, and one can then obtain high-resolution spectra of isolated, cooled molecules. Richard Smalley (now at Rice), Len Wharton and Don Levy successfully demonstrated[9] this method at the University of Chicago.

Figure 4 shows the vibrational overtone spectra for a cooled naphthalene crystal. Napthalene has 48 vibrational modes and benzophenone has 66. The C–H and C=O stretches are at about 3,000 cm^{-1} and 1,700 cm^{-1}, respectively. Because the dissociation energy of the C–H bond in the naphthalene molecule is about 44 000 cm^{-1}, it requires about 15 photons with the energies of the C–H mode to dissociate. To examine the C–H.and C=O bond locality we measured the spectra of the modes and the relaxation time of the excited mode by all other modes in the molecule.

From our naphthalene and benzophenone experiments[6] we found that:

▶ The overtone spectrum (that is, the spectrum of states with vibrational quantum number $v = 2, 3, \ldots$) is a simple progression of C–H or C=O bands (especially at high energies) with energies obeying a simple anharmonic rule:

$$E_v / v = A + Bv \qquad (4)$$

For naphthalene $A = 3086$ cm^{-1} and the diagonal anharmonicity, B, is -55.8 cm^{-1}; the fit of the data to equation 4 is excellent, as figure 5 shows.

▶ At about 15 000 cm^{-1} 9about 1.9 electron volts) in the molecule, the stretches of the C–H$_\alpha$ and C–H$_\beta$ bonds (figure 3) in naphthalene are distinguishable.

▶ The vibrations of the C–H$_\alpha$ and C–H$_\beta$ bonds in naphthalene have different apparent relaxation times (.075 and .11 picosec) even when the other modes (the "bath") are cooled to 1.3 K.

▶ The relaxation time (from the spectra) gets shorter as the energy increases in the C=O stretching mode of the benzophenone molecule.

What do these findings mean? If we think of naphthalene's C–H bonds as not communicating with each other, we could represent (rather naively) the molecule as simply the sum of the different bond deformations:

naphthalene
$$= 4C–H_\alpha + 4C–H_\beta + \cdots$$

And in fact, experimentally, in both the napthalene C–H stretches and the benzophenone C=O stretch, a simple relationship, like that in a diatomic molecule between the vibrational energy and the vibrational quantum number v, holds up very nicely and accurately.

Can we conclude from this that the energy is localized in the C–H or C=O bond? In my opinion, the answer is no. All it means is that our results are consistent with a local excitation in the C–H or C=O sub-systems. But to prove the locality we need further evidence.

From the experiments we do know that the energy stays in the carbon-hydrogen system for fractions of picoseconds or longer. Now we are faced with a dilemma. On one hand the spectra are consistent with a local-bond picture, but on the other hand the relaxation time of these excited modes is extremely short. Eventually lasers may be developed that can break a bond at this speed, but for the moment it might seem that we would have to give up.

However, the other channel of relaxation, dephasing, and inhomogeneous broadening could affect our previous measurement of the energy-relaxation time. (As we mentioned, dephasing describes the loss of phase coherence; for

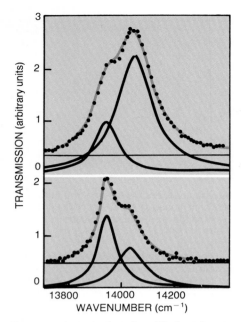

Transmission spectra of naphthalene (upper graph) and naphthalene deuterated in the α positions (lower graph). The experimental curves show the low-temperature (1.3–2 K) spectra for the fifth vibrational overtone; the theoretical curves show the fixed peak due to β hydrogen and the shifting peak due to α hydrogen or deuterium.   Figure 4

illustration we can think of the modes are dancers in a *corps de ballet*—if one dancer misses a step and gets out of phase with the others, there will be a disturbance in the routine but this will not immediately affect the *number* of dancers on stage. Similarly, the C–H oscillations can dephase without any net loss of energy.) Simple spectroscopic experiments will not measure dephasing rates directly;[6] the line width, for example, has contributions from both relaxation rates and from inhomogeneous spreading.

For successful selective chemistry we must know both T_1 and T_2: T_1 tells us how fast the deposited energy is flowing to all modes, and T_2 tells us what kind of lasers we should use. At the moment one does not know the contribution of each of these two relaxation times to the overall relaxation time of all overtones. It may be that the bond-locality time is much longer than the measured 0.1 picosecond overall relaxation time—even long enough for us to get at the bonds and break them with current laser technology. One also does not yet know how to describe theoretically the spectra of high-energy transitions. This is in contrast to our rich knowledge about the spectra of low-energy states. We shall return to this point later.

Which state do we excite?

Essential to the problem of selectivity is the question of state preparation. For any molecule, there are certain exact eigenstates that one may obtain, in

principle, by using just symmetry and quantum mechanics. But, does light excite these stationary states—or does it excite a non-stationary state? If the states are non-stationary are they normal or local modes? These questions are not fully answered yet. However, we may be able to shed some light on these questions by considering the results of the following experiment.

Recently, we have studied[6,10] the high-energy vibrational overtones of the durene molecule (tetramethyl benzene) and a partially deuterium-substituted analog (see figure 3). The durene molecule has two types of C–H stretches, those in which the hydrogen atoms are in the methyl groups remote from the benzene ring, called "aliphatic-C–H" and those in which the hydrogen atoms are directly attached to the ring, called "aromatic-C–H". Figure 6 shows the spectra of durene and of deuterated durene. Two remarkable observations are noteworthy. First, when the hydrogen atoms attached to the benzene were replaced by deuterium the spectrum of region I in figure 6 (between 0 and 800 cm^{-1}) is relatively unchanged while the band of region II (above 800 cm^{-1}) is completely missing in the spectrum of deuterated compound even at high energy. Second, the bands of the "triplet" observed in region I are very sharp (as small as 20 cm^{-1} wide—the narrowest band observed in any large molecule's overtone region) in contrast with all known aromatic-C–H bands, which have a typical width of 100 cm^{-1} or larger. The bands of region I are due to the aliphatic-C–H bonds, while that of region II is from the aromatic bonds. The implications of these results are quite interesting.

The fact that the absorption band of the aromatic-C–H bond disappears when hydrogen is replaced by deuterium indicates that the stretches of the aliphatic and aromatic bonds are not interacting with each other even in the quasicontinuum region. Furthermore the narrowness of the aliphatic bands in comparison with aromatic bands indicates that the coupling of the aliphatic-C–H vibrations to the intramolecular bath is much weaker than the coupling of the aromatic-C–H vibrations to the bath. In some sense the aliphatic-C–H bonds form a subsystem of the large molecule. This subsystem behaves like a small molecule in which the states are well separated from each other in energy. The aromatic-C–H bonds must, on the other hand, behave as if they were a part of a large molecule, where the bath is quite big and the dephasing or energy-relaxation rates are very large. This is consistent with the theoretical picture advanced by Joshua Jortner's group. An important point about the prepared states emerges from the observations. Because aliphatic-C–H stretches in the large molecule durene belong to the small- or intermediate-three molecule coupling limit, and because we observe three prominent lines in the overtone spectra, we conclude that the coupling among the aliphatic-carbon–hydrogen bond excitations of the methyl groups is larger than the coupling of the methyl groups to the bath. That is, the methyl groups behave as a strongly coupled system weakly coupled to a bath. We are still investigating the exact nature of the coupling, using polarized-light absorption spectroscopy.

A useful measure of the relation between the internal coupling of a subsystem of states and the coupling to the bath of other states is the ratio of the line splitting, Δ, to the line width, ϵ.

If $\Delta/\epsilon \ll 1$, we should observe only one band that in part is inhomogeneously broadened. What we excite under this condition is a statistical mixture of states (perhaps local), whose coherent character is ill-defined. On the other hand, if $\Delta/\epsilon > 1$, then the normal modes (or, more precisely, the anharmonic modes) can be excited. The important point here is that normal mode progressions in the quasicontinuum are possible, and that exciting a *single* bond in a molecule crucially depends on the magnitude of the intramolecular coupling and relaxation rate and on the laser source used in the experiment.

Laser-induced selective reactions.

Because of the extensive study of laser-induced chemistry in many laboratories, it is not possible to give here a full account of all the research. Rather we shall highlight the new ideas that relate to selectivity by discussing the results of three different experiments done at the University of Pennsylvania, at Allied Chemicals and at Exxon.

Robin Hochstrasser and David King[11] in an elegant experiment, have induced selective chemical reactions in the molecule tetrazine, $C_2N_4H_2$, which has a cyclic structure, very much like benzene. In their experiment the tetrazine molecules were dispersed in a host crystal at 1.6 K. Enrichment factors of at least 10^4 for carbon-13 and nitrogen-15 were obtained in the photodissociation products, hydrogen cyanide and nitrogen. Hochstrasser and King achieved this selectivity by tuning the laser to the absorption band of molecules with a particular isotopic composition. Such molecules will be lost because of dissociation, while molecules containing other isotopes are enriched. Even though the molecule is large and one expects the spectral levels to overlap, cooling the system to 1.6 K isolates the spectral bands of the different isotopes. In these low-temperature experiments, the reactions take place from excited electronic states and not from excited vibrational levels of the

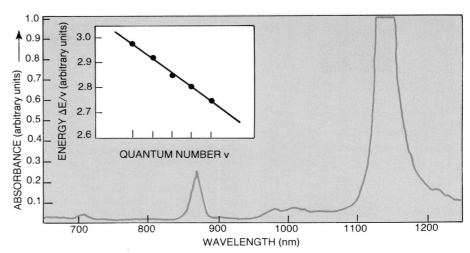

Overtone progression of the carbon–hydrogen bond stretch in naphthalene. The peaks are for vibrational quantum numbers $v = 5,4,3$ (left to right). The inset shows the simple anharmonicity of the spectrum: $\Delta E/v$ is a linear function of v. Figure 5

ground state. Furthermore, the selectivity is achieved because of the slow communication between different isotropically labelled molecules (molecular selectivity) and not between different vibrations in the same molecule (bond selectivity).

In a different type of experiment, Kammalathinna V. Reddy and Berry[12] have induced reactions in the gas phase by selectively exciting certain vibrational states. The molecule allyl isocyanide ($H_2C{=}CHCH_2NC$) undergoes photoisomerization when excited by a laser into the different C–H stretches. It was concluded that in this system, direct one-photon overtone photoactivation leads to bond-selective excitation that ultimately leads to nonstatistical photoisomerization reactions. Reddy and Berry conclude from their experiments that the time scale for complete randomization of the internal energy

due to intramolecular vibrational energy redistribution must be longer than or comparable to the isomerization time scale (10^{-7}–10^{-9} sec), that is, the behavior here is not according to the RRKM theory. If this conclusion turns out to be truly general, the outlook for bond-selective chemistry is very good.

The above experiment uses single photons from a dye laser. Multiphoton absorption has also been used to search for the possibility of selective chemistry. Richard Hall and Andrew Kaldor[13] excited cyclopropane, $(CH_2)_3$, by laser either at 3.22 microns exciting the C–H asymmetric stretch, or at 9.50 microns, exciting the CH_2 wag. The excitation is in the fundamental region but the reactions are induced by multiphoton infrared absorption. The reaction induced by the absorption is isomerization to propene, $CH_2{=}CH{-}CH_3$, in addition to some products from a fragmentation

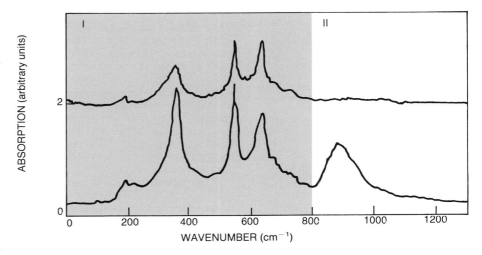

Spectra of durene and partially deuterated durene, obtained at low temperatures. The lack of absorption in region II for the deuterated molecule (upper graph) indicates that this absorption is due to hydrogen bonded to an aromatic carbon atom and that region I of the spectrum is due to the hydrogen atoms of the methyl groups. Figure 6

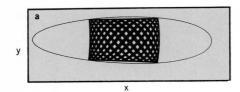

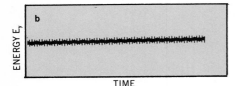

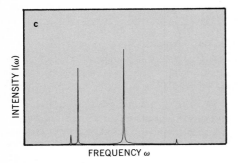

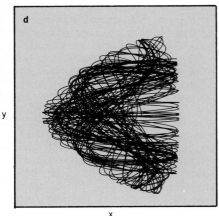

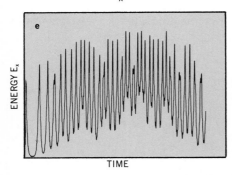

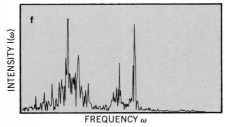

Two-mode systems in the quasiperiodic (upper three graphs) and ergodic regimes. We show the phase-space trajectories (**a** and **d**), the spectra (**b** and **e**) and the energy (**c** and **f**) of a computer model.[14]   Figure 7

reaction. The "branching ratio" of the yield of isomerization to the yield of fragmentation depends on the frequency of the laser used; at low pressures, almost no fragmentation occurs for the 3100 cm^{-1} excitation, whereas the yields of fragmentation and isomerization are roughly equal for excitation at 1050 cm^{-1}. The fact that in each case the products are typical of those formed with thermal excitation whereas the relative yields are not, led Hall and Kaldor to conclude that their results are consistent with a non-statistical distribution of the states that lead to a reaction. More experiments of this type are obviously needed before we can generalize the conditions under which selectivity can be achieved. It is quite interesting that the above experiments indicate that selectivity may be achieved by pumping the molecule with a laser directly into the fundamental region (multiphoton experiments) or into the overtones of modes in the quasicontinuum (single photon experiments). The question now is: can we relate the experimental findings on energy localization and selectivity to a theory?

Theoretical Descriptions

We are interested here in the interaction of molecules with the radiation field and in the subsequeent intramolecular energy redistribution. The former problem is usually formulated phenomenologically in the semiclassical nonlinear optical theory: the laser field is simply described as a classical field and molecules are described as quantum systems with phenomenological energy levels. The other theoretical problem, concerning the intramolecular energy redistribution, is not at all trivial.

The theory of intramolecular energy redistribution is difficult because of the many interactions among a large number of anharmonic modes, and because accurate calculations require precise knowledge of the multi-dimensional potential-energy surface. However, several simplifications have been advanced both in classical and quantum calculations. To illustrate these, let us consider perhaps the simplest case—a system containing two anharmonically coupled modes[14,15] (such as the symmetric and antisymmetric stretches in a linear triatomic molecule). Extension to large molecules might still be a difficult, if not impossible task.

Theorists at Irvine, Chicago, Caltech, UCLA, Colorado Tel Aviv and Hebrew University and other places have considered this challenging theoretical problem. Armed with chemical intuition and the mathematical tools of nonlinear mechanics, these groups have developed some criteria for describing the energy redistribution and in some cases for the onset of chaotic, or "ergo-

dic," behavior in "simple" molecules. One expects that as the molecule's energy increases, the vibrations will become more and more chaotic because of the large density of states at high energies. But what determines the onset for this chaotic behavior? And at what energy can classical and quantum calculations yield the same results?

Before we discuss the recent theoretical results, let us briefly review the classical theory of two coupled oscillators. For a system with two vibrational degrees of freedom, such as a triatomic molecule (neglecting bending), we can write the Hamiltonian as

$$H = T(p_1, p_2) + V(q_1, q_2)$$

where T is the kinetic energy and V is the potential energy. To describe oscillations one expands the potential around the equilibrium position

$$V = \tfrac{1}{2}k_1 q_1^2 + \tfrac{1}{2}k_2 q_2^2 + k_{12} q_1 q_2 + \Sigma k_{ijl} q_i q_j q_l + \cdots$$

where the q_i are generalized coordinates, the p_i are the conjugate momenta and the k's are constants. The normal modes of the system are those for which the quadratic terms of T and V are diagonalized. If we denote generalized coordinates for these "normal modes" by x and y we have

$$V = \tfrac{1}{2}(K_x x^2 + K_y y^2) + \Sigma K_n x^n y^{3-n} + \cdots$$

The exact nature of the vibrational motion is given by quantum-mechanical solution of the above Hamiltonian. Even for simple anharmonic potentials that is not a simple task, so one often starts with a classical solution.

The classical behavior of the oscillator are the solutions to Hamilton's equations, $\dot{x} = \partial H / \partial p_x$, and so forth. One can, for example, obtain the classical trajectories—the "orbits" in the x-y plane—by integrating Hamilton's equations on a computer, starting with some representative initial conditions. Marcus and his collaborators, first at Illinois and now at Caltech, have obtained approximate energy levels for coupled anharmonic oscillations by applying the semiclassical quantum conditions

$$\oint p\,dq = (n + \tfrac{1}{2})h$$

(with the integrals calculated along topologically independent paths) to the classical orbits. The classical trajectories also permit one to calculate the absorption spectrum for the resonance associated with each normal mode:

$$I(\omega) = \frac{1}{2\pi} \int_{-\infty}^{\infty} \langle x(0)x(t) \rangle e^{-i\omega t}\,dt$$

for the x-mode, for example. (Of course, to obtain the actual infrared spectra one must use $\mu(t)$ instead of $x(t)$ in the correlation function.) Figure 7 illustrates the trajectories and spectra [assuming that $\mu(t) = x(t) + y(t)$] obtained with these methods by Marcus's group for two important cases: the "quasi-periodic" and the "ergodic" limits.

What do we mean by quasiperiodic and ergodic motions? For a single oscillator (for example, the vibration of a single diatomic molecule) the motion is always periodic: the oscillation will be some combination of the fundamental and its overtones. For a system with two modes, the energy can go from one mode to another, but if the modes are not too highly excited, the motion will shift smoothly from one mode to the other and the motion is said to be quasiperiodic. In a phase-space diagram, such as figure 7a, the trajectory is confined to a limited region. At sufficiently high energies this quasiperiodic behavior is lost and the motion is said to be ergodic. Because of the many interfering Fermi resonances and combination bands, the behavior becomes quite chaotic and the trajectory essentially fills the entire phase space, as in figure 7c.

The next problem is to find out how accurate these classical results are by confronting them with quantum mechanical calculations on the same system. If possible, the comparison will allow us to find if quantum behavior correspond to classical behavior in the quasi-periodic and ergodic regimes.

Stuart A. Rice and his group at Chicago examined[15] quantum ergodicity of certain nonseparable, two-dimensional potential surfaces in an anharmonic oscillator basis. They pointed out that the microscopic ergodic theories of John von Neumann and Eberhard Hopf are not relevant to the problem of ergodic behavior in molecules. They then advanced their own idea, which can be summarized as follows: For a given Hamiltonian there exist a set of eigenfunctions which can, in principle, be obtained from computer diagonalization methods. The projections of these wavefunctions on the unperturbed harmonic-oscillator basis functions give an idea about the "degree of chaos": when a wavefunction has a substantial component over a number of basis functions it is termed "global", otherwise it is "local". The question is: Are the global states also ergodic (or statistical, stochastic, chaotic)? It is argued in the literature[14] that states can be global in a quantum sense without necessarily being ergodic or statistical in nature. Moreover, for certain system Hamiltonians, quantum and classical methods yield essentially the same answer in the quasiperiodic regime, but not necessarily in the ergodic limit.[14]

William Miller's group[16] at Berkeley made use of the nodal pattern of wavefunctions and concluded that many of the higher eigenstates can apparently be classified as ergodic. These particular states seem to correspond quite well to classical ergodic trajectories. Marcus's group detected the onset of chaos by searching for "overlapping avoided crossing" in the plot of eigenvalues as a function of parameters in the Hamiltonian, such as the coefficients k_{ijl}. At UCLA Eric Heller's group has found[17] a strong correlation between spectra, wavefunctions, and classical motion for certain model Hamiltonians. In more recent work Rice's group has used the Kolmogorov entropy to suggest that there is an inherent difference between classical chaos and quantum mechanical chaos. These results leave open the question of what precise relationship, if any, exists between classical chaos and quantum mechanical motion. Clearly, we still do not know the precise meaning of quantum ergodicity.

Problems for the future

Our understanding of selective bond-breaking in molecules bonds by lasers is in its infancy. There is a massive effort by chemists around the world to unravel many of the theoretical and experimental problems that stand in the way of successful laser selective chemistry. In my view we must know a lot more about intramolecular dynamics. Lasers are now available in almost laboratory, but the understanding of large-molecule dynamics is not yet fully available.

What we must expect from theory and experiments in the future can best be summarized, perhaps, by asking the following questions:
► Can we predict theoretically the yields of chemical reactions following the laser excitation into a selected quantum state?
► In large molecules, what is the precise nature of the state that we excite?
► Can we induce chemical reactions from truly local modes?
► To what extent does selective chemistry with lasers depend on the properties of the source of the excitation?
► Are the experiments on selectivity more likely to be successful in other systems than in those currently used?
► Under what conditions is the randomization of intramolecular energy slow or rapid relative to chemical reactions such as isomerization or fragmentation?
► Is our description of the statistical behavior in real isolated molecules correct?
► Can quantum calculations be done to predict intramolecular rates in real systems?
► What is the quantum origin of optical relaxation times T_1 and T_2 in large systems, such as large molecules?
► What exactly do we mean by quantum ergodicity in complicated systems?

Opportunities in this field of study are numerous. Lots of exciting questions are at hand, and if they can be answered—and if their answers are encouraging—selective chemistry, with lasers might revolutionize many fields such as photochemistry, analytical chemistry and catalysis.

With the advances made recently in lasers with ultrashort (picosecond to subpicosecond) pulses and ultrahigh resolution, the selective, and extremely fast, heating and breaking of bonds in large molecules will perhaps be no longer a chemist's dream.

* * *

I wish to thank many of my colleagues for their critical comments and for communicating their results and ideas: S. Rice, R. A. Marcus, M. Berry, R. Bernstein, A. Kaldor, M. El-Sayed, R. Hochstrasser, H. Rubalcava, E. Stechel, E. Heller and W. Miller. This work was supported in part by the National Science Foundation.

References

1. P. J. Robinson, K. A. Holbrook, *Unimolecular Reactions*, Wiley–Interscience, New York (1972).
2. For a review, see: J. Jortner, S. Mukamel, in *The World of Quantum Chemistry*, Riedel Boston (1974), page 145.
3. I. Oref, B. S. Rabinovitch, Acc. Chem. Res. **12**, 166 (1979).
4. E. Thiele, M. Goodman, J. Stone, Opt. Eng. **19**, 10 (1980).
5. N. Bloembergen, E. Yablonovitch, PHYSICS TODAY, May 1978, page 83.
6. A. H. Zewail, Acc. Chem. Res. **13**, 360 (1980); J. W. Perry, A. H. Zewail, J. Chem. Phys. **70**, 583 (1979); D. D. Smith, A. H. Zewail, J. Chem. Phys. **71**, 540 (1979); J. W. Perry, A. H. Zewail, Chem. Phys. Lett. **65** 31 (1979).
7. A. C. Albrecht, in *Advances in Laser Chemistry*, A. H. Zewail, ed., Springer Series in Chemical Physics, Springer, New York (1978); B. Henry, Acc. Chem. Res. **10**, 207 (1977).
8. R. G. Bray, M. J. Berry, J. Chem. Phys. **71**, 4909 (1979); D. Heller, S. Mukamel, J. Chem. Phys. **70**, 463 (1979); M. Sage, J. Jortner, Chem. Phys Lett. **62**, 451 (1979).
9. R. Smalley, L. Wharton, D. Levy, Acc. Chem. Res. **10**, 139 (1977).
10. J. W. Perry, A. H. Zewail, J. Phys. Chem., to be published.
11. R. M. Hochstrasser, D. S. King, J. Am. Chem. Soc. **97**, 4760 (1975).
12. K. V. Reddy, M. J. Berry, Chem. Phys. Lett. **66**, 223 (1979).
13. R. Hall, A. Kaldor, J. Chem. Phys. **70**, 4027 (1979).
14. R. A. Marcus, D. W. Noid, M. Koszykowski, in *Advances in Laser Chemistry*, A. H. Zewail, ed., Springer, New York (1978), page 298. D. Noid, R. Marcus, J. Chem. Phys. **67**, 559 (1977); Chem. Phys. Letts. **73**, 269 (1980); J. Chem. Ed. **57**, 624 (1980).
15. S. A. Rice, *Advances in Laser Chemistry*, A. H. Zewail, ed., Springer, New York (1978), page 2, and references therein.
16. R. Stratt, N. Handy, W. Miller, J. Chem. Phys. **71**, 3311 (1979).
17. M. Davis, E. Stechel, E. Heller, "Quantum Dynamics in Classically Integrable and Nonintegrable Regions," Chem. Phys. Lett., in press; E. J. Heller, J. Chem. Phys. **72**, 1337 (1980). □

Laser-induced chemical processes

Isotope separation, which continues to be an active field of laser photochemistry, is joined now by methods for producing very pure substances and by studies of biologically important molecules.

V. S. Letokhov

PHYSICS TODAY / NOVEMBER 1980

Proposals for controlling or inducing chemical reactions with laser light have been around almost as long as lasers themselves. Immediately after Theodore Maiman's invention of the ruby laser[1] in 1960 and Ali Javan, William Bennett and Donald Herriott's invention of the helium–neon gas laser[2] in 1961 suggestions for using this new type of light source in chemical processes followed. The proposals resulted essentially from adapting already well-known principles of photochemistry. Yet almost no research employing lasers in chemistry was done for ten years. Not until tunable lasers and high-powered infrared lasers arrived at the end of the 1960's and in the early 1970's did laser chemistry become experimentally feasible, and the period since then has seen intense and systematic study of atomic and molecular processes. Of particular interest are those selective processes applicable to laser isotope separation,[3] such as the work on carbon isotopes carried out in the apparatus shown in figure 1 and discussed later in this article. Selective processes also appear to be useful for the production of highly pure substances and for the study of certain photobiochemical compounds. In this article I will be able to give only a brief survey of the great amount of work currently in progress, concentrating particularly on those aspects being studied at laboratories in the Soviet Union.

Types of molecular photoexcitation

With laser radiation new methods for molecular photoexcitation have become available that are not possible with "classical" light sources. Figure 2 compares the old and new techniques for selective molecular photoexcitation in schematic form. The classical, or "pre-laser," photochemical method (figure 2a) is based on single-step excitation of an electronic state of an atom or molecule. It has the serious disadvantage for selective photochemistry that most molecules, particularly polyatomic ones, have broad structureless electronic absorption bands at room temperature; the scheme cannot be used for, say, isotopically selective excitation. Only for a few simple molecules is the absorption line narrow enough for isotope selection. On the other hand, excitation of electronic states does give a high quantum yield for the photochemical reaction.

Single-step excitation of a molecular vibrational state (photochemistry in the ground electronic state) features rather high excitation selectivity for

V. S. Letokhov is the head of the laser spectroscopy laboratory and deputy director for research at the Institute of Spectroscopy, of the Academy of Sciences of the USSR, and also a professor at the Moscow Physico-Technical Institute.

both simple and complex molecules. The main disadvantage of this method is that the fast relaxation of vibrational excitation to heat leads to a low quantum yield of the subsequent photochemical process. In any case, the method can only be used for photochemical reactions with low activation energy.

Two-step excitation of a molecular electronic state through an intermediate vibrational state by joint action of infrared and ultraviolet radiation (figure 2b, left side) combines the advantages of single-step ir and uv processes and removes their disadvantages.[3] In two-step excitation by a two-frequency laser field (ir + uv) it is possible to separate the functions of selective excitation, when the molecule acquires a rather low energy from the ir photon, and the absorption of a much greater amount of energy, from the uv photon, by the selectively excited molecule. Thus this kind of two-step photoexcitation combines sufficiently high selectivity with high quantum yield for photochemistry.

Two-step excitation of molecules through an intermediate electronic state (figure 2b, right side) is not so widely used as joint ir + uv excitation. Its advantage over single-step excitation of electronic states is that it offers the possibility of exciting states with specific properties and of reaching high-lying states (with large quantum yields for photochemical reactions) without requiring radiation in the extreme ultraviolet. The process is most important for laser excitation of biologically important molecules in solution.

For polyatomic molecules, a powerful source of infrared radiation acting alone may suffice for isotope-selective excitation to high vibrational or even excited electronic states.[4] Figure 2c shows how, by multiple absorption of infrared photons of the same frequency, a molecule can derive an energy comparable to the typical energy of electronic excitation. Therefore with this process we can simultaneously realize the excitation selectivity for separating isotopes and a rather high quantum yield of the subsequent photochemical process.[4] The method is, however, limited in that it can be applied only to polyatomic molecules with a high density of excited vibrational levels in the ground electronic state.

The methods of selective photochemistry illustrated in parts b and c of figure 2 can only be carried out with laser radiation, because the intermediate quantum levels need to be highly populated. "Conventional" (incoherent) light sources have too low a radiation temperature for all processes except the single-step one of figure 2a, and they have a higher efficiency for electronic states than for vibrational ones in this process.

Types of selectivity

In laser chemistry generally, an important feature of each different approach is the "selectivity" it offers, which of course determines the kind of work it is suitable for. As an example by which we can discuss different kinds of selectivity let us consider infrared multiphoton laser chemistry. This field provides an ideal example because of its recent rapid development (about 500 papers so far) and good prospects for further expansion.

Resonant multiphoton excitation of molecular vibrations in a strong infrared field forms the basis for several essentially different approaches to laser chemistry. They can be classified according to the relations between the various relaxation times for an excited vibrational level interacting with the infrared field:

$$\tau_{v-v'}{}^{\text{intra}} \ll \tau_{v-v'} \ll \tau_{V-T} \qquad (1)$$

Here $\tau_{v-v'}{}^{\text{intra}}$ is the time for intramolecular transfer of vibrational energy between different vibrational modes of the molecule being excited;

$\tau_{v-v'}$ is the time for intermolecular transfer of vibrational energy between molecules of different kinds in a mixture (for example, molecules of different isotopic composition), and

τ_{V-T} is the relaxation time for molecular vibrational energy to be transferred to translational degrees of freedom—which is the time for complete thermal equilibrium to be reached in the molecular mixture.

The first of these processes is intramolecular by definition, and it may therefore occur without collisions; the other two are naturally collisional, so that the times $\tau_{v-v'}$ and τ_{V-T} vary with the experimental conditions such as gas pressure.

Let us call the rate of vibrational excitation of a molecule by multiphoton absorption W_{exc}. The value of this rate W_{exc} depends on radiation intensity and vibrational transition cross sections. From a consideration of inequality 1 we can distinguish four different conditions for the relations between W_{exc} and the relaxation rates of vibra-

Carbon-isotope separation system operated jointly by the Kurchatov Institute for Atomic Energy and the Institute of Spectroscopy of the Academy of Sciences, at the industrial Institute of Stable Isotopes in Tbilisi. Separation is by multiphoton dissociation of CF_3I with pulsed CO_2 laser radiation of average power 1 kilowatt. See also figure 4. **Figure 1**

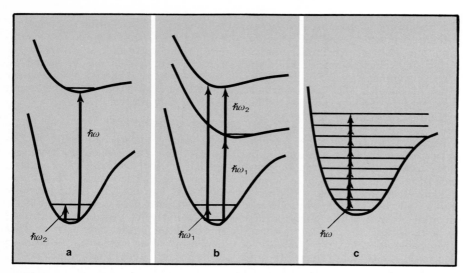

Selective molecular photoexcitation. **a**: Single-step excitation of electronic or vibrational state; **b**: Two-step excitation of an electronic state through intermediate vibrational or electronic states, and **c**: Multiple-photon excitation by infrared radiation.   Figure 2

tional energy in different circumstances. Accordingly, there are four different approaches to infrared multiphoton laser photochemistry depending on how far from equilibrium the vibrational excitation in the molecule and molecular mixture lies. Let us consider all four:

When $\quad W_{exc} > 1/\tau_{v-v'}{}^{intra}$ (2)

mode- or bond-selective excitation of molecules is possible when there is no vibrational equilibrium even inside the molecule interacting with the strong infrared field. A certain mode or a functional group of a polyatomic molecule must have a higher vibrational temperature compared with the remaining modes or functional groups to bring about this condition, in which mode- or bond-selective photochemistry is quite feasible.

When $\quad 1/\tau_{v-v'}{}^{intra} > W > 1/\tau_{v-v'}$ (3)

which is a more moderate condition on the multiphoton vibrational excitation rate, molecular-selective excitation becomes possible; the requirements are that there is no vibrational equilibrium among the molecules in the mixture but there is intramolecular vibrational equilibrium for those that interact with the infrared field.

The molecules in resonance with the infrared field acquire a higher vibrational temperature than all the other molecules. (The translational temperature of all the molecules in the mixture remains, of course, constant.)

The field of molecular-selective infrared photochemistry that becomes possible under condition 3 has been very well explored since its origins in 1974 with experiments on laser isotope separation.[4] Almost all the published work in this general area relates to this case, because of the important practical applications of selective excitation and dissociation of molecules in a mixture.

At the much more moderate condition:

$1/\tau_{v-v'} > W_{exc} > 1/\tau_{V-T}$ (4)

the vibrational equilibrium among all the molecules in a mixture is stronger, but there is still no relaxation to general heating. The condition is possible only if no component in the mixture has a fast V→T relaxation. Because the vibration temperature differs from the translational one in the mixture, nonselective vibrational infrared photochemistry is possible when reactions with a minimum energy barrier take place in a time no greater than about τ_{V-T}. When the fourth and final condition,

$1/\tau_{V-T} > W_{exc}$ (5)

applies, equilibrium thermal excitation of all the molecules in the reaction cell is established (provided no heat is supplied to the walls). In this case the vibrational and translational temperatures of all the molecules in the mixture are equal, or almost equal, because of the vibrational excitation relaxation to heat through collisions. We will therefore be dealing with thermal heating of the molecular gas as observed[5] as long ago as 1966 with continuous-wave CO_2 laser radiation. This field of infrared thermal chemistry is of interest only when the reaction-cell volume needs to be heated while the walls remain cool.

Having stated these different conditions, let us now review the various methods of infrared laser chemistry that follow them.

Nonselective laser photochemistry

When condition 4 and especially condition 5 apply, no selective photoprocesses are possible in molecular mixtures. Yet even in those circumstances multiphoton excitation of any component in the mixture can be of some practical interest despite the fast transfer of vibrational energy to other molecules. Suppose that certain molecules in the mixture with a high dissociation energy absorb the infrared energy and transfer it to other molecules that have a lower dissociation energy but do not interact with the infrared field. There will of course always be some vibrational-to-thermal relaxation, but if τ_{V-T} is sufficiently greater than $\tau_{v-v'}$ in condition 4 we can generate a vibrationally excited molecular mixture in which, for a limited time, photochemical reactions (particularly molecular dissociation at a minimum energy barrier) will occur.

For most mixtures of complex molecules there will be a channel of rapid V→T relaxation via which total heating of the mixture will inevitably occur. Each mixture needs to be studied in detail, comparing rather complicat-

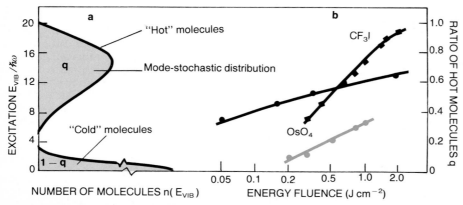

Generation of two molecular ensembles ("hot" and "cold") at multistep excitation by intense infrared laser pulses (part **a**); part **b** of the figure shows the relative fraction of highly excited ("hot") molecules as a function of energy fluence in the CO_2 laser pulse. (Data taken from publications in reference 8.)   Figure 3

ed V→V and V→T relaxation pathways. In the literature one therefore often meets discussions about the difference between vibrational and translational molecular temperatures in experiments at high pressure, when the degree to which condition 5 is fulfilled is unknown.

At first sight is seems unprofitable to use expensive laser systems to realize the processes of nonselective thermal chemistry. But in applications where remote or localized heating of matter is essential the laser may provide the best solution. Examples are thermal chemical surface treatment by infrared laser radiation, and highly localized thermal destruction by a focused laser beam in medical applications.

Intermolecular selective processes

Molecular-selective laser chemistry is the best developed and most promising field. In a wide range of problems we are tempted to use laser light for selective action on molecules of a particular type in gaseous mixtures, in solutions, on surfaces, and so on. Three major applications are laser isotope separation, laser production of very pure substances, and laser-induced photo-biochemical processes. Let us consider these three briefly.

Laser isotope separation. Many methods have been proposed and studied[6] for using lasers to separate isotopes. The best developed processes involve multistep selective photoionization of atoms, and photodissociation of molecules. These are both monomolecular processes, which are very amenable to control by laser radiation, so their prominence is not altogether surprising. In this article we are considering only molecular processes, and in any case the atomic process is relatively well understood from the physics standpoint and currently offers mainly technical problems. For isotope selection by multi-step or multiphoton dissociation of molecules, on the other hand, we still have physics problems to solve. Let us consider some of them.

Our qualitative understanding[7] of multistep photodissociation of molecules by infrared radiation is clear, and the work of the last two or three years has concentrated on obtaining experimental data for true (that is, nonaveraged) characteristics of the process and their correlation with theoretical models. Detailed checks of various theoretical models would be possible if we could measure the vibrational distribution produced in molecules by multistep infrared excitation; however, until very recently, we could only measure the average energy absorbed per molecule in the interaction region of the infrared field and the gas. Such measurements are rather uncertain, because the radiation pulse excites only a

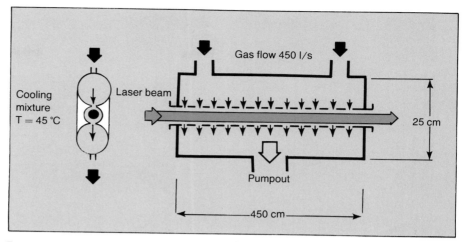

Fast transverse flowing separation cell for carbon isotope separation by high-power, high-repetition-rate CO_2 TEA lasers. (See also the photograph, figure 1.) A productivity of about 1 gram per minute of highly enriched carbon-13 is expected.

Figure 4

fraction q of the molecules, leaving a fraction $(1 - q)$ unexcited. (See figure 3a.) To measure the true characteristics of the process, not averaged over these two ensembles ("hot" and "cold") we need to know the value of q and its dependence on the intensity or energy flux and the laser frequency. In some recent work[8] various methods have been used to measure q and its dependence on energy fluence for the molecules OsO_4, SF_6 and CF_3I. Figure 3b shows some of the results—the variation of q with energy fluence for a 100 nanosec laser pulse in these three molecules.

Last year[9] we took an important step forward when we elaborated some theoretical models to make a determination of q by comparing theoretical data with experiment for the CF_3I molecule. The next step is to calculate q directly, starting from spectroscopic information about vibrational–rotational molecular transitions between the low lying vibrational levels responsible for molecular "leakage" into the vibrational quasicontinuum.

This type of isotope separation is very promising for light and moderate-mass isotopes in polyatomic molecules. For example, our experiments[9] in CF_3I (for carbon-isotope selection) show that theoretically limiting parameters for the multiphoton selective-dissociation process can be reached. Specifically:
▶ The minimum energy consumption is equal to the dissociation energy of the weakest molecular bond, C-I (2 eV);
▶ The dissociation yield for molecules in the irradiated volume reaches almost 100% for only moderate energy flux;
▶ The dissociation has high selectivity (> 30) at comparatively high gas pressure (5–10 torr);
▶ Significant extraction of the desired rare isotope carbon-13 is possible (about

50%) without substantial decomposition (< 2%) of the remaining molecules containing carbon-12, and
▶ Secondary photochemical processes that could destroy the high selectivity of the primary dissociation are completely eliminated.

Our group, together with the Kurchatov Institute for Atomic Energy, therefore proceeded with a scaled-up version of the original system to investigate carbon-isotope separation in macroscopic quantities. We use a CO_2 TEA laser with a high pulse repetition rate (up to 200 Hz) and average power 1kW. This is the system shown in figure 1. The rate of enrichment in CF_3I or CF_3Br irradiated in the laser-separation cell (figure 4) was so high that it required fast transverse pumping of the gas. We expect to attain a productivity of about 1 gram/minute of highly enriched carbon-13. The simplicity and technological advantages of this method should ensure its early industrial application.

Separating light and moderate-mass isotopes with abundances of about 1% or more is relatively easy, compared with the problem of extracting rare isotopes (for example, deuterium, with an abundance of about 1/5000) and isotopes of heavy elements from molecular compounds with a small isotopic shift (for example, the uranium-235 isotope in UF_6). To separate hydrogen and deuterium we should have a high separation selectivity—at least above 1000. Recent successful experiments[10] on multiphoton dissociation of some deuterium-containing molecules by CO_2-laser radiation encourage us to expect progress in this area.

Multiphoton molecular dissociation by a single-frequency high-power infrared pulse is a "brute-force" method of exciting a multilevel system with regularly spaced levels. High excitation levels and dissociation yields are

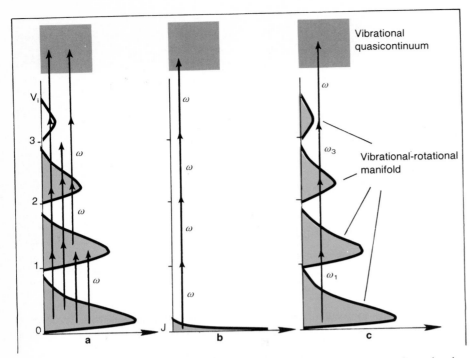

Multistep selective excitation of vibrational–rotational levels of polyatomic molecules, shown in three schemes. Part **a**: Low selective excitation at room temperature by single-frequency intense infrared pulses; **b**; Highly selective excitation at low temperature by single-frequency intense infrared pulses, and **c**: Highly selective excitation at room temperature by multifrequency infrared pulses of moderate intensity.   Figure 5

achievable, but with the inevitable reduction in isotopic selectivity—especially for heavy isotopes. This is because, although the frequency dependence of the rate of multistep molecular excitation by intense single-frequency radiation is resonant by its very nature, the resonance width is usually hundreds of times larger than the width of individual vibrational–rotational absorption lines. Indeed, a polyatomic molecule usually has very many vibrational–rotational levels (figure 5a). Also, at room temperature the molecules are originally distributed over many vibrational–rotational states. Because the anharmonicity of vibrations and isotopic shifts for heavy isotopes are usually less than the width of a vibrational–rotational band, several vibrational–rotational transitions turn out to be in resonance with fairly intense monochromatic infrared radiation. Cooling the molecules reduces the extent of the original distribution and so increases the sharpness of the resonance; this is particularly important for separating heavy isotopes and is illustrated in figure 5b. Yet the need to use an intense field makes the maximum selectivity impossible to realize, even in this single-frequency case. So we have to abandon multiphoton excitation and turn instead to two-step ir + uv photodissociation (figure 2b). To get some increase in selectivity we can photodissociate in a two-frequency infrared field.[11]

These techniques lead us naturally to the possibility of separating the functions of isotopically selective excitation and subsequent dissociation. Selective excitation needs radiation of finely tuned frequency ω_1 and intensity; for dissociation of the excited molecules a much less well-controlled frequency ω_2 suffices. Isotopic selectivity in such a system is enhanced because strong nonresonant radiation at frequency ω_2 does not perturb the resonant transition at frequency ω_1.

However, only by using multifrequency infrared radiation with its individual frequencies tuned in precise resonance with successive allowed vibrational–rotational transitions (figure 5c) can we achieve a high selectivity of excitation for molecular gases at room temperature with moderate requirements on laser radiation intensity. In my opinion, multistep excitation of high-lying vibrational levels by multifrequency infrared laser pulses of moderate intensity (10^2–10^5 watts cm^{-2}) is one of the most interesting approaches in fundamental and applied infrared laser photochemistry.

Purification of substances. Selective molecular photodissociation can be used to clear molecular impurities from a substance in the gas phase when standard chemical or physical methods are not suitable. The method depends on the presence of differences between the physical or chemical properties (or both) of the substance to be purified

and those of the impurity products after dissociation, so that standard separation techniques apply after irradiation. In this way purification of $AsCl_3$ (with infrared radiation) has been studied at the Institute of Spectroscopy in Moscow[12] and purification of SiH_4 (with vacuum ultraviolet radiation) has been studied at Los Alamos.[13]

Such techniques for purification by laser may have important industrial applications in the future; among the most interesting of these would be a method for removing toxic impurities from nuclear-reactor waste. But I should stress that the appropriate laser engineering today is, of course, in an embryonic state as far as such large-scale applications are concerned.

Chemical reactions and synthesis. With a beam of intense infrared radiation from a laser we can deposit energy—up to several electron volts—in molecules of a single type in a mixture without affecting the other molecules. This is in strong contrast to thermal excitation, in which all the molecules in the mixture store approximately the same amount of vibrational energy. We therefore have the prospect of activating chemical reactions in new ways, with highly excited molecules or their dissociation products.

Excitation of molecules under transient reaction conditions, and the formation of dissociation products before excitation thermalization, can make these unusual ways of chemical reactions studied under equilibrium more competitive. Indeed, under stationary conditions the composition of initial and final reactants is of course governed only by thermodynamics. For pulsed excitation under transient conditions, temporal factors become as essential as thermodynamic ones; for example, the ratio between the reaction rate along a particular pathway and the relaxation time for the nonequilibrium state of the molecular mixture.

An example of infrared photochemical synthesis is the work[14] on selective dissociation of CF_3I and CF_3Br in the presence of acceptors. This is part of a large study of a possible closed chemical cycle of laser separation of carbon isotopes on a practical scale. It is shown that photochemical conversion of CF_3Br molecules mixed with I_2 to CF_3I is highly effective (almost 100%) at pressures above 10 torr. The photochemical reaction in this example proceeds as follows:

$$CF_3Br + n\hbar\omega \rightarrow CF_3 + Br$$
$$CF_3 + I_2 \rightarrow CF_3I + I$$
$$CF_3 + I \rightarrow CF_3I$$
$$Br + Br \rightarrow Br_2 \qquad (6)$$

An analogous set of reactions occurs in multiphoton dissociation of CF_3I in the presence of Br_2. Analysis of the dissociation products of the $CF_3I + Br_2$

mixture shows that by changing the acceptor concentration and the temperature of the (cooled) walls it is possible to attain any ratio between the end dissociation products. When the bromine pressure is zero the only final product is the C_2F_6 molecule formed by recombination of CF_3 radicals. When CF_3I pressure is 0.2 torr and bromine pressure is greater than 3 torr, almost 100% of the dissociated CF_3I molecules are converted to CF_3Br by reactions analogous to those in 6 above.

The conclusion from this work, and other published studies, is that laser photochemical synthesis based on radical reactions will make possible the production (for short times) of any concentration of various radicals. The synthesis reaction can therefore be directed along the shortest pathway to provide the maximum yield of the desired product for the smallest quantity of initial materials. In the case of infrared multiphoton dissociation it should be remembered that the decay of the excited molecules occurs by the breaking of the weakest bond. Nevertheless, in practice this does not restrict the possibilities of laser chemical synthesis.

Photobiochemistry and photomedicine. In my opinion the prospects for laser-molecular interactions and their applications are far wider than the isotope-separation and gas molecular mixture work we have discussed so far in this article. I think the most promising possibilities lie in photobiochemistry and photomedicine. In my previous PHYSICS TODAY article[7] I briefly discussed several problems in this area, particularly photochemistry of nucleic acid bases, where the selective abilities of laser light may prove very useful for selective mutagenesis and physical sequencing of bases in DNA.

For high selectivity it is natural to use selective excitation of vibration followed by transfer of the vibrationally excited molecules to those excited electronic states that have a high quantum yield for photoreaction (figures 2a, 2b). But absorption by solvents in the infrared region, the abundance of vibrational levels in biomolecules, the low energy limit for the vibrational quasicontinuum and the very fast intramolecular and intermolecular vibrational relaxation make this approach hard to realize with present-day tunable infrared lasers.

The first successful data[15] on selective excitation of nucleic acid fragments have been obtained in a less obvious way—by two-step excitation through an intermediate electronic state. Despite the almost complete overlap of the ultraviolet spectra of electronic absorption—so that there is no selectivity at all at the first excitation stage—the total photoreaction

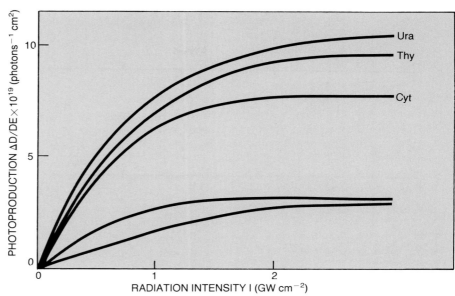

Photoproduct yield versus irradiation intensity for all five-nucleic-acid bases (from top: uracil, thymine, cytosine, guanine and adenine) in neutral aqueous solution. These data were obtained by two-step excitation. (From reference 15.)   Figure 6

yield differs substantially for various nucleic-acid bases. For example, figure 6 shows normalized dependences of photoreaction yield on picosecond pulse intensity (the pulse duration is 20 picosec) at a wavelength of 265 nm for the five nucleic-acid bases. The photoreaction yield is here measured from the relative value of irreversible decrease in the ultraviolet absorption bands for these bases, with the total decrease over many pulses normalized to the total energy fluence of irradiating pulses (in photons cm^{-2}), that is, to the value of the ultraviolet radiation dose.

When the intensity is below saturation for the first electronic transition, the photoreaction yield scales as the square of the pulse intensity, proving that we are here dealing with two-step photoexcitation. When the electronic transition becomes saturated, the photoreaction yield becomes linearly dependent on intensity, because the second electronic transition from excited states produces a very fast photoreaction; in other words, it is an unsaturable reaction.

The mechanism of these photoreactions may be ionization, dissociation or a chemical reaction of the bases after they have absorbed two ultraviolet photons with energy totalling about 8 eV, and our studies are directed toward revealing which it is. Selectivity in such multistep reactions may arise from:
▶ differences in the lifetimes of intermediate states
▶ differences in cross sections for the subsequent (second) electron transition
▶ differences in the yields of the last photochemical step of the process.

When we use shorter picosecond pulses and independently tunable picosecond radiation at the second electron transition we will be able to gain insight into the mechanism that causes selectivity.

One more important feature of ultraviolet-laser induced photobiochemical reactions is the very high potential reaction yield at small irradiation doses. Figure 7 shows schematically three potential schemes for photoexciting biomolecular reactions. Under

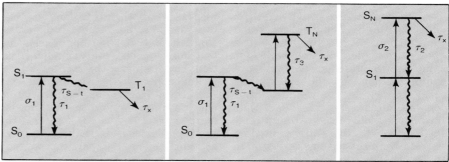

Quantum yields of photoreactions induced by single-photon (a) and two-photon excitation through the triplet (b) and singlet (c) electron excited states of biomolecules. The yield from scheme c can reach 100% in, say, thymine.   Figure 7

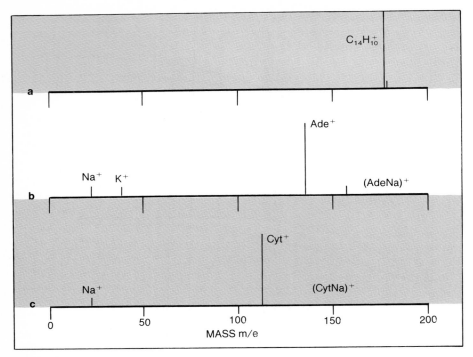

Mass spectra of photoions formed during irradiation of the surface of molecular crystals of anthracene (a), adenine (b) and cytosine (c) by an ultraviolet pulse from a KrF laser. (From reference 16.) Figure 8

low-intensity radiation (no more than 10^6–10^7 watts per cm^2) only single quantum photoreactions from the first singlet S_1 or first triplet T_1 states are possible (figure 7a). As the intensity increases, the excitation of the molecule from the triplet state becomes possible (figure 7b). The photoreaction yield for molecules that have absorbed two ultraviolet photons may be as great as 100%. Hence the ultimate yield is limited by the probability of molecular transition to the triplet state,

$$\eta_{ST} = \left(1 + \frac{\tau_{ST}}{\tau_1}\right)^{-1}$$

The value of η_{ST} is much less than unity because the singlet-state lifetime τ_1 is usually in the range 10^{-9}–10^{-11} sec.

If the high-power ultraviolet laser has a pulse length no longer than τ_1, the molecule can be excited in two steps through the singlet state (figure 7c). Such processes are responsible for two-quantum photoreactions of nucleic-acid bases, such as those shown by their yields in figure 6. The yield can reach essentially 100% in this case. The photoreaction yield for thymine, for example, through the scheme of figure 7a (dimerization from the triplet state) is only 0.5×10^{-3}. The yield for intercombination conversion is 10^{-2}, thus limiting the yield of the two-step reaction in figure 7b. But at the same time the yield from the scheme in figure 7c can reach 100%.

Thus we hope for new methods of phototherapeutics in molecular photo-medicine that will have very high efficiencies for small ultraviolet exposure.

Mode- and bond-selective effects

When multiphoton molecular processes were first studied, around 1975 to 1977, some scientists hoped to attain mode-selective or bond-selective excitation and photoreactions. Experiments quickly showed that all vibrational modes of polyatomic molecules were involved in the multiphoton-excitation process. Typical pulse lengths of the infrared radiation would be around 10^{-6} to 10^{-8} sec. Today, with our better understanding of the processes, we can identify the experimental conditions (large molecules and ultra short pulses) where we can hope to see (and subsequently use) mode- or bond-selective effects in polyatomic molecules in a strong infrared field.

The rate of intramolecular transfer of vibrational excitation among the modes depends on many factors, including the excitation level of an individual mode and its coupling with other modes, or whether collisions are permitted, and so on. Under collisionless conditions the time for excitation transfer between weakly coupled vibrational modes (when the excitation level is sufficiently high, amounting to a substantial fraction of the dissociation energy) can be placed in the approximate range 10^{-11} to 10^{-13} sec. If the excitation rate exceeds the rate of vibrational-energy equipartition we may hope to see mode-selective effects. Current experimentation has not yet reached this stage.

The most interesting molecules for studying bond-selective infrared photochemistry are large, and they have well-separated functional groups. Clearly strong vibrational modes associated with the different functional groups will be relatively weakly coupled. So by exciting vibrations in one functional group it is possible to achieve a substantially nonequilibrium energy distribution inside a large molecule during a short time interval, and hence nonstatistical behavior in the photochemical processes.

Assume, for example, that we are conducting an experiment to observe nonstatistical behavior of a long-chain polyatomic molecule with different functional groups at its two ends. Let one of the functional groups—say CF_3—be excited by a high-power CO_2 laser pulse. Because this group is not directly bound to the motion of the atoms at the other end of the chain, vibrational perturbation will reach that end after a characteristic time $\tau \approx n/2\pi c X_0$, where n is the number of bonds in the chain and X_0 is the characteristic anharmonicity constant responsible for the vibrational interaction (that is, for the anharmonic coupling of adjacent bonds in the chain). If the molecule is excited by a high-power infrared pulse of duration $\tau \lesssim 10^{-9}$ sec, then with $X_0 \approx 1\ cm^{-1}$ and n around 100 we can expect an essential difference between the results and the predictions of the statistical model. We could see the differences by observing hot bands in the ultraviolet or infrared spectra of another functional group at the opposite end of the molecule or (more directly) by observing the molecular photodissociation by breaking of a different bond from the one with the smallest dissociation energy.

Some experiments have shown mode-selective effects even when the laser-pulse duration is as large as 10^{-8} or 10^{-7} sec.

For example, a group at Exxon Research Laboratories has reported[16] multi-photon dissociation of a rather complex gaseous compound of uranium whose simplified name is $UO_2(hfacac)_2THF$. They used a CO_2 laser under both continuous and pulsed operation. The photodissociation occurred in a molecular beam, free of collisions and subsequent thermal heating even under continuous irradiation. The CO_2 laser frequency was in resonance with the asymmetric O-U-O vibration of the uranium-containing part of the molecule, which is sensitive to the isotopic composition of the uranium and oxygen.

The multiphoton absorption and dissociation properties of this molecule are those of a molecule with a very low vibrational quasicontinuum limit. Excitation of the $0 \rightarrow 1$ vibrational transition of the asymmetric O-U-O vibration quickly relaxes, transferring excitation to a few tightly coupled modes, and the asymmetric mode returns to its ground state where it can continue to absorb infrared photons. Even a pulse energy density as low as 0.1 J/cm^2 yields high values of dissociation. A theoretical model reported in the same paper[16] assumes that the absorbed energy is distributed among only five vibrational modes. Shorter laser pulses (about 1 nanosec or less) appear to be necessary to prove this conclusion.

Let us turn now to electronic transitions. Excitations of molecular electronic states appear to be more promising for bond-selective photochemistry than does the work with vibrational states summarized above; in fact, I believe this is the only possibility for complex biomolecules in a condensed medium. It may even be possible eventually to study, selectively, individual bases in a long chain of DNA molecules by this method. Success with that problem would allow us to develop a new method of selective mutagenesis and also give us a direct method for direct sequencing of bases in the DNA chain—that is, for reading the DNA code. This, I suspect, is the most complicated problem in selective laser photochemistry.

After the successful experiments on solutions with nucleic-acid fragments which we mentioned above (figure 6) we undertook experiments on selective action on uracil in the ApU dinucleotide (a part of a single-stranded RNA). These very preliminary experiments have already shown[15] some selectivity.

Bond-selective photoionization of molecules on metal surfaces is the concept behind a possible photoion projector for direct visualization of molecular bonds. I discussed the idea of this device, which combines wave and corpuscular microscopy, in my previous PHYSICS TODAY article.[7] To realize the device we have several problems to solve:

▶ Investigate molecular ion photodetachment from the surface under laser light

▶ Investigate the photodetachment of these molecular ions from a macromolecule

▶ Make these photodetachment operations photoselective.

Although these are major and difficult tasks we have made some progress already on the first one.[17] With a KrF ultraviolet laser at 249 nm, having radiation power less than 10^5 W/cm^2 (fluence less than 10^{-3} J/cm^2), we could observe molecular ions originating from various crystal surfaces—anthracene and all five nucleic acid bases. Figure 8 shows the spectra recorded with a time-of-flight mass spectrometer. We know the ions are photophysical in origin because they appeared at intensities much lower than the threshold of thermal action on the surface. Assuming that the mass of the molecular ion is a clue to the type of nucleic-acid base, we can simplify the solution of the difficult third problem, above, to a simple analysis of the mass of the detached photoions.

References

1. T. H. Maiman, Nature **187**, 493 (1960).
2. A. Javan, W. R. Bennett Jr, D. R. Herriott, Phys. Rev. Lett. **6**, 106 (1961).
3. V. S. Letokhov, Science **180**, 451 (1973). (The ideas presented in this article were stated in my report to the Lebedev Physical Institute in 1969, titled "On possibilities of isotope separation by the method of resonant photoionization of atoms and photodissociation of molecules by laser radiation." The report was published as a preprint of the Institute of Spectroscopy ten years later.)
4. R. V. Ambartzumian, V. S. Letokhov, E. A. Ryabov, N. V. Chekalin, Pis'ma Zh. Eksp. Teor. Fiz. (Russian) **20**, 597 (1974). [Sov. Phys.,—JETP Lett. 20 273 (1974)].
5. C. Bordé, A. Henry, L. Henry, C. R. Acad. Sci. Paris. **B262**, 1389 (1966).
6. V. S. Letokhov, C. B. Moore, in *Chemical and Biochemical Applications of Lasers*, Volume 3, (C. B. Moore, ed.), Academic Press, New York (1977): pages 1–165; V. S. Letokhov, Nature **277**, 605 (1979).
7. N. Bloembergen, E. Yablonovitch, PHYSICS TODAY, May 1978, page 23; V. S. Letokhov, PHYSICS TODAY, May 1977, page 23.
8. R. V. Ambartzumian *et al.*, Pis'ma Zh. Eksp. Teor. Fiz. (Russian) **28**, 246 (1978); V. N. Bagratashvili *et al.*, Zh. Eksp. Teor. Fiz. (Russian) **76**, 18 (1979); V. N. Bagratashvili *et al.* in *Laser-Induced Processes in Molecules*, (K. L. Kompa and S. D. Smith, eds.), Springer Series in Chemical Physics, Vol. 6, Springer-Verlag, Heidelberg (1979); page 179.
9. V. N. Bagratashvili, V. S. Doljikov, V. S. Letokhov, A. A. Makarov, E. A. Ryabov, V. V. Tyacht, Zh. Eksp. Teor. Fiz (Russian) **77**, 2238 (1979); V. N. Bagratashvili, V. S. Doljikov, V. S. Letokhov, E. A. Ryabov, Appl. Phys. **20**, 231 (1979).
10. I. P. Herman, J. B. Marling, Chem. Phys. Lett. **64**, 75 (1979), and J. Chem. Phys. **72**, 516 (1980).
11. R. V. Ambartzumian *et al.*, Pis'ma Zh. Eksp. Teor. Fiz. (Russian) **23**, 217 (1976). [Sov. Phys.—JETP Lett. **23**, 194 (1976)].
12. R. V. Ambartzumian *et al.*, Kvantovaya Elektronika (Russian) **4**, 171 (1977).
13. J. H. Clark, R. G. Anderson, Appl. Phys. Lett. **32**, 46 (1978).
14. G. I. Abdushelishvili *et al.*, Pis'ma Zh. Techn. Fiz. (Russian) **5**, 849 (1979).
15. D. A. Angelov, P. G. Kryukov, V. S. Letokhov, D. N. Nikogosyan, A. A. Oraevsky, Appl. Phys. **21**, 391 (1980).
16. D. M. Cox, R. B. Hall, J. A. Horseley, G. M. Kramer, P. Rabinowitz, A. Kaldor, Science **205**, 390 (1979).
17. V. S. Antonov, V. S. Letokhov, A. N. Shibanov, Pis'ma Zh. Eksp. Teor. Fiz. (Russian) **31**, 471 (1980). □

Laser annealing of silicon

Scanned and pulsed laser beams rapidly heat silicon surfaces, revealing basic mechanisms of crystallization and amorphization, while paving the way for the fabrication of novel semiconductor devices.

John M. Poate and Walter L. Brown

PHYSICS TODAY / JUNE 1982

Silicon is one of the best understood of all materials because it lies at the heart of one of the central technologies of the twentieth century: integrated circuits. How much more can we learn about silicon and ways of handling it? Quite a lot, apparently, if the remarkable developments in laser annealing over the past five years are any indication.[1–4]

A quick look at the process by which integrated circuits are made reveals the reasons for the great interest in the new technology of laser annealing and points to areas in which we need more knowledge about silicon. To make integrated circuits, manufacturers slice very pure crystals of silicon into thin wafers, such as those that appear in the cover photograph and in figure 1. Then they use complex fabricaton techniques to produce electrically doped regions close to the surface of the silicon and to produce insulating and conducting films that overlay the surface. These doped regions and films can be as thin as a few hundred atomic layers.

One can introduce dopants by diffusion, which requires heating the wafer to about 1000 °C for half an hour. Ion implantation is an alternative method of introducing dopants, but even this requires high-temperature heating to remove damage caused by the implantation process. Other currently practiced processing steps, such as oxidation to produce silicon dioxide films, also entail heating the whole wafer for extended periods.

Heating the entire wafer is not ideal because it limits the range of surface structures that one can produce. Because they permit heating only in the region of interest, focusable sources such as lasers, electron beams and even ion beams are attractive alternatives. Furthermore, with lasers, one can melt and solidify surface layers in times as

John M. Poate and Walter L. Brown are members of the radiation physics research department at Bell Laboratories, in Murray Hill, New Jersey.

short as nanoseconds. The initial studies, performed in Russia, showed[1] that pulsed or continuous-wave laser radiation removes ion-implantation damage in semiconductors. Thus the term "laser annealing" was coined. It now applies generically to a much broader class of phenomena ranging from crystal growth to alloying.

Laser annealing is being used in basic solid-state and materials-science research and in the investigation of new processing technologies for silicon integrated circuits. In this article we will highlight some of the more novel features and applications of silicon crystal growth and solidification that laser annealing techniques have revealed. We will look at the trapping of unusually large concentrations of dopants, the formation of amorphous silicon from the melt with very fast solidification, as well as the exciting subject of silicon crystal growth on amorphous substrates, something that raises the possibility of constructing three-dimensional integrated circuits. We will also discuss the basic question of the mechanism of electronic-to-thermal energy transfer, which has generated controversy over the past several years.

Fast crystal growth

Ion bombardment renders surface layers of silicon amorphous. Although amorphous silicon is thermodynamically unstable in the presence of crystalline silicon, it does not crystallize at room temperature because of kinetic barriers to atomic motion. However, at quite modest temperatures the amorphous portion crystallizes layer by layer, or "epitaxially," on the underlying crystal. In epitaxial growth, atoms form a new crystal plane by registering with the atoms of an existing plane. Figure 2 shows the exponential rate at which the velocity of the crystallizing interface increases with temperature. The continuous line is an extrapolation of furnace measurements of crystallization rates in the temperature range

500–600 °C. At temperatures of about 1100 °C, a crystalline layer 1000 Å thick will grow from the amorphous material in about 1 msec. It is not possible to heat for such short times in conventional furnaces.

At Hughes Research Laboratory in Malibu, California, researchers are using[5] an elegant technique in which they heat the surface with a single, long pulse from a continuous-wave argon laser and observe the crystal growth directly by monitoring the reflectivity of light from a probing laser. Interference between light reflected from the surface and light reflected from the amorphous–crystal interface causes the reflectivity to oscillate as a function of time. Thus they measure directly the time for the interface to travel one wavelength. It appears as though one can extrapolate the low temperature furnace rates to temperatures of about 1000 °C. The Hughes measurements indicate solid phase regrowth of amorphous silicon to within 50 °C of the 1423 °C melting temperature of crystalline silicon, as shown in figure 2. However, the curve above 1100 °C is dashed because there are predictions that amorphous silicon melts at temperatures considerably beneath the crystalline melting temperature. We will discuss the thermodynamics of silicon later.

Clearly, continuous lasers can heat silicon until it melts. But pulsed lasers can melt thin surface layers of silicon. As figure 2 shows, there is a sharp jump in the crystallization velocity above the solid phase values because of the much greater mobility of atoms in the liquid and because of the very high temperature gradients present in pulsed laser melting. Scientists at the University of Catania, in Sicily, and at Oak Ridge National Laboratory have used heat-flow calculations to estimate these velocities. The velocity depends on the rate at which the latent heat of crystallization is extracted from the interface; it therefore depends on the thermal con-

Laser annealing of a silicon wafer by localized melting restores the material's crystallinity after ion-implantation. The radiation is from a Q-switched frequency-doubled Nd-YAG laser. (Photograph courtesy of George Celler, Bell Laboratories.)

Figure 1

ductivity of crystalline silicon and the temperature gradient in the solid.

Recently, a collaborative group from Cornell University and Los Alamos and Sandia National Laboratories measured[6] the interface velocity directly. They melted surface layers with a Q-switched ruby laser, using 30-nsec pulses with energy densities of 2 J/cm². By monitoring the conductance of the silicon sample they determined the position of the liquid–solid interface as a function of time. This was possible because the conductance depends on the thickness of the metallic liquid layer, whose conductivity is much greater than that of the semiconductor. They measured an interface velocity of 2.8 m/sec, in excellent agreement with the calculated value of 2.7 m/sec based on heat flow. This dynamic conductivity technique will undoubtedly emerge as an important tool for measuring melting and freezing rates.

At velocities of about 3 m/sec the melt crystallizes with a high degree of epitaxial perfection. It is interesting to contrast this type of crystal growth with conventional growth from the liquid phase, where the growth rates are typically 10^{-5} m/sec. In the conven-

tional methods, the growth rate is the rate at which the seed crystal is mechanically drawn from the melt. What ultimately determines the velocity of crystallization is the degree of supercooling, or undercooling, of the growing interface with respect to the melting temperature: The greater the undercooling the greater the velocity.

There have been no definitive measurements or calculations of the relationship of undercooling to interface velocity in silicon; the boundaries of the hatched region in figure 2 represent plausible estimates. The upper limit to the recrystallization velocity occurs when solidification is so fast that an amorphous layer forms instead of a crystalline one, because the atoms do not have time to find equilibrium locations at the moving interface. Figure 2 shows that the range of crystallization velocities, from the amorphous solid and the liquid phases is quite remarkable—one can vary them experimentally over 14 orders of magnitude.

Trapping dopants

One of the most striking manifestations of the rapidly moving liquid–solid interface is the incorporation, or trap-

ping, of dopants in the solid at concentrations in excess of equilibrium solubilities. Experiments usually begin with ion implantation of impurities in crystalline silicon, followed by melting of the resulting implanted amorphous layer and part of the underlying crystalline silicon with a pulsed laser. The final step is to use Rutherford backscattering and ion-channeling techniques to observe the depth distribution and lattice-site location of the impurities in the recrystallized surface layer. The impurity depth distributions are fitted to theory by using known silicon liquid-state diffusivities and by assuming a unique interfacial segregation coefficient k', which is defined as the ratio at the interface of impurity concentration in the solid to that in the liquid. The fits are made assuming that once the solid traps the impurities their motion is negligible; this assumption is quite valid for most impurities because of the high cooling rates and low diffusivity in the solid phase.

The table, from Woody White and colleagues at Oak Ridge,[7] compares segregation coefficients k_0 at slow solidification with segregation coefficients k' at 4 m/sec solidification. It also compares solid solubilities C_0 at slow solidification with the maximum concentration C_{max} that can be trapped on lattice sites by laser annealing with 4 m/sec solidification. The maximum solid solubility of arsenic exceeds its equilibrium solid solubility by a factor of four; for bismuth, C_{max} exceeds C_0 by nearly three orders of magnitude.

We are collaborating with Pietro Baeri and Nuccio Foti of Catania and Tony Cullis of the Royal Signals and Radar Establishment in Malvern, England, to investigate the physical basis of these supersaturation phenomena. We model the trapping effects in terms of competition between the speed of the moving liquid–solid interface and diffusion of the impurities away from the interface. Impurities are trapped if they reside at the interface for times greater than the intrinsic recrystallization time of the interface. Clearly, supersaturation effects depend on the very rapid motion of the interface. To demonstrate the effect of this motion we varied the regrowth velocities by changing the laser pulse length and by changing the temperature of the substrate, which determines the rate of thermal conduction through the silicon.[8] At present we can vary the regrowth velocity from 1 to 20 m/sec.

Segregation coefficients. Figure 3 shows segregation coefficients k' for indium as a function of velocity and substrate orientation.[8] The strong effect of velocity on trapping is evident. Moreover, k' appears to saturate at values less than unity at high velocities. The other notable effect that figure 3 shows is the strong dependence of segregation on the crystal orientation of the substrate. The curve for the (111) plane is broken above 5 m/sec because many extended defects remain in the lattice and it is not possible to determine segregation coefficients. As velocities approach 20 m/sec, epitaxial crystal growth fails completely and amorphous silicon forms at the moving interface—a subject we discuss later.

These segregation data present a real challenge to the modern theory of crystal growth. Authors with various perspectives have presented[2] a number of kinetic analyses that predict the main features of the velocity dependence but appear unable to handle the orientation dependence in anything but an *ad hoc* fashion. At Bell Laboratories, George Gilmer is simulating the process of crystal growth in silicon using Monte Carlo calculations in the framework of the kinetic Ising model. He shows that at a growth velocity of several meters per second the undercooling at a (111) interface should be about a factor of 2 greater than that at a (100) interface; undercooling at a (110) interface should be about midway between these two values. The in-

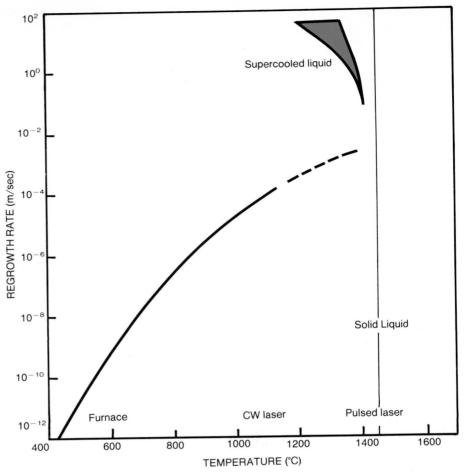

Solid-phase epitaxial regrowth rate is the speed at which amorphous silicon becomes crystalline when heated. The interface between these two solid forms of silicon advances into the amorphous region at a rate that increases exponentially with temperature. Temperatures characteristic of various heating schemes are indicated. At furnace temperatures of the order of 600 °C the interface moves 1000 Å in about 1000 seconds; with a continuous-wave laser heating to temperatures of the order of 1100 °C, this time drops to about 1 msec; and with a pulsed laser heating to the melting point, the time decreases to about 100 nsec. The blue region represents possible crystal growth rates from the liquid phase. Figure 2

creased undercooling at a (111) interface should give rise to increased values of k', because of the greater chance of impurity condensation.

Scientists at the National Bureau of Standards have considered[9] the thermodynamic limits to the maximum amount of impurity that silicon can incorporate. The Oak Ridge group's experimental results for arsenic impurities (see the table) are close to the thermodynamic limit, but those for indium are still considerably beneath the limit. The reason for this difference is that at sufficient concentrations of indium the rapidly moving planar interface becomes unstable before the thermodynamic limit is reached. The high concentration of impurity segregating ahead of the interface lowers the melting point and an instability develops that is known as "constitutional supercooling." A forward fluctuation of the freezing interface rejects impurity laterally and this material crystallizes locally faster than the adjacent materi-

al. In this way a cellular pattern of microsegregation develops. The lateral wavelength of the segregation is, to first order, D/v, where D is the liquid-state diffusivity and v is the interface velocity. Figure 4 shows transmission electron micrographs of the cellular structure produced by the laser annealing of (100) silicon implanted with a high concentration of impurity—3×10^{15} atoms of indium/cm². The dark cell walls are laterally segregated indium. The cell dimensions increase with decreasing regrowth velocity. Jagdish Narayan[10] of Oak Ridge and we[11] have been able to predict the stability and size of the cells remarkably well using morphological stability theory.

Amorphous silicon

Scientists at Bell Labs and Harvard University recently investigated[12] some of the thermodynamic properties of amorphous silicon in light of the possibilities offered by laser heating.

Trapping and solubilities in silicon

| Dopant | Dopant concentration ratio: solid/liquid | | Maximum concentration of dopant in solid | |
	at very slow solidification k_o	at 4m/sec solidification k'	at very slow solidification C_O (10^{18}/cm^3)	at 4m/sec solidification C_{max} (10^{18}/cm^3)
B	0.80	~1.0		
P	0.35	~1.0		
As	0.30	~1.0	1500	6000
Sb	0.023	0.7	70	2000
Ga	0.008	0.2	45	450
In	0.0004	0.15	0.8	150
Bi	0.0007	0.4	0.8	400

From reference 7

Their ideas are demonstrated schematically in figure 5, which shows the Gibbs free-energy diagram of amorphous, crystalline and liquid silicon. The intersection of the free-energy curves of crystalline and liquid silicon is, by definition, the equilibrium melting temperature T_c. Amorphous silicon certainly has a higher free energy than crystalline silicon, so if the shapes of the free-energy curves of these two forms are similar, as in figure 5, one can argue that the amorphous phase will melt at a considerably lower temperature (T_a) than the crystalline phase (T_c). This argument rests on the very plausible premise that the phase transition from the tetrahedral amorphous phase to the close-packed metallic liquid is first order.

Recently we measured[13] the heat of crystallization of amorphous silicon by using electron-beam heating, and, in collaboration with the Harvard group, by using calorimetry. Our results indicate that the differences in free energy are sufficient to imply a depression ($T_c - T_a$) in melting temperature of some 250 °C. Such a large depression in melting temperature is remarkable. One cannot observe this depression in furnace heating because the amorphous silicon will crystallize in the solid phase before such high temperatures are reached. But if amorphous silicon is heated fast enough the melting temperature depression should show up. Our pulsed electron heating experiments[13] did, in fact, give some evidence for a substantial reduction in melting temperature; this evidence was based in part on energy calculations and on an analysis of the unusual microstructures formed during solidification.

However, as we mentioned earlier, the Hughes group has evidence[5] that the amorphous phase can be heated to within 50 °C of T_c without melting. David Turnbull of Harvard has suggested that if the Hughes measurements are correct, the data are evidence for considerable superheating of amorphous silicon. More experiments are required to resolve this controver-

sy. We should note that neither of the experiments described here measures the temperature directly.

Related to this matter are experiments showing that amorphous silicon forms from the melt when the quench rates or regrowth velocities are fast enough. Groups at Harvard and IBM made[14] the first of these important observations using picosecond or nanosecond pulses of ultraviolet light. The group at Malvern, using 2.5-nsec pulses

of light from a high-power frequency-doubled ruby laser, recently produced amorphous spots some 3 mm in diameter, thus making materials analysis much easier. Heat flow calculations indicate that the velocities required to form the amorphous phase on the (100) plane of silicon are in the vicinity of 20 m/sec; lower velocities are sufficient on the (111) plane. The Harvard group has pointed out that it is thermodynamically possible to form the amorphous phase if the melt is undercooled to a temperature below T_a. Therefore if we knew the exact dependence of undercooling on interface velocity, we could determine T_a from these experiments. Estimates by Ken Jackson of Bell Labs indicate that at about 10 m/sec the undercooling may be several hundred degrees.

Crystals on amorphous substrates

One of the most important results of research into the transient thermal processing of semiconductors is the technology for forming crystalline silicon on amorphous substrates. Manufacturers have produced small-grain

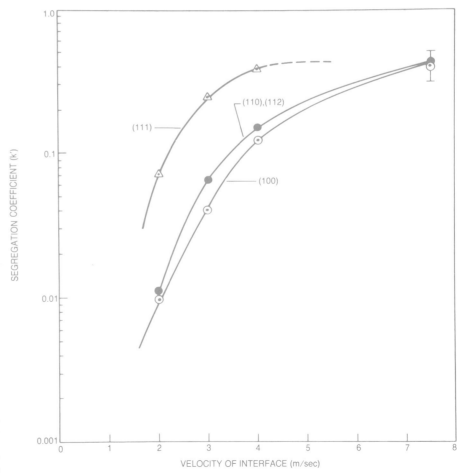

Segregation curves for indium impurity in silicon show that there is less segregation, or more trapping, at higher crystallization rates and for certain substrate orientations. The segregation coefficient (k_o) for asymptotically slow recrystallization is 4×10^{-4}. The (110) and (112) orientations showed no differences within experimental error. (Reference 8)   Figure 3

polycrystalline silicon on SiO_2 for many years by methods involving chemical vapor deposition. However, the resulting material has an exceedingly fine grain size of about 500 Å, so that while it is useful for making electrical interconnections, its electrical transport properties are much poorer than single-crystal silicon and it is uninteresting as a material in which to fabricate active devices.

At Stanford University, Jim Gibbons and his associates achieved[15] major increases in grain size by scanning a continuous laser spot over such a small-grain polycrystalline film. This method requires a laser intensity high enough to produce a region of molten silicon. The silicon recrystallizes in the trailing edge of this molten spot, as shown schematically in figure 6a. This process yields sliver-shaped grains that are typically 25 microns long and 2 microns wide, with their long dimensions in the direction of the motion of the molten spot. Although this material is still not ideal in devices, it compares quite favorably with the best known silicon-on-insulator material, epitaxial silicon-on-sapphire.

The sliver-shaped crystallites in the Stanford experiments grew laterally over a layer of amorphous SiO_2, seeded by the leading edge of the thin silicon layer being formed. Thus, we call this type of crystallization lateral crystal growth or lateral epitaxy. The fact that the material grown consists of many grains is a clear indication that the nucleation and growth of new grains is uncontrolled—that not all solidification takes place on a single seed. When the Stanford group used lithography to pattern the vapor-deposited polysilicon in small islands before laser scanning, they found it possible to grow each island as a single crystal, so long as the islands had a size comparable to the sliver-shaped grains grown in the continuous film.

Gaining control of crystal nucleation and growth is of great importance to semiconductor manufacturing. To increase the grain size in laterally grown material, research workers at Xerox Corporation in Palo Alto, California, investigated[16] the importance of the shape of the molten hot spot and concluded that a circular spot is particularly poor. With such a spot, random nucleation of refreezing occurs first at the edges of the path of the molten zone and grows toward the center of the path, which is the last to freeze. The Xerox group proposed using a spot with a kidney bean shape, concave at its trailing edge. In this case most of the trailing edge of the molten zone has a concave isotherm, so the center of the trailing edge is the first region to freeze and crystal growth propagates outward instead of inward. Using this technique they were able to grow single-

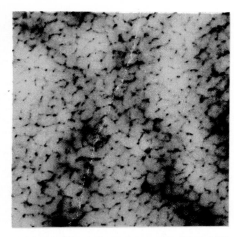

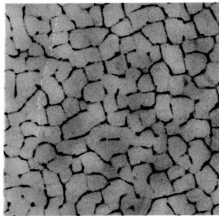

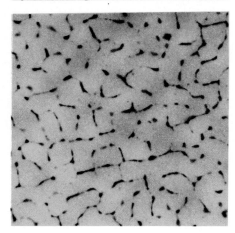

Silicon layers implanted with indium and resolidified show different cellular patterns of microsegregation for different interface velocities. From top to bottom the layers were resolidified at velocities of 4 m/sec, 3 m/sec and 2 m/sec. Each of these transmission electron micrographs shows an area about ¾ micron wide. (From reference 11.) Figure 4

crystal islands as wide as 20 microns. Workers at Bell Laboratories in Holmdel, New Jersey, have shown that tailoring the thermal expansion coefficients of the substrates is also important for the formation of large single crystals.

The success of lateral crystal growth

led the Stanford group to construct a novel two-level, high-rise complementary metal-oxide semiconductor device. A single gate drives an enhancement-mode p-channel metal-oxide semiconductor in n-type substrate silicon as well as its n-channel complement, which is in silicon that was laser recrystallized on an upper level. This is the first example of a vertically integrated circuit.

Controlling crystal orientation

Neither sliver-shaped crystallites in a continuous film nor single-crystal islands on an amorphous substrate have a unique orientation, though there may be a substantial preference for a particular surface normal. Researchers have tried to use an underlying crystalline substrate in conjunction with a scanning-beam technique, to direct or "seed" the crystal orientation of the solidifying surface layer. Figure 6b shows the idea: Specific areas of the deposited amorphous or polycrystalline silicon melt through to the crystalline substrate. Solidification from these points then defines the crystal orientation of the surface layer.

At Texas Instruments, Al Tasch and his associates report seeding crystal growth to distances of 30 microns over a SiO_2 substrate. At Bell Laboratories, Harry Leamy, George Celler and their associates have successfully seeded overgrowth for hundreds of microns by using a planar geometry with pads of vapor-deposited silicon recessed into thick SiO_2 films. A group at Hitachi has had success in forming crystalline bridges over silica by using a pulsed Q-switched laser. Because the thermal conductivity to the substrate is high where the molten silicon puddle is in contact with the crystalline silicon substrate, that region crystallizes first and seeds the subsequent crystallization over the oxide. Although the crystallization distances achieved in this way are rather short—about 3 microns—they are sufficient to form a bridge of epitaxial silicon over a silicon gate.

A laser is not the only means of producing a moving molten zone for lateral crystal growth on amorphous substrates. At MIT Lincoln Laboratories, John Fan and his coworkers used[17] graphite-strip heaters to melt a layer of silicon of SiO_2. One of their heaters is narrow and defines the molten zone. It moves parallel to the surface of the film at a speed of about 5 mm/sec. Striped openings through the SiO_2 to the underlying crystalline silicon seed the epitaxial growth of the overlayer. With this method single crystals grow to cover large areas—square centimeters in size—if the molten zone moves either parallel or perpendicular to the seeding stripes. With parallel motion lateral epitaxy may persist millimeters beyond the last seeding stripe.

Figure 7 shows an optical micrograph of a single-crystal overlayer grown by moving the molten front parallel to the stripes. Although graphite heaters provide a wider molten zone than does laser heating, they subject the whole substrate to near-melting temperatures. This may make them unsatisfactory for use in building high-rise device structures. But these heaters remain extremely attractive as a way of producing thin crystalline materials.

Energy transfer controversy

In all of the experiments that we have discussed, the laser introduces energy into the semiconductor's electronic system. Laser photons of energy greater than that of the band gap generate electron–hole pairs. Free electrons and holes may also absorb energy from photons and be promoted to states of higher kinetic energy in the conduction and valence bands. In all cases, the input energy flows into electronic excitation of the solid.

On the other hand, the experimental observations associated with laser annealing are remarkably well explained by models that simply involve heat, that is, by excitation of the phonon system not the electronic system. Sufficient heat melts a layer at the surface, with all of the various consequences discussed above.

The central issue is the *rate* at which the energy moves from the electrons to the phonons, thus becoming heat. If the transfer is fast compared with the time over which excitation occurs (the length of a laser pulse, for example) the electronic system will never be far out of equilibrium with the lattice; if it is slow, the electrons may become much hotter than the lattice. Jim Van Vechten of IBM made the controversial assertion[18] that a highly excited electronic system would retain its excitation for tens or hundreds of nanoseconds. He suggested that many of the results from annealing with a Q-

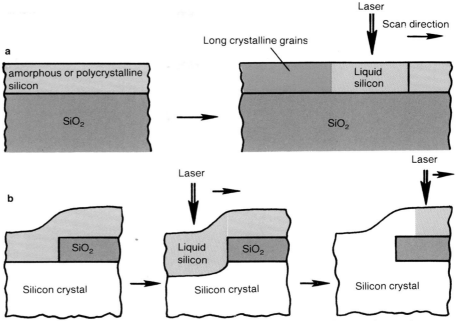

Lateral epitaxy. Schematic diagram shows two methods by which a continuous laser can produce a thin layer of single-crystal silicon on a silica substrate: **a**, Without seeding; **b**, With seeding. In **b**, an opening in the silica layer puts the molten silicon in contact with a silicon crystal, which defines the crystal orientation of the solidifying liquid. Figure 6

switched laser could be due to the effect of this "hot plasma" on atomic rearrangements in the solid.

There have been a number of experiments that bear on this question. These include measurements of time-dependent reflection and transmission during and following an excitation pulse,[19] time-dependent Raman scattering and transmission,[20] time-dependent x-ray Bragg scattering,[21] time-dependent electrical resistivity,[6] the velocity distribution of evaporated atoms[22] and thermionic electron emission.[23] Some of the experiments used laser pulses as short as 20 picoseconds.

Nicolaas Bloembergen and his co-workers at Harvard, using reflectivity and thermionic electron emission, found[23] the energy-conversion time in silicon to be less than about 10 psec. This is in agreement with the general theoretical considerations of Ellen Yoffa at IBM. Only the time-dependent Raman and transmission measurements from a group at Kansas State University have given experimental support[20] to a nonthermal model. The most provocative of these measurements—Raman scattering during the high-reflectivity state of the surface and selective transmission at 1.04 microns through the "molten" surface layer—were recently retracted.[24] The only remaining puzzle in the Raman results is the relatively low temperature they indicate shortly after the high-reflectivity molten phase.

The original and relatively simple picture advanced in 1977—that the mechanism of pulsed laser annealing is thermal heating and melting—is generally believed to be correct. It not only explains the overall changes in the

material properties of semiconductors subjected to pulsed lasers, but also the results of a wide array of detailed probes of the annealing process.

In this article we have dealt with several aspects of fast crystal growth and solidification, the main features of which appear to be qualitatively understood. It is important, however, to quantify the process. Quantification should enable us to correct the concepts of undercooling at rapidly moving interfaces with such diverse phenomena as impurity segregation and the melting temperature of amorphous silicon.

We have not been able to discuss all areas where laser annealing is likely to have an important impact. In surface science, for example, workers at Oak Ridge National Laboratories have demonstrated the feasibility of using pulsed lasers to produce atomically clean and selectively doped surfaces with metastable surface structures and novel distributions of electronic states. Researchers around the world are using laser annealing—in its broadest sense—to explore[1-4] other areas of science and technology, and the end is not in sight.

References

The progress of the field is nicely detailed in the four Materials Research Society annual symposia from 1978 to 1981, references 1–4.

1. *Laser–Solid Interaction and Laser Processing—1978*, S. D. Ferris, H. J. Leamy, J. M. Poate, eds., AIP Conf. Proc. No. 50 (1979).

2. *Laser and Electron Beam Processing of Materials*, C. W. White, P. S. Peercy, eds., Academic, New York (1980).

3. *Laser and Electron-Beam Solid Interaction and Materials Processing*, J. F. Gibbons, L. D. Hess, T. W. Sigmon, eds., vol.

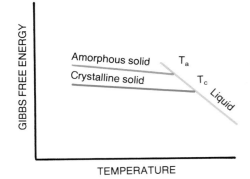

Free-energy diagram of amorphous, crystalline and dense metallic liquid silicon. The higher free energy of the amorphous phase indicates that it melts at a lower temperature (T_a) than the crystalline phase (T_c). Figure 5

Single-crystal film of silicon on a silica substrate, the result of seeded epitaxy using a scanning graphite heater. Beneath the substrate is crystalline silicon. This crystalline material seeded the growth of the single-crystal film through striped openings in the silica. The stripes, which are visible in this optical micrograph, are $\frac{1}{20}$ mm apart. (From reference 17.)　Figure 7

1, Mater. Res. Soc. Symp. Proc., North Holland, New York (1981).

4. *Laser and Electron Beam Interaction with Solids*, B. R. Appleton, G. K. Celler, eds., vol. 4, Mater. Res. Soc. Symp. Proc., North Holland, New York (1982).

5. S. A. Kokorowski, G. I. Olson, J. A. Roth, L. D. Hess, Phys. Rev. Let. **48**, 498 (1982).

6. G. L. Galvin, M. O. Thompson, J. W. Mayer, R. B. Hammond, P. S. Percy, Phys. Rev. Lett. **48**, 33 (1982).

7. C. W. White, S. R. Wilson, B. R. Appleton, F. W. Young Jr, J. Appl. Phys. **51**, 738 (1980).

8. J. M. Poate, in ref. 4.

9. J. W. Cahn, S. R. Coriell, W. J. Boettinger, in ref. 2, page 89.

10. J. Narayan in Inst. Phys. Conf. Series, No. 60, *Microscopy of Semiconducting Materials 1981*, A. G. Cullis, D. C. Joy, eds., IOP, London (1981). Page 101.

11. A. G. Cullis, D. T. J. Hurle, H. C. Webber, N. G. Chew, J. M. Poate, P. Baeri, G. Foti, Appl. Phys. Lett. **38**, 642 (1981).

12. B. G. Bagley, H. S. Chen, in ref. 1, page 97; F. Spaepen, D. Turnbull, in ref. 1, page 73.

13. P. Baeri, G. Foti, J. M. Poate, A. G. Cullis, Phys. Rev. Lett. **45**, 2036 (1980); E. P. Donovan, F. Spaepen, D. Turnbull, J. M. Poate, D. C. Jacobson, submitted to Phys. Rev. Lett.

14. P. L. Liu, R. Yen, N. Bloembergen, R. T. Hodgson, Appl. Phys. Lett. **34**, 864 (1979); R. Tsu, R. T. Hodgson, T. Y. Tan, J. E. E. Baglin, Phys. Rev. Letter. **42**, 1356 (1979).

15. A. Gat, L. Gerzberg, J. F. Gibbons, T. J. Magee, J. Peng, J. D. Hong, Appl. Phys. Lett. **33**, 775 (1978).

16. D. K. Biegelsen, N. M. Johnson, D. J. Bartelink, M. D. Moyer, in ref 3, page 487.

17. J. C. C. Fan, M. W. Geis, B-Y Tsaur, Appl. Phys. Lett. **38**, 365 (1981).

18. J. A. Van Vechten, in ref. 2, page 53.

19. D. H. Auston, J. A. Golovchenko, A. L. Simons, R. E. Slusher, P. R. Smith, C. M. Surko, T. N. C. Venkatesan, in ref. 1, page 11.

20. A. Compaan, H. W. Lo, A. Aydinli, M. C. Lee, in ref. 3, page 15; H. W. Lo, A. Compaan, Phys. Rev. Lett. **44**, 1605 (1980); A. Aydinli, H. W. Lo, M. C. Lee, A. Compaan, Phys. Rev. Lett., **46**, 1640 (1981).

21. B. C. Larson, C. W. White, T. S. Noggle, D. Mills, Phys. Rev. Lett. (1982), in press.

22. B. Stritzker, A. Pospieszczyk, J. A. Tagle, Phys. Rev. Lett. **47**, 356 (1981).

23. J. M. Liu, R. Yen, H. Kurz, N. Bloembergen, Appl. Phys. Lett. **39**, 755 (1981); N. Bloembergen, H. Kurz, J. M. Liu, R. Yen, in ref. 4.

24. A. Compaan, A. Aydinli, H. W. Lo, H. C. Lee, in ref. 4.　□

Infrared-laser-induced unimolecular reactions

The absorption of 30 or more infrared photons can raise a molecule into a highly excited vibrational state, leading to some intriguing problems in physical chemistry, quantum electronics and statistical mechanics.

Nicolaas Bloembergen and Eli Yablonovitch

PHYSICS TODAY / MAY 1978

Photochemistry, which deals with the way in which chemical reactions are induced or altered by the presence of photons, has been a very active branch of science for many years. Although visible and ultraviolet photons, by giving rise to excited electronic states, have the most pronounced effects, much attention in recent years has gone to infrared photochemistry. Molecules irradiated with a powerful electromagnetic wave in the infrared remain in the electronic ground state (in the Born–Oppenheimer sense), but their vibrational modes may become highly excited. As a result, infrared-laser radiation often profoundly affects chemical reaction rates.

Several reviews of this topic have appeared recently.[1,2] In this article we will restrict the subject further by the condition that the molecule absorbing the infrared radiation undergoes no collisions of any kind. Therefore our "universe" will be restricted to one single molecule and a large bath of infrared photons, supplied by one or more monochromatic lasers. This limited physical system is suitable for the study of intriguing questions in statistical mechanics, quantum electronics and physical chemistry.

Laser chemistry

It has been firmly established during the past few years that many polyatomic molecules can absorb sufficient energy from an intense infrared laser pulse that a true unimolecular reaction takes place. Because a chemical activation energy is typically about 3 eV and the infrared photon for the CO_2-laser wavelength near 10 microns is about 0.1 eV, the molecule is absorbing 30 or more photons.

Figure 1 illustrates the most studied

unimolecular reaction induced by infrared radiation, the dissociation of SF_6; this molecule is becoming the "hydrogen atom" for multiphoton chemistry and vibrational spectroscopy. The table on page 26 presents an incomplete list of the many polyatomic molecules that exhibit infrared-induced unimolecular reactions. The list is being augmented with each passing day. Large organic molecules are particularly favorable candidates for unimolecular ir photochemistry.

The interest in the field was greatly enhanced by the discovery of R. V. Ambartsumian and Vladilen Letokhov that the laser-induced dissociation reaction is isotope selective.[3] This, of course, has whetted the appetite of photochemists. They are excited by possible applications in the dissociation, cleavage, isomerization and synthesis of complex molecules, let alone the question of isotope selectivity. They are also led to the hope for bond-selective chemistry. Consider, for example, the halogen-substituted methane molecule CHFClBr. With an infrared wavelength exciting the C–H vibration they might hope to eject the H atom and obtain the CFClBr radical; with infrared wavelengths corresponding to the C–F, C–Cl and C–Br bonds, the other three respective dissociation channels might be activated. However, the realization of such dreams has proven more difficult than expected; we will discuss the reasons for this below.

From the physicist's point of view the system of one isolated molecule and a bath of monochromatic infrared photons is an intriguing one. What type of nonlinear optical effect could be responsible for the absorption of 30 or more infrared quanta? Does the energy remain in one mode, or will it be transferred to and shared with other modes? If so, will the distribution be statistical, and could the molecular behavior be ergodic—and on what time scale?

That these questions are nontrivial can be seen from the fact that a collisionless polyatomic molecule has only a few degrees of freedom, hardly the size of the systems to which statistical mechanics is usually applied. Nevertheless, we shall find below that statistical theory may give a good description of molecular behavior. How few degrees of freedom may a molecule have and still be capable of a statistical description? Will the dividing line occur for tri-atomics or quadra-atomics? This paper discusses these questions, but it remains for future work to provide definitive answers.

Experimental methods and results

Before the advent of powerful infrared lasers, highly excited vibrational states could only be reached by thermal-equilibrium heating, by chemical activation or by electronic excitation. With short CO_2-laser pulses from TEA lasers, we can now excite individual molecules in a time that is short compared to any collision time.

Figure 2 shows the frequency dependence of the dissociation probability as determined by Ambartsumian and Letokhov.[1] The isotope selectivity follows immediately from the 17 cm^{-1} shift in the infrared spectrum due to the mass difference between $S^{32}F_6$ and $S^{34}F_6$.

A key parameter of the multiphoton excitation process is the mean number of photons absorbed per molecule, $\langle n \rangle$. This quantity may be deduced from the direct attenuation of the laser beam[4] or by the heat deposition, as determined with an acoustic calorimeter.[5] If the excitation is strong enough, a fraction of the molecules may gain sufficient energy for dissociation. The reaction products may be analyzed by infrared spectroscopy, gas chromatography or mass spectroscopy.

The reaction yield is a rough measure of the tail of the molecular energy distribution, the fraction of molecules with

Nicolaas Bloembergen is the Rumford Professor and Eli Yablonovitch is an associate professor in the Division of Applied Sciences of Harvard University, Cambridge, Massachusetts.

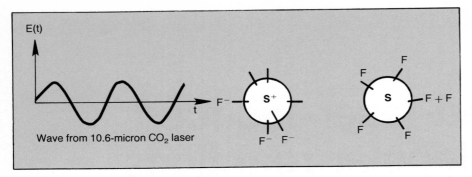

Dissociation of sulfur hexafluoride by absorption of infrared radiation in the ν_3 vibrational mode, in which the positively charged sulfur atom moves with respect to the octahedron of the negatively charged F atoms. The structure of SF_5 is schematic.   Figure 1

sufficient energy for dissociation. Figure 3 plots the reaction yield versus the mean excitation $\langle n \rangle$. Because $\langle n \rangle$ corresponds to the "center of gravity" of the population distribution over energy, this figure is probably the best available experimental measure for the shape of the energy distribution function. It compares the experimental points for two pulse durations with the theoretical curve calculated for a thermal distribution.

Collisionless dissociation occurs most convincingly at the intersection of a molecular beam with the laser beam;[6] it is also firmly established by experiments with very short pulses in cells at low pressure.[7] Alternatively, the reaction products may be probed, with good time resolution, by resonance fluorescence excitation by a frequency-doubled dye laser.[8] Figure 4, obtained by this technique, shows the population distribution over internal vibrational energy of the CF_2 fragments produced in the dissociation of CF_2Cl_2.

Some of the other important experimental methods for studying these multiphoton reactions are:

▶ Ultraviolet and visible luminescence from electronically excited radicals created in the unimolecular reaction.[9]

▶ Infrared luminescence from bimolec-ular reaction products of dissociation fragments with buffer-gas molecules.[10]

▶ Center-of-mass distribution of the kinetic energy of fragments created by dissociation in a molecular beam.[7,11]

▶ Intramolecular isotopic branching ratio.[12]

These reactions also may be studied with double or multiple frequencies, and as a function of such parameters as laser frequency, laser intensity, pulse duration, gas pressure and delay time between pulses. Rather than compile and critically review this large body of experimental results here, we shall discuss the theoretical concepts that are consistent with the dominant experimental features of collisional dissociation induced by infrared radiation.

The quasicontinuum

A diatomic molecule may be excited from the ground state to its first vibrational level by a resonant infrared photon. In general, the vibrational anharmonicity will necessitate a lower frequency for the absorption of a second photon, a still lower one for the third photon, and so on. To reach a level, $\nu = 30$, corresponding to a typical dissociation energy, 29 different frequencies therefore appear to be required. While available evidence suggests that indeed many selected infrared frequencies would be necessary to dissociate such a *diatomic* molecule, the situation in *polyatomic* molecules is qualitatively different.

One way to approach the problem of the polyatomic molecule is to imagine making an exact solution of the time-dependent Schrödinger equation for it. The initial condition would simply be the room-temperature canonical ensemble: a distribution of angular-momentum quantum numbers, internal vibrational energies and center-of-mass velocities. A solution of the time-dependent Schrödinger equation would require a knowledge of the energy, matrix elements, selection rules and exact wave function for each level up to the dissociation energy—an all-but-impossible task. Account would have to be taken of the octahedral splittings,[13] rotational energy changes and Coriolis coupling, which tend to compensate the anharmonicity.[1,2,14]

Fortunately, a simplification occurs. A schematic picture of the energy levels is shown in part **a** of figure 5. Each of the levels represents a specific vibrational–rotational state in the ground electronic manifold. The most important feature is the existence of several vibrational degrees of freedom.

In polyatomic molecules the density of states grows rapidly with energy E, due to the rapid expansion in the volume of accessible phase space. The number of possible permutations and combinations of vibrational modes increases rapidly with available energy.

The key point is that at a sufficiently

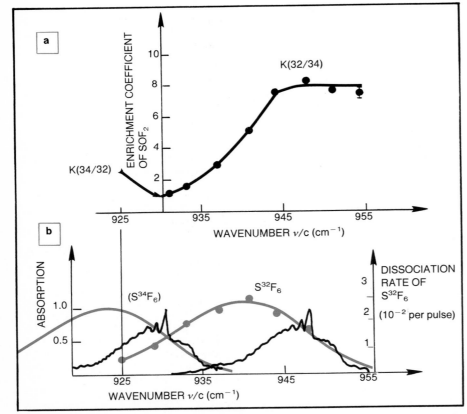

Laser isotope separation, based on the frequency dependence of the multiphoton dissociation of $S^{32}F_6$ and $S^{34}F_6$ by a CO_2-laser pulse. Curve **a** shows the variation of enrichment coefficients $K(32/34)$ and $K(34/32)$ as points ● and ▲ respectively. Curves **b** show (in color) the dissociation probability for the two isotopes and for comparison (in black) the respective linear absorption spectra. (After R. V. Ambartsumian and V. S. Letokhov, reference 1.)   Figure 2

high density of states the "Golden Rule," as Enrico Fermi called it, becomes valid. In it the transitions are described in terms of constant rates, and the full Schrödinger equation reduces to a set of rate equations. This criterion determines the energy region we call the *quasicontinuum* to distinguish it from the *discrete levels,* where the full machinery of the Schrödinger equation is required. The two regions are labelled in figure 5 (part **a**). The quasicontinuum has now become accessible to investigation by infrared laser spectroscopy.

The basic condition for the validity of the Golden Rule is that the transition rate should be neither too fast nor too slow:

$$[\hbar\sigma(E)]^{-1} \ll transition\ rate \ll T_2^{-1} \quad (1)$$

where $\sigma(E)$ is the density of available states and $\hbar/T_2$ is the width of the distribution of oscillator strength. When $\sigma(E)$ is large enough, a rate-equation description is possible. This conclusion is especially useful in large polyatomics, where the thermal energy at room temperature is sufficient to boost most of the molecules into the quasicontinuum. In that instance, from the initial condition onward, we may use a rate-equation description[15]:

$$\frac{dW_n}{dt} = K_n{}^a I W_{n-1} + K_n{}^e I W_{n+1}$$
$$- (K_{n+1}{}^a + K_{n-1}{}^e) I W_n - K_n{}^{diss} W_n \quad (2)$$

where W_n is the probability of being in that group of stationary states $n\hbar\omega$ above the starting level, $K_n{}^a$ and $K_n{}^e$ are absorption and stimulated-emission coefficients into that group, and $K_n{}^{diss}$ is the reaction rate, which differs from zero only for those states above the activation energy. Because the rates, except for $K_n{}^{diss}$, are all proportional to the light intensity I, it may be divided from the right-hand side of equation 2 to show explicitly[16] that the temporal evolution depends only on *intensity × time = energy fluence* (joules/cm2). This remarkable conclusion is confirmed by data of Paul Kolodner and his co-workers on the fractional dissociation of SF_6 as a function of pulse duration.[7] The laser energy is held fixed as the peak power increases by a factor 200. The fraction dissociated increases only 30%, showing that it is almost independent of peak power. The reaction yields in infrared multiphoton dissociation depend very strongly on energy fluence but only weakly on peak power.

When a significant number of molecules have energy in excess of the chemical activation energy, the reaction rates on the right side of equation 2 becomes important. Even when a molecule has enough energy to react, it will not do so immediately. Only a small fraction of the available volume of the phase space leads to immediate chemical reaction—the molecule must wander through phase space for some time before encountering

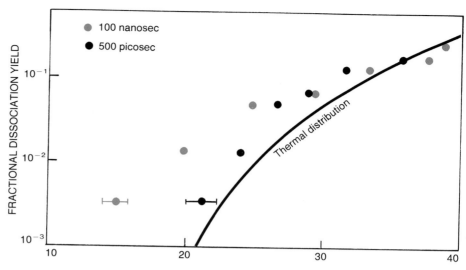

The probability of dissociating a molecule with a laser pulse is a function of the mean energy of excitation per pulse, $\langle n \rangle \hbar\omega$. This graph gives information relating to the shape of the energy-distribution function for multiple photon absorption. The experimental points for two pulse durations, 100 nanosec and 500 picosec, are compared with a theoretical curve based on the assumption of a thermal equilibrium distribution of the same energy.

Figure 3

that small region. This is the essence of the RRKM (Rice–Ramsperger–Kassel–Marcus) theory of chemical reactions.[17] The reaction rate is given by the fractional volume of phase space multiplied by a typical vibrational frequency ω_0. In the simplest version, the Kassel model, the reaction rate is given by

$$\omega_0 \frac{n!(n-m+N-1)!}{(n-m)!(n+N-1)!}$$

where m is the minimum number of absorbed photons required for chemical reaction and N is the number of normal modes. The factorial coefficient estimates the fractional volume of phase space leading to immediate reaction. Typically, $m = 30$ and excess energy $n - m = 10$ leads to reaction within 10^{-7} sec. The excess kinetic energy of the reaction fragments has been measured,[6] by Yuen Lee and Ron Shen and shown to be consistent with the RRKM theory.

Before leaving this aspect, notice that n is *not* the quantum number of the driven normal mode, for example the ν_3 mode in SF_6. While the quantum states in the discrete region tend to be readily identifiable in terms of specific normal modes of the molecule, each individual state of the quasicontinuum is generally a superposition of all the modes. An up transition in n could mean either an up or down transition in ν_3, depending on the individual states involved.

The intramolecular heat bath

The simplified picture implicit in equation 2 suggests that a quantum-mechanical basis set different from the energy eigenstates of the exact vibrational Hamiltonian might be appropriate. It is clear from the experimental observations that the energy originally fed into the ν_3

mode is at least partially distributed to other modes.

If we start with a very large molecule, we find a situation basically similar to that of spin relaxation in magnetic-resonance experiments. The radiation feeds energy into a transition between two (spin) energy levels, and a transfer to the continuum of energy levels of the lattice takes place with a characteristic relaxation rate T_1.

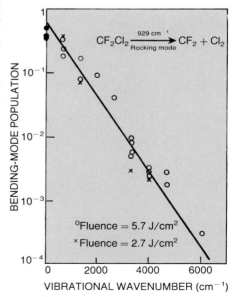

The fraction of CF_2 photofragments $P(E_v)$ formed initially with vibrational energy E_v in the ν_2 bending mode from dissociating CF_2Cl_2. The pulses came from a CO_2 TEA laser operated at the frequency of the ν_8 rocking motion on CF_2Cl_2. The data points agree with the Planck distribution function for a vibrational temperature of 1050 K, shown by the straight line. (After D. S. King, J. C. Stephenson, ref. 8.)

Figure 4

Single out the vibrational transition at resonance and consider all other modes as the lattice or heat bath; this is certainly permissible in a large molecule at room temperature. The anharmonic coupling between the selected mode and all others provides a relaxation mechanism. It is perfectly feasible to heat a lattice by means of electronic paramagnetic resonance. Similarly, a large molecule may be heated by absorption on one infrared-active vibration. As the molecule heats up, the resonance frequency shifts and the relaxation time becomes shorter, just as an epr line may shift and broaden with increasing temperature. This description is consistent with the data of figure 6, showing the measured absorption spectrum of SF_6 that has been boosted into the quasicontinuum of energy levels by thermal heating in a shock tube.[18] (For example, at 1000 K, SF_6 has thermal vibrational energy equivalent to 8000 cm^{-1},

high enough to put it into the quasicontinuum.)

It may seem surprising that the rather sharp resonance of figure 6 would occur in an energy region we have been calling the quasicontinuum. The energy levels of the exact Hamiltonian are indeed closely spaced. Nevertheless, the oscillator strength for transitions from any given level of the quasicontinuum is not uniformly distributed, but tends to cluster near the frequencies of the normal modes. Since the ν_3 absorption band does remain reasonably narrow, the quasicontinuum transitions do indeed contribute to the isotopic selectivity.

Away from the ν_3 frequency of SF_6, the absorption cross section falls off, but not to zero. The many combination and overtone bands produce a weak overlapping background absorption over broad regions of the infrared. The heating of the molecule makes available new com-

binations, further increasing the background absorption cross section, which is generally above 10^{-20} cm^2. This explains the results of two-frequency irradiation experiments[19] in which a heating pulse could be detuned from the main absorption line and still produce multiphoton excitation. At a cross section of 10^{-20} cm^2 an energy fluence of 100 J/cm^2 is more than sufficient to generate a chemical reaction. This also explains why the primary dissociation fragments[6] are able to absorb more energy and undergo further fragmentation.

The internal molecular degrees of freedom are divided into two groups, the system and the heat bath. The driven, infrared-active, vibrational mode should be regarded as the *system*, with Hamiltonian $H_s(x)$ and wave functions $\psi_s(x)$. The other $N - 1$ vibrational modes should be regarded as the *heat bath*, with Hamiltonian $H_b(y_1, y_2, \ldots y_{N-1})$ and wave

---

Unimolecular reactions induced by infrared multiphoton excitation

	K. Nagai, M. Katayama, Chem. Phys. Lett. **51**, 329 (1977)
$CClF_3 \longrightarrow CF_3 + Cl$	D. F. Dever, E. Grunwald, J. Am. Chem. Soc. **98**, 5055 (1976)
	R. B. Hall, A. Kaldor, Bull. Am. Phys. Soc. **23**, 73 (1978)
$CF_2Cl_2 \longrightarrow$ fragments	A. T. Lin, A. M. Ronn, Chem. Phys. Lett. **49**, 255 (1977)
$SeF_6 \longrightarrow SeF_5 + F$	J. J. Tiee, C. Wittig, Appl. Phys. Lett. **32**, 236 (1978)
$CH_3NC \longrightarrow CH_3CN$ (not strictly ir-laser induced)	K. V. Ready, M. J. Berry, Chem. Phys. Lett. **52**, 111 (1977)
$CH_3NC \longrightarrow$ fragments	M. L. Lesiecki, W. A. Guillory, J. Chem. Phys. **66**, 4317 (1977)
$BH_3PF_3 \longrightarrow$ fragments	K. R. Chien, S. H. Bauer, J. Phys. Chem. **80**, 1405 (1976)
$C_2H_3Cl \longrightarrow C_2H_2 + HCl$	F. M. Lussier, J. I. Steinfeld, Chem. Phys. Lett. **50**, 175 (1977)
	A. Yogev, R. M. J. Benmair, Chem. Phys. Lett. **46**, 290 (1977)
	J. M. Preses, R. E. Weston Jr, G. W. Flynn, Chem. Phys. Lett. **46**, 69 (1977)
$CH_3-\overset{\displaystyle O}{\overset{\|}{C}}-OCH_2CH_3 \longrightarrow CH_3COOH + C_2H_4$	W. C. Danen, W. D. Munslow, D. W. Setser, J. Am. Chem. Soc. **99**, 6961 (1977)
	Y. Haas, G. Yahov, Chem. Phys. Lett. **48**, 63 (1977)
$\longrightarrow H-\overset{\displaystyle O}{\overset{\|}{C}}-CH_3 + C_2H_4$ or radicals	R. N. Rosenfeld, G. I. Brauman, J. R. Barker, D. M. Golden, J. Am. Chem. Soc. **99**, 8063 (1977); D. M. Brenner, private communication
$NH_3 \longrightarrow NH_2 + H$	J. D. Campbell, G. Hancock, J. B. Halpern, K. H. Welge, Opt. Commun. **17**, 38 (1976)
$CF_3I \longrightarrow CF_3 + I$	S. Bittenson, P. L. Houston, private communication

functions ψ_b. The system and heat bath are coupled by anharmonic terms in the molecular potential described by $V(x, y_1, \ldots y_{N-1})$. The total vibrational Hamiltonian is:

$$H = H_s(x) + V(x, y_1, \ldots y_{N-1}) + H_b(y_1, \ldots y_{N-1}) \quad (3)$$

where x and the y's represent the phase-space coordinates of the system and the heat bath, respectively.

The new basis set consists of product wave functions of the form $\psi_s \psi_b$. The density-matrix equation of motion reduces to:

$$\frac{d\rho_s}{dt} = -\frac{i}{\hbar}[H_s, \rho_s]$$
$$-\frac{i}{\hbar}[\mu\mathcal{E}, \rho_s] + \text{damping terms} \quad (4)$$

where ρ_s is the system density matrix and the "damping terms" refer to the T_1- and T_2-type relaxation rates, which are familiar from nuclear magnetic relaxation theory. A T_1-type relaxation process is indicated in figure 5, part **b**. The driving Hamiltonian is the product of the dipole-moment operator μ and the laser field $\mathcal{E}$. The conditions for the validity of equation 4 are similar to those for equation 2:

$$\frac{1}{\hbar\sigma_b(E)} \ll \frac{1}{T_1} \leq \frac{1}{T_2} \ll \frac{1}{\tau_c} \quad (5)$$

where $\sigma_b(E)$ is the heat-bath density of states and τ_c is the autocorrelation time of the fluctuating anharmonic potential, which is typically about one vibrational period. The anharmonic potential V determines T_1 and T_2, and for $\tau_c \ll T_2$ the broadening is homogeneous. In nonrigid molecules, where slow internal motions are possible, τ_c could be much longer and the intramolecular damping could be partially inhomogeneous.

Intramolecular damping

Inhomogeneous contributions to the lineshape in addition to the intramolecular broadening are those from Doppler broadening, rotational structure and hot bands. Notice that these three inhomogeneous broadening mechanisms are associated with momentum, angular momentum and internal energy, respectively—the three conserved parameters of isolated systems. For $\hbar^{-1}|\mu\mathcal{E}| \ll T_2^{-1}$, the off-diagonal elements are small due to rapid dephasing, and the operator equation may be reduced to a set of rate equations for the populations:

$$\frac{dW_s}{dt} = [s\sigma(W_{s-1} - W_s) + (s+1)\sigma$$
$$\times (W_{s+1} - W_s)]\frac{I}{\hbar\omega} + \text{damping} \quad (6)$$

where σ is the absorption cross section and $\hbar\omega$ is the infrared quantum. Although equation 6 resembles equation 2, the interpretation is different. The index s represents the quantum number of the

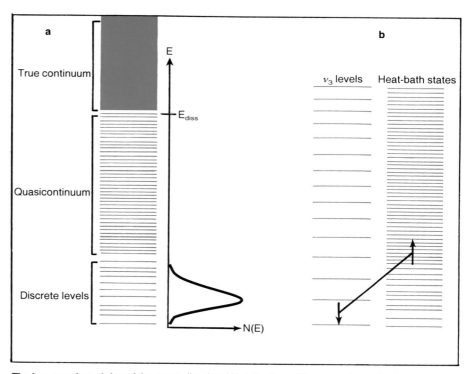

mode being driven, while n represents the total number of quanta in the molecule as a whole. For large molecules at sufficiently high temperature, most of the transitions take place between $s = 0$ and 1. This is because the energy is rapidly leaking out of the driven mode into the heat bath.

If the molecule is initially at absolute zero of temperature—that is, if there is no vibrational excitation—the concept of the heat bath is not valid for a one-photon transition. The molecule then must first undergo a multiphoton transition to a highly excited state with a sufficient admixture of other mode functions.

The SF_6 molecule at room temperature represents a borderline case. The discrete level transitions saturate, but for increasing intensity $|\mathcal{E}|^2$, multiphoton transitions to higher vibrational levels take place. Nature finds a ladder of nearly equally spaced energy levels with sufficient strength of connecting matrix elements to a level of excitation where conditions 1 or 5 are met. Thus the discrete levels represent a possible bottleneck, which becomes less important the higher the initial thermal excitation and the higher the laser intensity.

The subsequent vibrational heating of the molecule often causes a drop in the absorption cross section. Figure 6 shows that for SF_6 this is due to a shift and broadening of the ν_3 absorption band in

The true quantum states of the exact vibrational Hamiltonian. Diagram **a** shows the three regimes of vibrational excitation: discrete levels, quasicontinuum and true continuum. The initial population-distribution function $N(E)$ is the product of the vibrational density of states $\sigma(E)$ and the room-temperature Boltzmann factor. In the intramolecular-heat-bath picture, **b**, the energy levels of the infrared-active ν_3 mode are shown separately from the energy levels of the remaining modes, which form a quasicontinuous heat bath. Anharmonic coupling terms cause phase-fluctuation damping (with time constant T_2) and lifetime broadening (T_1) of the ν_3 mode. An energy relaxation process of the T_1 type is shown in black.   **Figure 5**

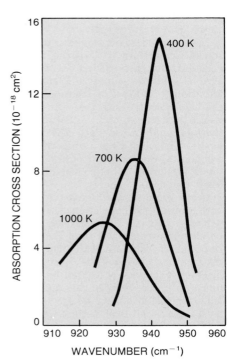

Absorption cross section per molecule in a thermal ensemble of SF_6. At 1000 K the molecule is boosted into the quasicontinuum, but the ν_3 absorption feature remains well defined. The surprising sharpness of the resonance is due to the nonuniform distribution of oscillator strengths. Not shown is the background absorption of about 10^{-20} cm^2. (A. V. Nowak, J. L. Lyman, ref. 18.)   **Figure 6**

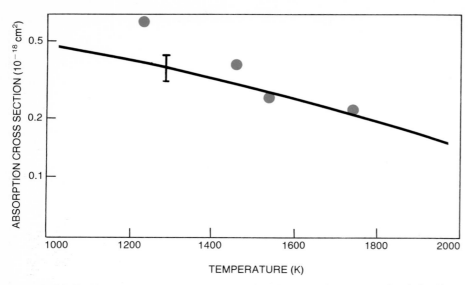

Shock-tube absorption data (in color) are compared with the absorption cross section derived from CO_2-laser heating with 500-picosecond pulses as a function of vibrational excitation (black line); both are at 944 cm^{-1}. The two heating methods give similar results.   Figure 7

the quasicontinuum—as opposed to a true saturation effect, which would occur between isolated levels in discrete regions. By employing a 500-picosec laser pulse of high peak power, the molecule may be boosted directly into the quasicontinuum, avoiding the saturation and bottlenecking effects in the discrete levels. Thus, the quasicontinuum absorption cross section should be similar for both laser-heated and shock-heated SF_6. The comparison, made in figure 7, shows good agreement in the temperature range above 1000 K.

In general, rate equations 2 or 6 can be solved numerically.[15] For the case of constant cross section there exists an analytical solution—simply a Planck distribution.[20] If the T_1 relaxation rate to the heat bath is faster than the rate of heating, the heat bath will come into thermal equilibrium with the driven mode. The molecule as a whole will have a canonical thermal distribution, which tends, of course, to be peaked at a relatively well defined energy.

If the cross section decreases as the bath heats up, the population distribution will tend to pile up at higher energies and be somewhat narrower than thermal.[15] In a realistic estimate of the population distribution, however, the effects of the discrete level barrier for entering the quasicontinuum must be accounted for. Inhomogeneous effects in the discrete level region tend to slow the energy deposition and smear out the population distribution.

As discussed earlier, a plot of reaction yield versus $\langle n \rangle$ is probably our best experimental measure of the shape of the population distribution. Figure 3 compares the experimental results with a thermal model. A 100-nanosec CO_2 laser pulse produces a significant yield even when the mean number of photons absorbed $\langle n \rangle$ is relatively low. The distri-

bution therefore tends to have a long high-energy tail and be broader than thermal. The distribution produced by a 500-picosec pulse, on the other hand, tends to be much closer to thermal, because the discrete level barrier plays a smaller role.

Another theoretical approach to the multiphoton chemistry problem is to treat the motion of the atoms in a purely classical manner, by making assumptions about the internuclear potential surface and effective charges as a function of internuclear distances. A Monte-Carlo type of calculation for various initial conditions of nuclear positions and velocities yields the required distributions and dissociation rates.[21]

Ergodicity

Chemists dream of being able to heat one part of a molecule while leaving the rest of the molecule relatively cold. Then a unimolecular reaction could be produced via a channel different from the thermodynamically favored one. Such a process was described above for the model molecule CHFClBr. In principle, this process would be possible if the rate of heating of a given vibrational mode could exceed $1/T_1$, the rate of relaxation to the rest of the molecule. This might be called mode-selective or bond-selective heating. In this situation, T_1 can be thought of as the time required to establish ergodicity.

Much fundamental work in statistical mechanics has concentrated on establishing the criteria under which ergodicity holds. (Inequality 5 is not such a condition; it merely determines whether the damping may be regarded as occurring at a constant rate, but it does not say whether the final steady state will be ergodic.) Perhaps the theoretical emphasis should shift to concentrate more effort on

$1/T_1$, the rate at which the equilibrium condition is approached.

Very little is known about the time constant T_1. Numerical classical-trajectory calculations[21] appear to show that T_1 can be less than 1 picosec. On the experimental side, there is also very little information. In an ingenious set of chemical-kinetics experiments, the intramolecular relaxation time was measured[22] indirectly, by varying a buffer-gas pressure. For the cases studied the upper limit was estimated to be on the order of picoseconds.

In another approach, the excitation of the infrared-active ν_4 mode by multiphoton pumping of SF_6 in the ν_3 line was detected[23] by means of a weak ir fluorescence in the ν_4 band of frequencies. Due to the technological constraints it was only possible to establish an upper limit of 1 microsec on the time required to transfer energy into the ν_4 mode.

In an important experiment J. P. Maier and collaborators[24] employed picosecond infrared pulses to probe the redistribution of energy within the huge coumarin-6 molecule. They placed an upper limit of 4 psec on the T_1 lifetime of the molecule in the vapor phase at 305°C. It would be especially interesting to repeat this type of experiment in a smaller molecule with less thermal energy. Because a knowledge of T_1 under a variety of different conditions is essential for mode-selective chemistry, more quantitative answers will likely be forthcoming in the next few years.

Related questions are addressed by overtone spectroscopy in polyatomic molecules, where high vibrational levels are excited by means of a very weak single-photon absorption rather than multiphoton absorption. Figure 8 shows the spectrum of the fifth overtone band of the C–H stretch in benzene in the gas phase,[25] as obtained by a laser-acoustic technique, which is also illustrated by the photograph on the cover of the May 1978 issue of PHYSICS TODAY.

To find such a sharp structure in the quasicontinuum should be no surprise to us by now. This has been understood by chemists in terms of a local-mode picture,[26] which is somewhat analogous to the heat-bath approach we have been using.

For the conditions of excitation in figure 8, the Doppler, rotational and hot-band broadening effects are negligible, and the broadening must arise mainly from the intramolecular anharmonic potential V. At present, the relative contributions of T_1 lifetime broadening and T_2 phase fluctuation broadening are unknown. The fifth overtone is a rather precise Lorentzian, while the fourth overtone has some substructure.[25] It is still too early to interpret this in terms of the onset of ergodicity at a definite energy level.

The infrared lineshapes in large polyatomic molecules are an interesting problem in their own right. These line profiles are usually observed in the liquid phase, in which strong solvent interactions obviously contribute to the lineshape. But the lines have similar broad shapes in the gas phase, as shown by figures 6 and 8. Until now only the three inhomogeneous effects, Doppler, rotational and hot-band broadening, were held responsible. The significance of fluctuations in the intramolecular anharmonic potential V for line broadening is now beginning to be recognized. Future developments will be quite fascinating as we learn to distinguish T_1 lifetime effects, T_2 phase-fluctuation effects, and in nonrigid molecules the effects of slow fluctuations.

Opportunities in chemical physics

We have already alluded to the potential applicability of infrared multiphoton chemistry in producing mode-selective chemical reactions. The pumping rate must be made high enough to compete with T_1. In addition, the rise time of the pulse must be short enough to prevent a premature nonselective reaction.[27] With T_1 as short as it is, these conditions will be difficult to achieve. The 30-picosec CO_2 laser pulses produced by optical free-induction decay[28] will probably play a role.

Mode-selective nonstatistical heating is only one possible application, but it may be difficult to achieve. A number of other unique attributes of multiphoton-induced chemistry present themselves, even under ergodic conditions:

▶ Multiphoton heating can be ultrafast, leading to a higher temperature than can be achieved otherwise. This would permit a high-energy reaction channel to compete more effectively with a low-energy channel. The relative yields of the different reaction products could be controlled in this way.

▶ Infrared multiphoton heating is collisionless, so unimolecular reaction channels can be made to compete more effectively with bimolecular or collisional channels.

▶ Most important, ir laser heating can operate on one component of a mixture, leaving the other components cold.

The latter, of course, is the essence of the isotope-separation capability, but it has exciting chemical applications as well. For example, consider two sides, A and B, of a chemical reaction:

$$A \rightleftharpoons B$$

By heating one of the components of A, it is possible to shift the equilibrium to the right, and vice versa for B. This might be rather useful in organic synthesis. Very frequently the challenge to the synthetic chemist is not in creating a molecule of a given formula, which is relatively easy, but in producing that molecule in a desired structure or isomer, which can be

much more difficult. The two isomers A and B will have rather different ir spectra. Irradiation at a wavelength at which the difference in absorption cross section is large can convert the molecules to the desired isomer.

There are many other ways to exploit the remarkable capability of heating one component of a mixture.

▶ Ultrapurification should be possible by selectively causing the undesired molecules to react.[29]
▶ A high density of ground- or excited-state radicals can be generated by infrared photochemistry and employed as reagents.
▶ The laser can produce homogeneous catalysts, enhancing the speed of subsequent collisional reaction steps in a mixture.

One of the problems of infrared multiphoton heating is that is has been restricted mostly to the gas phase, although much interesting biological chemistry occurs in solution. The main hindering influence of the solvent is the strong relaxation, which tends to damp out the excitation and conduct away the heat. Again, ultrashort laser pulses should be able to overcome this effect and open the liquid phase to the prospect of ir photochemistry.

Challenges ahead

The infrared multiphoton excitation process will provide problems and opportunities for some time to come. In this review we have attempted to organize a conceptual framework upon which a quantitative understanding can be built. For example, at the present time the contributions of the various line-broadening mechanisms are unknown, even for SF_6, the most studied case. Different types of molecular systems clearly will have to be studied in their own right; this will represent an important opportunity for specialists in nonlinear spectroscopy. The direct measurement of reliable T_1 relaxation times is an important challenge to picosecond spectroscopists.

The area of chemical applications is wide open and being pursued by many different groups. The results are still somewhat confusing, and it may take a while before we can make intelligent use of the tools that have become available.

In the area of fundamental statistical mechanics, the rate of approach toward ergodic behavior may turn out to be a more interesting question than whether true ergodicity is ultimately achieved. Theoretical calculations of T_1 in different situations would be very welcome. It will be interesting to learn whether an isolated triatomic molecule still falls within the realm of statistical theory!

* * *

The authors acknowledge receipt of a preprint of reference 2.

References

1. R. V. Ambartsumian, V. S. Letokhov, in *Chemical and Biochemical Applications of Lasers*, volume 3 (C. B. Moore, ed.), Academic, New York (1977), chapter 2; Comm. At. Mol. Phys. **6**, 13 (1976); V. S. Letokhov, PHYSICS TODAY, May 1977, page 23.

2. C. D. Cantrell, S. M. Freund, J. L. Lyman, in *Laser Handbook* (M. Stitch, ed.), North-Holland, New York (1978), volume 3.

3. R. V. Ambartsumian, V. S. Letokhov, E. A. Ryabov, N. V. Chekalin, JETP Lett. **20**, 273 (1974); J. L. Lyman, R. V. Jensen, J. Rink, C. P. Robinson, S. D. Rockwood, Appl. Phys. Lett. **27**, 87 (1975).

4. D. O. Ham, M. Rothschild, Opt. Lett. **1**, 28 (1977).

5. T. F. Deutsch, Opt. Lett. **1**, 25 (1977).

6. M. J. Coggiola, P. A. Schulz, Y. T. Lee, Y. R. Shen, Phys. Rev. Lett. **38**, 17 (1977); E. R. Grant, M. J. Coggiola, Y. T. Lee, P. A. Schulz, Y. R. Shen, Chem. Phys. Lett. **52**, 595 (1977).

7. P. Kolodner, C. Winterfeld, E. Yablonovitch, Opt. Commun. **20**, 119 (1977).

8. D. S. King, J. C. Stephenson, Chem. Phys. Lett. **51**, 48 (1977).

9. N. R. Isenor, R. V. Merchant, R. S. Hallsworth, M. C. Richardson, Can. J. Phys. **51**, 1281 (1973).

10. J. M. Preses, R. E. Weston Jr, G. W. Flynn, Chem. Phys. Lett. **48**, 425 (1977); C. R. Quick Jr, C. Wittig, Chem. Phys. Lett. **48**, 420 (1977).

11. K. L. Kompa, in *Tunable Lasers and Applications* (A. Mooradian, T. Jaeger, P. Stokseth, eds.), Springer, New York (1976).

12. S. W. Benson, A. J. Colussi, R. J. Hwang, J. J. Tiee, Chem. Phys. Lett. **52**, 349 (1977).

13. C. D. Cantrell, H. W. Galbraith, Opt. Commun. **18**, 513 (1976).

14. N. Bloembergen, C. D. Cantrell, D. M. Larsen, in *Tunable Lasers and Applications* (A. Mooradian, T. Jaeger, P. Stokseth, eds.), Springer, New York (1976).

15. E. R. Grant, P. A. Schulz, Aa. S. Sudbo, Y. T. Lee, Phys. Rev. Lett. **40**, 115 (1978).

16. J. G. Black, E. Yablonovitch, N. Bloembergen, S. Mukamel, Phys. Rev. Lett. **38**, 1131 (1977).

17. P. J. Robinson, K. A. Holbrook, *Uni-Molecular Reactions*, Wiley, New York (1972).

18. A. V. Nowak, J. L. Lyman, J. Quant. Spec. Rad. Trans. **15**, 1945 (1975).

19. R. V. Ambartsumian, G. N. Makarov, A. A. Puretsky, in *Laser Spectroscopy III* (J. L. Hall, J. L. Carlsten, eds.), Springer, New York (1977), page 76.

20. E. W. Montroll, K. E. Shuler, J. Chem. Phys. **26**, 454 (1957).

21. D. W. Noid, M. L. Koszykowski, R. A. Marcus, J. D. Macdonald, Chem. Phys. Lett. **51**, 540 (1977).

22. J. D. Rynbrandt, B. S. Rabinovitch, J. Phys. Chem. **75**, 2164 (1971).

23. D. S. Frankel, J. T. Manuccia, Chem. Phys. Lett., to be published.

24. J. P. Maier, A. Seilmeier, A. Laubereau, W. Kaiser, Chem. Phys. Lett. **46**, 527 (1977).

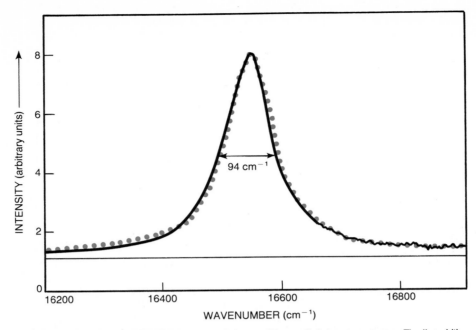

The fifth-overtone absorption spectrum of the C–H stretching mode in benzene vapor. The linewidth is associated with intramolecular energy relaxation (T_1) and dephasing effects (T_2); the experiment is depicted on the cover. (After R. G. Bray, M. J. Berry, ref. 25.) Figure 8

25. R. G. Bray, M. J. Berry, in *Proceedings* of the Conference on the Chemical Application of Lasers, University of North Carolina, Chapel Hill (1977).

26. R. L. Swofford, M. S. Burberry, J. A. Morrell, A. C. Albrecht, J. Chem. Phys. **66**, 5245 (1977).

27. E. Yablonovitch, Optics Lett. **1**, 87 (1977).

28. E. Yablonovitch, J. Goldhar, Appl. Phys. Lett. 25, 580 (1974); H. S. Kwok, E. Yablonovitch, Appl. Phys. Lett. **30**, 158 (1977).

29. J. H. Clark, R. G. Anderson, Appl. Phys. Lett. **32**, 46 (1978). □

Laser-induced thermonuclear fusion

Can focused laser pulses in the gigawatt range
be used to compress hydrogen droplets by a
thousand-fold to create energy-producing reactions?

John Nuckolls, John Emmett and Lowell Wood

PHYSICS TODAY / AUGUST 1973

Laser-induced fusion has recently joined magnetic-confinement fusion as a prime prospect for generating controlled thermonuclear power. During the past three years, the Atomic Energy Commission has accelerated the national laser-fusion program more than tenfold, to about $30 million annually,[1] and the Soviet Union has a program of comparable size.

In contemporary nuclear power plants, uranium is the primary fuel and fission reactions provide the nuclear energy. Fusion reactions between the heavy isotopes of hydrogen are another source of nuclear energy, one whose utilization for electricity generation lies in the future. Fusion power is important because, as Richard F. Post pointed out in his recent PHYSICS TODAY article,[2] the deuterium contained in the oceans is a virtually inexhaustible, low-cost, relatively clean fuel.

Fusion reactions are demonstrated in thermonuclear explosions and are thought to be the source of stellar energy. For the past twenty years, the thrust of controlled fusion research has been toward magnetic confinement of plasmas heated sufficiently (to about 10^8 K temperatures) to achieve fuel ignition. Laser-induced fusion, which exploits inertial confinement, has received increased public interest recently with the AEC declassification of important concepts and calculations. The key idea is to use laser light to

The authors are physicists at the University of California Lawrence Livermore Laboratory. John Nuckolls is associate leader of "A" Nuclear Explosive Design Division, John Emmett is leader of "Y" Laser Fusion Division and Lowell Wood is a member of the Director's Office and Physics Staffs.

isentropically implode pellets of deuterium and tritium to approximately 10 000 times liquid density and thereby induce efficient thermonuclear burning. Fusion energies 50–100 times larger than laser input energies of 10^5–10^6 joules have been achieved in sophisticated computer simulation calculations. There is as yet no experimental confirmation, but 10 000-joule lasers are being planned at the Livermore and Los Alamos Laboratories and at the Lebedev Institute in the USSR to explore laser-induced fusion.

The laser fusion implosion system consists of a tiny spherical pellet of deuterium–tritium surrounded by a low-density atmosphere extending to several pellet radii, located in a large vacuum chamber and a laser capable of generating an optimally time-tailored pulse of light energy. Before the main pulse occurs, the atmosphere may be produced by ablating the pellet surface with a laser prepulse. Most of the dense pellet is isentropically compressed to a high-density Fermi-degenerate state and thermonuclear burn is initiated in the central region. A thermonuclear burn front propagates radially outward from the central region igniting the dense fuel.

With 10^5–10^6 joule laser pulses initiating 10^7–10^8 joule fusion pulses, gigawatt electrical power levels may be generated by initiating 100 pulses per second, possibly ten per second in each of ten combustion chambers. The combustion chambers would have a diameter of a few meters and walls wetted with lithium to withstand the nuclear radiation and debris. For economic operation the pellets would have to cost less than a cent each, and they could be fabricated in a drop tower.

Electricity would be generated via neutron-heated lithium blankets as in conventional controlled thermonuclear reaction schemes, or by direct conversion in advanced power-plant schemes.

The implosion of bubbles in water was considered by William Henry Besant in 1859[4] and Lord Rayleigh in 1917.[5] A self-similar solution to an imploding shock wave was developed by G. Guderley who made some relevant calculations in 1942.[6] Early work on fission weapons at Los Alamos began by Seth Neddemeyer, John Von Neumann and Edward Teller and others explored spherical implosion systems driven by high explosives.[7] Subsequently, moderately high compressions were experimentally demonstrated and utilized. However, compressions approaching 10 000 fold (relative to liquid or solid densities)—which are required for practical laser fusion power reactors—have not been experimentally achieved.

The invention of the pulsed laser in the early 1960's stimulated further implosion calculations and a proposal of a laser fusion engine for CTR and propulsion applications. Nearly all laser-fusion implosion calculations remained classified until reported recently.[3,8] Experimental work began in 1963, and by the mid-1960's, Ray Kidder and S. W. Mead constructed a twelve-beam implosion-oriented laser at Livermore.[9] In 1972, Nikolai Basov and his colleagues reported implosion of a 100-micron diameter CD_2 microsphere with a few hundred joule, few nanosecond, nine-beam laser pulse.[10]

Implosions and thermonuclear burn

Conditions involving pressure, symmetry and stability must be satisfied to implode a DT sphere to a state at 10^4 times liquid density, in which both Fermi-degeneracy and thermonuclear propagation can be exploited to achieve maximum gain.

The optimum laser pulse shape generates an initial shock, which is near-sonic $(1/2 \times 10^6$ cm/sec) in the outer part of the pellet. This shock produces an entropy change sufficiently small that subsequent compression to a Fermi-degenerate state is possible. As this shock converges toward the center of the pellet it becomes sufficiently strong to produce significant heating. The pulse shape also generates a maximum implosion velocity of about 3.5×10^7 cm/sec, corresponding to the required average energy density of 6×10^7 joules/gram (see box). The implosion velocity is increased from the initial to the maximum value at such a rate that the hydrodynamic characteristics in the compressing pellet coalesce to form a strong shock near maximum compression, at a distance from the center approximately equal to the range of 10-keV alpha particles in DT. By numerous computer calculations of laser implosions, we know that the optimum pulse shape is approximately

$$E = E_0 \tau^{-s}$$

where E is the laser power, $\tau = 1 - (t/t')$, t is time, t' is the collapse time, and s is approximately 2.[11] No satisfactory analytic derivation of this equation is known. Figure 3 shows how this pulse shape may be approximated with sufficient accuracy by a histogram of 5–10 pulses.

The compression and burn processes that have been described are illustrated in figure 4 by results of a typical computer-simulation calculation of the implosion of a fusion pellet to 10 000 times liquid density, and of the resulting thermonuclear microexplosion. This calculation was carried out at the Livermore Laboratory by Albert Thiessen with a program developed by George Zimmerman.[12] The program includes the following physical processes:

▶ Hydrodynamics—Lagrangian; real and generalized Von Neumann artificial viscosities; ponderomotive, electron, ion, photon, magnetic and alpha-particle pressures.

▶ Laser light—absorption via inverse bremsstrahlung and plasma instabilities; reflection at critical density.

▶ Coulomb coupling of charged-particle species.

▶ Suprathermal electrons—multigroup flux-limited diffusion with self-consistent electric fields; non-Maxwellian electron spectra determined by results of plasma-simulation calculations for laser-light absorption by plasma instabilities; inverse bremsstrahlung elec-

tron spectrum for classical absorption.

▶ Thermal electrons and ions—flux-limited diffusion.

▶ Magnetic field—includes modification of all charged-particle transport coefficients, as well as most of the equilibrium MHD effects described by S. Braginskii.[13]

▶ Photonics—Multigroup flux-limited diffusion; LTE non-LTE average-atom opacities for free–free, bound–free, and bound–bound processes; Fokker–Planck treatment of Compton scattering.

▶ Fusion—Maxwell velocity-averaged reaction rates; the DT alpha particle is

transported by a one-group flux-limited diffusion model with appropriate energy deposition into the electron and ion fields; one group transport of the 14 MeV neutron.

▶ Material properties—opacities, pressures, specific heats, and other properties of matter are used, with nuclear, Coulomb, degeneracy, partial ionization, and other significant effects taken into account.

In this implosion-burn calculation, a 10kJ, short-wavelength (1/2 micron), frequency-modulated, pulse of laser light is focussed symmetrically onto a 1200-micron radius low-density atmo-

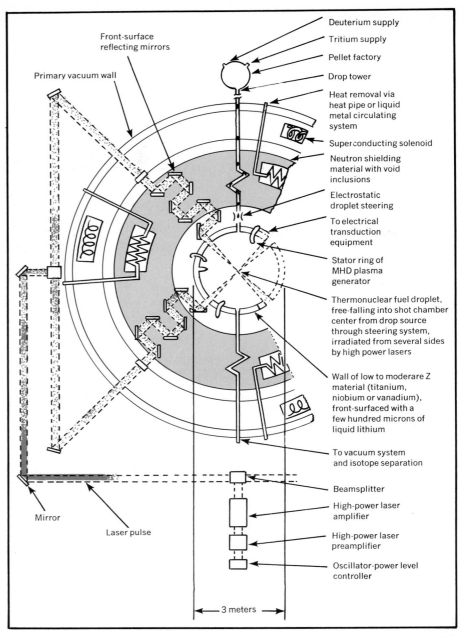

Conceptual design of a laser-fusion power-plant. One hundred laser-induced micro-explosions per second will produce 1000 megawatts of electrical power. The pulse of laser light, which is shown here in deep color, is shaped in time to yield optimum implosion and thermonuclear burning of the pellet. Several processes can be used to convert the explosion energy into electricity, such as thermal and MHD plasma conversion. Figure 1

sphere (generated by a laser prepulse) surrounding a 400-micron radius spherical pellet of liquid deuterium–tritium. The applied laser power is increased in eight pulses from about 10^{11} to about 10^{15} watts in 10 nanoseconds. These eight pulses closely approximate the ideal pulse shape described earlier. The laser light is absorbed via inverse bremsstrahlung near the critical density (the density where the laser light and electron plasma frequencies are equal) in the atmosphere, at a radius of approximately 600 microns, generating hot electrons. The pellet and atmosphere are seeded with small amounts of material of Z greater than 10 and short-wavelength laser light is used to make possible efficient absorption by inverse bremsstrahlung and in order to increase the thresholds of plasma instabilities. Frequency modulation of the laser light also increases the instability thresholds.[14] When these effects are accounted for, the peak laser intensity is less than a factor of ten above the threshold for instabilities—so that generation of strong non-Maxwellian electron distributions is avoided. The atmosphere is heated by the hot electrons to electron temperatures that increase in time from about 3×10^6 to 10^8 K at the absorption radius. The surface of the pellet is heated and ablated by electron thermal conduction through the hot atmosphere, generating implosion pressures that optimally increase from about 10^6 to about 10^{11} atmospheres. This increase in implosion pressure by five orders of magnitude occurs at an optimal rate during transit of the initial shock to the center. Consequently the outer part of the pellet is isentropically compressed into a high-density spherical shell ($p > 100$ gm/cm^3) while at the same time this shell is inwardly accelerated to velocities that increase in time from 10^6 to 3.5×10^7 cm/s.

As the internal pressure becomes larger than the ablation pressure the rapidly converging shell slows down and is compressed nearly isentropically, at sub-Fermi temperatures, to densities greater than 1000 gm/cm^3. The inner region is compressed by the outer shell to densities approaching 1000 gm/cm^3, and heated to ion temperatures greater than 10^8 K, initiating thermonuclear burn. A thermonuclear burn front then propagates outward. About 1200 kilojoules of fusion energy are produced in less than 10^{-11} seconds. The energy gain is about 20-fold.

There are, however, other effects that may reduce this gain; this is especially so for long-wavelength laser light such as the 10.6-micron CO$_2$ line. The pellet compression and energy gain might be strongly degraded by electron preheat[15] and decoupling.[16] Preheat occurs when the laser-heated electrons

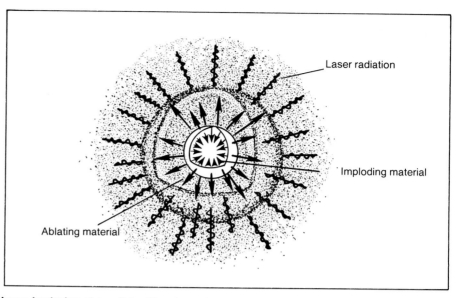

Laser implosion of a pellet. The atmosphere extends to several pellet radii and is formed before the main laser pulse by a prepulse that ablates some of the pellet surface. Absorption of the laser light in the outer atmosphere generates hot electrons. As the electrons move inward heating the atmosphere and pellet surface, scattering and solid-angle affects greatly increase the spherical symmetry. Violent ablation and blowoff of the pellet surface generates the pressures that implode the pellet; the effect is similar to a spherical rocket. The pellet core then undergoes thermonuclear burn.   Figure 2

have a range that is a significant fraction of the pellet radius; these electrons then preheat the fuel, making it more difficult to compress. Decoupling occurs when the electrons have a large enough range to cross the atmosphere, re-enter the absorption region, and are reheated to still higher energies with longer mean free paths until the pellet is effectively decoupled from the laser-heated electrons. Decoupling can be compensated for if the volume of the pellet is increased by making it hollow.

Efficient absorption of CO$_2$ light is not possible via inverse bremsstrahlung because the light absorption length is too long ($\gg 1$ cm at 10 keV).[17] Absorption is possible via plasma instabilities.[18] However, if the thresholds for these instabilities are greatly exceeded, then plasma-simulation computer codes indicate that non-Maxwellian electron spectra may be generated, with high-energy tails extending beyond 100 times the thermal electron energy.[19] Experiments are needed to determine the electron spectra reliably in such situations. If excessive numbers of superthermal electrons are not generated, then long-wavelength lasers may be suitable for CTR applications, provided that the hollow pellet can be constructed cheaply enough.

In compression of a sphere by 10^4-fold, the radius decreases somewhat more than 20-fold. If, after compression, spherical symmetry is required to within half of the compressed radius— or 1/40 the initial radius—then the implosion velocity (and time) must be

spatially uniform and synchronized to about one part in 40, or a few percent. The outer atmosphere may be heated uniformly to 10% to 20% by a many-sided irradiation system, consisting of beam splitters, mirrors, lenses, and other optical elements. This error is then reduced to less than 1% by physical processes occurring inside the atmosphere.[3,20] Asymmetries are reduced during electron energy transport through several scattering mean free paths of atmosphere to heat the surface of the pellet. In addition, since the atmosphere has a large radius compared to the pellet, each point on the pellet surface is heated by electrons coming from almost 2π steradians of the hot absorbing region in the outer atmosphere. Finally, during most of the implosion, the electron mean free path in the absorbing region is a significant fraction of the absorption radius.

The implosion of the pellet by diffusion-driven pressures generated by ablation is hydrodynamically stable, except for relatively long-wavelength surface perturbations.[3,20] Fortunately, these perturbations grow too slowly to be damaging if the pellet is imploded in one sonic transit time. In part, ablative stabilization occurs because the peaks of surface perturbations are effectively closer to the heat source than are the valleys, so that the ablation-driving temperature gradient is steeper. Consequently, the amplitude of the perturbation is reduced, both because the peak is more rapidly ablated and because the ablation pressure is higher on the peak.

Laser-imploded pellets

Thermonuclear micro-explosions scale as the density–radius product ρR. The rates of burn, energy deposition by charged reaction products, and electron–ion heating are proportional to the density, and the inertial confinement time is proportional to the radius of the pellet. Consequently, the burn efficiency, self-heating, and feasibility of thermonuclear propagation are determined by ρR. If ρR is very much greater than 0.3 gm/cm^2, then only 0.3 gm/cm^2 in the central region of the pellet need be heated to approximately 10 keV to initiate a radially propagating burn front that ignites the entire pellet. In this case, 1.6×10^{10} joules/gm of fusion energy will be released from the central region; one fifth of this energy is in alpha particles, sufficient to heat three times more DT to 10 keV. The alpha particles will deposit their energy in approximately this mass since their range at 10 keV is about 0.3 g/cm^2.

When ρR is approximately 3 gm/cm^2 the fusion energy released is about 10^{11} joules/gm. If we assume propagation of the thermonuclear reaction from the 10-keV central region and compression of the remaining DT to a Fermi degenerate state, all at 1000 gm/cm^3, the minimum average energy of ignition and compression is about 6×10^7 joules/gm. (If the degeneracy condition is not satisfied the compressional energy will exceed the ignition energy.) The gain is then about 1500, but approximately 95% of the laser energy absorbed by the pellet during implosion is lost to kinetic and internal energy blowoff. Consequently, the energy gain relative to the laser energy employed is about 75 fold. This is sufficient for CTR applications with a 10% efficient laser, a 40% thermal-to-electric efficiency and about 30% of the electrical energy circulated internally to pump the laser.

The figure shows the variation of gain (relative to laser light energy) with compression and with laser-light energy.[3,11] The curves have been normalized to computer calculations of the implosion and burn. Gains approaching 100 are predicted for laser energies of 10^6 joules. The calculations indicate that less than 1 kilojoule of laser light may be sufficient for breakeven (gain ≈ 1) and 10^5 joules may be sufficient to generate net electrical energy with a 10% efficient laser. These predicted gains are probably upper limits to what can be achieved. Similar gain curves may be generated for D_2 and DHe^3 pellets seeded with a small percentage of tritium to facilitate ignition.

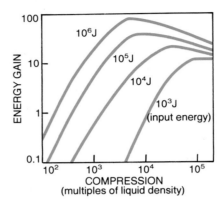

Laser technology

The development of lasers for high-density laser-fusion application poses a special set of problems that have not previously received much attention in the laser R&D community. In the optimum pulse shape, about half the total energy is produced in the final 100 picoseconds. Thus, in terms of a single pulse, a CRT laser has to be capable of producing at least 50 kilojoules in 100 picoseconds. In addition, the optimum plasma heating process may require short-wavelength lasers. Such lasers do not exist at present; however, development of short wavelength devices is receiving increasing attention in the US and several other countries.

The salient characteristics of four laser systems presently under consideration are indicated in Table 1. These systems are representative of the diversity that exists in the laser world. What is interesting to note about these laser systems is the almost total lack of overlap in the technologies required for the development of each. Thus, extensive development effort applied to one of these systems is not usually applicable to another. Also shown in Table 1 are the desired characteristics of a hypothetical laser system that matches the requirements of laser fusion as presently envisioned. The primary characteristics of this hypothetical laser system are high efficiency, high average power, short wavelength and high energy. It is clear that none of the real lasers in

Table 1 demonstrates all of these characteristics.

Neodymium–glass lasers develop gain at 1.06-microns wavelength with Nd^{3+} ions in a glass matrix pumped by xenon flash lamps. They have the best developed technology for operation in the sub-nanosecond region. In addition, the high second-harmonic conversion efficiency (60–80%) already demonstrated offers great potential for operation at 0.53 micron. Fourth-harmonic generation (0.265 micron) and stimulated Stokes–Raman scattering (1.9 micron) offer potential for additional wavelengths with efficiencies greater than 20%. Thus, the neodymium–glass laser system provides the best laboratory tool for near-term laser-fusion experiments. However, the extremely low energy of efficiency (0.1%) and the low average power capability (limited by the low thermal conductivity of glass) prevent any consideration of neodymium–glass systems for eventual laser fusion power-generation applications.

The carbon-dioxide laser develops gain at 10.6 microns between vibrational energy levels of the ground state of the CO_2 molecules. This system (actually a CO_2–N_2–He mixture) is pumped by relatively low-power electrical discharges in the gas. This laser has demonstrated efficiencies of approximately 5% for one-nanosecond duration pulses. Operation in the nanosecond regime has yet to be demonstrated. With the addition of high-speed gas flow, CO_2 lasers have the capability to generate high average powers. The major liability of the system is the 10.6-micron emission wavelength. Efforts are currently underway to convert the 10.6-micron energy efficiently to shorter wavelengths, although success has yet to be achieved. Development of high-energy, short-pulse CO_2 lasers continues because of their high efficiency and high average power. The ultimate usefulness of this system remains to be determined.

The iodine laser has recently come under consideration as a possible lower-cost replacement for neodymium–glass. It develops gain at 1.315 microns between the electronic levels of the neutral iodine atom, which is produced in an excited state by photodissociation of molecules such as CF_3I. The emission wavelength and efficiencies are similar; however, the cost of CF_3I is much below that of neodymium–glass. The stimulated-emission cross section of I*, even in the presence of a few atmospheres of a line-broadening buffer gas, is much larger than neodymium–glass. This necessitates an entire different approach to the laser design, in order to control parasitic oscillation within a single laser amplifier section. With the eventual solution to the parasitic oscillation problem and more detailed understanding of the pumping requirements of the system, it is possible that iodine lasers may be built in the 1–10 kilojoule regime.

Recent interest in the xenon laser stems from the short wavelength (1722 Å) and predicted high pumping efficiency (25%).[21] Little detailed information is available on this system, and

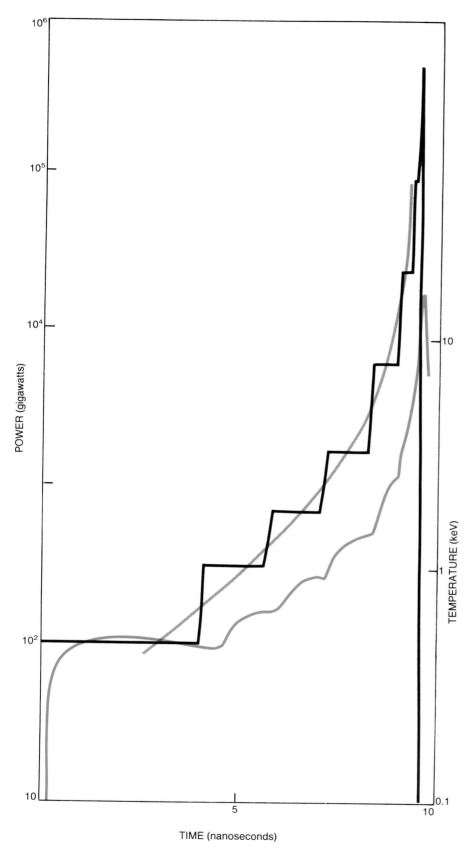

Laser pulse power rise in steps to approximate the ideal shape, which is shown in grey. The corresponding electron temperature in the critical density region is shown in color. These curves were generated by computer calculations and the assumption of a 10 000-fold compression of a fusion pellet. This time-tailored pulse is composed of eight subpulses that form the steps. The first step is a 500-joule pulse that lasts for 4200 picoseconds, and the remaining steps are formed by the following subpulses: 500 J-1600 psec, 1 kJ-1400 psec, 24 kJ-1200 psec, 4kJ-650 psec, 8kJ-330 psec, and finally 30kJ-65 psec.   Figure 3

the technology for the generation of subnanosecond pulses has yet to be developed in this region of the spectrum. Two comments are, however, appropriate. First, the stimulated-emission cross section is approximately 3×10^{-18} cm². Thus, a large high-energy amplifier will be severely limited in performance by superfluorescence (amplified spontaneous emission) and parasitic oscillation. Second, lasers useful for fusion applications must operate at flux levels of 10^{10} watts/cm² (1 J/cm², 100 picosec) or greater. At this flux level and 1722 Å wavelength, all transparent materials (window, lenses, coatings) from LiF to Al_2O_3 will exhibit two-photon absorption coefficients in the 1–25 cm⁻¹ range. Thus, to use such a short wavelength, a new optical technology of gas lenses and aerodynamic windows will have to be developed.

From the foregoing discussion we may draw several conclusions. Clearly, neodymium–glass laser systems provide the best technology base for the near-term laser-fusion experiments. The wide range of pulse widths obtainable (20 picosec–20 nanosec) and the range of wavelengths (0.265–1.9 microns) render it an almost ideal laboratory tool. For these reasons large multi-aperture neodymium–glass laser systems with energies of 10 kilojoules are in design or construction stages both in the US and the USSR. At the Lawrence Livermore Laboratory, a 10-kilojoule subnanosecond facility is being designed for spherical irradiation. Funds for it have been requested in the President's fiscal 1974 budget submitted to Congress. With this instrument, the important milestone of significant thermonuclear burn and scientific breakeven (fusion energy equals laser energy) will be achieved.

The high efficiency and high average-power capability of CO_2 lasers warrant their further study. The technology of high-energy, short-pulse CO_2 lasers is being aggressively pursued by the Los Alamos Scientific Laboratory. A ten kilojoule, one-nanosecond multiaperture device is being developed to answer the fundamental questions associated with the use of such long wavelengths for high density laser fusion.

The iodine laser system is under intensive development at the Institut fur Plasma Physik, Garching, where the objective is a one-kilojoule, one-nanosecond device for laser-fusion experiments. Additonal low-level investigations of this system are being carried out in many laboratories, both in the US and abroad. However, the only flexibility currently provided by this system is a lower-cost laser medium. This may be more than offset by the increased cost of capacitor banks and

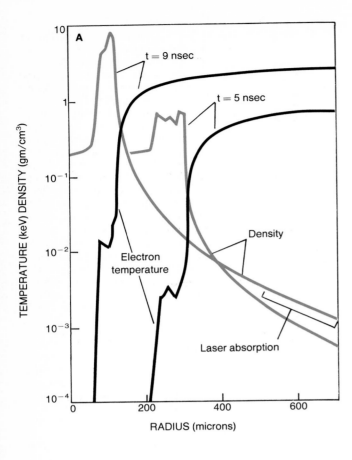

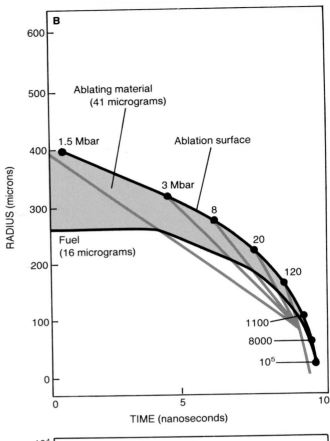

Compression of the fusion pellet from the beginning of the laser pulse and through absorption until thermonuclear ignition. These curves were generated by a computer and the assumption of a 10 000-fold compression. Part A shows the density and electron temperature versus the radius at times, five and nine nanoseconds after the start of the laser pulse but before ignition. In part B, the radius is plotted against time. The colored lines in this figure represent weak shocks from the steps of the pulse power. The inner curve represents that part of the pellet which undergoes thermonuclear burning, and the outer curve represents the outer region of the pellet; all the material in between is ablated away and is indicated by the colored region. The pressures at different stages of the compression are shown in megabars. In part C, the density and electron temperatures are plotted at the time of ignition.
Figure 4

flashlamps for the short pulse excitation required.

The xenon laser system is too new to make any meaningful projections as to its ultimate usefulness for laser fusion. It will obviously have significant applications to high-density plasma diagnostics where its short wavelength may be essential.

The development of new lasers is required if laser-fusion power production is to become a reality. The xenon laser represents a class of possible la-

sers based on the weakly bound or van der Waals molecules. Laser action has already been achieved from Xe_2^* and Kr_2^*, and it may be expected from some of the similar dimer systems of mercury, cadmium and zinc. Other non-dimer systems such as LiXe or HgXe also look attractive. These systems are pumped by efficient, high-current relativistic electron-beam machines, which have been extensively developed during the last decade. The availability of an efficient pump source

and short wavelength of emission makes these systems of great interest for laser fusion. However, it is clear that stimulated emission cross sections smaller than those in the xenon system will be necessary, as will extensive development of a means of reducing parasitic oscillation and superfluorescence.

Probably the most important characteristic of any new laser system developed for laser-fusion applications will be the ability to use energy efficiently.

Table 1. Lasers for Fusion

Laser	Wave-length (microns)	Effici-ency (%)	Energy storage (J/liter)	Pulse width (nsec)	Max. output short pulse (joules)	Average power capability	Wavelength convertability	Laboratory
Nd:glass	1.06	0.2	500	≥0.02	350 (0.1 nsec) 350 (1.0 nsec)	Very low	0.26–1.9 microns (40% eff.)	University of Rochester; Naval Research Laboratory
CO_2	10.6	5	15	≥1.0	17 (1 nsec)	High (flow)	Not demonstrated for high powers	Los Alamos Scientific Laboratory
Iodine	1.32	0.5	30	0.6	12 (10 nsec)	High (flow)	Similar to Nd:glass	Institut für Plasmaphysik, Garching
Xe_2	0.17	<20	300	≈10	0.01 (10 nsec)	High (flow)	Not required	Lawrence Livermore Laboratory
Desired characteristics of a new laser	0.3–0.5	>5	100–1000	0.1–1.0	10^4–10^6	High (flow)	Not required	

In this context, the future development of chemical lasers can be expected to influence the laser-fusion problem strongly. However, a significant amount of basic research on the detailed energetics and kinetics of chemical reactions will certainly have to precede the development of efficient chemical lasers operating in the visible or near-ultraviolet region of the spectrum.

Fusion-fuel combustion chamber

The fusion-fuel combustion chamber of a laser-fusion power plant must not only serve to admit the fuel pellet and direct the laser beams upon it, but must also endure perhaps as many as 100 multimegajoule thermonuclear pulses per second for of the order of ten years, and be technically and economically feasible to contruct and maintain. The fusion effects consist of an x-ray pulse, a neutron pulse, and blast and thermal effects from the plasma explosion debris (see figures 5 and 6). The x-ray pulse is fortunately heavily attenuated in the softest (10–1000 eV), most wall-threatening, portion of the spectrum by inverse bremsstrahlung in the superdense fireball. The neutron spectrum is dominated by a 14-MeV peak. Calculations involving x-ray opacities, neutron cross sections, specific heats, thermal expansion coefficients and compressibilities indicate a chamber of about 3-meter radius with a wall of a few layers of properly chosen low-to-moderate atomic-number materials (for example about 0.01 cm of beryllium backed by titanium, niobium or vanadium) will endure the x-ray and neutron pulses of a 10^7-joule microexplosion. If surfaced with a thin, low-Z liquid layer (for example lithium a few hundred microns thick) by continuous exudation, the plasma pulse of a 10^7-joule explosion may also be repetitively endured by the combustion chamber.[22]

The impulse associated with an explosion determines the size and material strength of a chamber that must contain it. This impulse is proportional to the square root of the product of the explosion energy and the mass of the explosion debris. Relative to a chemical explosion of the same energy, a fusion pulse involves about six orders of magnitude less explosive debris mass and thus about three orders of magnitude less impulse, provided that the surface of the wall is not vaporized. Then a 10^7-joule fusion micro-explosion produces no more impulse than a large firecracker.

If the combustion chamber is too small, the wall will be ablated by the thermonuclear debris. Then the peak pressures imposed on the wall may be multiplied as much as a thousandfold and may be unacceptably high (greater than one kilobar). A crucial advantage of not vaporizing the lithium on the wall is that the chamber pumpdown time does not severely limit the pellet burn repetition rate.

About one joule/cm² of thermonuclear plasma energy may be directed against a chamber wall "moistened" with a several-hundred-micron layer of liquid lithium before significant blowoff is produced. The suprathermal ion fluence (for example 3.5-MeV alpha particles, knock-on deuterons and tritons) associated with a one joule/cm² thermal plasma fluence poses no blowoff hazard, since it penetrates the moist layer relatively deeply and deposits its energy in a large amount of matter. Moistened-wall combustion chambers, rated for ten-megajoule pulses of approximately 3-meter radius would thus be satisfactory from a plasma wall-loading standpoint.

If the combustion chamber wall is shielded from the pellet debris by a minimum-**B** magnetic field, the surface-area requirement for the explosion chamber is determined by x-ray loading considerations. The combustion chamber radius may then be reduced

by approximately a factor of two.

The chamber wall might also be satisfactorily shielded from the plasma pulse, as well as from a portion of the x-ray pulse by pulsed injection of gas through the dry walls of the microexplosion chamber. However, the required mass injection rates are uncomfortably large, and the firing rate is limited by the chamber pumpdown time.

A very important problem for laser-fusion reactor design is how to arrange for input of the laser light and the target pellet while at the same time maintaining adequate neutron and x-ray shielding. Laser beams might be admitted through cheap, replaceable windows in the outer vacuum wall, passed through the neutron shield in neutronic-baffling dogleg tunnels on mirror trains, and focused onto the pel-

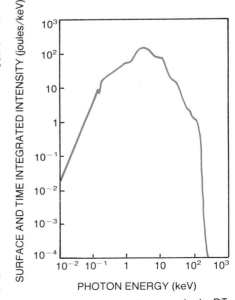

X-ray pulse spectrum of a megajoule DT-fusion microexplosion as calculated by a computer code; note the large self-absorption of the superdense fireball at low photon energies.

Figure 5

let atmosphere by aspheric mirrors facing into the explosion chamber through apertures in the inner wall. Continuous, low-Z liquid-metal exudation-surfacing of the mirrors would prevent degradation of the reflectivity by the thermonuclear environment for laser wavelengths greater than 0.2 microns. The fuel pellet, several millimeters in diameter, would free-fall or be electrostatically projected into the combustion chamber.

For tritium breeding and recovery necessary for pure DT burning, a lithium-rich neutron blanket similar to those being considered for magnetic confinement fusion power-plant designs would surround the combustion-chamber wall. A 1% void fraction in the lithium blanket is probably required to permit impulsive neutron heating of the lithium without mechanical damage to the wall.

Fusion energy conversion

Fusion energy pulses, as extremely high-grade energy sources, apparently admit of several very different means of converting their energy into electricity, depending on pellet-fuel composition, the ρR value at which the fuel is burned (product of density and pellet radius, see box) and the combustion-chamber system design and operation. Three types of systems have been identified so far: They are ordinary thermal conversion, MHD hot-gas-generator conversion and MHD plasma conversion.

For first-generation laser-fusion power plants, which would burn DT pellets, ordinary steam-thermal conversion of fusion energy deposited by neutrons in the lithium blanket appears preferable. Such systems would have capital costs of several hundred dollars per kilowatt and energy-conversion efficiencies of up to 40%, which is comparable to conventional and fission-reactor systems.

If the combustion-chamber wall is shielded from the plasma pulse by injected gas, the heated gas might be exhausted from the chamber through a relatively inexpensive, pulsed MHD hot-gas generator. Several atmospheres stagnation pressure at a few thousand degrees temperature could be produced. Such a system might permit electricity generation with higher total efficiency (approximately 60%) for moderate ρR (approximately 10 gm/cm^2) DT or DD pellet burning, or for a 5 gm/cm^2, high-charged-particle-fraction (for example D–He^3) pellet burning. Such ρR's may be obtained with a few-hundred-kilojoule lasers if the pellet is compressed to 10^4 gm/cm^3 densities. However, at these high ρR's, 10–30% of the fusion energy is radiated as x rays. Hence the ultimate efficiency of this approach is limited by the efficiency with which the x-ray energy may be converted to electricity.

The rapidly expanding fusion fireball may be made to do magnetohydrodynamic work on a magnetic field imposed from outside the combustion chamber, transforming its energy into that of a compressed magnetic field. Induction coils suspended from the combustion-chamber walls might be used to transform the compressed-field energy directly into electricity, in a manner basically very similar to the way an ordinary power transformer works. The basic feasibility of such fireball-to-electricity energy conversion has already been demonstrated.[23] Low capital cost, high efficiency (greater than 70%) electrical energy generation may thus be ultimately attainable, in advanced laser-fusion CTR systems. Various estimates also indicate that laser-fusion power plants will be economically feasible.[24]

In the mid 1970's crucial superhigh density laser-implosion experiments will be carried out with lasers now being designed. Edward Teller has recently emphasized the importance of these experiments when he said, "A third of a century ago liquids were considered incompressible for all practical purposes. We are talking now about at least a thousand-fold compression if laser fusion is to be practical. This is a challenge we cannot afford to ignore. I believe that we shall succeed and that the effort will profoundly change our views on how man and matter can interact." Practical power production also depends on the success of programs now underway to develop pulsed lasers with

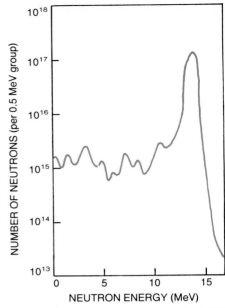

Neutron pulse spectrum of a megajoule DT-fusion microexplosion calculated by a computer code; note the neutron energy peaks produced by single and double 14-MeV neutron scatterings from deuterons and tritions and the 3.5 MeV peak. Figure 6

sufficiently high power, energy, frequency and efficiency, and on the engineering of economic reactors. We are excited by the challenge of these difficult and complex tasks, and by the prospect that the mastery of fusion may be more important to Man than the harnessing of fire.

* * *

This work was supported by the USAEC.

References

1. R. Hirsch, New Scientist **12**, 86 (1973).
2. R. Post, PHYSICS TODAY, April 1973, page 30.
3. J. Nuckolls, L. Wood, A. Thiessen, G. Zimmerman, Nature **239**, 139 (1972).
4. W. Beasant, *Hydrostatics and Hydrodynamics*, Cambridge U.P. (1859).
5. Lord Raleigh, Phil. Mag., **34**, 94 (1917).
6. G. Guderley, Luftfahrtforschung **19**, 302 (1942).
7. F. Hawkins, "Manhattan District History, Los Alamos Project," LAMS 2532 (1961).
8. K. Boyer, Astronaut. Aeronaut. **11**, 28 (1973). W. Daiber, A. Hertzberg, C. E. Wittliff, Phys. Fluids **9**, 617 (1966). J. S. Clarke, H. N. Fischer, R. J. Mason, Phys. Rev. Lett. **30**, 89 (1973). K. A. Brueckner, Trans. IEEE **PS1**, 13 (1973).
9. S. W. Mead, Phys. Fluids **13**, 1510 (1970).
10. N. G. Basov and others, JETP Lett. **15**, 417 (1972).
11. J. Nuckolls and others, Livermore report UCRL-74116 (1972).
12. G. Zimmerman, Livermore Report UCRL 50021-72-1, 107 (1972).
13. S. Braginskii, Rev. Plasma Physics **1**, 205 (1965).
14. S. Bodner, Livermore Report UCRL 74074 (1972).
15. J. Nuckolls, Livermore Report UCRL 74345 (1972).
16. R. Kidder, J. Fink, Nucl. Fusion **12**, 325 (1972).
17. J. W. Shearer, J. J. Duderstadt, Livermore Report UCRL 73617 (1972).
18. P. Kaw, J. Dawson, Phys. Fluids **12**, 2586 (1969). W. Kruer, J. Dawson, Phys. Fluids **15**, 446 (1972).
19. J. Katz, J. Weinstock, W. Kruer, J. Degroot, R. Faehl, Livermore Report UCRL 74334 (1972).
20. L. Wood and others, Livermore Report UCRL 74115 (1972).
21. B. Freeman, L. Wood, J. Nuckolls, Livermore Report UCRL 74486 (1971).
22. L. A. Booth, LASL Report LA 4858MS (1972).
23. A. Haught, D. Polk, W. Fadr, Phys. Fluids **13**, 2482 (1970).
24. R. Hancock, I. J. Spalding, Culham Report CLM-P310 (1972). □

Comments

When this article was published in 1973, the most promising approach to inertial fusion was classified. This "indirect" x-ray driven approach was discussed in my article in the Sept. '82 issue of PHYSICS TODAY, "The Feasibility of Inertial Confinement Fusion."

The discussion of target physics was obscured not only by classification, but also by uncertainties in the physics of laser pellet coupling and fluid instabilities in ablatively driven implosions.

Subsequently the coupling issues have been resolved—ICF lasers must have short wavelengths. However, the fluid instability issue is not yet resolved. There is experimental evidence that in ablatively driven implosions instability growth rates are sufficiently small to permit successful implosions of hollow targets with moderate aspect ratio shells, but we do not know how thin a shell can be successfully imploded.

Because hollow targets can have a much larger volume and area per mass than the bare drop targets discussed in the original article, the peak laser power and pulse shaping requirements can be greatly relaxed. However, the limiting target gain vs compression calculations shown in the article are unchanged since these limits depend only on conservation of energy and material properties and not on aspect ratio.

In retrospect our proposal to demonstrate laser fusion's feasibility with relatively small short wavelengths, high peak power lasers having unprecedented capabilities for precision pulse shaping and illumination symmetry contained unforseen difficulties. So far it has not been possible to construct such a laser. Consequently we still do not know whether or not breakeven can be achieved at the 10 kJ level. Instead it has proven easier, faster, and cheaper (since laser construction costs are a small fraction of the total program costs) to construct much larger lasers and implode larger targets which have relaxed requirements for pulse shaping, symmetry, etc. Furthermore, it is now clear that the smallest target implosions are the most difficult and sensitive, and that their failure would not prove the infeasibility of the much larger MJ scale targets which can achieve the gains of order 100 required for practical applications.

John Nuckolls
May 1986

Shiva moves closer to laser fusion

Bertram M. Schwarzschild

PHYSICS TODAY / NOVEMBER 1979

In recent months, the 20-armed Shiva laser system at the Lawrence Livermore Laboratory has attained a significant milestone on the road to the development of a laser-fusion reactor. The Livermore group has reported that with target pellets of classified design Shiva has driven the deuterium–tritium fuel inside the pellets to between 50 and 100 times its liquid density. (With unclassified ablative targets they report achieving 10 to 20 times liquid density.) One hundred times liquid density is only an order of magnitude short of the densities that will be needed to achieve "scientific break-even." This goal, namely the release of as much fusion energy as the lasers deliver to the target (or the somewhat more modest goal of thermonuclear ignition), may well be achieved by Nova, the next generation laser system at Livermore, on which construction began in May.

The Shiva laser, which has been in operation at Livermore for over a year (see PHYSICS TODAY, April 1978, page 17), can direct 10 to 15 kilojoules of 1.06-micron light at a target pellet in a pulse shorter than a nanosecond. Although Shiva can deliver higher energy than any other fusion laser currently in operation, it is still an order of magnitude below the energy that will be needed to achieve breakeven fuel densities at the 10-keV ion temperature required for deuterium–tritium fusion. One is therefore faced with a trade-off. According to John Nuckolls, who heads the target-design group at Livermore, one can design targets and shape laser pulses either to achieve high temperature or high densities, but not both simultaneously—at Shiva energies. (See figure.)

Temperatures near 10 keV, resulting in about 3×10^{10} fission neutrons, had already been achieved at Shiva some time ago, with an "exploding pusher" target-pellet design. But this temperature was attained at much lower densities than those recently attained with "ablative" targets and longer pulse durations. The exploding pusher targets at Shiva, about 0.3 mm in diameter, have very thin glass shells. With these targets, the entire laser pulse is delivered in about 100 picoseconds, roughly the time it takes a sound wave to traverse the thickness of the shell, thus initiating a supersonic shock wave in the fuel. Such a short laser pulse, reaching a power level of about 30 terawatts, generates very high temperatures in the fuel before it reaches high density.

To achieve high fuel densities with a 15-kJ laser, one must avoid the shock wave that prematurely excites the high temperature that hinders further compression. This slower, "isentropic" compression has now been accomplished at Livermore with ablative targets and targets of classified design. Much of the information about target pellets and density monitoring is classified. The ablative targets have about the same overall size as the exploding-pusher pellets, but their shells are significantly thicker, in several concentric layers. With such targets the laser pulse from Shiva is of longer duration than with exploding-pusher pellets, of the order of a nanosecond.

Rather than delivering the entire pulse in the time it takes a sound wave to traverse the thick ablative shell, the temporal shape of the pulse is designed to double the pressure of the fuel in that characteristic time. This compression is accomplished by the smooth inward acceleration of the inner shell in reaction to the outward ablation of the outer shell surface by the intense laser light. In contrast to the supersonic shock wave of the exploding-pusher case, the ablative-pellet fuel is driven to higher pressures more slowly and smoothly, by a subsonic thermal wave. The temperatures thus attained at Shiva are only a few keV, resulting in a lower neutron yield, about 10^8 per pulse. But even at these lower temperatures, the achievement of fuel densities fifty to a hundred times liquid density is, according to Lawrence Killion of DOE's Office of Inertial Fusion, an important step on the way to a laser-fusion reactor.

Because densities must be monitored by indirect techniques, the final densities are not known with great precision. The Livermore group quotes a maximum density of 70 plus or minus about 30 times liquid density for classified targets.

Nova. To achieve fusion-energy outputs of the order of the laser input energy to a deuterium–tritium target will neces-sitate fuel densities about a thousand times the liquid density, at a temperature of 10 keV. This will require a new generation of lasers, capable of delivering at least an order of magnitude more energy than Shiva. In May construction began at Livermore on the Nova Laser, whose first phase (Nova I) is expected to achieve about a hundred kJ early in 1983. Like Shiva, Nova will be a multi-armed neodymium-glass laser system. But each arm of Nova is expected to produce 15 to 20 times as much laser energy as a single arm of Shiva. Nova I will have 10 arms, directing a total energy in excess of 100 kJ at the target area in 1- to 3-nanosecond pulses. Phase II of Nova, on which a final funding decision from DOE is expected in 1981 (see PHYSICS TODAY, September 1979, page 106), anticipates 10 additional high-power laser arms.

Nova will achieve its twenty-fold increase in output per laser arm by two techniques: the use of improved optical materials in the neodymium laser, capable of tolerating the higher power flux densities, and an overall increase in the cross-sectional area of each arm. The host material in Shiva is silicate glass, doped with neodymium atoms, which provide the 1.06-micron lasing transition. At the power flux levels envisaged for Nova, the optical susceptibility of most glasses exhibits serious nonlinear behavior; that is, the effective index of refraction increases with the light intensity. This causes the beam to fragment into thin filaments, undermining the coherence of the laser beam and generating hot spots that can damage the laser material.

In Shiva this difficulty has been partly resolved by placing spatial filters in the optical line, which filter out this filamentary structure generated by the nonlinear optics. But at the higher Nova flux levels this remedy will not suffice. The recently completed 6-armed Zeta laser at the University of Rochester employs a phosphate glass whose nonlinear properties are somewhat better than those of traditional silicate glasses.

Nova will employ fluorophosphate glass, a new kind of glass with a very small nonlinear refractive index, developed by Livermore in cooperation with Owens-

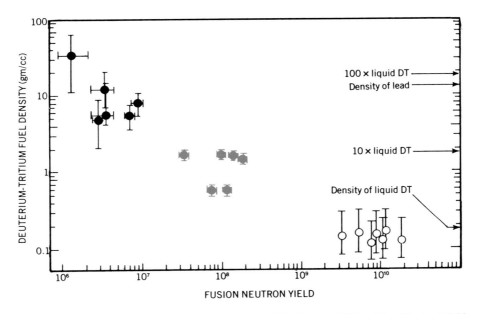

Trade-off between neutron yield and fuel density is evident in recent Shiva data.   Neutron yield increases rapidly with fuel temperature.   Open circles show compression of exploding-pusher targets, which achieve high temperature at low densities.   Colored points show ablative-target results.   Highest densities are obtained at lowest temperatures with classified targets (black).

Illinois, Schott, Hoya Optics and the National Bureau of Standards.   This glass, which consists mostly of fluorine, aluminum and calcium, with a bit of phosphate for stabilization, has a nonlinear susceptibility only a third that of the silicate glass in Shiva.   Earlier this year there was some concern that because neodymium ions sitting at different lattice sites have slightly different lasing frequencies, glass lasers might not achieve adequate energy extraction.   John Holzrichter, who heads the laser group at Livermore, told us that recent energy-extraction measurements satisfy him that Nova will achieve its design goal of 100 kJ in 3 nanoseconds.

The Gekko-XII neodymium-glass laser in Osaka, Japan is currently using phosphate glass in its two operational arms. When completed, its twelve arms are expected to deliver 40 terawatts.   Plans call for fluorophosphate glass lasers later.

With the new glass and improved coating technqiues, Nova is designed to tolerate at least a factor of two increase in power flux over Shiva levels.   But each Nova arm must accommodate power levels up to twenty times those of its predecessor.   The additional factor of ten comes from an overall tenfold increase of the cross-sectional area of the Nova arms, from a typical diameter of 20 cm to 74 cm.

Because the target volume should scale with the total laser-energy input, the pellets for Nova I will have a diameter of about a millimeter.   Nuckolls explained

to us that with increasing target size one wants a smaller incident wavelength, to optimize the coupling of the laser energy to the target.   The laser light couples to the ablated plasma from the pellet's shell primarily at the so-called critical surface, where the light frequency equals the characteristic frequency of the plasma. Theoretical calculations indicate that at smaller wavelengths the coupling efficiency of the laser light is enhanced, for several reasons: At smaller incident wavelengths the critical surface finds itself at higher densities, increasing the collision frequency in the ablated plasma. Furthermore, at shorter wavelengths the problem of electron preheating of the fuel is expected to be lessened and the efficiency of ablation should improve.

Color.   For shorter wavelengths one wants to avail oneself of the higher harmonics of the infrared neodymium lasing frequency, "to put color into Nova," as the jargon goes.   The second harmonic of the neodymium lasing frequency, 0.53 microns, is green; the third, 0.35 microns, is blue.   Experiments are now in progress at Livermore's two-armed Argus laser to generate these harmonics in the laser output by means of KDP (potassium-dihydrogen-phosphate) crystals.   By properly phase matching the 1.06-micron laser beam, one can use the nonlinear response of these crystals to intensify the higher harmonic components.   The Argus group and KMS Fusion (Ann Arbor, Michigan) have already succeeded in extracting the second harmonic with better than 50% efficiency in large-aperture

systems.   Plans are being formulated to include KPD crystals in the Nova beam in order to make use of the second, third, and perhaps even fourth harmonics of the neodymium line.

To achieve scientific breakeven one needs a target gain (fusion output over laser energy) of one.   But to offset losses in the lasers and other parts of the system, the target gain would have to be of order 100 in a useful laser-fusion reactor. Holzrichter told us that for a useful reactor the laser would need to be at least 5% efficient.

Although the neodymium glass laser has already proven itself a very useful research device, current designs could not approach this required efficiency level, and they could certainly not operate at the high repetition rates necessary for a practical fusion reactor.   At these high rates one needs a laser whose heat can be rapidly dissipated.   In this regard gas lasers appear to be more promising than solid-state lasers.   The CO_2 laser used in the 10-kJ Helios system now operating at Los Alamos (and in the 100-kJ Antares system under construction there) is the most obvious candidate.   But there is some question about the coupling efficiency of the long-wavelength (10-micron) CO_2 laser light.

The Department of Energy's Inertial Confinement Fusion Program is supporting the development of several other inertial-fusion driver systems, including KrF gas lasers, with a wavelength of 0.25 microns, and light- and heavy-ion bombardment techniques.

CHAPTER 7

X RAYS AND TOMOGRAPHY

It is sometimes stated that most of the dramatically new discoveries in physics are made by bright young researchers—usually still in their 20s—who are largely unencumbered with the traditional training and experience, and who are therefore able to see and interpret physical phenomena in a new and different way from the more mature but prosaic scientist. But Wilhelm Röntgen was in his middle-fifties, happily studying electrical discharge phenomena in vacuum tubes, when in 1895 he observed a curious glow that seemed to be caused by his experiment. He worked busily for three or four months before publishing a magnificent series of papers on a new, highly energetic sort of radiation that he named x rays. Other workers immediately turned also to this field, but it is interesting to note that Röntgen had managed in his initial papers to observe almost all of the fundamental properties of this short-wavelength radiation; the next generation of researchers largely only added another

decimal point to some of Röntgen's measurements. Before a year was out doctors were examining x rays of broken bones, and hosts of medical applications were developed. (Some early workers suffered serious damage from overexposure to x radiation. However, William Coolidge, who perfected a practical x-ray tube around 1915, lived to the ripe old age of 101.)

The x-ray image is really a shadow of some opaque or semi-opaque object, further blurred by the nonplanar radiation from the x-ray tube. But in the 1960s Allan Cormack (and others) realized that if a series of x rays were recorded at a sequence of angles, a computer could be used to reconstruct a well-defined image of a single slice through an object (such as a human patient). So the computer has been linked to x-ray optics to produce the new field of computerized tomography, with (at present) the most dramatic results in medical applications. The papers of this section sketch some of these recent developments.

John N. Howard

CONTENTS

Computerized tomography: taking sectional x rays

A source and detector moving around the patient yield data that, processed by a computer, make visible a two-dimensional slice of the living human body—at 2-millimeter resolution.

William Swindell and Harrison H. Barrett

PHYSICS TODAY / DECEMBER 1977

An important new diagnostic technique is making its appearance in our major hospitals. The technique, which uses x rays to render visible thin slices through any section of the human body, has been so dramatic in its development that no general agreement has yet been reached on its name—it has been called "computed tomography," "computerized axial tomography," "transaxial tomography" and "reconstruction from projections." Figure 1 shows one version of the apparatus in a clinical setting.

How is it possible to construct such an image, and what computations are needed to do this? What has computed tomography accomplished, and what future improvements are in the offing? We will consider these questions in this article, but first let us trace the origins of the new technique.

A useful shadow

For sheer rapidity of practical exploitation, few scientific discoveries can rival Wilhelm Konrad Röntgen's marvelous x rays. Within nine months of that day in 1895 when he observed that strange fluorescence in barium-platinocyanide powder, photographic records of the interior of the human body were produced in several laboratories. The potential power of these startling images was immediately obvious to even the least imaginative of observers.

The field of diagnostic radiology matured rapidly over the next two decades. Practical x-ray tubes, antiscatter grids, radio-opaque contrast media and intensifying screens of calcium tungstate, all mainstays of a modern radiology department, were developed in that period. Even x-ray movies, progenitors of today's ciné-fluoroscopy, were produced before the turn of the century.

After 1917, major advances in the art and science of medical radiology were slower in arriving. The rotating-anode x-ray tube came on the scene in 1939, and electronic image amplification became available in 1949. Although these two developments had considerable practical importance, they could scarcely be counted as revolutionary. If Röntgen had walked into any hospital radiology department in 1970, he would undoubtedly have been impressed by the equipment he saw there, but he would have had no difficulty in recognizing the underlying principles. Indeed, for all its technological sophistication, this modern equipment produced images that differed remarkably little from those produced by Röntgen himself. From a physicist's viewpoint, the field of radiology had reached senescence.

To understand this situation, it is worthwhile to consider briefly what physical mechanisms the designer of radiographic systems has at his disposal. The options are drastically restricted by the requirement that the radiation be sufficiently penetrating to pass through a human body, yet not so penetrating as to make detection of contrast in the images impossible. If photons are to be used, their energy must be near the 10–100 keV range. Energetic charged particles, such as protons at several hundred MeV, could be used, but the equipment is a little cumbersome for a dentist's office. Ultrasound is a reasonable alternative for some applications, but again the requirement for adequate penetration must be considered. Because the acoustic attenuation increases rapidly with frequency, the sound must have a wavelength of nearly a millimeter, so that the image quality is severely limited by diffraction.

Therefore for high-resolution imaging we are limited in practice to rather hard x rays—so hard, in fact, that refraction, reflection and diffraction effects are quite negligible. The only possible interactions that can be used to form images are absorption and scattering. The radiation can be blocked and diverted away from the detector, but it can not be controllably deflected in the way a lens deflects light; radiographic images are therefore, in a sense, shadows.

An instant success

Although such a shadow conveys much valuable information, it is by no means a complete, unambiguous representation of the x-ray transmission characteristics of the body being imaged. Instead, it is a two-dimensional projection of a three-dimensional structure—many planes are collapsed onto one. Not only is all depth information lost, but more importantly the confusion of overlapping planes makes detection of subtle abnormalities difficult or impossible. Details in soft tissue simply can not be discerned on a conventional radiograph.

A partial solution to this problem is the type of "tomographic" imaging shown in figure 2, in which a selected plane of the body is imaged in sharp focus. But this type of tomography (tomo is derived from the Greek word for slice) only blurs the out-of-focus planes without eradicating them; the image contrast is still very low. The ideal solution would be an image of only the selected plane of interest without interference from detail in other planes.

The mathematical basis for a solution to this problem existed as early as 1917 in the work of the Austrian mathematician Johann Radon,[1] but the first practical,

The authors both hold joint professorships with the Optical Sciences Center and the Department of Radiology, Arizona Health Sciences Center at the University of Arizona, Tucson.

In transaxial tomography the patient remains stationary while a gantry housing the x-ray source and detector rotates. In apparatus such as this Varian Three-second Whole Body CT Scanner the beam pivots about the desired plane to image a slice through the body. Figure 1

clinically oriented solutions did not begin to appear until the early 1960's. In that period Allan Cormack[2] of Tufts University began to popularize and extend Radon's work, while David Kuhl[3] and his co-workers at the University of Pennsylvania built a practical "transverse section scanner" for nuclear-medicine applications. Kuhl's device, unlike all earlier tomographic devices, isolated a single plane transverse to the long axis of the patient's body (whence its name), completely eliminating information from the other planes.

But the spark that ignited the widespread excitement in this field came from a commercial laboratory, EMI Ltd, in Great Britain. In 1971 they announced the development of the EMI Scanner, a symbiotic marriage of x-ray-scanning and digital-computer technology.[4] This remarkable system, largely the brainchild of Godfrey Hounsfield, generated images of isolated slices of the brain with exquisite discrimination of very small differences in x-ray absorption coefficient.

With these new techniques, radiologists for the first time could distinguish between different types of brain tissue, and even between normal and coagulated blood. The cover of this issue of PHYSICS TODAY shows such a sectional view through a human head, in which false colors are used to enhance the discrimination even further. This computerized tomographic image of a section of the human head taken horizontally through the eyes shows the nasal sinuses in black, brain tissue in blue and green, the bones of the skull in white and the left optic nerve in yellow. The scan, with a spatial resolution of approximately 2 mm, was made on an Ohio-Nuclear Delta Scan 50.

With such images radiologists can now visualize easily the ventricles of the brain,

the repositories of the cerebro-spinal fluid. This capability alone would have made the new device an instant success because it immediately made obsolete a rather unpleasant procedure, called "pneumoencephalography," in which air is pumped into the ventricles to displace the fluid and provide radiographic contrast.

Many other companies soon entered the market, and in short order machines

capable of this type of imaging in any section of the body became available.

To help appreciate the significance of this development, some typical specifications might be useful. Most machines have a spatial resolution around 2 mm and, almost without exception, the manufacturers claim a noise level corresponding to a 0.5% change in x-ray absorption coefficient. To achieve this level of discrimination in a normal radiograph

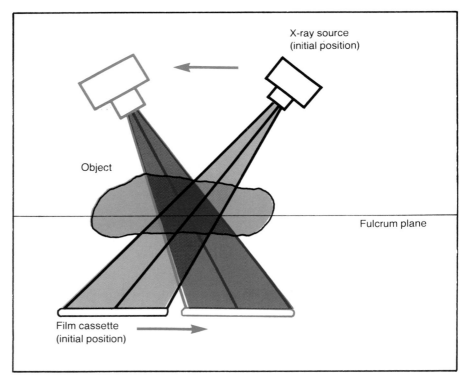

Shadows of points in the fulcrum plane remain stationary on the moving film because of the simultaneous motion of the x-ray source; hence they are sharply imaged. The image of a structure located away from this plane is blurred by the motion. Although this scheme, "conventional tomography," facilitates the x-ray examination of an isolated region deep within a patient's head or chest—or other object—the contrast it achieves is poor. Figure 2

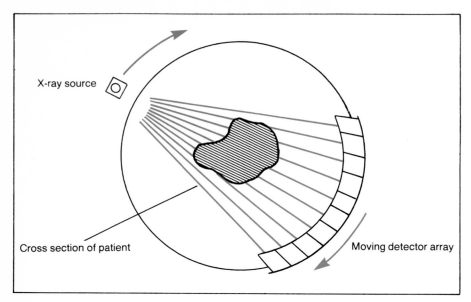

A third-generation machine. The x rays are formed into a fan of pencil beams that encompasses the section of interest. With the source–detector assembly rotating uniformly about the patient, the required set of exposures is obtained by flashing the x-ray tube on at the appropriate angular positions. In some recent versions only the source moves.   Figure 3

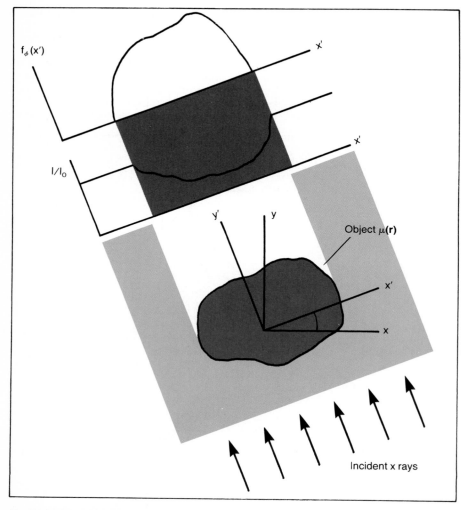

To obtain data for computer processing, the object—a two-dimensional distribution of the linear attenuation coefficient $\mu(r)$—is fixed with respect to the unprimed reference frame (x,y). The direction of the x-ray beam, as it rotates during the sequence of exposures, defines the y' axis. Projection data are gathered along the x' axis; ϕ is the angle of rotation. The projection function $f_\phi(x')$ is given from intensity data by $-\ln(I/I_0)$.   Figure 4

would require x-ray-intensity differences of about 0.02% to be rendered visible on the film. (The linear absorption coefficient of soft tissue is about 0.2 cm^{-1} and the change of intensity would thus be 0.2 cm$^{-1} \times 0.2$ cm $\times 0.5\% = 0.02\%$.) With conventional projection radiography an intensity change of 0.02% is two orders of magnitude smaller than the minimum detectible change possible with that modality. Furthermore, because the output of computed-tomography machines is a quantitative map of the x-ray absorption coefficient in a single slice, it is readily possible to display the subtle differences measured by the device.

Collecting the data

The principles of computerized tomography as applied to medicine were recently reviewed by Rodney Brooks and G. DiChiro,[5] and other recent advances were reported[6] at a 1975 conference in Puerto Rico. The wider applications of three-dimensional reconstruction are treated in the proceedings volumes resulting from a Brookhaven workshop[7] and a meeting sponsored by the Optical Society of America.[8]

To reconstruct the two-dimensional map of x-ray absorption coefficients, a large number of transmission measurements must be made through the slice of interest. By restricting the measurements to paths contained only in that slice, information from other parts of the body is automatically excluded from the data. The normal arrangement is for the patient to remain stationary and be positioned within a gantry that houses an x-ray tube and one or more detectors, as figure 1 shows. By suitably translating and rotating the source–detector assembly around the patient, the required series of measurements is obtained.

Three generations of machines have already appeared, the generic distinctions being mainly in the details of the scanning systems. The early machines used a single detector and a pencil beam of x rays. Actually, two detectors are to be found in these machines, but each one takes data for a different section of the patient and the system produces two output images for each patient scan. Because of the serial way in which the data are collected, these first-generation machines need approximately four minutes to complete the scan; they take, typically, about 200 measurements at each of 180 different angular positions.

Second-generation machines are somewhat faster, with scan times ranging from ten seconds to two minutes. Many projections are recorded simultaneously with several pencil beams and an array of detectors. The faster machines in this group can collect the data during the time a patient can hold his or her breath. This greatly facilitates imaging of the abdomen and chest, because patient motion during the scan can introduce severe artifacts

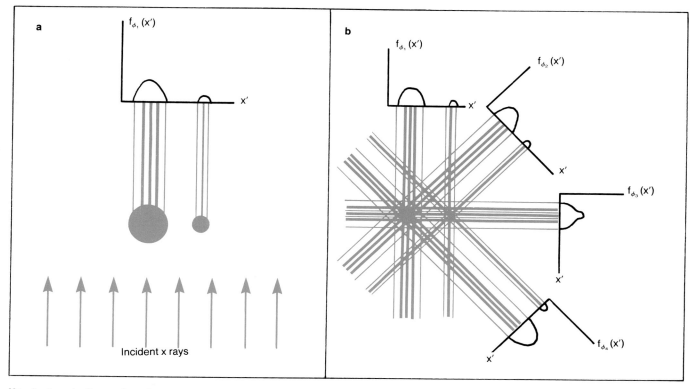

How back projection works. Consider as the object two absorbing disks of different diameters. At the **left** we see the first projection, at angle ϕ_1, and the corresponding function defined in the previous figure. Back projection, shown on the **right,** consists of smearing each projection back along the direction in which the original projection was made, where they sum to form a reconstruction of the object. With a sufficiently large number of projections (only four are shown here), points in the object become cusp-like patches (as suggested by the spoke pattern in the figure). This "point-spread function" has the effect of degrading the quality of the resulting image.

Figure 5

into the reconstructed image. It is still necessary to translate the assembly containing the x-ray tube and the detector, but the number of angular positions at which scans are made is reduced approximately in proportion to the number of detectors (per slice) used. Generally, these faster body-scanning machines take data from only a single slice of the body.

Third-generation machines are even faster, with data collection typically taking between five and ten seconds for the whole operation. They employ a wide fan-shaped beam covering the whole section, eliminating the need for translation; see figure 3. Because the method permits the angular motion to be continuous, the mechanical arrangements are much simpler. Inevitably a large number of separate detectors is required; several hundred are to be found in many of the recently announced systems. These systems often employ xenon-filled ionization detectors, because they are smaller and cheaper, although they are not as efficient as the scintillation detectors used in earlier systems.

The scanner output is processed by a digital computer and converted into useful form by peripheral devices. A basic system consists of a scanner unit, the central processing unit, and a control and display console. Options include magnetic data storage on tape or floppy disk, line printers for numerical printouts, and remote and off-line display devices.

The central-slice theorem

Imagine an object with a property that is distributed in the form of only a single spatial cosine wave, which is one Fourier component of a more general distribution. When a projection of this object is formed in an arbitrary direction, the projecting rays will cross an equal number of positive and negative crests of the cosine with a net result of zero. However, when the projection is parallel to the crests of the cosine wave, that is, in the y' direction, the projection will also be a cosine. A one-dimensional Fourier transform of the projection then reveals the amplitude and the spatial frequency of the two-dimensional cosine wave, while the projection direction reveals its orientation.

A mathematical derivation of this result is not difficult. Let the object distribution be denoted by $\mu(\mathbf{r})$ where $\mathbf{r} = (x, y)$. Consider a projection along the y' axis given by

$$f_\phi(x') = \int_{-\infty}^{\infty} \mu(\mathbf{r}) dy'$$

where the x'–y' axes are rotated an angle ϕ from the fixed x–y system. The one-dimensional Fourier transform of the projection is

$$\mathcal{F}_1\{f_\phi(x')\} = \int_{-\infty}^{\infty} f_\phi(x') e^{i2\pi\xi'x'} dx'$$

where ξ' is the spatial-frequency variable conjugate to x'. Substituting in the definition of $f_\phi(x')$, we find

$$\mathcal{F}_1\{f_\phi(x')\}$$
$$= \int_{-\infty}^{\infty} dx' \int_{-\infty}^{\infty} dy'\, \mu(\mathbf{r}) e^{i2\pi\xi'x'}$$
$$= \left[\int_{-\infty}^{\infty} dx' \int_{-\infty}^{\infty} dy' \times \mu(\mathbf{r}) e^{i2\pi(\xi'x'+\eta'y')} \right]_{\eta'=0}$$

The latter form is simply the two-dimensional Fourier transform $\mathcal{F}_2\{\mu\}$ evaluated along the line $\eta' = 0$ in a two-dimensional frequency plane. Because ξ' and η' are always conjugate to x' and y', respectively, the line $\eta' = 0$ rotates as the projection angle is changed. A set of projections over a 180-deg range in ϕ is therefore sufficient to determine $\mathcal{F}_2\{\mu\}$ at all points.

The theorem has wide utility. For example, in radio astronomy it provides the basis for Earth-rotation aperture synthesis, and it permits a determination of the modulation transfer function of an optical system from a series of transforms of *line-spread* functions. If the system is known to be rotationally symmetric, one line-spread function is sufficient.

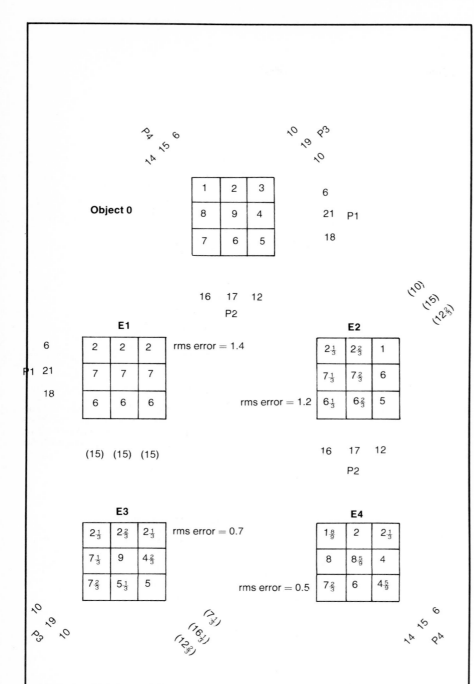

The many iterative algorithms differ in the manner in which the corrections are calculated and reapplied during each iteration. They may be applied additively or multiplicatively; they may be applied immediately after being calculated; alternatively, they may be stored and applied only at the end of each round of iteration. The order in which the projection data are taken into consideration may differ as well.

The simple example shown here illustrates additive immediate correction. Four three-point projections are taken through a nine-point object, O, giving rise to projection data sets P1 through P4. Taken in order, these are used successively to calculate estimates E1 through E4 of the original object.

The initial estimate is obtained by allocating, with equal likelihood, the projection data P1 into the columns of E1. Subsequent corrections are made by calculating the difference between the projection of the previous estimate and the true projection data and equally distributing the difference over the elements in the appropriate row of the new estimate. For example, the difference between the projection of the first column of E1 shown in parentheses (15), and the true measured value, 16, is 1. In creating the first column of E2, one third of this difference (1/3) is added to each element of the first column of E1. The first iteration is completed with the calculation of E4.

That the process converges in this numerical example is demonstrated by calculating the rms deviation of elements of E1 through E4 from the true values in O. As the figure shows, these rms errors decrease monotonically.

Data-processing time has kept pace with data-acquisition time. With some systems the picture is available immediately; others require from a few seconds to a few minutes after the scan is completed. Rapid processing expedites the flow of patients, and if a retake is found to be necessary it can be done at once, before the patient has left the area.

Constructing the image

In practice, discretely sampled data and fan-beam geometries are used. However, for clarity this article will describe mainly the reconstruction algorithms appropriate to continuous data sets; we will assume parallel rather than fan-beam projections.

When a monochromatic pencil beam of x rays of intensity I_0 passes through a uniform medium, it is attenuated in accordance with Beer's law

$$I = I_0 e^{-\mu x} \qquad (1)$$

For inhomogeneous objects the exponent is replaced with the line integral of the linear absorption coefficient, $\mu(x,y)$.

$$I = I_0 \exp\left[-\int_{SOURCE}^{DETECTOR} \mu(x,y)dl\right]$$

To linearize the algebra we take negative logarithms of the raw data, which gives integral equations of the form

$$f_\phi(x') = -\ln(I/I_0)$$
$$= \int_{SOURCE}^{DETECTOR} \mu(x,y)dy' \quad (2)$$

The geometry is shown in figure 4.

The problem is to obtain an estimate of $\mu(x,y)$, designated $\hat\mu(x,y)$, from a sufficiently large set of projection data $f_\phi(x')$. We can obtain only an estimate at best, because of noise considerations and because we are dealing in practice with sampled data.

A theorem that is fundamental to the understanding of reconstruction from projections is called the *central-slice* or *projection-slice theorem*; it is described and illustrated in the Box on page 303. In brief, it says that a one-dimensional Fourier transform of a one-dimensional projection of a two-dimensional object is mathematically identical to one line (a slice) through the two-dimensional Fourier transform of the object itself. Thus knowledge of all one-dimensional projections is sufficient to synthesize the two-dimensional transform of the object from which the object is readily obtained by an inverse two-dimensional transform. Although this method of reconstruction has been used commercially, it is not used extensively.

The most common reconstruction algorithm includes an operation called back projection, illustrated in figure 5. The result of performing this operation, $B[\ ]$, on the projection-data set is to generate an estimate $\hat\mu$, albeit a badly degraded one, of the absorption-coefficient distribution of the original object, as given by

$$\hat{\mu}(\mathbf{r}) = B[f_\phi(x')] = \mu(\mathbf{r}) ** \frac{1}{r}$$

The double asterisk represents the two-dimensional convolution integral. The statement simply says that, if we think of the reconstruction as an optical image, it resembles the object—except that each point in the object has been replaced by a cusp-like patch of light. The irradiance of this patch falls off inversely with the distance from its geometrical center. This patch of light, known as the "point-spread function," serves only to degrade the quality of the image to such a degree that it becomes useless for medical diagnoses and most other purposes.

Image processing

To overcome this defect, we employ the tools of the image-processing trade. Because the point-spread function is known and because it is everywhere the same, the degraded image can be improved by convolving it with an appropriate correcting function. (Equivalently, this correction can be done by taking the Fourier transform of the degraded image, multiplying that transform by a correction factor and then performing the inverse Fourier transform on this intermediate result.) Whether the correction is applied in real space by convolution or in Fourier space with frequency filters, the process is necessarily two-dimensional in nature. A considerable simplification is gained by filtering the projection data *before* they are back projected, when they are still in a one-dimensional format. That this interchange in the order of filtering and back projection is permissible follows immediately from the fact that both of the operations involved are linear and shift-invariant.

This reconstruction algorithm, known as the filtered back projection, can now be written

$$\hat{\mu} = B[h(x') * f_\phi(x')] = B[\hat{f}_\phi(x')]$$

The process is illustrated in figure 6. The one-dimensional convolution, designated by the single asterisk, is between each set of projection data $f_\phi(x')$ and the appropriate filter function, $h(x')$. It is this algorithm that is most commonly used in commercial machines. Not only is the algorithm structurally simple, its implementation is also straightforward; the whole process can be coded with less than 25 Fortran statements.

Another type of commonly used reconstruction algorithm is based on iterative techniques. Figure 7 shows how the projection data may be related to the discrete distribution of absorption coefficients μ_n through a set of linear simultaneous equations

$$f(\phi,x') = \sum_{n=1}^{N} \alpha_n(\phi,x')\mu_n$$

with coefficients $\alpha_n(\phi,x')$. With ap-

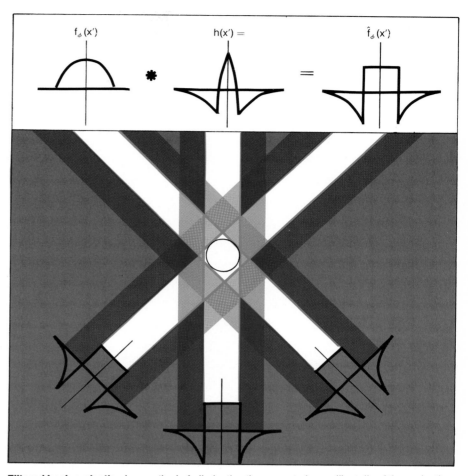

Filtered back projection is a method of eliminating the unwanted cusp-like tails of the projection shown in figure 5. The projection data are convolved (filtered) with a suitable processing function $h(x')$ before back projection, as shown at the **top** of the figure. The method is illustrated for a small absorbing disk, with only three back projections shown (**bottom**); in practice there would be at least 100 back projections. The filter function has negative side lobes surrounding a positive core (shown exaggerated in width for clarity) so that, in summing the filtered back projections, positive and negative contributions cancel outside the central core. With the tails thus eliminated, the reconstructed image will resemble the original object quite closely. Figure 6

proximately 200×200 equations in the set, the solution is best obtained iteratively. Various techniques have been proposed and employed. With names such as Algebraic Reconstruction Technique (ART), Simultaneous Iterative Reconstruction Technique (SIRT) and Iterative Least-Squares Technique (ILST), they differ in the details of the way the correction terms are calculated and reapplied to subsequent iterations—the Box on page 36 illustrates the method with a simple example. Reference 9 provides a review of these methods. These algorithms are capable of producing satisfactory reconstructions, as the success of the commercial instruments employing them attests.

Current research, future prospects

There is a trade-off in tomography between the signal-to-noise ratio SNR, the patient dose D, the spatial resolution ϵ and the slice thickness h, given by

$$D \propto (SNR)^2/\epsilon^3 h$$

in simplified form.[10]

This fundamental relationship is based on the Poisson statistics of the photon flux. The value of the constant of proportionality is relatively independent of the particular processing algorithm. Indeed, modern processing algorithms are sufficiently good that all commercial systems exhibit essentially the imaging performance implied by the above equation. This suggests that future improvements in resolution or signal-to-noise ratio will be achieved only by increasing the patient dose. Because this is normally unacceptable, significant improvements in resolution should not be expected.

Several trends, however, can be discerned in the current development of computed tomography. One of the most striking is the "horsepower race" to achieve even more rapid scans and image reconstructions. The fan-beam geometries described previously were an important step towards this end, yielding scan times under 10 sec. Some manufacturers now use as many as 600 separate detectors to speed up data acquisition.

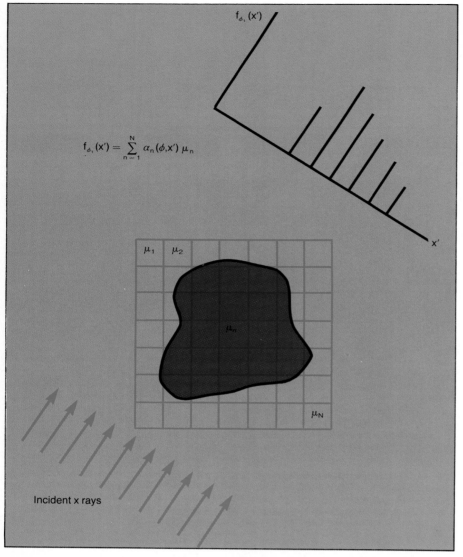

$$f_{\phi}(x') = \sum_{n=1}^{N} \alpha_n(\phi, x') \mu_n$$

In iterative-reconstruction techniques a matrix of N cells represents the object being imaged by x rays. The line integrals for the projection function $f_{\phi}(x')$ reduce to a set of linear equations in the absorption coefficients μ_n, with $\alpha_n(\phi, x')$ representing the path length traversed by the x-ray beam in the nth cell (n runs from 1 to N).

Figure 7

Reconstruction times have been decreased through the application of special-purpose computing hardware and more efficient software.

A somewhat extreme example of special-purpose computing hardware is the incoherent optical computer being studied by the authors.[11] In this approach the projection data are recorded on x-ray film and viewed with an analog optical system that mimics the more conventional digital machines, but in a parallel rather than serial fashion. Among the advantages being hoped for are high spatial resolution, low cost and rapid reconstruction.

A dramatic increase in scan speed is promised by the Dynamic Spatial Reconstructor now under development by Earl Wood and his collaborators at the Mayo Clinic.[6] This system uses 28 separate x-ray sources, each with its own image intensifier, low-light-level television camera and video recording channel.

Twenty-eight separate two-dimensional projections of the three-dimensional object are recorded in less than 10 msec—fast enough to freeze the beating heart! Unfortunately, the 28 projections are seldom enough for adequate image reconstruction. Recognizing this, the Mayo researchers have made provision to rotate the assembly through a small angle to fill in the missing projection angles; however, in so doing they have at least partially negated their speed advantage. Their machine promises nevertheless to be an extraordinarily useful research tool.

Another recent innovation is the use of software to reshuffle the reconstructed image planes to create new planes arbitrarily oriented with respect to the original set. For example, sagittal planes (dividing right from left) and coronal planes (dividing front from back), parallel to the long axis of the patient's body, are easily formed. These are often quite

useful techniques in clinical applications.[6]

Strictly speaking, the first equation of this article is correct only for monochromatic x rays. Because practical sources provide x rays with a wide energy range, and because the absorption coefficient μ is also a function of photon energy, there should be also an integration over the spectral energy distribution. When this effect is ignored, the contributions to μ from Compton scattering and photoelectric absorption are scrambled together. The resulting image is subject to errors called "beam-hardening artifacts" (so called because the lower-energy, "soft," photons are more rapidly absorbed). This defect may be remedied by forming two images, each with its distinct x-ray spectrum.[12] Suitable processing then yields maps of the absorption coefficient for the two constituent processes. Much valuable information regarding average atomic number and chemical composition can be gleaned from this new type of data. Lesions that could not be seen at all in a map of total attenuation coefficient stand out clearly when this two-energy technique is used.

Inner and outer space

Not surprisingly, the dramatic success of x-ray computed tomography has spawned a search for other applications of the same basic principles. One promising area of current research is, somewhat ironically, nuclear medicine. The irony is that medical computed tomography really originated in nuclear medicine with David Kuhl's early work, but the field foundered and might never have reached commercial viability had it been confined to this area.

One impetus to the development of tomographic techniques in nuclear medicine is the increasing interest in positron-emitting radionuclides.[6] Isotopes such as C^{11}, N^{13}, O^{15} and F^{19} are easily incorporated into almost any biologically interesting molecule. Imaging of the 511-keV annihilation radiation, produced within a few millimeters of the site of positron emission, thus provides valuable information about the distribution of these molecules in the body.

Ultrasonic applications of computed tomography are also receiving increasing attention. Maps of the ultrasonic attenuation or velocity can be produced by passing a set of sound beams through the body in direct analogy to the x-ray beams used in computed tomography.[13] On a still higher level of sophistication, ultrasonic Doppler shifts can map the velocity of blood flow.

Yet another new application area of medical interest is zeugmatography,[14] the marriage of computed tomography and nuclear magnetic resonance. In this case the output is a map of either the nuclear spin concentration or the relaxation time in the body.

Nonmedical applications abound as

well. Much of the theoretical basis for reconstruction was developed by Ronald Bracewell[15] in the context of radio astronomy. A strip-scan radio antenna—one with very poor spatial resolution in one dimension—measures basically a one-dimensional projection of the two-dimensional radio sky. Projections at other angles are recorded as the Earth rotates until enough data are accumulated to reconstruct the entire sky. The mathematics is virtually identical with that of x-ray computed tomography—although the physics could hardly be more different.

Three-dimensional refractive-index fields produced, for example, by temperature variations in a gas can be mapped by related techniques. In this case the projection involved is the phase shift accumulated by light as it traverses the field, and the measurement is made interferometrically[16] or holographically.

As one final application, consider the possibility of mapping the interior of planets by detecting the flux of solar neutrinos passing through them. This suggestion comes from a recent science-fiction novel,[17] but then who knows . . .

References

1. J. Radon, Ber. Verh. Sachs. Acad. Wiss. **69**, 262 (1917).
2. A. M. Cormack, J. Appl. Phys. **34**, 2908 (1964).
3. D. E. Kuhl, R. Q. Edwards, Radiology **80**, 653 (1963).
4. G. N. Hounsfield, Br. J. Radiol. **46**, 1016 (1973).
5. R. A. Brooks, G. DiChiro, Phys. Med. Biol. **21**, 689 (1976).
6. *Proceedings* of the 1975 Workshop on Reconstruction Tomography in Diagnostic Radiology and Nuclear Medicine (San Juan, Puerto Rico), University Park, Baltimore (1977).
7. *Proceedings* of the International Workshop on Techniques of Three-Dimensional Reconstruction, BNL 20425 (E. B. Marr, ed.), Brookhaven National Laboratory, Upton, N.Y. (1976).
8. *Technical Digest* of the 1975 Optical Society of America Technical Meeting (Stanford University), OSA, Washington, DC.
9. R. Gordon, G. Herman, Intl. Rev. Cytol. **38**, 111 (1976).
10. H. H. Barrett, S. K. Gordon, R. S. Hershel, Comput. Biol. Med. **6**, 307 (1976).
11. H. H. Barrett, W. Swindell, Proc. IEEE **65**, 89 (1977).
12. R. E. Alvarez, A. Macovski, Phys. Med. Biol. **21**, 733 (1976).
13. J. F. Greenleaf, S. A. Johnson, National Bureau of Standards Special Publication 453, Washington, DC (1976), page 109.
14. P. C. Lauterbur, in reference 8.
15. R. N. Bracewell, A. C. Riddle, Astrophys. J. **150**, 427 (1967).
16. D. W. Sweeney, C. M. Vest, Appl. Opt. **12**, 2649 (1973).
17. Hal Clement, "Star Light," Analog Sci. Fict. Sci. Fact **85**, 55 (1970).

The growth in the applications of physics to medicine is perhaps best illustrated by the developments in medical imaging. When AAPM was founded, medical imaging consisted primarily of x-ray fluoroscopy (direct images) and radiography (exposures on film). Although these techniques still dominate the work of a radiology department, there are a number of new ways of making images of the interior of the living body. (See, for example, the cover of this issue and figure 1.) Many of these new imaging methods depend on computers to perform the enormous amounts of data-analysis required—computed tomography or nmr imaging, for example, would not be possible without computers—and in many other cases computers serve to provide enhanced images that may allow more accurate diagnoses.

In this article we will discuss several of these new medically useful imaging techniques. Digital subtraction angiography and computed tomography both make more sophisticated use of the information available from x rays. Nuclear magnetic resonance was a well-known research field in 1958; in the last few years it has become useful in medical imaging. The techniques of computed tomography have made it possible to make images of the distribution of radioactive tracers in the living body. The use of ultrasound to form images, in its infancy in 1958, is now a fairly common diagnostic tool.

While some of the contributions that physicists have made to medical imaging are eminently newsworthy, other, more mundane contributions—such as the development of quality-control devices to improve medical images—may have a more significant effect on health care. Thus, for example, unnecessary x-ray exposure has received considerable attention in the press; however, a poor x ray that causes a radiologist to miss an early diagnosis of a breast cancer may cost the patient her life, and the amount of radiation is of minor concern.

Digital subtraction angiography

One of the drawbacks of conventional film radiography is that material other than bone shows up poorly. Even when one introduces an absorbing material into the organ or vessel one wants to examine (to enhance the

Paul Moran, Jerome Nickles and James Zagzebski are in the department of medical physics at the University of Wisconsin in Madison; Moran is currently visiting professor of radiology at Bowman Gray School of Medicine, Lake Forest University, Winston-Salem, N.C.

The physics of medical imaging

PHYSICS TODAY / JULY 1983

Images computed from x-ray absorption, from positron emission, from nuclear magnetic resonance and from reflections of ultrasound are showing tissues and details unimagined 25 years ago.

Paul R. Moran, R. Jerome Nickles and James A. Zagzebski

contrast with surrounding material) the additional absorption may be hard to see against the background of bone or other tissues. The conventional technique also fails to use all the information that may be available in the transmitted x-ray beam, such as the absorption spectrum of the material and time variations in its behavior.

The technique of "subtracting the background" is well known in other areas of physics. It can be applied to diagnostic x-ray images to provide information not available from single broadband projection images. For example, from an image obtained using an absorbing marker one can subtract an image obtained without the marker, thus making clear the contribution of the absorbing material. Alternatively, one can subtract images obtained at two different energies from each other to bring out the presence of materials whose absorption changes between these two energies—that is, materials that have an absorption edge in that energy range.

Because the subtraction technique is most commonly used to examine blood vessels and related structures, and because the images are generally obtained using digitized fluoroscopy apparatus, the technique is referred to as digital subtraction angiography.

The subtraction of background features from an image allows the clinician to work with much smaller amounts of absorbing material. One common application (shown in figure 2) is the examination of arteries; iodine injected into the bloodstream is the absorbing medium. The sensitivity to changes in absorption allows one to inject the iodine into a vein and follow its progress into the arterial system as the blood is pumped out of the heart again after passing through the lungs.

Not only is venous injection much less invasive than arterial injection, but the concentration of iodine is typically twenty times lower than in direct angiography. Typically, one follows the flow of the injected iodine as it makes its first pass through the arterial system by making a sequence of images—about 1 to 30 images per second—and subtracting the pre-injection background from each. Because the iodine signals are weak, the image must be amplified, and because noise is amplified as well, the system designers must take care to provide for adequate x-ray exposure, and to use tv cameras and amplifiers with high signal-to-noise ratios. Depending on the degree of patient motion one can expect, one may also use signal-integration techniques to reduce the effects of noise.

Digital subtraction angiography is just one example of a generalized technique[1] in which data obtained under slightly different conditions are subtracted to obtain clinically useful information about the attenuation x rays. In its passage through the body (or whatever else is being examined) the intensity I of the transmitted beam is a function of its energy, of time, and of position. By examining the changes in the function $I(E,t,\mathbf{r})$ with respect to small changes in E, t, or $\mathbf{r}$, one is, in effect, looking not at I itself but at its partial derivatives—or rather, at the first few terms in the Taylor series for the function, because the intervals one uses in practice can never be truly infinitesimal.

We have already mentioned energy subtraction, which may be used to provide images selectively displaying particular organs or tissues. Iodine, for example, has an abrupt discontinuity in its absorption spectrum at 33 keV (its K-edge). Subtraction images formed from data obtained with nearly monoenergetic beams with average energies above and below 33 keV will emphasize tissues containing iodine—whether naturally or artificially introduced—while suppressing the effects of bone and other tissues. Because one is not relying on a change in iodine concentration with time, one can allow the iodine to reach a temporary equilibrium distribution before one obtains the image. Heavily filtered beams from conventional x-ray sources as well as beams from monoenergetic sources have been used to investigate this technique, and it is currently being studied with finely tuned synchrotron radiation.[2]

First-order spatial derivatives yield not only the data for traditional image-processing techniques such as edge enhancement, smoothing, and various operations on the spatial-frequency content of the image, but also for more recently introduced techniques of computed tomography.

Image processing of single films is severely limited by the fact that many variables have already been averaged over in the production of the image. They are also constrained by the fact that the trained eye and brain are already an excellent image-processing system, so that one must take care not to degrade subtle features that an observer may use to analyze an image. Because of this, computer processing of conventional radiograms has had little impact on clinical practice. However, there is considerable information in a single x-ray film that can be brought out by appropriate processing. Spatial filtration, for example, can eliminate large variations in overall brightness in different areas of the image; alternately, it can pick out features having a particular spatial frequency. A group at Wisconsin has used the latter in an attempt to show coronary arteries (high spatial frequencies) against the left ventricle (low spatial frequency); the technique, in fact, involves a second-order derivative, in that the spatial filtration is performed on a subtraction angiogram of the sort we described earlier. So far this technique has not proved as useful as the more invasive cardiac catheterization techniques.

Other second-order techniques have also been investigated. These include:
▶ hybrid subtraction—the derivative of the intensity with respect to time and energy
▶ tomographic subtraction angiography—the derivative of the intensity with respect to spatial variables and time
▶ tomographic energy subtraction—the derivative of the intensity with respect to energy and depth
▶ multiple K-edge subtraction—the second derivative of the intensity with respect to energy
Each of these techniques combines some of the advantages of the separate subtractions involved.

The hybrid subtraction technique, for example, was introduced[3] to reduce the effects of patient motion in digital subtraction angiography. Instead of taking single exposures before and after injecting the absorbing material, one takes a pair of exposures at widely separated energies. Subtraction of the energy pair produces images in which no tissues except the carriers of the opaque medium show up; subtraction of the resulting time-separated images then removes the remaining effects of bone and displays clearly the effects of the injected contrast material.

Some of the arteries carrying the injected iodine may well overlap in the two-dimensional image obtained in a digital subtraction angiogram; this overlap often makes diagnosis difficult. One possible solution[4] to this problem is to use tomographic techniques in conjunction with subtraction angiography. To obtain such three-dimensional information one can either move the source rapidly or use multiple sources to illuminate the patient from several different angles. The resulting data can then be processed to provide images of tissues containing iodine viewed from several different angles, giving the clinician an effectively three-dimensional view of the arteries.

Energy subtraction combined with tomography can provide information

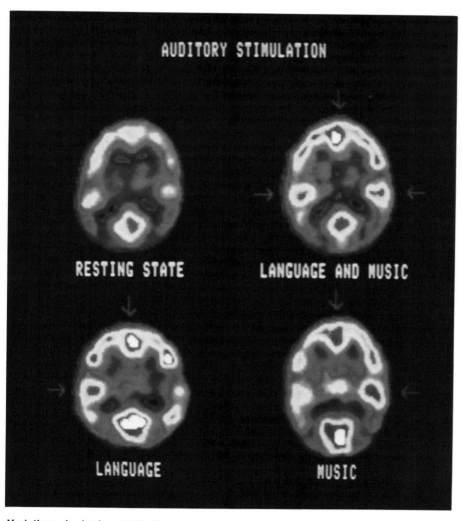

Variations in brain metabolism as a function of stimulation. These positron-emission tomograms show the distribution of F[18]-labeled 2-deoxyglucose in a transverse slice of a living brain with various kinds of auditory stimulus. (The eyes are open for all four images.) In the resting state the ears are plugged; in the other images the stimulation consists of a segment of the radio program "The Shadow" (language) and a Brandenburg Concerto by Johann Sebastian Bach (music). The frontal lobes are indicated by the vertical arrows; the auditory cortex (temporal lobes) are indicated by horizontal arrows and the visual cortex (occipital lobes) is at the bottom of the images. (Courtesy of Michael Phelps, UCLA.)   Figure 1

on the chemical composition of various regions of the body. One can also obtain information about materials by making use of the different variations with energy of the photoelectric and Compton-effect contributions of x-ray scattering. The ratio of these two contributions at widely separated energies can be used to get some idea of the effective atomic number. In a tomogram, one can thus get more information about the chemical composition of each region than from just the absorption data at a single energy.

The first-order techniques, specifically computed tomography (equivalent to depth subtraction) and digital subtraction angiography (time subtraction) have thus far clinically had the most importance. Their impact is probably due less to the computer processing, which can enhance the display of data, than it is due to the interesting data that can be generated by subtraction processes in general. By combining data from several images obtained under slightly different conditions, one can exploit much more fully the physical, geometrical and physiological information obtainable from what would otherwise be "just an x ray."

Computed tomography

Computed tomography consists of producing an image of a slice (*tomos* in Greek) of a three-dimensional object by combining a large set of scans through the object at different angles, each scan producing the equivalent of strip of a standard x-ray image. The information from all these scans is then used to reconstruct the two-dimensional variation of x-ray absorption through the entire slice.

The basic techniques of producing and detecting the x-rays is over 30 years old. What is new is the information-processing capacity. Allan Cormack and Godfrey Hounsfield—as well as others—independently developed[5] the mathematical techniques for reconstructing tomograms from the integrated absorption of beams passing through the absorber. Hounsfield and Cormack won the Nobel Prize for Physiology and Medicine in 1979 for this work (see PHYSICS TODAY, December 1979, page 19). The commerical development of scanners for computed x-ray tomography followed rapidly, and most major hospitals by now have access to CT scanners.

Modern x-ray CT scanners in general incorporate a large number (500 to 1500) highly collimated detectors mounted on a circular gantry that surrounds the patient in the region from which one wants the tomogram. An x-ray tube is also mounted on a gantry and rotates around the patient, producing some 500 to 2000 views, each a different angular projection of the

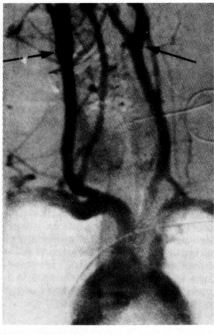

Digital subtraction angiogram of the carotid arteries that provide blood to the brain. The photo shows an x-ray image of the aorta branching, among others, into the two carotid arteries (indicated by arrows). The images are read off a fluoroscope by a tv camera, digitized and then subtracted. Note the absence of bones or other structures in the resulting image.   Figure 2

patient, in the course of a 360° rotation. Each view produces about 500 measurements of the x-ray transmission across the subject's diameter, and a typical view takes about 5 to 10 millisec. The average power input to the x-ray tube required for this speed is about 20 to 30 kW. The x-ray beam is collimated axially and is variable between 1.5 and 10 mm; the thinner the illuminated slice, the better is the axial resolution. To remove the effects of variations in x-ray output and detector sensitivity, the individual measurements are normalized to reference detector readings and scaled to a reference material. The result is the total attenuation along the path the x-ray beam takes through the sample.

One way to understand the computations involved is to consider each scan as consisting of parallel rays. The Fourier transform of a single angular view is then, it turns out, basically the two-dimensional Fourier transform of the x-ray attenuation density in the slice, for values of the wavevector perpendicular to the view angle. (See the articles by William Swindell and Harrison Barrett, PHYSICS TODAY, December 1977, page 32, and by Rowland Redington and Walter Berninger, PHYSICS TODAY, August 1981, page 36.) The complete scan thus produces a complete Fourier transform, for wavevectors in all directions. (It is, essentially,

the Fourier transform given in polar coordinates.) To reconstruct the tomogram, the data for a complete set of views are appropriately filtered ("apodized") and entered into a Fourier inversion algorithm. Other geometries of illumination and other ways of looking at the problem lead to different computational schemes, with other integral transforms replacing the Fourier transform.

The output of the calculations is typically displayed on a tv monitor divided into an array of 512×512 pixels, the brightness at each point being the result of an inversion computation that involves on the order of a million computer operations. The display is typically supported with image-array buffers and electronics so that the operator can select the mid-level brightness and the contrast range of the display. The computations are sufficiently sensitive that the x-ray quantum statistics limit the precision. Image noise, at spatial resolutions of 5 line pairs per centimeter, is typically less than $1/2\%$ the attenuation of an equal volume of water. The range of attenuations from air to very dense bone is about 8000 times as much. Figure 3 shows a computed tomogram.

Images for nmr

The frequency and decay rate of an nmr signal depend on the local magnetic—and hence chemical—environment of the nucleus being probed. One can thus produce images—of the interior of a body, for instance—that display the variations in nmr signals and thus display anatomical features. As in x-ray computed tomography, the measured data are given as a Fourier transform of the spatial image desired.

The initial work for producing nmr images was also done in the early 1970s. (See PHYSICS TODAY, May 1978, page 17.) Paul Lauterbur announced[6] the underlying principles of nmr "zeugmatography" in 1973, about the same time that the work on x-ray computed tomography was taking place. However, while CT scanners are now in place at many hospitals and are being actively used in diagnoses, nmr scanners are only now moving into research medical centers for clinical evaluation. The chief reason for the difference is probably historical: X-ray technology has had a long history in medicine, and computed tomography can be viewed as a more efficient use of information for which the interpretation was clear, the clinical efficacy was obvious and the engineering and technological problems were known and soluble. For nmr no such history exists, and clinicians must now begin to establish a tradition of interpreting nmr images.

Because nmr signals are intrinsically weak, data must be collected over long times to obtain reasonable signal-to-noise ratios. Thus, for example, while it takes about 2 seconds to complete an x-ray scan of a head with sufficient detail to study the morphology of the ventricles in the brain, a proton-nmr scan requires several tens of minutes to produce data for a comparable image. (Protons produce the largest intrinsic nmr signal and are the most abundant nuclei in the body; other nuclei, with smaller gyromagnetic ratios and smaller abundances, would require even longer times for scans.)

The physical appearance of the apparatus to produce an nmr scan is very similar to that used for x-ray computed tomography: A large circular gantry surrounds the patient, with the axis of the gantry usually along the long axis of the patient. The gantry houses coils to produce a strong static axial magnetic field, coils to produce gradients in this field, and an rf coil to transmit the excitatory pulses and to pick up the nmr signal.

The strong static, homogeneous base magnetic field—ranging from 0.07 T up to 1.5 T in some applications—produces a net equilibrium nuclear magnetization. From a semiclassical point of view, one can think of the nuclei as precessing when they are displaced from the equilibrium; the resonant frequency—the Larmor frequency—depends on the magnetic moment of the nuclei and on the magnetic field at the nucleus. The decay of the precession depends on local dephasing processes as well as thermodynamic relaxation processes. When one applies a short pulse of rf at the Larmor frequency, the spins are made to precess, and one can observe the resulting "ringing" as an induced emf at the Larmor frequency, which can be picked up by the same tuned coil that transmitted the excitation pulse. The signal is amplified and demodulated by phase-coherent single-sideband detectors, which use the exciting oscillator for the reference signal.

To obtain information about spatial position in the nmr signal, one adds a gradient to the applied field. The spatial variation in the field gives rise to a frequency modulation of the rf signal. A single measurement cycle with a particular gradient applied thus gives a spatial Fourier transform of the excited nuclear resonance. A series of such cycles with the gradient varied in direction from cycle to cycle generates the full two- or three-dimensional Fourier transform of the nmr signal. Computer algorithms then select particular delay times (to get consistent intensities), apodize the data and perform the Fourier transforms that yield the sort of images shown on the cover and in figure 4.

When a proton scan is made sensitive only to density, a superbly uninterest-

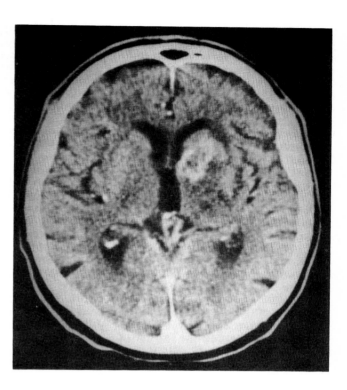

An x-ray computed tomogram of a head. The brain tissue is clearly differentiated from the surrounding bone as well as the (darker) ventricles, filled with cerebrospinal fluid. The light area to the right of the ventricle indicates a diseased condition. The image was obtained with a Siemens DR3 scanner at the University of Wisconsin.   Figure 3

ing image results. All tissues have just about the same nmr-observable density of hydrogen. However, the decay times of the nmr signal turn out to depend strongly on physical and chemical factors that differ from tissue to tissue.[7] The biology of these differences is not as yet well understood.

As we mentioned, there are several kinds of processes that affect the decay of the nmr signal:

▶ Processes that contribute to the establishment of thermal equilibrium of the spins with the rest of the body; one can think of these processes as analogous to the slow collapse of the spinning top, except that here, because the spin itself never decreases, the decay is to the aligned state. The characteristic time for these processes is called T_1.

▶ Processes that contribute to the loss of coherence among the various precessing spins. This, considerably shorter, relaxation time is generally called T_2.

By appropriately timing sequences of excitation pulses and selecting delay times for making the measurements, one can thus modulate the density-dependent nmr intensity by factors that depend on the local physical and chemical environment. Because the timing sequences as well as the field gradients can all be controlled, the operator not only acquires the data to produce an image, but in a profound sense also produces the specific subject to be imaged.

It even appears possible to observe directly the shifts in the Larmor frequency that arise from changes in the chemical environment, and thus obtain direct information about metabolic processes—for example, the relative levels and distribution of metabolic products incorporating phosphorus-31, say—in a living subject. The physician thus will be able to use nmr images to study normal and diseased tissues from a wide variety of viewpoints, anatomical, morphological, physiological or functional. Each of these uses of nmr data will of course be guided by—and contribute to—an already substantial knowledge of normal and pathological states. Thus, for example, while x-ray tomography yields highly detailed images of morphology and sensitivity to high-Z elements, an nmr scan can yield images that depend on physiology or functional properties. Most x-ray studies provide no distinction between a living subject and a cadaver; nmr scans can look totally different in the two cases.

Positron tomography

The information obtained from the imaging techniques we have discussed so far leads only indirectly to information about the chemical composition of the tissues imaged. Much disease, however, has a chemical origin; and changes in metabolism and biochemistry often come early, while changes in organ size, position or density generally occur late in the disease process. The use of radioactively labeled compounds to trace biochemical pathways has been a well-known research tool for quite some time. Applying the techniques of computed tomography allows one to map the distribution of the radionuclides.[8] To do this successfully, one must use radionuclides that meet two requirements: They must mimic the behavior of elements in the metabolic pathways and they must produce emanations whose paths one can determine. Positron emitters provide the natural choice for localization because the two 511-keV photons produced when the positron annihilates allow for easy reconstruction of the path. Metabolic equivalence then leads one to the short-lived positron emitters C^{11}, N^{13}, O^{15} and F^{18}.

Most positron tomography systems today look from the outside very much like x-ray tomography or nmr-imaging systems. Their discrete scintillators are arranged in rings; typically, there are about one hundred detectors per ring, with up to five rings in the gantry. A coincidence event between two detectors across the axis of the subject defines a line along which the positron annihilation must have occurred. The collection of millions of coincidence counts along the thousands of possible projection rays permits the reconstruction of the positron distribution through the now-standard backprojection techniques we described above. One can improve the resolution by spacing the detectors closer together. A group at UCLA, for example, is studying closely packed bismuth germanate detectors, with each detector indentified by cleverly adapted mercurous iodide photodiodes. Determining where along the ray the positron decayed can provide another improvement in resolution. This can be done with time-of-flight techniques and very fast counters.[9] Fast scintillators such as CsF and BaF_2 can produce time resolutions better than 300 picosec and can greatly help in suppressing image noise.

The amount of radioactive tracer material required to produce useful images is minuscule. One can use C^{11}-labeled carbon monoxide to trace blood flow to detect, for example, motion abnormalities of the cardiac walls by observing the heart contractions in a cine loop. A few millicuries of radiocarbon (its halflife is 20 minutes) will suffice for this purpose. This amount of radioactivity corresponds to 2×10^{11} molecules of carbon monoxide, about $\frac{1}{3}$ of a picomole. Carbon monoxide is a potent poison, but a factor of 10^{10} times as much as required to produce a detectable physiological response.

Nuclear medical procedures—that is, those that depend on introducing radioactive tracers—differ from other radiologic imaging techniques in that they are fundamentally dynamic. They are also sensitive to extremely small concentrations and can pinpoint highly specific receptors. The measurement of cardiac volume we mentioned above is not the best example of the sensitivity of positron emission tomography. The fluid volume of the heart is a simple mechanical property that can be obtained by many other imaging techniques; the receptors for carbon monoxide are simply hemoglobin, and are ubiquitous in the bloodstream.

An example of the sort of image unobtainable by other means is the use of rubidium-81 to trace the flow of blood to the heart muscle. In a normal heart this "myocardial perfusion" is about 50 ml/min for every 100 g of heart tissue. Rubidium chloride follows potassium pathways in the heart and is distributed by the blood flow. A dose of 16 mCi of Rb^{81}-labeled RbCl administered intravenously can provide a clear picture of the myocardial perfusion. Similarly, regional metabolic demands can be mapped through the kinetics of labeled fluorosugar analogs and fatty acids, probing the biochemical states of the heart and its fuel needs. Only nmr imaging can approach this type of information.

It is in the study of the brain, however, that positron tomography has made its greatest contributions. In the brain only oxygen and glucose serve as metabolic substrates. One can label glucose with carbon-11, but because the metabolic products of glucose (lactate and CO_2) are also mobile, the interpretation of an emission tomogram is difficult. However, a fluorinated analog of glucose, 2-fluoro-2-deoxyglucose can cross the blood–brain barrier but cannot be metabolized completely; it remains trapped[10] as the phosphate, 2-FDG-6-phosphate. Labeling 2-FDG with fluorine-18 allows one to make positron-emission maps of glucose uptake by the brain. Figure 1 shows such a set of maps for different kinds of activity. The differences in brain metabolism associated with differences in information processing are evident.

Oxygen does not have an analog capable of being trapped. One can, however, set up a steady state in which the subject comes into equilibrium with O^{15}-labeled oxygen inhaled over a period of 5–10 minutes. (The halflife of O^{15} is 2 min.) The regional input–output differences are balanced by decay and the activity thus reflects local oxygen needs. Regional blood flow can simply be mapped with O^{15}-labeled water. An even clearer picture can be obtained by observing the decay of O^{15} activity after the subject inhales a brief pulse of labeled oxygen. Such maps can show detail comparable to that obtained with labeled glucose analogs.

Diagnostic ultrasound

The last of the techniques we would like to discuss offers perhaps the least invasive method of probing the soft tissues of the body to produce images or detect disease states: the use of brief pulses of ultrahigh-frequency acoustic waves as a sort of sonar to map tissues within the body. As it travels through the body, an ultrasound pulse is reflected and scattered by changes in density and elasticity at tissue interfaces. Plotting the echo strength as a function of time allows one to visualize tissues as a function of distance within the body. For one-dimensional information one can simply plot the echo strength as a function of delay ("A-mode" display). For two-dimensional information, obtained by sweeping the source and detector across the patient, one generally just plots a point on a screen for echoes above a threshold, the distances along the line corresponding to delays

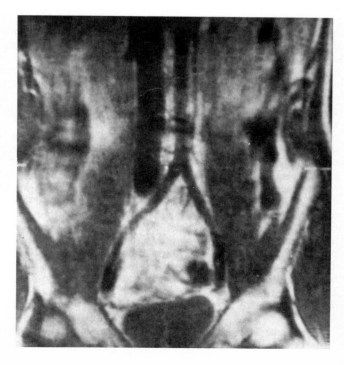

An nmr image of a coronal slice of an abdomen. Differences in brightness of the image correspond to differences in the decay times of the proton spin resonances and thus reflect different chemical environments— differentiating, for example, fatty tissue from muscle. (Photo courtesy Anne Deery, Siemens)   Figure 4

Radiological imaging systems

	Digital subtraction angiography	Computed tomography	Nuclear magnetic resonance	Positron computed tomography	Ultrasound
What is detected	transmitted x rays	transmitted x rays	rf radiation from induced emfs	511 keV annihilation photons	acoustic echoes
What is imaged	electron density of object	electron density of object	induced nuclear magnetization	in vivo positron emitters	discontinuities in speed of sound
What causes structure	variation of electron density with composition and density	variation of electron density with composition and density	variation of strength and decay of magnetization with composition	differential uptake of labelled compounds	variation of density and elasticity of tissues
What is inferred	location of tissues filled with absorbing material	sizes and shapes of organs	physical and chemical variations among tissues	flow and metabolism of tracer material	sizes and shapes of organs; acoustical properties
Typical application	examination of arterial narrowing	detection of brain tumors	imaging of brain tumors	mapping of glucose metabolism in brain	fetal growth; tumor detection; cardiology
Signal source	x-ray tube	x-ray tube	precession of nuclei	ingested labelled compounds	piezoelectric transducer
Signal detector	image intensifier	x-ray detector	rf pickup coil	scintillation detectors	piezoelectric transducer
Image plane	longitudinal	transverse	any	transverse	any
Spatial resolution	$\frac{1}{2}$ mm	1 mm	2 mm	10 mm	2 mm
Temporal resolution	10^{-2} sec	1 sec	10^{-1} sec to 10^2 sec	10^1 sec to 10^3 sec	10^{-2} sec
Typical radiation dose	2 rad (imaged field)	1 rad (imaged slice)	(not applicable)	10^{-2} rad (whole body)	(not applicable)
Cost (approximate)	$0.2 million	$1–2 million	$1–2 million	$1 million	$0.1 million
Chief use	clinical diagnosis	clinical diagnosis	physiological and clinical research	physiological research	clinical diagnosis

of the echo ("B-mode" display).

The ultrasound beam is produced and echoes are detected by a piezoelectric transducer. Manual-scanning instruments have the transducer attached to an electromechanical scan arm. The arm constrains the motion of the transducer to a plane. It also provides signals used to compute the position and orientation of the transducer for determining the path of the ultrasound beam, and thus the reflector locations. More commonly, the beam is swept automatically by a small hand-held scanning assembly; the operator positions and orients the scanner in the desired plane. Most instruments have an image memory or scan converter that presents the image on a tv monitor, and one can usually observe the image as it is produced. There are several techniques available for performing the automatic scans: mechanical motion of the transducer, rotation of a sound-deflecting mirror, electronic switching between elements in an array, or phased excitation of a multielement sectored array.

The rapid scanning speed and the "real-time" images of the hand-held automatic scanners offer several advantages to the clinician. By providing flexibility and more or less instantaneous feedback, the sonographer can move the scanner to find landmarks, such as blood vessels, that can help pinpoint the location of the tissue or organ under investigation. Such landmarks are especially important when there is only subtle acoustic contrast between the structure of interest and the background. Once the structure is found, the sonographer can position the scanner to provide optimum contrast and clarity.

The maximum possible scanning speed of acoustic imaging systems is limited by the speed of sound in tissue (about 1540 m/sec), the maximum depth of tissues of interest and the detail required for the image, which determines the number of lines per scan. Rapid scanners are very useful in providing the real-time images and the flexibility we have mentioned. They are also essential in viewing moving objects such as the heart. One can now obtain ultrasound images of the heart at a rate of more than 30 scans per second. The images are clear enough that one can diagnose cardiac problems such as mitral valve stenosis and other valve diseases as well as congential heart defects. Even faster scans are possible for examining superficial anatomic structures or for producing images with limited resolution.

One can add a further dimension to ultrasonic scans by coupling measurements of Doppler shifts with the detection of echoes so that one can, for example, measure blood flow at the same time that one is imaging the vessels through which the blood is flowing. Such "duplex" transducers could be very valuable in examinations of the heart, to detect abnormally high blood velocities or abnormal flow directions through defective valves.

A significant fraction of all ultrasound examinations performed involve studies of the abdominal organs (figure 5) or examinations of the fetus. In the abdomen, the images may help determine whether abnormal masses are likely to be malignant. Ultrasound provides a method of rapidly detecting stones and other obstructions in the gall bladder and related structures. Variations in the overall texture of the

image of an organ are sometimes an indication of a disease condition. For example, early signs of rejection of kidney transplants may be detectable with ultrasound.

Fetal maturity studies—using measurements of head size or other anatomical dimensions from ultrasound scans to indicate the development of a fetus—are routine in most obstetrics departments and even in some physicians' offices. The resolution available with some devices allows one to distinguish fetal structures with sufficient detail permit diagnoses of many types of fetal abnormalities.[13] Most obstetricians now perform amniocentesis—in which a sample of the amniotic fluid is withdrawn for laboratory analysis, particularly to detect genetic problems—under ultrasonic guidance. By following the position of the probing needle on a screen, the obstetrician can avoid puncturing the fetus, placenta or umbilical cord.

The spatial resolution in an ultrasound image is related to the volume occupied by the sound pulse as it propagates through tissue. Ordinarily this volume can be divided into two components: one that depends on the lateral dimensions of the transducer beam and a second that depends on the time duration of the pulse, that is, its axial dimension. A typical instrument may use a 3.5 MHz transducer with a circular aperture 20 mm in diameter, focused at a depth of 12 cm. The diffraction-limited lateral resolution in the transducer's focal region is about 2.6 mm. The axial resolution for such a transducer is usually on the order of 1 mm. For a scanner whose focal length is fixed, the lateral resolution is, of course, less for reflectors located away

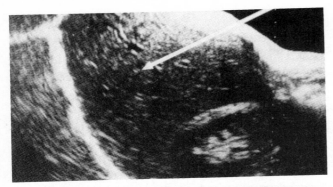

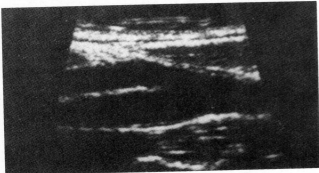

Ultrasound images.
Above, an image of the abdomen of an adult; the bright band at upper right indicates the skin; the curved band at the left and bottom is the diaphagm; the liver is indicated by an arrow. Below, an image of the carotid artery, showing its bifurcation (into the external and internal carotid arteries) in the neck; such an image can be combined with measurements of the Doppler shift to give further information about blood flow through the artery. (Lower image courtesy of J. Buschell, Diasonics.) Figure 5

"phantom." Such phantoms are built to have precisely known acoustic properties at all points and to mimic to some extent the behavior seen in living organs; they are valuable in testing imaging and signal-processing algorithms and for checking[18] theories on how beams of ultrasound interact with soft tissues.

* * *

We would like to thank John R. Cameron for assistance in coordinating the contributions to this article and Charles A. Mistretta for help in preparing the discussion on digital subtraction angiography.

References

1. C. A. Mistretta, Opt. Eng. **13**, March/April 1974, page 134.

2. E. Rubenstein, E. B. Huges, *et al.*, Proc. Soc. Photo-Opt. Instrum. Eng. **314**, 42 (1981).

3. M. S. Van Lysel, J. T. Dobbins III, W. W. Peppler, *et al.*, Radiology, to be published; G. S. Keyes, S. J. Riederer, B. J. Belandger, W. R. Brody, Proc. Soc. Photo-Opt. Instrum, Eng. **347**, 34 (1982).

4. R. A. Kruger, J. A. Nelson *et al.*, presented at 68th Ann. Mtg. RSNA, Chicago, Ill, 28 November 1982.

5. G. N. Hounsfield, Br. J. Radiol. **46**, 1016 (1973); G. N. Hounsfield, Med. Phys. **7**, 283 (1980); A. M. Cormack, Med. Phys. **7**, 273 (1980).

6. P. C. Lauterbur, Nature **242**, 190 (1973); P. C. Lauterbur, Pure App. Chem. **40**, 149 (1974).

7. R. Damadian, Science **171**, 1151 (1971).

8. M. E. Phelps, E. J. Hoffman, N. A. Mullani, M. M. Ter-Pogossian, J. Nucl. Med. **16**, 210 (1975).

9. M. M. Per-Pogossian, N. A. Mullani, D. C. Ficke, J. Markham, D. L. Synder, J. Comput. Assist. Tomogr. **5**, 277 (1981).

10. B. M. Gallagher, J. S. Fowler, N. I. Gutterson, R. R. MacGregor, C.-N. Wan, A. P. Wolf, J. Nucl. Med. **19**, 1154 (1978).

11. T. Jones, D. A. Chesler, M. M. Ter-Pogossian, Br. J. Radiol, **49**, 339 (1976).

12. G. D. Hutchins, R. J. Nickles, Proc. XI Int. Symp. Cereb. Blood Flow and Metab. (in press).

13. R. C. Sanders, A. E. James, eds., *Ultrasonography in Obstertics and Gynecology*, 2nd ed., Appleton-Century-Crofts, New York (1980).

14. W. J. Zwiebel, ed. *Introduction to Vascular Ultrasonography*, Grune-Stratton, New York (1982).

15. M. Linzer, ed., *Ultrasonic Tissue Characterization II*, National Bureau of Standards Special Publication 525 (1979).

16. J. F. Greenleaf, S. K. Kenue, B. Rajagopolan, R. C. Bahn, *et al.*, in *Acoustic Imaging*, A. F. Metherell, ed., Plenum, New York (1980), vol. 8, page 599.

17. P. N. T. Wells, *AAPM Medical Physics Monograph 6*, G. Fullerton, J. Zagzebski, ed., AIP, New York (1980).

18. E. L. Madsen, J. A. Zagzebski, M. F. Insana, T. M. Burke, G. Frank, Medical Phys. **9**, 703 (1982); D. Nicholas, Ultrasound in Med. & Biol. **8**, 17 (1982). □

from the focal region.

One can improve the resolution by using higher frequencies. Unfortunately, however, higher frequencies are more strongly absorbed; the attenuation of ultrasound in soft tissues increases approximately linearly with frequency. The short-wavelength path to improved resolution can thus be used only for superficial structures, where the beam need not penetrate deeply into the body. One can, for example, use sound frequencies as high as 15 MHz to examine the eye for tumors or detachments of the retina. The lower part of figure 5 shows an image of the bifurcation of the carotid artery (in the neck) obtained with a scanner operating at 8 MHz; the image was produced in "real time." Such images can show atheromatous placque (which can tear loose and cause a stroke by blocking a smaller artery in the brain) early in its formation. Small calcifications in the arterial walls cast a recognizable acoustic shadow over deeper structures. Measurements of Doppler shifts, as we have mentioned, can provide additional information on the blood flow through the arteries, showing evidence of higher flow speeds through narrowed sections or turbulence associated with protrusions into the vessel.[14]

Recent technical improvements in diagnostic ultrasound equipment include incorporation of digital processing and digital image memories, improvements in the design of single-element and array transducers, fabrication techniques that increase the sensitivity and reduce off-axis radiation, and the development of special-purpose scanners for the breast, for superficial examinations and for other applications. In spite of all the advances that have been made, ultrasonography currently uses only part of the potential information that might be extracted from echo signals. The contrast and detail on B-mode images are based on the magnitude of echo signals reflected by soft tissues. In the future it may be possible to extract additional diagnostic information by applying more sophisticated signal processing to the returning echo signals, using, for example, waves scattered in directions other than backwards to construct the image, or using more information from the backscattered wave in computerized reconstructions and to give quantitative characterizations of the tissues being examined.[15]

To produce such more sophisticated images, we will have to understand better the interactions of high-frequency sound waves and soft tissues. The human body is acoustically a very complex structure, with many sources of scattering, reflection, and refraction. The bulk of our knowledge of the quantitative transmission characteristics of human tissue, including speeds of sound, ultrasonic attenuation and ultrasonic scattering, stems from measurements carried out on excised human and animal organs.[16] However, it is not always clear that the behavior of sound waves *in vivo* is always the same, and experimenters are now developing methods to make measurements of ultrasound characteristics in living subjects. It may become possible to observe specific tissues by their specific ultrasonic signatures. One useful tool[17] in such studies is an acoustic

Nobel prize to Cormack and Hounsfield for medicine

Bertram M. Schwarzschild

PHYSICS TODAY / DECEMBER 1979

In the first year of this century Wilhelm Roentgen won the very first Nobel Prize in physics, for the discovery of a phenomenon which, in a remarkably short time, revolutionized medical diagnosis. On 11 October the Karolinska Institute announced that this year's Nobel Prize for Physiology and Medicine has been awarded jointly to a physicist and an engineer, for their contributions to what is widely regarded as the most revolutionary development in radiography since Roentgen's discovery of x rays. Allan Cormack, professor of physics at Tufts University (Medford, Mass.) and Godfrey Hounsfield, head of the medical-research division at EMI Ltd. (Middlesex, England) were awarded the Prize for their separate contributions to the development of computer-assisted tomography (see PHYSICS TODAY, December 1977, page 32).

Cormack was born in Johannesburg, South Africa in 1924. He is the only member of the Tufts physics faculty without a PhD. Curiously, neither of this year's laureates in medicine has a doctorate of any kind. Cormack left his graduate studies at Cambridge University in 1950 to take up a faculty position at Capetown, where he did experimental nuclear physics.

In 1956 he did part-time service as a radiological physicist at Capetown's Groote Schuur Hospital, to help out after the resignation of the hospital's physicist. Watching the x-ray treatment of tumors at Groote Schuur, Cormack told us, he was displeased with the imprecision with which the location of the tumors was known, and the consequent imprecision of the therapeutic irradiation. This set him thinking about the mathematical problem of using x-ray transmission information to localize internal structures accurately. The following year, while on a sabbatical at Harvard, he was offered a professorship at Tufts. He has been

there ever since, doing mostly experimental nuclear and particle physics till about 1970.

In his spare time Cormack stuck with the problem of reconstructing the spatial distribution of an x-ray absorber when one knows only the integrated absorption of beams passing through the inhomogeneous absorber. Having failed to find a solution in the mathematical literature, he published a Fourier series solution[1] in 1963, expressing the nth Fourier component of the original absorber distribution in terms of an integral of the nth Fourier component of the projected absorption data. He had solved the general problem of reconstructing a function in a two-dimensional domain in terms of the integrals of the function along every straight line through the domain.

Cormack learned only years later, he told us, that Johann Radon had solved the same problem in 1917. But Radon's closed-form solution, while equivalent to Cormack's, did not explicitly display

certain interesting properties of the solution explicit in Cormack's Fourier expansion. What Cormack calls "the hole theorem" follows at once from his form of the solution. The theorem states that in order to reconstruct the original distribution at a distance r from the origin, one needs only information about line integrals (x-ray beams) that come no closer than r to the origin.

Computer-assisted reconstruction. In a follow-up paper in 1964[2] Cormack provided a polynomial expansion of his solution useful for computer calculations, and put it to a successful test in what appears to have been the first computer-assisted tomographic reconstruction ever performed. He measured the attenuation of a beam from a gamma source passing in different directions through a configuration of aluminum disks packed in lucite, inside an aluminum ring. Applying the attenuation data to a computer program based on his polynomial expansion, Cormack was able to reconstruct the config-

CORMACK

HOUNSFIELD

uration of aluminum disks with surprising accuracy.

Cormack pointed out at the time that his reconstruction solution had applications beyond x-ray imaging. Reconstructing the distribution of positron annihilations in a subject from the resultant emerging gamma pairs is essentially the same mathematical problem. So is the problem of imaging with charged particles, to which Cormack referred pessimistically in 1964 but later took up experimentally. In 1976 he reported the results of a study of proton tomography, done with Andres Koehler at the Harvard cyclotron. Cormack believes that charged-particle tomography may eventually yield resolutions as good as those achieved with x-ray CAT scanners, at an order of magnitude less ionization dose to the patients.

Different approaches to the same mathematical problem have been applied over the years to an impressive array of imaging situations. As early as 1956 Ronald Bracewell of Stanford used a Fourier-transform image-reconstruction technique for solar radioastronomy. In recent years similar solutions have been applied to nmr zeugmatography (see PHYSICS TODAY, May 1978, page 17), ultrasonic imaging and electron microscopy. But as Rodney Brooks and Giovanni diChiro of NIH pointed out in a 1976 review article,[3] the history of this field is characterized by a striking lack of cross fertilization. This is illustrated by the fact that Hounsfield was quite unaware of Cormack's work when he developed the first clinical CAT scanners at EMI. Indeed, as of this writing the two laureates have never met.

Godfrey Hounsfield, whom the Karolinska Institute cites as the central figure in the development of computer-assisted tomography, was born in 1919. He grew up in rural Nottinghamshire, England. After serving as an RAF radar lecturer during the War, he studied at the Faraday House College of Electrical Engineering in London. Since 1951 he has been with EMI, working at first on radar and computer design.

His work on computerized pattern recognition led him to take an interest in x-ray imaging. When the Ministry of Health encouraged Hounsfield to test the practical feasibility of his ideas on x-ray tomography, he built the first simple laboratory CAT scanner in 1968. The device consisted of an americium gamma source and a crystal detector, with plastic and biological samples rotating and translating in the gamma pencil beam.

Hounsfield's original computer reconstruction program was an iterative algorithm, essentially a trial-and-error technique that converged slowly on the final reconstruction by the iterative solutions of a large set of linear equations. Nowadays most CAT scanners use filtered back-projection programs, developed independently by Hounsfield and others, to process the x-ray attenuation data significantly faster than did the interative technique. As far as Cormack knows, none of the clinical CAT scanners now in use employ his polynomial-expansion solution.

The first clinical CAT scanner was installed at Atkinson Morley's Hospital (London) in 1971, and the following year Hounsfield was granted a patent. The laboratory gamma source was replaced by an x-ray tube, which rotated synchronously with the opposing crystal detector about the patient's head, in the plane to be imaged. Since 1973 several thousand CAT scanners, produced by EMI and four other principal manufacturers, have been installed in hospitals all over the world.

The effect on neurological and other diagnostic arts has been momentous. X-ray CAT scanners can clearly delineate 0.5% differences in soft-tissue density, and they can achieve a spatial resolution of $3/4$ mm (with a matrix of 320×320 beam traversals of the plane). Pneumoencephalography, a painful method of increasing x-ray contrast for tumor location by displacing cranial fluid with air, has now been rendered obsolete by the CAT scanner.

Since their first introduction, Hounsfield has been closely involved with the further development of x-ray CAT scan-

ners, which are now said to be in their fourth generation of evolution. Single pencil beams have been replaced by a fan of multiple beams from a single x-ray source, impinging on several detectors simultaneously. A recently introduced EMI CAT scanner has about a thousand crystal detectors surrounding the scanning plane. In this configuration only the x-ray source moves, the detectors being fixed in the plane. The proliferation of beams and detectors speeds up the process of imaging, avoiding the problem of blurring due to breathing and heart beats. Whereas CAT scanners were at first used only for tomography of the head, more recent models can scan the entire body. Once the problem of image artifacts due to large differences in attenuation length had been surmounted, whole-body scanners became feasible.

The signal-to-noise ratio for x-ray CAT scanners at their present stage of evolution is largely limited by the Poisson statistics of the photon flux. Modern reconstruction algorithms are good enough that the noise approaches this theoretical limit. Therefore it is unlikely that resolution or signal-to-noise ratio can be significantly improved in x-ray scanners without unacceptable increases in the x-ray dose to the patient. This had led Cormack to investigate the alternative of charged-particle tomography, and Hounsfield is currently involved in the study of nmr imaging.

References

1. A. M. Cormack, Jour. Appl. Phys. **34**, 2722 (1963).
2. A. M. Cormack, Jour. Appl. Phys. **35**, 2908 (1964).
3. R. A. Brooks, G. diChiro, Phys. Med. Biol. **21**, 689 (1976).

Interest grows in synchrotron-radiation sources

Gloria B. Lubkin

PHYSICS TODAY / AUGUST 1973

The National Science Foundation has just funded a new synchrotron-radiation facility at Stanford University to be open to users throughout the US. The facility, which will use the electron–positron storage ring, SPEAR, is currently operating at 2.6 GeV at the Stanford Linear Accelerator Center. It will produce photons with a critical energy of 3.07 keV with some photon flux up to 20 keV.

Two other proposals were vying for NSF support, one from the University of Wisconsin, which has been operating its Tantalus 1 source for several years, and one from the Cambridge Electron Accelerator, which started to do synchrotron-radiation experiments in its waning days as a machine for high-energy physics. AEC support for CEA ended on 30 June.

Interest in synchrotron-radiation sources has been mounting throughout the world because these sources can be used for an enormous variety of experiments in solid state, atomic physics, chemistry, crystallography and biology.

Electrons circulating in a synchrotron or storage ring emit incoherent white light with a spectrum whose peak wavelength varies inversely as the cube of the electron energy and linearly as the radius of the electron orbit. The radiation comes out as a thin wedge-shaped "pancake" with a vertical angular divergence of about 1 milliradian for high-energy machines.

To do experiments at a specific wavelength, a monochromator is employed, and by adjusting or changing the monochromator, the source is continuously tunable. Although synchrotron radiation sources give high intensity at lower energies too, they are an excellent source in the extreme ultraviolet at photon energies of 15–500 eV, a region that is otherwise nearly inaccessible. The radiation is polarized. For storage rings, photon intensity is very stable. Typical synchrotron-radiation sources emit subnanosecond pulses. Within the vertically narrow light beam, the intensity is very high. As Richard Watson and Morris Perlman have noted,[1] for experiments that need high spatial and energy resolution this intensity is significantly greater than that obtainable from other sources.

At a Brookhaven conference on synchrotron radiation sources held last September, the following kinds of experiments were discussed: optical and electron emission experiments with solids and surfaces, photochemistry and atomic physics, optical biochemistry, photometric calibration, diagnostic x-ray radiography, x-ray scattering (both structural and nonstructural) and x-ray microscopy.

The Stanford facility will be symbiotic to the SPEAR operation and is expected to operate in 1974. It will be one large port that will then be broken into a number of beams for different experiments. One will cover the uv up to about 100 eV; a second will cover the soft x-ray region from 50–100 eV; the rest of the facility will be dedicated to x-ray experiments, for which the beam will be brought out through a beryllium window.

The NSF award is for $750 000 the first year; the following year, an additional $450 000 will go toward completing the construction and starting the operation. After that, NSF estimates that it will cost about $300 000 per year to operate. Construction will include a small building alongside the SPEAR ring to house the beam run and experimental areas, beam lines, vacuum equipment and protection for the vacuum lines.

Even with the Stanford source, the US will not have as strong a capability as is available at the DESY synchrotron in Hamburg, for photon energies above 8 keV. The Stanford intensity below 8 keV is 2–3 times the DESY synchrotron intensity. DESY is constructing a storage ring, DORIS, which will increase its capability. By late 1974 the SPEAR energy is expected to be raised to 4.5 GeV.

Sebastian Doniach and William E.

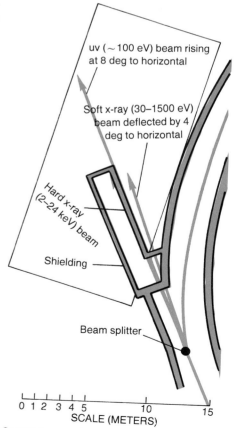

uv (~100 eV) beam rising at 8 deg to horizontal

Soft x-ray (30–1500 eV) beam deflected by 4 deg to horizontal

Hard x-ray (2–24 keV) beam

Shielding

Beam splitter

0 1 2 3 4 5   10   15
SCALE (METERS)

Stanford synchrotron radiation facility (preliminary layout). Beams come from port at SPEAR storage ring. The four or five hard x-ray beams will be both horizontally and vertically deflected.

Spicer are co-principal investigators on the Stanford project. Herman Winick, formerly with CEA, has joined Stanford as associate director of the synchrotron-radiation facility. Gerhard Fischer has served from the beginning as SLAC liaison with the Stanford synchrotron-radiation project. Spicer told us that Ingolf Lindau and Piero Pianetta have a pilot experiment under way. It will be used to do ESCA (Electron Spectroscopy for Chemical Analyses) studies with a resolution of 0.1 eV, which would be a factor of about five improvement over presently available ESCA resolution, Spicer said.

The next thing in the pilot project is to look at the fine structure in x-ray absorption edges, which would allow one to get structural information not available from conventional sources, and should be particularly useful for amorphous materials and complicated organic materials. This work is being done by Dale Sayers and Edward Stern (University of Washington), Farrel Lytle (Boeing Co.), Arthur Bienenstock, Mitchell Weissbluth and Doniach of Stanford.

Spicer noted that there is a large interest in biological studies—time-dependent studies of muscles, for example, should be possible. John Baldeschweiler (Cal Tech) is planning to use x-rays to study induced structural changes in retinal photoreceptor membranes. Frederick Brown (who recently left the University of Illinois to join the Xerox Research Center in Palo Alto) will be setting up the initial soft x-ray operation; he expects to extend to higher energy the optical absorption experiments he had been doing at Wisconsin. Victor Rheen of the Navy's Michelson Lab in China Lake, California, Don Baer (Stanford) and Spicer are developing a separate uv beam line for about 100 eV.

US facilities. The first far-uv synchrotron-radiation experiments were done at the 300-MeV synchrotron at Cornell University by D. H. Tomboulian and Paul Hartman in 1956. Since then the US has had three other synchrotron-radiation sources. One, at the National Bureau of Standards, under Robert Madden, is a 180-MeV synchrotron, which was used in pioneering atomic absorption experiments that began in 1961. This source is being converted to a storage ring and is expected to operate next year with an improved intensity of more than a factor of 100.

Several years later Tantalus I, a 240-MeV storage ring, then directed by Frederick Mills and now by Ednor Rowe, started doing synchrotron-radiation experiments with support from the Air Force. It has been used by many investigators from other agencies as well. Over the past three years, the Wisconsin group has developed a "wavelength shifter" that will give usable intensity (through one port) down to about 30 Å instead of the present cutoff of about 100 Å. Meanwhile, Wisconsin proposed to NSF to build a so-called "dedicated facility," known as Tantalus 2, that would use a microtron as an injector. It would have been a 1.76-GeV storage ring with a capability down to 1 Å. Anticipated construction cost was $1.5 million over a three-year period.

The third synchrotron-radiation source, at the Cambridge Electron Accelerator, has stored beams for synchrotron radiation up to 3.5 GeV. (Colliding beams were limited to 2.5 GeV.) Dean Eastman (IBM Research Center) and his collaborators have used CEA to extend uv photoemission studies begun at Wisconsin, studying the evolution of the photoemission spectrum as a function of photon energy. Paul Horowitz (Harvard), acting upon a suggestion of Edward Purcell, built at CEA an x-ray microscope. It is a scanning instrument with chemical discrimination, a large depth-of-field, resolution of about 1 micron and does not have to be operated in a vacuum.

CEA had proposed to NSF that it support a dedicated facility there that would have required relatively little construction. Up to 15 beam lines, each divisible into three parts, would have brought radiation into the experimental area that was used for high-energy physics. The final CEA proposal asked for about $2.5 million over a five-year period for installation of the beam lines and full operation of the accelerator, plus an additional $1.1 million over five years for an in-house laboratory.

Howard Etzel of NSF's Division of Materials Research told us that the three installations were visited by a select group charged with making recommendations to the Foundation as to whether or not the US should develop a major facility, and if so, which of the three should be recommended. The decision made by the NSF was primarily determined by factors such as cost, national diversification and long-range flexibility. Eventually, the US will probably have a second-generation source, Etzel told us, but not in the next five years.

What of the rest of the world? Rowe filled us in on the status of other machines. DESY has a 7.5-GeV synchrotron whose synchrotron-radiation spectrum reaches into the hard x-ray range, about 20 kV. By putting on an appropriate monochromator, a machine of the DESY class would give a mono-chromatic x-ray source one to three orders of magnitude more intense than the strongest x-ray tubes in the same energy range, Rowe told us. A group from the European Molecular Biology Organization plans to use DESY to study, among other things, muscles and long-chain protein molecules. Rowe notes that synchrotron radiation may be used also to help solve the phase problem in x-ray diffraction; with phase information a lot of ambiguity in large molecular structures would be removed. The rest of the DESY program is largely solid state and atomic physics. When the new storage ring, DORIS, starts operating, circulating currents of several amperes are expected; DORIS will be used as a synchrotron-radiation source and part of the building for these experiments is completed.

In Paris, ACCO, a 530-MeV storage ring, is by late next year expected to be completely converted to a dedicated synchrotron-radiation source, and go down to 10–20 Å. A 2.2-GeV storage ring now nearing completion will be used for synchrotron-radiation experiments.

At the University of Bonn, a 2.3-GeV synchrotron is doing some synchrotron-radiation work.

NINA, the 5-GeV synchrotron at Daresbury, UK, is now taking data. The Science Research Council has recommended construction of a dedicated storage ring, which would not use the existing synchrotron as injector.

At Frascati, Italy, the 1.1-GeV synchrotron has been used for some time.

In the Soviet Union, the Lebedev Institute in Moscow is using its 680-MeV synchrotron as a radiation source. A bigger synchrotron at Lebedev, now nearing completion, may be altered to operate as a storage ring and synchrotron-radiation source. In Yerevan, a Moscow group is using the 6.5-GeV synchrotron there as a synchrotron-radiation source. It was used to calibrate the x-ray telescopes for the Salyut space station. Sergei Kapitsa of the Institute for Physical Problems in Moscow would like to build a dedicated source using a 1.3-GeV storage ring.

At the University of Tokyo, the INS-SOR group is using the 1.2-GeV synchrotron for synchrotron-radiation studies. A 300-MeV storage ring is being added there; it is expected to be finished late this year.

Reference

1. Research Applications of Synchrotron Radiation, R. E. Watson, M. L. Perlman, eds., proceedings of a study-symposium at Brookhaven National Laboratory, 25–28 Sept. 1972, BNL 50381 (June 1973).

The renaissance of x-ray optics

New optical elements such as multilayer mirrors and zone-plate lenses allow the extension of imaging and holography techniques to the spectral region below 300 Å.

James H. Underwood and David T. Attwood

PHYSICS TODAY / APRIL 1984

Solar x-ray photograph made from original exposures at wavelengths ranging from 3 Å to 44 Å. The Skylab telescope that made the exposures used "Wolter's type I" x-ray optics, explained in figure 3. The photograph shows that the solar x-ray emission arises primarily from coronal regions over solar centers of activity. The large dark structures are coronal holes, regions where the coronal density is abnormally low. (Photograph courtesy of American Science and Engineering, Inc.).   Figure 1

There has been a spectacular resurgence of interest in the soft x-ray and extreme ultraviolet regions of the electromagnetic spectrum in the past few years. In part, this is due to the development of new x-ray optical devices that have given us an x-ray view of the universe on scales ranging from the microscopic to the astronomical (see figure 1).

Not long ago it was thought to be impossible to construct optical systems capable of forming x-ray images. Now, however, we have contouring and polishing techniques for producing supersmooth surfaces with accurate optical figures, for use as reflecting surfaces in x-ray microscopes and telescopes. Materials scientists have perfected methods for depositing layers of atoms to produce multilayer interference coatings that enhance the reflectivity of mirrors at selected wavelengths in the soft x-ray and extreme ultraviolet regions of the spectrum (together called the xuv region, which extends from about 3 Å to 300 Å). Finally, the efforts of the electronics

industry to build ever-smaller microcircuits have led to techniques that permit the fabrication of structures with dimensions as small as a few hundred angstroms, and thereby permit the construction of diffracting optics, such as zone plates and transmission gratings, for x rays. There have been parallel improvements in detectors, particularly in sensitive, high-resolution imaging detectors and fast detectors such as streak cameras.

At the same time, we are witnessing the development of new and interesting sources of xuv radiation, requiring new kinds of optical devices to diagnose their properties and exploit their characteristics. These sources include hot

dense transient plasmas generated by lasers or electrical discharges, modern synchrotron radiation generators that use multiperiod magnetic insertion devices in high-brightness storage rings, and magnetically confined plasmas for nuclear fusion. There is also considerable effort underway to provide coherent power at ever shorter wavelengths in the vacuum ultraviolet, through a variety of potential lasing or nonlinear upconversion schemes. Of particular interest, because of its assured success in making coherent x rays widely available, is the use of magnetic undulators with a hundred or more periods in specially designed low-emittance storage rings. If constructed, such stor-

James H. Underwood and David T. Attwood are staff members of the Center for X-ray Optics, at Lawrence Berkeley Laboratory, in California.

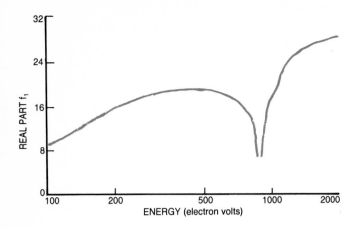
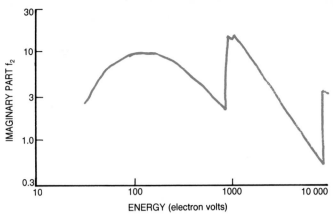

Atomic scattering factor for nickel. The wavelength or energy dependence of the element's index of refraction is a function of the scattering factor, whose real and imaginary parts are shown here. Nickel's complex index of refraction $\hat{n}$ is $1 - 4.1 \times 10^{-7} \lambda^2 (f_1 + i f_2)$, with λ in angstroms. Figure 2

age rings would provide a high-brilliance source of coherent x rays, tunable from 100 eV to several keV, an extremely important spectral range for x-ray microprobe, microscopy and holography studies in a wide variety of fields.

In this article, we look at developments in x-ray optics that will complement these new x-ray sources. We shall see that these developments make possible exciting new applications in fields as diverse as physics, astronomy, biology, medicine, chemistry and materials science—applications that could not have been seriously considered even a few years ago.

Refraction and absorption of x rays

Following his discovery of x rays in 1895, Wilhelm Roentgen himself attempted to reflect, refract and focus them. His lack of success was understood several years later, when it was demonstrated that x rays are short-wavelength electromagnetic waves whose frequencies are above most of the resonances of the bound electrons in all elements.

Thus, for x rays, the real part of the refractive index n is less than unity by only a very small amount δ, so refraction is negligible. In contrast, the frequencies of visible light are near but below important atomic transition frequencies, hence the refractive index for visible light is substantially greater than unity. The construction of optical elements for x rays is further compli-

cated by the fact that all materials are absorbing; this can be expressed by writing the refractive index as a complex number $\hat{n}$:

$$\hat{n} = 1 - \delta - i\beta$$
$$= 1 - (N_a r_e \lambda^2 / 2\pi)(f_1 + i f_2)$$

Here N_a is the number density of atoms, r_e the classical electron radius and λ the wavelength. The atomic-number dependence of the decrement δ and the absorption index β is contained in the complex atomic scattering factor $f_1 + i f_2$.

Burton Henke and his colleagues at the University of Hawaii recently compiled[1] tables of the real part f_1 and the imaginary part f_2 of the atomic scattering factor. Derived from a best fit to available measured and theoretical values for the elements hydrogen to plutonium, these tables represent a valuable resource for workers in the field of x-ray and xuv optics. The data for nickel appear in figure 2 as an example. From the figure and the above equation, we can calculate that both the decrement δ and the absorption index β are of the order of 10^{-4} for x rays of wavelength 10 Å, or energy 1240 eV. With the index of refraction so close to unity, a concave nickel refracting lens would have an impractically long focal length, and might in addition require such a strong curvature—be so thick at the edges—that it would absorb almost all of the x rays. This is the case for all materials.

Conventional mirrors are also im-

practical. The intensity reflection coefficient I/I_0 is given by the Fresnel equations. For a vacuum–material interface and a wave polarized so that its electric vector is perpendicular to the plane of incidence, the equation is

$$I/I_0 = |R|^2 = \left| \frac{\cos\phi - \hat{n}\cos\phi'}{\cos\phi + \hat{n}\cos\phi'} \right|^2$$

The angle of incidence ϕ is measured from the surface normal, and the complex angle of refraction ϕ' is formally given by Snell's law, $\sin\phi = \hat{n}\sin\phi'$, where $\hat{n}$ is the complex index of refraction. For near-normal incidence, the cosine terms approach unity, and the Fresnel equation above reduces to

$$I/I_0 = (\delta^2 + \beta^2)/4$$

Recalling the magnitudes of the decrement δ and the absorption index β, we see that the single-surface reflectivity for x rays is negligible.

Glancing incidence reflection. The situation is different at angles of incidence close to 90°, called "glancing" or "grazing" incidence. Because the x rays are going from a medium of higher refractive index (the vacuum) to one of lower index (any material for which $n < 1$), Snell's law gives imaginary values for the angle of refraction ϕ' when the angle of incidence ϕ is greater than the critical angle of incidence ϕ_c, given by $\sin^{-1} n$. Hence, no wave can propagate into the second medium; the energy is reflected back into the first medium in

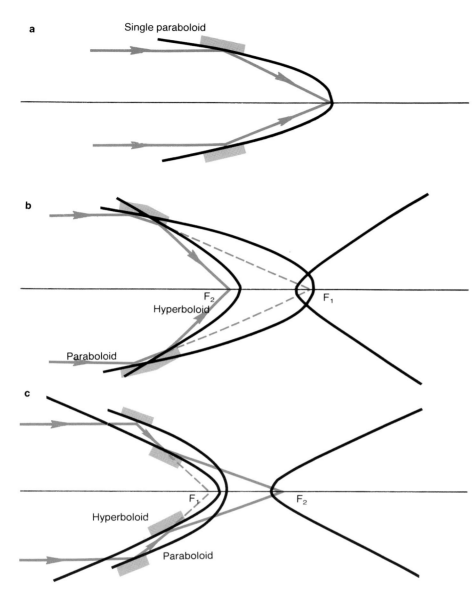

a Single paraboloid

b F₂

Hyperboloid

Paraboloid

F₁

c

F₁

Hyperboloid

F₂

Paraboloid

Glancing incidence optical systems based on conic sections. Hans Wolter proposed these systems for forming x-ray images. **a:** The single-paraboloid system, whose normal-incidence analog is the Newtonian telescope. **b:** Wolter's type I system, made up of an internal paraboloid and an internal hyperboloid. **c:** Wolter's type II system, which has an internal paraboloid and an external hyperboloid.

Figure 3

a manner analogous to the total internal reflection of visible light. For both the visible and the x-ray cases, Fresnel's equations are valid. The reflectivity for x rays is usually expressed as a function of the angle of glancing incidence θ, defined as $90° - \phi$. The term "total reflection" is somewhat misleading when applied to the case where the absorption index β is finite. The transition from high reflectivity, when the angle of glancing incidence θ is less than the critical angle θ_c, to very low reflectivity, when θ is greater than θ_c, takes place more gradually for high values of the ratio β/δ.

Arthur Compton first demonstrated the reflection of x rays at glancing incidence in 1922. Physicists soon realized that this phenomenon offered the possibility of using curved mirrors to make image-forming x-ray optics. However, many years passed before they overcame the severe aberrations—astigmatism in particular—of glancing-incidence optical elements. A concave spherical mirror at near-normal incidence forms a good image of a point object on the optical axis, but the image becomes progressively more elongated as the object moves further from the axis. At glancing incidence, this astigmatism is extreme: The image of a point becomes a line, and a single spherical mirror has limited use as an imaging device.

There are two principal methods for correcting this problem. In 1948, Paul Kirkpatrick and Albert Baez of Stanford University constructed[2] the first x-ray imaging system, an x-ray microscope, by using a pair of spherical concave mirrors in tandem, with their axes of revolution perpendicular so that each mirror corrects the astigmatism of the other. This method is used in laser fusion diagnostics, as we shall see below.

The second method of correction[3] uses a mirror with a radius of curvature much greater in one direction than in the perpendicular direction. The inner surface of a tube gently curved along its length has this form.

In 1952, Hans Wolter of Kiel University made an important contribution[4] to the theory and application of this kind of glancing-incidence x-ray optics. Working with surfaces of revolution based on conic sections, Wolter designed several systems for focusing x rays. Figure 3 outlines three of these systems. A single glancing-incidence paraboloidal mirror brings rays parallel to the axis to a focus and is evidently free from spherical aberration, as indicated in figure 3a. However, this mirror suffers badly from coma, the variation of magnification across the aperture. Wolter showed that coma could be alleviated by using two different conic-section surfaces of revolution, as shown in figures 3b and 3c. In the system of figure 3b, Wolter's type I system, the rays are first reflected from a paraboloid similar to that of figure 3a, and then from an internally reflecting hyperboloid having one focus coincident with that of the paraboloid. In Wolter's type II system, shown in figure 3c, the second element is an externally reflecting hyperboloid; it is the analog of the familiar Cassegrain telescope. Wolter showed[4] how the image quality of these systems could be improved

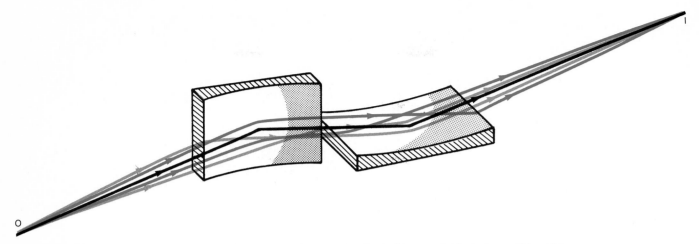

Microscope principle that corrects the astigmatism associated with glancing-incidence spherical mirrors. Kirkpatrick–Baez microscopes based on this method of crossed mirrors are used to make x-ray photographs of the implosions of laser-fusion capsules.   Figure 4

further by distorting the mirror surfaces from the exact conics into a form described by a theory of astromomer Karl Schwarzschild.

Applications of reflecting optics

Glancing-incidence reflection is the most mature technique of x-ray optics, and has found the most applications. Despite their severe aberrations, single mirrors have found important applications in a number of fields. Single spherical, cylindrical or toroidal mirrors, for example, are useful in xuv spectrometers. Single mirrors are essential to the utilization of xuv synchrotron radiation, for which they provide a means of beam extraction and deflection. Doubly curved optics are employed to form an image of the x-ray source—the circulating electrons—on the experimenter's instrument. The next generation of storage rings and insertion devices (discussed by Arthur Bienenstock and Herman Winick, PHYSICS TODAY, June 1983, page 48) will generate such intense radiation that x-ray heating and distortion of the optics will be severe, necessitating spectral filtering to reduce the flux, and special materials, such as silicon carbide, that can withstand high thermal loads.

Although physicists have built a number of Kirkpatrick–Baez microscopes, reflection x-ray microscopy has not yet achieved the spatial resolution obtainable with techniques that we will describe below. The performance of reflection microscopes is limited to about one micron by residual aberrations, scattering from the surfaces, and by the difficulty of figuring the mirrors to the high degree of accuracy required.

Recently, however, the inertial confinement laser fusion program has stimulated renewed interest in reflection x-ray microscopy. The objective of this program is to compress a small pellet containing deuterium–tritium

fuel to a high density at thermonuclear reaction temperatures, as John Nuckolls explains in his article (PHYSICS TODAY, September 1982, page 24). The spatial distribution, time history and spectrum of the x rays emitted by the hot plasma contain information valuable for diagnosing the compression. In this case, the required spatial resolution of a few microns is not as good as the 100 Å resolution desired for the x-ray microscopy of biological specimens, but on the other hand, the object is selfluminous, inaccessible and emits other radiations and debris, precluding the use of contact or projection radiography.

At Lawrence Livermore National Laboratory, Fredrick Seward, Thomas Palmieri and their collaborators constructed[5] a series of Kirkpatrick–Baez microscopes, including one that simultaneously obtained four separate images of the imploding target. By using different materials for each of the separate mirrors, setting them at different glancing angles and interposing various thin-foil x-ray absorption filters, they obtained sufficient x-ray spectral data to make images of the compressed plasma and determine its temperature with moderate resolution. Figure 4 illustrates the principle upon which the microscope is based.

Laser fusion workers have also used[6] Wolter type I microscopes to photograph imploding targets. And by combining x-ray imaging systems and ultrafast x-ray streak cameras, they have studied the implosion dynamics of laser-irradiated fuel pellets, achieving space and time resolutions of several microns and 20 picoseconds, respectively. These x-ray systems have led to significant insight[7] into the energy-transport mechanisms within the irradiated fuel capsules.

Wolter's x-ray optics have also seen extensive use in telescopes for x-ray

astronomy. The coronas of the Sun and many stars emit x rays, as do more exotic objects such as supernova remnants, galaxies, quasars, pulsars, neutron stars and possibly black holes. The mechanisms of emission include thermal radiation and a variety of nonthermal processes such as the inverse Compton process and synchrotron radiation. Riccardo Giacconi and Bruno Rossi of MIT first suggested[8] the use of glancing-incidence optics to study celestial x-ray emitters, and in the early 1960s, rockets carried small x-ray telescopes above the atmosphere to photograph the Sun. Gradual improvements in telescope technology and spatial resolution culminated[9] with the flight of two solar x-ray telescopes on Skylab in 1973. The numerous x-ray photographs they obtained revealed the solar corona to be complex and dynamic, and greatly enhanced our understanding of the physics of the upper solar atmosphere. Among the phenomena discovered or studied extensively for the first time were coronal holes, coronal loop structures and x-ray bright points. The x-ray photograph in figure 1 contains examples of all of these.

Non-solar x-ray astronomy requires larger, more sensitive, x-ray telescopes. The largest to date was launched in November 1978 as part of the second High Energy Astronomy Observatory HEAO-2, informally named the Einstein Observatory. The telescope, which operated until April 1981, consisted of four separate Wolter–Schwarzschild type I systems "nested" inside one another and aligned to have a common focus. The diameter of the outer mirror was 58 cm, the focal length 344 cm and the total soft x-ray collecting area 400 cm². A focal plane turret carried an imaging proportional counter, a high-resolution microchannel-plate imaging system, a crystal

Zone-plate form and diatom. The resist material in **a** was written on by an electron beam and etched to leave the pattern seen in this scanning electron micrograph. The form is ready to be electroplated with gold and dissolved to leave a free-standing gold Fresnel zone plate capable of focusing x rays. The spiked structure in **b** is a diatom a few microns in diameter. It was photographed with 44 Å synchrotron radiation and a zone-plate x-ray microscope. The smallest resolved features are approximately 1500 Å across. Elemental sensitivity, spatial resolution, and compatibility with unstained wet-cell speciments are major advantages of this new technique. (Zone-plate photograph courtesy of Dieter Kern, IBM; diatom photograph courtesy of Günther Schmahl, Dietbert Rudolph and Bastian Niemann.[22])   Figure 5

spectrometer and an energy-dispersive solid-state detector. To obtain low-resolution spectra, ground control could command the insertion of a free-standing transmission grating behind the telescope optics.

Although the Einstein observatory surveyed only about 1% of the sky, it has given us a new, fascinating and valuable x-ray view of the universe. One of its accomplishments was to show x-ray astronomy to be a powerful technique for studying the activity and evolution of relatively ordinary stars, as well as more spectacular objects. A comprehensive overview of the Einstein results is in reference 10.

The high-energy astrophysics community is looking forward to the launch, in the late 1980s, of an even larger and more elaborate NASA x-ray observatory, the Advanced X-Ray Astronomy Facility or AXAF. With a three-fold increase in collecting area, and a similar improvement in spatial resolution, AXAF will extend astronomical x-ray observations to much fainter and more distant objects, over much more of the sky.

Multilayer x-ray mirrors

New kinds of x-ray imaging and focusing devices now under development may replace or supplement glancing-incidence optics in certain laboratory and space applications. Already available are multilayer coatings that enhance the xuv reflectivity of mirrors over a wide range of angles and at the same time provide spectral selectivity. In their simplest form, these coatings consist of alternating thin layers of two materials with different refractive indices, and are analogous to the multilayer mirrors used at longer wavelengths. The high reflectivity of such mirrors is the result of the coherent addition of the weak reflections at the many interfaces. The single-surface intensity reflection coefficient I/I_0, given by the equations on page 45, is typically 10^{-4} at 45 Å, so that the amplitude coefficient $|R|$ is 10^{-2} and one can achieve a reflectivity approaching unity (even with absorption) with about 100 interfaces. The condition for constructive interference and thus peak reflectivity is given by the well-known Bragg equation

$$2d \sin\theta = m\lambda$$

In this equation, which is uncorrected for refraction or absorption, d is the thickness of a single layer pair, and m the order of reflection. Evidently, the thickness $2d$ must be comparable with the wavelength λ for angles θ greater than a few degrees, so that x-ray multilayer mirrors require layer thicknesses of about 10 Å, or just a few atomic diameters. One can compute the reflecting properties of the coatings by using optical multilayer theory, or by thinking of the coatings as synthetic crystals and using crystal diffraction theory.

Multilayer x-ray mirrors only recently became a practical possibility, with the development of techniques for the vacuum deposition of exceedingly thin, uniform layers on ultra-smooth substrates. These techniques[11] come primarily from the laboratories of Eberhard Spiller at IBM's Watson Research Center and Troy Barbee at Stanford University. A typical reflecting structure is made up of layers of a dense refractory element such as tungsten or molybdenum interleaved with a light element such as carbon. Such a combination of materials with very different indices of refraction yields a mirror

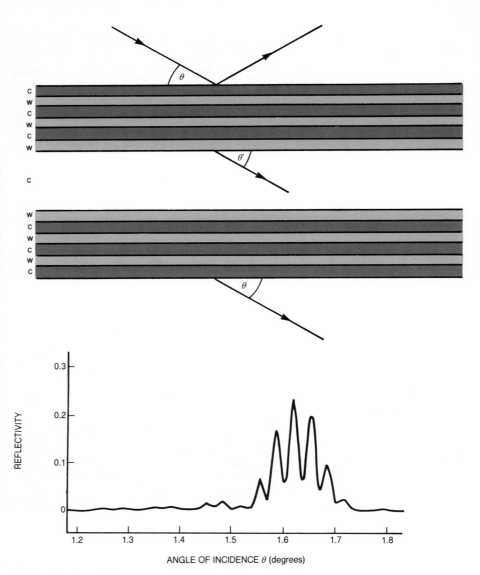

Etalon for x rays. The upper part of the figure shows the principle of a solid Fabry–Perot etalon. In this example, two multilayer mirrors made of tungsten and carbon are separated by a carbon spacer. The plot shows the calculated reflectivity of 1.54 Å radiation when each mirror is made from 15 layer pairs. The tungsten layers are 8.5 Å thick; the carbon layers are 19.1 Å thick; and the carbon spacer is 961 Å thick.   Figure 6

with a high reflectivity and a relatively broad wavelength bandpass: The ratio of wavelength to bandpass width, $\lambda/\Delta\lambda$, ranges from about 10 to 100. Combinations of light elements with more nearly equal indices of refraction yield mirrors with much narrower pass bands: here $\lambda/\Delta\lambda$ may be 1000. One must choose the material pairs carefully to give atomically smooth layers and to avoid problems such as interdiffusion and chemical reactions.

The performance of multilayer x-ray mirrors has now been measured over the spectral range of 0.5 Å to 120 Å, and in many cases the agreement with predictions is spectacular. With several laboratories obtaining excellent results, there is now much activity aimed at using multilayer mirrors in the

diverse fields of x-ray research and technology that we have mentioned in this article. The x-ray mirrors have possible uses as low-resolution dispersive elements for spectrometers and monochromators, as narrow-band reflection filters, beam splitters, polarizers, and in a variety of imaging or focusing applications. For example, they can be used to coat glancing-incidence optics of the Wolter kind to increase the reflectivity at a particular wavelength and glancing angle, and thereby increase the throughput of telescopes or microscopes. Because the technology of multilayer mirrors is less mature than that of glancing-incidence x-ray optics, much of the research has reached the stage where feasibility has been demonstrated, but where few

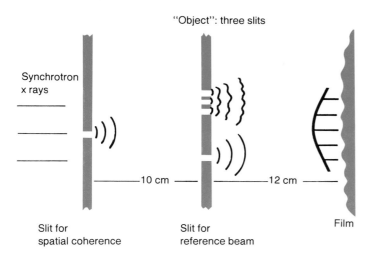

"Object": three slits

Synchrotron
x rays

10 cm — 12 cm

Slit for
spatial coherence

Slit for
reference beam

Film

Holography with x rays. This schematic diagram describes an experiment by a group in Japan. The "object" for the hologram consists of three 3-micron slits, 9 microns apart. The reference beam is introduced by diffraction from a 2.5 micron slit, 30 microns away, as part of a process called lensless Fourier-transform holography. The low-frequency interference pattern is chosen for subsequent visible-light reconstruction with a He–Ne laser. The Japanese group has used different x-ray sources and recording schemes to examine such objects as red blood cells and artificial fibers. **Figure 7**

working instruments are yet in use.

The feasibility of using multilayer coated mirrors at normal incidence to form images in the x-ray region below 100 Å has been demonstrated, raising the interesting and exciting prospect of extending normal-incidence optical techniques to the xuv spectral region. Patrick Henry of the University of Hawaii, working with Spiller and Martin Weisskopf of NASA's Marshall Space Flight Center, used a single multilayer mirror to achieve[12] spatial resolution of the order of one arc second at 68 Å, while Barbee and Underwood have obtained[13] clear images at 45 Å using a multilayer coated silicon wafer bent into a concave spherical mirror. Both experiments were directed toward the eventual construction of a soft-x-ray telescope of simple (single mirror) or compound configuration. Spiller and his colleagues at IBM have made[14] considerable progress toward the construction of a scanning x-ray microscope for operation at 68 Å, using compound multilayer optics, while Rolf Haelbich of Hamburg University has described[15] a similar instrument to work at around 200 Å.

Diffractive x-ray optics

In addition to mirrors and lenses, Fresnel zone plates can focus light and form images. Whereas mirrors and lenses focus radiation and form images by adjusting the phase at each point of the wavefront, zone plates block out those regions of the wavefront whose phase is more than a half period different from that at the plate center. In its simplest form, a zone plate is a diffracting mask consisting of concentric circular zones that are alternately transparent and opaque to the radiation of interest. The zones must be of equal area, so the radii of the boundaries between them increase in proportion to the square roots of the natural numbers. At a particular wavelength, the spatial resolution is proportional to the width of the thin, outermost zone.

Because drawbacks such as a diffraction efficiency of only about 10%, zone plates have limited applicability in the visible. However, in the xuv they may be the most direct route to the theoretical limit of resolution, the diffraction limit. Albert Baez was the first to demonstrate[16] the utility of zone plates for xuv radiation. Together with the Buckbee Mears Company, he developed free-standing plates with 20-micron-wide outer zones, and tested them with ultraviolet radiation at 2537 Å. As Baez predicted, the technology for making finer structures has improved dramatically. By using techniques of both photon interference and electron-beam pattern writing, several groups have formed patterns that represent improvements of several orders of magnitude. Groups at Göttingen, MIT and IBM have produced full zone plates whose smallest features are on the order of 1000 Å.

Figure 5a shows a scanning electron micrograph of an intermediate step in the formation of a zone plate. The pattern was written by an electron beam and etched in a resist material. The resulting object serves as a form for electroplating with gold, which is opaque to x rays. The resist material is then removed, leaving a free-standing gold structure. Although gold is favored by the microelectronics industry, it is not necessarily ideal for x-ray optics; the future development of partially transmitting phase plates promises to give significantly higher diffraction efficiency, while reducing the background of undiffracted radiation.

Although zone plates have a few applications in x-ray astronomy, they are potentially most useful in x-ray microscopy[17] and microholography. The short wavelength of x rays makes it possible to obtain much better spatial resolution with an x-ray microscope— 100 Å or less—than is possible with a light microscope. Because x rays can penetrate relatively thick, wet samples of biological material, in addition to a limited quantity of air, one avoids the need for the thin sectioning, staining and drying of specimens, as is required in electron microscopy. This raises the exciting prospect of dynamical studies on *living* biological specimens. Finally, the selective absorption of x rays by the chemical elements makes it feasible to map the distribution of elements within specimens. For biological applications, the spectral region containing the K absorption edges of oxygen (23.3 Å), nitrogen (31.2 Å) and carbon (43.6 Å) is most appropriate. These features of x-ray microscopy are clearly complementary to those of electron microscopy, which achieves higher spatial resolution on inanimate specimens, but has limited capability for distinguishing elements.

Several groups have embarked on programs to develop zone-plate microprobes and microscopes. Figure 5b shows a demonstration image of diatoms obtained by the group at Göttingen, using synchrotron radiation and an x-ray microscope based on zone plates. This group recently showed[18] a diatom resolved to 700 Å.

Zone-plate microscopy is not the only alternative to the reflection method. In fact, the most striking results to date come from *contact microscopy*. This simple but powerful technique uses no x-ray optical elements at all. One places the specimen in contact with a photosensitive material such as film or photoresist. After exposing and developing the material, one enlarges the resulting image in a light microscope if the detector is film, or in an electron microscope if the detector is a photoresist. Electron microscopists have obtained resolutions close to 100 Å with this technique. Transmission electron microscopy is preferred over scanning electron microscopy because it gives information that is more quantitative and easier to interpret. Although such techniques are simple to implement, they do not appear to provide a clear path to dynamical studies, so there is a continuing incentive to develop zone-plate microscopes.

Diffraction gratings. The same techniques used to make microscopic Fresnel zone plates—electron-beam and photon writing and optical and x-ray lithography—have also been used to make other x-ray diffractive elements. In 1974, Gary Bjorklund and his colleagues at Stanford formed a linear grating with a period of 836 Å in the photoresist PMMA from an interference pattern of harmonically generated coherent radiation at 1182 Å. In 1982, Andrew Hawryluk and his colleagues at MIT carried this a step further by fabricating free-standing gold transmission gratings, albeit with a 3000 Å period.

These new elements add a new degree of freedom to spectroscopic studies, not so much because of their resolving power, but rather because of the convenience with which they can be incorporated into existing recording devices, such as picosecond x-ray streak cameras or reflection x-ray telescopes and microscopes. Hawryluk, Natale Ceglio and their colleagues at MIT, Lawrence Livermore Laboratory and the University of Rochester's Laboratory for Laser Energetics have already used the resulting hybrid devices in a variety of experiments involving heating and energy transport in laser-irradiated targets.

Interferometry and holography

In 1965, Ulrich Bonse, now at the University of Münster, and Michael Hart of Bristol University invented[19] x-ray interferometry. Their monolithic single-crystal interferometers have become valuable and versatile tools for metrology and for studying imperfections in crystals and phase effects in materials at energies of a few keV. With the development of multilayer x-ray coatings and microfabrication, it now seems feasible to extend interferometric techniques to the xuv region of the spectrum.

An x-ray interferometer using multilayer reflectors has in fact been demonstrated,[20] in the form of a solid Fabry–Perot etalon. Like its optical counterpart, an x-ray etalon consists of two reflectors separated by a spacer—in this case two multilayer structures separated by a thick layer of the less dense material (figure 6). When one scans the etalon in angle at a fixed wavelength, the reflection curve shows the Bragg peak characteristic of either of the reflectors, modulated by a series of narrow dark fringes (the fringes of equal inclination). The angular position of these fringes depends upon the thickness of the spacer, its refractive index, and the phase shifts that occur on reflection at the interfaces. Figure 6 shows a schematic diagram

of an x-ray etalon, and the predicted reflectivity assuming tungsten–carbon reflectors separated by a 1000-Å-thick carbon spacer. The agreement with experiment is remarkably good, although the peak reflectivity and fringe contrast are somewhat less than predicted due to the roughness of surfaces and the divergence of the x-ray beam. One use for this type of interferometer may be to make high-precision measurements of the thickness of films, but perhaps it will be most useful for accurate measurement of the xuv refractive index of the spacer material in regions of anomalous dispersion.

There is active interest in using the new x-ray optical elements to extend holography to the x-ray region. There was significant progress[21] in this direction in the early 1970s in Japan, where Seishi Kikuta, Sadao Aoki and their colleagues demonstrated beautiful x-ray holographic recording and, in some cases, optical reconstruction. Objects of their studies extended from simple grids of micron spacing to artificial fibers and red blood cells. Figure 7 illustrates their technique for off-axis, lensless Fourier-transform x-ray holography. To get the required longitudinal coherence, the Japanese workers dispersed synchrotron radiation and used a narrow portion of its spectrum; to get the required lateral coherence, they put the beam through a slit, as figure 7 shows. They were able to make x-ray holograms with 60 Å radiation, and to view the holograms with He–Ne laser light at 6328 Å. Malcolm Howells is now doing related experiments at Brookhaven's National Synchrotron Light Source. Extension of these techniques to interesting problems of biological dynamics calls for a combination of developments, all of which are within the scientific community's collective ability to achieve.

Sufficient coherent power to produce off-axis x-ray microholograms efficiently will require intense quasi-coherent synchrotron radiation from advanced synchrotron facilities. In addition, efficient holography will require multilayer mirrors for undulator mode selection, as well as more complex microfabricated diffraction structures, such as a Fresnel lens side by side with an appropriate transmission grating to generate a reference beam. To resolve significant three-dimensional features will require fast x-ray optics, necessitating careful attention to sources of degradation in the imaging and recording processes.

All indications are that the current rapid evolution of the field of high-resolution x-ray optics will continue. And with the development of new sources of xuv radiation, we can expect to see even more exciting applications over a broad scientific spectrum.

The Lawrence Berkeley Laboratory Center for X-ray Optics is supported by the Department of Energy's Office of Basic Energy Sciences under contract DE-AC03-76F0098.

References

1. B. L. Henke, P. Lee, T. J. Tanaka, R. L. Shimabukuro, B. K. Fujikawa, Atomic Data and Nuclear Data Tables **27**, 1 (1982).

2. P. Kirkpatrick, A. V. Baez, J. Opt. Soc. Am. **38**, 766 (1948).

3. F. Jentzsch, Phys. Z. **30**, 268 (1929).

4. H. Wolter, Ann. der Phys. **10**, 94 and 286 (1952).

5. F. D. Seward, J. Dent, M. Boyle, L. Koppel, T. Harper, P. Stoering, A. Toor, Rev. Sci. Inst. **47**, 464 (1976).

6. R. H. Price in *Low-Energy X-ray Diagnostics 1981*, D. T. Attwood, B. L. Henke, eds., AIP Conf. Proc. **75**, American Institute of Physics, New York (1981), page 189; J. K. Silk, Proc. SPIE **184**, 40 (1979).

7. D. T. Attwood, J. Quant. Electronics **QE-14**, 909 (1978).

8. R. Giacconi, B. Rossi, J. Geophys. Res. **65**, 773 (1980).

9. G. S. Vaiana, L. VanSpeybroeck, M. V. Zombeck, A. S. Krieger, J. K. Silk, A. F. Timothy, Space Sci. Instr. **3**, 19 (1977); J. H. Underwood, J. E. Milligan, A. C. deLoach, R. B. Hoover, Appl. Opt. **16**, 858 (1977).

10. *X-Ray Astronomy with the Einstein Satellite*, R. Giacconi, ed., Proceedings of the High Energy Astrophysics Division of the AAS meeting, 28–30 January 1980, D. Reidel, Dordrecht (1980).

11. E. Spiller in *Low-Energy X-ray Diagnostics 1981*, D. T. Attwood, B. L. Henke, eds., AIP Conf. Proc. **75**, American Institute of Physics, New York (1981), page 124; T. W. Barbee, page 131.

12. J. P. Henry, E. Spiller, M. Weisskopf, Proc. SPIE **316**, 166 (1981); Appl. Phys. Lett. **40**, 25 (1982).

13. J. H. Underwood, T. W. Barbee, Nature **294**, 429 (1981).

14. I. Lovas, W. Santy, E. Spiller, R. Tibbetts, J. Wilczynsky, Proc. SPIE **316**, 90 (1980); E. Spiller in *Scanned Image Microscopy*, E. Ash, ed., Academic, New York (1980), page 365.

15. R.-P. Haelbich in *Scanned Image Microscopy*, E. Ash, ed., Academic, New York (1980), page 413.

16. A. V. Baez, J. Opt. Soc. Am. **51**, 405 (1961); J. Opt. Soc. Amer. **42**, 756 (1952).

17. J. Kirz, D. Sayre in *Synchrotron Radiation Research*, H. Winick, S. Doniach, eds., Plenum, New York (1980), page 277.

18. *Science*, 8 January 1983.

19. U. Bonse, M. Hart, Appl. Phys. Lett. **6**, 155 (1965).

20. T. W. Barbee, J. H. Underwood, Optics Comm. **48**, 161 (1983).

21. S. Kikuta, S. Aoki, S. Kosaki, K. Kohra Opt. Comm. **5**, 86 (1972); S. Aoki, Y. Ichihara, S. Kikuta. Japan. J: Appl. Phys. **11**, 1957 (1972); S. Aoki, S. Kikuta, Japan. J. Appl. Phys. **13**, 1385 (1974).

22. G. Schmahl, D. Rudolph, B. Niemann, Proc. SPIE **361**, 100 (1981); D. Rudolph, B. Niemann, G. Schmahl, page 103. □

CHAPTER 8

CONCEPTS AND EPILOGUE

Optics (including astronomy) is one of the oldest of the sciences, having begun when our most remote forebears gazed in awe at the starry firmament on high or marvelled at a rainbow. But today, far from being ancient or decrepit, optics is lively and vigorous as never before. The specific descriptions of developments given in this volume are good examples of the resurgence of applied optics. Most of the sensitive infrared detectors were not developed until after the Second World War; the use of high-speed computers to optimize optical designs, or to reduce interferometric data, is less than 35 years old; the discovery of the laser, and its application to holography, occurred just over 25 years ago. It is less than 30 years since the launching of the first satellite, and space vehicles are now available for the optical probing of distant planets. These recent efforts are brave beginnings, but they are still just beginnings.

It is interesting briefly to review this recent resurgence of optics: one has only to observe that we alive today overlap the lives of most of the greats of optics, spectroscopy, and astronomy. Although the burning glass and the mirror date from antiquity, it was only about 400 years ago that man fashioned the first lenses to make possible the eye glass, the microscope, and the telescope. Newton in 1666 discovered that white light could be dispersed into a spectrum when he introduced a prism into a beam of sunlight coming through a hole in a shutter, but over 100 years elapsed before Wollaston and Fraunhofer repeated that simple experiment using a slit, thereby discovering spectral lines. In today's scientific community that next step would have happened by the following day. The science of spectroscopy and spectrum analysis was begun by Bunsen and Kirchhoff about 115 years ago. Photography is also just over 120 years old, and 120 years ago Ernst Abbe was just beginning his research on lens design and image formation. The basic theory of geometrical optics and the aberrations of lenses were by then reasonably well understood and even the first textbooks on optics (e.g., by Parkinson) had appeared. Still, lens design was a painful and tedious procedure; Abbe worked for years developing the apochromatic lens, and he kept the composition of the fluorite lens component a secret for over two decades, calling it "glass x" on the circulation sheets. The splendid Tessar lens was also many years in development. The full flowering of the science of lens design began in 1917 when A. E. Conrady was appointed professor of optical design at Imperial College. Where the previous German (continental) school had used methods of ray tracing and of trial and error, Conrady, following Coddington and Airy, tried to use algebraic formulas for the aberrations. He instinctively understood the role of diffraction, and Rayleigh's quarter-wave limit for path differences became the foundation of his theory of optical

tolerances. By 1945 the art of optical design had advanced to the point where Arthur Hardy of M.I.T. said he visualized an era not too far away when machines could run rapidly through all possible optical designs and arrive at the best one for a given application. This prediction has long since come true: about 20 years ago some students at the Institute of Optics in Rochester, using a small, slow computer, in five minutes made 66 iterations of the computer program to calculate a three-element f/3.5, Tessar lens that had originally taken years to design. Today any optics amateur can do this on his personal computer.

Even more remarkable have been the recent developments of optical theory and of our knowledge of the nature of radiation. Just 115 years ago, Maxwell published his papers on the dynamical theory of the electromagnetic field, which provided a theoretical basis for the explanation of many optical effects, such as the Faraday, Zeeman, Kerr, Voight, Majorana, and Cotton-Mouton effects, which were observed and explained as quickly as improved spectral instrumentation permitted. Other optical effects remained puzzling, however, and Rayleigh recognized in early 1900 that classical theory was inadequate to explain blackbody radiation fully. Planck then considered this problem and within a year had postulated his quantum of action and a correct formula for blackbody radiation. The concept of quanta was extended by Bohr in 1912 to a model of the hydrogen atom and led in turn to the spectacular development of quantum mechanics and wave mechanics in the mid 1920s. But yet another tremendous development was being stimulated by optics during the same time period: just before the turn of the century 16-year-old Albert Einstein became intrigued by the question: "What would happen if a man should try to imprison a ray of light?" A few years later he tried to teach himself the use of Maxwell's equations and in rapid succession wrote his papers on Brownian motion, special relativity, and the photoelectric effect. He also reformulated the Planck equations for the absorption and emission of radiation in terms of spontaneous and stimulated emission and laid the groundwork for the more recent development of the maser and laser. Einstein perhaps more than anyone else has stimulated the present magnificent renaissance in optics, spectroscopy, and astronomy. Even his more abstruse general theory of relativity can only be tested by astronomical observations, mostly optical; he himself suggested the deviation of starlight by the sun as a test of his theory. Einstein's early postulate of the mass equivalence of energy also led to an explanation of how stars radiate: prior to this explanation there had been only the unsatisfying hypothesis by Lord Kelvin that the radiation energy resulted from a gravitational collapse of the sun. Now, a century later, the gravitational collapse hypothesis is sometimes revived in an effort to explain the behavior of quasars.

Within the past few years new sensors have extended the domain of optics from the visible region to the entire range of the electromagnetic spectrum; the linking of these sensors to modern electronic data processing and computers has opened up entire new fields of research, such as x-ray and infrared astronomy, Fourier spectroscopy, infrared interferometry, and holography. There is even a branch called space optics, using astronomical and spectroscopic sensors mounted on balloons, aircraft, rockets, and satellites to look back at the Earth or to probe remote planets. So accurate and sophisticated is our optical instrumentation today that the meter, our standard of length for all physical measurements, has been defined interferometrically in terms of the wavelength of an atomic spectral line, and the second, our unit of time, is defined in terms of a spectral transition (the cesium clock). More recently—in 1983—the velocity of light has been adopted as a universal constant, so that the meter is now defined as the distance traveled by light in a vacuum in 1/299,792,458 of a second. Large radiotelescopes on opposite sides of the world can now be linked together into a giant interferometer, synchronized by an atomic clock. Spectrometers mounted in powerful ground-based telescopes can measure both the Zeeman splitting and Doppler shift in the spectrum of sunspots, from which the magnetic field, temperature, and velocity of gases in the vicinity of solar active centers can be determined. We hope some day to be able to predict the occurrence of solar disturbances by optical techniques.

What advances can we expect in optics, spectroscopy, and astrophysics in the near future? When one reflects that the advances of the last 50 or 60 years have far outstripped all the thousands of years of previous effort, it is risky even to hazard a guess. Satellite-borne optical sensors monitor our weather and map cloud cover, temperature, water vapor, and ozone profiles. Efforts are underway to extend this optical sensing to the movements of winds and weather fronts. Others predict satellite-borne surveys of Earth's geological and timber resources and monitoring of crop development and movement of ocean currents. The atmospheres of distant planets are being probed radiometrically and spectrally. Astronomical observations will be made routinely from extraterrestrial platforms. It is interesting to observe that some of the traditional roles of physics and astronomy have now been reversed: whereas formerly physics tested theory by hands-on experiment and astronomy merely amassed observational data from far away, today much of physics digests remote observational data from bubble and sparkchambers, but optical probes and space platforms have made the entire universe a laboratory for *in situ* astrophysical experiments.

In spectroscopy itself tunable lasers are substituting for the spectrometer and bring to the optical region the resolving power now found in microwave spectroscopy. In astronomy many problems of cosmology lie almost within grasp: for example, are quasars immense radiant bursts at the periphery of space or perhaps not so distant accidental stellar lasers? Are there other solar systems around nearby stars? In physics the nature of the electronic charge still eludes us and the clue to this may lie in spectroscopy, in the theoretical determination of the fine structure constant $2\pi e^2/hc$, which remains a major unsolved problem of modern physics. We still argue about the nature of the photon. There are giant problems yet remaining, but happily there are yet giants on the Earth in these days.

John N. Howard

CONTENTS

The concept of the photon

It has its logical foundation in the quantum theory of radiation. But the "fuzzy-ball" picture of a photon often leads to unnecessary difficulties.

Marlan O. Scully and Murray Sargent III

PHYSICS TODAY / MARCH 1972

The idea of the photon has stirred the imaginations of physicists ever since 1905 when Einstein originally proposed the use of light quanta to explain the photoelectric effect. This concept is formalized in the quantum theory of radiation, which has had unfailing success in explaining the interaction of electromagnetic radiation with matter, seemingly limited only by the ability of physicists to perform the indicated calculations. Nevertheless, it has its conceptual problems—various infinities and frequent misinterpretations. Consequently an increasing number of workers are asking, "to what extent is the quantized field really necessary and useful?" In fact the experimental results of the photoelectric effect were explained by G. Wentzel in 1927 without the quantum theory of radiation. Similarly most electro-optic phenomena such as stimulated emission, reaction of the emitted field on the emitting atom, resonance fluorescence, and so on, do not require the quantization of the field for their explanation. As we will see, these processes can all be quantitatively explained and physically understood in terms of the semiclassical theory of the matter–field interaction in which the electric field is treated classically while the atoms obey the laws of quantum mechanics. The quantized field is fundamentally required for accurate descrip-

tions of certain processes involving fluctuations in the electromagnetic field: for example, spontaneous emission, the Lamb shift, the anomalous magnetic moment of the electron, and certain aspects of blackbody radiation. (The Compton effect also fits here, but see later under references 8b and c.) Here we will outline how the photon concept originated and developed, where it is not required and is often misused, and finally where it plays an essential role in the understanding of physical phenomena.[1] In our discussion we will attempt to give a logically consistent definition of the word "photon"—a statement far more necessary than one might think, for so many contradictory uses exist of this elusive beast. In particular consider the original coining of the word by G. N. Lewis:[2]

"[because it appears to spend] only a minute fraction of its existence as a carrier of radiant energy, while the rest of the time it remains an important structural element within the atom . . . , I therefore take the liberty of proposing for this hypothetical new atom which is not light but plays an essential part in every process of radiation, the name photon!"

(our exclamation point). Clearly the present usage of the word is very different.

From Maxwell to Schrödinger

Although the nature of light has been a subject of wonder since day one[3] (it was on a Monday), the conceptual basis for the understanding of radiative phenomena begins with James Clerk Maxwell and Heinrich Hertz. While it is true that Isaac Newton, Christian Huygens, Thomas Young, and many others contributed mightily to our understanding of optics, the Maxwell–Hertz demonstration that light is made

of the same stuff as electric and magnetic fields must be regarded as the first insight into the inner workings of the radiation field. According to their description, light is radiated by accelerating charges and is an electromagnetic excitation (of an "aether").

Among other things, it must have been the far-ranging success of Maxwell in explaining electromagnetic phenomena that led 19th-century physicists to state that there were really only two clouds on the horizon of physics at the beginning of the 20th century. Interestingly enough, both of these clouds involved electromagnetic radiation. The first cloud, namely the null result of the Michelson–Morley experiment, led to special relativity, which is the epitome of classical mechanics, and really capped things off in a logical way. The second cloud, the Rayleigh–Jeans catastrophe and the nature of blackbody radiation, led to the beginnings of quantum mechanics, which, of course, was a radical change in physical thought up to that point. Note that while both of these problems involve the radiation field, neither (initially) involved the concept of a photon. That is, neither Einstein nor Lorentz in the first instance nor Max Planck in the second called upon the particulate nature of light for the explanation of the observed phenomena. Relativity is strictly classical, and Planck only quantized energies of the oscillators in the walls of his cavity, not the field. Up to this point (before 1905) the discreteness of light quanta was never invoked.

The next chapter in the history of the photon concept came when Einstein applied Planck's quantization ideas to the photoelectric effect. The situation here was very different from that envisioned by Planck. Einstein invoked the existence of discrete bundles or

Marlan Scully and Murray Sargent are both at the University of Arizona in Tucson. Scully, professor of physics and optical sciences, is an Alfred P. Sloan Fellow and is supported in part by Professor Peter Franken. Sargent is assistant professor of optical sciences. Together with W. E. Lamb, Jr, they have written a text on quantum optics, which will appear shortly; this book will give a more complete account of the photon concept than can be presented here.

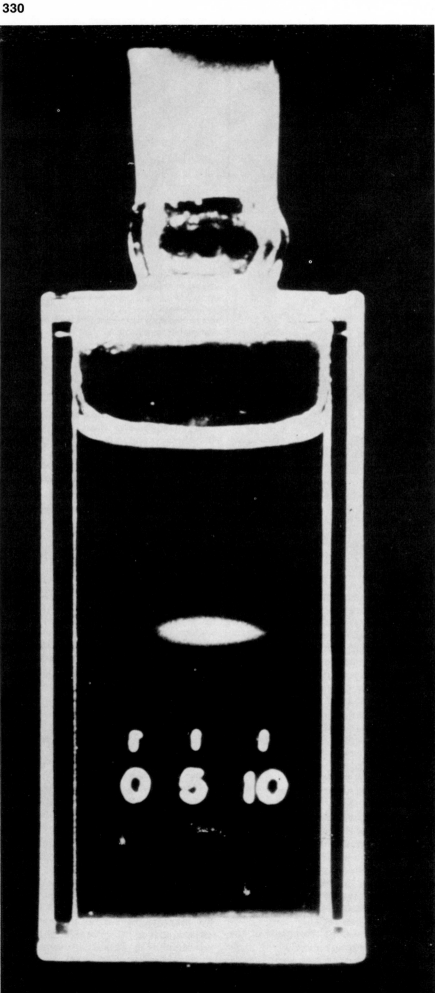

Laser pulse photographed in flight.
This is an ultrashort green pulse obtained as the second harmonic of 1.06-micron light from a neodymium-doped glass laser. To make the photograph the pulse was passed through a water cell; the scale on the cell wall is in millimeters. The camera shutter was a Kerr cell, triggered by an infrared pulse (1.06 micron) from the same laser; exposure time was about 10 picosec, the same period as the duration of these ultrashort pulses. During the exposure the pulse moved about 2.2 mm (right to left), the velocity of light in the cell being approximately 2.2×10^{10} cm/sec. (Photograph by Michel Duguay, Bell Laboratories.)

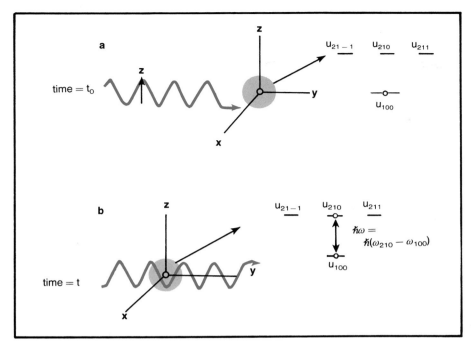

Radiation-induced transitions. This diagram shows how the electric field of equation (2) induces transitions from the energy levels u_{100} to u_{210}. The corresponding wave function is given by a superposition of states.
Figure 1

quanta of light energy (photons) to explain the ejection of photoelectrons from solids. This was distinct from Planck's idea in which only matter was quantized, as is illustrated in the following passage from George Gamow's delightful little book, *The Thirty Years That Shook Physics*[4]

"Having let the spirit of the quantum out of the bottle, Max Planck was himself scared to death of it and preferred to believe the packages of energy arise not from the properties of the light waves themselves but rather from the internal properties of the atoms which can emit and absorb radiation only in certain discrete quantities. Radiation is like butter, which can be bought or returned to the grocery store only in quarter-pound packages, although the butter as such can exist in any desired amount .... Only five years after the original Planck proposal, the light quantum was established as a physical entity existing independently of the mechanism of its emission or absorption by atoms. This step was taken by Albert Einstein in an article published in 1905, the year of his first article on the Theory of Relativity. Einstein indicated that the existence of light quanta rushing freely through space represents a necessary condition for explaining empirical laws of the photoelectric effect; that is, the emission of electrons from the metallic surfaces irradiated by violet or ultraviolet rays."

We shall return to the photoelectric effect later; however, we note that Planck's "butter-ball quantum" idea is not completely absurd, and, in fact, a modern version of it is being reconsidered by some modern theoretical

physicists.[8]

The next cornerstone is the realization that matter itself has a wave-like side to its personality. The first to put this in a concrete mathematical form was Erwin Schrödinger. He wrote his famous equation for the wave function of an atom, $\psi(r,t)$ in terms of its Hamiltonian as

$$i\hbar \frac{\partial}{\partial t} \psi(r,t) = \mathcal{H} \psi(r,t) \qquad (1)$$

and is responsible for demonstrating that the wave nature of matter is essential for its understanding.

Semiclassical theory

Atoms require quantum theory in the description of their behavior, because among other things, classical mechanics tells us that orbiting (therefore accelerating) electrons in atoms should radiate and spiral into the nucleus in contradiction of observed results! A surprisingly successful theory of the atom–field interaction can be obtained in which the atoms obey the laws of quantum mechanics and the electric field is treated *classically* according to Maxwell's equations—that is to say, without the concept of the photon. This semiclassical theory is important for our present purposes for two reasons: First it is important to understand which classical phenomena do not need or logically imply quantized fields for their explanation, and second, the semiclassical theory accounts quantitatively for most radiation–matter interactions. In this part of our article we will support this contention by reviewing the semiclassical description of:

▶ the response of an atom to a resonant, monochromatic field
▶ the self-consistent treatment of the atom–field interaction

▶ stimulated emission
▶ resonance fluorescence
▶ the photoelectric effect

Consider first a hydrogenic atom with energy eigenstates u_{nlm}, and suppose the atom is initially in its ground (1s) state, u_{100}, with energy $\hbar\omega_{100}$. We irradiate the atom by a light beam represented by the linearly polarized, plane-wave electric field

$$\mathbf{E}(y,t) = \hat{z} E_0 \cos(\nu t - Ky) \qquad (2)$$

where the (circular) frequency ν is nearly resonant with the 1s $\rightarrow$ 2p ($u_{100} \rightarrow u_{210}$) transitions, that is, $\nu \approx \omega \equiv \omega_{210} - \omega_{100}$. Radiation so polarized induces transitions from the u_{100} level to the u_{210} level, causing the wave function $\psi(\mathbf{r},t)$ to become a linear superposition of the two eigenfunctions as depicted in figure 1. The time development of $\psi(\mathbf{r},t)$ is determined by the Schrödinger equation (1) whose Hamiltonian includes the electric dipole interaction energy

$$\mathcal{H}_1 = -e\mathbf{r}\cdot\mathbf{E} \qquad (3)$$

The resulting z dependence of $\psi(\mathbf{r},t)$ varies in time as shown in figure 2a. There the probability density $\psi^*\psi$ oscillates back and forth across the (positively charged) nucleus with frequency $\omega = \omega_{210} - \omega_{100}$. Hence an ensemble of N such systems located in a volume (small compared to a cubic wavelength) about the position $\mathbf{R}_0$ produces an average oscillating dipole moment, namely

$$\mathbf{p}(\mathbf{R}') = N\left[\int d^3r\,\psi(\mathbf{r},t)e\mathbf{r}\psi(\mathbf{r},t)\right]\delta(\mathbf{R}' - \mathbf{R}_0) \qquad (4)$$

which depends on the detuning $(\omega - \nu)$, the strength of the atom–field interactions, and so on. We treat the expectation-value expression (4) as an ordi-

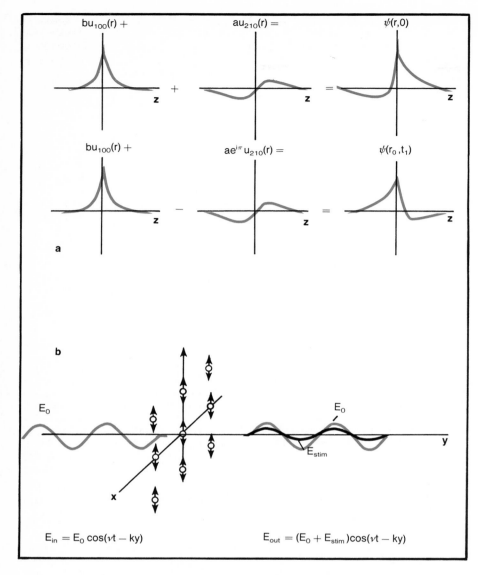

$bu_{100}(r) +$ $au_{210}(r) =$ $\psi(r,0)$

$bu_{100}(r) +$ $ae^{i\pi}u_{210}(r) =$ $\psi(r_0,t_1)$

a

b

E_0

E_0

E_{stim}

y

x

$E_{in} = E_0\cos(\nu t - ky)$ $E_{out} = (E_0 + E_{stim})\cos(\nu t - ky)$

Radiating dipoles. The z-dependence of the wave function $\psi(\mathbf{r},t)$ in part (a) are for $t = 0$ and $t = \pi/\omega$; the probability density $\psi^*\psi$ oscillates back and forth across the nucleus at frequency ω, so yielding an oscillating dipole. Part (b) of the figure shows how a sheet of dipoles radiates an electric field in phase with the incident field (equation 2). The individual fields become increasingly retarded the further off the axis one goes.
Figure 2

nary dipole density radiating, for example, the far-field electric field

$$\mathbf{E}(\mathbf{R},t) = [\omega^2/(4\pi\epsilon_0 c^2)]\,(n \times \mathbf{p}) \times$$
$$\hat{n}\,\frac{\exp[iK|\mathbf{R} - \mathbf{R}_0|]}{|\mathbf{R} - \mathbf{R}_0|} + \text{complex conjugate}$$
$$(5)$$

We see that the field induces a dipole moment in an ensemble of atoms and that this moment, in turn, contributes a field. So far we have neglected the effect of the *back reaction* of light emitted by the dipole back on itself. This reaction is included by requiring that the field be self consistent, that is, that the field which the atoms see be consistent with the field radiated, as outlined in figure 3. Solving the self-consistent set of equations in figure 3 simultaneously we account for the back action. This effect is one phenomenon sometimes said to require the quantum theory of radiation. We now apply the semiclassical method to several other problems.

Stimulated emission is the first of these. We wish to study a sheet of atoms in the x–z plane (figure 2b) sub-

ject to the incident electric field of equation (2). We suppose again that the atoms have two relevant levels, this time having atomic decay phenomena associated with, for example, collisions, and we carry out the appropriate time integrations to find the dipole density as in equation (4). We find the dipole moment density at the point $(x, 0, z)$

$$\mathbf{P}(x,0,z,t) \propto$$
$$\hat{z}E_0[\gamma\sin(\nu t) + (\omega - \nu)\cos(\nu t)] \quad (6)$$

in which the constant of proportionality depends on the number of atoms involved, the strength of the atom–field interaction, a Lorentzian involving detuning, the atomic decay rate γ, and so forth. The things to note from equation (6) are:
▶ the dipole oscillates at the driving frequency ν and not the atomic line center ω
▶ the magnitude of the dipole is proportional to the field amplitude E_0
▶ there are components "in phase" ($\cos \nu t$ term) and "in quadrature" ($\sin \nu t$ term) with the inducing field of

equation (2). The former modifies the index of refraction in the sheet; the latter acts as a source for gain due to stimulated emission.

Considering the second of these more closely, we note that on resonance the polarization, equation (6), is proportional to $\sin \nu t$. This is 90 deg out of phase with the applied field of equation (2), and in view of equation (5), one notes the radiated field (on axis) is not in phase either. In order to get back into phase with the incident field (indeed in phase with the textbooks!), we add up the contributions from a sheet of dipoles (integrate over x and z) to find a radiated field proportional to $\cos \nu t$ as is equation (2). This second radiated field has the same phase, frequency and direction as the incident field.

Resonance fluorescence[5] is our second application of semiclassical theory and is defined to be the emission of radiation by a ground-state atomic ensemble excited by an optical field. As depicted in figure 4, an incident field (spectral width Γ) and central frequency ν is absorbed by the ensemble (spectral width γ) which, in turn, emits into some new direction with the same spectrum as the incident field if Γ is small compared to γ. This follows from equation (7), which shows that the induced dipole has the same frequency as the inducing (driving) field. Alternatively, for a field whose spectral width is due to its finite duration $1/\Gamma$, we understand the atomic response as that of a driven oscillator with the frequency of the driving field. The atomic oscillator scatters for as long as it is driven, that is, for a time $1/\Gamma$. The spectral width of the emitted radiation therefore corresponds to the reciprocal of the lifetime, namely, Γ.

The photoelectric effect[5,6,8a] is our

The uncertainty principle violated

As an aside, we can note for example that the commutation relations for an atom damped by a quantized field are time independent, because of the presence of quantized Langevin noise sources associated with the atomic damping. These sources are quantized because the field is. Even if we could damp the atom with a classical field, as Michael Crisp and Ed Jaynes suggest, the associated Langevin sources would commute and allow the atomic commutation relations to decay to zero in time. (For further discussion see the paper by Melvin Lax, reference 10.)

This time dependence of the commutation relations then implies a violation of the uncertainty principle. This point was noted first (in a somewhat different way) by Niels Bohr and Leon Rosenfeld.

---

rate $\propto E_\mathrm{o}^2$),

▶ There is not necessarily a time delay between the instant the field is turned on and the ejection of photoelectrons.

To explain these three characteristics, we suppose the medium consists of ground-state ($|g\rangle$) electrons which can make transitions under the influence of an applied field, equation (2), to a quasi-continuum consisting of momentum states $|k\rangle$ as depicted in figure 5. With the electric-dipole energy, equation (3), and the philosophy of Fermi's Golden Rule, we find the probability for a transition from the ground state to the kth excited state within a time t to be

$$P_k = 2\pi[e|r_{kg}|/\hbar]^2 E_\mathrm{o}^2 t\delta[\nu - (\epsilon_k - \epsilon_g)/\hbar] \quad (8)$$

Writing energy $\epsilon_k - \epsilon_g$ as $mv^2/2 + \phi$ as in figure 5, we find that the δ function in equation (8) implies equation (7). This result conflicts with what is often taught, as the following quote from a well known text[7] illustrates:

"Einstein's photoelectric equation played an enormous part in the development of the modern quantum theory. But in spite of its generality and of the many successful applications that have been made of it in physical theories, the equation $h\nu = mv^2/2 + \phi$ is, as we shall see presently, based on a concept of radiation—the concept of 'light quanta'—completely at variance with the most fundamental concepts of the classical electromagnetic theory of radiation."

The second fact is also clearly contained in equation (8), since P_k is directly proportional to E_o^2. Finally the third point is accounted for, because

---

equation (8) is nonzero even for small times, a fact underlined by Peter A. Franken.[6] "As for the time delays [in the photoelectric effect], quantum mechanics teaches us that the *rate* is established when the perturbation is turned on [after several optical cycles]."

In fact, for the majority of quantum optical calculations the semiclassical theory proves most adequate. We note that in addition to those examples above, nonlinear optics,[7a] much of laser theory,[7b] pulse-propagation phenomena[7c] and even "photon" echo[7d] are all best explained without photons. That the list of successes of semiclassical theory is impressive is further illustrated by the following quote from the recent paper of Michael D. Crisp and Ed T. Jaynes,[8]

"Even though it is generally believed that a full quantum-electrodynamic treatment is necessary in order to obtain all radiative effects correctly, many calculations involving the interaction of radiation and matter were first done without quantizing the electromagnetic field. Thus is the case of the photoelectric effect,[8a] the scattering of radiation from a free electron (Klein–Nishina formula)[8b] stimulated emission and absorption of radiation by an atom,[8c] and vacuum polarization,[8d] the correct predictions were first obtained by semiclassical methods."

They continue with the assertion that, while spontaneous emission and the Lamb shift are generally conceded to require the quantized field, the self-consistent semiclassical theory does surprisingly well even here. In fact they derive a "Lamb shift" that is order-of-magnitude correct from their semiclassical calculation! However we

---

final example of semiclassical theory. Some readers may find this surprising, because the photoelectric effect provided the original impetus for ascribing a particle character to light. The three main facts of life, photoelectron-wise, are:

▶ When light shines on a photoemissive surface, electrons are ejected with a kinetic energy equal to Planck's constant times the frequency ν of the incident light less some work function ϕ, usually written as

$$h\nu = \frac{mv^2}{2} + \phi \quad (7)$$

▶ The rate of electron ejection is proportional to the square of the electric field of the incident light (ejection

---

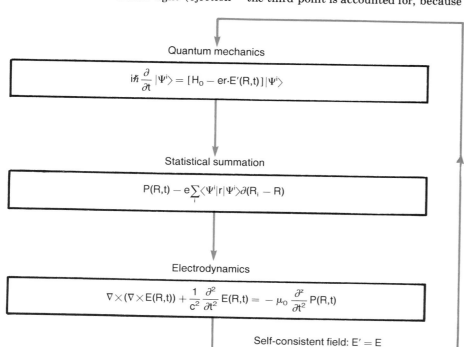

Self-consistent equations demonstrating that an assumed field $E'(\mathbf{R},t)$ perturbs the ith atom according to the laws of quantum mechanics and induces an electric dipole expectation value. Values for atoms localized at $\mathbf{R}$ are added to yield macroscopic polarization, $\mathbf{P}(\mathbf{R},t)$. This polarization acts as a source in Maxwell's equations for a field $\mathbf{E}(\mathbf{R},t)$. The loop is completed by the self-consistency requirement that the field assumed, E', is equal to the field produced, $\mathbf{E}$.
Figure 3

Quantum mechanics

$$i\hbar \frac{\partial}{\partial t}|\Psi^i\rangle = [H_\mathrm{o} - e\mathbf{r}\cdot E'(\mathbf{R},t)]|\Psi^i\rangle$$

Statistical summation

$$P(\mathbf{R},t) = e\sum_i \langle\Psi^i|\mathbf{r}|\Psi^i\rangle\partial(\mathbf{R}_i - \mathbf{R})$$

Electrodynamics

$$\nabla\times(\nabla\times E(\mathbf{R},t)) + \frac{1}{c^2}\frac{\partial^2}{\partial t^2}E(\mathbf{R},t) = -\mu_\mathrm{o}\frac{\partial^2}{\partial t^2}P(\mathbf{R},t)$$

Self-consistent field: $E' = E$

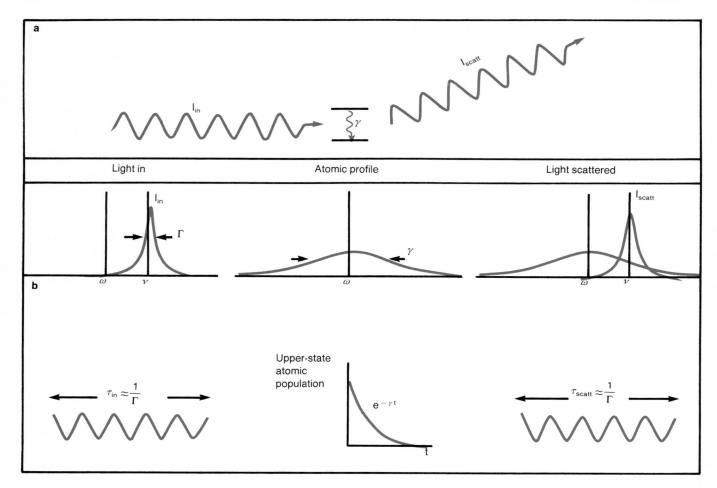

Resonance fluorescence. In part (a) incident light is scattered by an atom with a lifetime $1/\gamma$. In part (b) the incident light has a spectrum centered at frequency ν and corresponds to a wave train of duration $1/\Gamma$. Scattered light has width $1/\Gamma$, is centered at ν and lasts for time $1/\Gamma$; γ is much greater than Γ. Figure 4.

should note that their semiclassical explanation of spontaneous emission runs into conceptual difficulties for the case of atoms excited to a single eigenstate, because the initial atomic dipole **p** of equation (5) vanishes, resulting in infinite lifetimes. One can argue that this excited state is metastable much like a pencil standing on its point—that is, only a small fluctuation is required to get things started.[9] As we shall see, the quantized field readily provides such fluctuations. Furthermore, in the semiclassical theory, the electronic commutation relations[10] are not necessarily preserved in time, and hence the uncertainty principle for the matter can be violated. (See box on opposite page.)

Finally, and most importantly, the quantitative successes[11] of quantum electrodynamics are so impressive that we are virtually compelled to quantize the field as well as the atoms. In view of these facts, we now turn to the photon concept as it is embodied in the quantum theory of radiation.

The quantum theory of radiation

In 1927, P. A. M. Dirac[12] quantized the radiation field in addition to the atom, and the photon concept was for the first time placed on a logical foundation. We outline here the quantum theory of radiation in a form suitable for our purposes. (The present treatment, which uses E and B instead of the vector potential, follows that of reference 2.) For simplicity, we consider a one-dimensional cavity of length L that has perfectly reflecting mirrors. We take the electric and magnetic fields to be polarized in the z and x directions respectively and to be single modes of the cavity, as shown in figure 6. There we see that the electric and magnetic fields act as position and momentum coordinates. The corresponding energy in the cavity is given by the volume integral of the electric and magnetic field densities:

$$
\begin{aligned}
\mathcal{H} &= \frac{\int [\epsilon_0 E^2 + \mu_0 H^2]}{2} d(\text{volume}) \\
&= \frac{p^2 + \Omega^2 q^2}{2}
\end{aligned}
\tag{9}
$$

which is just the energy of a simple harmonic oscillator for a particle oscillating with frequency Ω, mass M and spring constant $M\Omega^2$. A more general multimode field is represented by a collection of such oscillators, one for each mode. To quantize the field (that is, to introduce the photon or particle nature of the radiation) we treat the electric-field "position" coordinate q and the magnetic-field "momentum" p according to the laws of quantum mechanics. We require the commutation relations $[q,p] = i\hbar$; $[q,q] = [p,p] = 0$.

Our single-mode field is then described by the quantum-mechanical wave function

$$
\psi(q,t) = \sum_{n=0}^{\infty} c_n(t)\, \phi_n(q)
\tag{10}
$$

where $|c_n|^2$ is the probability that the radiation oscillator is excited to the nth-energy eigenstate characterized by the eigenfunction $\phi_n(q)$ (the usual Hermite polynomial multiplied by a Gaussian) and having energy $\hbar\Omega(n + 1/2)$. This n-quantum state is said to be the "n-photon" state; that is, $\phi_0(q)$ has no photons (the vacuum), $\phi_1(q)$ has one photon, and so on. We note that the introduction of the wave function $\psi(q,t)$ (Schrödinger picture) or equivalently the noncommutativity of the operators p and q (Heisenberg picture) has the

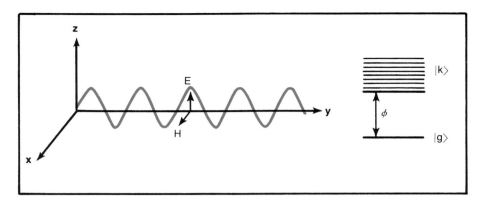

Photoelectric effect. An incident electromagnetic field interacts with a system in its ground state $|g\rangle$, causing transitions to occur to excited states $|k\rangle$, that is, ejecting an electron. Figure 5.

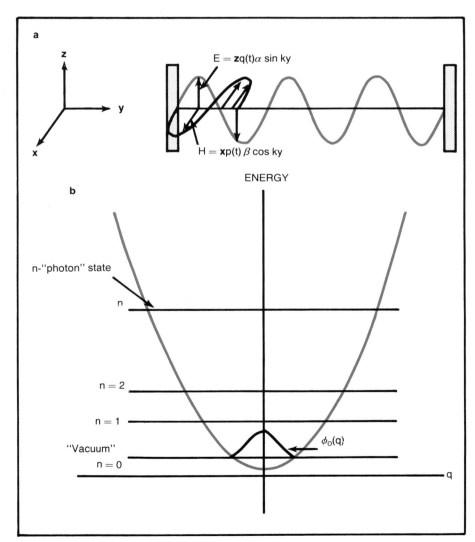

Electric and magnetic fields in a one-dimensional cavity, polarized in the z and x directions respectively. Part (a) shows single-mode standing waves proportional to coordinates q and p, with constants of proportionality α and β respectively. In part (b) we see the simple harmonic oscillator energy-level diagram resulting from quantization of the field. The nth level of the quantized oscillator, $\phi_n(q)$, corresponds to the state having n "photons," while the vacuum is associated with $\phi_0(q)$. Figure 6

effect of bringing out the *wave side* of "particles" (say, electrons) and the *particle side* of "waves" (say, electric waves).

The q and p operators serve to show that the single-mode electromagnetic field is dynamically equivalent to a simple harmonic oscillator. A more convenient and physically revealing set of operators is the annihilation operator $a \propto \Omega q + ip$ and its adjoint $a\dagger$, the creation operator. As their names suggest, these operators annihilate and create photons when acting on photon number states; in other words, $a(a\dagger)$ lowers (raises) ϕ_n to ϕ_{n-1} (ϕ_{n+1}). They are not Hermitian and hence do not themselves represent observables. However, the electric field is given by the Hermitian combination

$$E(y) = \mathcal{E}\,(a + a^+)\sin{(Ky)} \quad (11)$$

where $\mathcal{E}$ is the electric field "per photon" and the Hamiltonian is

$$\mathcal{H} = \hbar\Omega\{a^+a + [a,a^+]/2\} \quad (12)$$
$$= \hbar\Omega(a^+a + 1/2)$$

We emphasize that it is the introduction of the *commutation relations* $[q,p] = ih$, or equivalently $[a,a^+] = 1$, that leads to the *photon concept*.

The first thing to note about the quantized field is that it has fluctuations, *even in the absence of "photons."* In fact, denoting the vacuum state (0 photons) by $|0\rangle$, we find the Hamiltonian, equation (12), has the "zero-point" expectation value $\langle 0|\mathcal{H}|0\rangle = \hbar\Omega/2$, the electric field of equation (11) has vanishing expectation value, but that the vacuum average of the field squared is

$$\langle 0|E^2|0\rangle = \mathcal{E}^{\,2}\sin^2{(Ky)} \quad (13)$$

Thus the field has fluctuations about a vanishing mean in the vacuum. The zero-point energy $\hbar\Omega/2$ is given by a volume integral of $\langle 0|E^2|0\rangle$ and is therefore called the "energy" of the vacuum fluctuations. We shall outline the success of these considerations in

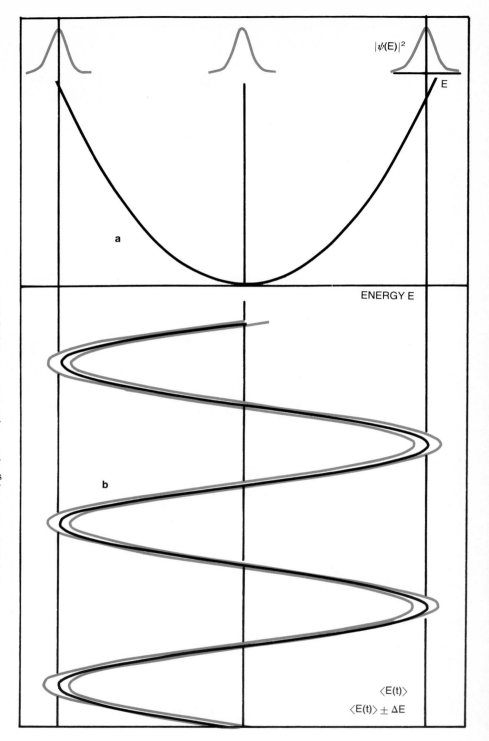

Time evolution of the expectation value ⟨E⟩ of the electric field operator, and variance ΔE indicated by error bars associated with the minimum uncertainty wave packet, are shown in part (a). Part (b) shows the time evolution of a wave packet with minimum uncertainties ΔE and ΔH.
Figure 7

accounting for the statistical fluctuations of light quanta, spontaneous emission, the Lamb shift, and so forth, in a few paragraphs below. However, let us first find out how we regain the classical field, equation (2), from the quantized (photon) field corresponding to the appropriate state vector in equation (10).

We often hear that large quantum numbers correspond to the classical limit. This is a misleading point of the view here, for the expectation value of the field in an n-photon state $|n\rangle$ vanishes, and this fact is true even for $n \rightarrow \infty$. The actual classical limit consists of a superposition of photon states, and this fact naturally leads us to a discussion of photon statistics. Essentially we desire a state of the field $|\psi(t)\rangle$ that yields the classical field of equation (2) for the expectation value $\langle E \rangle$, the square of equation (2) for $\langle E^2 \rangle$, and so on—that is, a field with precise amplitude and phase. But we must recall that the electric and magnetic fields correspond to position and momentum, which obey the uncertainty principle, so that

$$\Delta E \Delta H \geq \hbar/2 \times \text{(constant)} \quad (14)$$

The best we can do is to take the minimum uncertainty case (for all time) for which equality in equation (14) holds. This is described by the coherent (particle) packet[13]

$$|\alpha\rangle = \exp(-\alpha^*\alpha/2) \sum_{n=0}^{\infty} (\alpha^n/\sqrt{n!})|n\rangle \quad (15)$$

This state is the eigenstate of the annihilation operator a with eigenvalue α. (If statements such as this turn the reader off, we invite him to annihilate them from his copy.) We see in figure 7 that the probability density for this state, $|\langle q|\alpha\rangle|^2$, oscillates back and forth in the harmonic oscillator well without change in shape; that is, it coheres. The amplitude of the classical field, equation (2), is related to the complex constant α and the electric

field "per photon" $\mathcal{E}$ by $E_0/2 = \mathcal{E}|\alpha|$.

We see that this "most classical" state is *not* a single-photon number state, but rather a superposition with the Poisson probability of having n photons given by

$$P_n = \exp(-\alpha\alpha^*)(\alpha\alpha^*)^n/n! \quad (16)$$

The average photon number $\langle n \rangle$ is thus $|\alpha|^2$, from which we see that the intensity $(\mathcal{E}|\alpha|)^2 \propto \langle n \rangle \hbar\Omega$. Equation (16)

defines the photon statistical distribution for the coherent state. It is interesting to compare it with that for thermal radiation and that for a laser[14] as shown in figure 8.

It is perhaps worthwhile to note that the distinction between the thermal and coherent distributions is by no means merely academic, for the thermal implies a Hanbury-Brown–Twiss correlation, while the coherent does not. This

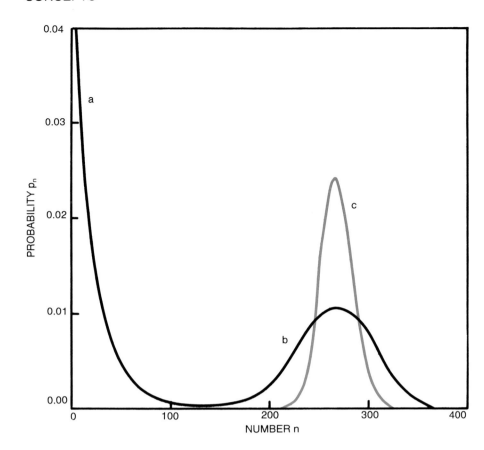

PROBABILITY p_n

NUMBER n

Photon statistical distributions compared for filtered blackbody (part a), laser (part b) and purely coherent (part c) light beams.
Figure 8

correlation is the measure of the excess probability of finding double photoelectron emission over that given by a purely random sequence of events such as raindrops on a roof. In fact, the probability of double photoelectron excitation is twice as large for the "single-frequency" blackbody distribution as for the purely coherent, Poisson, distribution. It is for this reason that photoelectrons produced by a purely coherent light beam are said to be completely uncorrelated—there is no bunching. In general a light beam is completely characterized not just by its spectral density, but by all higher order correlations as well.

For many problems of interest, it is convenient to expand the radiation state (or density operator) in terms of the coherent states $|\alpha\rangle$ instead of the number states. This is accomplished[13] in terms of the $P(\alpha)$ "distribution" defined by the density-operator expansion

$$\rho = \int d^2\alpha\, P(\alpha)\, |\alpha\rangle\langle\alpha| \qquad (17)$$

where integration is carried out over the complex α plane. For the coherent state $|\alpha_0\rangle$, $P(\alpha) = \delta^2(\alpha - \alpha_0)$. For "single-mode" thermal radiation

$$P(\alpha) = \exp\left(-|\alpha|^2/\langle n\rangle\right) \qquad (18)$$

$P(\alpha)$ gives one a measure of deviation from the classical state.

We now have at our disposal a simple method to show the validity of Einstein's[15] correct (but not immediately accepted[16]) interpretation of the fluctuations in the blackbody spectrum.

He claimed that, although the Planck law could be accounted for with a classical field (as Planck originally derived it), the energy fluctuations $(\Delta\mathcal{H})^2$ contained a part due to the wave nature of light plus a part due to its particle character. His formula is (in our notation)

$$(\Delta\mathcal{H})^2 = \bar{\mathcal{H}}\hbar\Omega + \mathcal{H}^2/g(\Omega) \qquad (19)$$

where the average energy

$$\bar{\mathcal{H}} = \hbar\Omega\langle n\rangle g(\Omega) \qquad (20)$$

and $g(\Omega)$ is the density-of-states factor. He identified the first term in equation (19) with the particle character and the second with the wave. Not everyone agreed. In fact, using the information then (1913) available, Wilhelm Wien[16] argued that there was no special reason to attribute the fluctuations to two separate causes (particle and wave).

We see that Einstein was indeed correct, by using the $P(\alpha)$ distribution of equation (18), to calculate the required averages as the following two-line derivation indicates. We note that $\langle\mathcal{H}\rangle = \hbar\Omega\langle n\rangle$ (dropping the $\hbar\Omega/2$ which ultimately cancels equation (19)) and find

$$\langle\mathcal{H}^2\rangle = (\hbar\Omega)^2 \int d^2\alpha \exp\left(-|\alpha|^2/\langle n\rangle\right) \times \langle\alpha|a\dagger\,aa\dagger\,a|\alpha\rangle$$
$$= (\hbar\Omega)^2 \int d^2\alpha \exp\left(-|\alpha|^2/\langle n\rangle\right) \times \langle\alpha|a\dagger\,a[a,a\dagger] + a\dagger\,a\dagger\,aa|\alpha\rangle$$
$$= \langle n\rangle\,(\hbar\Omega)^2 + 2(\hbar\Omega)^2\langle n\rangle^2 \qquad (21)$$

in which the first term resulted from the commutation relation $[a,a\dagger] = 1$, that is, from the quantum character of the field, and the second from the wave character of a classical average over intensity. Calculating the mean-square-deviation density, $(\Delta\mathcal{H})^2 = [\langle\mathcal{H}^2\rangle - \langle\mathcal{H}\rangle^2]\,g(\Omega)$, we find Einstein's formula, equation (19). We see that he correctly identified the particle and wave contributions, a noteworthy feat and a tribute to his insight inasmuch as the quantum theory of radiation was not developed until twenty years later!

As mentioned in our semiclassical discussion, the quantum theory of radiation accounts neatly for spontaneous emission.[17] To see this, we use the electric-field operator, equation (11), in the electric-dipole perturbation energy of equation (4). Using the Fermi Golden Rule we find the spontaneous transition rate (inverse of atomic lifetime)

$$\gamma = (er_{ab})^2\Omega^3/(\hbar\pi c^3) \qquad (22)$$

The emitted radiation is not perfectly monochromatic, for the exponential decay implied by equation (22) yields a Lorentzian frequency profile with width 2γ. We note that the sum over final states of the absolute value squared of equation (21), which enters the Golden Rule, is, in fact, proportional to the vacuum expectation value of E^2; that is, the vacuum fluctuations "stimulate" the atom to emit spontaneously.

Perhaps the greatest triumph of the photon concept is the explanation of the Lamb shift[18] between, for example, the $2s_{1/2}$ and $2p_{1/2}$ levels in a hydrogenic atom. According to the relativistic Dirac theory these levels have the same energy, in contradiction of the experimentally observed frequency splitting of 1057.8 MHz. We can understand the shift intuitively[19] by picturing the electron forced to fluctuate about its "Dirac" position because of the fluctuating vacuum field. Its average displacement $\langle\Delta\mathbf{r}\rangle$, is zero, but the squared displacement has a small positive value from this mean position, $\langle(\Delta\mathbf{r})^2\rangle$. This deviation may change the potential energy the electron experiences in the Coulomb field of the nucleus. To determine how much, we expand the energy in a second-order Taylor series. Noting that the first-order term vanishes in the average over fluctuations, that the fluctuations are isotropic, and that $\nabla^2(1/r) = \delta(\mathbf{r})$, we find the energy shift

$$\langle \Delta V \rangle_{\text{fluct}} = \langle V(\mathbf{r} + \Delta\mathbf{r}) - V(\mathbf{r}) \rangle_{\text{fluct}}$$
$$= (1/2)\nabla^2(V)(1/3)\langle(\Delta\mathbf{r})^2\rangle_{\text{fluct}}$$
$$= (1/6)e^2\delta(\mathbf{r})\langle(\Delta\mathbf{r})^2\rangle_{\text{fluct}} \quad (23)$$

We may now calculate the shift of the energy eigenvalues $\hbar\omega_{nlm}$ by calculating the matrix element $\int d^3r u_{nlm}{}^* \langle \Delta V \rangle_{\text{fluct}} u_{nlm}$. Because only s states have nonzero probabilities for being at $\mathbf{r} = 0$, only these states are shifted (in this approximation). Computation of $\langle(\Delta\mathbf{r})^2\rangle$ requires more discussion, and we refer the reader again to the texts for a complete discussion. However we emphasize that $\langle(\Delta\mathbf{r})^2\rangle$ is nonzero only because of the $(1/2)[a,a\dagger]\hbar\Omega$ vacuum fluctuations and is a direct consequence of the quantized field; that is $[a,a\dagger] \neq 0$. The changes in potential energy account for 1040 MHz of the 1057.8-MHz shift observed between the $2s_{1/2}$ and $2p_{1/2}$ states in atomic hydrogen. When various relativistic corrections and infinities are taken care of, the theory agrees beautifully with the experimental results and provides an impressive confirmation of the quantum theory of radiation.

The anomalous magnetic moment of the electron[20] is more difficult to interpret in simple physical terms because the origin of the spin is buried in the relativistic theory of the electron. Nevertheless, the origin is due to the modification of circulating electronic currents (and hence the magnetic moment), as they are affected by the fluctuating electromagnetic vacuum fields and vacuum polarization.

We conclude with a couple of remarks concerning the so-called wave-particle duality and its effect on interference phenomena. From Newton to Huygens to the present, this point has fascinated and often confounded scientists. Even such an outstanding optics text as Arnold Sommerfeld's[21] contains opaque remarks in this regard such as ". . . the photon theory, at leats in its present state of development, is unable to account precisely for polarization and interference phenomena." Indeed, the "photon theory," as embodied in the quantum theory of radiation, does very well even in these cases. For it is the normal-mode functions $U_{\mathbf{k}}(\mathbf{r})$ that describe interference phenomena in terms of nodal (dark) and antinodal (bright) regions of space. These functions are the same for both classical and quantum fields. Hence there is no need to switch from quantum to classical descriptions or to introduce a mysterious wave-particle dualism in order to explain interference and diffraction. This point is made clear in Fermi's article[22] on the quantum theory of radiation. He does a Lippman-fringe calculation in which light is emitted from one atom, strikes a mirror perpendicular to its direction of propagation, and is absorbed by a second atom. The calculation shows that the probability of excitation of the second atom varies periodically with its distance from the mirror because of interference between the incoming and outgoing light. Fermi comments that

"We may conclude that the results of the quantum theory of radiation describe this phenomenon in exactly the same way as the classical theory of interference."

Recent interference experiments[23,24] involving independent light beams have been made possible by the availability of coherent laser sources. These measurements were largely stimulated by Dirac's comment,[25] "Each photon then interferes only with itself. Interference between two different photons never occurs." The fact that interference between independent lasers *is* observed is not puzzling if we recall that the fringes are described by the normal modes of the system. Dirac's comment is consistent with this experiment in view of the fact that the photon is a quantized excitation of the normal modes of the entire system.

In conclusion: The photon concept as contained in the quantum theory of radiation provides the basis for explaining all known electromagnetic phenomena. However, the "fuzzyball" picture of a photon often leads to unnecessary confusion. Finally, most quantum and electro-optical physics is well understood and quantitatively explained semiclassically.

* * *

The authors wish to thank Willis E. Lamb Jr for a careful reading of the manuscript and valuable criticisms.

This work was supported by the US Air Force (Office of Scientific Research) and by the US Air Force (Kirtland).

References

1. M. Sargent III, M. O. Scully, W. E. Lamb Jr, *Quantum Electronics,* to be published.

2. G. N. Lewis, Nature **118,** 874 (1926).

3. *Genesis,* I. (For the purists only; it was really Sunday.)

4. G. Gamow, *The Thirty Years That Shook Physics,* Doubleday, Garden City, N.Y. (1966), page 22.

5. W. E. Lamb Jr, M. O. Scully, *Polarization Matter and Radiation* (Jubilee volume in honor of Alfred Kastler), Presses Univ. de France, Paris (1969).

6. P. A. Franken, "Collisions of Light with Atoms," in *Atomic Physics* (B. Bederson, V. W. Cohen, F. M. J. Pichanick, eds.) Plenum, New York (1969), page 377.

7. F. K. Richtmyer, E. H. Kennard, T. Lauritsen, *Introduction to Modern Physics,* 5th ed., McGraw Hill, New York (1955), page 94.

7a. N. Bloembergen, *Nonlinear Optics,* Benjamin, New York (1965).

7b. W. E. Lamb Jr, Phys. Rev. **134,** A1429 (1964). For more recent developments see M. Sargent III, M. O. Scully, "Physics of Laser Operation," Chapter 2 in *Laser Handbook,* (F. T. Arecchi, E. O. Schultz-DuBois, eds.), North-Holland, Amsterdam (1972).

7c. S. L. McCall, E. L. Hahn, Phys. Rev. **183,** 457 (1969). See also G. L. Lamb Jr, Rev. Mod. Phys. **43,** 99 (1971).

7d. I. D. Abella, N. A. Kurnit, S. R. Hartman, Phys Rev. **141,** 391 (1966). See also H. Pendleton, in *Proceedings of the International Conference on the Physics of Quantum Electronics,* (P. L. Kelly, B. Lax, P. E. Tannenwald, eds.), McGraw-Hill, New York (1965), page 822.

8. M. D. Crisp, E. T. Jaynes, Phys. Rev. **179,** 1253 (1969).

8a. G. Wentzel, Z. Physik **41,** 828 (1927).

8b. O. Klein, Y. Nishina, A. Physik **52,** 853 (1929).

8c. O. Klein, Z. Physik **41,** 407 (1927).

8d. E. A. Uehling, Phys. Rev. **48,** 55 (1935).

9. M. Crisp, private communication.

10. M. Lax, Phys. Rev. **145,** 110 (1966).

11. R. P. Feynman, Phys. Rev. **76,** 769 (1949); J. Schwinger, Phys. Rev. **73,** 416 (1948); S. Tomonaga, Phys. Rev. **74,** 224 (1948).

12. P. A. M. Dirac, Proc. Roy. Soc. (London) A, **114,** 243 (1927).

13. R. J. Glauber, in *Quantum Optics and Electronics,* (C. DeWitt, A. Blaudin, C. Cohen, Tannoudji, eds.), Gordon and Breach, New York (1965).

14. M. O. Scully, W. E. Lamb Jr, Phys. Rev. **159,** 208 (1967).

15. A. Einstein, Phys. Zeits. **10,** 185 (1909).

16. M. Jammer, *The Conceptual Development of Quantum Mechanics,* McGraw-Hill, New York (1966).

17. V. Weisskopf, E. P. Wigner, Z. Physik **63,** 54 (1930).

18. W. E. Lamb Jr, R. C. Retherford, Phys. Rev. **72,** 241 (1947).

19. T. A. Welton, Phys. Rev. **74,** 1157 (1948); see also V. Weisskopf, Rev. Mod. Phys. **21,** 305 (1949).

20. S. Koenig, A. G. Prodell, P. Kusch, Phys. Rev. **88,** 191 (1952).

21. A. Sommerfeld, *Optics,* Academic, New York (1964).

22. E. Fermi, Rev. Mod. Phys. **4,** 87 (1932).

23. L. Mandel, Nature **198,** 255 (1963).

24. A most useful collection of papers concerning these topics is found in *Coherence and Fluctuations of Light,* Vol. I and II (L. Mandel, E. Wolf, eds.), Dover, New York (1970).

25. P. A. M. Dirac, *The Principles of Quantum Mechanics,* Oxford U.P., London (1958), page 9. □

Light as a fundamental particle

**The question of whether the photon is "special" leads
to some remarkable conclusions about the interactions of matter
and about the underlying symmetry of nature.**

Steven Weinberg

PHYSICS TODAY / JUNE 1975

We take it pretty much for granted that the whole visible world of matter and radiation can be explained, if not in fact at least in principle, in terms of the interactions of a handful of so-called "elementary particles": the electron; the proton; the neutron; the quantum of light, the photon; the quantum of gravitational radiation, the graviton, and perhaps also the neutrino. We would like to know why these particles have the properties they have, and therefore why the world is the way it is. Or, if you do not believe that scientists should ask "why," you can restate the question in this form: What we want to know is the set of simple principles from which the properties of these particles, and hence everything else, can be deduced.

In our search for these principles we have uncovered a large zoo of other particles, some of which are listed in Table 1. The particles that we now include in the table of elementary particles are of varying familiarity. Among them, of course, are the photon, the graviton, the neutrino, the electron, the proton and the neutron. But there are other members of the list—the muon, the pi meson, the K meson—that are less familiar. Some of these, in fact, can be created only by cosmic rays or by artificial beams in accelerators. These various particles are distinguished from one another according to their masses, spins, charges and other properties, but the question of how familiar the particles are is basically a matter of their lifetime. The particles that are most

Steven Weinberg is Higgins Professor of Physics at Harvard University and Senior Scientist at the Smithsonian Astrophysical Observatory. He gave an invited paper at the Washington meeting of the Optical Society of America, Spring 1974, and this article is an adaptation of a tape recording of that talk.

familiar are naturally those of long lifetimes—photons, gravitons, electrons, neutrinos and protons, as well as neutrons—which are stable in nuclei although not in free space.

As you go to more and more unstable particles, they naturally become less and less familiar. But we really see nothing in this table to indicate that any one particle is fundamental in a way that other particles are not. This article originated as a talk at a meeting of the Optical Society of America while we know that there is no "Muon Society of America." The muon, of course, plays a much smaller role in everyday life than the photon but that is an accident. The photon happens to be stable while the muon is not, but we do not see any reason to suppose that the muon is in any sense a less fundamental particle than the photon—or any of the others.

But is that really true? Is it true that the photon is just another particle, distinguished from the others by a particular value of charge, spin, mass, lifetime and certain interaction properties? Or is there really something special about photons? Do they play in some sense a fundamental role; have they a deeper relation to the ultimate formulas of physics than the other particles?

I can—and will—argue both sides of this issue with great conviction. I will first present the case for the hypothesis that the photon is just another particle and that its properties—in fact, the whole of electrodynamics and optics—can be understood as flowing very simply from its particle properties, especially mass and spin, as given in Table 1. Then I will take the other side, and show why I think the photon really expresses something fundamental about the laws of nature. (If such a dialog appears to be rather confusing, I am sorry.) This "something fundamental"

is symbolized in figure 1. The Crab nebula is photographed in visible light, but the supernova that produced it may, as we shall see, have been caused by forces related to, but less familiar than, electromagnetism.

Light as a particle

The peculiar properties of the photon that allow us to go so far in deducing all of electrodynamics and optics on an *a priori* basis arise from the fact that it is a mass-zero particle with integer spin. In a 1939 paper, Eugene Wigner first explained how to analyze the states of a particle with a definite spin in the context of special relativity and quantum mechanics, in terms of the so-called "little group." This is the subgroup of the Lorentz group that leaves the state of motion unchanged. The little group is, very simply, the set of all the Lorentz transformations that do not change the velocity of the particle.

If we have a massive particle at rest, the little group is very well known: It just consists of all the rotations—obviously, the only Lorentz transformations that do not give motion to a particle at rest are rotations. These rotations are indicated schematically in figure 2 by small circles going around the x-axis, the y-axis and the z-axis. If we imagine the particle being given a velocity, say upwards along the z-axis, the Lorentz transformation, although it will do nothing to the little circles in the xy plane around the z-axis, will flatten out the circles around the x-axis and around the y-axis. As we go to higher and higher velocities, these circles become straight lines.

And so the little group, in the limiting case of a particle moving at the speed of light (as any massless particle does) is no longer the rotation group but something else: the product of a

HALE OBSERVATORIES

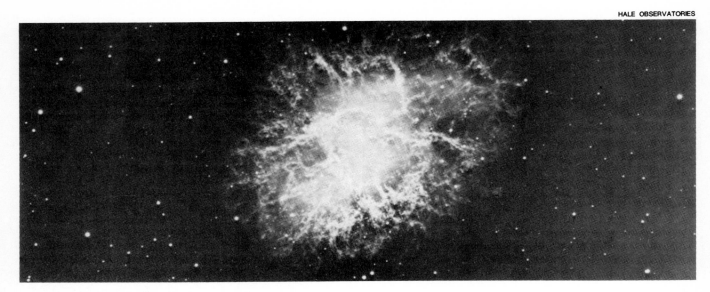

Visible light made this photo of the planetary nebula in Taurus, the Crab; but the supernova that created it in the year 1054 may well have involved other interactions. Despite their differences, do the basic forces of physics share similar invariances?   Figure 1

simple set of rotations only around the z-axis with the set of translations on the xy plane. Now it is precisely the fact that there is not only a J_z but also a J_x and a J_y—not only a component of angular momentum around the direction of motion, but also components perpendicular to it—that leads us to the famous conclusion of quantum mechanics that a particle with spin one and mass not equal to zero can exist in three states, with a z-component of spin equal to plus 1, zero and minus 1: three polarization states.

However, we do not have this conclusion for a particle moving at the speed of light, for the simple reason that the little group is not the rotation group. It only has one rotation, J_z. From a completely general point of view there is no reason why a particle with $J_z = +1$ has to be accompanied by a particle with J_z equal to zero or to any other value. In fact, parity conservation does require that it be accompanied by a particle with $J_z = -1$, and so we come to the conclusion that a particle with unit spin that moves at the speed of light has to exist in *two* polarization states, characterized by helicity +1 and −1, depending on whether its spin vector is parallel or antiparallel with its momentum. In optics, these states are known more familiarly as states of right and left circular polarization. Thus we arrive, without knowing anything about Maxwell's equations, at the first fact about photons: that they exist in only two polarization states. (This is necessary for my argument because I am trying to get Maxwell's equations!)

From the operators that create or destroy photons in these two polarization states one can attempt to construct various kinds of quantum fields, characterized by diverse Lorentz-transformation properties. One familiar example is the antisymmetric tensor formed from the six components of the electric and magnetic fields. In fact, there is a theorem that says this tensor field (along with its derivatives) is the most general Lorentz-invariant field that you can form from the operators for destroying and creating photons of helicity plus or minus one. In the absence of interactions with other particles or fields, the electric and magnetic fields formed in this way automatically satisfy Maxwell's free-field equations. Nothing could be more satisfactory.

Interactions

Now, what about interactions? If I imagine charged particles interacting by transmitting this kind of photon, and if I use the field-strength tensor mentioned in the previous paragraph to describe the interaction of the photon field with the charge, I will find that I get a factor of $\sqrt{p}$, where p is the momentum carried by the photon, at both vertices of the Feynman diagram:

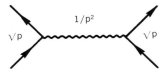

The propagation of the photon introduces a factor of $1/p^2$, and multiplying $\sqrt{p}$, $1/p^2$ and $\sqrt{p}$ again, I get $1/p$, which under a Fourier transformation gives $1/r^3$, the familiar dipole–dipole potential. That is, if we use only field strengths to describe the interactions of electromagnetic waves and charged particles, then we will find that the interactions between photons and other particles generate only $1/r^3$ potentials rather than the famous long-range Coulomb potential, $1/r$.

How does the $1/r$ potential come into physics? Well, as everyone knows, it comes into physics because the interaction of photons with charged particles takes place not only through the field-strength tensor, but also through the vector potential, the curl of which is the field-strength tensor. This statement appears to contradict the theorem I mentioned above, that says that the only fields that are formed from the creation and annihilation of photons with helicity plus or minus one are tensor fields and their derivatives, because here we seem to have a vector field!

However, if you remember that there are only two polarization states, you realize that the vector potential A_μ is *not* a vector. In fact, if you calculate how it transforms under a Lorentz transformation, you find that it picks up an additional gradient term $\partial\phi/\partial x^\mu$. This may be less surprising if I remind you that the only vector potential that could actually be formed in this way, without introducing spurious degrees of freedom, is what is usually called the "Coulomb-gauge" or "radiation-gauge" vector potential, defined to satisfy certain non-Lorentz-invariant constraints, such as, for instance, that the time component of the potential should vanish.

So then how can the Lorentz invariance be satisfied? It will only be satisfied if this vector potential interacts with a current J^μ in such a way that the extra gradient term in the Lorentz transformation law does not matter. In other words, the vector potential must interact with currents in such a way that the current is a four-vector and the divergence of the current is zero, so that when you integrate by parts, you find that the integral of $J^\mu\partial\phi/\partial x^\mu$ is zero.

Therefore we conclude that the only way of describing particles of mass zero and spin one that will give rise to the long-range forces, that is, $1/r^2$ forces, derived from $1/r$ potentials, is to use a vector potential that is not a four-vector; one, therefore, that requires the interaction with a conserved current—not just any old vector current!

Gauge invariance

Now, there is another way of saying all this, and that is that the field equations must be unchanged under a *gauge transformation*. This is a phase transformation in which the change in phase of any field is proportional to the charge q that is destroyed by that field times $\phi(x)$, an arbitrary function of space–time:

$$\psi(x) \longrightarrow e^{iq\phi(x)}\psi(x) \qquad (1)$$

In this gauge transformation, the vector potential undergoes a translation by the amount $\partial\phi(x)/\partial x^\mu$. The fact that A_μ couples to a conserved current means that this transformation will not in fact change the field equations.

This gauge invariance, then, which leads inevitably to Maxwell's equations, is a consequence of the requirements of Lorentz invariance, as applied to particles of mass zero and spin one. And with this we have the whole formalism of electrodynamics and optics: It follows inevitably if you believe that photons exist, of mass zero and spin one.

There is an interesting analogy between gauge invariance on the one hand and the general covariance of gravitational theory on the other. In fact, this whole line of argument could be repeated, starting with the idea that there exists a fundamental particle, the graviton, with spin two and mass zero. It may sound a little artificial to do it that way, because no one has detected any gravitons yet. There are, however, good reasons why no one has, and I think this is a perfectly defensible line of argument.

Turnabout

Having argued that electrodynamics and optics can be understood in terms of the photon as just another particle, but one that happens to have the particular quantum numbers of mass zero and spin one, I will now contradict myself and point out why that is not the most fruitful approach. Let us turn the argument around and say that gauge invariance is a fundamental symmetry of nature. From gauge invariance, now taken as an *a priori* assumption, we can derive the fact that there should exist a particle, the photon, with mass zero and spin one, and furthermore derive Maxwell's equations as well as all the other properties of electrodynamics.

Which way should we go, then? From the known quantum numbers to gauge invariance, or from gauge invariance to the known quantum numbers? Obviously questions like these do not make any sense in terms of predictions of experimental data, but they make a

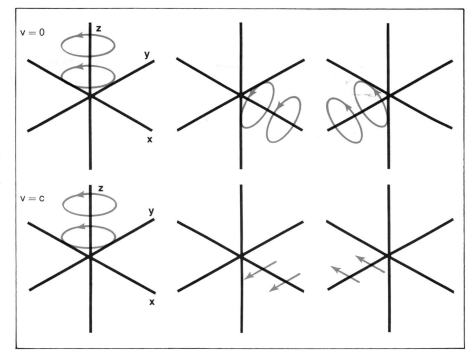

Why the photon has only two polarization states. Of the three circles (top) representing rotation components of a particle at rest, two (the ones normal to the motion) are squashed into lines for the massless photon, which travels at the speed of light. Figure 2

Table 1. Some elementary particles and their properties

Particle	Charge (e)	Spin ($\hbar$)	Mass (MeV/c^2)	Lifetime (sec)
Photon	0	1	0	∞
Graviton (?)	0	2	0	∞
Neutrino	0	1/2	0	∞
Electron	±1	1/2	0.511	∞
Muon	±1	1/2	106.66	2.199×10^{-5}
π meson	±1	0	139.576	2.602×10^{-8}
	0	0	134.972	0.84×10^{-16}
K meson	1	0	493.84	1.237×10^{-8}
	0	0	497.79	0.862×10^{-10}
η meson	0	0	548.8	2.50×10^{-17}(?)
Proton	1	1/2	938.259	∞
Neutron	0	1/2	939.553	935
Λ hyperon	0	1/2	1115.59	2.521×10^{-9}

lot of sense in terms of the direction theoretical research is likely to take.

I would like to argue that it makes sense to talk about gauge invariance as a fundamental symmetry of nature from which follows the existence of the photon and therefore also light and Maxwell's equations. The justification for this point of view is the fruitfulness of the idea—its fruitfulness, to be specific, in unifying electromagnetism with other areas of physics. To understand this, let us recall the way in which elementary-particle physicists these days categorize the various kinds of interactions enjoyed by these particles.

The properties of the four types of interactions are listed in Table 2. We have already discussed the gravitational and electromagnetic interactions, which

are the only ones felt in everyday life. They have ranges that, as far as we know, are infinite. They are characterized by strengths that are quite different, the gravitational being very much weaker than the electromagnetic. Gravitation, however, makes up for its weakness by affecting everything, whereas electromagnetism only affects charged particles. I have mentioned above that we now believe gravitational forces to be transmitted by the exchange of gravitons, and we know that electromagnetic forces are transmitted by the exchange of photons.

The strong interactions, which are supposedly responsible for holding the nucleus together, are much less familiar. They are two orders of magnitude stronger than the electromagnetic, and

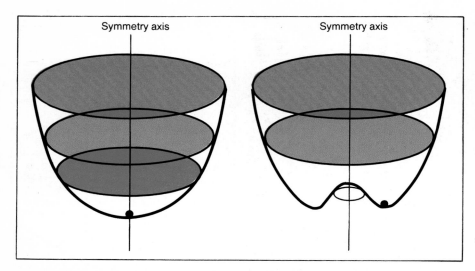

Spontaneous symmetry breaking is here illustrated by a mechanical analogy. The stable equilibrium condition of the ball in the bowl on the left is on the axis of cylindrical symmetry; when the bottom is indented (right), a circular trough becomes stable instead. In the analogy the position of the ball corresponds to the vacuum expectation value of a scalar field, the oscillation frequencies to scalar particle masses and the newly created rolling mode to a Goldstone boson of zero mass—the π meson can be regarded as one of these.   Figure 3

Table 2. The four interactions

	Gravitational	Electro-magnetic	Strong	Weak
Range	∞	∞	10^{-13}–10^{-14} cm	$\ll 10^{-14}$ cm
Examples	Astronomical forces	Atomic forces	Nuclear forces	Nuclear beta decay
Strength ($\hbar = c = m_p = 1$)	G_{Newton} $= 5.9 \times 10^{-39}$	$e^2 = 1/137$	$g^2 \approx 1$	G_{Fermi} $= 1.02 \times 10^{-5}$
Particles acted upon	Everything	Charged particles	Hadrons	Hadrons and leptons
Particles exchanged	Gravitons	Photons	Hadrons	?

Table 3. Gauge groups

Group	Vector fields	Physical application
O(2)	A_μ	Electromagnetism
O(3)	A_μ, $W_\mu\pm$	Yang–Mills theory of strong interactions
O(3) $\otimes$ O(2)	A_μ, $W_\mu\pm$, Z_μ	1967 model of weak and electromagnetic interactions

they act on a class of particles called hadrons, which includes the proton, the neutron and various mesons and hyperons. The strong interactions are believed to be transmitted by the exchange of hadrons.

The least familiar of all are of course the forces that have the shortest range; these are the weak interactions. While the strong interactions have ranges of the order of 10^{-13}–10^{-14} cm and so are only important inside the nucleus, the weak interactions have even a very much shorter range and, so far as we know, are not responsible for holding anything together. They are, however, responsible for nuclear beta decay, including some of the reactions that are responsible for producing the heat of the Sun. These forces are very much

weaker than the electromagnetic interactions. Although we know that they act upon hadrons and leptons (electrons, muons and neutrinos), we do not quite know what particles are exchanged in giving rise to these weak interactions.

There is a rule that the range of the force produced is inversely proportional to the mass of the exchanged particle:

range of force $\approx$

$(\hbar/c)$ /mass of exchanged particle

That is why we believe that there is a massless graviton in addition to the massless photon. It also suggests that a particle must be very massive if its exchange is to be responsible for the weak interactions.

It is, in fact, an old idea that, just as

Coulomb scattering, which is an electromagnetic process, proceeds by the exchange of photons, so beta decay proceeds by the exchange of a heavy kind of "photon." We can deduce the properties of the W particle, the supposed intermediate particle of beta decay, from the analogy between the electromagnetic and the weak interaction. Let us compare the Feynman diagram for Coulomb scattering

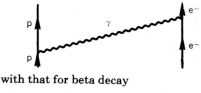

with that for beta decay

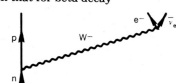

By analyzing the polarization states of the various particles involved in beta decay, we conclude that the spin of the W particle, if it exists, has to be one, the same as that of the photon. Integer spin makes it a boson and, because it has to be described in terms of the vector field, like the photon, it is called the intermediate vector boson. The W particle is, however, very different from the photon in that it has plus or minus one unit of charge. Furthermore, from the known strength of the interaction, we deduce that its mass is roughly 50 GeV.

These ideas go back to Enrico Fermi, with contributions since then by Hideki Yukawa, Julian Schwinger, Sheldon Glashow, Sydney Bludman, Abdus Salam and John Ward, and others.

Generalized gauge invariance

There are difficulties, however, with these ideas. One of the problems is that we are putting two radically different particles together into an analogous relationship: the photon, with zero mass, and the W particle, which has a mass of about 50 proton masses, much heavier than anything we have ever seen.

An equally important, although less obvious, problem is the appearance of infinities. Imagine the scattering, say, of a neutrino on a neutron. This is a process that normally would be expected to go by the exchange of a *pair* of W particles, since one of them alone can not do it. If you analyze this Feynman diagram:

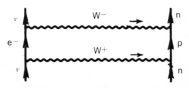

you find that it is infinite and that there

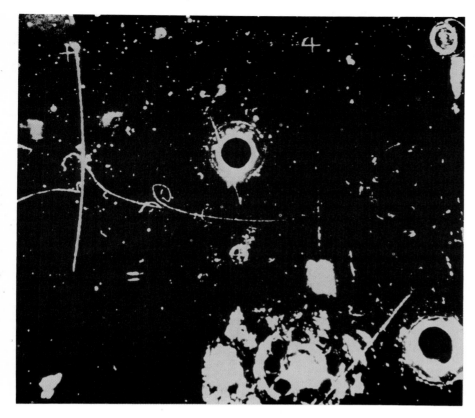

Detail of a photograph taken at the Gargamelle bubble chamber at CERN. A beam of muon antineutrinos enters the chamber from the right. One strikes an electron in an atom of the fluid, causing it to recoil to the left, spiralling clockwise in the chamber's magnetic field before it finally comes to rest. The tracks immediately to the left of the recoil-electron track are produced by an electron–positron pair that was created in the electric field of an atom by a photon emitted by the decelerating recoil electron. Figure 4

is no way of absorbing the infinity into a renormalization of coupling constants. When you further analyze this infinity, you find that it arises precisely from the exchange of W particles with helicity zero. For photons this diagram is not infinite because photons, as we saw, have no states of helicity zero.

In 1967 I made a proposal, which was subsequently also made by Salam, to derive the photon and the W as siblings from a general gauge-invariance principle—one that would be a generalization of the gauge-invariance principles I have described above in connection with electrodynamics. In this theory the mass of the W particle would arise from what is known in the theoretical trade as spontaneous symmetry breaking. This idea is that, on some fundamental level—on the level of the laws of nature, although not on the level of the predicted matrix elements of the scattering amplitudes—the photon and the W can be seen as members of the *same family* of vector particles.

A simple illustration of spontaneous symmetry breaking is provided by the two bowls of figure 3. In the analogy, the projection of the position of the ball onto the horizontal plane corresponds to the vacuum expectation value of a complex scalar field ϕ, and the oscilla-

tion frequencies to masses of a pair of scalar particles. When the bottom of the bowl is pushed in to form a dimple, the equilibrium point spreads out into a circle—the former symmetry is broken. The resulting new rolling mode of the ball about the axis of cylindrical symmetry represents, in this analogy, a new particle, called a Goldstone boson of zero mass. The pi meson may be regarded, at least approximately, as such a Goldstone boson.

Such an extended particle family would be required by a gauge-invariance principle broader than the gauge-invariance principle of electrodynamics.

The kind of generalized gauge invariance that I am referring to is this: Imagine a group, not just a simple group of phase transformations such as that of equation 1, but a group that has to be represented by two noncommuting matrices, M and $\mathfrak{M}$. The various charged fields transform according to the matrix transformations

$$\psi_n(x) \longrightarrow \sum_m M_{nm}[\phi(x)]\psi_m(x)$$

$$A_{\alpha\mu}(x) \longrightarrow$$
$$\sum_\beta \mathfrak{M}_{\alpha\beta}[\phi(x)][A_{\beta\mu}(x) + \partial\phi_\beta(x)/\partial x^\mu]$$

The $\phi_\alpha(x)$ are a set of arbitrary functions of position.

There would have to be one vector field for every one of the parameters that characterizes the group transformation, which would themselves form a nontrivial family. More than one member of this family would have to undergo a matrix transformation as well as the familiar gauge transformation.

This kind of generalized gauge invariance was first brought into physics by Chen Ning Yang and Robert Mills in the 1950's, and has been kicking around as a mathematical possibility for many years. The suggestion that was made in 1967 was in fact to see the photon as part of the family of vector fields required by a fundamental and *exact* gauge invariance of nature. The kind of gauge invariances that appear are shown in Table 3. There is, first, the gauge group 0(2) which, on the Argand diagram, is just a group of phase transformations. It is the same as the group of rotations in two dimensions, which would give rise only to a single vector field (there is only one way you can rotate in two dimensions) and, as I pointed out above, we identify this vector field with the photon.

The original suggestion of Yang and Mills really had to do with the strong interactions and isotopic spin. The idea was that the photon would be accompanied by a pair of charged vector fields, and the gauge group would be the group of rotations in three dimensions, 0(3).

New kinds of interactions

The particular proposal made in 1967, the simplest one possible, is that the fundamental gauge group is the direct product, 0(3) ⊗ 0(2), of 0(3) and 0(2). As a result, instead of three vector fields, there were four. This would imply that, in addition to the heavy charged intermediate vector bosons that would be responsible for the observed weak interactions such as beta decay and the nuclear reactions in the Sun, there would be an additional neutral vector meson, similar to the photon in having no charge, but unlike it in being heavy like the W's.

This 1967 proposal went nowhere for a long time, because, although it was suggested that it would eliminate the infinities from the theory of weak interactions, for a long time no one was clever enough to prove it. This was done in 1971 by Gerhard 't Hooft, then a graduate student at Utrecht, and his proof was later improved by other people including Benjamin Lee. As soon as it became clear that this theory was in fact a solution to the problem of the infinities in the weak interaction, one that had been with us since the mid-1930's, physicists began to put a tremendous amount of work into it, both in examining its applications and in looking for experimental tests.

Volume 46B, number 1   PHYSICS LETTERS   3 September 1973

OBSERVATION OF NEUTRINO-LIKE INTERACTIONS WITHOUT MUON
OR ELECTRON IN THE GARGAMELLE NEUTRINO EXPERIMENT

F. J. HASERT, S. KABE, W. KRENZ, J. Von KROGH, D. LANSKE, J. MORFIN,
K. SCHULTZE and H. WEERTS
III. Physikalisches Institut der Technischen Hochschule, Aachen, Germany

G. H. BERTRAND-COREMANS, J. SACTON, W. Van DONINCK and P. VILAIN *[1]
Interuniversity Institute for High Energies, V.U.B., U.L.B., Brussels, Belgium

U. CAMERINI *[2], D. C. CUNDY, R. BALDI, E. DANILCHENKO *[3], W. F. FRY *[2], D. HAIDT,
S. NATALI *[4], P. MUSSET, B. OSCULATI, R. PALMER *[5], J. B. M. PATTISON,
D. H. PERKINS *[6], A. PULLIA, A. ROUSSET, W. VENUS *[7] and H. WACHSMUTH
CERN, Geneva, Switzerland

V. BRISSON, B. DEGRANGE, M. HAGUENAUER, L. KLUBERG,
L. NGUYEN-KHAC and P. PETIAU
Laboratoire de Physique Nucleaire des Hautes Energies, Ecole Polytechnique, Paris, France

E. BELOTTI, S. BONETTI, D. CAVALLI, C. CONTA *[8], E. FIORINI and M. ROLLIER
Istituto di Fisica dell'Universita, Milano and I.N.F.N. Milano, Italy

B. AUBERT, D. BLUM, L. M. CHOUNET, P. HEUSSE, A. LAGARRIGUE,
A. M. LUTZ, A. ORKIN-LECOURTOIS and J. P. VIALLE
Laboratoire de l'Accélérateur Linéaire, Orsay, France

F. W. BULLOCK, M. J. ESTEN, T. W. JONES, J. McKENZIE, A. G. MICHETTE *[9],
G. MYATT *[6] and W. G. SCOTT *[6] *[7]
University College, London, England

Received 25 July 1973

Events induced by neutral particles and producing hadrons, but no muon or electron, have been observed in the
CERN neutrino experiment. These events behave as expected if they arise from neutral-current induced processes.
The rates relative to the corresponding charged-current processes are evaluated.

We have searched for the neutral-current (NC) and $CC (\nu_{\mu} N \rightarrow \mu^{-} N' + hadrons)$ (2)
charged-current (CC) reactions

$NC (\nu_{\mu} N \rightarrow \nu_{\mu} N' + hadrons)$ (1)   which are distinguished respectively by the absence
                                                                          of any possible muon, or the presence of one, and on-
                                                                          ly one possible muon. A small contamination of
*[1] Chercheur agree de l'Institut Interuniversitaire des $\bar{\nu}_{\mu} \nu_{\mu}$ events in the $\nu_{\mu}, \bar{\nu}_{\mu}$ beams giving some CC events
    Sciences Nucleaires, Belgium   which are easily recognised by the e^+e^- signature. The
*[2] Also at Physics Department, University of Wisconsin   analysis is based on 83 000 ν pictures and 207 000 $\bar{\nu}$
*[3] Now at Serpukhov   pictures taken at CERN in the Gargamelle bubble
*[4] Now at University of Bari   chamber filled with freon of density 1.5×10^3 kg/m³)
*[5] Now at Brookhaven National Laboratories   The dimensions of this chamber are such that most
*[6] Also at University of Oxford
*[7] Now at Rutherford High Energy Laboratory   ‡ A more detailed account of the analysis of this experiment
*[8] On leave of absence from Universita and INFN Pavia   appears in a paper to be submitted to Nuclear Physics.
*[9] Supported by Science Research Council grant

| 138

Fifty-five authors had a hand in writing this three-page paper from Gargamelle, in which neutral-current events were reported. Is this a record?   Figure 5

There are now a number of experimental reasons for believing that this theory is in some sense correct. One of them is that theories of this type—and there is more than one theory of this type—predict the remarkable fact that charge must come in quantized units.

Furthermore, these theories tend to predict that the weak interactions should have some effects comparable in strength to the effects of electromagnetism, but only of certain very limited types. For example, there is strong reason to believe that the mass difference between the neutron and the proton, which has usually been thought of as entirely electromagnetic (arising from the fact that the proton has a charge and the neutron does not) is in fact partly electromagnetic but mostly weak.

But there is a prediction, more accessible to direct test, of this kind of theory —there should be new kinds of weak interactions produced by the exchange of the heavy neutral intermediate vector particles; among these are the scattering of a muon type of neutrino by an electron, which has now apparently been observed in two events at CERN. A bubble-chamber photograph of one of these "neutral-current" events is shown in figure 4. Two events seem like a very small number, but the background is very low in this kind of experiment.

On the other hand, there are now hundreds of events in which a neutrino is seen to produce hadrons in a collision with a nucleon; in these, no exchange of charge takes place between the neutrino and the hadrons. Even though hundreds of events have been observed, both at CERN and at the National Accelerator Laboratory, the background in this case is a serious problem—so that this is also not conclusively settled. For its sociological interest, figure 5 shows what an experimental paper in high-energy physics looks like these days—there are 55 authors!

[Another recent encouraging development is the discovery of new long-lived vector particles at SLAC and Brookhaven. These new particles may be bound states of a new kind of "charmed" quark and the corresponding antiquark; such a charmed quark is needed in unified theories of weak and electromagnetic interactions to suppress certain unobserved neutral-current process.]

There has also recently been a breakthrough in the work of Daniel Freedman and James Wilson and others on the question of how supernovas manage to explode. The key idea that is involved is that the neutral currents produce a coherent interaction between the neutrinos and the iron nuclei in the outer core of a supermassive star. For the first time, this interaction provides a computer model that is actually capable of blowing the star apart and producing the observed phenomenon of a supernova, such as the one that occurred in the year 1054, the remains of which is the Crab nebula shown in figure 1.

The symmetry of nature

How then do we answer the question, What is light? The answer in which I now have the greatest faith is: The photon is the most visible member of the family of elementary particles required by a generalized gauge group that mediates the electromagnetic, weak and perhaps also the strong interactions. As far as I have seen, no one has claimed to be able to include gravitation in this scheme.

If these theoretical ideas, and the experiments that are going on, pan out, we will begin to understand in a fundamental way what light is, and that the photon (as well as other particles that are far less familiar because we live on a much longer time scale than they do) forms the manifestation of a symmetry principle of nature that describes the interactions of matter. This principle is about as fundamental as anything we know about the world.

Bibliography

- S. Weinberg, in *Lectures on Particles and Field Theory* (S. Deser, K. W. Ford, eds.), Prentice-Hall, Englewood Cliffs, N.J. (1965), vol. 2, page 405.
- E. S. Abers, E. W. Lee, Phys. Reports **9**, 1 (1973).
- J. Bernstein, Rev. Mod. Phys. **46**, 7 (1974).
- S. Weinberg, Rev. Mod. Phys. **46**, 255 (1974).
- S. Weinberg, Scientific American, July 1974, page 50.   □